CEREAL GRAIN CROPS

CEREAL GRAIN CROPS

NEAL C. STOSKOPF

Professor of Crop Science and
Director of The Diploma Program
Ontario Agricultural College
University of Guelph
Ontario, Canada

RESTON PUBLISHING COMPANY, INC.
A Prentice-Hall Company
Reston, Virginia

Library of Congress Cataloging in Publication Data

Stoskopf, Neal C.
Cereal grain crops.

Includes bibliographies.
1. Grain. I. Title.
SB189.S865 1985 633.1 84-18193
ISBN 0-8359-0733-3

Editorial/production supervision and
interior design by Norma M. Karlin

10 9 8 7 6 5 4 3 2 1

PRINTED IN THE UNITED STATES OF AMERICA

Contents

Preface

This was an exciting book to write. Excitement came with the fact that cereal crops are central to a major issue of our time, that of world food supply. To research and develop a book dealing with the key crops of agriculture and the world food system in the belief that education is a step in helping reduce the threat of hunger is an exciting prospect indeed. It cannot be denied that every aspect of agriculture offers excitement, for agriculture is both the oldest and the largest of the world's industries. Agriculture is a science that touches every part of our daily lives.

Agriculture is global in its scope. Modern transportation and communication have linked all areas of the world and all segments of agriculture into what is popularly known as a global village. Any real impact on cereal grain production, desirable or undesirable, environmental, political, or technological, whether in the fields of Kansas or Saskatchewan, the steppes of Russia, the plains of India or Africa, will have a ripple effect around the globe. Every cereal grain producer, whether large scale or small, regardless of the crop grown, and irrespective of the level of sophistication will be affected. Every consumer will feel the impact of a change in production. Understandably, then, this book was written with a global perspective and with the hope that a positive impact on world cereal production will result.

Changes in world cereal production are evident. Among the new and changing aspects covered in this book are the development of a network of International Agricultural Research Centers, high fuel costs and concerns about the availability of fuel to agriculture, intensification of production and associated problems of soil depletion and deterioration, hybrid crops, biotechnology and genetic engineering, gene banks, a systems approach to all aspects of agriculture, development of Breeders' Rights, crop-residue management, and implementation of a universal system of measurement in SI units. Exciting changes are occurring with the successful development of a new crop called triticale as well as initial steps in the domestication of wild rice.

This book is written with a broad perspective for a broad audience. It will appeal to the student whether interested primarily in production, plant breeding, crop physiology, or industrial and marketing aspects. It will appeal to instructors at both the diploma and degree level. It is my hope that the many challenging and promising aspects of cereal grain crops will stimulate students to new heights in their particular area of interest.

This book is not an updated version of a standard text. Statistical data, which are so subject to dating and which never seem to be complete or the most recent, are kept to a minimum. Replacing cold, hard, statistical facts are interpretations of trends, perspectives, insights, concepts, and ideas, some subjective, some new, and some not so new, but all presented in straightforward prose. The aim is to provide a sense of continuity among topics and to develop a connection between basic theory and application.

The ten cereal grain crops considered have much in common, and in Part I these commonalities are discussed in 14 chapters covering general aspects. The specifics of each crop are presented in the nine chapters of Part II. The order of presentation of the two parts is deliberate. The approach is to present first a general understanding and broad practical applications and then the specific details. By providing a challenge and motivation first, the presentation becomes compatible with the way students learn.

Gratitude is extended to those who contributed to the preparation of this book beyond the call of friendship or financial return. To my associates at the University of Guelph, especially those in the dean's office and the Crop Science Department of the Ontario Agricultural College, I am indebted for their individual encouragement and for generally creating an environment inducive to undertaking the research and the writing of a manuscript.

The following individuals contributed items that added greatly to the quality of the text, and their help is acknowledged and sincerely appreciated: Ev Everson, Michigan State University, East Lansing; Kevin Falk, former graduate student, University of Guelph; Thomas Hargrove, IRRI Philippines; Bryan Harvey and Les Henry, both of the University of Saskatchewan, Saskatoon; Bill Hopper, Sparrow Lake; Ken Kasha, University of Guelph; Karl Lucken, North Dakota State University, Fargo; Soon Park, Harrow; Denny Ranson, Edmonton; W. K. Robertson, Ottawa; Gloria Rosenberg, Hyderabad; Bea Ryan, Guelph; Morris Sagriff, Chatham; Frank Scott-Pearse, Chatham; Doug Wagner, Kitchener.

The following organizations were more than generous in their response to specific requests: Canadian Seed Growers' Association, Ottawa; Ciba-Geigy Ltd., Basel; Food and Agriculture Organization, Rome; International Crops Research Institute for the Semi-Arid Tropics, Hyderabad; King Grain Ltd., Pain Court; Ontario Ministry of Agriculture and Food, Guelph and Toronto; Pioneer Hi-Bred Ltd., Chatham and Des Moines; Ranson Photographers Ltd., Edmonton; The International Rice Research Institute, Manila.

It is not possible for me to list all who contributed by answering questions, directing my attention to sources of information, or providing specific services. The typists at Words Plus, who persevered with my handwriting and under the capable guidance of Sandy Sanders met all deadlines and made after-hour deliveries, have more than earned my respect. Special thanks are extended to the staff in the University of Guelph library, all of whom have been consistently helpful and courteous. Many of my associates have unknowingly helped me, simply by stimulating new ideas or concepts.

An undertaking of a textbook this size cannot be done without considerable effort and sacrifice. To my family, my wife Nora, children Joanne and Mike, I extend my appreciation for their tolerance.

The deepest debt of all, however, is owed to my wife Nora, for her unfailing support and continued encouragement. During the four-year effort, Nora faithfully looked after all my necessities, relinquished a room in the house, and sacrificed holiday time and other social functions. I am indeed grateful.

NEAL C. STOSKOPF

PART ONE

GENERAL ASPECTS OF CEREAL CROPS

1

Cereal Crops—The Heart of World Agriculture

INTRODUCTION

Botanists estimate that there are about 350,000 plant **species**[1] of which about 300,000 are vascular land plants (Burger, 1981). The dominant and economically most important class of plants are the 195,000 species of flowering plants, most of which produce edible parts potentially useful to man. The world's food supply is obtained either directly or indirectly from this abundance of plant species, but—surprisingly—less than 0.1%, or fewer than 300, are used for food. Worldwide, about 50 species are cultivated actively, and as few as 17 species provide 90% of man's food supply and occupy about 75% of the total tilled land on earth (Biswas and Biswas, 1976; Harlan, 1976). They include wheat, rice, corn,[2] potato, barley, sweet potato, cassava, soybean, oat,[3] sorghum, millet, rye, peanut, field bean, pea, banana, and coconut.[4]

Eight **cereal grains**—wheat, rice, corn, barley, oat, rye, sorghum, and millet—provide 56% of the food energy and 50% of the protein consumed on earth. By far the major part of the world's population lives on two crops, wheat and rice. Corn is a major crop for both direct and indirect human consumption, as it forms a major energy feed for animals.

In general terms, wheat, rice, and corn together make up three-fourths of the world's **grain** production (Table 1-1). Sorghum, barley, millet, rye, and oat make up the remaining cereal grain production on earth.

These facts underscore the worldwide importance of cereal crops as a food source for mankind. The domination of these crops serves as a focal point to direct research and production efforts to increase them to meet world food needs. The centrality of cereals among world crops is the very basis on which this international text on cereal crops has been prepared.

HISTORICAL CHOICE AND DEVELOPMENT

Archaeological evidence suggests that during the hunting and gathering period, which extended

---

[1]Words or terms that are defined or explained in the Glossary are set in boldface type the first time they appear in the text.

[2]Corn, often referred to as maize, and in England referring to all grain, normally will be used in this text to delineate *Zea Mays*.

[3]Note the use of singular crop names. Traditionally some crop names seem to be used always in the singular (e.g., wheat, rice), while some others are often in the plural (e.g., oats). In keeping with the suggestion of V. E. Green, Jr. [*Agronomy Journal* 74 (1982):64], the singular form for all crops is adopted.

[4]Botanical names for crops are given in Appendix A.

Table 1–1
World Production from Eight Major Grain Crops Averaged over the 1979–1981 Period and Compared to Production a Decade Earlier

Crop	Production, million tonnes (million tons)[a]				Percent of Total, 1979–1981	Percent Change
	1969–1971		1979–1981			
Wheat	327.9	(361.5)	443.9	(489.4)	28.0	+35.4
Corn	283.0	(312.0)	421.4	(464.6)	26.6	+48.9
Rice	309.8	(341.6)	396.3	(436.9)	25.0	+27.9
Barley	125.2	(138.0)	158.5	(174.8)	10.1	+26.6
Sorghum	55.6	(61.3)	64.3	(70.9)	4.1	+15.6
Oat	54.5	(60.1)	43.5	(48.0)	2.8	−20.2
Millet	33.2	(36.6)	28.5	(31.4)	1.8	−14.2
Rye	30.8	(34.0)	24.7	(27.2)	1.6	−19.8
World total	1231.7	(1353.5)	1592.8	(1750.3)	100.0	+29.3

Source: *FAO Production Yearbook 1981.*
[a]Refer to Appendix B for information on metric measurements and conversion factors.

over a million or more years, humans tried perhaps 3000 plants as food sources. The record of how and why a particular crop was selected is vague, but by the end of the Stone Age, 10,000 years ago, a much smaller number of plants had been selected as the most useful. Almost certainly, crops that were the most productive and most rewarding in terms of labor and capital invested were favored. Wheat and barley were recognized as valuable food sources for Old World and corn for New World agriculture, and probably these crops formed the basis for the transition from a hunting and gathering to an agricultural society. Choice of crops apparently was not directed to providing a varied or exciting diet. Basic food needs came first, and eventually variation in diet was achieved by methods of preparing the same basic food products.

Perhaps animals aided humans in their selection. Animal preference for the palatable and nutritious grain may have been evident as well as the value of cereal grains in providing a forage roughage for ruminants.

In selecting plants to meet their food needs, humans caused substantial changes to occur among the cereal crops. Cereal plants with a nonshattering seed habit were selected through ease of harvest over those which readily shattered their seed as maturity was reached. The development of a nonshattering cultivar is the conventional indicator of cereal domestication. By this criterion, wheat and barley were domesticated in the Fertile Crescent of the Middle East about 10,000 years ago (Evans, 1980). Until Gregor Mendel discovered the laws of **genetics** in the second half of the nineteenth century, improvement in cereal crops occurred through the process of unwitting or conscious selection, intuition, detailed observation and choice, or by chance.

The knowledge of genetics and development of scientific methods of crop improvement has not changed the crop species on which modern humans depend. Triticale, a man-made crop species, as well as other new crops, must earn a place under the sun by competing successfully with more ancient crops. The successful establishment of triticale as a world crop offers a tremendous challenge, for it represents an additional crop where the trend has been for the most important to increase while lesser crops disappear. Note that in the ten-year period between 1969–1971 and 1979–1981, wheat, corn, rice, barley, and sorghum production increased, while oat, millet, and rye production decreased (Tables 1-1 and 1-2). The ability of the latter three crops to produce in environmentally limited areas means they will continue to be important world crops. Their harvested area is reduced, however, by the expansion of more widely adapted cultivars of the other, superior, cereal crops.

WORLD CEREAL GRAIN PRODUCTION

Because cereal crops dominate world food production and because their low moisture content reduces the risk of deterioration in storage, total world output of only eight cereal grains serves as a reliable measure of world food production and adequacy.

Total world cereal grain output exceeded 1500 million tonnes (1792 million tons) for the 1979–1981 period (Table 1-1). The increase over the 1969–1971 period was 22.6% or a 2.3% per year increase. World population increased during the same decade at an average annual rate of 1.8%. Although total food production appeared to exceed population growth, inadequate distribution resulted in food shortages in some areas and surpluses in others.

Per capita grain use serves to emphasize further man's dependence on the cereal grain crops as well as to indicate the adequacy of world food supplies. Consumption in the world varied from less than 200 kg (440 lb) per year in Pakistan, the Philippines, Indonesia, India, and Nigeria to over 700 kg (1543 lb) in the United States (Table 1-3).

In countries highly dependent on direct grain consumption, average grain use equals about 150 kg (331 lb) per person per year or 411 g (14.5 oz) per day. Where average grain consumption falls below that figure, as in Nigeria (Table 1-3), alternate starchy, nongrain crops such as cassava or sweet potato often are eaten. Where per capita grain consumption is above 200 kg (440 lb) per year, the excess is usually consumed in the form of animal products. Of the 700 kg (1543 lb) consumed per person in the United States, about 100 kg (220 lb) is consumed directly as bread, pastries, and breakfast foods, with the remaining 600 kg (1323 lb) fed to livestock (Brown, 1978).

Constantly rising world demand necessitates an increase in world production of about 23 million tonnes (25 million tons) of grain crops every year or about 2% annually. The projected average annual population increase from 1975 to 2000 for the world is 1.8%, with a 0.6% increase predicted in the less developed regions compared to 2.1% in the developed regions of the world (Barney, 1980). Demand for food in developing countries is expected to average 2.2% over the same period.

Table 1–2
Total Area of Cereal Crops Harvested (Grain Only), Averaged over the 1979–1981 Period and Compared to the Area a Decade Earlier

	Harvested Area, millions of hectares (acres)		
Crop	*1969–1971*	*1979–1981*	*Change, %*
World total	682.1 (1684.8)	727.8 (1797.8)	+6.7
Wheat	212.9 (525.9)	235.1 (580.6)	+10.4
Rice	132.9 (328.2)	143.2 (353.6)	+7.8
Corn	114.5 (282.8)	129.5 (319.9)	+13.1
Barley	66.8 (164.9)	80.7 (199.4)	+20.8
Sorghum	48.0 (118.7)	46.3 (114.4)	−3.5
Millet	46.8 (115.6)	43.0 (106.3)	−8.1
Oat	30.4 (75.0)	25.9 (64.0)	−14.8
Rye	20.0 (49.5)	15.3 (37.8)	−23.5

Source: FAO Production Yearbook 1981.

Table 1–3
Annual Per Capita Grain Consumption in the 20 Most Populous Countries, 1975[a]

Country	Kilograms	Pounds
United States	708	1561
USSR	645	1422
Spain	508	1120
France	446	983
Federal Republic of Germany	441	972
Turkey	415	915
Italy	413	910
United Kingdom	394	869
Mexico	304	670
Egypt	286	630
Japan	274	604
Brazil	239	527
Thailand	225	496
People's Republic of China	218	481
Bangladesh	203	448
Pakistan	171	377
Philippines	157	346
Indonesia	152	335
India	150	331
Nigeria	92	203

Source: Data from Brown (1978).
[a]Includes both that consumed directly and that consumed indirectly in the form of meat, milk, and eggs.

A PERSPECTIVE ON CEREAL CROPS: REASONS FOR UNIVERSAL IMPORTANCE

Cereal crops have not achieved worldwide dominance and acceptability by chance or because they were among the first crops successfully cultivated by early agriculturists. Neither do cereal grains possess distinctive botanical features nor eye-catching superiority. Instead a number of distinct features can be identified that collectively account for the prominence of cereal grains worldwide. These features are considered here.

High Yield Performance

Cereal grains provide more grain for less effort than any other class of crop plants. Grain crops yield comparatively well under conditions that are unbelievably poor or under inadequate cultivation and management. Cereals produce under practices as ancient as the stick plow or under highly mechanized modern farming systems.

It is difficult to make valid comparisons as to which cereal crop yields the highest because of different responses to environmental variables. Record wheat and rice yields are similar, yet because much of the wheat is grown under dryland conditions, the average worldwide yield of wheat is much lower than that of rice (Table 1-4). As well, the yield potential of sorghum is as high as that of corn, but sorghum is grown primarily under arid conditions too dry for corn. Both sorghum and millet are capable of producing large quantities of grain under conditions of drought that would render the production of other crops impossible. The importance of this capability comes into perspective when it is realized that 36% of the cropland of the world receives less than 76 cm (30 in.) of rainfall per year and is classified as arid or semiarid. Ninety percent of the total agronomic crop production of

Table 1-4
Comparisons of Average World Grain Yields among Cereal Crops for the Years 1969–1971 and 1979–1981

Crop	Grain Yield kg/ha (lb/ac) 1969–1971	1979–1981
Corn	2472 (2207)	3252 (2904)
Rice	2331 (2081)	2767 (2471)
Barley	1875 (1674)	1965 (1755)
Wheat	1540 (1375)	1888 (1686)
Rye	1536 (1371)	1612 (1440)
Oat	1795 (1603)	1682 (1501)
Sorghum	1157 (1033)	1386 (1237)
Millet	709 (603)	662 (591)
World average	1806 (1613)	2188 (1953)

Source: FAO Production Yearbook 1981.

semiarid regions in the world is derived from wheat, barley, sorghum, and millet. The remaining 64% of the world's cropland suffers from drought sometime during the growing season (Duncan, 1982).

The relatively high yield performance of corn, sorghum, and millet, even under less than ideal conditions, is explained by an efficient photosynthetic system called the C_4 dicarboxylic acid pathway. C_4 plants produce well under temperatures of up to 35°C (95°F) and high-light-intensity conditions in a very efficient manner. Many other crops, including wheat, rice, barley, rye, and oat, have a less efficient photosynthetic system called the C_3 Calvin cycle pathway. C_3 plants are less responsive to high light, carbon dioxide and temperature conditions than C_4 plants.

After centuries of development directed toward cereal crops, many genetically superior **cultivars** exist within each crop. The best cultivar adapted to specific environmental conditions can be grown to ensure high yield. In addition, careful evaluation of specific cultivars and their combination to produce first-generation crosses with **hybrid vigor** has allowed yield to be increased substantially in crops such as corn, sorghum, and millet.

World cereal crop grain yields are more than double those of the grain legume and oilseed crops (Table 1-5). Direct yield comparisons, however, must be made with caution, since cereals are considered as energy-producing crops. Yield differences are expected on the basis that, because of their high energy content, production of protein and vegetable oils requires more energy from photosynthesis than does carbohydrate production. From comparisons of the known energy requirements of the various metabolic pathways, equivalent seed yields would be about 60 for rapeseed, 70 for soybean, and 90 for pea, when wheat and corn are rated as 100 (Evans, 1980).

Inevitably, low-yielding, less responsive and unadapted crops are replaced by higher-yielding ones. In the decade from 1970 to 1980, a slower rate of yield increase has been observed in the grain legumes and oilseed crops than in cereal crops (Tables 1-4 and 1-5). During this period, wheat, rice, and corn increased 22.6%, 18.7%, and 31.6%, respectively (Table 1-4), compared to soybean and broad bean crops, which exhibited the highest increases of 13.5% and 17.3%, respectively (Table 1-5). Minor increases or decreases in yield were observed in other legume or oilseed crops.

The low yield increase in the traditional and cherished nitrogen-fixing and high-protein grain legume crops may result in these crops' being displaced by cereal crops, a trend reported in India and Africa (Evans, 1980) where the demand for food is intense. Although such a displacement might appear to reduce the nutritional protein value of local diets, the extra calories produced by the higher-yielding cereal crops appear to be of more value. It is generally accepted that the energy needs will be met first through metabolic processes and before protein needs are satisfied.

Worldwide Dependability

Wide **adaptation,** flexibility, and reliability are criteria of cereal crops that make them dependable as a food source. Cereal crops can be grown over a wide range of environmental conditions—

Table 1-5
Grain-Yield Comparisons among World Cereal, Pulse, and Oilseed Crops for the Years 1969–1971 and 1979–1981

Class and Crop	Grain Yield, kg/ha (lb/ac) 1969–1971	Grain Yield, kg/ha (lb/ac) 1979–1981
Cereals (from Table 1-3)	1806 (1613)	2188 (1953)
Grain legumes[a]	674 (601)	642 (574)
Soybeans	1487 (1328)	1688 (1507)
Pea (dry)	1233 (1101)	1146 (1023)
Broad bean (dry)	961 (858)	1128 (1007)
Peanut (groundnut in shell)	916 (818)	945 (844)
Chickpea	655 (585)	615 (549)
Lentils	613 (547)	571 (510)
Bean (dry)	520 (464)	539 (482)
Oilseed crops[b]		
Sunflower	1173 (1047)	1178 (1052)
Rapeseed[c]	811 (724)	977 (872)
Safflower	707 (631)	692 (618)
Castor bean	589 (526)	576 (514)
Linseed (flax)	514 (459)	454 (405)
Sesame	323 (288)	297 (265)

Source: FAO Production Yearbook 1981.
[a]Value shown includes other pulse crops such as cowpeas and vetch.
[b]Value not available because crops such as olive oil and coconuts are not expressed on the same dry-weight basis.
[c]Rapeseed includes canola derivatives.

unlike citrus fruit crops, which cannot tolerate frost and require sun conditions; cassava, which has a high temperature requirement; cranberries, which thrive in acid, marshy, organic soils; or artichokes, which respond to foggy, cool climates.

Grain crops are grown in countries around the globe. To illustrate this, the area, yield, and total production are shown in Table 1-6 for countries exceeding 7 million hectares (17.3 million acres). One or more cereal grains are a basic crop in seven major areas of the world. Adapted cereal crop species and cultivars within each species are found in all latitudes from 60° north to 50° south, on soils slightly alkaline to acid, and in arid and well-watered regions. Rice is only slightly more confined to middle latitudes. The productiveness of rice is impressive under rainfed or dryland conditions as well as under irrigated or flooded conditions.

With the exception of taro, a minor crop, rice is the only major food crop that can be grown in standing water. Other crops are killed by flooded conditions, but even under conditions of continuously submerged roots, rice thrives. Like all crops, the roots of rice require oxygen for respiration and nutrient absorption. Oxygen requirements in the rice crop are met because of connecting tissues in the leaves, stems, and roots that permit air to move from the leaves to the roots. The semiaquatic nature of the rice plant means that the many great, fertile river basins and deltas of tropical and subtropical Asia can be utilized effectively.

Spring rye and oat are adapted to areas with a very short growing season. Winter rye is capable of surviving severe winter conditions, thereby providing a protective winter cover on fragile soils, as well as performing reliably on droughty and acid soils.

Cereal crops produce food grain without an abundance of commercial fertilizer, yet remain responsive to high-fertility conditions or heavy fertilizer applications. Corn is highly responsive to the use of commercial fertilizers, but for the predominantly cereal crop producing and extensive dryland areas of the Great Plains of North America and in similar areas of the world, fertilizer use is a relatively recent development. The response of cereal crops to fertilizer in dryland areas is variable, because soil moisture deficiencies have a greater effect on grain production than does fertilizer. When cereal crops are grown on unirrigated land, nitrogen fertilizer use is minimal. When cereal crops are irrigated, they respond to added nitrogen.

The dependability of cereal crops on marginal lands often means that tender or higher-value fruit and vegetable crops are grown on more preferred land. Cereal yields therefore are often low. Worldwide, and based on the 1979 to 1981 period, grain yield averaged an outstanding 2188 kg/ha (1953 lb/ac) and ranged from a yield in excess of 4600 kg/ha (4162 lb/ac) in France to 720 kg/ha (643 lb/ac) in Nigeria (Table 1-6).

In the densely populated area of Asia, cereal crops accounted for 43.5% of the area harvested (Table 1-6) because they are dependable crops. Crop failure in Asia could result in unimaginable human misery.

A possible reason for yield dependability is

Table 1–6

World Production Statistics for All Cereal Crops for Countries Exceeding 7 Million Hectares (17.3 Million Acres) of Production over the 1979 to 1981 Period

Region and Country	*Area Harvested, million ha (ac)*	*Yield, kg/ha (lb/ac)*	*Total Production, million tonnes (tons)*
Africa	72.4 (178.9)	984 (879)	71.3 (78.4)
Nigeria	13.4 (33.0)	720 (643)	9.6 (10.5)
South Africa	8.5 (21.0)	1612 (1440)	13.9 (15.3)
North Central America	104.8 (258.9)	3530 (3152)	370.0 (406.6)
Canada	19.4 (47.9)	2170 (1938)	42.4 (46.6)
USA	72.6 (179.4)	4162 (3716)	302.3 (332.2)
Mexico	97.1 (24.0)	2109 (1883)	20.7 (22.8)
South America	37.7 (93.2)	1784 (1593)	67.3 (74.0)
Argentina	11.1 (27.4)	2196 (1959)	24.6 (27.1)
Brazil	20.6 (50.9)	1500 (1339)	30.8 (33.9)
Asia	316.8 (702.4)	2095 (1870)	639.6 (703.6)
Bangladesh	10.7 (26.4)	1963 (1752)	21.0 (34.1)
China	96.1 (237.3)	2977 (2658)	286.0 (314.6)
India	103.5 (255.5)	1341 (1197)	138.8 (152.7)
Pakistan	10.7 (26.3)	1607 (1435)	17.1 (18.8)
Turkey	13.6 (33.6)	1855 (1656)	25.2 (27.7)
Europe	70.3 (173.6)	3525 (3147)	247.7 (272.5)
France	9.8 (24.2)	4661 (4162)	45.7 (50.3)
Poland	7.9 (19.5)	2344 (2093)	18.5 (20.4)
Spain	7.3 (18.0)	1984 (1771)	14.5 (16.0)
Oceania	16.3 (40.3)	1339 (1196)	21.9 (24.1)
Australia	16.1 (39.8)	1304 (1163)	21.0 (23.1)
USSR	121.0 (299.0)	1445 (1290)	175.0 (192.5)
World	727.8 (1797.7)	2188 (1953)	1592.8 (1750.3)

Source: FAO Production Yearbook 1981.

the fact that cereal crops have a synchronous flowering habit (**determinate**) which may be superior under certain environmental conditions to the asynchronous flowering habit (**indeterminate**) of other crops. A determinate characteristic clearly delineates growth and development from seed to maturity into vegetative and reproductive periods. With indeterminate crop plants, vegetative and reproductive periods overlap.

In climates with a highly predictable dry season, a determinate crop growth habit may be a useful feature to ensure rapid development, early maturation, and uniform ripening. These features allow for complete and mechanical harvesting and assure a reliable crop yield. With determinate crops, irrigation water can be precisely timed, and **transpiration** losses are reduced by the associated restricted **tillering** period (Blum, 1973; Jones and Kirby, 1977).

When the determinate habit is combined with a nontillering corn cultivar, plant population can be manipulated and, with precision planting, leaf area can be effectively managed. When plant population is properly managed to suit the environmental conditions, corn is recognized as a highly dependable crop. Adequate moisture prior to pollination and during the grain filling period is important.

Related to crop dependability is the annual growth habit of all grain crops, although many have a winter annual counterpart. Annual crops produce a yearly grain yield without risk of winter injury.

The determinate and annual features of cereal grains often make it possible to harvest two or more crops each year, at least in areas where light and temperature do not limit growth and where adequate moisture or irrigation water is available. In the Philippines, annual grain yields of up to 26 tonnes per hectare (12 tons per acre) have been obtained from three rice crops per year (Evans, 1980). Further shortening of the rice life cycle may permit four crops per year.

Annual grain crops that raise no concern about survival from one year to the next offer ample scope for ingenuity in modifying high-input agriculture to maximize grain yield. Under circumstances of two or more crops produced per year, yield per day rather than a yield per crop becomes the most important feature. Using this measure, it may be best to maximize the number of crops, rather than the yield of each crop.

Another factor related to worldwide dependability is that cereal crops generally are not limited by the variation in natural daylengths found at different latitudes because of the genetic insensitivity to daylength that is available in cereal crop species. Where sensitivity to daylength is desired to assure early maturity and double cropping, the genetic variability is available.

The flexibility associated with cereal crops has led to innovative and economical production practices. Broadcasting winter wheat seed into a standing crop of soybeans has resulted in a system of **double cropping.** In short-season areas, a full-season crop of soybeans cannot be harvested sufficiently early to allow fall seeding of winter wheat conventionally seeded into a prepared seedbed. Broadcasting the winter wheat seed before the soybean leaves drop off the plant, timed to precede a rainfall and where weed control is adequate, provides short-season areas with a flexible rotation system.

In areas where winter cereals can be harvested before the **critical planting date** for a second crop, double cropping is possible. To reduce the time to plant a second crop following the harvest of an early-maturing cereal crop, a system of minimum or no tillage may successfully be employed (Touchton and Johnson, 1982). Where maturity of the cereal crop cannot be reached before the critical planting date of a second crop, surface seeding into the standing first crop may be practiced without significantly damaging the yield of a second crop (Chan, Johnson, and Brown, 1980).

Worldwide Food Source

For about half of the human population, rice is a staple diet. At least a third of the world's population depend on rice for more than half their food. Rice is synonymous with food for the people in Asia, where more than half the people on earth live. Asian countries have become so dependent on rice that, throughout history, a rice crop failure has caused widespread famine.

About 93% of the rice produced in the world is consumed directly (Chandler, 1979). The apparent average annual rice consumption in the principal rice-consuming countries of Asia is mostly over 100 kg (220 lb) or an average of 274 g (9.7 oz) per day. Rice has an enduring palatability, and one can eat it daily for a lifetime, often at consecutive meals, without tiring of it (Chandler, 1979).

In other areas of the world, despite the few crops that supply humans, monotony in diet is avoided by variation in the preparation of cereal grains. Cereal grains can be processed readily into a number of attractive food commodities with wide acceptance among the many cultural groups and tastes of the world's people.

Bread, produced from wheat or rye flour, is the very symbol for food in many countries of the world. Bread has been a basic food for humans throughout recorded history and was given further importance with the advent of the Industrial Revolution. Industrial workers who left the farms for the factories required a change from their usual food form, which was perhaps a thin broth. Factory workers needed an inexpensive, abundant, reliable, palatable, storable, and transportable form of food. The answer was bread. By 1800, bread had become the established food of the industrialized countries.

Wheat bread was almost unknown in Japan when U.S. occupation forces began importing wheat as postwar relief into that nation in 1946. Bread baked from U.S. wheat provided food for civilians and school children in the war-torn, food-deficient country. Because the Japanese liked bread, its place in the Far East was established.

For other peoples of the world, corn has been portrayed as the staff of life. Corn has influenced life styles; has provided a culture, art, and craft; and has united families to a single purpose which in turn became a national driving force.

In Mexico, close to 98% of the corn crop is consumed in the form of **tortillas,** the daily bread of the Mexican people (Wellhausen, 1976). Tortillas can be made also from wheat, and in northern Mexico, where wheat is abundant, the tortilla is made of wheat instead of corn. The tortilla is similar to the **chapatti** of the Near and Middle East, except that the tortilla is made from white flour rather than whole wheat.

The people of Brazil depend on a combination of corn, wheat, and rice for their main food and protein source.

Durum wheat is made into yellow pasta flour to make products such as spaghetti and macaroni that are believed to have originated in Italy. In Algeria, durum wheat flour is used to produce **couscous.**

Soft or low-**gluten** wheats are preferred for cookies, cakes, Korean biscuits, and pastries. Although soft wheat flour is not regarded as suitable for making leavened bread, the delicious and distinctive-tasting French bread is made from soft wheat. French bread stales quickly and must be made fresh daily—accounting for the 40,000 decentralized bakeries in France (Morgan, 1979).

Scottish people have come to be portrayed as consumers of oat grain products. Dr. Samuel Johnson in his 1755 dictionary defined oat as a grain which in England is generally given to horses, but in Scotland supports the people. Perhaps the burly, robust, and healthy Scotsman is a clear example of the product of a cereal grain diet.

Recognition of bread and other cereal products came about as the result of the Irish potato famine in 1844. Both the politicians of the time and the average citizen came to realize that an alternate food was needed to reduce their dependence on the potato. The result was that the "Corn Laws," which had prevented the importation of cereal grains, were repealed. Millions of Irishmen readily switched to eating bread and other cereal products.

An Ideal Package for Storage and Handling

Cereal grains are a concentrated form of food that contain little water and hence require little space. Cereal grains are not damaged by freezing or dry conditions, are durable, and can be handled mechanically without severe damage by conveyors or elevating equipment (Figure 1-1). An important feature is that dry grain flows readily under the influence of gravity, thereby facilitating handling (Figure 1-2). These attributes allow for repeated mechanical handling from the field to the on-farm storage bin or drier (Figure 1-3), to a truck or railcar for transport and dumping in a primary elevator, where the grain is graded. After grading and possible cleaning, the grain can be sent to a transfer or terminal elevator for loading into the hold of a saltwater vessel for shipment to an international market. After delivery to its destination, the grain again may be handled several times. Repeated handling of grain does not result in damage by bruising, and deterioration from cracking is minimal.

Usually, a 15% moisture level is safe for storage, and it is the standard moisture level and weight for trading purposes. At this level grain can be readily stored if reasonable caution is exercised.

Moisture levels above 15% usually can be dried down in the field or by forced-air mechanical driers without serious kernel damage. Mature, harvested grain remains as a living entity, but respiration occurs at such a slow rate, compared to that of fresh fruit and vegetables with a higher water content, that elaborate controlled-atmosphere storage units are not needed. The slow deterioration rate of cereal

Figure 1–2. Cereal grain flows readily under the influence of gravity, a feature that greatly assists in handling. (Ranson Photographers Ltd., Edmonton)

Figure 1–1. Cereal grain is an ideal package that can be handled mechanically without damage. In this photo corn is being machine picked by a single operator. The mechanical corn picker removes the ear from the stalk, removes the husks, and separates the grain from the cob. (Courtesy of Ontario Ministry of Agriculture and Food)

Figure 1–3. Heat drying to remove excess water in cereal grains is required for safe storage, especially with full-season crops such as corn and sorghum produced in short growing seasons. The tower on the right is a continuous-flow drier. Dried grain can be elevated and distributed to selected storage bins for shipment or on-farm use. (Courtesy of Ontario Ministry of Agriculture and Food)

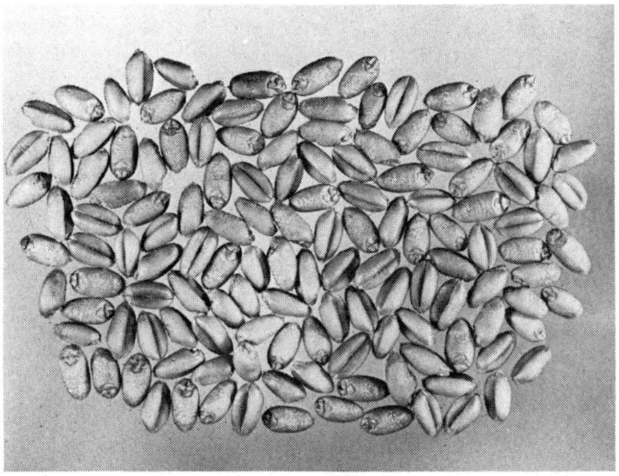

Figure 1–4. Kernels of wheat. The naked, dry kernel has a protective but edible seed coat that makes it a concentrated food form ideal for transport and storage. (Courtesy of Ontario Ministry of Agriculture and Food)

Figure 1–5. Kernels of barley showing the tightly adhering hull that accounts for 12% of the weight of barley. The hull protects the barley kernels, serves a useful purpose in the malting and brewing trade, and does not detract from the value of barley in shipping. (Courtesy of Ontario Ministry of Agriculture and Food)

grains means that storage over several years is possible.

The homogeneous nature of cereal grains, partly due to repeated handling and mixing, allows for accurate standardization and grading for entry into the international grain trade. This feature has resulted in the domination of cereal grains in the international market system.

Wheat and sorghum have a naked kernel with no adhering **hulls,** making these crops economical to transport (Figure 1-4). Rice grain needs some light, intermediate processing to remove the hull and clean the kernel. Hulled barley has a tightly adhering lemma and palea that accounts for 12% of the weight of barley (Figure 1-5). In the case of corn, the grain adheres to a cob but the kernels can be removed readily to facilitate handling (Figure 1-6). Only oat kernels, which have a hull comprising 30% of their weight, are considered to require too much space in proportion to their weight to be moved economically to distant markets. The adhering hull on oat kernels is illustrated in Figure 1-7.

Quality Food Source

Cereal grains enjoy a high status on the food acceptability scale and are free from any historical or cultural stigma. As a result, diets are dominated by cereal grains.

Generally, cereal grains have been regarded as a starch source of **carbohydrate** to supply food energy to the diet. As such, cereal grains

Figure 1–6. Two ears of dent corn lying on a bed of kernels. The grain of corn adheres to a cob, but the kernels can be removed readily to facilitate handling and to concentrate storage space. (Courtesy of Ontario Ministry of Agriculture and Food)

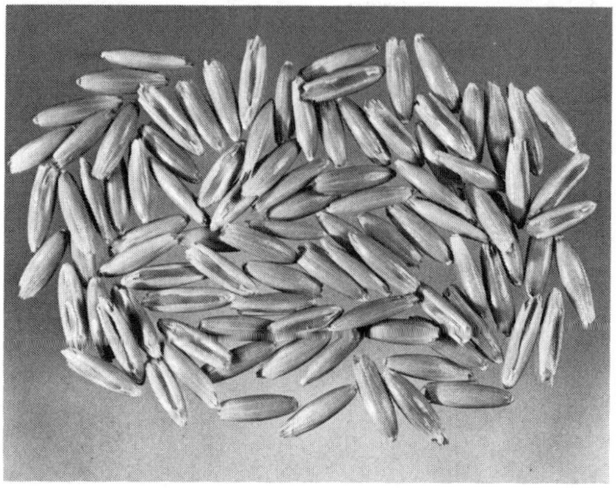

Figure 1–7. Oat kernels showing the adhering hull. About 30% of the weight of oat consists of hull, making oat too light and requiring too much space to be moved economically to distant markets. (Courtesy of Ontario Ministry of Agriculture and Food)

provide the bulk of energy consumed on earth, with rice and wheat the predominant sources (Table 1-7).

The value of cereal grains as a protein source often has been overlooked. Because of the vast volume of cereal grains consumed, cereal grains provide 60% of the protein consumed, the largest source in the human diet (Wittwer, 1977). The share of protein coming from cereals, however, varies widely among geographic regions, from a low of 24% in North America to a high of 74% in Asia.

Recognition of the importance of cereal grains as a protein source, as well as of the fact that cereals normally are deficient in the essential **amino acids** lysine and methionine, led to efforts to improve the amino acid composition. Successful breeding and selection programs have substantially increased the **biological value** of cereal protein. Prior to this breakthrough, a varied diet consisting of animal products or protein from several plant sources was needed to obtain the essential amino acids. Cereal grain products can yield more essential amino acids per unit area than that from animal products.

An ideal diet may consist of an annual intake of not less than 454 kg (1000 lb) of grain equivalent per person. Of the 20 most populous countries, only three exceed 454 kg (1000 lb) cereal grain intake annually (Table 1-3). A suitable diet might consist of a simplified diet rich in whole-grain cereals, fresh fruit, and vegetables and low in fat-rich livestock products.

Protein deficiencies in the world are sufficiently severe that all sources should be examined carefully. Protein comparisons between cereal grains and staples such as the potato favor the cereal grains. Because of their high water content, potatoes appear to have a low protein value and have acquired the undeserved stigma of being a poor man's starch food. The potato contains 70% to 80% water, which increases its bulk compared with such foods as wheat, rice, and corn. The reported 2% to 3% protein content of the potato makes it appear inferior to the 7% to 15% protein of the cereal grains. But few cereal grains are eaten uncooked. When rice, for example, is boiled, water is absorbed and the bulk more than doubles. The protein content decreases proportionately.

Cereal grain starch is easily digested in a cooked, raw, or semiprocessed form. An exception, however, occurs in people suffering from a malabsorption problem known as the sprue syndrome or as **celiac disease** in children. In such

Table 1–7
Sources of Food Energy in the Human Diet

Food	*Energy Supplied, %*	
Cereal grains	56	
Rice		21
Wheat		20
Corn		5
Other cereals		10
Roots and tubers	7	
Potatoes and yams		5
Cassava		2
Fruits, nuts, and vegetables	10	
Sugar	7	
Fats and oils	9	
Livestock products and fish	11	
Total	100	

Source: Brown (1974).

instances, gluten, a form of protein common in wheat, has a toxic effect on the small bowel of patients. Strict elimination of gluten from the diet may restore bowel function.

In addition to being a source of carbohydrates and proteins, cereal grains provide fats, minerals, and vitamins to the human diet.

Major grocery chain outlets in developed areas of the world feature from 5000 to 12,000 food products and brands, three-quarters of which are produced from no more than eight crop species. Corn, rice, wheat, and oat, for example, are processed into more than 150 brands of breakfast cereals.

Hard wheat has the type and combination of proteins, specifically known as gluten, required to produce a high-quality leavened bread. Rye and triticale are the only other grain crops that will yield a true bread flour. The flour of rye, however, is darker, heavier, and coarser than flour from wheat. Cereal crops that lack gluten and hence the ability to produce a leavened product have found wide acceptance in unleavened breads, tortillas, chapattis, and pancakes.

The starch carbohydrate stored in cereal grains can readily be fermented into alcohol for use in beverages or as a fuel source. The abundance and high yields of corn in North America make this the preferred source for alcohol production. Rice can be used to produce a distinctive product called saki. Alcohol produced from cereal grains can be flavored and colored to produce a number of products and tastes such as bourbon, scotch, or rye whiskey.

The technology exists for converting starch grains to ethanol for combustion, and considerable capacity could be developed to convert surplus grains to fuel. Generally, the principal use of ethanol is to extend deficient fossil-fuel supplies by blending. In the United States, corn has a high national production and availability, an important consideration for large distilleries. Grain sorghum has an ethanol yield per unit of grain similar to that of corn and probably is second choice among the grains. World demand for wheat and rice is likely to limit their use as energy crops. The area seeded to oat is too limited, and the yield of ethanol per unit of grain and per unit area is low (Hodgson, 1982).

Most cereal grains can be processed into malt to be used in a number of products and in the brewing process. Malt is produced by germinating the seed. In the germination process, the enzyme amylase is produced and converts the starch, comprised mainly of amylose and amylopectin, into sugar.

Livestock Feed

Since 1950 there has been a very rapid increase in the use of cereal grain for livestock feeding (Abercrombie, 1982). As affluence in a country increases, the North American pattern emerges of eating more animal products from grain-fed beef, dairy, hogs, and poultry and less bread, rice, or potatoes. Under conditions of surplus grain production the consumption of animal products can occur.

Livestock feed from cereal grain crops may take the form of whole grains blended with protein to produce a well-balanced energy ration, of by-products from the processing industry, or of whole-plant forage either as dry feed, ensiled wet, or as a grazing crop.

In the southern Great Plains of the United States, fall-seeded cereal crops are grazed during the winter months when the plants are in the seedling stage. Controlled grazing allows efficient utilization of the cereal crop as a forage, yet allows for the same fields to be harvested later for grain. This dual-purpose cropping system can reduce grain yield and is not recommended where winter survival is critical and in short-season areas.

Corn and sorghum are used extensively as whole-plant silage because the high carbohydrate level found in the grain and vegetative parts allows for ready fermentation into organic acids that serve as a preservative. Silage production is fully mechanized and results in a minimum of spoilage and waste; also, because cereal crops are annuals, complete removal is done at the proper stage of development without the concern about subsequent regrowth or winter survival associated with perennial forages. Dry-matter yields of forage cereal crops are competitive with perennial forage yields in which several cuts may be taken each season.

Corn offers livestock producers considerable versatility in producing feed commodities such as dry shelled corn, high-moisture grain, corncob meal, center-cut forage, whole-plant silage, or dry stored feed.

Livestock numbers of the earth exceed by two to three times the human population (Wittwer, 1977). Livestock constitute a food reserve and, when viewed as converters, harvesters, and gleaners of products that might otherwise go to waste, they add to man's food supply.

Genetic Variability

The wide range of genetic variation that exists in the cereal crops not only has provided humans with an opportunity to select plants that have met past needs but also offers considerable scope for future plant improvement. The seed shattering habit, useful in nature but of no value to humans and a hindrance to efficient harvesting, has been eliminated because genetic variability existed for nonshattering cultivars.

The **perennial** habit, perhaps once found in cereal crops, mainly has been eliminated except for some degree of tillering and the **winter annual** habit. Presumably constant selection pressure for annual growth habit emphasized cultivars that matured each season with reduced risk of loss by winter kill, seasonal flooding, or other peril.

Genetic variability for cross- and self-pollination existed, and, because early agriculturists favored plants with a full seed set, a shift to self-fertilization may have been favored. Cross-fertilization has been retained in some cereal crops such as corn and rye, perhaps because more emphasis was placed on vigor than on reliability.

Selection for heavy grain was high on the list of priorities. Genetic variability allowed for a fivefold increase in weight per grain in the evolution of wheat from a wild grass (Evans and Dunstone, 1970). An increase in excess of 3.6 times in seed size from a primitive and to a cultivated wheat can be observed in Figure 1-8.

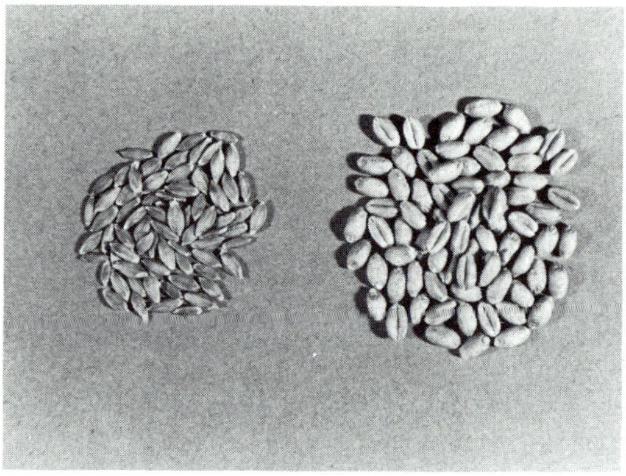

Figure 1–8. Breeding and selecting for seed size resulted in an increase of over 3.6 times from a wild form of wheat. Einkorn is on the left and a present-day common wheat cultivar on the right. (Courtesy of Ontario Ministry of Agriculture and Food)

The vast amount of variation that exists among the cereal crops means that cultivars can be found to suit many conditions around the world. Tall or short, early or late maturing, winter or spring growth habit, wide or narrow leaves, and numerous other variables can be found. Rice has the genetic variability to be adapted to a wider range of environmental conditions than any other crop.

Although sorghum is widely distributed in the tropical and subtropical regions of the world, cultivars adapted to temperate zones have been developed. The full potential of sorghum is yet to be realized.

Likewise, the first man-made cereal crop, triticale, has added new genetic variation to cereal crop resources. The role of triticale among world crops is yet to be determined, but it appears promising on poor soils or at high elevation.

The genetic diversity required to provide cultivars for worldwide distribution over an enormous range of environments is impossible to calculate.

Protective Seed Cover

All cereal crops are included in a botanical grouping called **angiosperms** characterized by seed produced with an ovary. Angiosperms are considered to be a more advanced evolutionary form than **gymnosperms,** a grouping of plants with naked seeds. Seed of cereal crops is enclosed to a varying extent in a protective cover of **glumes** or husks while growing on the plant. Rainwater is prevented from weathering or spoiling the seed, direct attack by pests is reduced, and protection is provided against contamination by **pesticide** applications.

Primitive cultivars of corn exist with glumes over each kernel as well as protective outer husks. Selection has substantially reduced glume size, but protection after maturity against adverse weather is afforded by a change of the ear from an upright to a suspended position and by the tightly adhering husks.

Except in oat and barley crops, cereal crops are free-threshing; that is, the glumes and husks are readily removed. Although it is possible to select hullless oat and barley cultivars, the adhering hull has come to be accepted or even preferred for feed and malting purposes, respectively.

Ecologically Safe Crops

A large portion of cereal grains are consumed directly as human food. From a health standpoint it is important, therefore, that they be free of pesticide contamination. Cereal crops are, for a number of reasons, ecologically safe crops.

Although cereal crops are plagued by their share of diseases and pests, extensive chemical pesticide use on cereal crops is restricted because it cannot be justified economically. In contrast, high-value fruit and vegetable crops require frequent and timely spray applications to reduce or prevent pest attacks and the resulting deformities, blemishes, and loss of cosmetic appearance and yield. Often, too, cereal crop products undergo processes that remove protective outer contaminated layers.

Genetic variability found in cereal crops allows for selection and use of genetic resistance. This, coupled with an annual growth habit (as compared to the perennial habit of fruit trees, such that resistant cultivars cannot be easily seeded), has reduced the need for pesticides.

The observation that cereal crops are ecologically safe crops is important in the context of increasing crop yields to meet future food needs. In general, as crop production is intensified to increase yield, the biological constraints (such as pest and lodging) to yield increases become stronger. Additional inputs are required to overcome these constraints. Increases in cereal crops are expected through more intensive production without approaching limits set by biological constraints. Cosmetic appearance of cereal grain is of minor importance, and this removes many constraints.

When cereal grains or their by-products are used for livestock production, animals serve as a **biological buffer.**

IMPROVEMENT OF CEREAL CROPS

The perfect plant species or cultivar does not exist and probably cannot be expected to be developed, because changing situations provide continuing shifts in emphasis, priorities, or problems. A constant reexamination of each crop is needed so that breeding, selection, production, and management objectives and procedures can be upgraded. Crop improvement has been going on since plants were first domesticated, and further progress is anticipated, because technology has provided exciting new genetic variability.

This section describes improvements that are now being developed; they will be described in more detail in subsequent sections. The objective in describing possible future improvements is to stimulate ideas and not to enter into the realm of science fiction.

Improved Bioconversion of Solar Energy

The fact that approximately 95% of all plant dry matter is produced through the process of photosynthesis, with 5% being contributed from plant nutrients, emphasizes the need for improved plant conversion of light energy through the photosynthetic process. Present conversion rates seldom exceed 1% even under the best of conditions. There are many options for improvement, for this is what agriculture is all about.

One option is to transfer the C_4 photosynthetic mechanism into C_3 plants. There must be a means by which the enzymatically controlled C_4 mechanism can be regulated chemically, genetically, or by both methods.

Biological Nitrogen Fixation

Ranking second in importance to photosynthesis is the naturally occurring process of gathering nearly inert nitrogen from the atmosphere and producing nitrogen that is available to plants. The contribution to agricultural productivity worldwide by nitrogen fixation, found mainly in legumes, probably exceeds by four or five times all the nitrogen fertilizer currently fixed by chemical means utilizing high pressures and temperatures and a massive fossil-fuel input. The high costs of producing nitrogen fertilizer have turned attention toward biological nitrogen fixation in grasses and specifically wheat, rice, and corn. In areas of the tropics where soluble nitrate nitrogen is quickly leached out or where shifting cultivation depletes soil fertility, biological nitrogen fixation is an attractive thought.

Preliminary evidence does exist for the development of successful nitrogen-fixing relationships between bacteria and cereal grasses (Hardy, 1975; Neyra and Dobereiner, 1977).

If biological nitrogen fixation in cereal grasses could be achieved satisfactorily, hopefully it would not entail a significant loss in grain yield. Like the industrial process of nitrogen manufacture, biological nitrogen fixation is an energy-intensive process. In the legume plant, 10% to 20% of energy produced in photosynthesis is directed toward nitrogen fixation (Hume, 1977). It may be incorrect, however, to conclude that yield could be increased by a comparable amount if legumes did not fix nitrogen. Plants do not obtain nitrogen without an energy cost. An amount of energy equal to that used in fixation may be required to assimilate nitrogen supplied as commercial fertilizer.

Perennial Cereal Crops

The extensive use of annual cereal crops means that soil preparation, seeding, weeding, and seed storage is an annual feat and one that leaves the soil exposed and unprotected between crops. The availability of perennial cereal crops would minimize erosion problems and greatly reduce production costs.

Modern humans have accepted the decision made by early agriculturists 10,000 or more years ago to produce annuals. Perhaps this decision was based on the annual deposition of soil in river deltas that obliterated perennial crops. Given another set of circumstances, early agriculturists might have emphasized perennial cereal plants. Surely perennial forms existed, as evidenced by the tillering habit considered as a remnant form of **perennation,** by the winter annual habit associated with most spring forms, and by wild perennial cereal relatives.

A serious modern agricultural problem is the reduction in soil productivity associated with intensive cropping systems using annual cereal crops. Frequent tillage for reseeding cereals, coupled with complete residue removal to facilitate tillage, has reduced soil organic matter, fertility, and water penetration. Soils are exposed, and unprotected by a cover crop over winter in temperate zones. The use of machines leads to soil compaction. The result is that greater inputs are required to overcome biological restraints. The use of perennial cereal crops may lead to a more sustainable agriculture (Jackson, 1981).

Increased Lodging Resistance

The potential **lodging** hazard (Figure 1-9) in most cereal crops, especially the fine-stemmed cereals such as barley and oat crops, has limited cereal yields. In areas where use of commercial fertilizer has increased, the need for lodging resistance has intensified. Even in the large-stemmed cereal crops such as corn, lodging problems have increased, especially as corn is moved into short-growing-season areas, by the build-up of root worm and stalk rots associated with intensive production and by the

Figure 1–9. A barrier to yield increases is the inability of the plants to support the ever-increasing weight of economic yield; they fall over or lodge. The oat crop in the photo has lodged, and a yield reduction can be expected because of poor light interception and reduction in photosynthesis. (Courtesy of Ontario Ministry of Agriculture and Food)

development of thinner stalks associated with high plant populations.

Despite progress toward stronger-stemmed cereal crops through breeding for mechanical strength, often coupled with reduced plant height that reduces the wind torque on the lower stem portion, the lodging problem has not diminished.

Improved Pest Resistance

Intensive agriculture, often associated with large geographical areas seeded to a single cultivar containing specific genetic resistance to a specific pest or strain of a disease, has increased the vulnerability of these areas. The constant threat of a new virulent strain of a new disease pest means that more durable disease resistance is of high priority. Destruction of a crop in a region could spell famine for millions of people.

With the exception of rice, many small cereals cannot be grown on the same land year after year without a serious build-up of soil-borne pests and diseases. Multiple cropping, necessary to increase food output, has placed severe pressures on such cropping systems. New forms of resistance are required at a pace that rivals the rate at which new genetic resistance can be obtained and incorporated into suitable cultivars.

Durability of existing cultivars may be achieved by combining several or all means at our disposal. These include cultural, general, and specific genetic resistance, chemical pesticides, **multilines,** diversification of crops, sanitation, and rotational measures.

Conservation of Plant Diversity

Genetic improvement is dependent upon diversity of plant types. Gene banks to assure conservation of plant diversity must be made a worldwide project if collections are to be thorough and if **germplasm** depletion is to be avoided.

Quality Aspects

The discovery of wheat, rice, and corn with improved amino acid composition has awakened the world to opportunities for improvements in quality aspects. Balancing amino acid composition in the major cereal crops remains a challenge in terms of finding specific genotypes and management practices to avoid a yield reduction.

Problems associated with the "control" of germination prior to harvesting need attention. Preharvest rain may cause sprouting in wheat, sorghum, rye, barley, oat, triticale, and even in rice. Unwanted germination is detrimental to the producer and the processing industry, as it causes yield losses, reduced test weight, and unwanted chemical changes that reduce quality.

The new crop, triticale, is one of the most sensitive cereals to preharvest sprouting. Triticale may sprout prior to harvest maturity without any rain (Derera, 1980).

Despite extensive interest in it, this problem is so complex that the fundamentals still are not understood.

Quality losses occur when the vitreous texture of hard wheats changes to a more "mealy" one, as sprouting stimulates alpha-amylase that converts starch to sugar.

A third quality concern is celiac disease, whereby afflicted people cannot digest the protein gluten found in wheat, rye, barley, and oat. One in 3000 Canadians suffers from this disease, while in the United States and Australia it is practically nonexistent. In the British Isles the incidence is one in 300 (Rickerd, 1982). Sufferers from celiac disease must abstain from such foods as bread and all products made from wheat, rye, barley, and oat.

Hard red spring wheat is the dominant wheat grown in Canada, and hard red winter wheat dominates in the United States. Both wheats can be used to produce bread. Observed differences between Canada and the United States in terms of the incidence of celiac disease may relate to the different genetic sources. The lineage of winter wheat cultivars in North America relates to Turkey Red; the lineage of spring wheat relates to Marquis (Buller, 1919; Quisenberry and Reitz, 1974). Use of electrophoretic techniques to measure protein gluten heterogeneity to identify genetic differences between Marquis and Turkey Red may be useful in determining factors associated with the celiac syndrome.

Questions

1. *a.* Speculate on events surrounding the domestication of plants. Give consideration to how seed for food and seed for replanting occurred; the problems associated with seed shattering; the impact of the sickle on production; and seed storage for next year's crop.
   b. Consider the statement that humans are dependent on crops, but domesticated crops are dependent on humans.
2. As few as 17 plant species provide nearly 90% of the world's food supply and occupy about 75% of the total tilled land on earth. Name additional crops; consider their importance in providing food and why they have not gained greater prominence.
3. It is suggested that humans are not likely to increase the number of food crops on which they are dependent, but rather they are likely to continue improving the few plant species on which they now depend. On the basis of this statement, speculate on the future of triticale, the first man-made crop. What features might triticale have to exhibit as a grain crop in order to become an important food crop? What special features might help it reach importance?
4. *a.* Consider the oat crop as one of the eight major grain crops in terms of past, present, and future importance.
   b. If triticale were to become an important crop, what impact might it have on the oat crop or the rye crop?
5. *a.* Name cultivars of grain crops commonly grown in your area.
   b. Describe and compare features of commonly grown cultivars of the same species. How distinctive and identifiable are individual cultivars? Can a specific cultivar be identified by the seed alone, in the vegetative stage, or in the reproductive phase?
6. *a.* Consider your food sources for a single day. How many of the foods you consume are produced directly or indirectly from cereal crops?
   b. Does a day ever pass in which you do not consume some cereal crop products?
7. Speculate on the potential of biological nitrogen fixation in cereal crops.
8. Compare cereal crop yields for your region with those shown in Table 1-4.

9. *a.* Consider factors that influenced yields to increase or decrease as shown in Table 1-4.

   b. What factors are most important in yield changes in your area?

10. Consider the merits and problems of crop production if perennial cereal crops were developed.

11. *a.* What is the average life span of a cereal crop cultivar?

    b. Why are older cultivars replaced?

References

Abercrombie, K. 1982. "Intensive Livestock Feeding: Reckless Folly or an Emergency Food Reserve?" *Ceres* 15 (1): 38–41.

Biswas, A. K., and M. R. Biswas. 1976. "Energy and Food Production." *Agro-Ecosystems* 2: 195–210.

Blum, A. 1973. "Components of Analysis of Yield Responses to Drought of Sorghum Hybrids." *Experimental Agriculture* 9: 159–167.

Brown, L. R. 1974. *By Bread Alone*. New York: Praeger Publishers. 272 pp.

———. 1978. *The Twenty-Ninth Day*. New York: W. W. Norton & Company, Inc. 200 pp.

Buller, A. H. R. 1919. *Essays on Wheat*. New York: The Macmillan Company. 339 pp.

Burger, W. C. 1981. "Why Are There So Many Kinds of Flowering Plants?" *BioScience* 31: 572, 577–581.

Chan, L. M., R. R. Johnson, and C. M. Brown. 1980. "Relay Inter-Cropping Soybeans into Winter Wheat and Spring Oats." *Agronomy Journal* 72: 35–39.

Chandler, R. F. ,Jr. 1979. *Rice in the Tropics: A Guide to the Development of National Programs*. Boulder, Col.: Westview Press. 256 pp.

Derera, N. F. 1980. "The Audit of Sprouting." *Cereal Research Communications* 8 (1): 15–22.

Duncan, R. R. 1982. "Water Use Efficiency." *Crops and Soils* 35 (1): 19–20.

Evans, L. T. 1980. "The Natural History of Crop Yield." *American Scientist* 68: 388–397.

———, and R. L. Dunstone. 1970. "Some Physiological Aspects of Evolution in Wheat." *Australia Journal of Biological Science* 23: 725–741.

FAO. 1982. *FAO Production Yearbook 1981*. Vol. 35. *Rome*. Food and Agriculture Organization of the United Nations. FAO Statistics Series, No. 40

Hardy, R. W. F. 1975. "Nitrogen Fixating Research: A Key to World Food?" *Science* 188: 633–643.

Harlan, J. L. 1976. "The Plants and Animals That Nourish Man." *Scientific American* 235 (2): 88–97.

Hodgson, H. J. 1982. "Fuel From Crops: Part I, Grain and Sugar Crops." Editor's Forum. *Crops and Soils* 34 (7): 5–7.

Hume, D. J. 1977. "Nitrogen Fixation: Free Fertilizer?" *Information for Industry Personnel*. Ontario Ministry of Agriculture and Food. 141/544.

Jackson, W. 1981. "Investigations in Sustainable Agriculture." *The Land Report*, No. 13: 14–15.

Jones, H. G., and E. J. M. Kirby. 1977. "Effects of Manipulation of Tillers and Water Supply on Grain Yield in Barley." *Journal of Agricultural Science* 88: 391–397.

Morgan, D. 1979. *Merchants of Grain*. New York: The Viking Press. 519 pp.

Neyra, C. A., and J. Dobereiner. 1977. "Nitrogen Fixation in Grasses." *Advances in Agronomy* 29: 1–38.

Quisenberry, K. S., and L. P. Reitz. 1974. "Turkey Wheat: The Cornerstone of an Empire." *Agricultural History* 48: 98–112.

Rickerd, J. P. 1982. "Catering to Special Diets." *Science Dimension* 14 (1): 2.

Touchton, J. T., and J. W. Johnson. 1982. "Soybean Tillage and Planting Method Effects on Yield of Double-Cropped Wheat and Soybeans." *Agronomy Journal* 74: 57–59.

Wellhausen, E. J. 1976. "The Agriculture of Mexico." *Scientific American* 235 (3): 128–150.

Wittwer, S. H. 1977. "Alternatives Available for Improving Plant and Animal Resources." *Proceedings of the World Food Conference of 1976*. Ames, Iowa: The Iowa State University Press. pp. 429–448.

2

Botany and Physiology of Cereal Crops

GENERAL DESCRIPTION AND TERMINOLOGY

All true cereals are members of the grass family, one of the largest and economically the most important family in the plant kingdom. Grasses are found everywhere plants grow; from sea level to the upper limits of vegetative growth on mountains; from the tropics to the farthest polar regions; from wet, humid areas to dry, desert regions.

Cereal crops are grown for their edible fruit, botanically called a caryopsis but popularly referred to as a grain, a kernel, or a seed. The seed is a single fruit consisting of two major components, the **endosperm** and the **embryo** or germ. The endosperm is stored food and comprises the bulk of the seed. It may be used for human food purposes or by the embryo as an energy source for germination. It is not surprising, then, that cereal crops are of universal importance in meeting humans' food needs.

Endosperm development is initiated by the fusion of the sperm nuclei in the pollen with two polar nuclei in the ovary. Essentially, double fertilization is required for cereal seed development.

The endosperm is encased by an outer wall called the **pericarp** that develops from the ovary wall. Under the pericarp is a semipermeable layer called the **testa** that surrounds the embryo and is derived from the inner ovary wall. The testa is permeable to water but not to dissolved salts and is important for germination.

A third layer of thick-walled cells called the **aleurone** layer is usually present. Unlike the starchy endosperm cells, aleurone cells are free of starch. In barley, the aleurone layer is three or four cells thick as compared to one cell thick in other grasses. The aleurone cells in barley are the source of diastatic **enzymes,** which enable the young plant to utilize the starch present in the endosperm. Barley is a preferred grain for malt production because the more abundant aleurone cells supply enzymes to convert starch to sugar.

The pericarp, the testa, and the aleurone layer are collectively called the **bran.**

The major true cereals are wheat, rice, corn, barley, oat, rye, sorghum, millet, and the manmade crop triticale. Minor cereals include wild rice, Guineagrass, mannagrass, and other tropical grasses from which the grain may be used as food. A few plants that are not grasses, such as buckwheat, flax, and amaranth, also produce seeds similar to the true cereals.

Grain crop is a collective term including the true cereals, as well as buckwheat, flax, amaranth, and even soybean and other seed-producing legume crops. The term *cereal,* derived from the Greek goddess Ceres, or giver-of-grain, is applied to the entire plant, to the grain, and to grain products. In some areas of the world the term *food grains* includes wheat and rice and the term *feed* or *coarse grains* is applied to corn, barley, oat, sorghum, rye, and millet. There is no firm criterion for this loose classification, and triticale may be termed as either a food or feed grain. "Coarse" implies that the seed has a hull, that it is intended primarily for feed, or that it has one or more coarse or undesirable features. In India and China, sorghum is regarded as "coarse" and to be consumed by those who cannot afford wheat or rice.

Included among coarse or feed grains is a mixture of oat and barley produced by growing these crops simultaneously together. The inclusion of wheat in a mixture does not improve the status above that of a coarse grain.

The term *large grains* may refer to corn and sorghum because of their large plant size, with the remaining cereals being referred to as *small grains*.

When cereal crops are underseeded with forage grasses or legumes, the cereal crop is termed a **companion crop. Ratoon cropping,** or ratooning, implies more than one harvest from a single planting, a situation possible in tropically grown rice and sorghum.

Cereals grains should not be confused with grain legumes such as cowpeas, peas, beans, lentils, or soybeans. Oilseed crops such as peanut, canola, rapeseed, flax, sunflower, and safflower are not cereal grains. Oat is not an oilseed crop, even though it has a fat content of 4.3% on a whole-grain basis.

BOTANICAL ASPECTS

All cereal crops are **monocotyledonous** plants having one **cotyledon** in the seed in contrast to **dicotyledonous** plants, such as soybean, flax, buckwheat, peanut, or bean, which have two cotyledons. The cereal cotyledon is sometimes called a seed leaf and botanically is a **scutellum** rather than a cotyledon.

Cereal crops are classed as **annuals,** as they complete their cycle of growth from seed in one year. Those seeded in the spring are spring cereals. Most spring cereals have a winter annual counterpart. These cereals are seeded in the fall, pass through the winter in a dormant state, and complete their cycle of growth the following year. In regions where the winters are mild, typical spring cultivars may be grown as winter annuals.

The pattern of cereal growth is typically determinate; i.e., after a period of vegetative growth the plant switches to reproductive development, and vegetative growth ceases. During the vegetative growth period, internodes elongate to form a stem. The lower internodes are short, but as growth through cell division and cell elongation gets progressively faster, upper internodes are extended. With the onset of **anthesis,** reproductive growth begins, with fertilization and grain development continuing until **physiological maturity** is reached.

The stem of the cereal crop, typically called a **culm,** is cylindrical and, except in corn and sorghum, is hollow, except for the solid nodes or **joints.**

Leaves of grasses consist of two parts, the sheath and the blade. The sheath arises from each node, and a flat elongate or straplike blade with parallel *venation* extends from the sheath. At the junction of the sheath with the blade, a distinct membranous fringe called the **ligule** may be found (Figure 2-1).

At the juncture of the leaf blade with the sheath, the edges of the blade may extend into distinctive structures known as **auricles.** Large, clasping auricles characterize the barley plant (Figure 2-2).

The head or **inflorescence** of cereal crops consists of flowers grouped on a central axis, called a **cob** in corn and a **rachis,** an extension of the stem, in other cereals. The grouping of flowers is called a **spikelet.** These spikelets are variously arranged; if they grow from a unbranched main stem, the inflorescence is a **spike** (Figure 2-3). A **panicle** consists of a main stem with branches and subbranches arising from a central axis and with spikelets borne on a **pedicel.** In corn the seed-bearing organ is an **ear.**

The inflorescence of wheat, triticale, rye, and barley is a spike; of oat, rice and sorghum, a

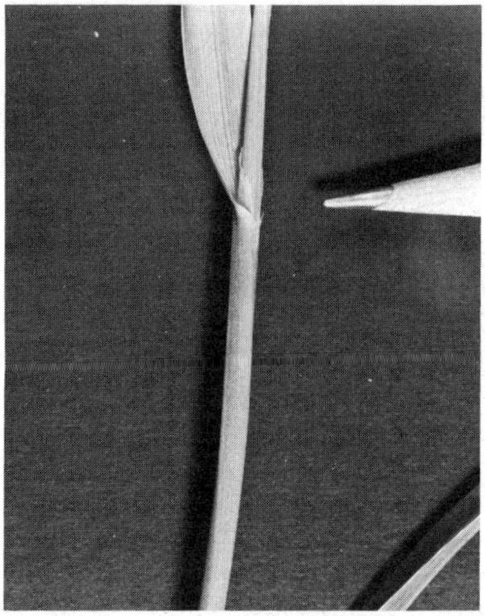

Figure 2–1. Section of a grass shoot showing the sheath and leaf blade arising from the sheath. The pencil is pointed to the ligule, a membranous fringe at the inner junction of the sheath. The size of the ligule may serve as a feature for identification. Note the parallel veination of the leaf blade, typical of members of the grass family to which cereal crops belong. (Courtesy of Ontario Ministry of Agriculture and Food)

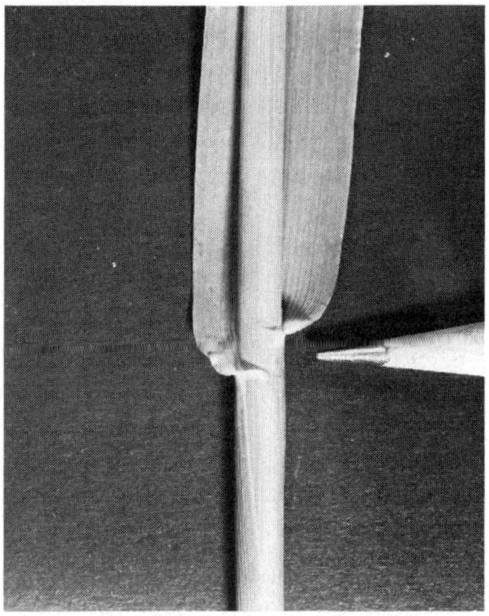

Figure 2–2. Section of a barley shoot showing the large, clasping auricles that characterize this crop. Auricles appear as extensions of the edges of the leaf blade. The ligule is hidden behind the culm, but a small portion is visible on the right-hand side immediately above the right auricle. (Courtesy of Ontario Ministry of Agriculture and Food)

panicle; of millet, mostly a panicle, but in some a spike. In barley, both six-row and two-row cultivars are found. In six-row cultivars, all three spikelets at a rachis node are fertile. Groups of three spikelets arranged alternately at opposite sides of the rachis nodes give the appearance of having six rows of grain on the spike. In two-row cultivars, only the central spikelet is fertile, giving the appearance of having two rows of grain, one on either side of the rachis.

Spikes may be dense, mid-dense, or lax, depending on the distance between rachis internodes. A code number may be used for reporting spike laxness ranging from 0 to 9 (Figure 2-4). A compact or dense spike having nine rachis internodes 2.5 mm apart will have a spike length approximately 22.5 mm (1 in.) and is given a 1 coding. A mid-dense spike with 1.8 rachis internodes each 5.0 mm apart is coded 5 and has a spike 90 mm (3.5 in.) in length. A lax spike with

Figure 2–3. Two views of the inflorescence of wheat called a spike. The spike of wheat consists of a series of spikelets arising from an unbranched main stem or rachis. (Courtesy of Ontario Ministry of Agriculture and Food)

Spike laxness

Enter code number which best describes spike laxness.
Rachis internode length of some classes is given as a guide.

Code		Guide
0, 1, 2, 3	dense	(2·5 mm)
4, 5, 6	mid-dense	(5·0 mm)
7, 8, 9	lax	(10·0 mm)

Figure 2–4. Documentation of plant genetic stocks includes information relative to the identification, origin, and genetic makeup. Standard data recording methods might include guides for recording the important features of each specific crop. An example of the type of recording guide was suggested by Konzak and Dietz (1969) and is shown here. The figure also serves to illustrate the marked range in variability for spike laxness found in wheat.

12 rachis internodes each 10 mm apart is coded 9 and has a spike length of 120 mm (4.75 in.) in length (Konzak and Dietz, 1969).

Spikelets are enclosed in bracts. The two lowermost are called **glumes** or **chaff** (Figure 2-5). The next bract above the glumes is the **lemma.** Conspicuous prolongations on the glumes or the lemmas are called *awns*. The lemma may be fully awned, partly awned, tip-awned, or awnless. The glumes in Figure 2-5 are partly awned. Several degrees of awn development are shown for barley in Figure 2-6.

On the vertical part of the seed is the **palea.** The lemma and palea form two bracts that enclose the reproductive organs to comprise a **floret.** Such a structure forms a **perfect flower** because it contains both male and female reproductive organs. The spikelet in Figure 2-5 had four fertile florets and an infertile fifth floret. Second-order florets arise from the first-order florets; third-order florets arise from second-order florets. The number of seed-bearing florets per spikelet is an indication of the size of the developing grains (the *sink*), which may reflect yield capacity. The number of spikelets per inflorescence and the number of fertile florets per spikelet depends on the cultivar, the environmental conditions affecting photosynthesis, and the supply of photosynthate to the sink.

Individual spikelets arise from rachis nodes (Figure 2-5) on a short axis called the **rachilla.** The lemma arises from the rachilla and the palea from a short pedicel or flower stalk. In some cereals, such as wheat, rye, and triticale, the lemma and palea readily separate from the grain when threshed. In rice, the lemma and palea are removed by a mild rubbing action. In oat and barley, the lemma and palea, as well as the rachilla, may adhere to the grain. Distinctive features of the rachilla may be useful in identifying a specific cultivar from the seed alone.

Within the floret are **stamens** or the pollen-

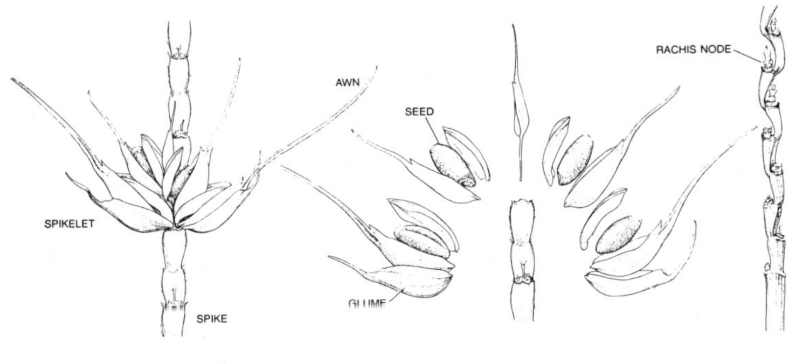

Figure 2–5. Detail of a wheat spikelet. On the left is a single spikelet attached to an internode on the rachis. The center drawing shows the spikelet exploded to show two outer glumes or chaff, the free-threshing lemma and palea encasing each of the four kernels, and a central sterile floret for a total of 5 florets contained in one spikelet. Note the conspicuous prolongations on the glumes or the lemmas known as awns. A naked rachis on the right shows the alternate nodes. (From M. Feldman and E. R. Sears (1981), "The Wild Gene Resources of Wheat," © January 1981 by Scientific American, Inc. All rights reserved.) *Scientific American* 244 (1):105.

Figure 2–6. Degree of awning in barley. On the left the barley spikes have an awn arising from each lemma that adheres to the barley caryopsis. These spikes are said to be fully awned. The spikes in the center are awnless. On the right, the fully awned spikes exhibit an extreme amount of awn development because the rudimentary outer glumes each have produced an awn. All spikes shown are six-row. Six-row barley seed can be distinguished from two-row barley seed by the fact that two-thirds of six-row barley seeds are lateral seeds that usually exhibit a typical shape relative to the main seed. In two-row barley all seeds are main seeds. (Courtesy of Ontario Ministry of Agriculture and Food)

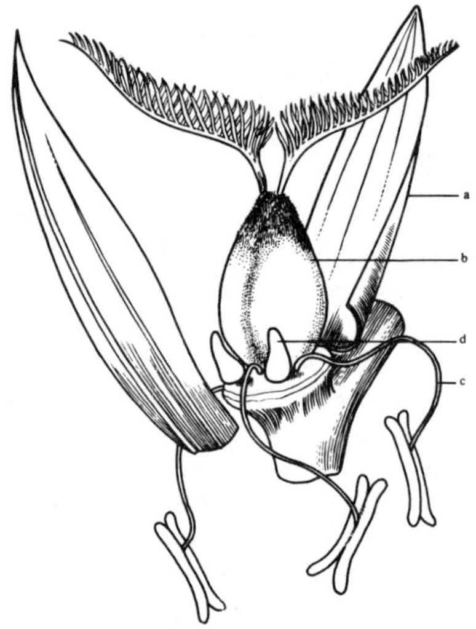

Figure 2–7. Stamens, pistil, and lodiculae are enclosed by the lemma and palea to form a floret in cereal grasses. Stamens produce pollen, which falls on the feathery stigma of the pistil. Swelling of the lodiculae (d) at the base of the floret will push open the glume (a) to facilitate cross-pollination. Stamens consist of a pollen-bearing anther held on a filament (c) or stalk. The stigma, style, and ovary (b) make up the pistil. (Courtesy of The Canadian Seed Growers' Association)

producing male organs, a **pistil** or female organs, and two **lodiculae** (Figure 2-7). Stamens in each floret of cereal grasses are found generally in threes and sixes, never in fives. Rice, for example, has six stamens; wheat has three.

Each stamen is composed of a filament or stalk which bears an anther. When pollen is ready, drying and shrinking causes slits to form on the sides of the anthers, and pollen is shed. At this point the filament or stalk commonly undergoes rapid elongation to extrude the anthers out of the floret. This action may be useful to effect cross-pollination and may be a remnant of the era when cereals were all cross-pollinated.

The single pistil is composed of an ovary, two styles, and two feathery stigmas. Pollination occurs when pollen lands on the feathery stigmas; fertilization is accomplished when the nucleus of the pollen grain unites with the egg. The pollen grain contains a tube nucleus and two sperm nuclei. The tube nucleus develops a pollen tube that elongates down through the style to the embryo sac. One sperm nucleus unites with the egg in the ovule while the other fuses with the two polar nuclei (termed double fertilization). This mechanism accounts for **xenia,** or the immediate observable effect of foreign pollen on the maternal endosperm tissue. The effect can be seen in the various endosperm colors often produced by foreign pollen when the outcrossed grains ripen.

In cereal crops grown for their grain, successful seed set is important. Without pollination and fertilization, no seed set will occur. In breeding programs and especially in hybrid seed development, the mode of pollination is significant for the plant breeder as well as for seed producers. Cereal plants are normally classed as either **cross-pollinated** or **self-pollinated.** Corn, rye, sorghum and pearl millet are naturally cross-pollinated; wheat, oat, barley, and rice are naturally self-pollinated. Triticale, produced by combining wheat and rye genomes, plants which are self- and cross-pollinated, respectively, is largely a self-pollinated crop.

All cereal crops, however, can be induced to either cross-pollination or self-pollination, a situation useful for inbreeding for selection and eventual F_1 hybrid development. The natural occurrence of selfing or crossing depends on a number of factors including flower morphology. The perfect flowers of wheat, rice, oat, barley, and triticale are easily self-pollinated by the falling of pollen from the anthers onto the stigma under the influence of gravity. Although cross-pollination occurs in these crops at a level of only 1 to 4%, depending on the cultivar and environmental conditions, male sterility can be used to prevent self-pollination, and cross-pollination may occur.

Imperfect flowers lack either stamens or a pistil and require cross-pollination for seed set. Corn is described as **monoecious** because it bears both pistillate and staminate flowers on the same plant and can be either cross-pollinated or self-pollinated. In nature, cross-pollination, also called **allogamy,** is common. Although rye and pearl millet have perfect flowers, they are cross-pollinated crops because of mechanisms that discourage or prevent selfing from occurring. In millet, the female reproductive structures develop and mature before the male structures. In this manner, self-pollination is prevented. This mechanism is known as **protogyny.**

Triticale, sorghum, and pearl millet are described as often cross-pollinated cereals. Pollination and fertilization may occur with the floret of the same plant, from another flower on the same plant, or from another plant. If all plants in the field are similar, progeny will be similar whether fertilization was as a result of self- or cross-pollination.

Cross-pollination in cereals occurs by wind transfer of the pollen rather than by insects' carrying of the pollen from the male to the female organs. Cross-pollination can occur only if the lemma and palea are separated to provide an opening. Swelling of the lodiculae at the base of the floret will push open the floret to facilitate cross-pollination. If self-pollination does not occur because of male sterility or because the anthers have been removed by hand, the lodiculae will become enlarged and open the floret for cross-pollination.

CEREAL ROOT SYSTEMS

Two sets of roots normally support the growth of cereal plants. The **seminal,** primary, or seedling roots are produced from the seed at germination. The coronal, permanent, or **adventitious** roots arise from a basal or crown internode just below the surface of the soil. The same crown internode also is the place of origin of the shoots or above-ground parts.

Seeding depth can influence the depth of the seminal roots as they arise from the seed. Seeding depth should not be so great that the primary shoot or **plumule** cannot emerge. The crown from which adventitious roots arise is formed near the surface. Under conditions of shallow seeding, the crown is formed at or a little above the seed, possibly at about the same location as for seeds planted more deeply. Thus deep seeding does not increase the depth of the secondary, coronal, or adventitious root system. The part of the plant between the seed and the crown is the **mesocotyl,** subcoronal internode, or underground stem. The mesocotyl may vary in length from 1 to 10 cm (½ to 4 in.) depending on seeding depth.

Seedlings are nourished by the food energy stored in the endosperm, eventually from water and nutrients secured by the seminal root system and until the adventitious crown or secondary roots are developed. This latter root system greatly surpasses the seminal or seedling roots in size. In corn, Weihing (1935) found the coronal roots to be 15 to 20 times as numerous as the seminal roots.

Seminal roots are often regarded as serving the plant until the adventitious or secondary roots become established. To regard the seminal roots as temporary may be incorrect. In wheat, seminal roots were found to support the main stem of the plant (Locke and Clark, 1924) and may function until the plant matures (Weaver and Zink, 1945). In corn, it is believed that the seminal roots degenerate and cannot be found six weeks after germination begins (Brouwer, 1966; Weihing, 1935). Under dryland conditions, seminal roots may be important in seedling establishment (Locke and Clark, 1924).

Adventitious or coronal roots initially tend to grow horizontally before turning downward. Interrow cultivation may cut these developing roots, reduce nutrient and water uptake, and lower yield. Strong evidence of this in cereals is lacking, however, and interrow cultivation that does cut some coronal or adventitious roots may be beneficial in causing greater branching and proliferation of root growth (Russell, Fehr, and Mitchell, 1971).

The majority of the root system in cereal crops consists of fine, fibrous roots which remain almost the same diameter throughout their length. In corn, adventitious roots may arise from the internodes above the subcrown internode; these are known as brace or anchor roots and are useful in supporting the plant in an upright position.

Roots serve several functions. Major and obvious ones are to anchor the plant to prevent root lodging and to secure water and nutrients. Roots also serve in the synthesis of growth substances, in the partitioning and redistribution of stored photosynthate, and as a site for **nitrate reductase** activity in some species.

Root density, duration, and rate of extension and depth may vary among cereal species and cultivars. Density influences the efficiency with which available soil is explored. Duration and rate of extension determine the ability of the plant to explore new pockets of soil. Depth of penetration is important in exploiting soil moisture reserves, particularly if water becomes limited in a period of drought. The rooting pattern may not be associated with crop height. Dwarf cultivars, particularly in wheat, may have more root deep activity than some of the taller cultivars (Lawes, 1978).

Root characteristics of a tall and semidwarf wheat are illustrated in Figure 2-8. These plants were grown indoors, side-by-side, in a container with fertile black soil throughout the entire depth and were given adequate water daily. Identical environmental conditions suggest that observed differences are genetic.

MEASURING CEREAL YIELD AND QUALITY ASPECTS

For many years the bushel was a standard measure for cereal grain weight. The bushel weight was developed in Anglo-Saxon times, based on the Winchester bushel, so named after a cylinder

Figure 2–8. Root characteristics of a tall wheat (right) compared with a semidwarf wheat (left). Both roots appear to penetrate to the same depth, but the density of roots for the tall wheat was maintained at greater depth. Note the fine, fibrous structure of the root system. (Courtesy of Ontario Ministry of Agriculture and Food)

kept in the town hall of Winchester, England. The weight of grain when the cylinder was filled was converted to bushel weight, and all measurements had to conform. Small and shrunken seed had a low bushel weight compared to plump seed. The weight also varied according to the amount of moisture in the grain. Levels established varied among 14.0%, 14.5%, and 15.0%.

Metric information in Appendix B recognizes the hectoliter (Tables B-3 and B-5) as the SI equivalent to bushel weight. Grams per half liter replaces the Winchester cylinder for determining hectoliter weight. The grades of wheat (Table 2-1) reflect the range in hectoliter weight for five grades of wheat.

Test weight is a function of kernel packing volume and kernel density. Variations in test weight can be found among cereal grain cultivars depending on grain shape and surface characteristics such as humping, dorsal, or lateral depressions and wrinkling (Yamazaki and Briggle, 1969). Traditionally, a low test weight in wheat has been associated with low flour yield, and this remains true when environmental variables reduce test weight. Kernel density and packing volume that reduce test weight may not reduce flour yield.

The historical use of bushel weight led to the adoption of this measure to reflect yield per unit area. Bushel weight varied among crops, however, and this led to difficulties in yield comparisons among cereal crops. Wheat and rice, for example, become standardized at 60 lb/bu; corn, sorghum, and rye at 56 lb/bu; barley at 48; and oat at 32 in the United States and 34 lb/bu in Canada.

Acceptance of SI units has eliminated this problem. Yields can be expressed in kilograms or tonnes per unit area (hectare). Quintals are metric, based on kilogram, but are not SI units.

Table 2–1
Grade Specifications for Wheat Based on Test Weight

Wheat Grade	*Minimum Weight, g/half liter*	*Hectoliter Weight kg/hl*	*Bushel Weight pounds*
No. 1	368	73.6	60
No. 2	358	71.6	58
No. 3	342	68.4	56
No. 4	316	63.2	52
No. 5	290	58.0	48

Source: Ontario Wheat Producers' Marketing Board (1981).

PHYSIOLOGICAL ASPECTS OF CEREAL CROPS

The basis of all crop growth and yield is photosynthesis, the process whereby sunlight energy is converted to chemical energy. Photosynthesis is the most important biological phenomenon on earth, the source of all primary energy for mankind. Photosynthesis, therefore, is the very essence of agriculture, the cornerstone of food production.

Although the fact that all food comes from photosynthesis is intuitively obvious, photosynthesis needs to be front and center when crop production and yield aspects are considered. It is a common assumption that all crop activity is under genetic control, and this view may obscure observations of crop responses that are primarily physiological or environmental and that can modify photosynthesis substantially. High-yielding cereal crop cultivars, for example, are superior because they either produce more total photosynthate or partition a greater portion into grain yield. Cereal crops that are resistant to pests produce a higher yield because photosynthetic rate is not reduced by pests that destroy photosynthetic tissue or complete for photosynthate. Crop cultivars tolerant to stress conditions such as drought or heat yield more because the photosynthetic rate is higher than that of stress-susceptible cultivars. Weeds are removed so that they do not shade or compete for sunlight energy needed in photosynthesis. Multiple-cropping systems are designed to increase the amount of photosynthate produced per unit area. Seeding rates and early planting dates are designed to develop a crop canopy that intercepts sunlight to best advantage in photosynthesis. Fertilizers are applied to promote leaf and root growth so that photosynthesis is increased.

Yield must be viewed as the outcome of the integration of many variables affecting the plant. Photosynthesis can take place only in a genetically adapted plant, but environmental variables may override genetic influences. When the importance of photosynthesis is kept in the forefront, then steps to increase yield through production or breeding practices are put on a positive basis. Cultural practices change from an art to a science based on scientific principles.

From a plant-breeding standpoint, attempts to increase grain yield by increasing the relative contribution of the different components of grain yield (ears per plant, grains per ear, grain size) without attention to the amount of photosynthate available generally have failed. An increase in one component tends to be accompanied by a reduction in another (Adams, 1967). Improving the partitioning of available photosynthate by reducing vegetative plant parts has generally led to an increase in grain yield (Mohamed, Okiror, and Rasmusson, 1978; Lawes, 1977).

THE PROCESS OF PHOTOSYNTHESIS

The site of photosynthesis is the **chloroplasts.** These are small pigmented bodies which stream around in the **cytoplasm** of the cells of leaves and other green tissue. The pigments are **chlorophyll.** Genetic and environmental differences alter the relative amounts of chlorophyll pigments and produce leaves which are dark green to yellow (no chlorophyll) in color. Dark green leaves are preferred, but there is no evidence that dark green leaves have a higher photosynthetic ability than leaves less green in color. Immature leaves may be yellow-green, presumably because chlorophyll synthesis is incomplete. If mineral nutrition is adequate, the leaf will become a darker green as it ages.

In addition to chloroplasts, only three components are required—carbon dioxide (CO_2), water (H_2O), and sunlight energy—with glucose sugar ($C_6H_{12}O_6$) being produced. In chemical symbols the balanced photosynthetic equation can be written:

$$6\,CO_2 + 12H_2O \xrightarrow[\text{chloroplasts}]{\text{sunlight energy in presence of}} C_6H_{12}O_6 + 6\,H_2O + 6\,O_2$$

Note that fertilizer nutrients are not a part of the equation, but they are essential for the development of chloroplasts and plant structures needed to conduct water to the site of photosynthesis, or to allow carbon dioxide to enter the leaf, or to allow glucose sugar to be transported to and transformed into parts including grain. Carbon dioxide is the main ingredient in the photosynthetic process. Products of carbon dioxide comprise 90% to 95% of the dry weight of a plant. Nutrients taken up from the soil account for the remaining 5% to 10% of a plant's dry weight.

During photosynthesis, water is separated into hydrogen and oxygen ions, and the oxygen in carbon dioxide is separated from carbon. Hydrogen and carbon recombine to form sugars. Oxygen is released to provide the primary source of atmospheric oxygen. Plants, therefore, provide not only food, but also oxygen to breathe.

PHOTOSYNTHETIC EFFICIENCY

It may come as a surprise to many to learn that photosynthesis proceeds with an efficiency of about 1%. This means that of the total visible radiation striking a field of plants, only 1% is converted to chemical energy. Other causes are related to poor sunlight interception, carbon dioxide deficiencies, water stress, deficiencies in mineral nutrition, extremes in temperature, poor cultural practices, inadequate pest control, and other factors. An aim of crop production is to reduce or eliminate the limiting factors of photosynthesis.

Higher photosynthetic efficiency is possible but very difficult to obtain on a field basis, where many factors are beyond our direct control. An upper limit of photosynthetic efficiency of 20% was calculated on a purely theoretical basis (Bonner, 1962), but an efficiency of greater than 6% is not expected under field conditions. Efficiencies of 2.0% to 6.0% have been reported for wheat and rice crops.

PHOTOSYNTHESIS AND CEREAL CROP YIELDS

Since carbon is the main constituent of plant dry matter, the logic is straightforward that increasing the assimilation rate of carbon dioxide will result in increased yields. Since about 1960, the rate of CO_2 assimilation has been a popular starting point for considering improvements in crop yields. Considerable research work has been directed to finding cultivars with high photosynthetic rates or finding techniques to increase leaf photosynthesis.

Paradoxically, no relationships appear to exist between the rate of net CO_2 exchange per unit of leaf area and crop yield (Moss and Musgrave, 1971; Evans, 1975 and 1980). Atmospheric enrichment with carbon dioxide to increase photosynthetic rate seems to increase yield more from improved plant-water relations and water-use efficiency than from increased photosynthetic rates (Kramer, 1981).

An understanding of why increased leaf photosynthetic rates may not increase grain yield forms the basis of an understanding of crop physiology relating to crop growth and development. Not all the photosynthate produced is directed into marketable or grain yield; part is directed to the development of roots, leaves, or stems and part is used in respiration and in keeping vital functions of the plant operating. Certainly, the duration of leaf photosynthetic rate appears to be as important as its magnitude. We must recognize, too, that measurements of photosynthetic rate in an individual leaf of an isolated plant may not reflect the total rate of that plant or of the plant population. It might be concluded that the marketable yield of a crop depends more on sink or kernel capacity than on the supply of photosynthate from the source (Evans, 1980).

In a review of the many reports that photosynthetic rates had no relationship to crop yield, Zelitch (1982) found that instantaneous measurements of photosynthesis were misleading and that crop yield is closely related to the net photosynthetic assimilation of CO_2 when calculated over an entire season. As well, experiments on CO_2 enrichment provide the best evidence available that increasing net photosynthesis will increase yield.

It is interesting to note that although plant improvement has been going on for centuries, there is no evidence of any increase in the maximum photosynthetic rate per unit area since grain crops were first domesticated (Evans, 1980).

VARIATIONS IN PHOTOSYNTHETIC EFFICIENCY

It is well known that some crop species exhibit different photosynthetic rates; sorghum and corn, for example, have a capacity to utilize sun light energy at a rate approximately 50% to 66% higher than tobacco (Hesketh, 1963). A major part of the difference can be explained by differences in respiration. If the photosynthetic rate is considered in relation to the amount of photosynthate left over after respiration, the resulting **net assimilation rate** or NAR is a more useful value in determining the relationship to total dry matter or **biomass** production when considered over a period of time.

C_3 Versus C_4 Plants

When the first product of photosynthesis in a plant is a three-carbon compound formed via the Calvin-Benson pathway, the term C_3 plant is applied. When the first product of photosynthesis is a four-carbon compound formed via the Hatch-Slack pathways, the term C_4 is applied. A variation of the C_4 type of photosynthesis is the *crassulacean acid metabolism type* or CAM. Plants characterized by this latter type are adapted to arid regions.

C_4 plants under warm, high light conditions are more efficient than C_3 plants; their magnitude of efficiency overall is about 40% (Monteith, 1978). Of the cereal crops, wheat, rice, barley, rye, and oat are C_3 types; corn, sorghum, and millet are C_4 types.

Some of the yield advantages of C_4 over C_3 plants can be explained through differences in respiration and the net assimilation rate. All plants require energy to exist, and this energy is obtained through respiration. Two types of respiration are known. One type reconverts sugars and carbohydrates produced in photosynthesis back to carbon dioxide, water, and energy. This form of respiration is independent of light and proceeds at the same rate in both light and darkness. The classic agronomic view is that respiration represents a loss that should be minimized, but this, of course, is not correct, as respiration is required to provide energy for the growth and survival of the plant. But respiration over and above that needed for crop yield production is wasteful. Net assimilation rate rather than the rate of photosynthesis and respiration is therefore important (McCree, 1982).

A second type of respiration, stimulated by light and termed **photorespiration,** is found in C_3 plants but apparently not in C_4 plants. Photorespiration has no known essential function and may be regarded as an evolutionary error. By eliminating photorespiration substantial yield increases could be expected.

Environmental responses of cereal crops can be explained on the basis of their C_3 and C_4 classification. The differences can be listed as follows:

1. C_4 plants respond to higher light intensities than C_3 plants. C_4 plants will respond to light intensities double that of C_3 plants (Table 2-2).
2. C_4 plants are more efficient users of carbon dioxide. Under a constant and relatively high light intensity, C_4 plants are capable of reducing the carbon dioxide concentration in the surrounding air to 0–10 ppm of CO_2. This is the **compensation point**—the level at which the rate of photosynthesis equals the rate of respiration. C_3 plants have a compensation point of 50–150 ppm of CO_2. A low compensation point indicates a high level of photosynthetic efficiency.
3. A major and economically important difference is that C_4 plants have a higher net assimilation rate than C_3 plants. This is in part due to the absence of photorespiration.
4. C_3 plants are adversely affected by high temperatures, whereas C_4 plants are not. Within the temperature range of 0° to 35°C (32° to 95°F) respiration rate increases about two to four times for each 10°C rise. The rate of photorespiration increases with temperature faster than gross photosynthesis. Many C_3 plants become unproductive at temperatures from 25° to 35°C (77° to 95°F), whereas C_4 plants increase in productivity at these temperatures. Because rice is the most common tropical food cereal, it is generally considered to be adapted to high-temperature regions. Studies have shown, however, that optimum tempera-

Table 2-2
Photosynthetic Rate (in Milligrams of Carbon Dioxide per Gram on Dry Weight) for C_3 and C_4 Cereal Crops Grown under Four Light Intensities in the Field

Type of Crop	*Light Intensity during Growth, percent of full sun*			
	100	*70*	*50*	*30*
C_3 species:				
Wheat	36	38	27	38
Oat	35	34	31	40
Rice	35	42	41	39
C_4 species:				
Sorghum	177	176	203	227
Corn	183	191	190	215

Source: Data from Singh, Ogren, and Widholm (1974).

tures exist for each growth stage, and that temperatures exceeding the optimum probably occur under field conditions (Owen, 1971). Rice is a C_3 plant, hence its yield will be reduced by high temperatures. Under high-temperature conditions during reproductive growth, sterility may be induced (Satake and Yoshida, 1978).

5. A greater water-use efficiency is found in C_4 plants than in C_3 plants. When water-use efficiency was averaged over a number of C_3 and C_4 species, the mean dry matter produced for each 1000 g of water used was 3.29 g for C_4 plants compared to 1.54 g for C_3 plants (Begg and Turner, 1976).
6. *Translocation* rates are often correlated with photosynthetic rates, as slow translocation or storage capacity can limit yield. In C_4 leaves, translocation is about twice as fast as in C_3 leaves (Zelitch, 1982).
7. C_3 plants suffer an oxygen stress and are adversely affected by the oxygen level of 21% found in nature. Oxygen stress is closely related to photorespiration, and C_3 plants perform better at low oxygen levels than at normal levels (Table 2-3). It is recognized that although dark respiration is independent of oxygen concentrations and proceeds at about the same rate in oxygen concentrations ranging from 2% to 20%, photorespiration in C_3 plants increases as oxygen concentrations increase from 2% to 20%. (Zelitch, 1982).
8. Maximum crop growth rates are, with few exceptions, much greater for C_4 than C_3 crops (Zelitch, 1982). When averaged over the entire season, the maximum crop growth rate for C_3 stands was 13 g/m^2/d and for C_4 stands 22 g/m^2/d (Monteith, 1978). By enclosing C_3 and C_4 plants in a sealed, illuminated container, differences in photorespiration can be demonstrated. Because the C_4 plants have a lower CO_2 compensation point than C_3 plants, the C_3 plants will die when the C_4's exhaust the CO_2.

Table 2-3
Inhibition of Net Photosynthesis Caused by Oxygen Stress; Comparison of Rates in Normal Air (21% O_2) with 1%-3% O_2

Crop	*Photosynthesis Rate, mg CO_2/dm^2/hr*		*Percent Inhibition at 21% O_2*
	1%-3% O_2	*21% O_2*	
C_4 species:			
Corn	37	37	0
Sorghum	45	45	0
C_3 species:			
Wheat	37	24	34
Bean	22	15	32
Sugar beet	58	25	57
Tobacco	25	16	56

Source: Data obtained from summaries presented by Zelitch (1982).

PLANT DEVELOPMENT AND PHOTOSYNTHESIS

The endosperm of cereal seeds serves as the food or energy supply from the time the seed begins to germinate until the **coleoptile** emerges from the soil and the first leaf begins to photosynthesize. Under ideal warm, moist conditions, the coleoptile will emerge within four or five days after seeding, but under cool or dry conditions, two weeks or longer may be required. The sooner the seedling emerges from the soil, the sooner photosynthesis can begin to feed the plant.

Depth of planting influences the length of time from planting to emergence. Seeding deeper than the recommended depth for a region may delay seedling emergence, subject the seeds to cooler temperatures, and increase the risk of emergence failure if soil crusting occurs in the interval between seeding and emergence. Unless local conditions dictate deeper seeding, seeds of most cereals should be placed 2 to 4 cm (1 to 2 in.) in the soil to promote rapid emergence.

After emergence, rapid growth of the photosynthetic area will occur if the primary or seminal roots contact a starter fertilizer placed in a band on either side and slightly below the seed. Too much fertilizer in direct contact with the seed may result in osmotic injury.

The juvenile or vegetative phase of plant development covers the period from germination to flower development. During the **juvenile phase** the internodes elongate into a stem and branches or tillers arise from the lowermost nodes. Stem elongation occurs by cell division in the **meristem regions** above each successive node and by cell elongation. The number and size of cells, the number of tillers, and the length of each **internode** are influenced by the duration of the juvenile phase, which is influenced by temperature and daylength. Increase in stem diameter occurs through cell enlargement, not cell division, after the essential structures have been formed.

The reproductive phase begins with the formation of flower initials and lasts until fertilization of the flower is completed. Because of the determinate nature of cereal crops, no new leaves are initiated after **anthesis** or reproductive development begins, although leaf blades may continue to enlarge owing to cell expansion.

The ripening phase is the time from fertilization of the flower to maturation of the seed. It is critical to grain development, and applications of water, the presence of a leaf area to intercept sunlight effectively, or the addition of nutrients during this phase may stimulate photosynthesis for grain development without stimulating vegetative growth.

The rate of dry-matter accumulation in the seed can be divided into three phases (Johnson and Tanner, 1972). The *lag period* occurs when cells are dividing in the embryo to form the storage or sink area. When the sink capacity has been established, photosynthate is transported very rapidly into the developing seed during the *linear period*. In the third or *grain-filling period* the rate of filling levels off and ceases. Final grain dry-weight yield is the product of the rate of grain filling and the duration of the grain-filling period.

When no further increase in dry weight of the grain occurs, the crop is said to have reached physiological maturity. At this stage, drying of the grain may occur, but harvesting could take place without a yield loss. The ability to simply or visually identify physiological maturity is therefore important. Procedures based on a periodic kernel or ear sampling to determine when dry-weight increases cease have been used with varying degrees of success. These procedures, however, are time-consuming, cumbersome, and subject to considerable error. Efforts to find a simple means of detecting physiological maturity led to the discovery of the kernel black layer in corn (Daynard and Duncan, 1969). Formation of this black layer represents a visual signal that all translocation to the kernel has stopped. A period of two to four days is required from the time of first appearance until the layer has completely developed.

The black layer occurs in the **placentochalazal** region of the kernel, and we can see it by removing the kernel from the cob and removing the pedicel. Black-layer development has been reported in sorghum and pearl millet (Rench and

Shaw, 1971; Eastin, Hulquist, and Sullivan, 1973; Fussell and Dwarte, 1980). The black layer appears to block the pathway of assimilate into the kernel and hence coincides with physiological maturity. In rice, wheat, and oat, less precise visual indicators exist (Hanft and Wych, 1982).

In corn, the following observations should be noted:

1. Black-layer formation is a common characteristic of all corn types including pop, sweet, flint, dent, and South American cultivars.
2. A black layer is formed at the base of all fertilized corn florets when their translocation activities cease. In fully developed kernels, black-layer formation occurs at maturity; in aborted tip kernels, it forms within one to three weeks after fertilization. At harvest the black layer (or its absence) serves as an indicator as to whether undeveloped tip florets aborted before or after fertilization, as unfertilized florets do not develop black layers.
3. Within an ear, tip kernels will develop black layers first, the large basal kernels will mature last, but most of the kernels in the central part of the ear will mature (form black layers) within a period of two to four days.
4. At any single location, a given number of days are required from silk emergence to maturity. Within a field of corn, ears will mature at different dates because of differences in silking dates. The length of the grain-filling period varies among cultivars and is considerably affected by environment.
5. At physiological maturity, grain moisture content varies considerably among cultivars and from one environment to another. Date of planting can affect kernel moisture content at maturity.

SINK-SOURCE RELATIONSHIPS

In the successful production of a cereal crop for grain, consideration must be given to seeding at a suitable calendar date and at an appropriate photosynthetic capacity to produce grain yield. In cereal crops, the time interval for vegetative growth exceeds that of reproductive growth. Whether the time span from seeding to maturity could be more advantageously divided into vegetative and reproductive growth periods is a matter of considerable speculation. Resolution of this question depends on whether grain yield under favorable conditions of light, temperature, water, and nutrients, and freedom from pests and diseases is more likely to be limited by the capacity for photosynthesis (the source) or by the capacity for storage (the sink).

In barley, neither the source nor the sink appear to present an overriding limitation to grain yield, and feedback interactions suggest that the photosynthetic rate may adjust to the requirements of the grain (Evans and Rawson, 1970).

Sink size in most cereals is dependent on the number of heads per unit area, spikelets per head, grains per spikelet, and grain volume and weight. A study of wheat and rice suggested that storage capacity may be a major limitation to grain yield (Evans, 1975).

CEREAL CROP ADAPTATION

Plants respond to variations in the duration of light and darkness which can influence crop adaptation. In equatorial areas, daylength is approximately of 12 hours duration year round. At latitudes north and south of the equator daylength varies according to the season of the year. At 60° north latitudes, for example, the longest day is 6 minutes short of 19 hours and the shortest day is less than 6 hours (Table 2–4).

Reproductive responses in plants to daylength are called **photoperiodism** or photoperiodic reactions. For long-day plants to reach reproductive development, a daylength of approximately 14 hours is required. Wheat, oat, barley, and rye are characteristically long-day plants and under short-day conditions these plants remain vegetative.

Short-day plants require a short day for reproductive development to be induced. Short-day crops include corn, rice, millet, and sorghum. Short-day crops remain vegetative when grown under conditions of long-days.

Photoperiodic sensitivity provides the explanation for why some crops do not reach reproductive development or why crop response is considered unusual when seed of daylength sensitive plants is exchanged among diverse lati-

Table 2–4
Approximate Daylight Hours at Eleven Latitudes North of the Equator

Degrees Latitude North	March (Spring Equinox)		June 21 (Summer Solstice)		Sept. 21 (Autumn Equinox)		Dec. 21 (Winter Solstice)	
	hr.	min	hr.	min	hr.	min	hr.	min
0	12	7	12	7	12	7	12	7
10	12	7	12	43	12	8	11	33
20	12	9	13	19	12	10	10	56
25	12	10	13	42	12	10	10	35
30	12	10	14	4	12	12	10	11
35	12	12	14	32	12	12	9	48
40	12	12	15	2	12	13	9	20
45	12	14	15	38	12	15	8	46
50	12	15	16	24	12	17	8	4
55	12	17	17	24	12	19	7	10
60	12	18	18	54	12	21	5	52

tudes. Short-day corn cultivars, for example, which respond to a short 12-hour photoperiod, will not be triggered into reproductive development by the long 15-hour day at 40°N. latitude. When the daylength becomes less than 14 hours about August 21, these corn plants may begin reproductive development. A long vegetative period may result in these corn plants being twice the usual height. By the time pollen is ready to be released, fall frosts at 40° latitude can be expected and no seed is produced. A plant breeder would have obvious problems crossing this corn with his usual breeding material. Corn adapted to 40° N. latitude will normally respond to a shortening daylength following June 21 and will reach reproductive development by late July.

A corn cultivar adapted to daylength conditions at 40° latitude and planted near the equator will be triggered into reproducive development as soon as the juvenile stage is completed. The plants may achieve about half their usual height.

Photoperiodic sensitivity in cereal crops can severely limit the successful exchange of germplasm among breeding programs or for commercial production at contrasting latitudes. The short-day photoperiodic response of tropical sorghum cultivars for many years restricted the universal use of much of the germplasm in this crop. Development of sorghum cultivars less specific in their photoperiodic response has allowed the range of sorghum to be greatly extended northward (Miller, Barnes, and Cruzado, 1968).

In some crops, and in some crop cultivars, daylength insensitivity is a major objective. In barley, insensitivity was found to be a multiple genetic trait. A plant breeder, however, can effectively select for insensitivity under a specific daylength because heritability values were found to be high despite the fact that major genes are not associated (Barham and Rasmusson, 1981).

The most desirable level of photoperiod insensitivity in crop species is unknown. In rice, for example, photoperiod insensitivity is important in the second crop of the two-crop pattern common in the summer season in areas such as the Yangtze River Valley in China. The second crop of rice must mature early enough to permit the establishment of a winter crop such as wheat before cold weather begins. Some degree of photoperiod sensitivity may be desirable as well, to lengthen the juvenile phase of growth which may lead to greater sink capacity and possible higher yields (Wilson, 1981).

Daylength is the term used to describe plant reactions to photoperiod even though it was subsequently learned that the length of the dark period, rather than the length of the light period, is the critical factor in the photoperiodic response. Interrupting the dark period with light, reverses the effect of a long night. This is a useful piece of information for inducing plants to reach reproduction in the growth room or on a small-scale basis for handling long-days crops in tropical and semi-tropical regions. By interrupting the dark period for about one hour in the middle of the night with flood lights, reproductive development can be triggered. A period of light during the night for short-day plants will delay flowering. In effect, a short-day plant should be referred to as a long-night plant and a long-day plant as a short-night plant.

Knowledge of the optimum photoperiod for a crop species would be useful for growing plants in controlled environmental rooms to obtain seed in the shortest possible time. Because the optimum photoperiod is not known for most crop species, the daylength commonly used in controlled environmental rooms is 12 and 20 for short- and long-day plants respectively (Major, 1980).

VERNALIZATION REQUIREMENTS

High temperatures have the effect of hastening flower initiation and development and therefore daylength does not exert exclusive control over flowering. Temperature in this case is said to have an indirect effect.

Temperature, daylength, and light intensity exert an influence on plants requiring a period of cold temperatures and short-days and is called **vernalization.** Winter annuals require a vernalization period to initiate reproductive development. When seeded outdoors in the fall, these crops are vernalized by natural low temperature, short days, and low light intensity over the winter months. In controlled environmental units, vernalization of winter annual seedlings can be achieved by temperatures maintained at 5° to 8°C (35° to 45°C), (Trione and Metzger, 1970), with 8 to 10 hours of low light intensity per day for a period of 6 weeks. Crops such as rye that are strongly winter hardy have a greater vernalization requirement than less winter-hardy crops such as winter oat or barley. In order of increasing vernalization requirements cereal crops can be listed in the following order: winter oat, winter barley, winter wheat, winter triticale, and winter rye. Winter annuals seeded in the field in the spring will remain in the vegetative phase all summer unless subjected to temperatures low enough to cause vernalization. Spring cultivars will reach reproductive development when planted in the spring in the more extreme latitudes or when grown during winter in the tropical regions.

Fully vernalized plants can revert back to the original nonflowering condition if exposed to a high temperature. The process is called **devernalization.**

TRANSPIRATION

Transpiration is the emission of water vapor from the aerial parts of plants, chiefly through the stomata. The term water requirement or **transpiration ratio** indicates the ratio of weight of water absorbed by the plant during its growth period to the weight of dry matter harvested. The ratio usually is expressed as units of water required to produce one unit of crop dry matter. The transpiration ratio varies from 200 to 1000 for different crops, i.e., for each unit of dry matter produced, 200 to 1000 units water are utilized. Transpiration ratios may vary among locations according to temperature, humidity, wind, and soil fertility.

The transpiration ratio is based on the weight of the above-ground dry matter harvested but without the roots. Transpiration ratios from the classical work of Shantz and Piemeisel (1927) conducted during the years 1911 to 1917 at Akron, Colorado, are accepted as reflecting standard values and are shown for cereal grains in Table 2-5.

Based on transpiration ratios shown in Table 2-5 and comparing the two extremes of rice and millet, the production of one tonne of dry matter in rice requires 682 tonnes of water compared to 280 tonnes for millet. Agriculture must be viewed therefore as a large consumer of water. Water

Table 2–5
Water Requirements of Cereal Crops Determined at Akron, Colorado 1911–1917

Crop	*Mean Transpiration Ratio*	*Units of Dry Matter Produced for Every 1000 Units of Water Consumed*
Millet (8 cultivars)	280	3.64
Sorghum (9 cultivars)	305	3.46
Corn (15 open-pollinated)	349	2.88
(8 single-cross hybrids)	350	2.87
Barley (4 cultivars)	517	1.93
Wheat–Emmer	517	1.93
Durum (6 cultivars)	520	1.93
Common (16 cultivars)	557	1.81
Oat (4 cultivars)	583	1.72
Rye (1 cultivar)	634	1.58
Rice (1 cultivar)	682	1.47

Source: Data from Shantz and Piemeisel (1927).

used for plant growth cannot contain excessive levels of salts and must be available at the proper times. Many plant species cannot cope with excess water. Management of water to meet crop needs in terms of availability, quantity, and quality is a major challenge in crop production.

Less than 1% of the water taken up by the root system of a plant is retained in the plant. The other 99% is lost through transpiration, by water vapor passing from the leaves through the numerous **stomatal openings** and evaporating. As water loss occurs, a low water potential draws the water taken up by roots to the leaves by a network of tubes called the **xylem.** Under conditions inducive to high transpiration rates, plant water losses may exceed the water uptake by the roots and wilting occurs. In addition to environmental variables affecting water loss, the size of the stomatal openings also exerts a strong influence. The opening and closing of the stomata is a mechanical process regulated by the turgidity of the **guard cells.** Plants transpire according to the demands of the atmosphere and not according to the demand of the plant for water for functioning.

Sugars produced in photosynthesis with the aid of the very small percentage of water taken up by the plant are transported chiefly in the **phloem.** A crop efficient in the use of water, as indicated by a low transpiration ratio, will not necessarily produce a high yield nor is a high yield necessarily a result of efficient water use. A low transpiration ratio does indicate a high degree of adaptation to dry conditions. Transpiration is a measure of a physical plant process to supply the water needs of the plant, and is not a measure of the yielding ability of a crop under drought conditions. Under conditions of equal dry-matter production, crops with a low transpiration ratio may require less water than those with a higher transpiration ratio.

If insufficient soil water is available to meet the water needs of the plant, yield will be restricted. Transpiration is sometimes viewed as a wasteful process but reductions in transpiration rate are usually associated with reduced yields, possibly because of a reduction in flow of plant nutrients needed for cell growth and development, or by high temperatures related to a reduction in evaporative cooling. Development of a thickened **cuticle,** or waxy bloom on the cuticle or cereal leaves and stems can reduce transpiration.

Rainfall or irrigation water applied to a crop may be used up by transpiration, may be lost through evaporation, runoff, or through percolation. Plants with a large and deep root system are able to extract more water from the soil and survive a period of drought better than plants with a limited and shallow root system. Water loss from a field through transpiration and evaporation is termed **evapotranspiration.** This total consumptive value offers a practical measure of the water requirements needed for crop production in a given area.

LODGING IN CEREAL CROPS

The single most important barrier to increased yields of many cereal crops is their susceptibility to lodging (Figure 1-8). Lodging is the bending and breaking over of plants from their normal upright position. The morphology of most cereals, whereby the grain bearing organs are found at the top of the stem, places a strain on the stalk to support the weight of grain produced, especially under conditions of high winds or wind driven rain. In corn, the ear is borne mid-way up the plant and from a morphological viewpoint, this is a superior location. Corn, however, is not exempt from the lodging problem.

The potential lodging hazard precludes intensive management such as high fertilizer applications and other production inputs to increase grain yields, especially in the small-stemmed cereals.

Lodging results in a direct yield loss if it occurs when the crop is actively producing photosynthate for vegetative growth or reproductive development. Lodging that occurred at heading resulted in a 10% greater average yield reduction than lodging that occurred 15 to 20 days after heading (Table 2-6) (Pinthus, 1973). Lodging disrupts the highly organized plant leaf canopy, resulting in greater mutual shading and reduced air movement and carbon dioxide supply. Any factor that reduces the rate of photosynthesis can lower grain yields. The severity of photosynthetic reductions is increased if lodging results in a sharp bending of the stems that disrupts the flow of water and nutrients needed for photosynthesis as well as the flow of assimilates to the roots or developing grain.

Indirect yield losses occur by harvesting difficulties of lodged grain at harvest time, by the resting and free feeding of birds on the lodged grain, by the creation of a suitable habitat for rodents such as mice, by molding associated with restricted air movement, and by premature seed sprouting and decay when lodged heads dry improperly or come in contact with the soil. Lodging may result in wet grain that cannot be stored without supplemental drying and that has a lower value due to poor quality.

When lodging occurs while the plants are metabolically active, **apical dominance** may be

Table 2–6
Percent Reduction of Grain Yield in Selected Cereal Crops at Several Locations by Induced 90-Degree Lodging at Two Stages of Development

		Lodging Induced at:	
Crop	*Location*	*Heading*	*Day 15 to 20 after Heading*
Winter wheat	Kansas	27	22
Winter wheat	Illinois	31	20
Spring wheat	Manitoba	34	24
Fall-sown barley	Arizona	40	39
Oat	Iowa	36	23
Oat	Illinois	37	17
Mean		34	24

Source: Data from Pinthus (1973).

destroyed and late tillering may occur, especially if the disrupted leaf canopy allows sunlight to strike basal nodes. Tillering that is induced by lodging and that occurs after the normal tillering period has ceased is not desirable, as late tillers do not mature with the rest of the crop and contribute little to grain yield.

Use of cereal grains as a companion crop can result in forage seedling failure if lodging occurs and sunlight energy, essential for photosynthesis in the forage seedlings, is reduced. Companion crops should either be seeded at a very low rate, be removed as a green forage rather than as mature grain, or be removed if lodging occurs. The high risk of lodging and the severe competition offered by companion crops generally means that direct seeding of forages accompanied by chemical weed control is the most reliable means of establishing a forage stand.

In addition to the tangible direct and indirect yield losses, the intangible yield losses may be even greater because small-grain producers are preconditioned to limit fertilizer or manure applications or to avoid small-grain production following a legume crop because of the lodging hazard. Lodging was described by Pinthus (1973) as an "abundance disease," meaning that it may occur under highly productive environmental conditions, thereby restricting the exploitation of otherwise yield-promoting factors. Greater straw strength in semidwarf wheat cultivars is strained by an accompanying increase in grain yield.

Productive practices must aim at a balance or compromise between yield-promoting inputs and the prevention or minimizing of lodging. Considering the number of interacting factors associated with crop production, and the climatic and weather variables that are beyond human control, producing a high-yielding crop without lodging is a very real challenge. Intensive cereal production systems can be used successfully to reduce the problem.

TYPES OF LODGING

Three types of lodging can be identified:

1. Bending of stems at the base.
2. Breakage of stems at any point along their length.
3. Root lodging.

Lodging may be promoted by physical forces exerted by wind, rain, or hail; by insects' damaging and weakening the roots and stems; by diseases; by fertility imbalance such as inadequate potassium relative to nitrogen; by high yield in association with high plant populations; by herbicide damage; or by genetically weak straw. The degree of stem lodging—that is, the angle at which stems deviate from the perpendicular—can vary from slight slanting to lying flat on the ground. Lodging may occur at any stage of plant development with serious negative effects (Gardener and Rathjen, 1975). If lodging in the seedling or vegetative stages occurs, plants may recover their upright position—a negative **geotropic response**—resulting in goosenecking or curvature of the lower stems. Reduced light on the underside of actively growing stems causes cell elongation, pushing the stems upright.

In corn and sorghum, large and vigorous root systems and sturdy young stalks prevent lodging in the early stages of plant development. Lodging in senescing cultivars of these crops is most prevalent late in the season, after physiological maturity has been reached, when the weight of the grain is greatest and plant stalks have been weakened because carbohydrates have moved from the source to the sink and are not being replaced because photosynthesis has ceased. Diseased, damaged, and weathered stalks may promote lodging in these crops after physiological maturity, as they are often allowed to dry down in the field.

Lodging creates a very real barrier to the production of high grain yield and represents a severe limitation to increasing the yield potential. The development of cereal crops with reduced plant height has represented a giant step forward. Owing to a reduction in wind resistance, short plants have less torque pressure on the roots or lower parts of the stem. Production practices must aim at proper row width, so that sunlight penetration to the lower internodes induces cell differentiation rather than cell elongation. Breeding and management practices must be combined to help reduce the lodging problem. Lodging is a problem that is not likely to be eliminated. As population pressure increases the demand for food, and as advances in technology increase the grain-yield potential, lodging is likely to remain as a permanent consideration.

TILLERING IN CEREAL CROPS

A feature common to all cereal crops is the arising of branches that resemble the main stem, known as tillers, suckers, or stools, from buds on nodes below or at the soil surface. The process of tillering is defined as the development of new shoots from the base of the original plant, as contrasted to branching, in which new shoots may arise from auxiliary nodes throughout the plant. Under spaced-planted conditions, small grains can produce from ten to 50 tillers per plant and under exceptional conditions up to 150 tillers per plant, each capable of developing reproductive organs and contributing to grain yield. Under field conditions of close spacing, fewer tillers are initiated and only a portion reach maturity and produce grain. The remainder die, largely during the period from jointing until flowering; this occurrence is known as tiller mortality (Table 2-7). Under competitive conditions in small grains associated with recommended seeding rates, the genetic tillering potential is substantially reduced, often to as few as one to three productive tillers per plant with several unproductive tillers. Crop production practices therefore have a tremendous impact on tillering. Minnesota tests with high- and low-tillering cultivars did not give a consistent grain yield, suggesting a possible association between tillering capacity and yield stability (Simmons, Rasmusson, and Wiersma, 1980).

In corn, tillering capacity varies from one to 12 (Kuleshov, 1933). Genetic selection against tillering virtually has eliminated tillering in commercial cultivars, allowing producers to control plant population through seeding rate.

Tillering occurs during the vegetative period of plant development and is influenced by temperature and light. Cool temperatures result in slow growth and in good sunlight penetration to the basal nodes that favors tiller bud differentiation. Early seeding generally favors tillering. Late-seeded cereals that develop rapidly under warm growing conditions and have a reduced period of vegetative growth may not tiller at all.

Tillering is also subject to hormonal influences. If apical dominance is destroyed by removal of the lead shoot or by lodging, late dormant tiller buds may be stimulated into development. Once apical dominance is expressed by the lead or primary shoot, tillering ceases. An unvernalized winter annual can produce a profusion of tillers, because shoots that exert apical dominance never develop. In dwarf cereal crops, apical dominance appears to be incomplete, and such plants exhibit a high level of tillering over an extended time period. In extreme cases, late-

Table 2–7
Average Tillering Characteristics of Five Spring Barley Cultivars Grown at St. Paul, Minnesota, in 1976, 1978, and 1979

Cultivar Number	*Shoots*		*Spikes*		*Percent Shoot Mortality*
	/m^2	*/ft^2*	*/m^2*	*/ft^2*	
1	446	41	312	29	30
2	463	43	296	27	34
3	552	51	325	30	39
4	651	60	405	38	39
5	763	71	459	43	40
Mean	575	53	359	33	36

Source: Data from Simmons, Rasmusson, and Wiersma (1982).

formed tillers may be green and immature when the first tillers are mature, thereby complicating harvesting and lowering grain quality (Stoskopf and Fairey, 1975). The full expression of apical dominance is associated with a more favorable synchronous tiller development.

Genetic differences in tillering capacity among species and among cultivars within a species can be readily observed. The full genetic tillering potential, however, is influenced strongly by environmental factors.

Winter cereals have a high tillering capacity as tillering occurs in the fall as well as over an extended spring period.

Two-row barley has a higher genetic tillering potential than six-row barley cultivars. Under similar environmental conditions the two types produce comparable yields, because differences in productive tillers account for differences in number of kernels per head.

The discovery of a genetically controlled nontillering or uniculm barley led to a proposal to develop commercial uniculm barley cultivars to reduce intra- and interplant competition. Uniculm cultivars yielded less than normal tillering cultivars.

A high degree of tillering is exhibited in sorghum and rice, and these crops are ratoon cropped because of this feature and because of their adaptation to tropical conditions.

From an evolutionary viewpoint, tillering can be viewed as a form of perennation. Tillering provides the mechanism for regrowth from season to season in perennial forage grasses. When perennial plants migrated out of the favorable tropical regions, the law of survival of the fittest favored annuals from among the perennial populations. A remnant form of perennation or tillering was retained. Winter annual cereals more closely resemble the perennial ancestor and retain a greater tillering habit than most annual cereals. The discovery of a perennial ancestor of corn (Nault and Findley, 1982) could result in a perennial corn that could regrow from tillers and be harvested for several years without replanting. Although perennial cereal crops are a possibility, tiller buds development and associated root reserves must be viewed as a competitive sink for photosynthate that could result in reduced grain yields compared to annual crops.

In winter cereals, tillering is desirable under conditions that reduce plant stand. Tillering can compensate for winterkill damage.

Plausible arguments can be made for or against the virtues of tiller production in annual cereal crops. Successful selection for nontillering grain corn is viewed as a positive step forward, but the lack of a commercial uniculm barley cultivar suggests that tillers have a positive value in barley production. Roots that develop from basal tiller nodes lend physical support to the plant and reduce root lodging, a problem with uniculm plants. Unproductive tillers often are viewed as being parasitic on the main stem because they use photosynthate, water, and nutrients that might be used to better advantage. Unproductive tillers increase shadings and use water through transpiration.

Questions

1. Consider any effect spike density might have on grain yield.
2. If you were asked to estimate the grain yield of a field of wheat prior to harvest, what items would you look for in the crop?
3. Consider the value of awns in crop production.
4. Of what botanical, economic, and nutritional value are the pericarp, the testa, and the aleurone layer which together make up the bran?
5. *a.* Consider the evolutionary evidence that self-pollination followed the development of cross-pollination.
   b. From an evolutionary standpoint what are the advantages of cross-pollination?
   c. How and why do you suppose self-pollinated crops developed?

6. Consider factors that influence depth of seeding of cereal crops around the world. In general, what do you conclude about the adequacy of seeding methods in terms of depth? Is seed in the main seeded at the right depth or is it seeded too deeply or too shallowly?
7. Review the problems with seeding too deeply; too shallowly.
8. How important are prop roots in corn?
9. Compare a 50-bushel-per-acre wheat yield with a 50-bushel-per-acre oat yield by expressing both in pounds per acre. Calculate the same yields in kilograms per hectare. Are there any similarities between the expressions of the yields in pounds per acre and in kilograms per hectare? What do you conclude about the expression of wheat and oat yields in SI terms as compared with 50-bu/ac terms?
10. *a.* Consider the advantages and disadvantages of tillering versus a uniculm cereal cultivar in:
    (1) spring cereals.
    (2) winter cereals.

    b. Consider why uniculm cultivars of spring barley have not achieved any commercial success.

    c. Compare features of uniculm barley cultivars with nontillering hybrid corn cultivars. Why has uniculm corn proven to be a commercial success and uniculm barley less of a success?
11. *a.* Consider the importance of "the components of grain yield" (ears per plant, grains per ear, grain size) in grain yield. What happens if tillers per plant are increased?

    b. How and why can a two-row barley cultivar equal the yield of a six-row cultivar?
12. Why are spring crops such as oat, barley, or wheat referred to as "cool-loving" crops?
13. Explain why the process of photosynthesis is considered to be the most important biological phenomenon on earth.
14. In terms of a sink-source relationship, what might you conclude about a field of a cereal crop that had only one or two developed kernels per spikelet, or a field of corn with poorly developed ears?
15. Consider the evolutionary significance of the black-layer development in corn.
16. Describe the growth and reproduction of a corn cultivar adapted to conditions at a high latitude and grown at the equator. Explain why the corn cultivar performed in this manner.
17. *a.* How could daylength insensitivity be developed?

    b. Consider the merits and disadvantages of crop cultivars that are insensitive to daylength and hence have worldwide adaptation.
18. Speculate on how an unproductive tiller may be "parasitic" on the main plant.
19. Which type of lodging do you consider to be the most serious? Explain your answer.
20. *a.* If transpiration appears as a wasteful process, consider the effect of reducing transpiration on crop growth and yield.

    b. What effect might the evaporative cooling in a transpiration have on C_3 and C_4 plants?

    c. C_4 plants are reported to have a higher water-use efficiency. To what extent does Table 2-5 support this view?

REFERENCES

Adams, M. W. 1967. "Basis of Yield Component Compensation in Crop Plants with Special Reference to the Field Bean, *Phaseolus vulgaris.*" *Crop Science* 7: 505–570.

Barham, R. W., and D. C. Rasmusson. 1981. "Inheritance of Photoperiod Response in Barley." *Crop Science* 21: 454–456.

Begg, J. E., and N. C. Turner. 1976. "Crop Water Deficits." *Advances In Agronomy* 28: 161–217.

Bonner, J. 1962. "The Upper Limit of Crop Yield." *Science* 137 (3523): 11–15.

Brouwer, R. 1966. "Root Growth of Grasses and Cereals." (In) F. J. Milthorpe and J. D. Ivins (eds.), *The Growth of Grasses and Cereals.* London: Butterworth & Co. (Publishers), Ltd. pp. 153–166.

Daynard, T. B., and W. G. Duncan, 1969. "The Black Layer and Grain Maturity in Corn." *Crop Science* 9: 473–476.

Eastin, J. D., J. H. Hulquist, and C. Y. Sullivan. 1973. "Physiological Maturity in Grain Sorghum." *Crop Science* 13: 175–178.

Evans, L. T. 1975. "The Physiological Basis of Crop Yield." (In) L. T. Evans (ed.). *Crop Physiology.* Cambridge: Cambridge University Press. pp. 327–355.

———. 1980. "The Natural History of Crop Yield." *American Scientist* 68: 388–397.

———, and H. M. Rawson. 1970. "Photosynthesis and Respiration by the Flag Leaf and Components of the Ear During Grain Development in Wheat." *Australian Journal of Biological Science* 23: 245–254.

Fussell, L. K., and D. M. Dwarte. 1980. "Structural Changes in the Grain Associated with Black Region Formation in *Pennisetum americanum.*" *Journal of Experimental Botany* 31: 645–654.

Gardener, C. J., and A. J. Rathjen. 1975. "Juvenile Lodging in Barley: A Yield-Depressing Phenomenon." *Australian Journal of Agricultural Research* 26: 231–242.

Hanft, J. M., and R. D. Wych. 1982. "Visual Indicators of Physiological Maturity of Hard Red Spring Wheat." *Crop Science* 22: 584–588.

Hesketh, J. D. 1963. "Limitations to Photosynthesis Responsible for Differences among Species." *Crop Science* 3: 493–496.

Johnson, D. R., and J. W. Tanner. 1972. "Calculation of the Rate and Duration of Grain Filling in Corn (*Zea mays* L.)." *Crop Science* 9: 473–476.

Konzak, C. F., and S. M. Dietz. 1969. "Documentation for the Conservation, Management, and Use of Plant Genetic Resources." *Economic Botany* 23: 299–308.

Kramer, P. J. 1981. "Carbon Dioxide Concentration, Photosynthesis and Dry Matter Production." *BioScience* 31: 29–33.

Kuleshov, N. N. 1933. "World's Diversity of Phenotypes of Maize." *Journal of The American Society of Agronomy* 25: 688–700.

Lawes, D. A. 1977. "Yield Improvement in Spring Oats." *Journal of Agricultural Science* 89: 751–757.

———. 1978. "Physiological Aspects of Yield in Cereals." *National Institute of Agricultural Botany Fellows Conference Report* No. 3. pp. 16–18.

Locke, L. F., and J. A. Clark. 1924. "Development of Wheat Plants from Seminal Roots." *Journal of The American Society of Agronomy* 16: 261–268.

Major, D. J. 1980. "Environmental Effects on Flowers." (In) W. R. Fehr and H. H. Hadley (eds.), *Hybridization of Crop Plants.* Madison, Wis.: American Society of Agronomy and Crop Science Society of America, Publishers. pp. 1–15.

McCree, K. 1982. "The Role of Respiration in Crop Production." *Iowa State Journal of Research* 56: 291–306.

Miller, F. R., D. K. Barnes, and H. J. Cruzado. 1968. "Effect of Tropical Photoperiods on the Growth of Sorghum When Grown in 2 Monthly Plantings." *Crop Science* 8: 499–502.

Mohamed, A. M., S. O. Okiror, and D. C. Rasmusson. 1978. "Performance of Semidwarf Barley." *Crop Science* 18: 418–422.

Monteith, J. L. 1978. "Reassessment of Maximum Growth Rates for C_3 and C_4 Crops." *Experimental Agriculture* 14: 1–5.

Moss, D. N., and R. B. Musgrave. 1971. "Photosynthesis and Crop Production." *Advances in Agronomy* 23: 317–336.

Nault, L. R., and W. R. Findlay. 1982. "Update on Perennial Corn Discovery." *Crops and Soils* 34(6): 10–13.

Owen, P. C. 1971. "The Effects of Temperature on the Growth and Development of Rice." *Field Crop Abstracts* 24: 1–8.

Pinthus, M. J. 1973. "Lodging in Wheat, Barley and Oats: The Phenomenon, Its Causes and Preventive Measures." *Advances in Agronomy* 59: 155–159.

Rench, W. E., and R. H. Shaw. 1971. "Black Layer

Development in Corn." *Agronomy Journal* 63: 303–305.

Russell, W. J., W. R. Fehr, and R. L. Mitchell. 1971. "Effects of Row Cultivation on Growth and Yield of Soybeans." *Agronomy Journal* 63: 772–774.

Satake, T., and S. Yoshida. 1978. "High Temperature-Induced Sterility in Indica Rices at Flowering." *Japan Journal of Crop Science* 47: 6–17.

Shantz, H. L., and L. N. Piemeisel. 1927. "The Water Requirements of Plants at Akron, Colorado." *Journal of Agricultural Research* 34: 1093–1190.

Simmons, S. R., D. C. Rasmusson, and J. V. Wiersma. 1982. "Tillering in Barley: Genotype, Row Spacing and Seeding Rate Effects." *Crop Science* 22: 801–805.

Singh, M., W. L. Ogren, and J. M. Widholm. 1974. "Photosynthetic Characteristics of Several C_3 and C_4 Plant Species Grown under Different Light Intensities." *Crop Science* 14: 563–566.

Stoskopf, N. C., and D. T. Fairey. 1975. "Asynchronous Tiller Maturity—A Potential Problem in the Development of Dwarf Winter Wheat (*Triticum aestivum* L.)." *Plant Breeding Abstracts* 45: 467–472.

Trione, E. J., and R. J. Metzger. 1970. "Wheat and Barley Vernalization in a Precise Temperature Gradient." *Crop Science* 10: 390–392.

Weaver, J. E., and E. Zink. 1945. "Extent and Longevity of the Seminal Roots of Certain Grasses." *Plant Physiology* 29(3): 359–379.

Weihing, R. M. 1935. "The Comparative Root System of Regional Types of Corn." *Journal of the American Society of Agronomy* 27: 526–537.

Wilson, D. 1981. "Breeding for Morphological and Physiological Traits." (In) K. J. Frey, (ed.), *Plant Breeding II*. Ames: The Iowa State University Press. pp. 233–290.

Yamazaki, W. T., and L. W. Briggle. 1969. "Components of Test Weight in Soft Wheats." *Crop Science* 9: 457–459.

Zelitch, I. 1982. "The Close Relationship Between Net Photosynthesis and Crop Yield." *BioScience* 32: 796–802.

3

Variability in Cereal Grains

INTRODUCTION

To the uninitiated, most cereal crop plants look alike, and certainly during the vegetative growth phase it may be hard to tell one cereal species from another. Likewise, difficulty may be experienced in identifying seed of various cereal crops. The extent of crop diversity and the very real differences within and among cereal crops are, however, substantial.

The more experienced find differences among crops and cultivars within a crop apparent and realize that there are no monotypic species. In wheat, for example, Vavilov[1] (1949–1950), a famous Russian plant explorer, estimated that 3000 subdivisions exist, all of which are distinguishable morphologically and physiologically. Vavilov determined more than 400 heritable characters, which in most cases can combine freely, making possible millions of recombinations from crosses. Each combination could become a cultivar having a different genetic combination from either parent and hopefully having superior commercial value.

Some of the readily apparent variations in wheat include: a range in height from dwarf to tall; winter and spring habit; seed color ranging from purple, red, and amber to white; seed texture ranging from hard and **vitreous** to soft and starchy; glume color ranging from white, grey, and red to black; glumes having awns, no awns, semiawns, or tip-awns; glumes that are hairy or smooth; leaf blades that are wide or narrow, long or short, hairy or smooth; spikes that are lax, intermediate, **clavate,** or compact. Basic chromosome number in wheat can vary from 7 to 14 to 21 (Appendix A), with each set of chromosomes providing additional variation. Similarly thousands of forms exist in barley, oat, millet, and sorghum and hundreds of forms in rye, corn, and rice. Awns in barley, for example, may be smooth or barbed, long and slender, or **hooded.**

Despite the enormous list of differences that could be listed, similarities exist among all the cereal crops, because they presumably evolved from a common ancestor. Vavilov carefully observed the enormous range of **genetic diversity** found within and among crops, and his findings form the basis for this chapter.

PLANT CLASSIFICATION

The numerous differences and similarities within and among cereal crops make plant classification

[1] N. I. Vavilov was a geneticist and the director of the All-Union Institute of Plant Industry in Leningrad from 1920 to 1940. He and his colleagues observed and collected masses of genetically diverse crop cultivars on a global scale. For information on Vavilov refer to Leppick (1969).

an absolute necessity. Crop plants can be classified by taxonomists on the basis of morphological features to give an indication of their biological and evolutionary relationships or **phylogeny.** Crop plants also can be classified according to their chromosomal composition by plant breeders or agronomists who wish to know whether specific plants can be used for crossing purposes. Early classifications were based primarily on **morphology,** and fortunately this is usually a reasonable approximation of genetic and evolutionary relationships.

However, increased knowledge of experimental genetics has resulted in some notable exceptions to the rules of classification. It generally is believed that the level of chromosome pairing in crosses between two species reflects the degree of relationship between parental species. Whenever morphological and genetic boundaries fail to coincide, taxonomic disagreements or new insights into classification have resulted in two or more names existing for the same species (Appendix A). Additional changes—hopefully improvements in the classification—can be expected.

Plants have acquired common names that may vary among locations or lack precision. Believing that he had arrived in India, Columbus referred to the common native crop as Indian corn. In North America, Indian corn commonly is called corn and in Europe maize. Barley in Europe may be called corn. Rice native to Africa is distinct from Asian rice, and both differ from the wild rice native to North America. Some crops such as triticale and timopheevi wheat have no common name and are referred to by a version of their scientific name. To clarify ambiguities developed by the use of common names, a system of scientific names was developed that is used in scientific publications instead of the common name.

The following scheme for wheat illustrates the classification system:

Kingdom: Plantae (plants)
  Division: Spermatophyta (plants bearing seed)
    Subdivision: Angiospermae (seeds enclosed)
      Class: Monocotyledonae (seeds with one cotyledon, scutellum, or seed leaf)
        Order: Graminales
          Family or Tribe: Triticae (includes wheat, barley, and rye)
            Genus: *Triticum* (the wheats)
              Species: *Aestivum* (bread wheat)
                Cultivar: Columbus

Bread wheat is classified by the **binomial system** as *Triticum aestivum* L. The double name includes the name of the **genus,** which is capitalized, and the species name, written in lower-case letters. Both words are italicized (or underlined). It is the practice to follow the binomial name with the name or the abbreviation of the name of the botanist who gave the plant its scientific name. The scientific name for bread wheat is followed by the letter L., which stands for the immortal Carolus Linnaeus (1707–1778), who established the binomial system of nomenclature. The citation of the authority allows a scientist to find the original description of the plant and, in the case of name duplication, be certain of the exact identity of the plant referred to.

The genus may be considered as a group of species which includes plants that are essentially alike in the majority of their fundamental structural characters, have a common ancestry, interbreed freely, and maintain themselves through subsequent generations.

The species name is a specific designation of the larger genus grouping. Species from the genus *Triticum* may be grouped according to the chromosome number (Appendix A). In corn, several subspecies are included as part of the species.

The term *cultivar,* a contraction of the words "cultivated variety," is an inclusive term for lines, varieties, hybrids, or selections of crops. To be classed as a cultivar, each must exhibit a distinct feature(s). The term cultivar has relevance to the individual producer as it refers to the commercial crop produced on farm fields. The term cultivar is used to avoid confusion with a trade, brand, or variety name. A cultivar will breed true and will reproduce itself from generation to generation. *Hybrid cultivars* refer to a particular combination of inbred lines that can be maintained to reproduce the hybrid cultivar each generation.

Table 3–1
Variability for Heritable Characters of the Cereal Inflorescence[a]

Heritable Variable Character		*Rye*	*Wheat*	*Barley*	*Oat*	*Millet*	*Sorghum*	*Corn*	*Rice*
1. Articulation of spikelets and florets (tendency to self-sowing)	Spikelets and florets breaking on ripening (rachis brittle)	+	+	+	+	+	+	+ (in Teosinte)	+
	Spikelets & florets not breaking apart (rachis not brittle)	+	+	+	+	+	+	+	+
2. Hulledness	Grain hulled (tightly enclosed in glumes); not shattering on maturity	+	+	+	+	+	+	+	+
	Grain naked (easily separated from hull on threshing); easily shattering on ripening	+	+	+	+	+	+	+	+
3. Compactness	Compact	+	+	+	+	+	+	+	+
	Loose	+	+	+	+	+	+	+	+
	Intermediate	+	+	+	+	+	+	+	+
4. Sexuality	Dioecious	+	0	0	0	0	0	+	0
	Monoecious	+	+	+	+	+	+	+	+
5. Awnedness	Spikelets awned	+	+	+	+	0	+	0	+
	Spikelets awnless	+	+	+	+	+	+	+	+
	Spikelets short-awned and semiawned	+	+	+	+	+	+	0	+
	Spikelets with deformed awns	+	+	+	0	0	0	0	0
	Spikelets with awn-like appendages	+	+	+	0	0	0	0	0
6. Character of awns	Coarse	+	+	+	+	0	+	0	+
	Fine	+	+	+	+	0	+	0	+
	Barbed	+	+	+	+	0	+	0	+
	Smooth	0	+	+	+	0	0	0	+
7. Number of florets in a spikelet	One floret	+	+	+	+	+	+	+	+
	Two florets	+	+	+	+	+	+	+	+
	Several florets	+	+	0	+	0	0	0	0
8. Color of glumes	White (straw yellow)	+	+	+	+	+	+	+	+
	Red	+	+	+	+	+	+	+	+
	Brown	+	+	+	+	+	+	+	+
	Gray (black)	+	+	+	+	+	+	+	+
	Violet (anthocyanin)	+	+	+	+	+	+	+	+
9. Hairiness of glumes	Hairy	+	+	+	+	0	+	+	+
	Smooth	+	+	+	+	+	+	0	+
10. Rachis	Simple	+	+	+	0	0	0	0	+
	Branched	+	+	+	+	+	+	+	+
11. Hairiness of rachis	Very hairy	+	+	+	+	+	+	+	+
	Smooth	0	+	+	+	+	+	0	0
	Slightly hairy	+	+	+	+	+	+	+	+
12. Waxy layer on glumes	Present	+	+	+	+	+	+	+	+
	Absent	+	+	+	+	+	+	+	+

Source: Data from Vavilov (1949–1950).

[a] A "+" indicates the occurrence of the character; a "0" indicates it was not described by Vavilov or not present.

GENETIC DIVERSITY AND THE LAW OF HOMOLOGOUS SERIES

It would be repetitious and a demanding task to describe and list the numerous morphological variations found in each cereal crop species discussed in this book. Thousands of variations could be described, and hundreds to thousands of cultivars exist in each species. Although the botanical classification of wheat is divided into three groups consisting of 13 species (Appendix A), it is readily apparent that a similar pattern of variability exists among the three chromosome groupings.

Vavilov (1949–1950) showed that among genera, many common features may be found; this he called **parallel variation.** Based on the common features associated with parallel variation, Vavilov developed the *Law of Homologous Series*. This law may be stated as follows:

1. It is possible to predict that variation found in one plant will also be found in plants of other species and genera because of the occurrence of parallel forms. The more closely related the species, the greater will be the resemblance among the series of variations.
2. Definite cycles of variability occur among whole families of plants, and common traits develop through all genera and species making up the family.

Went (1971) concurred that parallel development, or convergence, is a very common evolutionary phenomenon and that similar forms, shapes, structures, and functions occur in a large number of species, related or unrelated.

Several general tendencies in heritable variability were observed by Vavilov that apply to all genera of the family. For example, cereals can be divided according to the structure of the inflorescence (Table 3-1) into loose, compact, and intermediate types. These divisions are readily apparent in rye, wheat and barley, but more subtle differences may be found (Figure 3-1). The wheat spike may be densely compacted, clavate, or **fusiform** in shape. Apical extensions on the lemma midrib, or the lack of them, results in spikes that are awned, tip-awned, or bald (Figure 3-1).

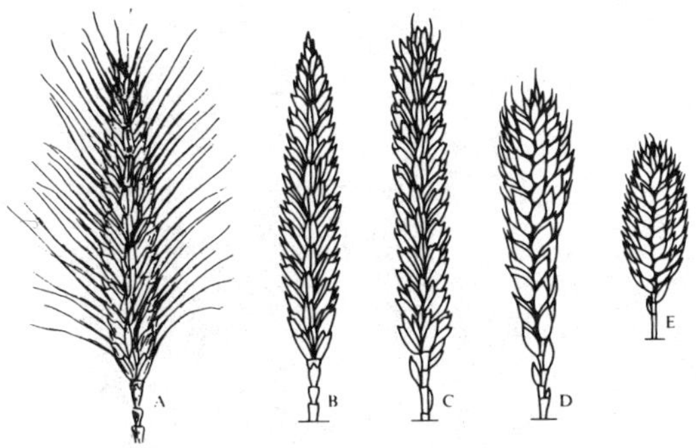

Figure 3–1. Genetic diversity is expressed in head shape and through awns. Shown here are spike forms of wheat: A, bearded; B, bald and somewhat fusiform; C, tip-awned and elongate; D, clavate; E, club. (Courtesy of The Canadian Seed Growers' Association)

In corn and rice, loose to compact inflorescence may be less evident but nevertheless can be divided according to the degree of compactness of the ear or panicle. Millet displays an inflorescence ranging from lax or **ramose,** more compact, drooping, to very compact or balllike. In sorghum, similar divisions can be found. Oat panicles are divided into forms ranging from compact with short branches and internodes, to lax or ramose types (Figure 3-2). On the basis of such observations, Vavilov therefore indicated the three levels of inflorescence compactness as existing in the eight cereal species observed (Table 3-1).

Eleven other phenotypic characteristics of the inflorescence are reported in Table 3-1, the majority of them commonly present in all cereal species observed. The shape of the rachis internodes is an example of an often missed but unmistakable form of genetic diversity (Figure 3-3).

In terms of grain color, form, size, and consistency, the eight species likewise exhibited many common features (Table 3-2). Kernel color ranged from white to black. Consistency of the endosperm ranged from hard and vitreous to soft and mealy.

Kernel and glume forms may serve to identify cultivars. Figure 3-4 shows a typical wheat kernel as well as variations that range from ellip-

Figure 3–2. Oat panicle shapes showing the common bilateral and symmetrical panicle of A, the unilateral or "mane shaped" panicle of "side oat" B, and the lax or ramose panicle of wild oat C. (Courtesy of The Canadian Seed Growers' Association)

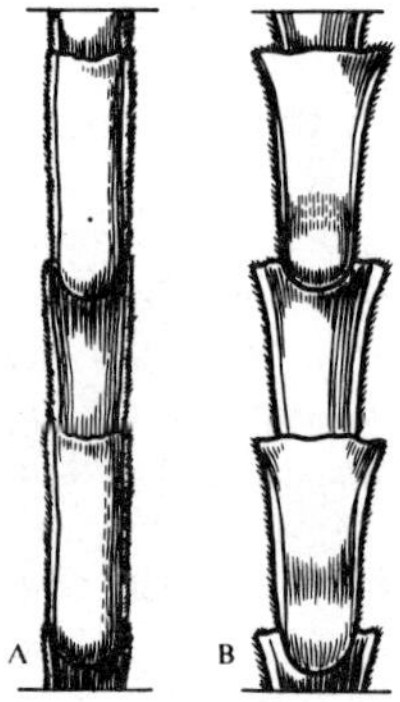

Figure 3–3. Genetic variation in rachis internodes of barley. A, parallel sided; B, flared. Internode length of the rachis can also vary. The shorter the internodes, the more compact the ear, and the more dense it is. Node and internode shape may vary, and differing degrees of hairness may be exhibited. (Courtesy of The Canadian Seed Growers' Association)

tical to ovate or oval. Wheat glumes may display characteristic shoulder widths and beak shapes and be glabrous or pubescent (Figure 3-5).

Barley seed variations can be expressed in the marginal nerve of the lemma as being smooth, barbed, or semismooth as well as by identifying differences in the structure of the rachilla (Figure 3-6). Basal markings on the kernels of barley may serve as an aid in the identification of individual barley cultivars (Figure 3-7).

Table 3–2
Variability for Characters of the Grain in Cereals

Heritable Variable Character		*Rye*	*Wheat*	*Barley*	*Oat*	*Millet*	*Sorghum*	*Corn*	*Rice*
1. Color	White	+	+	+	+	+	+	+	+
	Green (gray-green)	+	+	+	+	0	+	+	+
	Black (dark gray)	+	+	+	+	+	0	+	+
	Violet (anthocyanin)	+	+	+	0	0	+	+	+
2. Form	Round	+	+	+	+	+	+	+	+
	Elongate	+	+	+	+	+	+	+	+
3. Size	Large	+	+	+	+	+	+	+	+
	Small	+	+	+	+	+	+	+	+
4. Consistency of grain	Vitreous	+	+	+	+	+	+	+	+
	Mealy	+	+	+	+	+	+	+	+
	Waxy (reacting with iodine in contrast to preceding types)	+	+	+	0	+	+	+	+

Source: Data from Vavilov (1949–1950).

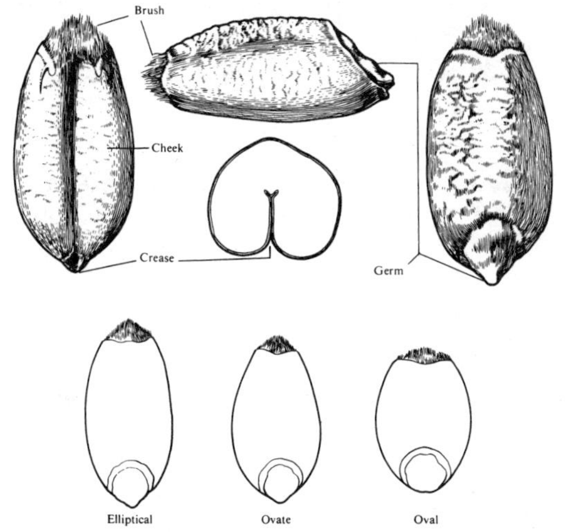

Figure 3–4. Ventral side and dorsal view of a wheat kernel showing the germ, brush, crease, and cheek. Wheat kernel shape varies from long and narrow (elliptical) to wide at the germ end and narrowing toward the brush (ovate). (Courtesy of The Canadian Seed Growers' Association)

Spikelet arrangements vary among wheat, oat, barley, and rye plants (Figure 3-8), yet features common to these four crops are readily apparent. Although genetic diversity is strong among the cereals, this does not preclude similarities from being found as suggested by the Law of Homologous Series.

An appreciation and awareness of the Law of Homologous Series is of value when the existence of a specific feature is in question. For example, Lambert and Johnson (1978) reported that plants with upright leaves gave a 9.8% grain-yield advantage. Plants lacking a ligule at the juncture of the leaf blade and sheath had pronounced upright leaves and under weed-free conditions and reduced row width outyielded flat- or droopy-leaved cultivars. A plant breeder wishing to test this observation in other crops could be assured that genetically liguleless types could be found (Table 3-3) with the possible exception of millet.

Likewise, items 5, 6, and 7 in Table 3-3 indicate that smooth or **glabrous** plant parts can be

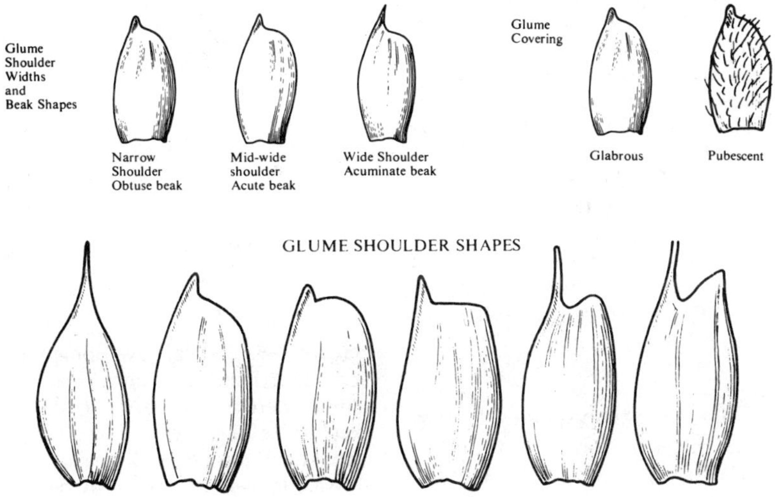

Figure 3–5. The size, shape, and color of outer glumes of wheat exhibit genetic diversity that can serve as a useful feature in cultivar identification. Variation is found in shoulder width and beak shapes, and the glumes may be smooth (glabrous) or pubescent (hairy). (Courtesy of The Canadian Seed Growers' Association)

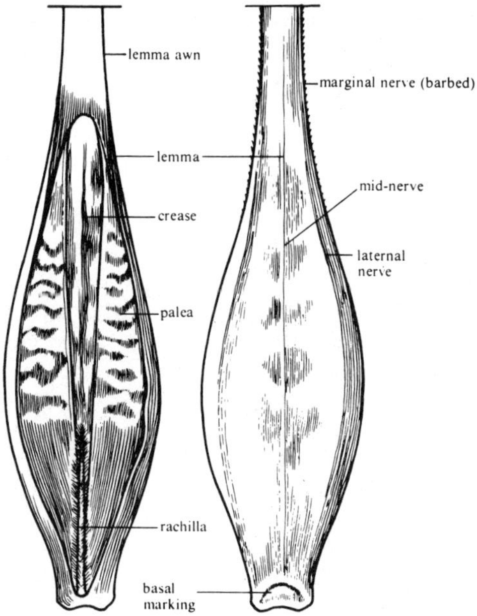

Figure 3–6. **Barley kernels exhibit features useful in cultivar identification and description. Barley may have an adhering lemma or palea, or may be naked. In this photo the lemma and palea are adhering and exhibit distinct features. Note the rachilla that arises at the basal end of the kernel and lies in the ventral furrow. The rachilla may be long or short; they may vary in degree of hairiness; and the hairs themselves may be long or short. In two-row barley, only the central spikelets are fertile and the grains are uniformly symmetrical. In six-row barley, all three spikelets at each node are fertile. The median grains, one-third of the total number, are symmetrical, but the remaining lateral grains are unsymmetrical, each with a right- or left-handed bias. The kernels shown here are symmetrical. (Courtesy of The Canadian Seed Growers' Association)**

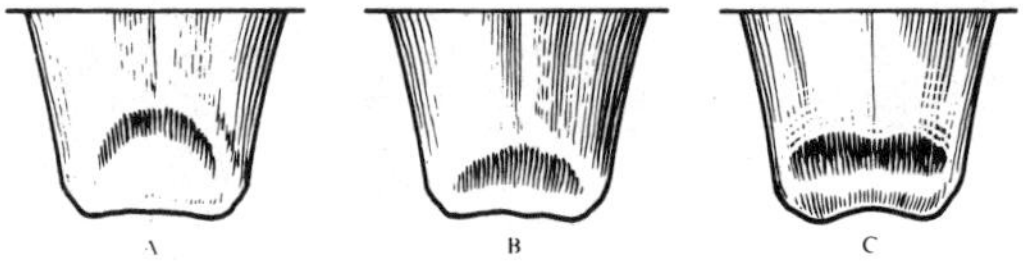

Figure 3–7. **Basal marking of barley kernel: A, horseshoe; B, incomplete; C, transverse crease. (Courtesy of The Canadian Seed Growers' Association)**

found in most of the cereal crop species. Leaves having **pubescence** or **trichomes** were found to be of practical value in terms of resistance to the cereal leaf beetle[2] that was introduced into North America in 1962 (Leisle, 1974; Ringlund and Everson, 1968). Plants with highly pubescent leaves are largely avoided by the insect. Density of pubescence is also a factor in avoiding insect attacks. Leaf pubescence was reported by Vavilov (1949–1950) to be present in all cereal plants but sorghum (Table 3-3). Hairiness of stems, leaves, glumes, and nodes all exhibit genetic diversity and may serve as important distinguishing characteristics (Figures 3-9 and 3-10) useful for identifying rogues in a field.

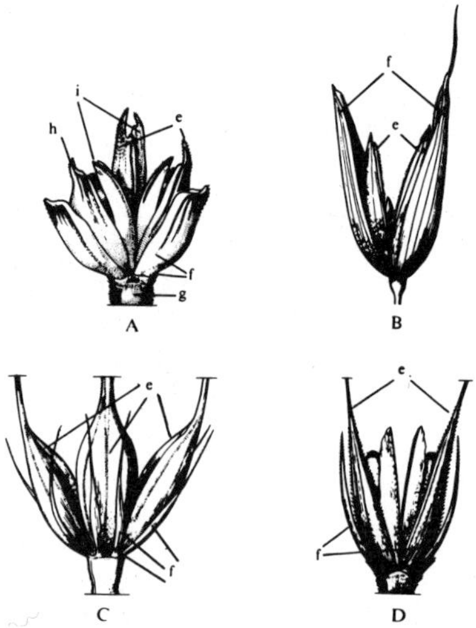

Figure 3–8. **Parallel development is evident from the common features found in spikelets of cereals as shown here by wheat, A; oat, B; barley, C; and rye, D. Each spikelet is enclosed by outer glumes, f, which in wheat and oat are well developed but in barley and rye are lanceolate structures that may carry fine awns of varying lengths. Spikelets contain one or more inner florets, e, and arise from a rachis, g, which varies in shape. The lemma, h, and palea, i, may be free-threshing as in wheat and rye or adhering as in oat and barley. Variability for free-threshing oat and barley cultivars can be found. (Courtesy of The Canadian Seed Growers' Association)**

[2]Cereal leaf beetle—*Oulema melanopus* L.

Table 3–3
Variability for Vegetative Characters in Cereals

Heritable Variable		*Rye*	*Wheat*	*Barley*	*Oat*	*Millet*	*Sorghum*	*Corn*	*Rice*
1. Leaf structure	Leaves with ligule	+	+	+	+	+	+	+	+
	Leaves without ligule	+	+	+	+	+	0	+	+
2. Stem structure	Hollow	+	+	+	+	+	+	0	+
	Solid	+	+	0	(thick-walled)	0	+	+	0
3. Color of seedlings	Violet	+	+	+	+	+	+	+	+
	Green	+	+	+	+	+	+	+	+
	Variegated	+	+	+	+	0	+	+	+
4. Form of growth	Upright	+	+	+	+	+	+	+	+
	Sprawling	+	+	+	+	+	+	+	+
5. Hairiness of stem below the inflorescence	Smooth	+	+	+	0	+	+	0	0
	Hairy	+	+	+	0	0	0	+	0
6. Hairiness of leaf sheath	Smooth	+	+	+	+	+	+	+	+
	Hairy	+	+	+	+	+	0	0	+
7. Hairiness of leaves	Smooth	+	+	+	+	+	+	+	+
	Hairy	+	+	+	+	+	0	+	+
8. Waxy layer on stems and leaves	With	+	+	+	+	+	+	+	+
	Without	+	+	+	+	+	+	+	+
9. Thickness of straw	Thick	+	+	+	+	+	+	+	+
	Thin	+	+	+	+	+	+	+	+
10. Nodes	Hairy	0	+	0	+	0	+	0	0
	Smooth	0	+	0	+	0	+	0	0
11. Height of plant	Tall	+	+	+	+	+	+	+	+
	Dwarf	+	+	+	+	+	+	+	+
	Medium	+	+	+	+	+	+	+	+
12. Color of straw	Yellow	+	+	+	+	+	+	+	+
	Violet (anthocyanin)	+	+	+	+	+	+	+	+
13. Color of leaves	Dark green	+	+	+	+	+	+	+	+
	Light green	+	+	+	+	+	+	+	+
14. Width of leaf	Narrow	+	+	+	+	+	+	+	+
	Medium	+	+	+	+	+	+	+	+
	Wide	+	+	+	+	+	+	+	+
15. Size of leaf	Large	+	+	+	+	+	+	+	+
	Small	+	+	+	+	+	+	+	+

Source: Data from Vavilov (1949–1950).

All cereal crops exhibit spring growth habit, and usually their winter counterpart is present (Table 3-4). Winter growth habit was not observed by Vavilov in millet, sorghum, or corn, but, according to the Law of Homologous Series, winter forms should be expected to occur. On this premise, perennial forms of an ancestor of corn, known as teosinte, were sought. In 1910 a perennial cultivar of teosinte had been found, but, because the chromosome number was twice that found in corn, the cross could not be made. In 1977 a perennial form having the same chromosome number as corn was found (Barney, 1980). It may be possible to obtain desirable germplasm from this plant and specifically to develop a perennial form of corn.

Stem-structure comparisons in Table 3-3 indicate that hollow stem occurs in all cereals

Table 3–4
Variability of Biological Characters in Cereals

Heritable Variable Character		*Rye*	*Wheat*	*Barley*	*Oat*	*Millet*	*Sorghum*	*Corn*	*Rice*
1. Growth habit	Winter type	+	+	+	+	0	0	0	+
	Spring type	+	+	+	+	+	+	+	+
	Semi-winter type	+	+	+	+	+	+	+	+
2. Earliness	Late	+	+	+	+	+	+	+	+
	Early	+	+	+	+	+	+	+	+
3. Ecological type	Hydrophytic	+	+	+	+	+	+	+	+
	Xerophytic	+	+	+	+	+	+	+	0
4. Cold resistance	Low	+	+	+	+	+	+	+	+
	High	+	+	+	+	+	+	+	+
5. Response to fertilization	High	+	+	+	+	0	0	+	0
	Low	+	+	+	+	0	0	+	0
6. Character of blossoming	Open	+	+	+	+	+	+	0	+
	Closed	+	+	+	+	0	+	0	+
7. Formation of albinos		+	+	+	+	0	+	+	+

Source: Data from Vavilov (1949–1950).

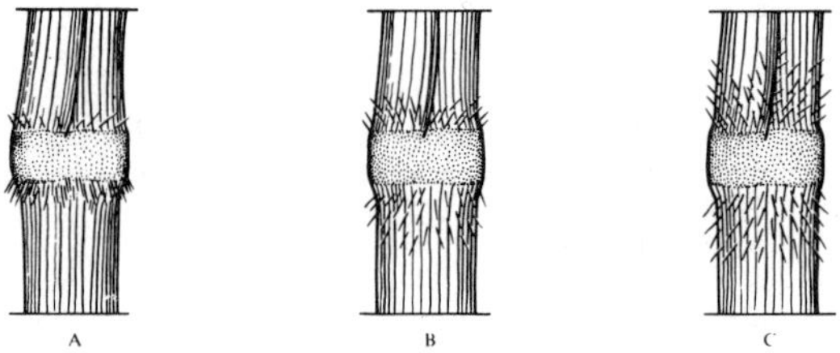

Figure 3–9. Hairiness or pubescence of stem nodes is found to vary in the oat crop and may be an important distinguishing characteristic of cultivars as well as an indication of genetic diversity. Shown here are oat nodes having a crown of pubescence below the node, A, scattered below the node, B, and scattered above and below the node, C. (Courtesy of The Canadian Seed Growers' Association)

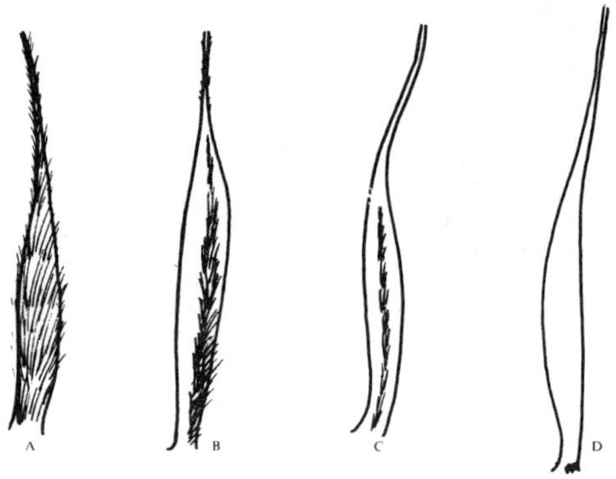

Figure 3–10. Hairiness or pubescence on barley glumes: A, covered; B, in band; C, on midline; D, without hairs. (Courtesy of The Canadian Seed Growers' Association)

but corn. In contrast, solid stem, common in corn and sorghum, is present also in rye and wheat. Solid-stemmed wheat cultivars are known to possess resistance to the wheat stem sawfly[3] (Kemp, 1934). Should the solid-stem feature be desired in barley or millet, an exhaustive search should reveal such a type, as assured by the Law of Homologous Series. Unfortunately, solid-stemmed cultivars of wheat are characteristically lower-yielding than hollow-stemmed wheat cultivars, possibly because the development of solid pith requires photosynthetic energy that might otherwise have gone into grain development (McNeal et al., 1965).

[3]Wheat stem sawfly—*Cephus cinctus* Nort.

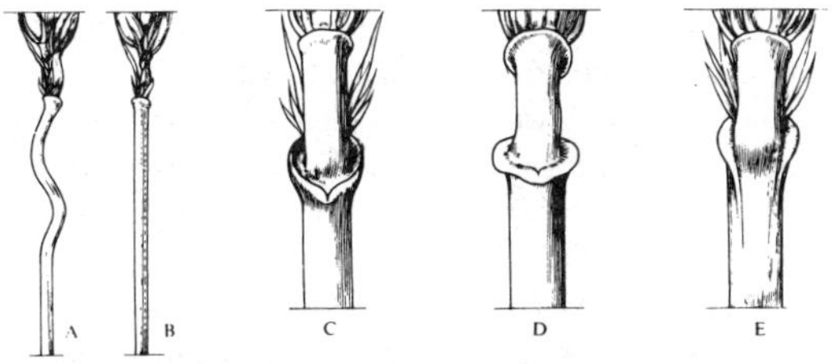

Figure 3–11. Genetic diversity expressed in neck shapes and collar forms found in barley. A, snaky neck; B, straight neck; C, V-shaped collar; D, closed collar; E, open collar. (Courtesy of The Canadian Seed Growers' Association)

Additional stem-structure features can be expressed in diversity of neck shapes and collar forms. Variations found in barley are shown in Figure 3-11.

Of use in identifying cereal crops in the seedling or vegetative stages are characteristic structures at the junction of the leaf blade and sheath (Figure 3-12). In general, these structures among the cereals are similar, only the degree of expression varies.

Tables 3-1 to 3-4 show a total of 38 morphological features with a total of 92 variations present in the majority of the cereal crops listed. In rice, Holcomb, Tolbert, and Jain (1977) listed 41 morphological features, but these lists are far from complete. Detailed analysis of enzymes or other physiological traits undoubtedly would reveal many more. A morphological trait not mentioned by Vavilov is the common feature of tillering in cereal crops. Development of nontillering corn cultivars provided a means of leaf-area control through plant population. Because nontillering corn plants exist, it was not surprising

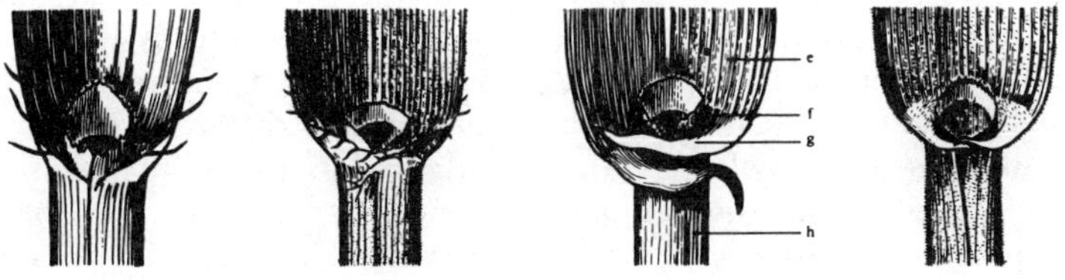

Figure 3–12. Distinguishing characteristics of cereals found at the junction of the leaf blade and sheath in oat, A; wheat, B; barley, C; and rye, D. Structures shown include the leaf blade, e; the ligule, f; auricle, g; and leaf sheath, h. In these four crops the following should be noted:

Oat—auricles absent, leaf sheath and blade mostly glabrous; a few long hairs on the margins of lower leaf blades and on lower leaf sheath of some cultivars. Ligule medium length. Blades with about 12 veins, usually twisted anticlockwise.

Wheat—auricles short, with blunt tips, hairy; leaf sheath and blade covered with short, fine hair (slight to very hairy); ligule medium length. Blades with about 12 veins, twisted clockwise.

Barley—clasping auricles long, slender, and with pointed tips. Leaf sheath and blade usually glabrous; ligule medium length. Blades with about 20 veins, twisted clockwise.

Rye—auricles very short, glabrous; leaf sheath and blade covered with short, fine hair (slight to very hairy); ligule short. Blade with about 12 veins, twisted clockwise.

Rice—not shown, can be distinguished from other grasses by the presence of a prominent ligule and long, slender hairy auricles. (Courtesy of The Canadian Seed Growers' Association)

that a genetically controlled nontillering or **uniculm** barley cultivar was found (Donald, 1968).

Another feature not described by Vavilov—that of response to daylength or photoperiod—is a dramatic example of effective use of germplasm variability. On the assumption that long- or short-day response could be found in all cereals, selections were made for wide adaptation to permit the exchange of cultivars between tropical and temperate zones. By crossing germplasm with widely differing maturities and selecting them under widely differing daylengths, it has been possible to select for wide adaptation and to permit the free exchange of cultivars around the world.

Parallelism is so conspicuous that it strongly points to a universal law of **homologous variation.** In addition to providing clues as to whether a specific form of variation may exist, such a law provides taxonomists with a rational expedient means of studying and describing plant variation. Plant breeders are able to predict the possibility of finding corresponding forms in nature or obtaining them through genetic manipulation, crossing, and selection. Homologous variation therefore is a useful piece of knowledge. The variation or genetic diversity that exists among and within species is of great importance also.

CULTIVAR IDENTIFICATION

Despite the enormous diversity that exists among cereal crops, it is extremely difficult to visually identify specific cultivars in each crop. Among the reasons are that breeding programs developing cultivars for a specific region select new cultivars from among a relatively narrow gene base of well-adapted material and that a complete key of morphological features seldom exists. A new cultivar may possess genetic resistance to disease but may be morphologically very similar to other susceptible cultivars. When morphological features associated with the seed, the seedling, and the mature plant are considered, specific cultivars might be identified, but this is an elaborate and time-consuming process.

But just as no two people have the same fingerprints, so, too, no two cultivars of a crop are the same, and the uniqueness of each cultivar appears to lie in protein patterns. Because the protein in each cultivar appears to vary according to its genetic constitution, it is possible to unequivocally "fingerprint" a cultivar for identification.

The ability to differentiate among crop cultivars becomes important when plant breeders attempt to patent a new cultivar for which royalties may be collected. Sometimes it may be necessary to prove that a new plant has been developed.

ORIGIN OF GENETIC DIVERSITY

Most of the similarities among plants can be attributed to a common ancestry. Genetic diversity emerged from natural evolutionary forces. The phenomenon of survival of the fittest led to the emergence of adapted genotypes. Genetic diversity occurred through natural **mutation, hybridization** caused by wind or insects, and natural selection. Gene recombination led to many genetically different plants. Those that were adapted to a specific area, soil type, or climatic condition survived; those that were not adapted were lost. The processes of mutation, hybridization, selection, and reselection that went on over several millennia led to an enormous range of genetic diversity.

When cereal plants were domesticated about 10,000 years ago, natural evolutionary forces were supplemented by artificial pressures exerted by humans. Selection that favored the non-shattering seed habit, large seed, lodging-resistant stalks, productivity, and other desirable traits led to rapid change and commercial progress. Rapid change and improvement was possible only because genetic diversity existed.

Application of the laws of genetics in the early 1900s led to controlled matings, and selection began in earnest. Instead of an increase in genetic diversity with modern breeding programs, a reduction occurred. The reasons for this decrease can be listed as follows:

1. Development of wide adaptation through photoperiodic insensitivity in the 1960s and 1970s in

selected cereal crops has resulted in the free exchange of seed among plant breeders worldwide. A genetically superior cultivar often was found suitable for production over geographically large areas.

2. The continued and increasing demand for food has forced the production of only the best-performing cultivars over an extensive geographical area. Widespread use of a limited number of genotypes has increased the risk of and vulnerability to a destructive epidemic.
3. The production of first-generation or F_1 hybrids, based on the combining of two inbred cultivars, each with a high degree of genetic uniformity, has reduced genetic diversity by comparison with the era when open-pollinated corn was produced. The risk of a trait-specific epidemic also increases, because a pest can spread unchecked.
4. The need for continued and rapid yield increases has resulted in an ever-increasing replacing of old cultivars with new, improved ones. Less productive cultivars are discarded and, unless a deliberate effort is made to preserve them, may be lost forever.
5. The habitat of wild cultivars, a source of new genetic resistance, is being destroyed by the extension of urbanization, intensive farming, or widespread grazing by cattle and goats. As a result, not only are existing wild relatives of cereal crops destroyed, but genetic diversity is diminished by a reduction in the range that exists for their development and in the opportunity for crosses among wild relatives. Destruction of the natural habitat for cereal crop relatives means the elimination of the primary natural source of genetic stocks.
6. For some time it was believed that common and widespread fungus diseases could be eliminated from farm fields by developing genetically specific disease-resistant cultivars. As a result, plant breeders spent 50% to 90% of their time in the mid twentieth century locating and incorporating genes for resistance. But genetic resistance has not proved to be lasting. Disease organisms are evolving constantly to overcome it, and a continuous battle is required to keep pests in check. Despite the excellent cultivars that were produced with genetic resistance to a specific race or races of the pathogen, these cultivars lasted only ten to 15 years before a genetic change to a more virulent form of the pathogen occurred. The process of finding and incorporating genetic resistance was begun all over. The rate at which new sources of genetic resistance are required threatens to deplete the readily available genes for specific resistance.

IMPORTANCE OF GENETIC DIVERSITY

Genetic diversity is desirable to reduce **genetic vulnerability** of crops and to provide an effective gene pool for crop improvement through breeding.

When large geographical areas are seeded with the same crop cultivar or with cultivars having the same genes or **cytoplasmic** factors, vulnerability increases. This genetic uniformity in commercial fields may lead to a widespread and unchecked outbreak of a disease or pest. In 1970, southern corn leaf blight[4] swept through the corn belt of North America because most of the corn produced contained the Texas male sterile factor in the cytoplasm. Corn containing this factor was acutely susceptible to the fungus, and the epidemic spread unchecked (Wade, 1972). In subsequent years breeders checked the fungus by producing hybrid cultivars that did not have the male sterile trait.

Genetic diversity is essential to the plant breeder, not only to provide new genetic sources for resistance to new or changing pests, but also to provide new morphological features with changing agricultural practices, such as narrow row spacings or changes in mechanization. To find the desired gene(s), plant breeders may sort through thousands of cultivars. Finding a desired gene may depend on the quality of the available seed banks that have been assembled through plant exploration, collection, and preservation.

PLANT EXPLORATION

In 1768, Sir Joseph Banks, the director of Kew Gardens, the famous Royal Botanic Gardens in London, accompanied Captain Cook on the ship *Endeavor* to the Pacific Islands to collect plants and make drawings. In those days plant explora-

[4]Southern corn leaf blight—*Helminthosporium maydis* Nishikado and Migake, Race T.

tion was conducted to survey little-known regions of the world in respect to plant life for agricultural and botanical purposes.

Since the initial explorations by Banks, plant explorers from many nations have searched for plants, primarily as a source of germplasm for crossing purposes. Unfortunately, early collections were not well preserved, because the concept of germplasm conservation had not occurred to plant breeders.

Historically, plant explorations have provided valuable material for the development of agricultural crop plants. Since the arrival of the first colonists in North America, agriculture has been dependent largely upon introduced plants—with the notable exception of corn. Many early plant explorations were made, and thousands of plants were introduced.

By the 1940s, concern about loss of thousands of plant introductions—collected at great expense—because of inadequate storage facilities led to the concept of a national seed storage. In 1958 the United States opened the National Seed Storage laboratory at Colorado State University, Fort Collins, Colorado. This facility is a milestone in the conservation of germplasm. Equipped with controlled storage rooms, it is designed to preserve the viability of seeds during intervals between their periodic growth in the field. Inventories of stocks are published, and seed is available to any worker.

Plant explorations remain important for locating and preserving the remaining gene pools before these plants are obliterated forever by modern progress.

CENTERS OF DIVERSITY

Plant explorations by Vavilov and his colleagues revealed an astonishing genetic diversity for selected plants in certain areas of the world. Vavilov described geographical areas that were rich in wild forms as "centers of diversity," and he considered them to be "centers of origin" for the crops found in them. Vavilov postulated the existence of eight main centers with two or three subsidiary ones (Harlan, 1975a and 1976). Six centers were in the Old World and two in the New World.

There is no doubt that centers of diversity do exist, but Vavilov's "centers of origin" theory has been modified. Harlan (1976) recognized that a single crop may have originated in more than one center and that many crops did not originate in the centers identified by Vavilov. Harlan (1976) suggested that some crops may have evolved over vast regions, rather than having a precise center of origin. Harlan believed that humans have scattered some of their crops far and wide and have obliterated the origins of still other crops. The probable origins of cereal crops, based on intensive studies by Harlan (1975a), are shown in Table 3-5. The probable origins of these cereal crops serve as locations for plant explorations.

GENE BANKS

Despite the establishment of national seed storage laboratories in several countries, it became apparent in the 1960s and 1970s that germplasm conservation efforts were far from adequate. Cataloguings of existing collections were incomplete; entries were duplicated and some were lost. Criticisms of the system were numerous (Harlan, 1972 and 1975b; Creech, 1970; Creech and Reitz, 1971; Miller, 1973; Reitz and Craddock, 1969; Konzak and Dietz, 1969; Webster, 1976). These references provide a historical perspective on the difficulties associated with gene banks and germplasm conservation.

Although criticisms of past programs may be valid, it must be recognized that species preservation represents an enormous undertaking (McClintock, 1982). Seeds of self-pollinated cereals can be readily stored indoors, propagated, described, and distributed. Cross-pollinated or asexually propagated crops are more complex, but compare this problem to that of holding large, perennial crops alive outdoors, subject to disease, insect, or climatic damage when grown out of their region of adaptation.

Many collections historically have contained a preponderance of genetic stocks of old

Table 3–5
Probable Origins of Eight Cultivated Cereal Crops

Crop	Description	Area of Origin
1. Wheat (*Triticum*)	Seven-chromosome wild wheat known as einkorn [*T. monococcum* (L.)]	Southeastern Turkey
	Fourteen-chromosome wheat known as emmer [*T. dicoccum* (Shrank)]	Near East
	Fourteen-chromosome wheat known by its scientific name [*T. timopheevi* (Zhuck)]	Soviet Georgia
	Forty-two-chromosome bread wheat [*T. aestivum* (L.)]	Trans-Caucasia and Caspian Sea
2. Rice (*Oryza*)	African rice [*O. glaberrima* (Steud.)]	West African savanna
	Oriental rice [*O. sativa* (L.)]	South China to East India
3. Corn (*Zea*)	Indian corn or maize [*Z. mays* (L.)]	Mesoamerica
4. Barley (*Hordeum*)	Common barley [*H. vulgare* (L.)]	Near East
5. Sorghum (*Sorghum*)	Common sorghum [*S. bicolor* (L.) Moench)]	Savanna zones of Sudan—Chad
6. Rye (*Secale*)	Common rye [*S. cereale* (L.)]	Anatolian plateau of Northern Europe
7. Oat (*Avene*)	Common oat [*A. sativa* (L.)]	Northern Europe
	Fodder oat [*A. strigosa* (Schreb.)]	Mediterranean
	Ethiopian oat [*A. abyssinica* (Hochst.)]	Ethiopia
8. Millet	Guinea millet [*Brachiaria deflexa* (Schum.) Hubb.]	Guinea highlands
	Finger millet [*Eleusine coracana* (L.) Gaertn.]	Ethiopia
	Pearl millet [*Pennisetum americanum* (L.) Schum.]	Dry savanna, Sudan to Senegal
	Japanese millet [*Echinochloa frumentacae* (Link.)]	Eastern China
	Italian or foxtail millet [*Setaria italica* (L.) Beav.]	Northern China
	A millet [*Papalum scrobiculatum* (L.)]	Milgiris of South India
	A millet [*Digitaria Cruciata* Nees ex Hoff. f.]	Hills of North East India
	Slender millet [*Panicum miliare* Lam.]	Himalayas and Upper Burma

Source: Adopted from Harlan (1975, pp. 69–76).

cultivars from breeding collections rather than collections from the gene centers. For years, controversy existed as to the merits of widening the gene base to include wild species. Hawkes (1977) urged that wild species be included in gene banks and suggested that a plant breeder attempting to solve a plant breeding problem should look first at working collections or material of old and current cultivars and breeding stocks for the desired genetic trait. Failing this, a world-scale search could be made through genetic resources collections or gene banks for primitive cultivars. The virtues of gene resources of the wild relatives of wheat have been described by Feldman and Sears (1981).

Considerable progress appears to have been made in evaluating, storing, and retrieving stored seed (Hawkes, 1981). The problems of space and economic considerations have restricted gene bank development, but the realization that seed banks hold the world's most precious assets has encouraged activity. Plant scientists in developing and developed countries alike have realized that germplasm collection and conservation is an essential prerequisite for the continued survival of humans on the planet earth. A most promising aspect in gene bank development is the move toward global coordination of germplasm collection and conservation.

In Europe, **Eucarpia** took a leading role starting in 1960 to establish a European network of germplasm conservation. On Eucarpia's initiatives, a gene bank was founded at Bary, Italy, to cover Southern Europe including the Mediterranean region; one at Braunschweig, West Germany, to cover northwestern Europe; and one at Lund, Sweden, to cover Scandinavia (Hawkes and Lamberts, 1977).

The **Food and Agriculture Organization (FAO)** of the United Nations in the late 1950s and through the 1960s assumed a leadership role in developing ideas that eventually led to global coordination. The groundwork for the FAO headquarters in Rome to act as a clearing house for information on collections and for plant exploration was established at conferences in 1967 and 1973 (Frankel and Hawkes, 1975). In 1974 the International Board for Plant Genetic Resources (IBPGR) was formed to assist and encourage the collection, preservation, and exchange of plant genetic material (Appendix C). Eventually a network of international institutions was formed for the collection and preservation of crop germplasm, located more or less in the areas encompassed by the centers of diversity. International programs operate in addition to and in conjunction with the European network under Eucarpia, and national programs operate in the United States, the USSR, India, China, Canada, Brazil, Turkey, Great Britain, and Australia. Individual countries which cooperate in regional programs make independent decisions to explore, conserve, evaluate, and utilize genetic resources. The help of private industries interested in developing private breeding programs has been enlisted, and their cooperation is apparent.

Among the cereal crops, substantial collections of wheat, rice, corn, barley, sorghum, and millet exist, often consisting of tens of thousands of **accessions** (Bunting, 1979). Efforts have been made to increase the range of genetic variability by exchange of seed (Anonymous, 1982).

The effort required to handle so many accessions is great. Ford-Lloyd (1978) pointed out that, for corn, the collections from Mexico and South America total about 40,000 accessions and represent only 40% of the global corn collection. Collections of wheat, rice, and sorghum are already large, and much collection is yet to be done. For each accession, possibly ten to 50 pieces of descriptive data exist. It is clear that electronic computers are required to deal with the massiveness of the information bank.

Global cooperation in germplasm collection and preservation is part of a larger global experiment in agricultural development under the Consultative Group on International Agricultural Research (CGIAR). The work of the international agricultural research organizations, listed in Appendix C, will be described in a subsequent chapter; in the present chapter we shall describe the work of the International Rice Research Institute (IRRI) in germplasm collection and conservation. IRRI, established in 1960, was the first international agricultural research institute and had the following objectives in relation to germplasm conservation:

1. To assemble at one site the world's available rice germplasm for basic and applied studies to improve the crop.
2. To produce and preserve sufficient seed of each rice accession so as to conserve the world's dwindling stock of germplasm.
3. To develop a morphological and agronomic catalogue of descriptions of accessions.

4. To supply rice researchers with needed seed, plant material, and technical information.

IRRI began its gene bank in 1961 by systematically contacting national and international agencies in rice-producing countries. Included in the collection are commercially important cultivars, obsolete cultivars, exotic stocks, breeding material, types reported to have special merit, primitive cultivars, and wild forms (Chang, 1972). A modern gene bank storage, named the N. C. Brady Laboratory after IRRI's founder, was opened in 1982 and houses 60,000 accessions of rice (Brady, 1982).

Gene banks can be divided into two groups depending on whether they contain base or active collections. **Base collections** are used for long-term preservation. Even though seeds may have a storage life of more than 100 years, germination tests may be conducted every five to ten years to determine the decline in germination. When regeneration is required, samples should be sufficiently large to assure that representative genotypes are increased, hopefully under conditions similar to those in which they were collected (Hawkes, 1981). Base collections may be stored in large glass jars with silica gel to assure that seeds remain dry and viable and to prevent decay. **Cryopreservation** offers a technique whereby all metabolic processes essentially are stopped and held in suspended animation. Consequently, all sources of seed deterioration that are metabolically related are reduced greatly or stopped, thus providing "indefinite" preservation. How long the seed can be held in liquid nitrogen without loss of viability is unknown and will be unknown for years. The potential of cryopreservation is considerable, and it may provide a much better quality of seed and improve maintenance of plant genetic resources in gene banks (Stanwood and Bass, 1981). Cryogenic storage may be suited best to base collections, as the system may not be flexible enough to handle active collections where seed is needed frequently.

Active collections are used for medium-term storage, regeneration, multiplication and distribution, evaluation, and documentation and may be kept separate from the base collections. The National Plant Germplasm System of the United States involves Regional Plant Introduction Stations differing in climate zones and hence in the crops studied. Active collections of all seed stocks for all accessions held at the National Seed Storage are maintained at each regional laboratory. The base collection at the National Seed Storage Laboratory at Fort Collins, Colorado, does not distribute seed to users unless other supplies are unavailable.

ASSURING CROP DIVERSITY

It is impractical to attempt to collect every genotype of a particular crop, but it is essential that geographical areas be sampled. There are definite limits to the number of samples that can be handled effectively in gene banks. A measure of genetic diversity would be of value to ensure the preservation of the maximum amount of useful genetic variability while keeping the total number of samples within acceptable limits. Samples from a specific area having the same soil type, elevation, or other similarities may be bulked in base collections. This would help to reduce the numbers handled, yet hopefully would preserve the genetic diversity that allowed this type of plant to survive and flourish in that particular environment.

Representative variation within and among collections of genotypes is a key to the success of world gene banks. The Shannon-Weaver information index (H^1) provides a measure of phenotypic diversity based on frequency data (Marshall and Brown, 1975). The index can be used to evaluate geographical patterns of diversity and to determine specifically the relative contributions of various countries to the germplasm resources of an existing world collection (Jain et al., 1975). The estimates of diversity, H^1, can range from 0, which indicates theoretically no variation, to 1, for the maximum possible variation. Calculation of H^1 values for 3000 durum wheat accessions in the collection in the United States showed that accessions from Ethiopia and Portugal contributed the greatest diversity; H^1 values were 0.87 and 0.86, respectively. Accessions from Turkey, Bulgaria, and the United

States contributed substantially less diversity, and H[1] values were 0.50, 0.42, and 0.39, respectively. The conclusion reached was that diversity in durum wheat in the collection of the United States was limited and collections needed to be expanded.

A similar analysis of diversity in 1895 rice accessions randomly drawn from IRRI's catalogue of germplasm resources suggested considerable overlap, even though the entries analyzed represented nine different countries (Holcomb, Tolbert, and Jain, 1977). This conclusion was based on 41 phenotypic features similar to those reported in Tables 3-1 to 3-4.

A careful screening of over 17,000 cultivated sorghum accessions showed that genetic diversity could be represented well by approximately 2000 carefully chosen lines (Webster, 1976). It is evident that large numbers in a collection are not a reflection of a high level of genetic diversity. Gaps in gene banks may exist because some regions are relatively inaccessible. Easily obtained "roadside samples" may comprise the bulk of sampling from regions and may not represent the extent of the variability found in the species in the region. Often ten or fewer plants were taken to represent the population.

To ensure genetic diversity when little or no information is available on the natural variation found in a region, Marshall and Brown (1975) suggested that 50 to 100 individual plants be collected per site. It is also important to sample as many sites as possible and to ensure that sampling sites represent as broad a range of environments as possible.

CONCLUSIONS

For some time the value of genetic diversity has been appreciated as a source of germplasm so necessary for the development of cultivars having genetic resistance to pests or for morphological and physiological yield increases. Until the mid-1970s progress on the network of base collections was slow and inconclusive. Action occurred when it became evident that the genetic treasures found in the centers of diversity not only were immeasurably valuable but were vanishing rapidly. Interest has spread to including primitive cultivars because of the immense range of variation they contain and because they have as yet been scarcely explored and exploited. Widespread concern and awareness of the value of genetic diversity led to the development of a global network of gene banks.

The development of an elaborate world-scale collection and storage network should not convey the idea that germplasm conservation is solved. Considerable progress has been made, but the task is enormous. No longer, however, is plant exploration and collection viewed as a "rescue operation," whereby specific traits are searched out when the need arises.

Very large world collections do not assure that a sufficient genetic sample from a center of diversity is on hand. Gaps exist because some regions are relatively inaccessible, and the base sample may be comprised from easily obtained roadside samples.

We must recognize also that there are definite limits to the number of samples that can be handled effectively. Careful attention must be devoted to preservation of the maximum amount of useful genetic variability while keeping the total number of samples within acceptable limits.

It is fortunate that humans' major food crops are cereals, because they possess an enormous range of genetic diversity and are propagated by seeds that can be readily stored in gene banks. In the National Plant Germplasm System of the United States the small-grains collection is the largest working collection of germplasm and is deemed by many to be a "national treasure."

Questions

1. *a.* Consider possible morphological changes that may be desirable as a result of changes in agricultural practices.
   b. What evidence exists that some of these desired changes might be achieved?
2. *a.* Give examples of crop losses due to the outbreak of an epidemic.
   b. Can genetic resistance to a virulent strain of an epidemic be found reliably and readily?
3. *a.* What evidence exists that perennial or winter forms of millet or sorghum exist or could be developed?
   b. If winter or perennial forms of these crops are not found, does it indicate a breakdown of the Law of Homologous Series? Explain.
4. Speculate why a cultivar lacking ligules and having upright leaves may outyield a flat-leaved cultivar under certain environmental conditions.
5. Discuss the validity of the concern that natural reservoirs of germplasm resources are being destroyed or that genetic resources erosion is occurring at an accelerating rate.
6. How might genetic diversity be affected in a crop that evolved under conditions that varied in elevation, soil type, moisture level, or other climatic extremes?
7. *a.* Review the comment that genetic erosion has occurred because plant breeders have been so successful.
   b. Speculate on the agronomic advances that might be expected if plant breeders did not spend so much time locating and incorporating genes for pest resistance.
8. In addition to providing resistance to the cereal leaf beetle, what other possible functions might leaf pubescence serve? Consider both desirable and undesirable features.
9. Trace the rise and fall of common cereal crop cultivars grown in your region to determine their life span. Why are old, faithful cultivars replaced in your region?
10. *a.* Locate some of the centers of early agriculture and compare them to the probable origins of crops.
    b. What crops did these first agriculturists produce in these areas?
11. Chapter 3 has been restricted to genetic diversity and gene banks for cereal crops. The task of germplasm conservation, however, extends beyond these crops. Hawkes (1981) reported that collections exist for some 200 crop species with a total of about 7,300,000 accessions. Comment on the problems associated with such an extensive system of collection and gene preservation. Note that a comprehensive review of this broad problem is available in the book edited by Frankel and Hawkes (1975).
12. *a.* The numbers of accessions for selected crops were reported as follows: 36,000 wheat accessions at the United States Department of Agriculture Small Grain Collection at Beltsville, Maryland (Porter and Smith, Jr., 1981); 60,000 rice accessions in the Philippines (Brady, 1982); and 100,000 corn accessions worldwide (Ford-Lloyd, 1978). Compare the problems of regenerating new seed for these accessions among these crops.
    b. Explain why very large collections may not assure that a sufficient genetic sample from a center of diversity is on hand.
13. *a.* How can genetic vulnerability be reduced?
    b. Is genetic vulnerability a regional, national, or international problem?
14. List additional morphological items found in cereal crops (items not listed in Tables 3-1 to 3-4) that may be of interest to a plant breeder or a taxonomist.
15. Explain how greater genetic diversity existed in corn when open-pollinated corn was produced.
16. *a.* Speculate on how specific areas came to be centers of origin.
    b. On what basis do you suspect Vavilov concluded that a high degree of genetic diversity reflects the probable center of origin?

References

Anonymous. 1982. "U.S. Germplasm Collection Team Makes Visit to USSR." *Diversity* 1 (3): 12.

Barney, G. O. (study director). 1980. *The Global 2000 Report to the President of the U.S. Entering the 21st Century*. Vol. 1. *The Summary Report*. New York: Pergamon Press. 360 pp.

Brady, N. S. 1982. *A Global Experiment In Agricultural Development*. International Rice Research Institute, Los Banos, Philippines. 24 pp.

Bunting, A. H. 1979. "Genetic Conservation in Crop Plants." *Span* 22 (1): 9–11.

Chang, T. T. 1972. "International Cooperation in Conserving and Evaluating Rice Germ Plasm Resources." (In) *Rice Breeding*. Los Banos, Philippines: International Rice Research Institute. pp. 177–185.

Creech, J. L. 1970. "Crop Germ Plasm: A Resource for the Future." *Span* 13 (2): 112–113.

———, and L. P. Reitz. 1971. "Plant Germ Plasm Now and For Tomorrow." *Advances In Agronomy* 23: 1–49.

Donald, C. M. 1968. "The Breeding of Crop Ideotypes." *Euphytica* 17: 385–403.

Feldman, M., and E. R. Sears. 1981. "The Wild Gene Resources of Wheat." *Scientific American* 244 (1): 102–112.

Ford-Lloyd, B. V. 1978. "Data Storage and Retrieval Systems in Genetic Resources Information Exchange." (In) J. G. Hughes (ed.), *Conservation of Plant Genetic Resources*. Proceedings of Section K, British Association for the Advancement of Science. Birmingham: University of Aston. pp. 42–45.

Frankel, O. H., and J. G. Hawkes (eds.). 1975. *Crop Genetic Resources for Today and Tomorrow*. Cambridge: Cambridge University Press. 492 pp.

Harlan, J. R. 1972. "Genetics of Disaster." *Journal of Environmental Quality* 1: 212–215.

———. 1975a. *Crops and Man*. Madison, Wis.: American Society of Agronomy. 295 pp.

———. 1975b. "Our Vanishing Genetic Resources." *Science* 188: 618–621.

———. 1976. "The Plants and Animals That Nourish Man." *Scientific American* 235 (3): 89–97.

Hawkes, J. G. 1977. "The Importance of Wild Germplasm in Plant Breeding." *Euphytica* 26: 615–621.

———. 1981. "Germplasm Collection, Preservation and Use." (In) K. J. Frey (ed.), *Plant Breeding II*. Ames, Iowa: The Iowa State University Press. pp. 57–83.

———, and H. Lamberts. 1977. "Eucarpia's Fifteen Years of Activities in Genetic Resources." *Euphytica* 26: 1–3.

Holcomb, J., D. M. Tolbert, and S. K. Jain. 1977. "A Diversity Analysis of Genetic Resources in Rice." *Euphytica* 26: 441–450.

Jain, S. K., et al. 1975. "Geographical Patterns of Phenotypic Diversity in a World Collection of Durum Wheats." *Crop Science* 15: 700–704.

Kemp, H. J. 1934. "Studies of Solid Stem Wheat Varieties in Relation to Wheat Stem Sawfly Control." *Scientific Agriculture* 15: 30–38.

Konzak, C. F., and S. M. Dietz. 1969. "Documentation for the Conservation, Management and Use of Plant Genetic Resources." *Economic Botany* 23: 299–308.

Lambert, R. J., and R. R. Johnson. 1978. "Leaf Angle, Tassel Morphology, and the Performance of Maize Hybrids." *Crop Science* 17: 499–502.

Leppick, E. E. 1969. "The Life and Work of N. I. Vavilov." *Economic Botany* 23: 128–132.

Leisle, D. 1974. "Genetics of Leaf Pubescence in Durum Wheat." *Crop Science* 14: 173–174.

Marshall, D. D., and A. H. D. Brown. 1975. "Optimum Sampling Strategies in Genetic Conservation." (In) O. H. Frankel and J. G. Hawkes (eds.), *Crop Genetic Resources For Today And Tomorrow*. Cambridge: Cambridge University Press. pp. 53–80.

McClintock, D. W. 1982. "Politics and Economics of Genetic Diversity." *Diversity* 1 (3): 17–18.

McNeal, F. H., C. A. Watson, M. A. Berg, and L. E. Wallace. 1965. "Relationship of Stem Solidness to Yield and Lignin Content in Wheat Selections." *Agronomy Journal* 57: 20–21.

Miller, J. 1973. "Genetic Erosion: Crop Plants Threatened by Government Neglect." *Science* 182: 1231–1233.

Porter, W. M., and D. H. Smith, Jr. 1981. "Genetic Vulnerability." *Science* 214: 862.

Reitz, L. P., and J. C. Craddock. 1969. "Diversity of Germ Plasm on Small Grain Cereals." *Economic Botany* 23: 315–323.

Ringlund, K., and E. H. Everson. 1968. "Leaf Pubescence in Common Wheat, *Triticum aestivum* L., and Resistance to the Cereal Leaf Beetle, *Oulempa melanopus* (L.)." *Crop Science* 8: 705–710.

Stanwood, P. C., and L. N. Bass. 1981. "Seed Germplasm Preservation Using Liquid Nitrogen." *Seed Science and Technology* 9: 423–437.

Vavilov, N. I. 1949–1950. *The Origin, Variation, Immunity and Breeding of Cultivated Plants.* Translated from the Russian by K. S. Chester. Waltham, Mass.: Chronica Botanica 13. 364 pp.

Wade, N. 1972. "A Message from Corn Blight: The Dangers of Uniformity." *Science* 177: 678–679.

Webster, O. J. 1976. "Sorghum Vulnerability and Germplasm Resources." *Crop Science* 16: 553–556.

Went, F. W. 1971. "Parallel Evolution." *Taxon* 20: 197–226.

4

Cereal Crop Yield—A Systems Approach

INCREASING GRAIN YIELD

In the year 2000 the population of the world is projected to reach 6.35 billion people—an increase of 50% over the 1980 population. With a projected increase in arable land between 1980 and 2000 of only 4%, it is obvious that most of the increased output of food will have to come from higher yield per unit land area.

Agriculturists are not in agreement whether yield can increase at the rate required, or exactly how the required increases will be achieved. On the one hand, the magnitude of record yields compared to average yields in the United States has demonstrated the large genetic potential that exists in five cereal crops (Table 4-1). The fact that yields average only 17.3% of the record yields suggests that opportunity for increased yields exists and that emphasis should be directed to bringing productivity closer to the very high genetic potential demonstrated.

On the other hand, Jensen (1978) observed an apparent levelling off of grain yield in newly released wheat cultivars in New York State and suggested that a yield plateau may have been reached and that further yield increases might be difficult to achieve. Jensen argued that an already advanced yield base exists and that further improvement in grain yield will be slow and limited.

Lupton (1982), observing that wheat and barley yields in Great Britain had doubled since the end of World War II, suggested that about half the increase in yield was due to genetic improvement. Similar grain-yield increases were demonstrated for 20 barley cultivars released between 1880 and 1980 and tested simultaneously over a three-year period under conditions in which disease and lodging were prevented (Table 4-2). The genetic gain in gain yield was 0.39% per year during the 100-year period and 0.84% per year between 1953 and 1980.

If 50% of the improvement arose genetically, presumably the other 50% arose from improved agronomic practices. The plant breeder appears to have selected the highest-yielding **genotype** through **empirical** yield trials; the crop producer grew the resulting **phenotype** using cultural practices that produced the highest yield under empirical tests. The producer may have had to adjust cultural practices to suit the genotypes available to allow the full potential of each cultivar to be expressed. Perhaps neither the plant breeder nor the producer has addressed fully the fact that photosynthesis is the basis of all yield and that close cooperation in the breeding and production of crops using techniques that emphasize photosynthesis might together produce a higher yield.

Table 4–1
Record and Average Grain-Yield Comparisons for Five Cereal Crops in the United States

Crop	Record Yield kg/ha	Record Yield lb/ac	Average Yield kg/ha	Average Yield lb/ac
Corn	19,300	17,232	4600	4107
Wheat	14,500	12,946	1880	1679
Sorghum	20,100	17,946	2830	2527
Oat	10,600	9,464	1720	1536
Barley	11,400	10,179	2050	1830
Mean	15,180	13,553	2616	2336

Source: Adapted from Boyer (1982).

CHANGES IN PLANT ARCHITECTURE—A MODERN INNOVATION

By itself, **plant architecture** has no merit in a yield sense except as it is associated with or leads to improved physiological functions (Adams, 1982). Architectural changes that result in better sunlight interception and an increase in photosynthesis have the potential for increasing yield. An opportunity for improving plant architecture came about with the development of herbicides and led to the development of smaller plants with upright leaves. Both are desirable, because smaller plants require less photosynthate to develop vegetative plant structures, and their smaller size, along with upright leaves, allows light to penetrate more uniformly through the crop, allowing lower leaves to make a greater contribution to photosynthesis or extending the photosynthetic duration for these leaves. Smaller plants with upright leaves may be able to carry on photosynthesis at an acceptable level under stress conditions caused by a high plant population; and although the yield per plant may be reduced, yield per unit area may be increased.

For centuries, crops such as corn and sorghum were planted in rows to facilitate cultivation between the rows for weed control. The width of the row for these crops was commonly a meter (39 inches), established on the basis of the space required for a draft horse pulling a scuffler or cultivator to pass without damaging the plants. Under such wide-row spacings, the best-yielding cultivars were those that had many large, horizontal leaves which could intercept sunlight most effectively and compete well with weeds when rows could no longer be cultivated because of plant size. Cultivars that tillered well and suppressed weeds effectively, even though many late-formed tillers were unproductive, were considered to be best in cereal grains. Tall cultivars that produced many tillers with large, horizontal leaves, competed well with weeds, and outyielded shorter or upright-leaved cultivars.

In the late 1950s and early 1960s, selective herbicides became commercially widespread and resulted in virtually 100% weed control. No longer was it necessary to plant in wide rows,

Table 4–2
Grain Yield of 20 Barley Cultivars Released Between 1880 and 1980 and Grown at the Plant Breeding Institute, Cambridge, England, When Averaged over the Years 1978, 1979, and 1980

Cultivar	Year of Introduction	Grain Yield kg/ha	Grain Yield lb/ac
Chevalier	1880 (ca.)	4610	4116
Goldthorpe	1880 (ca.)	4820	4304
Plumage	1900	4530	4045
Kenia	1932	5090	4545
Spratt Archer	1933	5190	4634
Plumage Archer	1935	5440	4857
Freja	1942	5590	4991
Rikà	1951	5420	4839
Proctor	1953	5650	5089
Vada	1960	5700	5089
Zephyr	1966	5950	5313
Golden Promise	1966	5510	4920
Julia	1968	6200	5536
Maris Mink	1973	5930	5295
Sundance	1976	6320	5643
Georgie	1976	6300	5625
Ark Royal	1976	6350	5670
Egmont	1980	6930	6188
Koru	1980	6740	6018
Triumph	1980	6680	5964

Source: Data from Riggs et al. (1981).

because interrow cultivation was unnecessary. Reducing row width without an increase in plant population helped equalize competition among plants, but before an increase in plant population was possible, changes in plant architecture were required.

Although the importance of photosynthesis for increased crop yield is intuitively obvious, changes in plant architecture were slow to develop. Corn producers overly eager to obtain the theoretically higher yields from increased plant populations often had disappointingly low yields, because stress conditions such as mutual shading of plants having an architecturally large plant size with horizontal leaves resulted in **barren plants** or **nubbin** development (Smith, Mock, and Crosbie, 1982).

The most rapid changes in plant architecture occurred in short-season areas where small plants are synonymous with early maturity. Changes were slow and methodical but positive, as full-season crops such as corn and sorghum were introduced into areas with a shorter and shorter growing season. In time, in the short-season areas, corn plant populations using adapted, architecturally small hybrid cultivars were increased gradually from 44,000 plants per hectare (18,000 plants per acre) to 62,000 pph (25,100 ppa) with a grain-yield increase.

A reduction in the need for corn plants to compete with weed interference allowed for the development of tillerless or uniculm corn hybrid cultivars. Uniculm corn provided plant-population control as well as **leaf-area index** control, so that precise planting rates were possible.

The discovery and use of dwarfing genes in crops including sorghum allowed for dramatic changes in plant architecture. When dwarfing genes were incorporated into adapted cultivars, yield increases were experienced. The architectural change to a short plant structure helped reduce lodging, a feature important in all cereal crops.

In Asia, a similar revolution in rice plant architecture was begun with the establishment of the International Rice Research Institute (Appendix C), motivated by the desire to increase rice yields. Indica, the traditional tropical rice plant, grows to a height of 160 to 200 cm (64 to 80 in.) with weak stems and droopy leaves; it produces many tillers but lodges readily (Figure 4-1). It is well adapted to Asian conditions, re-

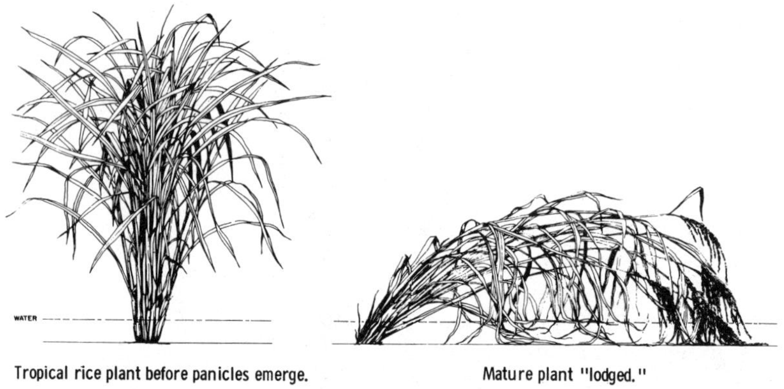

Tropical rice plant before panicles emerge.

Mature plant "lodged."

Figure 4–1. Traditional rice cultivars are tall, have weak stems, with wide, horizontal or droopy leaves. When fertilizers such as nitrogen are applied, traditional rice cultivars produce large heads but frequently lodge or fall over. Lodging disrupts light interception and photosynthesis as well as allowing the grain to be spoiled by the paddy water or be destroyed by rodents. The wide and droopy leaves at the top of the plant intercept most of the sunlight and prevent light energy from penetrating the leaf canopy. These rice cultivars are low-yielding. (Courtesy of International Rice Research Institute) quires little management input, competes well with weeds, and even under low-fertility conditions gives a reliable but low yield of 1 or 2 t/ha (900 to 1800 lb/ac).

Efforts to increase grain yield of the conventional tall indica rice cultivars through increased application of fertilizer, especially nitrogen, usually led to additional plant height, extra leaf growth, and increased lodging. Applications greater than 30 kg/ha (27 lb/ac) of nitrogen either do not increase yields or actually depress them owing to early lodging. The earlier in the season lodging occurs, the greater the yield reduction.

In contrast to indica rice, japonica rice cultivars grown in Japan are characterized by erect leaves (Figure 4-2), a low tillering capacity, and resistance to lodging, but they lack ability to compete with weeds and are susceptible to many diseases and insects. Under proper management, the best japonica cultivars are capable of yielding 6 to 8 t/ha (5400 to 6600 lb/ac) (Chandler, Jr., 1979).

Although previous attempts had been made to combine the indica and japonica cultivars, it

Modern semidwarf before panicles emerge. Mature semidwarf.

Figure 4–2. Modern semidwarf rice cultivars have strong lodging-resistant stems that are able to support high grain yields. The cultivars have a high rate of photosynthesis, because the upright leaves allow sunlight to penetrate the leaf canopy. Greater sunlight penetration promotes tillering and causes differentiation of the lower internodes to reduce the lodging problem. Many modern rice cultivars are insensitive to daylength, so that farmers can grow them at many latitudes and at any time of the year. (Courtesy of International Rice Research Institute)

Figure 4–3. Tall and short rice cultivar comparisons. Note the pronounced upright leaves of 1R8, one of the first cultivars developed at IRRI. Other successful rice cultivars developed at IRRI, including IR36 and IR54, have continued to give yield advances as well as providing genetic resistance to insects and diseases. (Courtesy of International Rice Research Institute)

was not until 1966 that cultivar IR8-288-3 (named IR8) from the IRRI program made the cross truly successful (Figure 4-3). IR8, an abrupt change from the traditional to the modern plant type, is characterized by seven features:

1. Semidwarf plants ranging from 90 to 100 cm (36 to 40 in.) in height (Figure 4-4).
2. Short, thick, sturdy stems that provide lodging resistance even under high nitrogen levels and conditions of strong winds accompanied by heavy rains.
3. A high photosynthetic rate related to improved sunlight interception by short, upright leaves of medium width (Figure 4-5).
4. High tillering capacity, which means more rice panicles per unit area of land.
5. An increase in the grain-to-straw ratio from about 30% of the total dry matter in the form of grain to about 50%.
6. Insensitivity to photoperiod, which means that reproductive development will occur regardless of the month seeded, allowing for more than one crop per year.
7. Increased grain yield under high levels of nitrogen fertility.

Figure 4–4. Rice test plots showing height variations. The measuring stick is calibrated in centimeters. Daylength insensitivity has allowed the modern semidwarf rice cultivar on the left to progress more rapidly than the tall, late-maturing cultivar in the center. Early maturity allows for more than one crop to be grown in the tropics. (Courtesy of International Rice Research Institute)

Figure 4–6. The semidwarf, upright-leaved wheat cultivar in the center of this replicated yield trial is an example of the modernization in the morphology of cereal crops. (Courtesy of Ontario Ministry of Agriculture and Food)

Similar changes in wheat occurred simultaneously (Figure 4-6). Following World War II, Japanese semidwarf wheat cultivars, named Norin 10 and Brevor, were introduced into North American wheat breeding programs. By 1953 the use of Japanese dwarf wheat cultivars had led to a reduction in plant height, an increase in the number of fertile florets per plant, an increase in tillering, and an unusual breadth of adaptation.

Exploitation of the outstanding features of modernized wheat and rice cultivars led to a worldwide acceptance of modern cereal crops. The result was an unprecedented increase in yield that became popularized as the **green revolution** (Borlaug, 1968, 1972).

Figure 4–5. Paddy rice plants with upright leaves and a high tillering capacity. (Courtesy of International Rice Research Institute)

SHORT-STATURE CEREAL CROPS

Changes associated with the modernization of cereal crops include: herbicidal weed control; a change in the status of straw from that of a valuable by-product for livestock bedding to that of a nuisance material; direct combine harvesting of cereal crops once considered too short to be harvested with a binder; and an increase in the lodging problem owing to irrigation and increased fertilization. Success of semidwarf cultivars of corn, sorghum, wheat, and rice have dramatized the opportunities associated with a height reduction in other cereal crops and introduced the concept of further height reductions as advantageous.

Change to true dwarfs in wheat and rice and semidwarfs in other cereals has, however, been slow and discouraging. This might be explained by the fact that a strong genetic base for height reductions is not readily available in many crops. The difficulty in developing dwarf cultivars caused some plant breeders to conclude that it is impossible to combine high yield with short straw (Briggle and Vogel, 1968) even in wheat, despite the initial success of the major dwarfing genes from Norin 10 and Brevor (Allan, Vogel, and Peterson, 1968).

From a production viewpoint, Army, Isleib, and Greer (1971) suggested that the proliferation of factors associated with crop production may have outstripped the ability of producers to incorporate them in the total management system required.

From a physiological standpoint, an increase in the harvest index has resulted in more grain being produced, even though little change in biomass has occurred. Lupton (1982) warned about the obvious limits to the extent to which this process can continue without an increase in photosynthate produced.

The problems associated with dwarf cereal crop development and production have permeated all areas of the world and merit consideration from the viewpoint of lessons to be learned in physiology, plant breeding, and crop production.

PLANT-HEIGHT AND GRAIN-YIELD RELATIONSHIPS

Although reductions in plant height have been a breeding objective for many years, success has been limited and slow (Table 4-3). Over a 100-year period, height reductions from about 100 to 75 cm (39 to 30 in.) or about 25% for spring wheat in Britain have occurred. In winter wheat, height was reduced over an 80-year period from about 112 to 75 cm (44 to 30 in.) or about 39%. Despite evidence that yield increases accompanied height reductions, these combinations were not easily obtained. In general, attempts to reduce plant height usually are accompanied by reduced grain yield in common and durum wheat, oat, barley, and sorghum (Law, Snape, and Worland, 1978). The shortest segregates following a cross of parents carrying dwarfing genes appear to be the lowest-yielding, with the "tall-dwarfs" having a more satisfactory yield performance (Fedak and Fejer, 1982). In durum wheats as well, Joppa (1973) reported the taller semidwarfs as having the best yield when cultivars differing by one height gene were compared.

Some of the reasons associated with yield-height problems may be as follows:

HARVEST INDEX *LIMITATIONS.* Short-stature plants may have a higher harvest index than tall-strawed cultivars (Table 4-4). A high harvest index may be a measure of the efficiency of partitioning photosynthate into grain and straw. Lupton (1982) and Riggs et al. (1981) warned that the extent to which grain yield can be increased under conditions of constant or declining total dry weight, as portrayed in Table 4-4, may be limited.

If grain yield of short-statured cultivars is to be increased, an increase in photosynthate produced may be essential. Harvest index values and total dry-matter weight for the spring barley and winter wheat cultivars in Table 4-3 are shown in Tables 4-5 and 4-6, respectively. In both crops, total dry matter or biomass yields remained relatively constant, but, because a better distribution of photosynthate into grain and straw occurred, as shown by a rising harvest index, grain yield increased. Austin and his co-workers (1980) argued that breeders may be able to increase harvest index values from the present 50% to about 60%, thereby achieving a gain in grain yield of about 25%. This could be achieved by a continuation of the trend toward reduced stem length, with no change in above-ground biomass.

An upper limit for the harvest index value is unknown. Fischer et al. (1981) suggested that at a harvest index of 50% and a height of 80 cm (32 in.) in wheat, no further grain yield increases can be expected without increasing total dry-matter production. Lukyanenko (1970), however, reported a dwarf wheat having a harvest index of 85% and felt that wheat plants with a height of 50 to 60 cm (20 to 24 in.) offer wide possibilities.

Table 4–3
Plant Height in Spring Barley Cultivars Developed from 1880 to 1980 and Winter Wheat Cultivars Developed from 1908 to 1980 in England

Spring Barley Cultivars[a]	Plant Height cm	Plant Height in.	Winter Wheat Cultivars[b]	Plant Height cm	Plant Height in.
Chevalier	98	39	Little Joss	127	50
Goldthorpe	102	40	Holdfast	112	44
Plumage	112	44	Cappelle Desprez	99	39
Kenia	96	38	Maris Widgeon	112	44
Spratt Archer	92	36	Maris Huntsman	96	38
Plumage Archer	98	39	Hobbit[c]	72	28
Freja	82	32	Mardler[c]	67	26
Rika	85	33	370/500[c]	70	28
Proctor	84	33	730/3637[c]	69	27
Vada	89	35	989/10[3]	74	29
Zephyr	79	31	Armada	87	34
Golden Promise	65	26	Benoist 10483	77	30
Julia	77	30			
Maris Mink	65	26			
Sundance	77	30			
Georgie	72	28			
Ark Royal	82	32			
Egmont	83	33			
Koru	81	32			
Triumph	70	28			

[a]*Source:* Riggs et al. (1981). Results averaged from crops grown over the three years 1978–1980 at Cambridge, England.
[b]*Source:* Austin et al. (1980). Experiments were carried out in 1978 at Cambridge, England, and averaged over two tests.
[c]Homozygous for a major drawfing gene.

SINK-SOURCE LIMITATIONS. A high harvest index value associated with plants with reduced stature cannot be regarded as a cause-and-effect relationship but, rather, as a final measure of distribution. The greater part of the carbohydrate utilized in filling the grain of determinate crops is synthesized during the period after ear emergence—that is, after the stems have achieved their final height. Some carbohydrate stored in the stem prior to grain development

Table 4–4
Total Dry Weight or Biomass, Grain Yield, and Harvest Index of Three Winter Wheat Cultivars Grown at Reading, England, in 1967

Cultivar Type	Total Dry Weight kg/ha	Total Dry Weight lb/ac	Grain Yield kg/ha	Grain Yield lb/ac	Harvest Index, %
Tall	9650	8616	3500	2917	37
Tall	9250	8259	3950	3527	43
Reduced in height	8600	7679	4500	4018	52

Source: Data from Drennan (1970).

Table 4–5
Total Dry-Matter Yields and Harvest Index Values for Spring Barley Cultivars Grown at Cambridge, England, and Averaged over the Years 1978–1980

Spring Barley Cultivars	Total Dry Matter		Grain Yield		Harvest Index, %
	kg/ha	lb/ac	kg/ha	lb/ac	
Chevalier	13,880	12,393	4610	4116	33
Goldthorpe	12,400	11,071	4820	4304	39
Plumage	12,160	10,857	4530	4045	38
Kenia	12,630	11,277	5090	4545	40
Spratt Archer	12,800	11,429	5190	4634	41
Plumage Archer	14,150	12,634	5440	4857	40
Freja	12,980	11,589	5590	5991	43
Rika	12,150	10,848	5420	4839	45
Proctor	13,590	12,134	5650	5045	42
Vada	13,400	11,964	5700	5089	42
Zephyr	12,800	11,429	5950	5313	46
Golden Promise	11,410	10,188	5510	4920	48
Julia	13,770	12,295	6200	5536	45
Maris Mink	12,490	11,152	5930	5295	47
Sundance	12,630	11,277	6320	5643	50
Georgie	12,670	11,313	6300	5625	50
Ark Royal	14,200	12,679	6350	5670	45
Egmont	14,450	12,902	6930	6188	48
Koru	14,700	13,125	6740	6018	46
Triumph	13,310	11,884	6680	5964	50
Mean	12,129	10,829	5748	5132	43.9

Source: Data from Riggs et al. (1981).

Table 4–6
Total Dry-Matter and Grain Yields with Harvest Index Values for Winter Wheat Cultivars Developed from 1908 to 1980 in England: Results Averaged over Two Tests Grown at Cambridge, England, in 1978

Winter Wheat Cultivars	Total Dry Matter		Grain Yield		Harvest Index, %
	kg/ha	lb/ac	kg/ha	lb/ac	
Little Joss	12,171	10,867	4260	3804	35
Holdfast	11,243	10,038	3935	3513	35
Cappelle Desprez	11,429	10,204	4800	4286	42
Maris Widgeon	11,869	10,597	4625	4129	39
Maris Huntsman	11,500	10,268	5290	4723	46
Hobbit	12,173	10,869	5965	5326	49
Mardler	10,708	9,561	5140	4589	48
370/500	10,806	9,648	5295	4728	49
730/3637	11,653	10,404	5710	5098	49
989/10	12,160	10,857	6080	5429	50
Armada	12,614	11,263	5550	4955	44
Benoist 10483	11,734	10,477	5515	4924	47
Mean	11,672	10,421	5180	4625	44.4

Source: Data from Austin et al. (1980).

can be drawn on to help develop the sink or grain yield (Lupton, Oliver, and Ruckenbauer, 1974). It appears that in short-stature cereal crops this source limitation may be a factor in the low yields of dwarf cultivars.

LESS YIELD RELIABILITY. The fact that short-stature cultivars store less of their photosynthate produced after grain development begins in the stem and more in the grain has not turned out to be an advantage, as initially believed (Fiddian, 1978). During periods of stress, tall cereal crops are able to transfer stem-stored photosynthate to the developing ear. If this source of dry matter is reduced because of a reduction in height, the development of short-stature cereal crops may result in cultivars less well buffered against stress.

BRITTLE STEMS. Short-stature crops are less subject to lodging, primarily because they have less torque on their lower stems because of reduced wind resistance. A unit length of stem from a tall or a short cultivar, however, may not differ in its resistance to breaking. It is possible that, because short-stature cereals store less dry matter in their stems, they may have brittle, dry, and inflexible stems that break under pressure caused by wind or rain.

POOR SEEDLING EMERGENCE. Drill-seeded short-stature cereal crops emerge more slowly and with less consistency than tall cereal crops seeded at the same depth. This appears to be explained by the fact that a height reduction in straw also is associated with a reduced length of coleoptile. When seed is planted, the coleoptile emerges from the seed and forms a spearlike shoot that is pushed through the soil by the elongating mesocotyl and serves to protect the first leaf or plumule. The depth of planting cannot exceed the total length of the coleoptile and mesocotyl. If seeded beyond this depth, the plumule will emerge from the coleoptile but may not be able to penetrate to the soil surface (Turner, Chen, and Bollich, 1982). Caution must be exercised not to plant short-stature cultivars too deeply. When semidwarf wheat cultivars were sown 10 to 13 cm (4 to 5 in.) deep, difficulty was encountered in obtaining proper stands (Peterson and Allan, 1980).

ROOT LIMITATIONS. Concern has been expressed that the roots of semidwarf cereals may not go deeply enough into the soil for dryland production (Briggle and Vogel, 1968). Considerable variation appears to exist in root structure, and short-stature cereals may not have an inadequate root system (Cannell and Drew, 1973; Lupton et al., 1974; Irvine, Harvey, and Rossnagel, 1980). When grown indoors in a 1.2-meter (4-ft) deep container, a tall and a dwarf wheat had roots equally as deep but differing in volume (Figure 2-8).

LOWER NET PHOTOSYNTHETIC RATES. Height reductions are achieved by a shortening of each internode. Since a leaf sheath with a leaf blade arises from each internode, dwarf plants have the same number of leaves as their tall counterparts and are called *brachytic*. Brachytic plants may have a reduced rate of photosynthesis because of mutual shading related to foliage density. A high tillering rate associated with semidwarf cultivars adds to this foliage density. Mutual shading may reduce the net photosynthetic rate. Changes in plant architecture to one with upright leaves may be an essential requirement with short-stature cereal crops.

C_3 cereal crops are subject to an increase in photorespiration rates with increases in temperature. The temperature near the soil under a crop is often substantially different from the temperature 1 meter (39 in.) above the soil surface. Short-stature C_3 cereal crops may be subjected to more extreme temperature differences associated with the **microclimate.** In dwarf spring wheat, a 1°C rise in temperature above the optimum was associated with a 4% grain-yield reduction (Fischer and Mauer, 1976). In barley, the microclimate effect on the developing spike resulting from incomplete emergence from the boot of short-stature barley cultivars adversely affected grain yield (Fedak and Fejer, 1982).

HEIGHT GOALS IN CEREAL CROPS

The development of semidwarf and dwarf cereal cultivars appears as a necessary objective to increase yield, reduce lodging, and facilitate harvesting. The ease with which plants with short stature can be identified in segregating populations allows for rapid shifts toward a specific height objective. Unfortunately, desirable short-stature plants may have poor yield performance until other limitations previously described are incorporated adequately into a single cultivar.

The image of an ideal crop height varies. Drennan (1970) suggested a scale for wheat might be in the following range:

Tall—100 cm (40 in.) or more
Semidwarf—70 to 90 cm (28 to 30 in.)
Dwarf—50 to 70 cm (20 to 28 in.)
Double dwarf—shorter than 50 cm (20 in.)

Continued intensive efforts in developing cereal crops toward true dwarfs will be required to develop a suitable cultivar.

YIELD AS A PRODUCT

Components of Yield

Efforts to describe small-grain yield often have been based on the number of ears per unit area, the number of grains per ear, and grain size. Grain yield can be expressed as the product of these three components. Values obtained among cultivars (Table 4-7) are useful for comparison purposes. Examination of the three components reveals considerable variation, but generally as one component increased or decreased, one or both of the other components compensated accordingly unless grain yield increased. Variations in yield are not determined by the components of yield but by the amount of photosynthate available for distribution and how it is partitioned.

The compensation among the components was described by Grafius (1965) as the geometry of plant breeding. Grafius worked with oat cultivars and expressed compensation values in mathematical terms in an effort to understand how a given environment might be used to advantage.

Under conditions of a constant rate of photosynthesis, components of yield cannot be manipulated successfully to increase yield. This fact has sometimes been overlooked, and cultivars with a high value for one component have been crossed to a cultivar with a high value for another component in an effort to combine the two high values. Only when the total photosynthate available for distribution is increased can the components of yield be increased simultaneously. Attention therefore should be directed to factors that increase photosynthesis or harvest index.

Photosynthetic Components of Yield

The amount of photosynthate available for distribution into the components of yield is a prod-

Table 4–7
Grain Yield and Its Components Averaged over the Years 1978, 1979, and 1980 for Nine Spring Barley Cultivars Grown at Cambridge, England

Cultivar	Grain Yield		Ears per Square Meter	Yield per 10 Ears, g dry wt	1000-Grain Weight, g dry wt
	kg/ha	lb/ac			
Chevalier	4610	4116	586	6.8	38.3
Plumage	4530	4045	481	8.6	40.4
Kenia	5090	4545	524	8.0	38.1
Plumage Archer	5440	4857	575	8.5	41.1
Freja	5590	4991	697	6.4	36.6
Golden Promise	5510	4920	759	6.6	31.9
Ark Royal	6350	5670	774	6.9	34.9
Egmont	6930	6188	566	9.9	43.6
Triumph	6680	5964	800	7.3	38.0

Source: Data from Riggs et al. (1981).

uct of the rate and duration of photosynthesis. Many factors may influence both rate and duration, including environmental factors, which will be considered in a subsequent chapter. Physiological factors are important also and include partitioning of plant development into vegetative and reproductive periods, leaf area, photorespiration, grain-filling duration and rate, and nitrogen absorbed.

A readily apparent factor influencing photosynthesis is photosynthetic area, which for convenience is termed leaf area. In the modernized rice cultivar IR8, Yoshida, Cock, and Parao (1972) noted that photosynthetic rate increased as leaf-area index (LAI) increased up to a value of 6. These workers noted that it is possible to produce an IR8 rice crop with an LAI value as high as 10 without inducing lodging. Such exceptionally high LAI values can be achieved because of the upright nature of IR8 leaves that allows for excellent penetration of sunlight into the crop, promoting photosynthesis in lower leaves, and helps reduce lodging. In crop cultivars with horizontal or droopy leaves, the reduced light energy at the lower internodes causes cell elongation and reduces **differentiation,** so that straw is weak and lodging occurs.

Plant populations that exceed the **optimum leaf area** index may result in detrimental mutual shading, especially at high nitrogen levels on fertile soils. In addition to leaf area and leaf angle, leaf shape and distribution also influence the absorption of sunlight energy by a crop canopy.

In the determinate cereal crops, crop producers attempt to have the optimum LAI coincide with the onset of grain development. This is attempted by manipulating seeding rate to suit leaf characteristics of the cultivar, the date of seeding, tillering, and fertility levels. In nontillering corn hybrids, precise seeding rates can be established for each area. For cereal crops with a variable tillering rate, a precise seeding rate cannot be given. For these crops, the yield response over a wide range of seeding rates can be depicted by a parabolic type of curve rising quickly to a maximum yield followed by a slow decline at high densities (Faris and De Pauw, 1980/81). Typically, there is a wide range of seeding rates that result in a near-maximum yield.

The shape of the yield response curve can be changed by several factors. In general, the greater the environmental resources, the higher will be the optimal seeding rate. Under conditions of adequate moisture, high soil fertility status, high light intensity, or early seeding, the optimum seeding rate of cereal crops can be increased.

Too high a seeding rate may result in a leaf area that exceeds a **critical leaf area** index, a level above which mutual shading of the lower leaves reduces the net photosynthetic rate. As respiration of lower leaves exceeds their gross photosynthetic rate, such leaves are said to be parasitic on the plant. Considerable controversy exists as to whether parasitic leaves or tillers exist. Tests with the rice cultivar IR8 and its consistent performance at high nitrogen levels in many parts of the world led Yoshida, Cock, and Parao (1972) to conclude that even at an LAI of 10, parasitic leaves did not exist.

The effect of nitrogen fertility on leaf-area index on 12 winter wheat cultivars measured two to three weeks prior to anthesis is shown in Table 4-8. Under high-fertility conditions, both the LAI and grain yield more than doubled over the low-fertility level. Substantial variations in LAI can be observed at both fertility levels. The cultivar 989/10 had the lowest and the cultivar 370/500 had the highest LAI at both fertility levels. A producer who knows the characteristics of each cultivar can manage his cultival practices accordingly. Cultivar 989/10, which had the lowest LAI, was generally the highest-yielding cultivar, which raises the question of whether additional fertility might be applied or if a higher seeding rate might be used to advantage.

In latitudes where daylength varies, LAI should be made to coincide with the high level of sunlight energy available around June 21, the longest day of the year in the northern hemisphere. Winter cereals, seeded in the fall, develop additional leaf growth early in the spring, and the maximum leaf area most closely coincides with the light-energy distribution curve (Figure 4-7) at Guelph, Ontario, Canada. Corn

Table 4–8
Grain Yield and Leaf-Area Indices in Winter Wheat Cultivars at Two Fertility Levels Grown at Cambridge, England, 1977–1978

	Low-Fertility Yield			*High-Fertility Yield*		
Cultivar	*kg/ha*	*lb/ac*	*LAI*	*kg/ha*	*lb/ac*	*LAI*
Little Joss	3300	2946	3.6	5220	4661	6.9
Holdfast	2910	2598	2.7	4960	4429	6.6
Cappelle-Desprez	3740	3339	3.1	5860	5232	7.4
Maris Widgeon	3570	3188	3.1	5680	5071	8.1
Maris Huntsman	4040	3607	3.1	6540	5839	8.2
Hobbit	4630	4134	3.0	7300	6518	6.3
Mardler	4070	3634	3.3	6210	5545	7.6
370/500	4250	3795	3.7	6340	5661	8.8
730/3637	4350	3884	3.1	7070	6313	8.1
989/10	4590	4098	2.8	7570	6759	5.9
Armada	4240	3786	3.1	6860	6125	7.6
Benoist 10483	3800	3393	3.1	7230	6455	7.5
Mean	3960	3536	3.1	6400	5714	7.4

Source: Data from Austin et al. (1980).

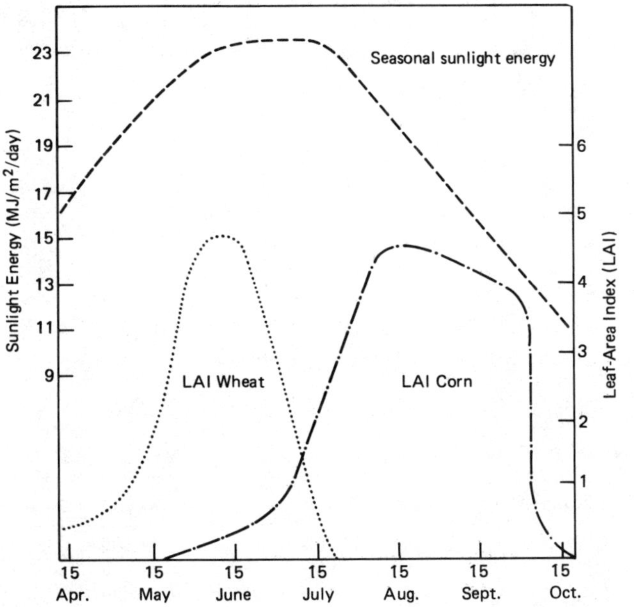

Figure 4–7. Average daily sunlight energy and leaf-area development of a wheat and a corn canopy during the growing season at Guelph, Ontario. (*Source:* Data from Tollenaar and Daynard, 1981.)

Table 4-9
Average Solar Radiation Received at Nine Locations in North America from May 1 to September 30

Location	Solar Radiation, 10^9 joule/m^2
Dayton, Ohio	2.90
South Bend, Indiana	2.94
Lexington, Kentucky	2.97
Minneapolis, Minnesota	2.98
Guelph, Ontario	3.08
Springfield, Illinois	3.22
Des Moines, Iowa	3.24
Lubbock, Texas	3.95
Sacramento, California	4.20

Source: Tollenaar and Daynard (1981).

at this location is seeded about May 1 and reached an LAI of 4 by about August 1, when reproductive development begins. At Guelph, complete or almost complete light interception by corn occurs at a LAI of 4. The duration of complete light interception by both the wheat and corn crop canopies constitutes a major limitation to seasonal sunlight interception, as becomes clear from the leaf development in wheat and corn at this location (Figure 4-7).

Optimum leaf-area indices may vary according to the average sunlight energy received during the growing season at various locations (Table 4-9). Hence the seeding rate or other management inputs also will vary. Potential crop dry-matter production was estimated by Tollenaar and Daynard (1981) for various scenarios using average daily sunlight energy, ranging from the unrealistic situation of complete light interception during the whole growing season to complete light interception during the month of August only (Table 4-10). Assuming a harvest index of 50% and a shoot-to-root ratio of 10, grain yield can be estimated.

Potential productivity, as estimated in Table 4-10, can be compared with record and average grain yields actually obtained in Table 4-1. Record yields of corn and sorghum compare favorably with the yield scenario estimated for complete light interception between August 1 to September 15, the period coinciding with reproductive development in these crops. The yield scenarios in Table 4-10 give an indication of the yield potential if sunlight could be effectively utilized over an extended period of the

Table 4-10
Scenarios of Potential Total Dry-Matter Production During a Growing Season Extending from May 1 to September 30 at a 40° N Latitude for Various Periods of Complete Interception of Solar Incident Radiation and Using an Average Photosynthetic Efficiency of 11.5 Quanta of Photosynthetically Active Radiation

Full Interception of Solar Radiation Extending from:	Estimated Yield			
	Total DM		Grain	
	kg/ha	lb/ac	kg/ha	lb/ac
May 1 to Sept. 30	112,000	100,000	44,800	40,000
July 1 to Sept. 30	79,000	70,536	31,600	28,214
July 15 to Sept. 15	59,500	53,125	22,800	20,357
Aug. 1 to Sept. 15	45,500	40,625	18,200	16,250
Aug. 1 to Aug. 31	36,500	32,589	14,600	13,040
Mean	66,500	59,375	26,400	23,571

Source: Data from Tollenaar and Daynard (1981).

growing season and emphasize the importance of crop management to capture sunlight energy. Other factors of competition, carbon dioxide level, moisture, pest control, and genetic constitution of the cultivar cannot be overlooked, as crop yield depends basically on the growth factor that is available to the plant in the smallest relative amounts. This basic fact, stated as the "law of the minimum" by the German chemist Justin Von Liebig in 1862 (Cannon, 1963), helps emphasize the need for a systems approach to crop production (Figure 4-8).

The CO_2 concentration in the atmosphere may limit grain yield. Concentrations higher than the normal 350-ppm level found in the atmosphere increase photosynthesis and grain yield. By increasing the CO_2 concentration to 900 ppm during the day around rice plants in the field, photosynthesis and grain yield (Yoshida, Cock, and Parao, 1972) were increased. The CO_2 enrichment before heading increased grain yield by 29% above the control and after heading by 21% (Table 4-11). The yield increase by the enrichment before heading was caused by increased grain number and grain weight. Enrichment after heading did not change grain number, as panicle size was fixed, but increased grain weight and filled-grain percentage. The plants that received CO_2 enrichment before heading had identical environmental conditions

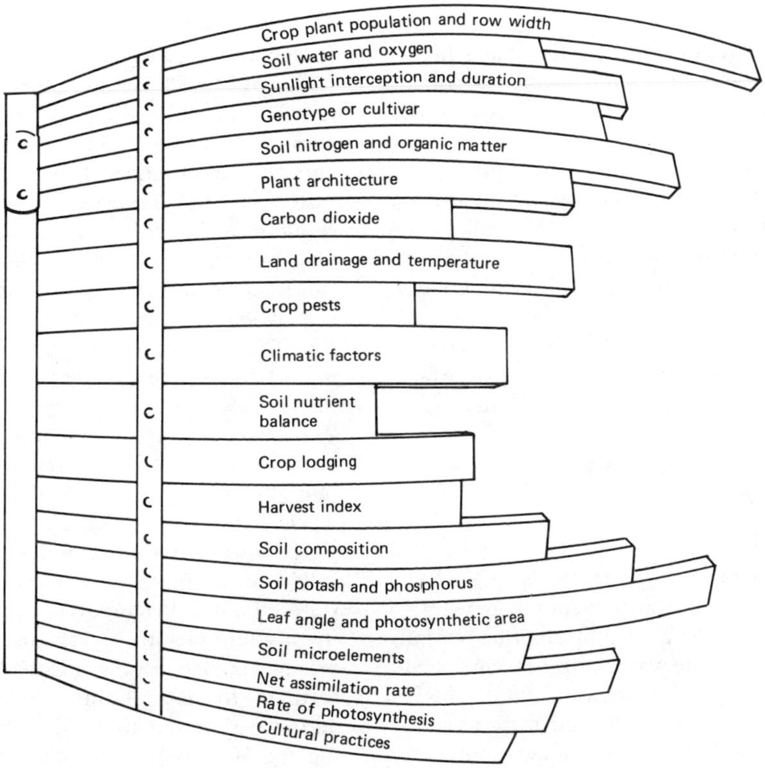

Figure 4–8. The "law of the minimum" can be likened to that of a barrel having staves of different lengths. The barrel cannot hold anything above the shortest stave, the most deficient or limiting factor. Crop yield will be held down by the one factor in least supply. The "law of the minimum" illustrates the need for a comprehensive systems approach to crop production.

Table 4–11
Effects of CO_2 Enrichment Before and After Heading on Grain Yield and Crop Growth Rate of IR8 Rice, When Grown in the Philippines in the 1971 Dry Season

Treatment	Grain Yield		Crop Growth Rate, $g/m^2/wk$	
	kg/ha	lb/ac	Before Heading	After Heading
Control	9,000	8,036	173	99
CO_2 for 30 days before heading	11,600	10,357	224	157
CO_2 for 28 days after heading	10,900	9,732	173	147

Source: Data from Yoshida, Cock, and Parao (1972).

to the control after heading, yet they had a higher crop growth rate. This can be regarded as an example of feedback interaction between sink and source.

The volume of air that the plant must process through photosynthesis to obtain the carbon needed for crop development, as well as the confounding factor of wind and air movements, make the logistics of direct CO_2 fertilization in the field so unmanageable as to be impractical. Some management of CO_2 levels may be achieved by the use of manure, crop residues, windbreaks, and the application of lime.

POSITIVE AND NEGATIVE YIELD FACTORS

The historical change in corn yield from 2010 kg/ha (893 lb/ac) in the pre-1930s to the 1980 level of 6290 kg/ha (5616 lb/ac) was used to allocate the relative contributions that various practices have made to Minnesota corn yield over a 50-year period (Cardwell, 1982). This increased yield can be attributed to a series of technological, cultural, and management practices adopted by farmers to produce results believed by Cardwell to be typical of the northern corn belt region. The example serves to illustrate a systems approach.

The genetic gain resulting from a switch from open-pollinated to hybrid corn accounted for 58% of the yield increase. This compares well with a 57% to 60% genetic gain estimated by Duvick (1977) for corn in the United States. Genetic gain in yield potential has been far more significant than adaptation of hybrids, and the transition from double-cross to three-way and to single-cross hybrids also has contributed. Three-way hybrids gave a 2.6% yield advantage, and single-cross hybrids contributed a 5% advantage when compared with double-cross hybrids. By 1979, Cardwell (1982) reported that in a survey of Iowan corn farmers, 75% planted to single-crosses and 17% to three-way crosses.

Applied commercial nitrogen fertilizer in balance with other fertilizer elements had a potential yield impact of 47%. When nitrogen needs were tempered by changes in other cultural practices that reduced mineral element availability, an increase in demand for nitrogen resulted. Cultural changes included a reduction in the amount of nitrogen-fixing forages used as a green manure crop. Substituting commercial fertilizer for natural sources was partly offset by an increase in the area grown in soybeans. During the 50-year period, soil organic matter had an absolute decline of about 1.35% of the total soil organic matter present in 1930. This reduced the amount of nitrogen available from organic-matter decomposition. These factors were estimated by Cardwell to be equivalent to a reduction of 28% in yield.

Improved weed control, mainly through the rise of herbicides, used on 93% of the area seeded to corn in 1979, increased yield 23%. Effective pest control represents a yield-saving as contrasted to a yield-promoting practice and will not increase yield potentials once effective control has been achieved.

Plant-density increases from about 30,000 plants/ha (12,000 ppa) under a **check-row planting** in 107-cm (42-in.) spacings, which permitted cross-cultivation to an average plant density in Minnesota in 1979 of 49,870 pph (20,190 ppa), increased yields 21% over the 50 years. Much of this increase in plant density occurred as a result of increased commercial fertilizer levels, earlier planting dates, narrower row spacing, and architecturally improved hybrids with increased stalk strength, tolerance to plant density, and more upright leaves.

The availability of hybrid seed with seedling cold tolerance, better herbicides, improved soil drainage, fall plowing, fewer spring-performed seedbed tillage operations, and large machinery that allows for timely field operations when field conditions are suitable, collectively permitted planting ten days earlier. This gave an 8% yield increase.

Average row spacing in Minnesota decreased from the 107-cm (42-in.) standard row in the 1930s to 102 cm (40 in.) in the 1950s, 96 cm in the 1960s, and 90 cm in 1979. Rows as narrow as 60 cm are used. Narrow row widths were estimated by Cardwell (1982) to increase grain yield 4% in Minnesota. The resulting change in planting pattern whereby interplant competition is reduced accounted for a yield gain of 8%.

An increase in the area devoted to corn production occurred at the expense of the area seeded to oat, barley, and flax. The benefits of a rotational system were reduced, and in some areas monoculture corn is produced. The effects of cropping sequence, when averaged over the total Minnesota corn-growing area, reduced grain corn yield 7%.

The European corn borer[1] was introduced into the United States in 1917 and spread through Minnesota by 1948. The resulting average annual loss in Minnesota since 1950 is 5%. When corn follows corn in the cropping sequence, corn rootworm[2] losses can be expected. Only a third of the area was treated with an insecticide specific for corn rootworm control in 1979. The estimated annual loss was 3%.

Mechanization which allowed for speed and timeliness of operation as well as for greater precision of spacing, depth, and fertilizer placement resulted in a 5% yield advance. Part of this increase in mechanization was related to removal of a previous crop to facilitate fall plowing with its accompanying yield advantage over spring plowing. Fall plowing increased soil erosion, estimated to reduce corn yields 8%. Other negative and unaccounted factors reduced yield 23%.

A summary of positive and negative factors that affected corn yields in the Minnesota study is shown in Table 4-12.

The Minnesota study is presented in some detail because it outlines the relative importance of many crop production factors and emphasizes the importance of a systems approach in crop production. Although individually the contributions of row spacing, date of planting, method of planting, and tillage practices are small, collectively they account for approximately 25% of the net yield gain.

The study also gives estimates of long-term negative aspects of changing cultural and management practices. Inorganic nitrogen has been substituted for manure and crop residues or green-manuring practices. Soil erosion and loss of organic matter has reduced the yield potential 21% in Minnesota. Close attention to the total system may reduce crop losses shown in Table 4-12. This in itself offers considerable promise for yield increases needed to match the demand for food by the exploding world population.

PLANT GROWTH REGULATORS FOR CEREAL CROPS

Synthetic or natural growth regulators are used routinely to control **suckering** in tobacco or to

[1]European corn borer—*Ostrinia nubilalis* Hubner.

[2]Northern corn rootworm—*Diabrotica longicornis* Say; western corn rootworm—*Diabrotica virgifera* Le Conte.

Table 4–12
Technological, Cultural, and Management Factors and Their Percentage Impact on the 4280-kg/ha (3822-lb/ac) Yield Increase in Minnesota from 1930 to 1979

Factor	*Percentage Influence*	
Genetic factor	+58	
Commercial fertilizer	+47	
Reduced manure and organic matter		−28
Herbicidal weed control	+23	
Increased plant production	+21	
Early planting	+8	
Improved spatial arrangement	+8	
Soil erosion		−8
Cropping sequence		−7
Mechanization	+5	
Corn borer damage		−5
Narrow rows	+4	
Corn rootworm damage		−3
Other negative and unaccounted factors		−23
Total change	+174	−74

Source: Data obtained from Cardwell (1982).

stimulate rooting in clonal cuttings of sugar cane. For cereal crops, however, the use of growth regulators to affect morphological or physiological change is considered as an emerging technology. An assessment of five new areas of technology in the United States that included photosynthetic enhancement, cell or tissue culture, plant growth regulators, genetic engineering, and biological nitrogen fixation placed plant growth regulators as having the greatest potential for increasing corn yield by the year 2000 (Menz and Neumeyer, 1982). The most likely mechanism for that contribution is an increase in harvest index. The use of growth regulators was regarded as a more expensive means of growth regulation than the other four technologies in the survey, as they are embodied in the purchased seed.

Plant growth regulators are nonnutrient chemicals that do not occur naturally in plants but when applied to plants can influence their growth and development. **Plant hormones** occur naturally in plants and act to control plant activities. There are five recognized groups of natural plant hormones: **auxins, gibberellins, cytokinins, ethylene,** and **asbscisic acid.** The action of these hormones has been described by Galston and Davies (1969).

In addition, a number of chemicals with herbicidal action may show quite unexpected and beneficial hormonal effects when applied in low doses (Thomas, 1970).

The behavior of the plant as a whole is controlled by the interaction of all the natural regulators. There are, however, some marked specific and desirable effects of applying single growth regulators to plants.

Before the development of semidwarf cereal plants, or where semidwarf cereal cultivars are not available, the application of CCC[3] to cereal plant foliage can produce a yield advantage (Brown and Earley, 1973). Typical effects include dwarfing, reduction in lodging, greater tolerance to drought, increased tillering, and in-

[3]CCC—(2-chloro-ethyl) trimethylammonium chloride.

creased yield. The use of CCC is an example of a commercial field application of a plant growth regulator that produces reasonable success.

Development of a plant growth regulator for use as a pollen suppressant is a much sought after possibility to facilitate hybrid cereal development.

YIELD STABILITY, ABILITY, AND QUALITY

The large and growing demand for food, economic considerations, and producer pride have shifted the emphasis in cereal crop production to **yield ability** rather than **yield stability.** First-time producers may choose wide adaptability associated with yield stability, whereas more experienced producers, prepared to gamble on maximizing yield, may choose a specific narrowly adapted cultivar that has performed well in some seasons or areas. A specific example is the selection of a late-maturing rather than an early-maturing corn hybrid cultivar in a medium-length-season area. Full-season cultivars with their potentially longer duration of photosynthesis have a higher yield potential than short-season hybrids but are subject to damage from an early fall frost. Corn growers prefer a single-cross hybrid over a four-way-cross hybrid because of the greater yield ability but lower yield stability.

When cultivars lacking yield stability are compared over a series of environments, the relative rankings differ. Such genotype-environment interactions are of major importance to the plant breeder in developing improved cultivars and cause the producer considerable difficulty in determining the best cultivar. Stratification of the environments based on temperature gradients, rainfall distribution, and soil types can be used to reduce these interactions but does not improve cultivar stability. The best method to improve cultivar stability is the development of stable genotypes. To find such stable cultivars, Eberhart and Russell (1966) proposed a method of ranking cultivar stability.

The use of genetic mixtures rather than homogeneous or pure-line cultivars was suggested as an alternative means to reduce genotype-environment interactions. Jensen (1952) suggested that a multiline cultivar of oat, as compared to pure-line cultivars, would possess greater stability of production, broader adaptation to environment, and greater disease protection. The multiline concept will be developed in a later chapter.

Allard and Bradshaw (1964) suggested that heterozygous and heterogeneous populations offer the best opportunity to produce cultivars which show small genotype-environment interactions. Allard and Bradshaw used the term "individual buffering" to describe individual members of the population that are well adapted to a range of environments so that "population buffering" assures yield stability.

The "purity concept" has been perpetuated in cereal crops by plant breeders, producers, and commercial seed suppliers because of pride in the uniformity of a cultivar and because of ease of perpetuating a pure-line cultivar generation after generation. The development of the haploid breeding technique, whereby completely homozygous cultivars are produced by chromosome doubling, presents a new concept in yield ability and stability. It is unlikely that both features can be combined in a single cultivar.

Just as high yield and wide adaptability do not go together, so, too, high yield and quality do not always go together. A generalization is that as dry-matter yields are increased, either through breeding or cultural practices, both crop quality and yield stability may decline. Quality is a complex entity, however, and cannot be described in simple, general terms. Quality may be determined by color, texture, gluten strength, digestibility, nitrogen content, or other physical and chemical properties.

Bread wheat, for example, favored for its strong gluten and high protein level, will have the best quality under conditions that favor differentiation, such as low soil moisture, high tissue temperature, and slow growth, conditions that restrict vegetative growth, photosynthesis, and high yield. In contrast, soft white wheat, favored for the manufacture of pastry flour, is characterized by a low protein level and a low level of soft gluten. Best-quality soft white wheat is produced in areas of adequate soil moisture and cool tissue temperatures that favor vegetative growth, the C_3 photosynthetic mechanism, and high grain yield.

In the production of grain for malt purposes, attempts to increase yield with high levels of soil nitrogen fertility often result in high protein levels that are undesirable in the **brewing process.**

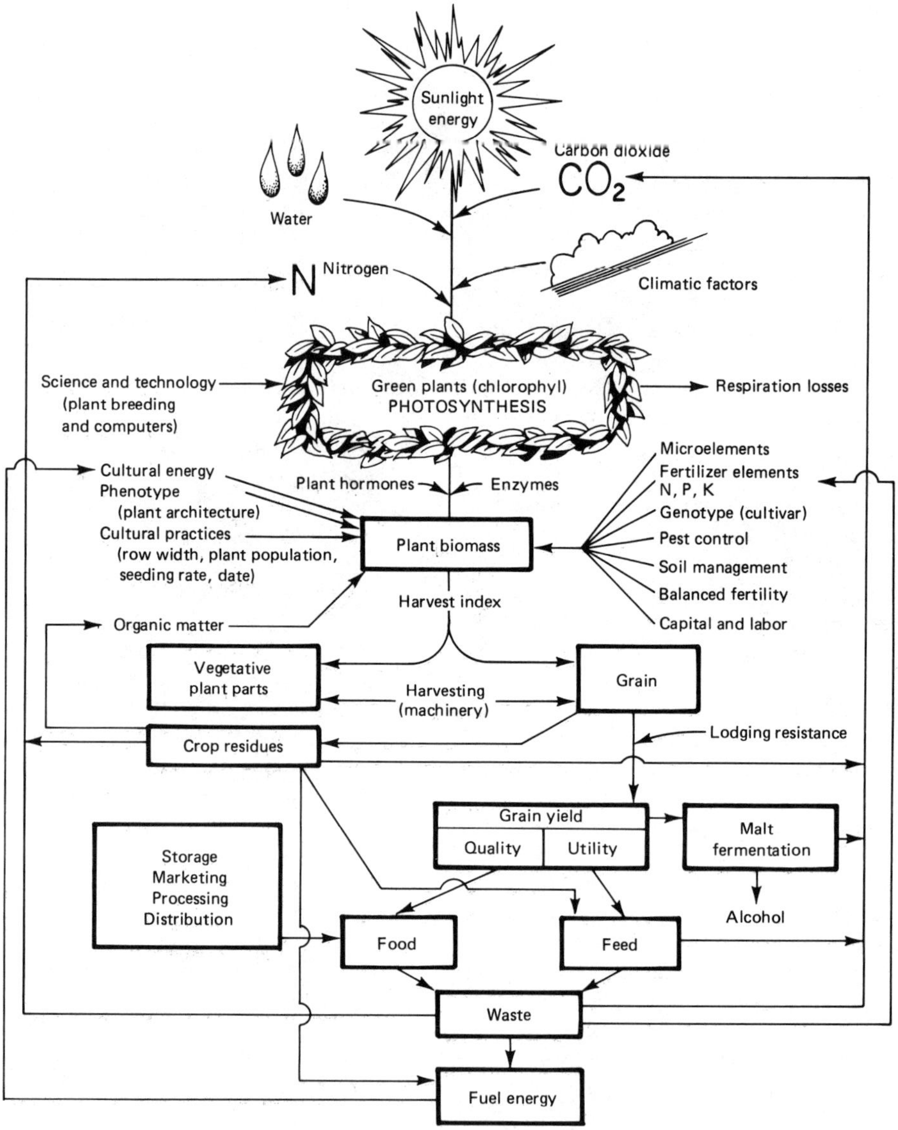

Figure 4–9. Diagrammatic sketch of the food-production system. Food production is successful when the appropriate factors are integrated into a viable system. The multiplicity of interacting factors in crop production and the ability to incorporate them into the total management system may represent the greatest challenge facing modern humans.

THE SYSTEMS APPROACH IN PERSPECTIVE

Advanced crop production requires a total management system—an integrated adaptation of phenotype, genotype, research findings, business management, and cultural inputs. A diagrammatic scheme of many general factors involved in the system is shown in Figure 4-9. Crop production involves the sciences of physics, chemistry, and biology along with the management of land, labor, and capital. To meet the projected 2.5% worldwide yearly increase in yield in the next two decades, an integrated systems approach is imperative.

The potential of a cultivar to produce high grain yield depends on a complex interaction of numerous factors, many of which are not completely understood. Confounding grain yield are factors related to quality and stability. An ideal phenotype would be one in which growth and development are closely tailored to the climatic conditions usually experienced, yet would be sufficiently flexible to produce a reliable yield, season after season.

Continued yield increases will be achieved by an understanding of the many factors comprising the production system. Leading the system and demanding close attention are factors that make a positive contribution to increasing the rate of photosynthesis and biomass production per unit land area.

QUESTIONS

1. *a.* Consider the evidence from your area that either cereal yields have reached a yield plateau or else opportunity exists for further yield increases.

   b. Consider the same question for the world as a whole with separate consideration for the developed and developing worlds.

2. *a.* Suggest specific inputs that could be used immediately to increase yield.

   b. Are these inputs now being deployed, and how successful are they?

3. If economic considerations were of no consequence, what would you do to increase yield and what magnitude of increase would you expect? Could you increase yield 10%, 100%, or 500%?

4. *a.* Consider specific architecturally superior features now existing in individual cereal crops.

   b. Consider architecturally inferior features now found in individual cereal crops.

   c. Construct an "ideotype" cereal crop.

   d. What possibilities exist for developing commercially your ideotype?

5. The songwriters Rodgers and Hammerstein immortalized the image of corn in the lyrics, "The corn is as high as an elephant's eye, and it looks like it's climbing up to the sky"—at least for Oklahoma. Consider plant size from a crop production viewpoint.

6. *a.* In addition to plant size, other morphological features must be integrated into a crop production system. Consider morphological aspects that impact on a crop production system.

   b. "The proliferation of factors associated with crop production may have outstripped the ability of producers to incorporate them in the total management system required." Has this statement any substance, or is it an expected part of a biological system? Could crop production systems be simplified without a grain-yield reduction?

7. *a.* Describe crop production factors that influence leaf-area index.

   b. To what extent can leaf-area index of cereals be manipulated by a cropping system?

   c. At what time of year and/or stage of plant development should the optimum leaf-area index be reached in the tropics and in temperate zones?

d. Consider the duration of the optimum leaf-area index in relation to yield as well as timing.

8. Discuss the importance of the "law of the minimum" in a systems approach to crop production.

9. List all the primary or essential features that must be considered in producing a crop. Recognize that photosynthesis is the basis of all yield.

10. *a.* Why are adequate carbon dioxide levels in the field of major importance in crop production?

b. What possibilities exist for the economic increase of carbon dioxide levels in the field for commercial crop production?

11. Successful crop production requires the effective manipulation of many interacting factors in a total system. Consider the application of a single growth regulator in terms of the entire plant enzyme and hormonal system. What do you conclude about the rate, timing, and effectiveness of growth regulators?

12. In 1974 it was suggested that within a decade, British wheat might stand a mere 30 cm (12 in.) high, owing to the introduction of dwarf, high-yield cultivars (Gale, 1974). This does not appear to have occurred. Explain.

References

Adams, M. W. 1982. "Plant Architecture and Yield Breeding." *Iowa State Journal of Agricultural Science* 30: 11–24.

Allan, R. E., O. A. Vogel, and J. C. Peterson, Jr. 1968. "Inheritance and Differentiation of Semi-Dwarf Culm Length of Wheat." *Crop Science* 8: 701–704.

Allard, R. W., and A. D. Bradshaw. 1964. "Implications of Genotype-Environment Interactions in Applied Plant Breeding." *Crop Science* 4: 503–507.

Army, T. J., D. R. Isleib, and F. A. Greer. 1971. "Integrating Production Systems to Advance Yields and Quality." (In) *Moving Off the Yield Plateau.* American Society of Agronomy Special Publication No. 20. pp. 83–89.

Austin, R. B., et al. 1980. "Genetic Improvements in Winter Wheat Yields Since 1900 and Associated Physiological Changes." *Journal of Agricultural Science, Cambridge* 94: 675–689.

Borlaug, N. E. 1968. "Wheat Breeding and Its Impact on World Food Supply." (In) K. W. Findlay and K. W. Shepherd (eds.), *Third International Wheat Genetics Symposium.* Canberra: Australian Academy of Science. pp. 1–36.

———. 1972. "Breeding Wheat for High Yield, Wide Adaptation and Disease Resistance." (In) *Rice Breeding.* Los Banos, Philippines: International Rice Research Institute. pp. 581–592.

Boyer, J. S. 1982. "Plant Productivity and Environment." *Science* 218: 443–448.

Briggle, L. W., and O. A. Vogel. 1968. "Breeding Short-Statured Disease Resistant Wheats in the United States." *Supplement to Euphytica,* No. 1: 107–130.

Brown, C. M., and E. B. Earley. 1973. "Response of One Winter Wheat and Two Spring Oat Varieties to Foliar Applications of 2-Cholorethyl Phosphonic Acid (Ethrel)." *Agronomy Journal* 65: 829–832.

Cannell, R. Q., and M. C. Drew. 1973. "Plant Root Systems and Crop Growth." *Span* 16(1): 38–40.

Cannon, G. C. 1963. "Baron Justus Von Liebig." *Great Men of Modern Agriculture.* New York: The Macmillan Company. pp. 148–159.

Cardwell, V. B. 1982. "Fifty Years of Minnesota Corn Production: Sources of Yield Increase." *Agronomy Journal* 75: 984–990.

Chandler, R. F., Jr. 1979. *Rice In The Tropics: A Guide to the Development of National Programs.* Boulder, Col: Westview Press, 256 pp.

Drennan, D. H. S. 1970. "Short Wheats for Britain?" *Span* 13 (2): 105–107.

Duvick, D. N. 1977. "Genetic Rates of Gain in Hybrid Maize Yields During the Past 40 Years." *Maydica* 22: 187–196.

Eberhart, S. A., and W. A. Russell. 1966. "Stability Parameters for Comparing Varieties." *Crop Science* 6: 36–40.

Faris, D. G., and R. M. DePauw. 1980/81. "Effect of Seeding Rate on Growth and Yield of Three Spring Wheat Cultivars." *Field Crop Research* 3: 289–301.

Fedak, G., and S. O. Fejer. 1982. "Progress in Breeding Semidwarf Barley." *Canada Agriculture* 27(1): 17–19.

Fiddian, W. E. H. 1978. "Basic Principles in Choice of Variety." *National Institute of Agricultural Biology, Fellows Conference Report* No. 3. pp. 1–3.

Fischer, R. A., et al. 1981. "Leaf Photosynthesis, Leaf Permeability, Crop Growth and Yield of Short Spring Wheat Genotypes under Irrigation." *Crop Science* 21: 367–373.

———, and R. O. Mauer, 1976. "Crop Temperature Modification and Yield Potential in a Dwarf Spring Wheat." *Crop Science* 16: 855–859.

Gale, M. 1974. "Towards Super Cereals." *New Scientist* 31: 248–251.

Galston, A. W., and P. J. Davies. 1969. "Hormonal Regulation in Higher Plants." *Science* 163: 1288–1297.

Grafius, J. E. 1965. *A Geometry of Plant Breeding.* East Lansing: Michigan State University. Research Bulletin 7: 59 pp.

Irvine, R. B., B. L. Harvey, and B. G. Rossnagel. 1980. "Rooting Capability as It Relates to Soil Moisture Extraction and Osmotic Potential of Semi-Dwarf and Normal-Statured Genotypes of Six-Row Barley (*Hordeum vulgare* L.)." *Canadian Journal of Plant Science* 60:241–248.

Jensen, N. F. 1952. "Intra-varietal Diversification in Oat Breeding." *Agronomy Journal* 44: 30–34.

———. 1978. "Limits to Growth in World Food Production." *Science* 201: 317–320.

Joppa, L. R. 1973. "Agronomic Characteristics of Near-Isogenic Tall and Semi-Dwarf Lines of Durum Wheat." *Crop Science* 13: 743–746.

Law, C. N., J. W. Snape, and A. J. Worland. 1978. "The Genetical Relationship Between Height and Yield in Wheat." *Heredity* 40: 133–151.

Lukyanenko, P. P. 1970. "The Breeding of Dwarf and Semi-Dwarf Winter Wheat Cultivars." *Acta Agronomica Academiae Scientarium Hungaricae* 19: 209–217.

Lupton, F. G. H. 1982. "Recent Advances in Cereal Breeding." *Netherlands Journal of Agricultural Science* 30: 11–24.

———, R. H. Oliver, and R. Ruckenbauer. 1974. "An Analysis of the Factors Determining Yields in Crosses Between Semi-Dwarf and Taller Wheat Varieties." *Journal of Agricultural Science, Cambridge* 82: 483–496.

Lupton, F. G. H., et al. 1974. "Root and Shoot Growth of Semi-Dwarf and Taller Winter Wheats." *Annals of Applied Biology* 77: 129–144.

Menz, K. M., and C. F. Neumeyer. 1982. "Evaluation of Five Emerging Biotechnologies for Maize." *BioScience* 32: 675–676.

Peterson, C. J., and R. E. Allan. 1980. "Production Problems after 18 Years of Growing Semi-Dwarf Wheats in the Pacific Northwest." *Proceedings of the Third International Wheat Conference.* Madrid, Spain. pp. 382–388.

Riggs, T. J., et al. 1981. "Comparison of Spring Barley Varieties Grown in England and Wales Between 1880 and 1980." *Journal of Agricultural Science, Cambridge* 97: 599–610.

Smith, C. S., J. J. Mock, and T. M. Crosbie. 1982. "Variability of Morphological and Physiological Traits Associated with Barrenness and Grain Yield in the Maize Population, Iowa Upright Leaf Synthetic #1." *Crop Science* 22: 828–832.

Thomas, T. H. 1970. "Plant Growth Regulators and Crop Production." *National Agricultural Advisory Service Quarterly Review* 89: 33–42.

Tollenaar, M., and T. B. Daynard. 1981. "Limitations to Yield of Maize in Southern Ontario." *Breaking the Climate Soil Barriers to Crop Yield.* Proceedings of a Symposium, University of Guelph, Guelph, Ontario, Canada. pp. 75–84.

Turner, F. T., C. C. Chen, and C. N. Bollich. 1982. "Coleoptile and Mesocotyl Lengths in Semi-Dwarf Seedlings." *Crop Science* 22: 43–46.

Yoshida, S., J. H. Cock, and F. T. Parao. 1972. "Physiological Aspects of High Yields." (In) *Rice Breeding.* Los Banos: Philippines: International Rice Research Institute. pp. 455–469.

5

Cereal Crop Yield— Improvement Through Genetics and Plant Breeding

INTRODUCTION

Without an understanding of the fundamentals of science, for centuries the improvement and production of field crops was purely an art. The state of the art was helped by empirical observations, but progress was slow, because biological phenomena often were steeped in mystery, superstition, and tradition.

With the development of science in the nineteenth century, progress improved, but the complexity of crop improvement and production and other biological aspects meant that the shift from an art to a science was slow, incomplete, and still underway. Despite scientific achievements such as splitting the atom and placing a man on the moon, scientists have not duplicated the process of photosynthesis outside of the plant, have not transferred the level of biological nitrogen fixation found in legumes to grasses, have not eliminated unnecessary C_3 photorespiration, have not brought climatic variables affecting crop production under control, and have not eliminated the limitations imposed by pests. Compared to the level of sophistication of science achieved in other applied fields, agriculture appears to remain more an art than a science.

In the biological sciences, however, it is in the area of the application of genetics to agriculture that some of the greatest scientific advances of the twentieth century are found. It is imperative, therefore, that cereal agronomists understand the basic principles of genetics, **heredity,** and **plant breeding.**

GENETICS AND CEREAL CROP PRODUCTION

High yields are obtained when crop producers match good management with genetically superior cultivars. However, the best cultural practices cannot overcome genetic deficiencies; the best genotype cannot overcome cultural deficiencies. The science of genetics therefore, is not reserved for specialists in the field of plant breeding, **cytology, physiology,** and **biochemistry** but is necessary to growers for the understanding of many aspects associated with cereal crops.

Seed growers who multiply the basic seed stocks developed by plant breeders need a knowledge of genetics and heredity to assure that agricultural, legal, and certification standards of the country are met. An understanding of how off-types or rogues emerge is essential. Hybrid-seed producers must understand how genetic traits are transmitted and all aspects of hybrid-seed production. Hybrid-seed growers must appreciate the implications of producing second-generation seed (see Chapters 9 and 10).

Plant Breeders' Rights, maintenance of the genetic constitution of plants, and genetic shifts are a part of modern-day agriculture. This chapter provides fundamental knowledge of genetics, heredity, and plant breeding that can be understood by crop producers and which has application to the scientific production of cereal crops.

Basic Genetic Principles

Gregor Mendel (1822–1884), an Austrian monk, is known as the father of genetics. Mendel experimented with hereditary traits in the common garden pea. Although he published his results in 1865 (Mendel, 1930), his work was overshadowed by the fiercely debated question of evolution. It was not until the rediscovery of Mendel's laws in 1900 that genetics attracted attention that led to the development of hybrid corn in 1917.

Mendel crossed pea plants having contrasting characters. When a cultivar that produced only red-flowering plants year after year was crossed with a similarly pure-breeding white-flowering cultivar, only red-flowered progeny were produced in the next generation. Mendel reasoned that, because only red-flowered plants were produced instead of pink or red-and-white-flowered plants, the red flower color was **dominant** and white flower color **recessive.** By convention, the capital letter *R* is used to represent the dominant red character; the lower-case letter *r* represents the recessive white color. In each case *R* or *r* represents one **gene** carried on separate **chromosomes.** Each plant has two genes for flower color—one received from the male parent and transmitted in the pollen grain and one from the female parent, carried in the ovary. Red-flowering plants, therefore, could have a genetic constitution of *RR* or *Rr*; white-flowering plants, *rr*.

Plants that are pure-breeding are **homozygous** (*RR* or *rr*). When reproductive germ cells or **gametes** develop, they contain half the normal chromosome number, each with the genes it carries. Pollen grains unite at random with egg cells. To determine the genetic combinations produced, the **Punnett square** provides a schematic method of presentation. The cross that Mendel made between pure-breeding red- and white-flowering plants is presented in Example A.

All plants produced from this cross are termed first-generation or **F_1 hybrid** progeny and in this case are uniformly red-flowered and **heterozygous.** When the first-generation plant is heterozygous, hybrid vigor or **heterosis** is often expressed.

Example A

		Gamates from Male Parent	
		R	*R*
Gametes from Female Parent	*r*	*Rr*	*Rr*
	r	*Rr*	*Rr*

Note: Capital letters representing genes are given first regardless of whether they come from the male or female parent.

Example B

		Gamates from Male Parent	
		R	*r*
Gametes from Female Parent	*R*	*RR*	*Rr*
	r	*Rr*	*rr*

Note: A standard ratio of three red to one white is normally produced from such a cross.

In the development of commercial hybrids, two parents are crossed with many homozygous but contrasting characters in each parent. Cross-pollination is essential to produce F_1 hybrid seed. If both parents are homozygous for the genetic traits transmitted, all progeny will have the same heterozygous genetic makeup, will be highly uniform, and may exhibit hybrid vigor. Only first-generation or F_1 progeny will be heterozygous, and so hybrid seed must be produced annually.

The consequences of producing a crop from seed obtained from F_1 parents can be illustrated when the red-flowering, phenotypically uniform but genetically heterozygous plants in Example A are allowed to intercross or become self-pollinated (Example B). Part of the progeny produce red flowers; part produce white flowers. This is known as **segregation,** which means that uniformity is lost. Both heterozygous and homozygous plants are present. For these reasons, crop producers do not use second-generation or **F_2 seed** if satisfactory performance based on hybrid vigor and uniformity is desired.

Example C, with two contrasting genetic characters, emphasizes F_2 segregation and the concept of **independent assortment** and **linked genes.** Some genes may be transmitted from one generation to the next independently of other genes, or they may be linked to other genes, which prevents random or independent assortment. In corn, the following genetic traits may be present:

S represents a gene for dominant starch endosperm

s represents a gene for recessive sugary endosperm

Y represents a gene for dominant yellow endosperm

y represents a gene for recessive white endosperm

Example C

		Gametes from Male Parent			
		SY	*sY*	*Sy*	*sy*
Gametes from Female Parent	SY	SSYY	S*s*YY	SSY*y*	S*s*Y*y*
	*s*Y	S*s*YY	*ss*YY	S*s*Y*y*	*ss*Y*y*
	S*y*	SSY*y*	S*s*Y*y*	SS*yy*	S*s*yy*
	sy	S*s*Y*y*	*ss*Y*y*	S*s*yy*	*ssyy*

1 SSYY 2 SSY*y* 2 S*s*YY 4 S*s*Y*y*	9 starchy-yellow endosperm
1 *ss*YY 2 *ss*Y*y*	3 sugary-yellow endosperm
1 SS*yy* 2 S*syy*	3 starchy-white endosperm
1 *ssyy*	1 sugary-white endosperm

When plants homozygous for starchy-yellow endosperm (SSYY) are crossed with sugary-white endosperm (*ssyy*), the F_1 progeny are heterozygous for starchy-yellow endosperm (S*s*Y*y*).

When F_2 progeny are allowed to self- or cross-pollinate, a typical 9 : 3 : 3 : 1 ratio results.

If independent assortment is restricted because the genes are close together or linked on the chromosome, then a ratio of three starchy-yellow endosperm to one sugary-white endosperm would have resulted.

Segregation of F_2 plants can occur for early or late maturity, tall or short plant height, wide or narrow leaves, adapted or unadapted genotypes, aggressive or noncompetitive types. The yield of such a heterozygous array of F_2 plants would be expected to be substantially lower because of **interference** among plants.

The term genotype refers to the genetic makeup of a plant. Thousands of genes are present in a plant. Each gene may have a large or a small effect. Genes are part of the chemical makeup of a chromosome. The number of chromosomes in the nucleus of each vegetative cell is the same throughout the plant and throughout that plant species. The basic chromosome number varies among plant species and is expressed as the 2*n* or *diploid* number. Half the chromosome number is transmitted through the sex cells and is termed the **haploid** or 1*n* chromosome number (Appendix A).

Genetic material is contained in the nucleus present in each cell, bases which collectively are called deoxyribonucleic acid or **DNA.** Associated with DNA is ribonucleic acid or **RNA,** which serves to translate genetic information into action through enzymes.

The genetic code is reproduced by nitrogen bases consisting of adenine, thymine, cytosine, and guanine. Just as eight musical notes can provide a seemingly unlimited number of musical sounds, so, too, combinations of four bases can make up a seemingly unlimited number of genes. Just as music can be loud or soft, harmonious or disharmonious, gene combinations can be effective or ineffective, beneficial or undesirable.

DNA occurs in double strands connected

by chemical bonds between specific base compounds. Adenine is always paired with thymine; quanine with cytosine. The sequence of one DNA strand serves as a model for the other, thereby providing for each gene to be replicated precisely and allowing for multiplication of cells and transmission from one generation to the next.

Chemical bonds between base compounds in the chromosome can be broken without damage and can be reassembled but perhaps in a somewhat different order. Chromosomal restructuring may take several forms, known as **translocation, inversion,** or **crossing over.** Failure of a segment of a chromosome to form a chemical bond may result in a **deletion** or loss of one or more genes. Changes in the chomosomal model means that such changes are perpetuated and passed to subsequent generations.

The number of genes controlling a specific character may vary from one to a number of genes. When two or more genes with similar and cumulative effects influence the same trait, they form a system of **multiple alleles.** Kernel color in wheat is controlled by several independent genes that have equal effects.

When genetic variation can be described in discrete classes, **qualitative genetics** is said to operate. Red and white flower color, sugary or starchy endosperm, and kernel color are examples of qualitatively inherited factors.

Quantitative genetics deals with the inheritance of genetic differences that are expressed in degree rather than kind. Yield, adaptation, vigor, and nearly every plant process and function are examples of quantitatively inherited traits. Rather than being characterized by discrete classes, quantitatively inherited traits form a continuous array of values from one extreme to the other.

When numerous genes are involved, with each expressing a small effect, they are known as **polygenes.** Red or purple plant pigmentation in corn is controlled by polygenes. The dark color that masks the normal green color is due to the presence of excessive amounts of anthocyanin pigmentation. The polygenes involved are influenced by light energy and soil nitrogen fertility.

Cell Division

Plant growth occurs in the meristematic regions of plants through division and expansion. A form of cell division known as **mitosis** occurs whereby both the chromosomes and the cell divide. At one stage in mitosis the chromosomes line up in the middle of the nucleus and split, longitudinally. The missing half of the chromosome is reconstituted from raw materials available in the nucleus, and **spindle fibers** then pull the duplicated or **homologous chromosomes** to opposite sides of the cell, and a wall forms between the two sets of chromosomes. The result of mitosis is two cells having exactly the same number and kind of chromosomes as the original cell. Other cell materials are divided about equally between the two cells, which then expand and redivide or differentiate into specialized tissues and organs. Differentiation into different tissues occurs despite the fact that each cell has the same genetic constitution. Because each cell contains all the necessary genetic material, it is possible to develop an entire organism from a single **somatic cell** by means of **tissue culture.** The formation of various plant parts from a single somatic cell is made possible by mitosis.

In the cells that form a pollen grain and an egg, a form of cell division called **reduction division** or **meiosis** occurs. Meiosis is similar to mitosis except that it consists of a sequence of two cell divisions, during which the chromosomes duplicate themselves only once. Meiosis therefore results in four cells, each with half the chromosome or haploid number present in somatic cells and each capable of giving rise to a gamete (pollen or egg).

As in mitosis, the two paired chromosomes divide in half longitudinally to form **chromatids** which double and **synapse** into sets of four. Crossing over of the synapsed chromosomes may occur, resulting in an actual exchange of segments, shown schematically in Figure 5-1. In the next stage, the attraction between the synapsed chromosomes changes to repulsion, and paired chromosomes move to opposite poles of the nucleus without separation. Immediately following this first division, a second division of the

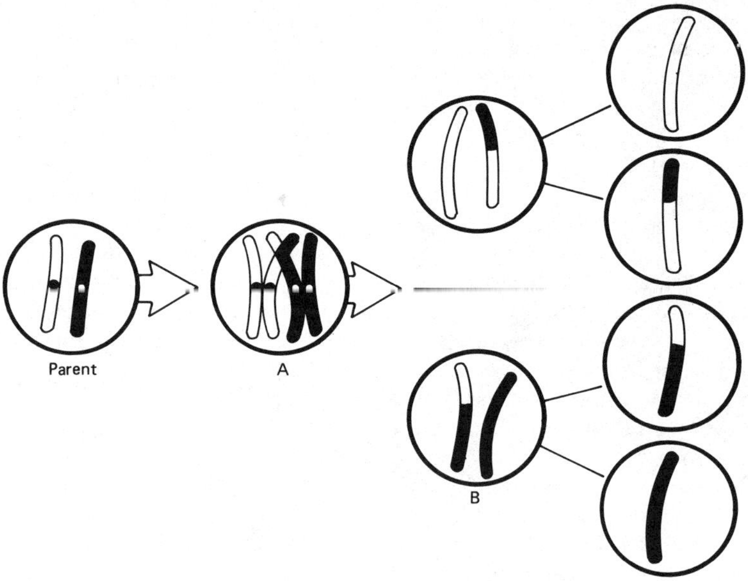

Figure 5–1. This illustration diagrammatically illustrates crossing over. During meiosis or reduction division, paired homologous chromosomes divide in half longitudinally to form chromatids which double and synapse into sets of four (A). Crossing over of the synapsed chromosomes sometimes occurs resulting in an actual exchange of segments (A). Crossing over results in an exchange of genes carried on the exchanged segments. The resulting chromosomes carry a different genetic message (B) than that of the parent. Any one of the gametes (C) may unite with another parent, resulting in a random genetic combination in the progeny.

paired chromosomes occurs, resulting in a random shuffling of chromosomes among the four gamete cells. This double division accounts for the four male and four female gamete cells in Example C. Genes SS, as in Example C, are located at the same points or **loci** on homologous chromosomes and are termed **alleles.**

GENE EXPRESSION

The phenotype of plants is influenced by both the environment and the genotype. Stress conditions may prevent the genotype from expressing its full genetic potential.

The term **heritability** is a measure of the effects of environment and gene complement on the phenotype of a plant. A phenotypically stunted or superior plant that was restricted or favored by environmental variables so that its true genotype was not expressed will continue to carry its true genetic complement, which can be expressed in subsequent generations under more suitable conditions. Phenotypic selection for superior performance of heterozygous plants will be successful only under environmentally similar conditions. Selection of plants with large ears or seed at the end of a row in the hope that similar progeny will be produced will be successful only if the plants are genotypically superior.

APPLICATION OF GENETICS TO INCREASING CROP DIVERSITY

Depletion of naturally occurring genetic variation has made it increasingly necessary to produce additional variation (Borlaug, 1971; Wittwer, 1974). Crop diversity, essential for crop improvement, can be developed by several methods as follows:

Genetic Recombination

By combining genetic traits from two parents, crop diversity can be increased. Independent assortment of genetic factors assures that new combinations can be achieved, but linked genes may restrict a particular genetic combination. Gene linkages may be broken but, because the chances of finding a desirable gene combination are reduced by linkages, large numbers of segregating progeny from many crosses may have to be screened. This can extend the breeding program and may require considerable effort and expense.

Crossing over results in new genetic combinations. If the original gene sequence was *ABC/abc* and crossing over occurred between *B* and *C*, then the new gene sequence would be either *ABc* or *abC* when these chromosomes appeared in the gametes.

Changes in genotype may occur through chromosome breakage and deletion of a chromosome segment if chemical bonding does not occur before the spindle fibers draw the chromosomes to their respective poles. If the chromosome segment changes position on the same chromosome or is transferred to a different chromosome, *translocation* is said to occur. If the segment breaks in the central portions of a chromosome by looping and is rebonded in a reverse position, the process is called *inversion*. Such changes are errors. Each error is called a *mutation*.

Mutation

Plants consist of thousands and thousands of cells, each of which has resulted from a previous cell division. It is not surprising that occasional errors occur when the hereditary factors are duplicated on such a grand scale. Mutations are important sources of genetic diversity. A mutation can affect a single gene and is called a *point mutation*.

Mutations may result in the deletion of a chromosome segment, the duplication of a segment, or the inversion of a segment. Many mutations are detrimental and lead to undesirable effects. The rate of mutation varies from gene to gene but generally is low.

In the early 1950s, experiments began in order to find techniques to induce mutations by using irradiation from **x-rays, gamma rays, thermal neutrons** or by using chemical **mutagens** such as diethyl sulphate or ethyl methane sulphonate. The ability to alter the heredity components in a sudden manner and to have this spontaneous change repeated in subsequent progeny aroused considerable interest in mutation breeding to increase genetic diversity. The desire to generate a large number of desirable or useful genetic changes led to the conclusion that mutation breeding could lead to superior genotypes more rapidly and economically than conventional crossing techniques. Mutation breeding, however, has not replaced traditional hybridization techniques for the following reasons:

1. The ultimate hope that mutation breeding could be controlled and that any given change might be induced at will has not materialized. Although efforts to find the optimum dosage rate and technique continue (Yonezawa and Yamagata, 1977), controlled mutation breeding does not appear on the horizon.
2. The impact of a single mutated gene may be so slight as to go unnoticed. Conventional techniques for screening and testing for possible mutations may not be adequate, as the plant breeder has to process a vast array of worthless mutations for the remote chance of finding a useful mutant. If the desired changes cannot be recognized on a single-plant basis and without effective screening techniques for particular desired characters, mutation breeding programs can be cumbersome.
3. Most mutations are deleterious. Some mutations associated with chlorophyll development occur with a high frequency and usually are undesirable. Many mutations result in sterility or reduced seed yields.
4. Mutations are frequently recessive, and a mutant will be expressed in the phenotype only in the presence of a similar recessive gene. Since this is

not likely to occur in cross-pollinated crops, mutation breeding is most applicable to self-pollinated crops.

5. Perhaps fewer mutations, especially desirable ones, were induced than expected. Some mutations are known to occur regularly, others infrequently. Mutation breeding may only repeat many known mutations.

Examples of desirable genetic diversity obtained through mutations can be cited: short-strawed variants of tall cereals were reported in Italian wheat; an awned mutant of an awnless but otherwise desirable wheat in India; a wanted kernel-color change in an Indian wheat cultivar; a mutant barley in Sweden with stiffer straw and improved lodging resistance; an oat cultivar in the United States with stem rust resistance; a rice cultivar that matures 60 days earlier than the parental form in Japan; and a mutant rice with higher protein levels. Mutation(s) are believed to have increased rice yields (Micke, Hsieh, and Sigurbjornsson, 1972). It should be noted that yield is the sum total of all genes acting together, not the product of one or a few genes, and mutation(s) that induce a yield increase may be a result of many interacting factors.

Although a few mutations have resulted in new and desirable genetic diversity, most examples of mutations are restricted to changes readily identifiable in the phenotype (Gregory, 1972).

Experience has shown that mutations, both spontaneous and induced, provide useful and exciting methods of obtaining genetic variation. Mutation breeding is a useful tool, but it is not an alternative to other techniques of obtaining genetic diversity.

Polyploidy

The diploid or $2n$ number (Appendix A) contains one set of chromosomes (**genome**) from the pollen parent and the other genome set from the female parent. Extra sets of chromosomes may be found in plants and are termed **polyploid.** A tetraploid barley, for example, has twice the diploid chromosome number and has the conventional symbolism of $2n=2x=28$. Polyploids can be produced with the use of a chemical called **colchicine,** extracted from the autumn crocus and discovered in mid-1930s (Kostoff, 1938). An interesting type of polyploidy is obtained by crossing two species and doubling the hybrid. Triticale is a polyploid of this kind and may be produced by combining seven rye chromosomes with 14 wheat chromosomes. The resulting combination with 21 chromosomes may abort, but, by doubling the 21 chromosomes, a hexaploid or 42-chromosome triticale is produced. By using a hexaploid wheat with $n=21$ chromosomes and combining diploid rye with $n=7$ chromosomes, an octoploid or 56-chromosome triticale can be produced. Octoploid triticale suffers from a high level of infertility.

When sets of unlike genomes are combined, such as wheat and rye genomes, the progeny are *alloploids.* When the genome is doubled, it is an *autoploid* (Figure 5-2).

Manipulation of the level of ploidy greatly increases the amount of genetic variability available, because many desirable genetic factors are found in species with different chromosome numbers. Crossing of plants at similar levels of ploidy is facilitated.

In comparison with diploids, polyploids often exhibit the **gigas** or enlarged features; leaves are wider and thicker; stems are larger and sturdier; and flowers and seeds are larger than the corresponding diploids (Simmonds, 1980). The overall effect in polyploid plants is of being lusher, more luxuriant, and a littler larger than diploids.

One of the earliest tetraploid rye cultivars, Tetra Petkus, was obtained by doubling the chromosome number of the diploid rye cultivar Petkus to produce an autoploid (Ross, 1953). Tetra Petkus and other rye and barley tetraploid cultivars commonly exhibit gigas features which are often associated with increased lodging resistance, but winter hardiness in winter annuals is often reduced. Increased water content of large polyploid cells may be associated with the reduced hardiness. In autotetraploids, seed fertility is often reduced because four sets of chromosomes ($4n$) exist and normal pairing may not occur. Random pairing within each group of

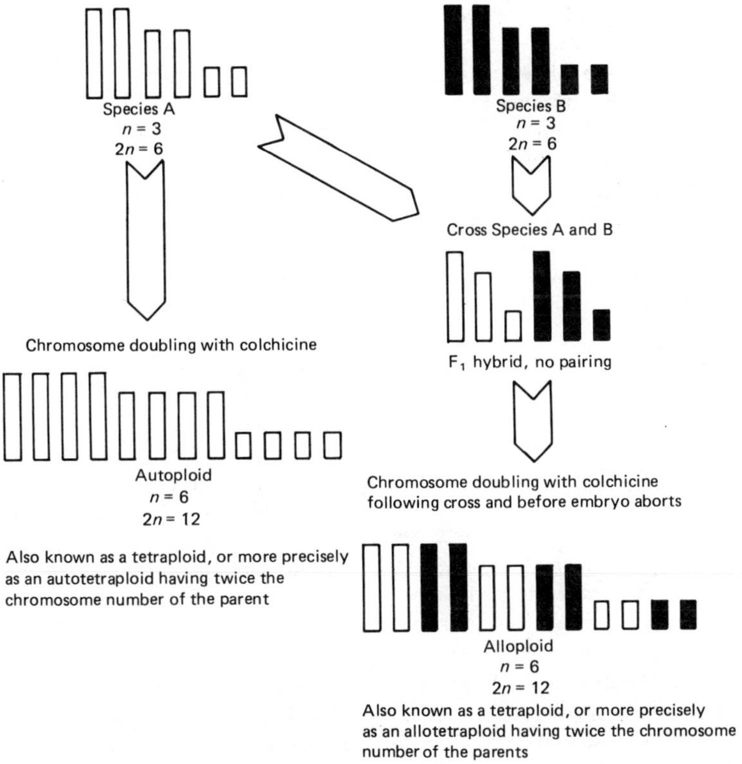

Figure 5–2. Diagram showing development of an auto- and an alloploid.

four homologous chromosomes is expected, but abnormalities may occur such as homologous chromosomes' fusing together in a group of three (**trivalent**) or a group of four (**quadrivalent**) at meiosis, resulting in **aneuploidy** and sterility (Reinbergs and Shebeski, 1959). Tetraploid corn has been developed with the typical gigas features, but seed set has been low.

So far, therefore, efforts to develop autopolyploid cultivars of the seed-producing cereal crops have failed because of sterility. Marginal economic success with tetraploid rye has encouraged long-term selection programs for improvement in rye, barley, and sorghum, but prospects for success are uncertain.

Autotetraploids have four alleles per gene, and the inheritance of traits is more complex than in diploids. Two types of homozygotes, *AAAA* and *aaaa*, are possible and give rise to three types of heterozygotes: *AAAa*, *AAaa*, and *Aaaa*. The genetic ratios of tetraploid progeny therefore differ from those of diploids.

Polyploids of cross-pollinated crops such as tetraploid rye or corn must be grown in isolation from their diploid counterparts, because tetraploid flowers fertilized with pollen from a diploid plant may produce shrivelled **triploid** seed, lacking in viability.

In contrast to the limited success with autopolyploids, allopolyploids have proven to be very successful and are grown widely. Some, such as wheat, developed through natural evolutionary processes. The 42 ($2n$) chromosomes of hexaploid common wheat were derived from three different diploid parents. Each contributed a distinct genome designated as A, B, and D.

Wheat falls naturally into three groups: the diploids (n=7), the tetraploids (n=14), and the hexaploids (n=21). Some wild wheat cultivars are diploid; durum wheat is tetraploid and may have been derived from two different diploid ancestors having the A and B genomes. Common hexaploid wheat has the A, B, and D genomes. Possibly the three genomes evolved from a common ancestor, making them mutually compatible but nonhomologous. Thus it is possible for the 21 haploid chromosomes to pair in groups of threes, each consisting of seven chromosomes. Sterility is not a problem.

Interspecific and Intergeneric Combinations

Crossing two plants may result in genetic recombination and lead to a desirable increase in plant variation. When parents to be crossed are both from the same biological species, the cross can be made readily to produce fertile offspring and presents little or no hindrance to genetic recombination. Sometimes, however, it is desirable to transfer useful genes from one species to another or even to a different genus. Often such a wide cross is difficult to make and, even if achieved, produces offspring that are inviable (die) or are sterile.

The failure of interspecific and intergeneric crosses in the cereals commonly is due to inviability of the newly formed embryo, embryo abortion, or endosperm abnormalities, all of which occur after fertilization. In addition, there are difficulties in obtaining fertilization because the pollen tube fails to reach the embryo sac (Heslop-Harrison, 1982).

To help understand the barriers to crossing different species and genera, consider a hypothetical situation in the animal kingdom. Suppose that all but some very large and some very small dogs died from an epidemic. If it were physically impossible for the two types to mate, and if artificial insemination techniques were unknown, taxonomists might classify the two types as different species. Taxonomic classifications, as suggested in Chapter 3, are arbitrary, often based on morphological features, and subject to modifications. A cross of the two dog "species" could be successful because they evolved within the same genus.

Likewise, within the wheat genus *Triticum*, several species exist such as *Triticum monococcum* and *Triticum dicoccum*. Combining these two species would be described as an *interspecific* cross. A successful cross could increase genetic variability.

An *intergeneric* cross is akin to crossing a dog and a cat. Triticale was produced from an intergeneric cross involving wheat (*Triticum*) and rye (*Secale*), although it should be noted that both are from the same tribe (*Triticeae*) (Dewey, 1982). Difficulties associated with wide crosses such as intergeneric ones are well illustrated by triticale,[1] which took 90 years of development for a commercial cultivar to emerge. Although the intergeneric hybrid leading to triticale was produced over a century ago (Wilson, 1876), triticale has yet to establish its place among world crops. Systematic efforts to develop triticale began in the 1930s, and it took nearly 40 years of intensive work by many plant breeders around the world to produce cultivars with commercial potential. Even though the progeny is an allopolyploid, a high degree of sterility had to be overcome and remains as a minor problem. It is apparent that intergeneric crosses may require a great deal of genetic adjustment to reach commercial value (Simmonds, 1980).

The protective mechanisms or barriers that exist in plants serve to maintain the integrity of species. Barriers to hybridization may act through morphological, physiological, or biochemical means to prevent fertilization, to induce embryo abortion, or to produce embryo-endosperm incompatibility. Plant breeders use many sophisticated techniques to circumvent some of the problems of sexual incompatibility. The barriers to wide crosses and the techniques to overcome them have been reviewed thoroughly by Hadley and Openshaw (1980). The magnitude of the problems, however, has been so great that relatively few viable wide interspe-

[1]The botanical nomenclature of triticale, as shown in Appendix A, is × *Triticosecale*, the × indicating the derivation of triticale as an intergeneric hybrid.

cific or intergeneric crosses have been produced for commercial application (Bates and Deyoe, 1973).

Despite seemingly insurmountable problems, considerable interest persists for combining the genomes from different species or genera—for the following reasons:

1. The promise of increased genetic diversity is great.
2. It is known that polyploids offer considerable promise and that allopolyploids offer greater promise than does doubling the chromosome number to produce autopolyploids.
3. The success of allopolyploid wheat, the world's most extensively grown crop, has provided an incentive for scientists to attempt additional interspecific crosses.
4. The potential of the intergenerically produced crop triticale under stress conditions has provided considerable encouragement.
5. Ingenious approaches associated with **biotechnology** offer new techniques to overcoming crossability barriers that have prevented wide crosses in the past. Areas of incorporating a genome from barley (*Horderum vulgare*) into wheat (*Triticum aestivum*) may be realistic (Mujeeb-Kazi and Bernard, 1982), considering they both are members of the tribe Triticea (Dewey, 1982).

Efforts to cross a wild barley (*Hordeum bulbosum*) with common barley (*Hordeum vulgare*) led to an exciting and unexpected development, namely the repulsion of chromosomes and a reliable mechanism to produce haploid barley plants (Kasha, 1974). By doubling of the otherwise sterile haploid, completely homozygous and fertile plants are produced. The first cultivar to be developed using the doubled-haploid method was named Mingo, licensed in 1979 by Ciba-Geigy Seeds Ltd. (Ho and Jones, 1980).

The use of wide crosses may be viewed as another tool of the plant breeder to develop new genetic diversity. Scientific opinion may vary as to the value of wide crosses, but success has shed new light on the future promise and potential. Problems and limitations of wide crosses and the low probability of obtaining the desired combination of genes from widely diverse parents make this a challenging avenue of research.

UTILIZING GENETIC DIVERSITY—PLANT BREEDING AND SELECTION PROCEDURES

The aim of the plant breeder is to improve agronomically important plants by producing cultivars that are physiologically more efficient and consequently more productive in relation to human need. Disease resistance, tolerance to climatic variations, morphological features, and quality considerations are major objectives. This section illustrates various procedures to achieve these ends, the steps involved, and the time frame required.

Although improvement by conscious or unconscious selection for superior phenotypes had proceeded for 8000 to 10,000 years, or since the development of agriculture, plant breeding began in earnest only with the rediscovery of Mendel's laws of genetics in 1900.

Plant breeding and selection involves several steps, including selecting parental material; making a cross between the parents; growing, observing, and selecting promising progeny; and hopefully releasing a superior new cultivar. A plant breeder does not make one cross and carry it through to completion, but makes many crosses every year to increase the chances of success. Despite careful attention to detail, there is no guarantee of success. Plant breeding is a long-term undertaking, requiring meticulous attention to detail; it is a very challenging but potentially rewarding task.

Many techniques have been developed to aid selection among the large number of progeny that emerge from crossing programs. A general objective of all selection and handling methods is to identify superior plants and to discard the remainder. There is no single, best technique to achieve this goal, and methods vary according to breeding objectives, the species, labor and facilities available, and individual preference. Several approaches will be outlined here to indicate the number of steps and associated problems.

In general, plant breeding can be described as a "numbers game." Independent assortment of numerous genes affecting cereal crops means that large segregating populations must be grown in easily segregating generations to increase the chances for a genetically superior combination to occur. With as few as 25 desirable yield genes involved in a cross—and this is only slightly more than one gene per chromosome in a hexaploid wheat—a minimum of 1330 plants would have to be grown in the F_3 segregating generation for the chance occurrence of only one plant that would contain all the desired alleles. Less than one plant per 100,000 of the F_3 population would have all the desired genes in either the heterozygous or homozygous condition.

Clearly defined objectives are a primary step before selection begins. In Example D, involving a cross between two heterozygous red wheats, it is essential that the desirable kernel color be established. In the F_2 generation in Example D, one plant in 16 will have white seed; if white seed is desired, it will be readily found, but it could be missed if not stated as an objective.

The need to distinguish the plant(s) having the desirable genetic combinations for subsequent performance testing and to discard genetically inferior ones, so as to reduce the number to be handled in subsequent generations, presents a difficult challenge. The necessity for identifying high-yielding genotypes in the earliest possible generation is obvious, for once superior genetic combinations have been discarded, they cannot be recovered in subsequent generations. Adding to the complexities of selection are environmental variables that influence the phenotype, the basis on which superior genotypes must be selected.

A successful plant breeder must exhibit a high level of perceptiveness to select parental material; an awareness of present and anticipated producer needs and desires; a keen sense of observation to identify superior segregates; the ability to make numerous decisions about retaining or discarding segregates; the courage and conviction to select those plants that approximate the "ideotype" in his or her mind's eye; a strong sense of dedication, as plant breeding is a long-term undertaking; and the ability to collect and present evidence of plant superiority. The final step is the licensing of a cultivar for commercial production. A sense of achievement is the main reward that accompanies commercial acceptability and performance.

Example D

Kernel color in wheat is a quantitatively inherited trait influenced by multiple gene action. A breeder wishing to develop a white wheat ($r_1r_1r_2r_2r_3r_3$) may cross two phenotypically red parents with the following genotypes: $R_1R_1r_2r_2r_3r_3$ and $r_1r_1R_2R_2r_3r_3$. All the F_1 progeny will be phenotypically red with the following genotype: $R_1r_1R_2r_2r_3r_3$. The following summary gives the genotype and kernel color of F_2 plants and breeding behavior in the F_3 generation.

F_2 Genotype	Kernel Color	F_3 Breeding Behavior
1 $R_1R_1R_2R_2r_3r_3$	Red	Breeds true for red color
2 $R_1r_1R_2R_2r_3r_3$	Red	Breeds true for red color
2 $R_1R_1R_2r_2r_3r_3$	Red	Breeds true for red color
4 $R_1r_1R_2r_2r_3r_3$	Red	Segregates; 15 red to 1 white
1 $R_1R_1r_2r_2r_3r_3$	Red	Breeds true for red color
2 $R_1r_1r_2r_2r_3r_3$	Red	Segregates; 3 red to 1 white
1 $r_1r_1R_2R_2r_3r_3$	Red	Breeds true for red color
2 $r_1r_1R_2r_2r_3r_3$	Red	Segregates; 3 red to 1 white
1 $r_1r_1r_2r_2r_3r_3$	White	Breeds true for white

BREEDING SELF-POLLINATED CROPS

Cereal crop cultivars are homozygous, hence are genetically stable and can be grown year after year and maintain their genetic identity. Following a cross, the F_1 progeny are highly heterozygous and in the F_2 maximum segregation occurs. A key step in breeding self-pollinated crops is the development of homozygosity. Four systems commonly are used to achieve this goal:

The **pedigree method.**
The **bulk method.**
The **backcross method.**
The **haploid method.**

The Pedigree Method

Routine steps are as follows.

STEP ONE: MAKING THE CROSS. Parents for crossing are selected, and, unless cytoplasmically inherited factors are involved, one is selected arbitrarily as the female and the anthers are removed (Figure 5-3) prior to pollination to prevent self-fertilization (**autogamy**). Some florets may be removed to facilitate **emasculation,** then heads are covered with **glassine bags** that prevent stray pollen from causing fertilization. Within 48 hours, pollen is collected from male parent plants by removing anthers about to shed pollen. Viable pollen is dabbed onto the stigma of each emasculated floret, and the glassine bag is replaced.

Figure 5–3. In preparation for making a cross between two parents, one parent is emasculated by removing the pollen-producing anthers. In this photograph, a technician is using forceps to remove the three anthers from each floret of this barley spike. Emasculation must occur prior to pollen release to prevent selfing. Note that the spike had not emerged from the sheath, which has been partially removed to facilitate emasculation. Caution must be used to minimize mechanical damage, however. (Courtesy of Ontario Ministry of Agriculture and Food)

Crosses can be made outdoors or under controlled environmental conditions provided the development of male and female reproductive organs can be made to coincide.

Crossed seed produced on female plants must be sufficient to provide enough F_1 plants for the desired number of plants in the F_2 generation. F_1-generation plants can be spaced to promote tillering and seed set; are heterozygous and provide no opportunity for selection; and can be compared to parent cultivars to determine dominant genetic traits and to assure that crosses have been made.

STEP TWO: THE GREATEST SELECTION OPPORTUNITY. Approximately 2000 to 10,000 individual F_2 plants are grown under spaced conditions to allow the plant breeders to thoroughly examine all plants at various stages of development. Maximum segregation occurs in the F_2 generation, and alertness for desirable phenotypes is essential. Weak, diseased and undesirable plants can be removed, but selection of 1000 or more heads for the F_3 generation should be based on positive selection criteria. Each selection is kept separate and given a number that establishes its pedigree.

STEP THREE: THE F_3 GENERATION. The 1000 selected heads are planted in individual progeny rows to provide for close subjective examination of each selection (Figure 5-4). Decisions to continue or reject each progeny row are based on the overall appearance, and, because segregation continues to occur in the F_3 generation, three or four heads for the F_4 generation are selected from the best plant within the row. Perhaps half the rows may be discontinued, but 1000 or more heads from individual plants may

Figure 5–4. In the "pedigree method" of cereal breeding, individual rows of segregating generations are grown in the field for subjective examination and selection. In this photograph rows are 2 meters (6 feet) long and 23 cm (9 in.) apart. Note the extensive area devoted to segregating populations. (Courtesy of Ontario Ministry of Agriculture and Food)

be selected. Hardiness, vigor, lodging resistance, disease reaction, maturity, and morphological features all serve as selection criteria.

STEPS FOUR AND FIVE. The three or four selected heads are grown in a three- or four-row plot, not only to provide an opportunity to more fully assess selections, but also because, by bulking the F_5 generation, sufficient seed can be obtained for a preliminary yield trial. Before bulking the F_5 seed, three or four heads from each continued row are taken to provide a seed source for the next generation. Bulked seed is used only for test purposes (yield and quality) and then discarded. Advancing only the best selections of each generation and maintaining the status of each is the very essence of the pedigree system. Perhaps 25 to 50 entries will be advanced to the F_6 yield trial from each cross.

STEP SIX: THE F_6 GENERATION AND SUBSEQUENT GENERATIONS. By the F_6, each selection has reached a reasonable level of homozygosity, and yield performance can be evaluated reliably (Figure 5-5). Replicated yield trial may be given various designations, but a descriptive terminology consists of the preliminary test, often with two **replications,** the screening test with four replications, and the cooperative test with six replications. If the breeding program is designed to serve a region or large geographic area, several widely scattered test sites should be selected to test for adaptability. Each entry may be tested a minimum of two years in each test, which means that the second cooperative test will be completed in the F_{11} generation.

On the basis of objective yield trials, two entries may emerge as having sufficient advantages over existing cultivars to merit licensing. One may be chosen for commercial production.

THE F_{12} GENERATION. A decision to license means that the plant breeder must provide basic seed stocks for multiplication and from which the genetic integrity and purity can be main-

Figure 5–5. Performance testing, involving replicated yield trials, is a sizable part of a plant breeding program. Plots in this photograph are 5 meters (16 feet) long, and may consist of four or five rows from which only the center two or three rows are cut to eliminate border effect. In addition to yield measurements, notes on height, disease, heading and maturity dates, lodging, or other morphological features may be recorded. The technicians in the background are recording these items. The size of the test area gives some indication of the scope of crop testing, and the need for careful record keeping is self-evident. (Courtesy of Ontario Ministry of Agriculture and Food)

tained. This will occur if basic seed sources are homozygous and uniform for phenotypic characters. The source of this seed is the three or four head rows from which 300 to 400 individual heads may be selected to be grown as individual head rows to verify phenotypic uniformity. Rows that do not represent the population may be rogued out. After an individual head has been collected from each remaining row, the bulked seed is known as *breeder seed.* Breeder seed can be reproduced at any future date as required. The breeder is responsible for maintaining the stock true-to-type to that originally released and to continue to produce breeder seed as long as the cultivar is in demand.

STEP 13: THE FINAL STEP. Breeder seed is distributed to farm producers who are "certified" as capable seed growers. Classes and terminology may vary geographically, but five pedigreed seed classes in self-pollinated cereal crops are common: **breeder, select, foundation, registered,** and **certified seed** (see Chapter 10).

Maintaining a record of each selection lends itself to computerization of recorded observations, necessary to reduce the cumbersome aspects of the pedigree system.

The Bulk Method

Self-pollinated plants are sufficiently uniform for quantitatively inherited characters by about the F_5 or F_6 generation, at which time yield trials can be meaningfully conducted. The bulk system involves advancing plants to the F_5 or F_6 generation without selection and by growing the segregating generations in bulk. All or a random portion of each generation bulk plot is harvested for the next-generation seed.

In the F_6 generation, individual plants are selected, seed of each plant is produced, and performance trials similar to those of the pedigree system are conducted.

The bulk system is useful when natural-selection pressure is exerted, such as winter hardiness, that may eliminate tender segregates. Under rigid natural-selection procedures, the remaining F_6 plants should be the most hardy within the population.

The ease with which bulk crosses can be handled up to the F_6 generation means that a greater number of crosses can be handled. A considerable amount of undesirable genotypes may be maintained that cannot be identified for elimination in the bulk plot.

Backcross Method

When a readily identifiable qualitatively inherited trait such as specific disease resistance is to be transferred to an otherwise desirable cultivar, the **backcross method** may be used. The method involves repeated backcrossing of the F_1 progeny to the recurrent parent. Selection for the recognizable single trait from the nonrecurrent parent will lead in six backcrosses to a genotype that has most of the genes of the recurrent parent plus the desired trait of the nonrecurrent parent. Fifty percent of the remaining heritable increment of the recurrent parent will be achieved in each backcross, as shown in Example E.

Example E

Recurrent Parent A × Nonrecurrent Parent B

The first backcross, B_1, will contain 50% of the genes of parent A.

B_2 will have 75% of the genes of parent A.

B_3 will have 87.5% of the genes of parent A.

B_4 will have 93.75% of the genes of parent A.

B_5 will have 96.88% of the genes of parent A.

B_6 will have 98.54% of the genes of parent A.

A cultivar differing by only one gene from the original parent is referred to as an **isoline.** Six backcross generations will not produce a true isoline, which should be indistinguishable from the recurrent parent except for the added trait. However, it is close enough for most practical purposes.

Haploid Breeding Method

In the three breeding methods just described, it was not until the F_5 or F_6 generation that a reasonable degree of homozygosity was achieved. The time required to develop a new cultivar through hybridization could be reduced if a shortcut to homozygosity could be achieved. In

1970, use of a wild, cross-pollinated barley species (*Hordeum bulbosum*) for producing haploids in common barley (*Hordeum vulgare*) was reported (Kasha and Kao, 1970). The technique involves removing the embryo shortly after fertilization and culturing the immature embryo on an artificial medium in a test tube (Figure 5-6). Although a fertilization rate of about 65% can be obtained between the wild and common barley, shortly after fertilization the embryo aborts, hence the need for embryo culture. As well, the chromosomes from the wild barley are selectively eliminated, thereby resulting in a haploid embryo that will not set seed unless the diploid chromosome number is restored. This is achieved by treating the haploid seedling with a 0.1% colchicine solution for five hours in the two- to three-leaf stage to double the haploid chromosome number (Kasha, 1974). Chromosome-doubled sectors on the original haploid plant produce seeds all of which are identical in genotype and which produce completely homozygous plants.

Figure 5–6. Haploid barley seedlings grown in test tubes on a nutrient agar medium. Two weeks have passed since the fertilized embryo on the left was removed from the parent plant, and the seedling on the right is three weeks of age. Doubled haploids will be produced by treating these plants with colchicine, so that the seed they produce has the diploid chromosome number. (Courtesy of Ontario Ministry of Agriculture and Food)

The haploid breeding procedures are as follows:

Step One. Select desired parents and cross them to produce 15 to 20 F_1 seeds. This procedure may require three or four months.

Step Two. Grow F_1 progeny in environmentally controlled rooms and cross F_1 plants with the wild barley to produce haploid embryos. This procedure may require up to six months.

Step Three. Grow the embryos into seedlings; treat with colchicine and grow plants to maturity. Requires four to six months.

Step Four. Increase seed from doubled haploids to go directly into replicated field trials. Requires a growing season.

Step Five. Begin replicated yield trials.

The haploid breeding method results in a time saving of nearly 50% and allows breeders to meet changing market demands or a crisis situation or to respond to changing economic and cultural conditions in crop production more rapidly than over the 12- to 14-year period required to produce a commercial cultivar using the pedigree or bulk breeding methods. Ho and Jones (1980) reported that only five years were required from the time the parents were crossed to the time the first commercial barley cultivar, Mingo, was licensed. The haploid technique gives a new incentive to commercial barley breeding programs, where profit accountability is required, and although the technique is applicable only to barley, the application of doubled haploids is secured and reliable haploid development in other crops is inevitable.

BREEDING CROSS-POLLINATED CEREAL CROPS

After centuries of cross-pollination, the genotypes of plants making up a crop such as open-pollinated rye and corn are highly heterozygous. Cross-pollinated crops can be selfed, but this likely leads to a marked decline in vigor due to inbreeding (Figure 5-7).

Cultivar development of open-pollinated crops can be achieved by selecting superior cross-pollinated plants and allowing cross-pollination to occur among these selections grown in isolation. The technique, known as **mass selection,** is similar to the bulk method in self-pollinated crops and often leads to an initial, rapid improvement, but because mass selection maintains heterozygosity, continued improvement becomes difficult.

If normally cross-pollinated plants are inbred, plants become homozygous, and superior inbreds can be identified, as deleterious recessive genes can be identified in the homozygous state. Selfing of cross-pollinated plants leads to a marked reduction in vigor. Recessive genes are not always undesirable, but they are easily masked or covered when in combination with dominant genes. Although inbreeding reduces vigor and inbreds may be reduced in size, restoring heterozygosity by crossing two unrelated inbreds restores plant vigor.

Figure 5–7. When naturally cross-pollinated, heterozygous corn is self-pollinated, it becomes homozygous with an accompanying loss in vigor. The tall vigorous plants in the background are hybrid progeny produced from a cross of two inbreds. The small plants in the foreground have been self-pollinated for six or more generations and are considered homozygous in their genetic composition. Superior inbreds will be selected to produce F_1 hybrids for commercial use. (Courtesy of Ontario Ministry of Agriculture and Food)

DEVELOPING F_1 HYBRIDS

The concept of utilizing F_1 hybrid vigor first came into prominence with corn, because this crop bears male and female flower clusters at separate locations on the same plant. Controlled crossing can be achieved in isolated field conditions by the removal of the pollen-producing tassel on the female parent plants.

Initial efforts in producing controlled hybrids involved planting two different corn cultivars in alternate rows and removing the tassels of one cultivar. The seed produced on the other cultivar was a cross. Quantitative characters such as yield, however, were not increased appreciably, because both parents were of mixed origin after centuries of out-crossing. A technique was needed to give a greater degree of control over the hereditary factors of corn.

The technique adopted was the development of inbred lines. There are at least three excellent reasons for producing inbreds:

1. To expose both superior and inferior recessive genotypes which may be masked by a dominant gene in the heterozygous state.
2. To develop a true-breeding inbred that can be crossed with other true-breeding inbreds to produce a consistent progeny year after year.
3. To provide an opportunity to evaluate the combining ability of true-breeding inbreds with other inbreds to produce an F_1 hybrid of superior performance.

If inbreeding is continued for five to ten generations and if weak and undesirable inbreds are eliminated, a remarkable degree of control over the heredity of corn can be exerted. Crossing two inbreds that combine well together to produce a truly outstanding inbred can result in a high level of heterosis or hybrid vigor and a yield increase. In addition to yield hybrid plants have superior stalk rot resistance, stronger stems, and better root systems to help resist lodging, F_1 plants are more uniform, which facilitates harvesting; the stalks remain green even when the ears are maturing; and controlled parental combinations provide for a wide range of maturities.

Hybrid corn is defined as the utilization of F_1 hybrid vigor following a cross of two unrelated parents. Mangelsdorf (1951) suggested that hybrid corn development during the first quarter of the twentieth century may prove to be the century's most far-reaching contribution of applied biology. The success of hybrid corn has led to interest in hybrid development in other crops, with success in sorghum. The opportunity to utilize hybrid vigor may some day exist in all crops (see Chapter 9).

The development of techniques to produce haploid corn plants, similar to that in barley, would be an asset in speeding up the development of homozygous lines.

APOMIXIS

Seed can be formed in some crop species without fertilization, a process known as **apomixis.** If apomixis could be completely influenced and controlled, a crop producer would have the efficiency of reproduction through vegetative organs, but with the convenience of propagation through asexually developed seed. The major accomplishment of controlled apomixis would be the ability to combine the desirable characteristics of two parents and perpetuate permanently F_1 hybrid vigor through true-breeding apomictic progeny. Commercial seed production would be simplified, since no isolation requirements exist as there is no danger of outcrossing.

The word apomixis is derived from the Greek meaning "without mixing." Apomixis is a form of **asexual reproduction;** seed develops by mitotic division of a nucleus in a somatic cell of the ovary, so that the resulting seed has exactly the same chromosome complement and genetic constitution as the mother plant. The offspring is a true clone, just as if it had been propagated asexually.

Apomixis has been reported in grain sorghum and pearl millet as well as in numerous perennial forage grasses. Some species, including sorghum, are *facultative apomicts,* which means that the same ovule is capable of either sexual or apomictic reproduction. *Obligate apomicts* reproduce only by means of apomictic seed set, and these can be detected through their progeny. Lack of variation among the progeny and a close resemblance to a heterozygous or cross-pollinated maternal parent suggests that apomictic seed set has occurred. Progeny of homozygous, self-pollinated crops are expected to lack variation and to resemble the maternal parent. Apomictic seed set can be detected in self-pollinated crops by a lack of F_2 segregation following a known cross to a diverse plant.

The Law of Homologous Series suggests that because apomixis exists in sorghum, it should exist or be developed in other cereal crops. Such a possibility offers plant breeders a challenge to develop this interesting form of reproduction.

BIOTECHNOLOGY—A FRONTIER OF SCIENCE

Advances in gene manipulation, called biotechnology, have opened new avenues in agricultural research and have added a new dimension to the future of cereal crop breeding and production. Biotechnology involves the manipulation of genetic materials through the application of engineering and technological principles to basic research at the cellular and molecular level. The ultimate benefits to agriculture production and society are substantial, perhaps equal to or greater than those of the "green revolution" of the 1960s, but the benefits may be one or two decades away (Bodde, 1982b). Biotechnology can be regarded as a frontier of science—an area where the boundaries of knowledge are being pushed most vigorously.

Biotechnology may make it possible to modify the genetic makeup of an individual by means other than normal sexual reproduction by using **in vitro** techniques. More specifically, biotechnology may provide a relatively quick and accurate method of incorporating specific, desirable genes into an individual, even between sexually incompatible species (Ingle, 1982). For third-world agriculture Swaminathan (1982) used the rice crop to demonstrate the potential that exists for integrating biotechnology with standard techniques to speed up rice breeding to overcome problems of salinity and alkalinity.

Aspects of biotechnology include genetic engineering, the fused-cell technique, and plant cell or tissue culture, discussed below.

Genetic Engineering

The ability to take selected fragments of DNA from a plant cell, from microorganisms, or from chemical synthesis and to join these fragments to another piece of DNA and then to transfer the reconstructed DNA to a microorganism is known as **recombinant DNA** or genetic engineering. A desirable gene to be inserted is carried by a **vector,** using in vitro techniques. The desired gene can be isolated by cleavage, using a specific enzyme. The same enzyme is used to "cut" an isolated vector molecule to produce identical severed ends, so that bonding occurs. The resulting DNA mixture is introduced into a bacterial host cell. To identify if the desired genetic combination exists in a random mixture of DNA fragments, the mixture is screened by introducing toxic compounds to which the desired gene combination is resistant.

The transfer of genes into plants uses similar procedures. When a desirable gene in a plant is isolated, the gene is manipulated into a bacterium using recombinant DNA procedures. It is then necessary to incorporate this gene into a plant cell, a tricky process because vectors equivalent to those in microorganisms do not exist in plants. Viruses may serve as potential vectors, or a suitable vector may have to be synthesized.

Plant cells are screened to select for the desired trait, using a drug that inhibits the growth of susceptible plant cells. The desirable survivors must be regenerated into a plant by means of plant cell or tissue culture, starting with a single cell or naked **protoplast.**

The successful application of genetic engineering or recombinant DNA technology to cereals should provide a precise tool for the transfer of a specific gene into a plant. The limitations caused by sexual incompatibility may be overcome, allowing for gene transfer between any two species. Perhaps a single gene transfer will cause minimal disruption. The time required to transfer a gene using recombinant DNA technology should be reduced over that of conventional crossing techniques.

Fused-Cell Technique

The ability to fuse together two cells from two different species through somatic hybridization may overcome incompatibility barriers. This parasexual hybridization may enable interspecific crosses to be obtained that might otherwise be impossible. The new entity contains nuclear material from both, and although combinations of unrelated genomes may be followed by elimination of chromosomes, total elimination may not occur. Novel new plants may therefore be produced (Abelson, 1983; Shepard et al., 1983).

Protoplast fusion between two related but sexually incompatible species promises a more immediate possibility. Primitive species with disease resistance might be incorporated into a commercial cultivar. Fusion is achieved by removing the cell walls with enzymes to expose the naked cell contents or protoplast. The newly fused cell must then be induced to divide by mitosis and to grow into a new hybrid plant. This phenomenon, called *totipotency,* can be achieved in tobacco[2] and petunia,[3] but is more difficult to accomplish in cereals. Totipotency is achieved by culturing the cells in vitro, using techniques similar to those involved in haploid plant production, but a mass of callus or undifferentiated tissue may develop instead (Cocking, 1977).

Some traits, such as **male sterility,** inherited through cell material outside the nucleus, are

[2]Tobacco—*Nicotiana tabacum* L.
[3]Petunia—*Petunia hybrida.*

known as **cytoplasmic inheritance.** Cell fusion may be used to transfer cytoplasmic features from one species to another. Cytoplasmic male sterility is useful in the development of hybrid crops.

Plant Cell or Tissue Culture

Tissue culture, the development of an entire plant from a single cell, includes the culture of embryos, protoplasts, somatic cells, anthers, or pollen. The objectives dictate the part of the plant used.

Plant tissue culture provides an opportunity for in vitro selection of cells containing attributes from a mass of growing cells in a cell culture, provided traits are expressed at the cell level. Often masses of cells respond to herbicides or plant toxins as mature plants would respond (Abelson, 1983). Salt tolerance, disease resistance, herbicide tolerance, and quality factors are expressed at the cell level.

The advantage of cell culture is that a single flask culture contains approximately 40 million cells, each a potential unit of selection (Tomes and Kasha, 1981; Tomes, 1982). While growing in a culture medium, plant tissue at the cellular level can be manipulated and exposed to various stress factors. Tissue that survives the stresses can be regenerated into a normal plant with its resistance to a specific trait already proven. Plant tissue-culture techniques may be applied effectively to produce virus-free plants or to multiply and preserve valuable genetic material (Holdgate, 1982). In strawberries,[4] for example, more than one million plants can be obtained from one donor plant in a year (Keller and Rajhathy, 1981).

Plants derived from haploid sex cells are themselves haploid, and, if they can be produced reliably, will allow for greater development of the haploid breeding method.

Predictable plant regeneration from tissue cultures of any crop is a prerequisite to the use of this technique in a plant-improvement program. Currently, all major cereal crops, including corn, oat, pearl millet, sorghum, rice, barley, and wheat, have been grown in cell-culture programs with various degrees of success. Limitations exist, including the fact that quantitative traits for yield cannot be screened using tissue culture.

[4]Strawberry (cultivated)—*Fragaria grandifora.*

Future Possibilities for Biotechnology

A potential nitrogen fertilizer shortage and rising fertilizer costs, both related to a dwindling world supply of fossil fuels, and the fact that cereal yields are critically influenced by the amount of nitrogen in the soil, have raised interest in alternate sources of nitrogen fertilizer. Biological nitrogen fixation in cereal crops is a possibility that has considerable appeal. Attempts to transfer a **symbiotic relationship,** similar to that which is formed between rhizobia bacteria and legume roots, to cereal crops has been attempted. Evidence exists that free-living nitrogen-fixing bacteria can be encouraged to associate with roots of cereals, enabling the cereal plant host to receive some nitrogen through bacterial nitrogen fixation (Barton and Brill, 1983). This association may be improved with time. The transfer of **genes from a bacterial complex[5]** is a possibility but is considered a giant step that may not be feasible. Some very difficult and perhaps insurmountable problems must be overcome (Barton and Brill, 1983; Bodde, 1982a; Ingle, 1982; Rennie, 1981).

The incorporation of enzymes associated with C_4 photorespiration into C_3 cereals (Ingle, 1982) is an exciting concept. Other hopefuls include improvement in quality of plant products, increased protein content, stress tolerance, and the obtaining of disease resistance from genetically incompatible species.

Large-scale exploitation of biotechnology offers exciting advances for plant improvement in cereal crops. Biotechnology is being heralded as the scientific answer to many problems in biology. Success in polypoid species such as wheat,

[5]A bacterium known as *Klebsiella pneumonial* contains a cluster of 17 contiguous *nif* genes that produces a complex of enzymes required for nitrogen fixation.

however, may require many years of development before techniques can be used to breed superior crop cultivars. Biotechnology should not be viewed as a substitute or replacement for classical breeding methods.

CONCLUSIONS

Genetics and environment together determine our food supply. An understanding of these two components is essential to successful crop production; the environmental aspects will be outlined in the next chapter. The promise of genetics lies in the fact that the genetic constitution of crop plants can be controlled and manipulated; unfortunately, much of the environmental impact cannot be controlled.

New techniques in genetic engineering promise opportunity for genetic control and manipulation in the future. Success in biotechnology will shift the biological aspects associated with plant breeding and inheritance more toward being an exact science and away from being an art.

Plant breeding represents the application of genetic principles. Plant breeding is a long-term undertaking, yet plants can be grown to reproduction much more rapidly than can large animals in animal breeding programs. Capable students often regard plant breeding as an exciting and challenging field in which to help meet our present and future food needs.

QUESTIONS

1. *a.* Consider the state of food production as an art or a science and compare it to scientific progress in other fields.
   b. Consider scientific progress in terms of the age of various fields of endeavor including agriculture.
   c. Why has progress in agriculture seemingly been so slow?
2. *a.* Dwarfing genes in sorghum have been useful because they favor mechanical harvesting and reduce lodging. Four nonlinked, brachytic dwarfing genes are known. Tallness is partially dominant to shortness, and five height classes are recognized. Tall sorghum may be as tall as 3 or 4 m (10 to 13 ft), and a dwarf plant may be as short as 1 m (3.3 ft). If no additional genes impact on sorghum height, describe the phenotype of the following genotypes and the appropriate terminology.
   (1) $Dw^1Dw^1Dw^2Dw^2Dw^3Dw^3Dw^4Dw^4$
   (2) $Dw^1Dw^1Dw^2Dw^2dw^3dw^3Dw^4Dw^4$
   (3) $dw^1dw^1Dw^2Dw^2dw^3dw^3Dw^4Dw^4$
   (4) $dw^1dw^1dw^2dw^2Dw^3Dw^3dw^4dw^4$
   (5) $dw^1dw^1dw^2dw^2dw^3dw^3dw^4dw^4$
   b. Consider the F_1 genotype and phenotype of a cross between (3) and (4) in question 2*a*.
   c. What precautions would have to be taken to ensure that F_1 progeny of a crop were of the dwarf genotype and all phenotypically short?
3. *a.* Develop a Punnett square to determine the phenotypes of a cross between two parents with the following genotypes:
   Parent A: *Dd ma ma*
   Parent B: *dd Ma ma*
   where D = tall, d = dwarf, Ma = late, ma = early.
   b. If a breeding objective were to select for a dwarf, early-maturing cultivar, how many true breeding progenies would be produced in the F_1 from the cross in 3*a*?
4. Consider the implications for a plant breeder wishing to develop a cultivar having the following features: semidwarf height, early maturity, adult resistance to mildew and rust, a high protein level, lodging resistance, strong gluten, red kernel color, upright leaves, deep-rooting ability, a long period of reproductive growth, seedling vigor,

sprouting resistance, and resistance to Hessian fly attack. How might a plant breeder undertake such a task?

5. At the time of gametic union, a recombination of genetic factors occurs. Explain.
6. *a.* Open-pollinated and F_1 hybrid corn are both highly heterozygous. Why then are F_1 hybrids superior to open-pollinated cultivars?

   b. In the heterozygous state a genetic combination may be *TtMm*. Assume that *TT* and *MM* represent genetic characters in corn that are desirable and *tt* and *mm* are undesirable. If the two characters are inherited independently and when the *TtMm* plants are selfed, what percentage of the plant population of the next generation could be eliminated as undesirable?

   c. If an undesirable recessive gene remains undetected in corn inbred, is it likely to be expressed in the F_1 hybrid population?
7. *a.* Virtually 100% of commercial corn grown in North America, as in many advanced nations of the world, is hybrid. But hybrid corn is not universally grown. Consider the disadvantages of hybrid corn and why it is not universally produced.

   b. In self-pollinated crops farm producers are able to use seed produced from a previous crop year after year. With hybrid crops, seed must be purchased each year. What are the consequences of using F_2 seed?
8. *a.* Would nature have improved upon crop production if all plants had evolved mechanisms to reproduce apomictically as the major mode of reproduction? Explain.

   b. If humans were able to induce a plant to produce apomictically whenever they wished, how would this help overcome the problems of hybrid crops as considered in question 7*a*?
9. *a.* Biotechnology promises exciting new advances in plant breeding and crop improvement in the next decades. Consider some of the possibilities in developing a crop "ideotype," using techniques in biotechnology.

   b. Explain the statement, "Biotechnology should not be viewed as a substitute or replacement for classical breeding methods."
10. *a.* Biotechnology is a sophisticated step forward in the science of crop production. Consider food production as an art or a science in the context of biotechnology.

    b. Is the sophistication of biotechnology so great that its advances are applicable only to the most advanced areas of the world? Review references by Swaminathan (1982) and Bodde (1982b).

References

Abelson, P. H. 1983. "Biotechnology: An Overview." *Science* 219: 611–613.

Barton, K. A., and W. J. Brill. 1983.. "Prospects in Plant Genetic Engineering." *Science* 219: 671–682.

Bates, L. S., and C. W. Deyoe. 1973. "Wide Hybridization and Cereal Improvement." *Economic Botany* 27: 401–412.

Bodde, T. 1982a. "Biotechnology in the Third World." *BioScience* 32: 713–716.

———. 1982b. "Genetic Engineering in Agriculture: Another Green Revolution?" *BioScience* 32: 572–575.

Borlaug, N. E. 1971. "The Green Revolution, Peace and Humanity." *Cereal Science Today* 16: 401–429.

Cocking, E. C. 1977. "Protoplast Fusion: Progress and Prospects for Agriculture." *Span* 20 (1): 5–7.

Dewey, D. R. 1982. "Genomic and Phylogenetic Relationships among North American Perennial Triticeae." (In) J. R. Estes, R. J. Tyrl, and J. N. Brunken (eds)., *Grasses and Grasslands: Systematics and Ecology.* Stillwater: University of Oklahoma Press. pp. 51–88.

Gregory, W. C. 1972. "Mutation Breeding in Rice

Improvement." (In) *Rice Breeding.* Los Banos, Philippines: International Rice Research Institute. pp. 551–572.

Hadley, H. H., and S. J. Openshaw. 1980. "Interspecific and Intergeneric Hybridization." (In) W. R. Fehr and H. H. Hadley (eds.), *Hybridization of Crop Plants.* Madison, Wis.: American Society of Agronomy and Crop Science Society of America. pp. 133–159.

Heslop-Harrison, J. 1982. "Pollen-Stigma Interaction and Cross-Incompatibility in the Grasses." *Science* 215: 1358–1364.

Ho, K. M., and G. E. Jones. 1980. "Mingo Barley." *Canadian Journal of Plant Science* 60: 279–280.

Holdgate, D. P. 1982. "Tissue Culture for Commercial Plant Propagation." *Span* 25(1): 24–26.

Ingle, K. 1982. "Genetic Manipulation and Plant Improvement." *Span* 25 (2): 50–53.

Kasha, K. J. 1974. "Haploids from Somatic Cells." (In) K. J. Kasha (ed.), *Haploids in Higher Plants: Advances and Potential.* Guelph, Canada: University of Guelph. pp. 67–87.

———, and K. N. Kao. 1970. "High Frequency Haploid Production in Barley (*Hordeum vulgare* L.)." *Nature* 225: 874–876.

Keller, W. A., and T. Rajhathy. 1981. "Crop Improvement: The Role of Biotechnology." *Agrologist* 10(3): 11–13.

Kostoff, D. 1938. "Colchicine and Acenaphthene as Polyploidizing Agents." *Nature* 142: 753.

Mangelsdorf, P. C. 1951. "Hybrid Corn." *Scientific American* 185 (2): 39–47.

Mendel, Gregor. 1930. English translation of 1865 presentation, *Experiments in Plant-Hybridization.* Cambridge, Mass.: Harvard University Press. Translation by the Royal Horticultural Society of London. pp. 313–352.

Micke, A., S. C. Hsieh, and B. Sigurbjornsson. 1972. "Rice Breeding with Induced Mutations." *Rice Breeding.* Los Banos, Philippines: International Rice Research Institute. pp. 573–580.

Mujeeb-Kazi, A., and M. Bernard. 1982. "Somatic Chromosome Variations in Backcross I Progenies from Intergeneric Hybrids Involving Some Triticeae." *Cereal Research—Communications* 10: 41–45.

Reinbergs, E., and L. H. Shebeski. 1959. "Fertility in Successive Generation of Four Autotetraploid Barley Varieties and the Effect of Selection for Fertility in the OAC 21 Autotetraploid." *Canadian Journal of Plant Science* 39: 98–107.

Rennie, R. J. 1981. "Dinitrogen Fixation Biotechnology at Work." *Agrologist* 10 (3): 20–21.

Ross, C. 1953. "New Lodge-Defying Tetraploid." *Southern Seedsman* 16: 16–17, 72.

Shepard, J. F., et al. 1983. "Genetic Transfer in Plants Through Interspecific Protoplast Fusion." *Science* 219: 683–688.

Simmonds, N. W. 1980. "Polyploidy in Plant Breeding." *Span* 23 (2): 73–75.

Swaminathan, M. S. 1982. "Biotechnology Research and Third World Agriculture." *Science* 218: 967–972.

Tomes, D. T. 1982. "Selection at the Cellular Level in Crop Plants: Potential for Agriculture." *Highlights of Agricultural Research in Ontario* 5 (1): 1–3.

———, and K. J. Kasha. 1981. "Tissue Culture Improving Crops Through Biotechnology." *Agrologist* 10 (3): 17–18.

Wilson, A. S. 1876. "On Wheat and Rye Hybrids." *Transactions and Proceedings of the Botanical Society of Edinburgh* 12: 286–288.

Wittwer, S. H. 1974. "Maximum Production Capacity of Food Crops." *BioScience* 24: 216–224.

Yonezawa, K., and H. Yamagata. 1977. "On the Optimum Mutation Rate and Optimum Dose for Practical Mutation Breeding." *Euphytica* 26: 413–426.

6

Cereal Crop Yield—Impact of Environment

INTRODUCTION

Crop yield reflects the combined effects of genotype, cultural practice, and **environment** on crop growth and development. **Climate, soil,** and **physiographic** and **biotic factors** each exert a direct effect on plant growth as well as an indirect effect through interactions with other environmental factors. Vulnerability of crops to adverse environmental conditions not only causes crop failure and harvest losses, but also prevents a significant extension of the global surface of arable land and a shift of more productive crops to new areas. Indeed, environmental factors exert a more powerful impact on food production than any other crop input. Environmental variables that cause yield fluctuations cannot be eliminated or controlled at the present level of technology. Agriculture, therefore, may be the only branch of economic activity that is subject to wide and irregular output fluctuations beyond human control. If environmental variables were to upset the food supply/demand situation in two or more major food-producing regions of the world, widespread famine could result.

Although humans cannot control the environment, they have done everything within their power to dominate it. Excessively large amounts of nonrenewable energy inputs are expended directly and indirectly on crop production. Direct inputs include fossil fuels, pesticides, fertilizers, machinery, and irrigation water; indirect inputs include energy expanded on manufacturing, transportation, crop testing, and advisory inputs. These energy inputs are supplemental to sunlight energy for photosynthesis and are called **cultural energy** inputs. Although outputs of crops per hectare and per man-hour have soared spectacularly, the **law of diminishing returns** applies, and the efficiency of energy use in food production in many technologically advanced areas of the world is low.

Mechanization has provided agricultural producers with an enormous capacity to dominate natural forces. Machinery is used to speed field operations during favorable **weather** conditions and to extend operations under unfavorable conditions. Rather than wait for freezing and thawing action or root forces to loosen soil, pulverizing equipment is used; to avoid moisture deficiencies irrigation water is applied; drainage systems are applied to remove excess water; and pesticides and fertilizers are applied at levels that dominate the soil environment. Domination of the environment has meant that producers become inflexible, and soil protection is placed at a low priority level.

Despite genetic manipulation and some success at modifying plants for specific environmental conditions, and despite expensive and time-consuming performance-testing programs, patterns of world crop distribution remain subject to basic principles of **crop ecology.**

Human efforts in crop production are little more than a buffering action against yield fluctuations. Environmental variations can override the best cultural conditions and the most suitable cultivar. Measured yield fluctuations over an extended time period are shown in Table 6-1 and were found to vary a substantial 10.3% to 23.1% in North America per year despite the most intense efforts to dominate the environment. The magnitude of various environmental limitations in the United States is expressed through the distribution of insurance indemnities from 1939 to 1978 (Table 6-2). Environmental factors of drought, excess water, cold, hail, and wind, all beyond human control, accounted for 89.3% of crop losses as measured by insurance indemnities. As agriculture around the world is pushed into less favorable areas through urbanization and the need for increased production, the adverse effort of environmental variables can be expected to increase. If environmental variables cannot be modified, improvements in the adaptation of plants to adverse environments can make major contributions to agricultural production. If the fundamental mechanisms of adaptation are understood, faster progress can be expected. Without environmental control, the best that can be hoped for is an understanding of the environmental effects, so that cultural practices can be modified to reduce the short-term impact and genetic factors modified for the long-term impact.

Table 6–2
Distribution of Crop Insurance Indemnities in the United States, 1939–1978

Cause of Crop Loss	*Percent of Payments*
Drought	40.8
Excess water	16.4
Cold	13.8
Hail	11.3
Wind	7.0
Insect	4.5
Disease	2.7
Flood	2.1
Other	1.5

Source: Data from Boyer (1982).

THE HOLOCOENOTIC ENVIRONMENT

Crop-environment relationships are many and exceedingly complex, with the result that little quantitative information exists that characterizes the response of a crop to the overall environment. Many useful studies have been conducted

Table 6–1
Measurements of Annual Crop-Yield Variations for Selected Cereal Crops in North America over Specific Time Periods

Crop Species and Country	*Period of Record*	*Mean Grain Yield*		*Percent Variation in Yield*
		kg/ha	*lb/ac*	
United States:				
Corn	1866–1975	2233	1994	12.5
Wheat	1866–1975	1109	990	10.3
Sorghum	1929–1975	1788	1596	13.3
Canada:				
Barley	1922–1975	1473	1315	16.4
Wheat	1922–1975	1250	1116	23.1

Source: Anonymous (1976).

on individual effects, such as temperature or light, but the yield of crops is the composite effect of all environmental, biological, and technological factors. To consider the environment in its entirety, as a **holocoenotic** environment, is complex because of its dynamic and ever-changing nature. The intensity of environmental factors varies with the hour, the day, and the season.

Approaches to the holocoenotic environment have become more urgent with the possibility of climatic change. The aim is to apply quantitative crop-environment relationships to crop observations obtained by remote sensing and to determine the impact of a specific environmental extreme. A holocoenotic approach on a global basis would aid in a rapid and reliable assessment of global crop yields to help those with surplus production match the demand before famine strikes.

Climatic Change

Various theories exist that cyclical climatic change has occurred in the past and that future changes can be expected. In the race between population and food, adverse climatic change presents a potentially dangerous factor. In the long term, changes in climatic factors can be determined; in the short term, minor climatic shifts are difficult to determine because they are obscured by short-term weather variations, by volcanic activity that spouts dust particles into the upper atmosphere, or by the release of carbon dioxide into the atmosphere by combustion of fossil fuels.

One possible threat is a cooling trend in the middle latitudes (Thompson, 1975) away from a period of benign weather of the 1940s, regarded as an agricultural optimum in terms of climate. In support of a cooling trend, Bryson (1975) extended cyclical temperature patterns, based on mean annual temperatures in Iceland over a 1000-year period as determined from historic records of the duration of sea ice on the shores of Iceland, to show a cooling trend in the northern hemisphere.

Small-cereal crop production is concentrated in the middle latitudes, where summer temperatures average 21° to 24°C (70° to 75°F). For C_3 cereal crops, cool growing conditions favor grain yield because of low respiration and because the effectiveness of limited rainfall is increased. A cooling trend may reduce the length of the growing season and render fringe areas unsuitable for full-season cereal crops, thereby reducing the area of cereal grain production.

Another possibility is that a warming trend may be generated by the increase in atmospheric carbon dioxide (Anonymous, 1982a; Hanson et al., 1981; Revelle, 1982; Rotty, 1978; Verma and Rosenberg, 1979; Wittwer, 1980). Potential effects include dramatically altered areas of agricultural production, duration of the growing seasons, and precipitation patterns (Hansen et al., 1981). Shifting climatic zones may lead to the development of drought-prone regions in North America and central Asia. A warming trend could reduce the area of Antarctic ice sheets by melting action and raise sea levels worldwide. Possible biological changes include increased photosynthetic efficiency with a carbon dioxide increase, reduced water requirements, accelerated growth, and increased productivity. The positive impacts on biological processes may be of greater significance for agriculture than climatic effects (Wittwer, 1980). A warming trend would favor C_4 plants (Patterson and Flint, 1980). Increases in total biomass will occur with increased carbon dioxide, only if other factors such as water, nitrogen fertility, and sunlight energy are not limiting (Kramer, 1981).

The lack of a holocoenotic understanding means that great uncertainties exist about the total impact of a global cooling or warming trend on food production. Many unknowns and uncertainties exist with regard to climatic measurements, except for carbon dioxide. Accurate and regular carbon dioxide monitoring dates back only to 1958, when a level of 315 ppm was recorded. By 1977 the level had reached 330 ppm (Rotty, 1978), and it appears to be increasing (Verma and Rosenberg, 1979). The level by the year 2025 may be near 600 ppm, double the estimated 298 ppm at the start of the industrial revolution.

Although the major effect of climatic change may not be felt for a century, preceding slow and subtle climatic changes will have a ripple effect throughout agriculture. In the face of climatic uncertainty, humans should safeguard food production through preservation of genetic diversity and a system of grain reserves.

Crop-Climate Relationships

A holocoenotic approach on a global basis would aid in a rapid and reliable assessment of global crop yields to help those with surplus production match the demand, and it would aid in monitoring production patterns that accompany climatic shifts (Schware and Kellogg, 1982). Understanding the impact of environmental variables could be of value for the judicious application of expensive inputs in terms of amount and timing. The technology for a holocoenotic approach to global crop-climate relationships may be aided by satellite monitoring.

Satellite technology allows for remote sensing of the reflected solar radiation to estimate the areas planted to specific crops. Accurate final harvest estimates can be made from climatological data, only if quantitative relationships can be applied meaningfully (Morain and Williams, 1975). Although knowledge of crop responses to temperature and moisture conditions is advancing steadily, the systems are highly complex, not yet fully understood, and subject to numerous complicating factors.

Based on 8100 surface weather observations as well as information from satellites, United States scientists are able to compare deviations in weather variables from the expected or long-term average (Richter, 1982). Survey data determine whether rainfall is lighter or heavier than normal; how soil moisture levels compare to those of previous years; heat units accumulated; and if weather occurrences in general are normal or represent departures from the expected for a given area. The area planted to a specific crop, the disease situation, the level of technology, and general agronomic practices can be used for yield determination and global crop estimates. The success of this program rests not only on weather information but on the assembling of high-quality, historic data bases for all the major regions of the world. The task is made easier by the fact that relatively few crops, including cereal grain, need to be monitored. When used for global crop forecasting, an accuracy within 10% of the true production 90% of the time was reported (Bauer, 1979). In the winter wheat regions of the world this level of accuracy was achieved 1.5 to 2 months before harvest.

Global forecasting requires a holocoenotic approach, although weather information is used indirectly. Whether the use of information is direct or indirect, a hierarchy of environmental variables must be determined in terms of primary, secondary, and tertiary effects. Views on the importance of each variable vary, reflecting individual interpretations and the ease of measurement and application. Temperature and moisture, for example, have been described as the two primary environmental determinants of crop yield (Idso, Jackson, and Reginato, 1977), possibly because both are discrete and finite entities. Light, water, and carbon dioxide, however, are the primary factors for photosynthesis and yield. Specific spectral bands of light needed for photosynthesis are less easy to measure in terms of intensity and duration and to relate to crop yields than are temperature and moisture.

Which environmental variable is of greatest importance can be debated, but such considerations may be purely academic when the holocoenotic environment is considered.

LIGHT ENERGY OR RADIATION

Sunlight energy is the primary source of energy for photosynthesis, hence it is absolutely essential for plant growth and development. Characteristics of light include intensity, spectral energy distribution or quality, and duration.

Intensity

Solar radiation is measured in terms of its total energy and is expressed as watts per square meter. Extraterrestrial sunlight has an intensity value of about 1400 W/m^2, but some of this energy is absorbed or scattered by the atmosphere and its constituents. Smoke, dust, and gases scatter light, and water vapor absorbs light.

Radiation intensity when the sun is directly overhead ranges from a high of 1200 on mountain tops to 1000 W/m^2 at sea level (Shirley, 1945). Because light is absorbed by water vapor, the intensity is less in humid than in arid climates; in areas where cloud and fog are prevalent, light intensity may be reduced to as little as 4% of its potential (Daubenmire, 1959). Plants are capable of using cloudlight energy for photosynthesis but at a reduced rate.

Plants growing beneath a canopy of tall plants, or basal leaves on a plant, receive less intensity because leaves absorb light energy. Some light can be transmitted through leaves.

Light intensity varies according to geographic location, season, and time of day. Each day the intensity of sunlight energy increases from zero to a peak near midday, depending on cloud cover, and diminishes to zero again at the end of the day. The intensity at any one location on the earth's surface depends on the distance of the sun because of the absorption and diffusion of light by the atmosphere. In temperate zones light intensity varies with seasons; in winter it is low because of the angle of incidence.

Hourly values and yearly totals of light are much higher in the tropics than at high latitudes. During the growing season at high latitudes, however, the greater length of day compensates for the lower intensity, and the amount of light received during a single day may be only slightly below that received during the short days in the tropics.

Spectral Energy Distribution or Light Quality

Not all the radiation a plant receives is useful in photosynthesis. About 99% of solar radiation lies between wavelength bands of 0.2 to 4 microns (200 to 4000 nanometers), but the region between 0.4 and 0.7 microns (400 to 700 nanometers) is most important in terms of plant photosynthetic efficiency and is called **photosynthetically active radiation** or PAR (Figure 6-1). PAR is measured in terms of **photon** content and is expressed in einsteins. At Hyderabad, India, on a clear day at noon on September 21, for example, PAR measured about 2000 microeinsteins per square meter second (2000 $\mu E/m^2/s$). Under the same conditions but on a cloudy day, PAR was less than 700 $\mu E/m^2/s$ (Sivakumar, 1979). These differences in PAR suggest that on a cloudy day,

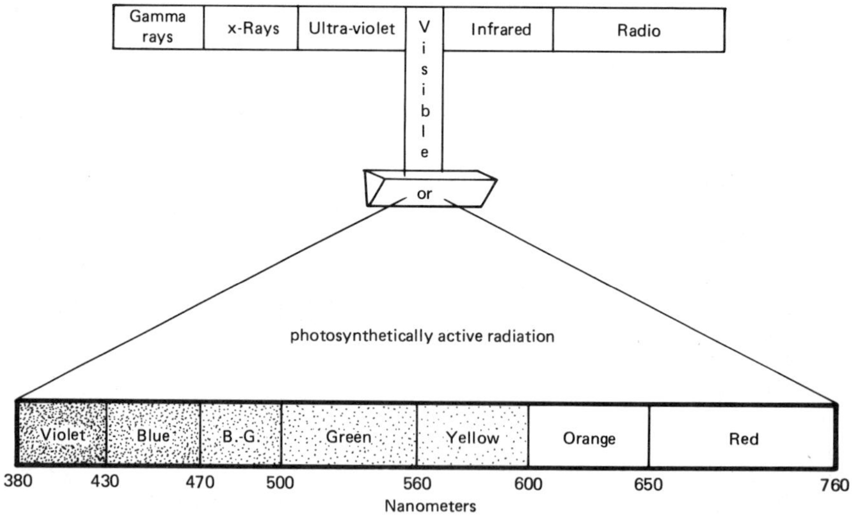

Figure 6–1. Visible light between 400 and 700 nanometers in the electromagnetic spectrum is used in photosynthesis.

photosynthetic activity is considerably less than on a clear day.

For plants to grow, the rate of photosynthesis during the light period must exceed losses by respiration during both the day and night. The compensation point is reached when photosynthesis equals respiration losses. The compensation point varies among plants; in cereals it is about 3% of full sunlight. Producers are careful to avoid plant densities at which many leaves are below the compensation point. Photosynthesis and growth is a function not of a single leaf but of the entire crop canopy and hence of factors that affect the distribution of light over the leaves of the canopy. Management practices attempt to distribute the available light over a maximum portion of the crop canopy for optimum plant productivity (Yoshida, 1972). Important factors include the number and size of leaves, their position relative to each other, whether leaves are disposed upright or horizontally relative to the vertical culms, the transmission and reflecting properties of the leaves, the ratio of diffuse and direct light, and the height of the sun (de Wit, 1965).

For leaves at the top of the plant, PAR is likely to exceed the level required for maximum levels of photosynthesis. Photosynthesis increases more or less linearly with increasing light intensity from the compensation point, until a maximum is reached. This maximum is about 30% of the average light intensity at a horizontal surface in the middle of the day (de Wit, 1965). PAR falling on a crop may strike the leaves and be absorbed and used in photosynthesis or used to produce heat; some may be reflected; some is transmitted through the leaves; some may strike the soil between the plants and be lost for photosynthesis. Highest crop yields are achieved when light is effectively intercepted. Plant density and the arrangement of leaf blades in cereal crops must be such that the lower leaves are at or above the compensation point.

The leaf-area index (LAI) of a crop is a measure of the leaf area in relation to a unit land area. A leaf-area index of 3 means that there are 3 square meters of leaves on each square meter of land area. The leaf area at which dry-matter production is maximized is regarded as the **optimum leaf area.** Below the optimum leaf-area index, PAR interception is incomplete; above it, lower leaves are shaded and are below the compensation point.

In cereal crops grown for grain, the optimum leaf-area index should coincide with the beginning of grain production and should coincide as closely as possible with the period when the highest levels of PAR are received (refer to Figure 4-7).

An absolute value for the optimum leaf area cannot be stated, but in general, complete or almost complete light interception by a corn canopy occurs at an LAI of 4. Generally it is impossible to plant at a density to achieve precisely the optimum LAI, which is influenced by the amount of light transmitted through a leaf, by absorption, and by leaf angle. Upright leaves are considered desirable, especially at high plant populations and under high levels of PAR. Leaf arrangements or dispersion can influence light interception, and dwarf and semidwarf plants may have a less desirable leaf dispersion than tall plants. In plants that tiller, the degree of tillering will affect leaf area, and abundant tillering can lead to excessive mutual shading.

Light Duration

The amount of light received by plants is a function of light intensity, the duration of the growing season, and the duration of sunlight hours per day.

At latitudes where the duration of sunlight hours varies over the season, crop producers should seed crops to benefit from the long days of the summer solstice (June 21 in the Northern Hemisphere and December 21 in the Southern Hemisphere). With determinate cereal crops, the more closely reproductive development coincides with long days, the greater the yield potential.

The impact of the relative length of light and dark periods in a 24-hour period on photoperiodically sensitive plants was outlined in Chapter 2.

Specific response of crop cultivars to photoperiod is important in regions where plant seeding dates vary widely. In South Africa, for example, wheat is grown between 23° and 34°S and may be sown in spring, fall, or winter. Climatic conditions vary from cool short days to warm long days. Generally, wheat is regarded as a long-day crop.

When three wheat cultivars were grown in controlled-environment chambers at 11- and 13-hour photoperiods and at a 15°/5° and 20°/10°C day and night temperatures, specific cultivar light-temperature interactions were observed (Hammes and Marshall, 1980). High temperatures speeded up reproductive development at both photoperiods by 4.5 days. High temperatures shortened the time to maturity by 27 days at a short and by 50 days at a long photoperiod.

LIGHT ASPECTS AND CONTROLLED-ENVIRONMENT CHAMBERS

All cereal plants can be grown successfully in controlled-environment chambers at light intensities considerably less than those found outdoors. The reason is that mutual shading is limited, chamber walls are highly reflective, constant air conditioning promotes a rhythmic leaf movement that results in **light flecking,** there is a concentration of light of suitable spectral quality, and the light has a controlled duration.

Controlled-environment chambers are suitable units for cereal crop research and are an important resource for environmental and physiological studies, as a tool in biotechnology research, as a supplement to field testing in plant breeding, and as an aid in seed multiplication of photoperiodically sensitive crops outside their area of adaptation. In climates where field research is seasonally limited, controlled-environment chambers are an important part of a research laboratory.

SUNLIGHT ENERGY IN PERSPECTIVE

Only an infinitesimal amount of the total solar energy striking the outer atmosphere is utilized as chemical or stored energy. Less than 1% of that total solar energy is captured in photosynthesis for use by humans (Figure 6-2).

TEMPERATURE

Temperature is a dominant, universal, familiar, and easily measured and recorded environmental aspect. Almost every plant process is influenced directly or indirectly by temperature, including respiration, translocation, cell growth, and plant development.

Specific plant responses to temperature can be identified. The minimum, optimum, and maximum temperatures for each activity are known as **cardinal temperatures.** Extensive research by Nuttonson (1958a) showed the universal impact of temperature on cereal grain development.

Temperature effects, however, cannot be considered in isolation from other environmental variables. Plant reactions to temperature vary with plant age, plant part, water and mineral nutrition, previous history, and radiant-energy levels. Cardinal temperatures are not absolute but relative values.

Temperature and photoperiod interactions were demonstrated with spring-seeded wheat, oat, barley, and millet crops when grown at two latitudes (Carder, 1957). Radiant energy was approximately the same at the two latitudes, but at 55°N latitude (Beaverlodge, Alberta), temperatures were 40% lower than at 43°N (Madison, Wisconsin). Photoperiods averaged 20.1 and 16.2 hours, respectively, over a 21-day period following emergence. Both temperature and daylength were considered responsible for substantial differences in morphology, heading date, and yield for the crops involved.

AIR TEMPERATURE

Latitude, altitude, proximity to the sea, the characteristics of the prevailing air mass, cloud cover, and radiant energy all influence air temperature.

Outside of the tropics latitude plays a dominant role in seasonal temperature cycles. Temperature is in effect the intensity aspect of heat energy received from the sun. Temperatures at a given latitude will generally determine what crop(s) can be grown successfully. The length of the growing season is determined by

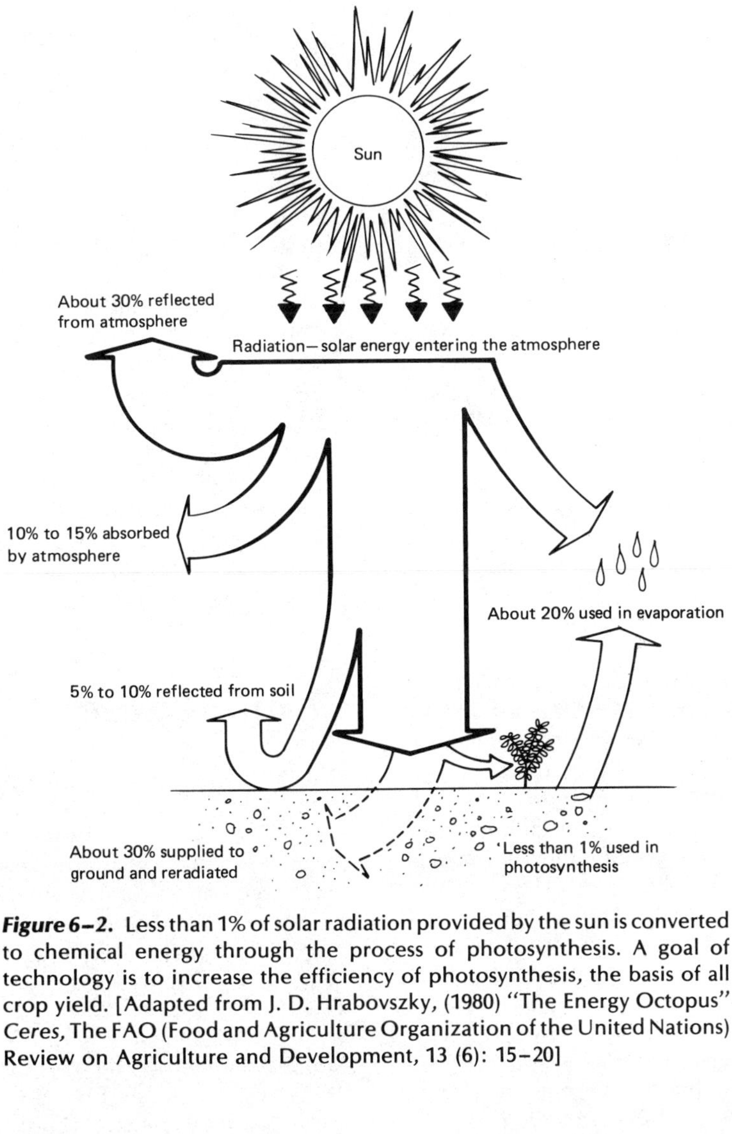

Figure 6–2. Less than 1% of solar radiation provided by the sun is converted to chemical energy through the process of photosynthesis. A goal of technology is to increase the efficiency of photosynthesis, the basis of all crop yield. [Adapted from J. D. Hrabovszky, (1980) "The Energy Octopus" *Ceres*, The FAO (Food and Agriculture Organization of the United Nations) Review on Agriculture and Development, 13 (6): 15–20]

annual temperature fluctuations. Plants can be successfully grown for grain only if their growth rate is rapid enough to complete the life cycle.

The critical effect of temperature on cereal production can be demonstrated by the impact of climatic change. If the climate of the Canadian prairies cooled by 3°C (5.4°F), the area suitable for maturing spring barley would be reduced by about 30% (Thompson, 1975).

Temperature fluctuations in the tropics are much smaller than at higher latitudes. The annual fluctuation of monthly mean temperatures seldom exceeds 5°C (9°F) and often is less than 1°C (1.8°F) in humid regions near the equator; altitude is responsible for greater fluctuations. Night frosts are a regular phenomenon at altitudes of about 2500 m (8200 ft) (Ferwerda, 1979). Near the tropics of Cancer and Capricorn, night frosts may occur during the coolest season at 1000 m (3281 ft) above sea level.

Heat stress is a major factor limiting the productivity and adaptation of crops, especially when temperature extremes coincide with critical stages of plant development. Extreme temperatures may cause damage to crops or prevent their cultivation entirely. The rate of temperature change and the duration and degree of high temperatures all contribute to the intensity of heat stress. Where heat stress occurs, it is important that crops possess a certain degree of heat tolerance to survive. It is difficult, however, to define adequately heat stress in plants, because response depends on thermal adaptation, the duration of exposure, and the stage of growth of the exposed tissue.

Diurnal temperature changes may reach 20° to 30°C (36° to 54°F) in hot areas under clear skies, which facilitate maximum radiation in daytime and rapid loss of heat all night. Plant responses to temperature changes are known as *thermoperiodism.* Plants may respond to seasonal as well as daily temperature cycles. Physiological processes affected include germination, stem elongation, reproductive development, and frost hardiness.

Overheating is common in the tropics but can occur elsewhere, as it is caused by the combined effects of high temperature and drought. Hot dry winds add to the stress. Heat damage is expressed by the inactivation of photosynthesis, while respiration is not inhibited, so that the net assimilation rate declines rapidly (Moss, Musgrave, and Lemon, 1961). Atmospheric temperatures that increase leaf-tissue temperatures affect transpiration. Overheating results in yellowing, growth inhibition, burns or lesions, and eventual death of the whole or a part of the plant (Ferwerda, 1979).

LEAF TEMPERATURE

Temperatures of plant leaves are governed by transpiration, convection, and radiation. Transpiration is an important leaf-cooling mechanism. When the rate of water supply to the leaves is inadequate for transpirational cooling, leaf temperatures increase rapidly. Leaf temperatures 5° to 10°C (9° to 18°F) above adjacent air temperatures are common (Sivakumar, 1979); in the tropics, tissue temperatures 20°C (36°F) higher than the surrounding air have been measured (Ferwerda, 1979). Sorghum grown under different moisture treatments in India had leaf temperatures closely related to available soil moisture (Sivakumar, 1979).

The difference between the temperature of the plant canopy and the temperature of the surrounding air can serve as an index of water stress. By using remotely acquired crop-canopy and adjacent-air temperature differences during the period of head emergence to the cessation of head growth in wheat, final crop yields can be estimated (Jackson, Reginato, and Idso, 1977). The difference between the temperature of the plant canopy and the temperature of the surrounding air provides a quantitative index of water stress and is useful for the scheduling of irrigation water.

SOIL TEMPERATURE

Soil temperature is an important factor in determining the growth rate and distribution of roots as well as their functional activity. Extremes in soil temperatures that limit these three functions can adversely affect the mineral and water needs of the plant, restricting the rate and duration of plant growth and increasing the occurrence and severity of plant diseases.

In temperate regions, soil temperatures may be too low for good germination and seedling establishment. Soil temperatures are influenced by soil moisture, and because water has a specific heat requirement about five times greater than that of solid constituents, a wet soil in the spring is a cold soil. Heat required to evaporate water to dry the soil is not available to warm the soil until drying has occurred. Tillage is effective and often required to help loosen and dry soils to aid warming. On heavy soils at high latitudes, **zero-tillage** is not practical. To promote early and rapid warming of the soil, drainage, tillage, a high level of organic matter, and removal of a surface mulch may be required.

Temperatures of bare-soil surfaces exposed to direct sunlight may rise considerably above air temperatures. Soil surface temperatures as high as 55° to 75°C (131° to 167°F) have been measured in the tropics (Ferwerda, 1979). To avoid damage to emerging seedlings, a mulch may be preferred, or zero-tillage practices may be favored. Plants are killed at temperatures at which protein coagulates.

Penetration of the temperature wave into the soil varies with soil type, tillage, and moisture content. Sandy soils, which do not retain as much water as clay soils, warm up more readily.

MATURITY RATING USING TEMPERATURE

Farmers need to know the time from planting to harvest of various crops to provide a reasonable assurance that harvest will be close to predetermined date. With full-season crops, it is especially important to know that the crop will mature before end-of-season frosts.

The hybrid-corn industry has developed a number of different methods of rating hybrids for maturity. One method gives each hybrid a "relative maturity" in days, but rate of growth per day varies greatly from cold seasons to warm seasons. A hybrid that is a 90-day corn at a warm, southern location may require 115 to 125 days to reach maturity in a cooler, northern location.

Another method widely used for corn is to compare new hybrid cultivars with old ones. This gives the farmer a good basis for comparison, but only if he is familiar with the old hybrid.

Heat units have been used extensively to develop various maturity indexing systems. The heat-unit concept relates to the amount of "effective growth heat" accumulated during the day (Holmes and Robertson, 1959). This value is obtained by subtracting the plant base temperature, or minimum temperature at which growth occurs, from the daily mean temperature. The resulting value is called the "growing degree-day" or "heat" accumulation for the day. This amount is added to that of the previous day until the plant matures.

The heat-unit concept assumes that there is only one base temperature throughout the life of the plant; that day and night temperatures are of equal importance to plant growth; and that the plant response to temperature is linear over the entire temperature range. Despite these limitations, the heat-unit concept has been enthusiastically received.

Growing degree days (*GDD*) ratings have been used as maturity ratings for full-season crops. GDD are calculated between two specified dates by accumulating temperatures on a daily basis. From the mean temperature for the day, 50°F (10°C) is subtracted. For all temperatures below 50°F (10°C) no heat units are added; for all temperatures above 86°F (30°C), the cardinal maximum for corn, a maximum of 80° is used.

A refinement of the GDD system is to accumulate heat units to physiological maturity, as indicated by black-layer development (Baker, 1970). This has increased the dependability of relative maturity comparisons.

Further refinements to the heat-unit system resulted in a rating called **corn heat units** (CHU) (Brown, 1978). This system rates both hybrids and geographic regions in the same terms, so that hybrids can be matched to the proper maturity zone. The corn-heat-unit system uses two relationships, one for daytime temperature and one for nightime (Figure 6-3). The daytime relationship uses 10°C (50°F) as the base and 30°C (86°F) as the optimum for maximum rate of development. The system is based on the assumption that corn hybrids will not grow until the daytime temperature exceeds 10°C and that they grow fastest at 30°C (86°F). The nightime relationship is 4.4°C (40°F) as the base temperature. The geographical growing season for corn is expressed in terms of the accumulated corn heat units from planting to a fall frost. The date coinciding with a 30-year mean temperature of 12.8°C (55°F) in the spring is used as the planting date. The date before which a killing frost can be expected one year in ten was selected as the fall frost date. (See Figures 6-4 and 6-5.)

Corn heat units are accumulated daily between these two dates using values from Table 6-3. For example, if the daily maximum temper-

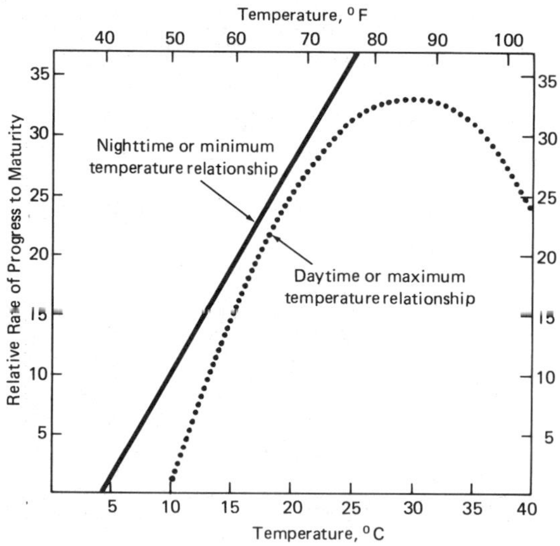

Figure 6–3. Relationships between daily rate of development of corn and nighttime or minimum and daytime or maximum temperatures used to determine corn heat units (CHU). (From Brown, 1978)

Figure 6–4. Photograph of corn seedlings damaged by a previous night frost. The seedlings are wilted, have a water-soaked appearance, and will turn brown in the sun. Unless the growing point has been damaged, however, regrowth of frosted plants may occur and reseeding may not be required. A spring frost is less serious, therefore, than a fall frost. (Courtesy of Ontario Ministry of Agriculture and Food)

Figure 6–5. Corn tissues damaged by frost as indicated by the light areas on the leaf blades. Frost damage is expressed through a loss of chlorophyll and a reduction in photosynthesis. Depending on the stage of development of the ear when frost damage occurs, a reasonable corn yield can be produced from photosynthate stored in the stalk, or from photosynthesis in remaining green tissue. Frosted corn plants can be ensiled for ruminant animal feed, but loss of green tissue is associated with a loss of vitamin A. (Courtesy of Ontario Ministry of Agriculture and Food)

ature is 20°C (68°F) and the minimum is 16°C (20.8°F), the respective values from the table are 24.9 and 20.8. Taking the mean of these two values gives 22.9 corn heat units for that day.

Hybrid corn cultivars are rated by comparing them with a rated standard check hybrid cultivar according to kernel moisture at harvest time. CHU ratings are regarded as being accurate to the nearest 100 corn heat units.

The CHU system has proven to be a reliable guide for producers in Canada. For the corn-growing regions of Canada, CHU ratings vary from 2200 to 3700 CHU. If soil conditions or any other factor delays planting from the normal date, 100 CHU are deducted for each one-week delay from the area rating.

Table 6–3
Corn-Heat-Unit Values for Minimum and Maximum (Night and Day) Temperatures as Determined from Figure 6–3

Temperature		*Corn Heat Units*	
°C	°F	*Minimum (Night) Temperature*	*Maximum (Day) Temperature*
4		0.0[a]	
6		2.8	
8		6.4	
10	50	10.0	0.0[a]
12		13.6	6.3
14		17.2	12.0
16		20.8	17.0
18		24.4	21.3
20	68	28.0	24.9
22		31.6	27.9
24		35.2	30.2
26			31.8
28			32.7
30	86		33.0
32			32.6
34			31.5
36			29.8
38			27.4
40	104		24.3

Source: Data from Brown (1978).
[a]Nighttime and daytime temperatures at or below 5°C (41°F) or 10°C (50°F), respectively, are given a value of 0.0.

Because solar radiation is the driving force behind agricultural production, it has been suggested (Bates, 1976) that the average solar radiation curve for a geographic region be related to crop development and be used instead of heat units. This might provide a more direct approach.

The classic work of Nuttonson, however, has shown the consistency of temperature requirements for various growth stages of small grains. His studies included data from European sites as well as sites in the United States (Nuttonson, 1958a, 1955, and 1958b). Extending Nuttonson's work, Ramirez and Bauer (1974) suggested that temperature units may suitably serve as a predictor of harvest dates for small grains, given the planting dates and expected mean temperature for the growing season. Temperatures in the range of 40° to 90°F (4.4° to 32°C) were suggested for hard red spring wheat, durum wheat, oat, and barley. Highest yields were obtained when temperatures were low, because C_3 cereals are cool-loving crops.

CARBON DIOXIDE (CO_2)

Photosynthesis is dependent on CO_2 in the atmosphere. Global atmospheric CO_2 concentration is increasing, as indicated by accurate and regular measurements taken since the mid 1950s. There is strong evidence that a major contributor is the combustion of fossil fuels. Indirect evidence suggests that CO_2 increases have occurred at an **exponential rate** for more than 100 years. If this trend continues, the CO_2 concentration could increase from the 1980 level of 330–360 ppm to 500–600 ppm by 2030 to 2050 (Hansen et al., 1981; Rotty, 1978; Woodwell, 1978).

Generally, an increase in CO_2 concentration will increase photosynthesis. Greenhouse operators attempt to increase production by CO_2 enrichment. The large air mass associated with field crop production and the resulting lack of control has limited outdoor application. Rapid gaseous exchange with the atmospheric mass cannot be managed (Begg and Turner, 1976). A 10% increase in atmospheric CO_2, however, could benefit crop production by an estimated 5% (Verma and Rosenberg, 1979; Wittwer, 1980), provided that other factors are not limiting. Under greenhouse conditions, yield responses peak at 1000 to 1200 ppm of atmospheric CO_2 (Wittwer, 1980).

CO_2 levels rise and fall in a diurnal cycle as photosynthesis proceeds or stops during the day and night. More specifically, comparisons at 1 and 16 meters (39 in. and 52 ft) above ground in an agricultural region showed that shortly after sunrise, as the rate of photosynthetic capture of CO_2 exceeded the rate of respiratory release, a sharp reduction in CO_2 concentration began at both levels. During the entire day, CO_2 levels remained relatively constant. The closer to the ground, however, the greater the daily CO_2 variation (Verma and Rosenberg, 1979). During the dark hours concentrations at 1 meter reached 400 to 500 ppm and at 16 meters 350 to 400 ppm.

Annual cycles in CO_2 levels are influenced also by photosynthetic activity. In Nebraska, early spring concentrations averaged 320 to 330 ppm over a 24-hour period. As crop size increased, CO_2 levels decreased to a minimum of 295 to 298 ppm in early August. As photosynthetic activity decreased, owing to plant aging and a shortening daylength, CO_2 concentrations increased to 315 to 325 ppm and in the winter stabilized at 325–335 ppm (Verma and Rosenberg, 1979).

At Ithaca, New York, the CO_2 concentration decreased to 260 ppm at noon within a corn crop canopy (Wright and Lemon, 1966).

In addition to CO_2 concentrations at any level within or above the crop, the transfer of CO_2 to or away from the ground, called the CO_2 flux, is important in crop production.

Measurements of CO_2 fluxes taken in Nebraska between the elevations of 5.6 and 16 meters (18 and 52 ft) so as to represent free CO_2 movement showed a flow from the air above to the photosynthesizing crop for about 11 to 12 hours daily (Verma and Rosenberg, 1979). As photosysnthesis increased during the growing season, CO_2 fluxes increased. Both daytime gains and nightime losses increased. CO_2 for photosynthesis is supplied by a downward flux from the atmosphere and an upward flux from the soil.

CO_2 fluxes are of interest to cereal producers, as decaying organic matter, crop residues, roots, or manure influence CO_2 levels. Techniques to increase downward CO_2 levels by air currents may be a step in increasing cereal production.

WIND

Wind exerts a major influence on CO_2 fluxes, as it causes air turbulence. The rate of photosynthesis increases with the supply of CO_2, which in turn is favored by turbulence. Greater turbulence and mixing of air can result if rows of cereals, especially wide rows of corn and sorghum, are planted at right angles to the prevailing wind.

In addition to influencing CO_2 availability, wind also affects plant growth through its influence on transpiration, through mechanical effects, and through leaf movement that results in flecking of light on lower leaves, thereby influencing photosynthesis.

Winds accompanied by high temperatures can increase water losses markedly through transpiration. Extended high wind speeds or whipping wind can cause severe dessication or mechanical injury to tender leaves and result in lodging, a common problem in cereal crops.

The obvious impact of wind is mechanical injury, and simplified statements about turbulence and transpiration portray wind as a simple, direct environmental factor. Wind is in fact most complex in relation to its effect on transpiration and is generally the least understood environmental variable. An increase in wind speed may increase, decrease, or have no effect on transpiration, depending on temperature differences between leaf and air, moisture differences between leaf and air, and leaf boundary-layer resistance. This latter may be influenced by pubescence or other factors and by alterations to external leaf resistance, such as irreversible damage caused by dessication (Montieth, 1965).

Generally, crops are more productive when protected by a windbreak, but the reasons for this increase may be attributed to a higher relative humidity, greater CO_2 fluxes, or increased air and soil temperatures due to reduced wind speed.

ATMOSPHERIC MOISTURE

Low air humidity may cause water deficits that are damaging to plant tissue even under conditions of sufficient soil water. High air humidity may limit the ability of leaves to cool by evaporation, and thermal stress may result.

When considering transpiration, **relative humidity** is a meaningless term. As air temperature changes, the relative humidity changes, even though the actual moisture content of the air remains the same. Likewise **vapor-pressure deficit,** often used as an indicator of atmospheric moisture content, is not very revealing, unless the leaf and air temperatures are identical. No matter how atmospheric moisture level is expressed, however, its influence on the growth of plants should not be underestimated. Transpiration is influenced by several environmental factors, including atmospheric moisture, temperature, wind speed, and moisture supply to the leaves. It is influenced also by plant factors, including leaf structure, leaf orientation, curling or rolling of leaves, leaf-area index, root-shoot ratio, plant diseases, and compounds applied to the leaves that may influence stomatal openings.

WATER

Water is the most abundant compound on earth, and water surplus or deficiency is a primary factor contributing to the stability or instability of agricultural production. Water makes up 85% to 90% of the fresh weight of actively growing plant parts, and extreme dehydration kills most plants. Water is as essential in photosynthesis as is CO_2. Worldwide, a deficit of water is the single most significant factor limiting crop yield (Begg and Turner, 1976). Physiological activities of plants decrease as water content is reduced. The reverse is also true; as physiological maturity is reached, water content decreases—a fact useful for field drying. A shortage or an excess of water is the most common limiting factor in world crop production. Drought is unquestionably the most important environmental factor influencing crop growth and production in the semiarid regions of the world.

In seeds, an increase in water content signals a substantial increase in respiration and other physiological activity.

Plant nutrients enter plants or move from cell to cell and tissue to tissue within the plant as solutes in water.

Plant turgidity is maintained by water, and insufficient water can lead to wilting and a loss of plant form. Turgidity is important in connection with the opening of stomata, necessary for gaseous exchange and photosynthesis.

Approximately 1% of the water absorbed by cereal plants is required to maintain turgidity and carry on plant functions. The other 99% is lost by transpiration (Table 6-4). A water deficit may lead to loss of turgidity, accompanied by an immediate reduction or cessation of growth and eventual death from dehydration. Water deficits occur in plants because the rate of water loss and the rate of absorption are only partly interdependent. Water absorption is controlled chiefly by soil factors, and water loss chiefly by atmospheric factors. When water loss exceeds absorption, a water deficit occurs. The rate of transpiration and the plant water supply are not direct measurements, therefore, of plant growth and yield. Rather, internal water balance is the most important factor in plant-water relations. It is possible that drought conditions in crop plants may occur even under conditions of adequate soil moisture.

Table 6–4
An Estimated Water Budget for a Corn Plant

Water occurring as constituent, solvent, and in maintenance of turgidity	1,872 g
Water used as a reagent	250 g
Water loss in transpiration	202,106 g
Total	204,228 g or 54 gallons

Source: Data reported by Kramer (1959).

Cereal crops such as sorghum and millet are better suited for cultivation under dry conditions than others. This feature, known as **drought resistance,** may arise from many sources. Russell (1959a) supplied a list of 30 factors affecting drought resistance, which serves to illustrate its complexity. Among the structural features affecting transpiration and drought resistance are:

1. Root-to-leaf ratio.
2. Conducting tissue.
3. Reduction of leaf surface.
4. Rolling, folding, or thickening of leaves.
5. Leaves that fall off.
6. Epidermal coverings.
7. Reductions in intercellular spaces.
8. Sunken stomata.
9. Size and number of stomata.
10. Stomatal regulation.
11. Surface hairs.

In many grasses, including sorghum and corn, **bulliform** or motor cells are commonly found and help to increase drought resistance. They are large cells, filled with cell sap, that shrink when the leaf wilts. This causes the leaf to roll or fold across its width. Rolling helps reduce water loss through transpiration, as a reduced leaf area is exposed to the sun. When adequate moisture conditions are restored, bulliform cells expand and return the leaf to its normal flat shape.

Some plants have a greater capacity than others to survive dessication; this is known as **drought hardiness.**

Despite the complexity of drought resistance, screening for this feature was possible in rice under artificially induced stress conditions (O'Toole and Maguling, 1981). Drought-resistant rice cultivars had a greater deep-root-to-shoot ratio and had the leaf-rolling ability. This latter feature is useful as a visual field-selection criterion (Loresto, Chang, and Tagumpay, 1976).

In regions where rainfall normally exceeds crop requirements, water deficiencies may occur because of:

1. Variations in total average rainfall from the long-term average. The mean annual rainfall at Hyberabad, India, for example, is highly erratic, and over a 77-year period varied from a high of 140 cm (55.1 in.) in 1917 to a low of 32.0 cm (12.6 in.) in 1972 (Sivakumar, 1979).
2. A distribution pattern that does not match crop water needs. Water stress at critical stages, such as immediately following seed set in cereals, can limit grain yield.
3. Excessive runoff resulting from high intensity of precipitation, steep slopes, or a low infiltration rate and crop cover.
4. Low soil moisture storage capacity for supplying crop needs between rains. This factor is influenced by soil depth and porosity.
5. A **potential evapotranspiration rate** or water need in excess of actual evapotranspiration rate, thereby developing a water deficit. In other words, high temperatures, low humidity, and winds may reduce rainfall effectiveness.

Since consumptive use of water is geared largely to solar radiation, moving a crop to a cooler region or to a cooler part of the growing season may, within limits, increase the quantity of plant product per unit of water used. Increasing the hardiness of winter cereals allowed this principle to operate as cereal production moved northward.

Developing corn capable of germinating at lower temperatures increased its efficiency of water use, as it permitted corn production at the northern fringe of the corn belt.

RAINFALL EFFECTIVENESS

Near the equator, one of the dominant factors in water use is the large amount of energy available for evaporation. In addition, therefore, to the total amount and seasonal distribution of precipitation, some measure of rainfall effectiveness is needed. Potential evapotranspiration is more or less constant from year to year, because solar radiation is quite constant from year to year. With increasing latitude from the equator, the degree of variation in potential evapotranspiration on a seasonal basis reflects variations in the sun's altitude, cloud cover, wind speed, and humidity.

At Beaverlodge, Alberta, located at 55°N latitude, spring wheat was found to produce the same yield produced at Madison, Wisconsin, at 43°N latitude with 30% less rainfall (Carder, 1957). The lower evapotranspiration rate at Beaverlodge, compared to that in areas with higher light intensity, meant that the meager 45.2 cm (17.8 in.) average rainfall at Beaverlodge was effective moisture for crop production.

Differences in rainfall effectiveness were illustrated by Stacey (1974) in spring wheat grown at three locations in 1965. At Beaverlodge, a yield of 3629 kg/ha (3240 lb/ac) was produced with a 28.2 cm (11.1 in.) rainfall; at Swift Current, 724 km (450 miles) south, a yield of 3360 kg/ha (3000 lb/ac) was produced under a rainfall of 40.4 cm (15.9 in.); and at Dickinson, North Dakota, 1207 km (750 miles) south of Beaverlodge, 46.0 cm (18.1 in.) of rainfall gave a yield of 2419 kg/ha (2160 lb/ac). Of course, differences in soil texture, soil fertility, seasonal distribution of rainfall, and management cannot be eliminated in these comparisons.

Variation in rainfall and potential evapotranspiration at Hyderabad, India, (Figure 6-6) suggests that there are three months in the year when water supply could meet the potential requirement at this site.

MOISTURE AVAILABILITY

Three types of soil water exist: hygroscopic, capillary, and gravitational. A film of hygroscopic water normally adheres so tightly to soil particles that it is of no practical value to plants. Gravitational water can be regarded as surplus water, is not taken up by plant roots and excludes oxygen from the soil. Gravitational water responds to gravity and helps replenish the subsoil water. Gravitational water may also be impounded on the surface until it returns to the atmosphere by evaporation. A high water table may limit the volume of soil in which good soil aeration exists and hence restrict rooting volume. A fluctuating water table may result in roots' being killed as the water table rises. When the water table recedes, the plant is left with insufficient root surfaces. A drainage system may be required to remove excess water so that oxygen can reenter the soil. The purpose of drainage in humid areas is to create (1) an aerated root zone and (2) a surface soil that is dry enough to enable farm machinery to be used whenever it is required. Surplus rainfall must be removed and the water table maintained at a level that is appropriate for the particular soil, crop, and climate.

Capillary water resists the pull of gravity and it is the water used by plants. Capillary water is replaced by rainfall, by irrigation water, or by capillary rise from the water table. Plants can use the moisture from the water table if it is near the surface so that the roots can reach the capillary fringe.

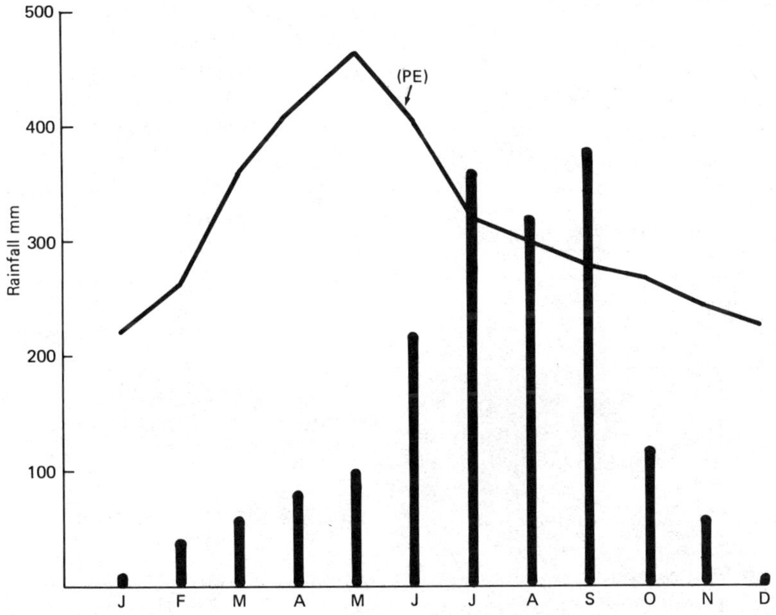

Figure 6–6. Monthly rainfall (vertical bars) and potential evapotranspiration (PE) at Hyderabad, India, with a mean annual rainfall of 761 mm (30 in.). (From Sivakumar, 1979)

Since the water content of the soil varies from time to time during the growing season, the status of soil water can be described in terms of **field capacity** and permanent wilting point.

Gravitational water moves downward in the soil, and after about one day in sand and four to five days in clay this movement virtually stops. Soil moisture in the wetted portion becomes constant and is at field capacity.

As capillary water is utilized by plants and as the soil dries out, plants have increasing difficulty obtaining soil moisture, and wilting may occur. The permanent wilting point is the level of soil water beyond which water is no longer available to plants.

The maximum amount of **available water** can be determined by subtracting permanent wilting percentage from field capacity. Field capacity and permanent wilting point may be expressed in terms of water weight as a percentage of dry soil weight. Available water may be expressed as centimeters of water per 30 cm soil depth and can be calculated using the **bulk density** of the soil. The following is an example of a calculation of available soil water:

A Field capacity, 40%

B Permanent wilting percentage, 20%

C Available soil water (*A* minus *B*), 20%

D Depth of soil considered, 30 cm

E Bulk density, 1.3 g/cm^3

F Available soil water, expressed as centimeters for 30 cm of soil depth at field capacity:

$$C \times E \times D = F$$

$$^{20}/_{100} \times 1.3 \times 30 = 7.8 \text{ cm}$$

Soils differ in bulk density and hence in the amounts of water they can hold. As well, roots of most annuals such as cereals do not penetrate more than 90 or 120 cm (3 to 4 ft) below the soil surface.

The amount of available moisture in five soil groups averaged from 40 different soils throughout Manitoba, Canada, varied considerably (Table 6-5). The amount of moisture available varied by as much as 56% among soils.

After decades of intensive water use for crop production in moisture-deficient areas of the world, and for industrial and human use, ground-water supplies are declining (Anonymous, 1982b; Walsh, 1980). The evidence for this statement is a reduced volume of water from wells, and the need to pump water from greater depths. Both factors require an increase in energy, and although actual water shortages may not occur for many years, an increasing problem is the shortage of easily obtained water. Depleted ground-water reserves are not readily replenished, as the total ground-water reserves and lakes in the United States are equivalent to about 13 years of precipitation (Russell, 1959b). Under limited irrigation systems, producers may have to turn to dryland farming and better water management and possibly to less water-intensive crops, such as wheat, sorghum, and millet.

***Table* 6–5**
Available Water in Five Soil Textural Groups in Manitoba, Canada

Soil Textural Group	*Available Soil Water to 120 cm (47 in.)*		*Kinds of Soil*
	cm	*in.*	
Coarse	14.5	5.7	Sand, loamy sand
Moderately coarse	24.0	9.4	Loam, very fine sand, sandy loam
Medium	25.5	10.0	Very fine sandy loam, loam, silt loam, silt
Moderately fine	26.0	10.2	Clay loam, sandy clay loam, silty clay loam
Fine	26.0	10.2	Sandy clay, silty clay, clay

Source: Data from Eilers et al. (1977).

EXCESS SOIL MOISTURE

Surplus soil moisture influences soil temperature, amount of air in the soil, and plant-food relationships. A wet soil is a cold soil, and, because additional heat is required to increase the temperature of a water-saturated soil, both the magnitude and rate of diurnal and seasonal changes in soil temperature are small.

Restricted air movement in a wet soil is detrimental to crop development, except in rice. Rice grows in standing water because of a unique mechanism whereby oxygen is supplied to the roots. In other cereals, waterlogged conditions will prevent economic crop production, because **anaerobic** soil conditions may lead to a reduction in oxidation and the availability of plant nutrients.

Excess water flowing over the land surface can result in water erosion that can lead to major problems of the agro-ecosystem. Impounded water and associated flooding, except for rice production, may delay seeding or cause losses of growing crops. Excess water percolating through the soil may cause leaching of soluble soil nutrients.

Heavy traffic on a wet soil can result in soil compaction and plastic deformation. Under very wet conditions, soils become fluid and lose their load-bearing properties. Compacted soils are low in oxygen and mechanically impede root growth. The result of restricted root growth is inadequate water and nutrient uptake for crop functions and growth.

SOIL

Soil serves as a root zone for plants and provides anchorage, a storage and exchange site for mineral nutrients, and a medium through which water and air can move. These seemingly simple functions do not lessen the importance and value of soil as an environmental variable upon which our food supply is dependent.

The rating of soil for agricultural purposes may vary from prime to unsatisfactory, depending on specific soil properties, topography, drainage, and the adequacy of other environmental variables. Soil productivity may be improved by management or decreased by soil depletion or soil deterioration.

Soil depletion is associated with a change in soil properties, such as the reduction in mineral nutrients, soil organic matter, texture, or density. These changes usually can be corrected through proper management or addition of fertilizers, manure, or crop residues.

Soil deterioration is more serious, as it is associated with the washing or blowing away of the soil. It is more difficult or even impossible to restore economically the productivity of an eroded soil. Soil deterioration therefore offers the greatest threat to world food needs.

Soil depletion and erosion, together with the removal of land from agricultural use for housing or industrial needs or by contamination, strain agricultural output. However, life on earth neither originates from the soil (as the term "mother earth" may convey) nor depends on soil for sustenance. The soil provides only about 5% of the dry weight of plants from the nutrients it supplies; the remaining 95% is produced from photosynthesis.

Plants can be produced without soil by means of **hydroponics,** a term used to describe the nutrient-solution culture of plants. Costs associated with hydroponics limit its use to high-value crops. Soil is the most convenient and economic medium for anchorage, mineral nutrients, and water supply.

PROPERTIES OF A PRODUCTIVE SOIL

Soil ratings for agriculture can be evaluated on selected soil properties. Structure, texture, bulk density, particle density, percent pore space, degree of compaction, percent organic matter, water infiltration rate, erodability, water-holding capacity, cation exchange capacity, and ***p*H** are important soil properties.

These properties and others such as temperature can be modified by tillage practices. *Tillage* means manipulating the topsoil to provide a suitable environment for agricultural purposes. The depth and extent of tillage depends on whether the objective is to prepare a rootbed or a seedbed. A *seedbed* in the case of row crops may be a 5-cm (2-in.) wide strip prepared to provide fine, firm soil to receive the seed; the seedbeds may be placed a meter (39 in.) apart. The seedbed for broadcast or narrow-row cereals may comprise the entire soil surface to be seeded. A fine, firm seedbed assures good contact between seed and soil, promotes capillary moisture movement from below, and reduces the risk of soil crusting. Depth of cultivation for seedbed preparation is about 5 cm (2 in.).

The rootbed is not a clearly defined zone, as roots may penetrate well below the cultivated zone. Conventional soil tillage systems include plowing, secondary tillage, and seedbed preparation. Less conventional systems involve stubble farming, minimal tilling and zero-tilling. Tillage is sometimes required to loosen the soil to aid warming, but where conditions are suitable for germination and crop growth without tillage, no primary or secondary tillage may be used either before or after sowing the crop.

ACID SOILS

When a considerable portion of the exchangeable cations are hydrogen, acidity develops. Hydrogen cations in soil water are produced from the formation by carbon dioxide and water of a weak acid; from ammonium fertilizers; or from sulfur. Leaching that removes calcium, magnesium, sodium, or potassium cations increases soil acidity.

Strongly acid soils are not productive for cereal crops, although sensitivity varies (Table 6-6). Barley prefers less acid conditions; corn, sorghum, and wheat a medium acidity; and millet, oat, rice, rye, and triticale will tolerate more acid conditions than the other cereals. Cereal cultivars vary in the soil *p*H values they will tolerate. Briggs (1978) reported on one barley cultivar that yielded well at a *p*H of 3.5. Values in Table 6-6 serve as a guide.

Table 6–6
Optimum, Upper, and Lower *p*H Limits for Cereal Crops

Crop	Minium	Optimum	Maximum
Barley	5.4	6.5–7.8	
Corn	5.0	5.5–7.5	
Millet	4.5	5.0–6.5	
Oat	4.5	5.0–7.5	
Rice		5.0–6.5	
Rye	4.5	5.0–7.0	8.0
Sorghum		5.5–7.5	
Wheat	5.0	5.5–7.5	
Triticale	4.5	5.0–7.0	

Source: Foth (1978) and Parks (1955).

SOIL COMPACTION

Soil physical factors such as compaction influence root growth and proliferation owing to the high pressure imparted by high soil density. Oxygen starvation from reduced gaseous exchange may slow root growth. Water infiltration into the soil profile is impeded by soil compaction. Efforts to improve the productivity of a compacted soil include tillage, incorporation of crop residues, and increased **soil fertility**. Soil compaction can be described by three parameters: bulk density, particle density, and **soil strength.**

Bulk density is the weight per unit volume of over-dry soil, commonly expressed as grams per cubic centimeter. Bulk density is calculated as follows:

$$\text{bulk density} = \frac{\text{weight of oven-dry soil}}{\text{volume of oven-dry soil}}$$

Volume of soil can be measured by means of core samplings, carefully extracted to preserve the natural structure of the soil.

Granulated clay surface soils commonly range in bulk density from 1.0 to 1.3 g/cm^3 (62.4 to 81.0 lb/ft^3). Coarse-textured soils range from 1.3 to 1.8 g/cm^3 (81.0 to 112.3 lb/ft^3). The greater the degree of compaction, the higher the bulk density. Robertson and Erickson (1978) suggested that bulk densities of 1.46 to 1.66 g/cm^3 (91.1 to 103.6 lb/ft^3) found in subsoil materials in Michigan are too compact for rapid root growth and high crop yields. A reduction in soil organic matter may allow soil compaction to become a serious and extensive problem.

Soil strength, resulting from the combined factors of bulk density, moisture, texture, organic matter, and management, was believed to account for 76% of the variability in cotton root growth (Gerard, Sexton, and Shaw, 1982).

Particle density of soil is determined from the solid particles only and does not vary with the amount of space between the particles. Particle density of any soil is a constant and for mineral soils will average about 2.65 g/cm³ (162.2 lb/ft³).

Particle density and bulk density are of value in determining **pore space** in a soil as follows:

$$\text{pore space (percent)} = 100\% - \frac{\text{bulk density} \times 100}{\text{particle density}}$$

At a bulk density of 1.3 g/cm³ and a particle density of 2.65 g/cm³, total pore space is 50%.

ENVIRONMENTAL VARIABLES AND AGRICULTURAL SYSTEMS

In order of priority, Geissbuhler, Brenneisen, and Fisher (1982) listed the major constraints for crop production in 1982 and for the future in the United States as soil deterioration by erosion, increasingly erratic water supplies, energy costs, and progressive water, air, and soil pollution. The objectives of agriculture, however, generally have been to achieve maximum yields, operate with a maximum profit, minimize year-to-year instability in production, and prevent long-term degradation of the productive capacity of the agricultural system (Altieri, Letourneau, and Davis, 1983). If a sustainable agriculture is to be achieved, the central issue is not maximum yield but, rather, long-term stabilization of production.

QUESTIONS

1. *a.* Give specific examples of output fluctuations in crop production caused by environmental variables on a worldwide scale.
   b. Do fluctuations occur at random, or are they confined to specific regions?
   c. Give an indication of the extent of crop output fluctuations and their impact on economic activity.
   d. Consider crop output fluctuations caused by environmental variables in your region.
2. Moisture exerts a major influence on crop production. Obtain long-term precipitation records for your region and determine:
   a. Total annual precipitation.
   b. Yearly variation in precipitation.
   c. Seasonal distribution on a monthly basis.
   d. Soil water-holding capacity.
   e. The possibility of a moisture deficiency.
3. Core samplings of a loam soil removed by the use of a metal cylinder 5.0 cm (1.97 in.) in height and 4.4 cm (1.73 in.) in diameter weighed, when oven-dried, 87.6 g (3.09 oz). Calculate the bulk density of this soil.
4. *a.* What bulk density is preferable for crop growth—1.3 or 1.69/cm³? Explain.
   b. Suggest how the adverse effects of bulk density can be reduced.
   c. What are the causes of an undesirable bulk density, and how can it be avoided?
5. Mechanization has allowed producers in many areas of the world to dominate the environment. Explain the statement that domination has made producers inflexible and soil protection is placed at a low priority level.
6. *a.* What are the major benefits of mechanization?
   b. Should a high level of mechanization be employed in all areas of the world?
7. "Only an infinitesimal amount of the total solar energy striking the outer atmosphere is utilized as chemical or stored energy" (see Figure 6-2). Consider cereal crop production in outer space in terms of:
   a. Advantages from a photosynthetic viewpoint.
   b. Limitations on the types of plants.
   c. Choice of crops and methodology.

8. "Less than 1% of the total solar energy striking the outer atmosphere is captured in photosynthesis for use by humans."

   a. What opportunities exist for increasing this level?

   b. Will an increase in photosynthetic rate increase grain yield?

   c. What are the limitations to utilizing light more efficiently to increase crop grain yields?

9. Classify the following in terms of soil depletion or soil deterioration:
   a. Soil salinity.
   b. Low fertility.
   c. Low organic-matter levels.
   d. Wind erosion.
   e. Water erosion.
   f. Poorly drained soils.
   g. Acid soils.

10. Describe cereal production under acid soil conditions.

References

Altieri, M. A., D. K. Letourneau, and J. R. Davis. 1983. "Developing Sustainable Agroecosystems." *BioScience* 33: 45–49.

Anonymous. 1976. "Scenarios for Major Crops in the U.S. and Canada." *Impact of Climatic Fluctuation on Major North American Food Crops.* The Institute of Ecology—The Charles F. Kettering Foundation. pp. 5-9.

———. 1982a. "Higher CO_2 Will Warm Earth, Change Climate, NRC Finds." *BioScience* 32: 717–718.

———. 1982b. "Irrigation Crisis." *Crops and Soils* 34 (7): 14–16.

Baker, R. F. 1970. "Relative Maturity Rating of Corn by Growing Degree Days." *Proceedings of the Twenty-Fifth Annual Corn and Sorghum Research Conference.* American Seed Trade Association, Chicago, Dec. 8–10. Publication No. 25: 154–160.

Bates, E. M. 1976. "Solar Radiation a More Accurate Guide Than Heat Units." *Crops and Soils* 28 (6): 12–15.

Bauer, M. E. 1979. "LACIE: An Experiment in Global Crop Forecasting." *Crops and Soils* 31 (9): 5–7.

Begg, J. E., and N. C. Turner. 1976. "Crop Water Deficits." *Advances in Agronomy* 28: 161–217.

Boyer, J. S. 1982. "Plant Productivity and Environment." *Science* 218: 443–448.

Briggs, D. E. 1978. *Barley.* London: Chapman Hall. p. 252.

Brown, D. M. 1978. *Heat Units for Corn in Southern Ontario.* Ontario Ministry of Agriculture and Food. Agdex 111/31. 4 pp.

Bryson, R. A. 1975. "The Lessons of Climatic History." *Environmental Conservation* 2: 163–170.

Carder, A. C. 1957. "Growth and Development of Some Field Crops as Influenced by Climatic Phenomena at Two Diverse Latitudes." *Canadian Journal of Plant Science* 37: 392–406.

Daubenmire, R. F. 1959. *Plants and Environment.* New York: John Wiley & Sons. 422 pp.

de Wit, C. T. 1965. *Photosynthesis of Leaf Canopies.* Agricultural Research Report No. 663. Wageningen, The Netherlands: Center for Agricultural Publications and Documentation. 57 pp.

Eilers, R. G., et al. 1977. "Soils." *Principles and Practices of Commercial Farming.* Faculty of Agriculture, The University of Manitoba. 5th ed., pp. 17–59.

Ferwerda, J. D. 1979. "The Ecology of Tropical Crops. Part 2. Climate." *Span* 22 (2): 58–60.

Foth, H. D. 1978. *Fundamentals of Soil Science.* New York: John Wiley & Sons. 6th ed., p. 210.

Geissbuhler, H., P. Brenneisen, and H. P. Fisher. 1982. "Frontiers in Crop Production: Chemical Research Objectives." *Science* 217: 505–510.

Gerard, C. J., P. Sexton, and G. Shaw. 1982. "Physical Factors Influencing Soil Strength and Root Growth." *Agronomy Journal* 74: 875–879.

Hammes, P. S., and R. J. Marshall. 1980. "Effect of Photoperiod and Temperature on the Development Rate of Three Cultivars of Wheat (*Triticum Aestivum* L.)." *Field Crops Research* 3: 121–128.

Hansen, J., et al. 1981. "Climate Impact of Increasing Atmospheric Carbon Dioxide." *Science* 213: 957–966.

Holmes, R. M., and G. W. Robertson. 1959. *Heat Units and Crop Growth.* Ottawa: Canada Department of Agriculture. Publication 1042. 35 pp.

Idso, S. B., R. D. Jackson, and R. J. Reginato. 1977.

"Remote-Sensing of Crop Yields." *Science* 196: 19–25.

Jackson, R. D., R. J. Reginato, and S. B. Idso. 1977. "Wheat Temperature: A Practical Tool for Evaluating Water Requirements." *Water Resource Research* 13: 651-656.

Kramer, P. J. 1959. "The Role of Water in the Physiology of Plants." *Advances in Agronomy* XI: 51-70.

———. 1981. "Carbon Dioxide Concentration, Photosynthesis, and Dry Matter Production." *Bio-Science* 31: 29–33.

Loresto, G. C., T. T. Chang, and O. Tagumpay. 1976. "Field Evaluation and Breeding for Drought Resistance." *Philippine Journal of Crop Science* 1: 36–39.

Monteith, J. L. 1965. "Evaporation and Environment." *The State and Movement of Water in Living Organisms*. Symposium of the Society of Experimental Biology 19: 205–234.

Morain, W. A., and D. L. Williams. 1975. "What Production Estimates Using Satellite Images." *Agronomy Journal* 67: 361-364.

Moss, D. N., R. B. Musgrave, and E. R. Lemon. 1961. "Photosynthesis Under Field Conditions II: Some Effects of Light, Carbon Dioxide, Temperature and Soil Moisture on Photosynthesis, Respiration and Transpiration of Corn." *Crop Science* 1: 83-87.

Nuttonson, M. Y. 1958a. *Rye-Climate Relationships and the Use of Phenology in Ascertaining the Thermal and Photo-thermal Requirements of Rye*. Washington, D.C.: American Institute of Crop Ecology. 219 pp. [Similar books for wheat (1955) and barley (1958b).]

O'Toole, J. C., and M. A. Maguling. 1981. "Greenhouse Selection for Drought Resistance in Rice." *Crop Science* 21: 325–327.

Parks, D. L. 1955. *Successful Crop Production in Eastern Canada*. McClelland and Stewart Limited. p. 13.

Patterson, D. T., and E. P. Flint. 1980. "Potential Effects of Global Atmospheric CO_2 Enrichment on the Growth and Competitiveness of C_3 and C_4 Weed and Crop Plants." *Weed Science* 28: 71–75.

Ramirez, J. M., and A. Bauer. 1974. "Growing Degrees for Many Crops." *Crops and Soils*. 26 (6): 16–18.

Revelle, R. 1982. "Carbon Dioxide and World Climate." *Scientific American* 247: 36–43.

Richter, J. 1982. "Weather Monitoring and Crop Forecasting." *Span* 25 (3): 112.

Robertson, L. S., and A. E. Erickson. 1978. "Soil Compaction: Symptoms, Causes, Remedies." Parts I, II, III. *Crops and Soils*. 30 (4, 5, 6): 11–14; 7–9; 8–10.

Rotty, R. M. 1978. "Atmospheric Carbon Dioxide. Possible Consequences of Future Fossil Fuel Use." *Resources and Energy* 1: 231–249.

Russell, M. B. 1959a. "Drought Tolerance of Plants." *Advances in Agronomy* 11: 70–73.

———. 1959b. "Management Practices Affecting Runoff and Water Yield." *Advances in Agronomy* XI: 115–131.

Schware, R., and W. H. Kellogg. 1982. "How Climatic Change Could Affect Food Production Patterns." *Ceres* 15 (2): 40–42.

Shirley, H. L. 1945. "Light as an Ecological Factor and Its Measurement." *Botanical Review* 11: 497–532.

Sivakumar, M. V. K. 1979. "Environmental Factors Affecting Plant Productivity." (In) S. Ahmed, H. P. M. Gunasena, and Y. H. Yang (eds)., *Adaptive Production Systems*. Proceedings of the Final INPUTS Review Meeting. Honolulu, Hawaii: East-West Resource Systems Institute. pp. 183-203.

Stacey, E. C. 1974. *Peace Country Heritage*. Saskatoon: Western Producer Book Service. p.72.

Thompson, L. M. 1975. "Weather Variability, Climatic Change, and Grain Production." *Science* 188: 535–541.

Verma, S. B., and N. J. Rosenberg. 1979. "Agriculture and the Atmospheric Carbon Dioxide Build-up." *Span* 22: 62–65.

Walsh, J. 1980. "What To Do When the Well Runs Dry." *Science* 210: 754–756.

Wittwer, S. H. 1980. "Carbon Dioxide and Climatic Changes: An Agricultural Perspective." *Journal of Soil and Water Conservation* 35: 116–120.

Woodwell, G. M. 1978. "The Carbon Dioxide Question." *Scientific American* 238: 34–43.

Wright, J. L., and E. R. Lemon. 1966. "Photosynthesis Under Field Conditions. IX. Vertical Distribution of Photosynthesis Within a Corn Crop." *Agronomy Journal* 58: 265–268.

Yoshida, S. 1972. "Physiological Aspects of Grain Yield." *Annual Review of Plant Physiology*. 23: 437–464.

7

Production Aspects: Winter Crops; Hardiness and Cold Tolerance; Summer Fallow; Soil Salinity

WINTER CEREALS

The temperate-zone C_3 spring cereals such as wheat, barley, oat, rye, and triticale each have a winter counterpart. Winter cereals are established in the fall; develop a limited amount of vegetative growth, usually four to six leaves, before growth is stopped by low temperatures; pass through the winter in a dormant state; resume growth in the early spring before spring cereals can be seeded; develop reproductive organs that effectively coincide with the energy-rich spring days; and mature early in the season, thereby avoiding the high temperatures, drought, and diseases that frequently occur in late summer.

Winter cereals are called winter annuals because they require a period of cold temperatures and short daylengths to become vernalized. Vernalization requirements must be met for winter annuals to develop reproductive organs.

YIELD PERFORMANCE AND ADVANTAGES

Winter cereals offer several advantages over their spring counterparts. Under conditions of satisfactory winter survival, winter cereals are capable of outyielding spring cereals by approximately 25% and offer greater yield stability because variations in spring seeding date are avoided. Superior yield is partly a consequence of fall establishment and an ability to benefit photosynthetically under the cool, moist conditions of early spring that favor C_3 plant development. Conditions that favor slow growth and surplus photosynthate produce plants that have well-differentiated cells, resulting in strong, lodging-resistant stems, profuse tillering, and a well-developed root system, which collectively allow the plant to respond to spring nitrogen applications. Superior yield performance reflects an ability to benefit from environmental conditions rather than a superior genotype.

Additional advantages of winter cereals include an opportunity for double-cropping practices because of the early maturity of winter cereals. Removal of winter cereals as whole-plant silage or as high-moisture grain to be used for livestock feed facilitates this concept (Chapter 8). Of very great and growing importance is the protective soil cover that winter cereals provide to prevent erosion (Figure 7-1). In dryland regions, winter cereals help capture and hold snow needed for soil moisture reserves. A favorable distribution of labor is provided by winter cereals because of fall seeding and early harvesting, although large, fast machinery has minimized this advantage. In favored, southern areas, winter

Figure 7–1. Winter cereals are valued for the protective soil cover they provide, especially on fields where erosion is a problem. In dryland regions, winter cereals help capture and hold snow needed for soil moisture reserves. (Courtesy of Ontario Ministry of Agriculture and Food)

cereals provide a winter grazing crop that can recover to produce a reasonable grain yield (Chapter 8). Although a spring application of herbicide may be required to control winter annual weeds, winter cereals compete well with annual weed species, and generally herbicide use is minimized. Generally, winter cereals should not receive a fall herbicide treatment.

IMPORTANCE OF WINTER CEREAL PRODUCTION

Because of the advantages just outlined, winter cereals are preferred over spring cereals wherever it is possible for them to survive the winter period (Hunt, 1980b; Salmon, 1917a). Winter cereals account for over 70% of the tilled area of Britain (Bullen, 1977).

Wheat is grown more extensively than any single crop, and winter wheat accounts for about 75% of world production. Almost all the rye produced in the world is of the winter type. In Europe, winter barley is supplanting spring barley in many areas, such as France (Hunt, 1980a) and England (Jenkins, 1970). Winter barley was described by Jenkins (1970) as a profitable cereal or plant form of rich genetic potential which has not been adequately exploited, but also as an irritating source of disease for the more important spring barley crop.

As plant breeders succeed in producing cultivars with greater winter hardiness, and as cultural practices secure greater reliability against winter injury, then the limits to winter cereal production will move into more extreme areas and displace spring cereals. In North America, for example, increased winter hardiness has allowed winter barley to become more firmly established around the Great Lakes region (Hunt, 1980a). Except in favored areas of the northern Great Plains region of North America, winter cereal production is unlikely to expand unless winter survival becomes more reliable (McLeod, 1980). In the past, quantum jumps in winter hardiness have not occurred because of the multiplicity of genetic factors involved in winter survival.

In temperate regions where midsummer drought severely damages full-season crops, and in areas where winter survival is acceptable, winter cereals may be an acceptable substitute.

WINTER HARDINESS

A serious obstacle in winter cereal production, and one that spring cereals avoid, is winter injury that can weaken or kill plants. Winter survival is the ability to withstand a series of environmental stresses which vary from one region to another, which vary in severity from one year to the next, and which may interact in numerous ways. Stresses include:

Freezing Temperatures

Geographic limits of winter cereals are determined largely by temperature extremes. Plants exposed to subzero temperatures may be killed as a result of mechanical injury, desiccation of the protoplasm, chemical effects, or suspension of metabolism (Salmon, 1917b). Whether plants are injured by low temperatures depends on a number of factors.

SITE AND STRUCTURE OF ICE CRYSTALS IN THE CELL WALL. Hardy winter cereals contain soluble polysaccharides which hinder ice-crystal formation in the cell wall and intercellular spaces. Ice-crystal formation can penetrate the cytoplasm and disrupt cell structure. Cells damaged by low temperatures have a typical water-soaked appearance, because membranes have been ruptured, allowing cell contents to flow out. Such cells become dehydrated and wilt when exposed to the sun.

MOISTURE CONTENT. Tender winter cereal seedlings can become conditioned to survive low-temperature stress conditions through an **acclimatization** process associated with cell differentiation. Acclimatization involves complex biochemical changes that are triggered by environmental signals such as low temperature (0–5°C; 32–41°F) and shortened daylength. The biochemistry of the plant cell is changed from promoting active growth to promoting high stress tolerance, involving gums and resins to resist freezing damage. The potential winter hardiness of any cultivar depends on the success of these various biochemical changes or the degree of acclimatization or cell differentiation.

External moisture content may influence also freezing susceptibility. Following a midwinter thaw, winter wheat and barley plants were more susceptible to low-temperature injury because of a high moisture content in the crown (Metcalf et al., 1970).

Thawing conditions may reduce the effect of acclimatization. Cold tolerance in winter wheat and rye was reduced an average of 5°C (9°F) after two thawing and freezing cycles (Gusta and Fowler, 1977). Reductions in hardiness, however, were variable.

FERTILITY LEVEL. Phosphorus fertilizer can promote winter hardiness by promoting cell differentiation and acclimatization. Efforts to increase cold hardiness in winter wheat and rye by increased levels of phosphorus fertilizer could not override the extreme conditions of the Northern Great Plains of America, however (Fowler and Gusta, 1982).

Salt stress associated with saline soils was observed to reduce the cold tolerance of winter wheat and rye (Fowler, 1981), although reductions were variable and not large enough to be of practical concern.

RATE OF FREEZING AND THAWING. A sudden drop in temperature before acclimatization has occurred or a sudden rise in temperature of frozen tissue will have a more detrimental effect than a slow drop or rise in temperature. Winter cereal crowns are the most vulnerable plant part to freezing injury. The crown contains the meristematic tissue needed for growth of new tissue. The crown is a diverse part of the plant, and different patterns of freezing occur simultaneously in its various tissues (Everson and Olien, 1975). A protective snow cover may prevent plant cells from experiencing rapid and wide temperature fluctuations. A light snowfall lodged in the crown of winter cereals can provide some protection against damage by low temperatures.

DURATION OF FREEZING. The longer the tissue is exposed to sublethal temperatures, the more severe the freezing stress becomes. McKersie (1981) reported that at −12°C (5°F) the ability for plants to survive began to decline after 12 hours, and at −16°C (1.5°F) after only one hour.

LEVEL OF FOOD RESERVES. Warm fall conditions that do not promote cell differentiation, followed by subfreezing temperatures on exposed plants, could damage cells. After a long winter, and when the protective snow cover has melted and before plants have had an opportunity to restore lost reserves, subfreezing temperatures likewise are detrimental.

PLANT HEAVING. Repeated freezing and thawing of soil and the associated expansion and contraction can lift plants out of the ground, breaking and exposing roots and promoting plant dessication.

SMOTHERING. Although plants may cease photosynthetic activity during the winter months and become dormant, they continue to be living, respiring plants that require energy. Plants encased in ice, under a cover of ice, or under dense, wet, or crusted snow that occurs when partially melted or waterlogged snow refreezes, may be deprived of oxygen needed for respiration and can die from oxygen deficiency. Without oxygen, sugars cannot be oxidized, and carbon dioxide and ethanol are produced. Levels of these compounds may become toxic to the plants or may deprive plants of an energy source so that death results.

PHYSIOLOGICAL DROUGHT OR DESICCATION. Under conditions of dry winter winds, the potential transpiration rate of leaf tissue may be enhanced despite subfreezing temperatures. If the soil is frozen, it is said to be physiologically dry, and plants cannot obtain sufficient moisture. Plants may be killed by excessive desiccation. Cereal seedlings with narrow leaves and prostrate growth habit may be less subject to injury than wide-leaved, upright plants.

DISEASE ORGANISMS. Although a protective snow cover is desirable to protect winter cereals in regions with low winter temperatures, continuous snow cover favors the development of a pathogen known as snow mold.[1] Under an early and continuous snow cover, when soil and tissue temperatures remain above freezing, conditions favorable to the disease exist. When spring melting conditions expose the infected plants, secondary infection occurs, and spores are produced on a white, moldlike mycelium. This mycelium is typical of snow mold but dries up quickly when exposed to the sun. Infected plants may wither and die.

FLOODING. Low-lying areas in a field may become flooded during a midwinter or spring thaw and result in injured plants. Injury is caused by smothering but without accompanying low temperatures.

The ability of a plant to withstand the above stresses or a combination of several is termed winter hardiness.

[1]Snow mold—*Fusarium nivale.*

VARIATION IN WINTER HARDINESS AMONG CEREAL CROPS

Genetic differences in winter hardiness can be found among cereal crop species and among cultivars within each species. Absolute ratings for species or cultivars cannot be provided because of the diversity of factors influencing winter hardiness. Cultivars of the most hardy cereal species may not withstand the rigors of one or more of the factors determining winter survival. Under favorable conditions, cultivars of a cereal species not considered as winter hardy may survive. Crop producers attempt to minimize the risk of winter injury by sound cultural techniques and careful selection of the site for winter cereal production.

Producers recognize that winterkill does not mean crop failure. A crop lost to winterkill has served the purpose of providing a protective soil cover and can be reseeded with a spring-seeded counterpart.

Winter rye is considered the most winter hardy of the cereal crops and usually can be grown under more severe conditions than other winter cereals. But despite the high level of winter hardiness, production of rye is well below that of other winter cereals, possibly a reflection of a smaller demand and a greater worldwide preference for wheat. Rye may find a greater role as a protective and reliable soil cover crop in soils subject to erosion.

Among the traditional winter cereal crops, wheat is regarded as second to rye in overall winter hardiness. As world food needs expand and as wheat production is pushed into fringe areas, the need for more reliable winter wheat production increases. Transfer of the outstanding winter hardiness genes from rye and wheat to other cereal crops through biotechnology is a hope for the future.

Winter triticale was produced to combine the hardiness of rye and wheat and to incorporate improved consumer acceptability through quality considerations. Although the first fertile triticale was produced as early as 1888, much of the effort toward developing triticale as a commercial crop began in 1954 and was directed primarily toward spring triticale. Research on winter triticale was initiated in Canada only in 1974 (Poysa, 1980). Improvements in winter hardiness of winter triticale may have only begun but look promising.

The first cultivar release of winter triticale appears to have hardiness equal to that of the most hardy winter wheat cultivars. However, this level of hardiness may not be sufficient for triticale to become widely accepted or to cause an expansion in the winter cereal growing area. Further improvement in both winter hardiness and agronomic characteristics is needed and expected (Poysa, 1980). Experience has shown, however, that rapid progress in a characteristic as complex as winter hardiness is as unlikely as a quantum jump in grain yield.

Winter barley is not as hardy as winter wheat. Determined and continued efforts to increase winter hardiness have been successful but very slow, because segregating material may suffer extensive but nonselective winterkill. Genetic variation for winter hardiness may be limited in this crop, or else selection techniques inadequately demonstrate the agronomic importance of individual physiological characters associated with winter hardiness (Mahon, 1983). Using conventional breeding techniques, therefore, the winter hardiness of winter barley cannot be expected to reach that of winter wheat.

Winter oat is the least winter hardy of the cultivated cereals. Its range of adaptation is restricted to favored southern regions. The vernalization requirement of winter oat is the lowest among those of the winter cereals, suggesting strongly that the limitation to improved winter hardiness is genetic.

IMPROVING WINTER HARDINESS

Winter hardiness in cereal crops has been investigated intensively (Dexter, 1956; Everson and Olien, 1975; Steponkus, 1978; Warnes and Johnson, 1972a and b), but despite extensive efforts, improvement has been slow in all the cereal crops. Perhaps research directed at improving overall winter hardiness is too general and too broad and should be defined more clearly and precisely. Many genes are associated with winter hardiness, thereby adding to its complexity. On the other hand, effort directed to one feature may improve that component, but winter survival is a combination of many factors.

The ultimate test for winter hardiness is survival in the field, but screening for winter hardiness under field conditions is difficult, for only rarely are environmental conditions suitable to obtain a clear ranking of the different genotypes. Indoor tests are used frequently as a selection aid, usually involve a single lethal temperature measurement, and hence determine only one trait involved with winter hardiness. Indoor screening techniques often are regarded as being of limited value for screening purposes, and sometimes screening tests appear to be more of an art than a science, as the period of cold acclimatization may determine the degree of frost hardiness in the seedlings.

Indoor screening tests nevertheless are an essential tool to breeding programs aimed at improving winter hardiness. Although the degree of frost hardiness is not the only factor that determines the winter survival of plants, it is the most significant factor in locations where plants are exposed to cold, dry winters, such as in the Ukraine or the northern Great Plains of North America. Continued research into techniques that offer small, but definite, differences in frost hardiness among cultivars is needed, as big differences in frost hardiness cannot be expected. Speed, ease, and reliability are essential attributes, so that large numbers of segregating materials can be screened. Cloutier (1982) announced a test in which seedlings 1 cm long were stressed by desiccation techniques before being subjected to low-temperature conditions.

In addition to cold tolerance, ability to grow under stress conditions may be a desirable attribute where full-season crops cannot be removed sufficiently early to provide timeliness of all seedlings but where a winter cover is needed.

Using additional freezing tests, Everson and Olien (1975) evaluated the world wheat and rye collections to determine the range of variation found in these two species (Figure 7-2). Wild grasses exhibited greater winter hardiness than rye, which was hardier than wheat. It is also interesting to note that in similar freezing tests,

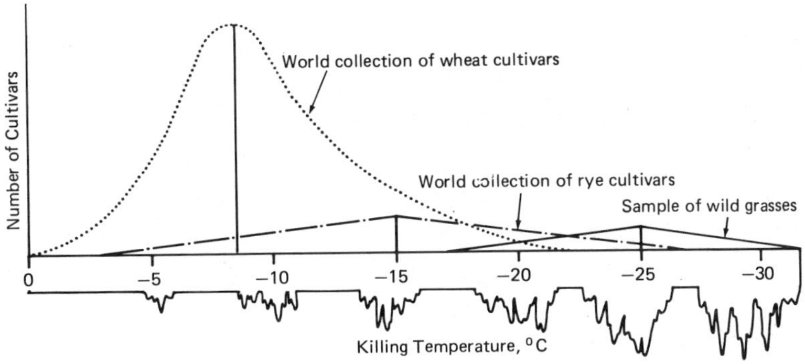

Figure 7–2. To determine the range of winter hardiness found in the wheat and rye crops, the world collections were subjected to artificial freezing tests. Although a wide range of hardiness was found to exist, relatively few winter wheat cultivars are highly resistant to low temperatures. A number of rye cultivars are more winter hardy than winter wheat. A sampling of wild grasses suggests that improvement in hardiness is possible if genes for hardiness can be obtained from these grasses by genetic engineering. (*Source:* Everson and Olien, 1975)

dormant seeds with relatively low moisture content survived at temperatures as low as −190°C.

Additional information on cold tolerance among and within crop species is shown in Table 7-1. Rye appears to be more cold tolerant than wheat or triticale. The three cereal crops are more cold tolerant than the forage legumes alfalfa and trefoil.

Small but real differences can be measured among crop cultivars (Table 7-1). The winter wheat cultivars listed are believed to be among the most hardy in the world. Less hardy wheat

Table 7–1

Comparative Cold, Icing, and Flooding Tolerances of Winter Cereals and Perennial Forages from Field-Grown Plants in November 1980

		LT_{50}[a]		*Percent Survival*	
Species	*Cultivar*	°C	°F	*7 Days Icing, −1°C (30°F)*	*14 Days Icing, 4°C (39°F)*
Wheat	Fredrick	−17	−1.4	ND[b]	ND
	Kharhov 22MC	−20	−4.0	97	58
	Lennox	−19	−2.2	95	38
	Moskovskaja	−19	−2.2	58	35
	Starke 2	−17	−1.4	58	17
Triticale	OAC Wintri	−17	−1.4	68	30
	AD 206	−19	−2.2	68	40
Rye	Frontier	−22	−5.6	80	ND
Alfalfa	Bonner	−08	17.6	ND	37
	Iroquois	−10	14.0	ND	52
Trefoil	Leo	−08	17.6	ND	80
	Empire	−07	19.4	ND	51

Source: McKersie (1981).
[a] LT_{50} is the temperature at which 50% of the plants survive.
[b] ND—not determined.

cultivars exist. Spring wheat cultivars, by comparison, may tolerate only brief exposures of −2° or −3°C (29° to 30°F) and cannot survive even moderate winter conditions if fall sown. Spring wheat cultivars may be damaged by a late spring frost following spring seeding (Single, 1975). At the end of the hardiness scale are winter wheat cultivars that can withstand temperatures in the range of −18° to −20°C (0° to −2°F).

PRODUCTION ASPECTS OF WINTER CEREALS

Except under ideal situations, winter conditions can be expected to cause some setback in winter cereal crops. Producers must assess winter damage and its severity and determine whether seedlings will recover or whether the field should be reseeded with a spring crop. Generally, a stand reduction of 25% or more should be replanted. Winter cereals, however, have a high tillering ability, and fields suffering a general overall stand reduction can produce a full stand through tillering. In other cases, sections of a field may be entirely winterkilled, and such areas will allow weed infestations unless reseeded. An application of nitrogen fertilizer will help promote photosynthetic area for energy production to promote tillering.

Following most winters, the chlorophyll produced in the leaves the previous fall will be destroyed and leaves will be brown, but such plants may not be dead. A positive indicator that plants are alive will be the appearance of new, white roots. These can be detected by pulling individual plants.

Successful winter cereal production begins with a proper fall seeding date (Fowler, 1982). Plants in the three- to four-leaf stage generally survive the winter better than plants with fewer or more leaves. Early fall seeding can result in excessive leaf growth that leads to smothering under a heavy snow load. Late fall seeding restricts root growth and food reserves, and the limited top growth is less effective in capturing a protective snow cover. A general recommendation is to seed winter cereals at least four weeks before the average date for the end of the growing season in areas where snow cover is heavy, and seven weeks in areas where snow cover is light. The end of the growing season is defined as occurring when the mean temperature reads 55°C (40°F).

Extending the fall growth period beyond seven weeks may subject winter wheat seedlings to Hessian fly infestations.[2] The Hessian fly is considered to be the most destructive insect pest of winter wheat throughout the world. Sometimes the insect damages winter barley and rye.

The Hessian fly is a mosquitolike insect that lays its eggs on wheat leaves. When the eggs hatch, the larvae move to the stem and feed on plant sap. Severe stunting in the early stages and stalk lodging in the later stages of wheat development are symptoms of fly damage.

Hessian fly eggs laid on the leaves of young wheat plants in the fall hatch, and larvae move down the plant and imbed themselves in the base of the stem before entering a dormant stage. The flies pass the winter as larvae inside a puparium, also called a flaxseed (Figure 7-3). In early spring the insect pupates, and flies emerge to lay the eggs for the spring brood. Once again, the larvae hatch out and move down the stem to the first joint above the ground. The larvae feed on cell sap by scraping the plant tissue, which results in severe stunting in the vegetative stage. Larval activity results in weakening the straw and interfering with the supply of nutrients to the head. In the reproductive stage, injured plants lodge and have shrivelled seed. In severe infestations, complete loss of crop may result, either from the fall or the spring attacks.

Several generations of the insect are produced during the summer if food plants such as volunteer wheat are available. The population of Hessian fly begins to decrease as the weather cools during the late summer and early fall, and unless wheat is available at this time for egg laying, the flies die without provision for the next generation. Fortunately, the date after the Hessian fly disappears is still early enough for wheat to become established before winter. Observa-

---

[2]Hessian fly—*Phytophaga destructor.*

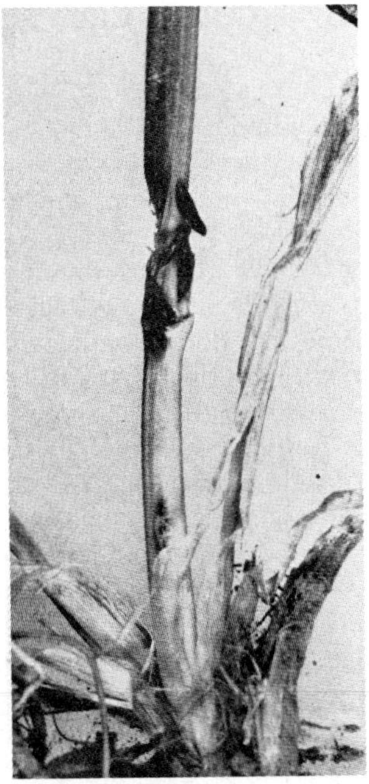

Figure 7–3. Hessian fly damage to winter wheat, rye, or barley can occur if seeded before the "fly-free date." At harvest time dark-colored "flaxseeds" or puparium form. The larvae that hatch during the early fall feed on the new wheat and change to the flaxseed stage before winter. Hessian fly damage can be avoided by seeding after the "fly-free date" and using genetically resistant cultivars. Flaxseeds are evident on this wheat stem. (Courtesy of Ontario Ministry of Agriculture and Food)

tion and experience have determined a "fly-free date" after which wheat can be safely seeded.

VERNALIZATION REQUIREMENTS

The production of winter cereals under greenhouses or controlled-growth chambers requires a vernalization period. Trione and Metzger (1970) found that vernalization proceeded most rapidly at 8°C (47°F) under a low light intensity and a short-day photoperiod. The duration of this period varies according to cultivar but will be completed for wheat in 42 to 56 days. Winter rye may require a full 56 days and winter barley 42.

Under field conditions, late planting may result in a delay in maturity. Reproductive development is a function of a vernalization requirement and a response to daylength. A 60-day delay in fall seeding (September 1 and October 30) will not delay maturity by 60 days, however, but by perhaps 14 to 17 days (July 24 to August 6 to 10).

Despite vernalization requirements, winter cereals lacking winter hardiness should be seeded earlier than winter-hardy species. Winter barley, for example, should be seeded two weeks earlier than winter wheat. Winter rye can be seeded two weeks following the recommended date for winter wheat.

HYBRIDIZATION OF WINTER AND SPRING CEREAL CROPS

The concept of utilizing winter × spring crosses to develop greater genetic variability is not new. The objective of such crosses is to transfer desirable traits—often specific genes for disease resistance (Kronstad et al., 1975). The semidwarf cultivars that gave rise to the green revolution were developed from winter × spring wheat crosses.

From a taxonomic standpoint, spring × winter types within the same species can be readily hybridized, provided that reproductive organs of the two types correspond. This, along with a concern about a loss of winter hardiness, generally has discouraged extensive programs whereby the winter and spring cultivars are crossed systematically.

In the early 1960s, however, the late Joseph A. Rupert, working with the Rockefeller Foundation Program in Chile, observed that conditions were suitable to provide simultaneous flowering in winter and spring wheat. He initiated a systematic program to exploit the desirable genes that had been isolated from one another by the vernalization requirement (Anonymous, 1968). By dividing the seed among three carefully chosen testing locations, it is possible to select for wide adaptation, spring types, winter types, and associated winter hardiness. If spring-planted, those with a vernalization requirement will not reach reproductive development and will be eliminated; if fall-planted, progeny lacking winter hardiness will be killed. Surviving plants can be backcrossed with winter-hardy cultivars in an attempt to increase winter hardiness.

Specific objectives of spring × winter crosses include transferring the superior disease resistance, cold tolerance, and possible drought tolerance from winter into spring cultivars. Spring wheat can contribute superior breadmaking quality and genes for leaf and stem rust resistance that differ from the genes carried by the winter cultivars. It is hoped that superior yield performance of both winter and spring cultivars may emerge from spring × winter crosses.

By 1980, more than 10,000 spring × winter wheat crosses had been made. Initial results were encouraging, but cultivar evaluation has not been evaluated adequately (Kern, 1980). In Turkey, winter hardiness comparable to that of standard wheat cultivars was obtained in two years of screening (Yakar and Altay, 1980). After three consecutive years of natural selection, spring × winter wheat crosses produced cultivars with a high degree of winter hardiness (Stoskopf, 1972).

A total of 360 wheat entries were tested in the first international spring × winter wheat screening nursery in 1973–1974 at 38 locations in 24 countries within the major winter wheat regions of the world. These and subsequent tests have revealed good yields, reasonable hardiness, and wide adaptation of spring × winter crosses. International tests of this magnitude, successfully conducted through international cooperation, are indicative of the progress that can be made on a worldwide basis (Kern, 1980).

A spring × winter barley hybridization program was begun at CIMMYT in Mexico in 1976–1977 (Anonymous, 1978).

FACULTATIVE CEREAL CROPS

Spring × winter crosses form the basis of efforts to develop **facultative cereals** that are to survive and produce under more than one specific type of environmental condition. Facultative cultivars, for example, may remain longer in the vegetative stage. Such a feature may be important to avoid frosts in high plateau regions that normally coincide with flowering. Facultative wheat cultivars may be useful in parts of central India if sown at the end of the monsoon period in late September. Residual moisture would encourage stand establishment, but plants would not head until late January or early February. Under such conditions, they would have a longer maturation cycle than shorter-season wheat cultivars planted in November. Areas of Argentina, North Africa, and the Near East need facultative wheats with some cold tolerance to avoid late spring frosts (Anonymous, 1978; Kern, 1980). Selection of intermediate-maturing types that will tolerate high temperatures early in the season are possible under proper environmental conditions and using spring × winter populations.

A spring × winter barley program aimed at developing a facultative type of barley has been undertaken (Anonymous, 1978, 1980).

COLD TOLERANCE OF SPRING CEREALS

In temperate regions of the world with short growing seasons, maximum grain yields of cereal crops are not achieved unless seed is sown in the early spring. Frequently early seeding is associated with cool soil and air temperatures. In high latitudes and altitudes of the world, the zone of adaptation could be extended in such crops as corn and sorghum if planting earlier than traditional planting dates were possible. This could be achieved by incorporating ability to germinate and grow under cool conditions. Ability to withstand freezing conditions may be desirable, as the reliability may be increased. Furthermore, early planting in the central latitudes followed by "normal" growth and development of the plants would result in near-coincidence of the grain-filling period of the crop with the period of highest light intensity for the growing season. Consequently, more photosynthate would be available for deposition in the grain.

Many studies have shown the existence of genetic variability for germination of crops at low temperatures. Effective fungicidal seed treatments, however, may have reduced the need for genetic resistance (Mock and Eberhart, 1972) to seed-rotting microorganisms. Ability to develop rapidly and reliably at low temperatures is important.

Sorghum is of tropical origin and is one of the more sensitive crops to low temperatures. A temperature of 18°C (64°F) was considered by Stoffer and Riper (1963) to be the minimum soil planting temperature. The near-minimum growth temperature for most sorghum cultivars was suggested by Rhykerd, Gross, and Sullivan (1960) to be 16°C (61°F). At 12°C (54°F) sorghum, most other tropical and subtropical plants, and some warm-season temperate plants grow poorly; below this level they are injured by nonfreezing temperatures (Lyons, 1973).

Cool-tolerance in sorghum would appear to be essential if full advantage of extending the adaptation of this crop is to be realized. Some genetic variation in sorghum cool tolerance has been observed (Stickler, Pauli, and Casady, 1962; Nordquist, 1971) and has allowed the crop to be grown in the southern parts of Canada. Greater tolerance to low temperatures, however, would be a welcome asset in cool areas.

A study to develop screening techniques was conducted at Lincoln, Nebraska, where seasonal night temperatures average 8°C (46°F) (Clegg, Eastin, and Nelson, 1983). Tolerant sorghum cultivars were generally early-maturing.

Tolerance to cool temperatures is not a well-defined criterion for crop improvement. In general, seedling cold tolerance and final yield are not clearly related. Cold tolerance may be similar to seed vigor (considered in Chapter 10).

The use of controlled-environment units has provided the necessary precision needed to screen crops for cold-tolerance genes. There is little doubt that improvement in cold tolerance is possible.

In corn, Eagles and Hardacre (1979) observed considerable genetic variation. Mock and Skrdla (1978) evaluated cold-tolerance response of 144 corn plant introductions and, based on the variation observed, suggested it should be possible to develop cold-tolerance genotypes to all latitudes. The same genetic factors controlled cold-tolerance responses of two Iowa corn breeding populations in both growth-chamber and field environments (Mock and Eberhart, 1972).

In mountainous regions of the tropics, and in the temperate rice-growing zones of the world, cold weather and cold water are problems for rice growers. Crops grown at high elevations may be subject to seasonal and diurnal temperature fluctuations.

Based on the premise that cold-tolerant rice cultivars exist and that these genes would be useful, a rice cold-tolerance workshop was held in 1979 (International Rice Research Institute, 1979). This workshop developed out of previous findings of cold tolerance in rice and partly out of research using the excellent cold-water facilities built by the Korean Government at Chuncheon, Korea. Reports from various countries at the workshop revealed a wide range in genetic variation for cold tolerance.

Cold-tolerant cultivars are considered one of the most important assets in rice in some regions of the world. Four aspects of cold tolerance have been identified in rice:

1. Low temperature at the seedling stage is a common limiting factor in the growth of rice in many low-temperature areas. Screening in the seedling stage to isolate genes for seedling cold tolerance is an important aspect.
2. Many areas with problems of low temperature have critically low temperatures at flowering. Panicle initiation and anthesis are the two stages most susceptible. Screening for cold tolerance at anthesis is a second important aspect.
3. Ability to produce roots rapidly when transplanted in soils with low temperature is impor-

tant, as it is related to plant vigor and earliness of flowering.

4. In low-temperature areas, blast,[3] a disease that produces spots on leaves of rice, is severe and can result in death of plants. The production of blast-resistant cultivars that are stable under low-temperature conditions is an important rice-breeding objective.

In wheat, some of the physiological mechanisms involved in cold hardiness may be associated with the plant's ability to emerge under soil moisture stress (Morrison et al., 1981). It may be possible to combine several cold-tolerance traits within a cultivar.

The large land area of cereal crops grown in the Northern Hemisphere suggests that cold tolerance could be an important asset on a world scale.

SUMMER FALLOWING

Over much of the land mass of the world, seasonal distribution of precipitation is not in phase with seasonal water use by crops. Crop demands and water supply can be brought more closely into balance by the use of winter cereals, which benefit from snow melt and spring rains that often delay seeding of spring cereals. Winter cereals are grown under conditions of a reduced evapotranspiration rate and are characterized by a deeper root system than spring cereals. Advancing the date of seeding in semitropical areas with a facultative cereal so that periods of greatest water use coincide with rainfall helps water-use efficiency (Spratt and Chowdhury, 1978). Complete weed control to reduce transpiration can improve crop-moisture relationships.

PURPOSE OF SUMMER FALLOWING

An extensively used system of moisture management in semiarid regions of the world involves **summer fallowing.** It allows crop production to be extended further into arid areas than would otherwise be possible. Its purpose is to store soil moisture in order to compensate for an annual precipitation regime that is barely adequate to permit a profitable agriculture on an annual basis. Traditional summer-fallowing practices involve leaving a field uncropped for an entire season, but the soil is cultivated as needed or sprayed with a herbicide to prevent weed growth and associated transpiration. In this manner, the period of time that soil water reserves can be replenished for a subsequent crop is extended.

WATER-CONSERVATION EFFICIENCY OF SUMMER FALLOWING

Summer fallowing cannot be regarded as an efficient method of conserving water, as about 50% of the summer precipitation on bare soil is lost to evaporation; about 30% to 35% percolates deeply into the soil below the crop root zone; and only 15% to 20% remains in the upper 2 meters (6.5 ft) of soil for use by the following crop. The actual percentage of rainfall stored depends on soil type, amount of rainfall, and evaporation rate and was suggested by Rennie (1982) to vary from 3% to 24% in efficiency.

Summer fallowing is used extensively in the production of both winter and spring cultivars. In the southern Great Plains of the United States, winter wheat alternated with summer fallow is considered a practical sequence when the total annual rainfall is less than 55 to 60 cm (22 to 24 in.); in the central Great Plains, spring wheat replaces winter wheat in the rotation where winterkilling is a problem, and summer fallow is used at an annual precipitation of 45 to 50 cm (22 to 24 in.), and in the northern Great Plains at 35 to 40 cm (14 to 16 in.).

Production and efficiency aspects of a summer fallow–wheat sequence are illustrated by the following considerations:

1. Assume the minimum amount of moisture required during the growing season to produce a harvestable crop is 26 to 27 cm (10 to 11 in.).
2. Assume that the total annual rainfall is 30 to 35 cm (12 to 20 in.) and that, because the effective rain-

[3]Blast is a fungus disease—*Pyricularia oryzae Cav.*

fall is somewhat less than this, insufficient moisture exists to produce a crop. The seasonal precipitation is 15 to 20 cm (6 to 8 in.), sufficient for minimum vegetative growth only (Figure 7-4).

3. For the location being considered, it is a fact that for every 2.5 cm (1 in.) of moisture over the minimum required for vegetative growth, 400 to 470 kg/ha (357 to 420 lb/ac) of grain is produced.
4. If 15% to 20% of the annual precipitation is stored, an additional 4.5 to 10 cm (1.75 to 4 in.) is available for the crop.
5. At this location it appears that only if summer fallowing is used is sufficient total moisture available for a cereal crop.

Under the dryland conditions at Swift Current, Saskatchewan, the need for summer fallowing is borne out (Table 7-2), but at Winnipeg the system is questionable. At Swift Current 21% of the annual precipitation was stored, compared to 10% at Winnipeg. Despite the inefficiency of summer fallowing, the system is widely used in areas with limited rainfall. From 1960 to 1980, 20%, 35%, and 40% of the cultivated land in Manitoba, Alberta, and Saskatchewan, respectively, has been summer-fallowed annually. Perhaps the area devoted to summer fallow was encouraged by surplus production and by subsidies to keep land in fallow. Since about 1972, the escalating costs of inputs, the price of grain, and the marketing quotas used in North America to control production have encouraged the continued extensive use of summer fallowing in dryland regions, to the point where the practice is considered detrimental.

Figure 7–4. This spring cereal crop is showing evidence of moisture stress. Heads are small, tillering is limited, and yield is unsatisfactory. Under dryland conditions a system of fallow is often used to store soil moisture to compensate for an annual precipitation regime that is barely adequate to produce a harvestable crop. (Courtesy of Ontario Ministry of Agriculture and Food)

SOIL SALINITY

A growing awareness of an accelerating loss of soil organic matter, soil erosion, and a major increase in the area of salt-affected soils has raised concerns about monoculture cropping systems and summer fallowing. The long-hallowed and treasured practice of summer fallowing has come to be regarded as a major form of mismanagement (Rennie, 1982) that has led to a condition of **saline soil.** The problem is so severe that major changes in summer fallowing and irrigation may be necessary to halt saline soil development.

Soil salinity is the accumulation of soluble salts and exchangeable sodium on the soil surface that is detrimental to plant growth. The source of these salts is the underlying bedrock material. When the prairie had a continuous natural cover of perennial grasses which used up moisture, the salts remained deep down. The introduction of a fallow period encouraged a rise in the salt-laden water table when unused water percolated below the root zone. Vegetation is needed to consume the rainfall, no matter how sparse and inadequate it may be for economic crop production. Transpiring plants pump the rainfall back into the atmosphere and leave the salt at a safe depth. Although thousands of years of weathering have produced a few localized saline areas, significant changes in the water cycle brought about by summer fallowing have propelled salinity into a major problem. Removal of natural vegetation in Western Australia resulted in a 10-m (32.8-ft) rise in the water table (George and Bull, 1979).

When water infiltrates into the soil below the root depth and when an impermeable soil

Table 7–2
Moisture Conserved by Summer Fallowing at Swift Current, Saskatchewan, and at Winnipeg, Manitoba, Canada

Annual Precipitation				Moisture Conserved			
Swift Current		Winnipeg		Swift Current		Winnipeg	
cm	in.	cm	in.	cm	in.	cm	in.
47.5	18.7	85.3	33.6	10.2	4.0	8.6	3.4
Total available moisture at seeding:				10.2	4.0	16.8	6.6

Source: Data from Eilers et al. (1977).

layer causes the water to move laterally, a saline seep develops (Halvorson and Black, 1974). Eventually, at some lower elevation on slopes, the water will seep laterally out at the soil surface by capillary action (Figure 7-5). Saline seeps may develop within the same field only at a short distance from higher ground or may form at a distance of several kilometers (miles). Subsurface movement of saline waters have caused salinity problems as far away as 25 km (15.5 m) (Rennie, 1982). As the water evaporates at the seep area, residues of soluble sulphate and chloride salts of sodium form a white surface crust (Figure 7-6) with a *p*H value of about 8.5.

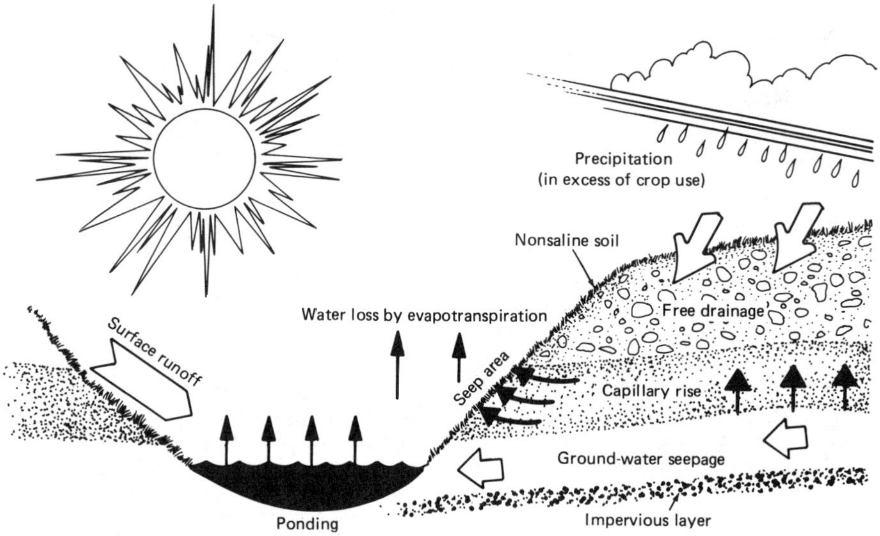

Figure 7–5. Illustration of the formation of a saline seep. A saline seep develops when water is carried below the root zone and when lateral movement occurs due to an impervious soil layer. Water may be carried to the surface by capillary rise, and, as the water is lost by transpiration and evaporation, it leaves soluble sulphate and chloride salts of sodium, calcium, and magnesium on the surface to form a white crust.

Figure 7–6. The white crust on the soil surface is salt accumulation, rendering this area totally unproductive. (Courtesy of Prof. Les Henry, Soil Science Department, University of Saskatchewan)

The concentrations of total soluble salts or of sodium are sufficient in concentrated areas to prevent agricultural production by altering plant metabolism through an increase in osmotic tension, by reducing root growth, and by altering soil properties.

SODIC SOILS

Sodic soils have a high sodium content and have a *p*H value ranging from 8.5 to 10. The negatively charged soil particles have a high proportion of exchangeable sodium ions which can lead to a puddled soil condition. Puddled soils are impervious to water, are difficult to till, are hard to drain, are low in oxygen, and restrict root growth. A concentration of neutral salts may interfere with water absorption and nutrient uptake, because a higher osmotic pressure exists in the soil solution than exists in the root cells.

Irrigation water containing soluble salts and applied too sparingly so that salts are not leached out may result in sodic soils. Good-quality irrigation water may contain a salt concentration up to 1000 parts per million (ppm). When irrigation water has from 1000 to 3000 ppm, it is considered marginal for irrigation use. Water in the Pacific Ocean has 35,000 ppm. Irrigation water applied to soils with poor drainage so that water cannot leach the salts away can result in sodic soil conditions.

EXTENT OF THE SOIL-SALINITY PROBLEM

On almost all irrigated cropland and in nonirrigated semiarid crop and rangeland, salt build-up is an existing or potential danger (Loomis, 1976). Although the area of saline soils is as yet somewhat limited, problems do exist in the Great Plains of North America, the steppes of the USSR, Western Australia, and dryland areas of Israel. In North America about 800,000 ha (2 million ac) are seriously affected by salinity (Cherry, 1980). In Canada, saline seepage has reduced crop yields by an estimated 50% on 1 million ha (2.5 million ac) of cropland (Rennie, 1979). In Saskatchewan, saline seepage has adversely affected 10% of the cropland (Kirkland and Keys, 1981).

In Western Australia, 170,000 ha (420,000 ac) or 1.3% of the total cleared area is significantly affected by rising salt levels in the 27.5 to 60 cm (10.8 to 23.6 in.) rainfall area. Of concern is the rate of encroachment, estimated by Rennie (1982) to be about 1% annually in the semiarid plains of North America.

The accumulation of salts on agricultural soils, however, is not a modern phenomenon but has plagued civilization for centuries. A major reason for the collapse of early agricultural societies dependent on irrigation water may have been soil salinity. Historically, humans have not been able to grow cereal grains in semiarid regions continuously without creating deserts.

The persistence of the problem suggests that there are no easy solutions. Saline and sodic soils can be reclaimed by the application of sufficient water to flush salts into the lower soil depths (2.5 to 3 m or 8 to 10 ft), provided that drainage is adequate. Without adequate drainage, the problem may be increased, as the addition of so much water will raise the water table and increase salt accumulations on the soil surface. Drainage can be improved by providing deep canals or by a system of underdrainage, provided that soil permeability is satisfactory. Without adequate drainage, however, reclamation may not be feasible.

A SYSTEM APPROACH TO DRYLAND PRODUCTION

At this point in time, efforts should be directed to preventing or slowing the spread of soil salinity rather than reclaiming saline soils. Perhaps a radically different system of semiarid farming that maximizes water-use efficiency needs to be developed. Perhaps no single approach is adequate to overcome dryland agriculture problems, but rather a combination of management inputs may be required.

The solution to the further development of saline soils is to discontinue the practice of summer fallow. It is not easy, however, to reverse nearly 100 years of fallow tradition through a textbook recommendation. But the fact that increasing salinity is attributed to inefficient water utilization by fallow in a two- to three-year rotation (Kirkland and Keys, 1981) clearly indicates a change is needed. An abrupt change without regard to crop moisture requirements is not feasible. Such a change can be accomplished only when accompanied by a method or methods that increase water available to crops. The change to continuous cropping may be gradual, as soil moisture conservation or water management techniques are developed.

A systems approach involves understanding the possible associated factors and integrating them into a system whereby summer fallowing can be eliminated. The following factors collectively may be effective in this regard. The list is presented in no particular order of merit and includes techniques to cope with saline soils where they now exist.

1. The efficiency of available soil moisture can be increased by sound fertility practices. More efficient water use can reduce the need for fallow. Fertilizer usage has been considered uneconomic under dryland conditions, and breakdown of organic matter during the fallow year was needed as a source of plant nutrients. The decline in soil organic matter to the point where soils have lost 50% to 70% of their nutrient restoring value has resulted in a need for an annual application of fertilizer.

The increased yield with fertility under dryland conditions was demonstrated clearly in North Dakota from 91 spring wheat tests grown at an average moisture level of 32.3 cm (12.7 in.). Fertilizer increased grain yield an average of 34.2% (Table 7-3) as a result of greater water-use efficiency.

2. The use of plants with a high transpiration capacity is being considered in Australia (Cherry, 1980). By strategically planting trees in the agricultural landscape so that farm cropping programs can proceed with a minimum of disruption, it is hoped that the soil water table can be lowered, thereby reducing lateral water movement and saline seep development. Tentative predictions suggest that 8% of the cleared land area may have to be planted to an adapted tree species with a high transpiration capacity for the plant to be successful. Although such an idea may not be pleasing to wheat producers, the merits of the method cannot be ignored.

3. By matching cropping practices to moisture distribution and not only to total soil moisture available, the need for fallow may be reduced. Crops may tolerate moisture stress periods without a serious yield reduction if these periods do not coincide with critical periods of development. With determinate crops, greatest sensitivity to moisture stress may occur at the time of floral initiation, during flowering, and to a lesser extent during seed development (Begg and Turner, 1976). A timely rainfall or application of irrigation water can influence crop yields favorably.

Under conditions in southern Arizona in 1966 and 1970, soil moisture for winter wheat at the **jointing** stage proved to be more critical than at the flowering or dough stage (Day and Intalap, 1970). Soil moisture stress may be increased if coupled with high temperatures, and the resulting reduction in photosynthesis may decrease grain yield, as it may reduce the development of both the sink and source.

Table 7–3
Effect of Soil Fertility on Spring Wheat Yields under Dryland Farming Conditions in North Dakota over a Range of Soil Moisture levels

Number of Treatments	Average Moisture Supply[a]		Average Grain Yield				Percent Increase Due to Fertility
			No Fertilizer		Fertilized		
	cm	in.	kg/ha	lb/ac	kh/ha	lb/ac	
7	16.3	6.4	1050	938	1430	1277	36.2
9	21.3	8.4	1180	1054	1450	1295	22.9
8	24.1	9.5	1390	1241	1990	1777	43.2
9	26.7	10.5	1530	1366	1990	1777	30.1
10	30.5	12.0	1700	1518	2360	2107	38.8
9	32.8	12.9	2050	1830	2570	2295	25.4
10	36.0	14.2	1940	1732	2620	2339	35.1
11	40.1	15.8	2000	1786	2630	2348	31.5
9	43.4	17.1	1780	1589	2470	2205	38.8
9	51.5	20.1	1960	1750	2740	2446	39.8
Mean	32.3	12.7	1658	1480	2225	1987	34.2

Source: Data from Bauer and Young (1966).
[a]Sum of growing-season precipitation and stored available soil water at seeding time.

These facts are of value when limited amounts of irrigation water are available, or if stages of crop development can be made to coincide with moisture availability.

4. The development of salt-tolerant crops produced under a system of moisture management could improve agricultural production on saline soils and extend agriculture to previously unsuited regions. Crops can be adapted genetically to conditions more saline than previously thought possible (Epstein et al., 1980). Barley is the most salt-tolerant cereal grain, a fact observed in ancient agricultural civilizations when increased soil salinity led to the eventual collapse of irrigated river flatlands. Screening of the world collection of barley genotypes has revealed barley cultivars capable of producing up to 1500 kg/ha (1340 lb/ac) at salinity levels of 35,000 ppm on highly permeable sandy soils (Epstein et al., 1980). Wheat is less salt-tolerant than barley, although variability exists. Screening of 5000 spring wheat accessions from the wheat collection has revealed 34 as being capable of producing grain at a salinity level 50% that of sea water, a level of stress lethal to most commercial wheat cultivars.

5. The inclusion of deep-rooted crops in a rotation that will use water at a greater depth than cereal crops and thereby reduce lateral water movement is suggested as a means of reducing saline seeps. Sweet clover[4] is one of the few legumes that is quite tolerant to salinity.

6. To control the development or spread of soil salinity, a modification in irrigation practices may be required. Less frequent applications that apply the same or even greater volumes of water to assure a uniform depth of wetting may be useful. Irrigation water free of salts is essential if salinity is not to be increased by irrigation practices.

7. Recognition and use of drought-survival mechanisms may help avoid drought stress. Plant survival under drought stress is conditioned by many complex traits, but they can be classed as avoidance, tolerance, or escape mechanisms. Drought-avoidance mechanisms help maintain a high tissue-water status under drought stress. Rolling of the leaf is one such mechanism (Figure 7-7).

[4]Sweet clover—*Melilotus* spp.

Figure 7–7. Corn plants showing drought stress as evidenced by the rolling of leaves. This action, caused by contraction of bulliform cells, is a form of drought avoidance. (Courtesy of Ontario Ministry of Agriculture and Food)

Drought tolerance allows the plants to survive and produce under conditions of a high internal water stress. Plants with the C_4 photosynthetic pathway, such as corn, sorghum, and millet, are able to tolerate greater drought stress than C_3 cereals.

Escape traits, such as early maturity, allow the plant to complete its life cycle before the onset of drought stress. Winter cereals are noted for their rapid, early development under long-day conditions, and the ability to escape subsequent hot, dry conditions is an important feature for C_3 plants.

Under dryland conditions in Oregon, *drought-resistant* winter wheat cultivars exhibited both drought-avoidance and drought-tolerance traits (Keim and Kronstad, 1981).

8. In areas where part of the annual precipitation falls as snow, snow management can be an effective means of reducing the need for fallow. At high latitudes, most of the precipitation in the winter months occurs in the form of snow that is blown off the fields by high winds. About 20% to 30% of the annual precipitation in the northern Great Plains falls as snow. Generally, 10 cm of snow is equivalent to 1 cm of rainfall.

A key to moisture management is to capture and hold the snow uniformly on the fields (Figure 7-8) to help replenish soil moisture at the beginning of the growing season. By retaining the snow on the fields, 7.5 to 15 cm (3 to 6 in.) of soil moisture could be captured in the northern Great Plains region. Surface stubble, crop residues, surface trash, or rough soil conditions are among the simplest method of holding snow.

A more elaborate system of snow control involves windbreaks. Windbreaks may range from rows of trees (Figure 7-9) or shrubs to tall perennial or annual plants systematically sown into fields of shorter crops (Rosenberg, 1977). These sown windbreaks may not restrict movement of a pivotal irrigation system or directional movement of field machinery and are easily removed, replaced, shifted, or established. Sown windbreaks do not occupy a large land area or reduce the yield of the sheltered crop adjacent to the windbreak.

More permanent windbreaks do not appeal to farmers, as the area between strips may have a moisture gradient that delays seeding, results in uneven development and maturity, and restricts machinery movement and flexibility.

The ideal windbreak should be dense enough to slow the wind by absorbing much of its momentum but be porous enough to allow a

Figure 7–8. In dryland areas, a key to moisture management is the capture and holding of snow uniformly over the field to help replenish soil moisture. Stubble, mulch, or a rough soil surface aid in this process. (Courtesy of Ontario Ministry of Agriculture and Food)

Figure 7–9. Permanent windbreaks have an established and documented history of successfully reducing wind speed and holding snow. Many established windbreaks have been removed by farmers who find the value of the land occupied by trees too great when considered in terms of the crop which might otherwise be produced on that land. The protection that windbreaks offer against wind erosion, mechanical protection for the sheltered crop, improved distribution of snow across the fields, and consequent improvements in moisture supply does justify their use in many regions. (Courtesy of Ontario Ministry of Agriculture and Food)

relatively smooth nonturbulent flow over and through the barrier. A solid wall does not make a good windbreak, as wind forced over the wall gains speed.

For eight years in Montana, Black and Siddoway (1976) studied a system of tall wheat grass[5] barriers to control wind erosion and to collect and hold snow over winter. They sought to determine whether increased soil water recharge can support more intensive cropping than crop-fallow rotations. The barriers nearly doubled overwinter soil water storage the first nine months compared with unprotected land. With nitrogen and phosphorus fertilization, average annual grain yields were 30% to 69% higher for intensive cropping systems than for conventional spring wheat-fallow. Continuous cropping systems used 71% to 80% of the average precipitation received between crop harvests; wheat-fallow used 30%. The amount of water unaccounted for between crops, some of which may contribute to saline seeps, averaged 55.4 cm (21.8 in.) for spring wheat-fallow and 7.9 to 11.4 cm (3.1 to 4.5 in.) for continuous cropping. In addition to controlling wind erosion, the grass barrier system provided the additional soil water needed, without fallow, to make intensive cropping feasible. Cropping intervals between grass barriers was 14.6 cm (48 ft). Grass barriers consisted of two rows, 91 cm (36 in.) apart, oriented at 45 degrees to the prevailing northwest winter winds.

Grass barriers a meter in height (3 ft) were found to reduce wind speed by as much as 20% at a distance of up to 18 m (60 ft), sufficient to cause a 32-km/hr (20-mph) wind to deposit half the snow it carried (Anonymous, 1979).

9. Herbicidal weed control can be used to reduce weed growth and associated transpiration, to increase moisture content through increased capture of winter snow, to reduce tillage costs, and to reduce destructive tillage on organic matter. This system has been used in Nebraska and is known as *ecofallow* (Fenster and Wicks, 1982; Meyer, 1983). The method consists of one year of winter wheat followed by fallow. After harvest the field is sprayed with a herbicide to remove weeds and volunteer wheat. Fewer cultivations reduce costs, and undisturbed surface stubble holds snow and reduces erosion losses. The result is an increase in storage and maintenance of soil moisture.

Ten years of tests at Akron, Colorado, showed that herbicides used in minimum-tillage fallow (ecofallow) suppressed an average 70% of potential weed and volunteer weed growth. Increased soil water storage increased winter wheat yields with 1.8 fewer tillage operations per fallow season (Greb and Zimdahl, 1980).

10. Zero-tillage allows crop stubble and residues to provide an effective cover to hold snow, thereby reducing the need to fallow. Zero-tillage is a system of crop production in which seed is planted in a seedbed that has not been tilled since the previous crop was harvested (Figure 7-10). Tillage operations destroy some of the de-

[5]Tall wheat grass—*Agropyron elongation.*

Figure 7-10. Zero-tillage is a system of crop production in which seed is planted in a seedbed that has not been tilled since the previous crop was harvested. (Courtesy of Ontario Ministry of Agriculture and Food)

sirable surface trash cover and—perhaps more important from the standpoint of winter snow accumulation—dislodge and disorient the stubble from its original upright, anchored position. Advantages of zero-tillage include protection against erosion, moisture conservation, and improved organic-matter levels. Disadvantages include changes in soil moisture and temperature regimes that can delay spring seeding, an important factor at high latitudes (Pelton, 1982). The rate of decomposition and the biological release of plant nutrients are affected also. An accumulation of trash may occur that hinders seeding equipment and complicates seeding. Despite problems, zero-tillage is used extensively in the United Kingdom, Australia, and the United States and on light, warm soils in Canada.

In a system involving zero-tillage in a winter wheat-corn-fallow rotation in Nebraska, standing wheat stubble trapped snow that increased corn yields as compared to plants where tillage had knocked stubble down (Hoefer, Wicks, and Burnside, 1981). Removal of winter wheat stubble after corn planting caused less crop root penetration, more lodging, and a 15% reduction in corn yields as compared to plots where stubble was left undisturbed.

11. Under dryland conditions, seeding rate may require modification to suit moisture availability. A 50% reduction in the seeding rate from that commonly used by the farmers in the semiarid region in southwestern Saskatchewan resulted in an increase in yield of spring wheat when measured over a nine-year period (Pelton, 1969). The seeding rate was reduced to 22 kg/ha (20 lb/ac). Perhaps fewer plants reduced water loss through transpiration and more closely matched plant population with moisture availability.

12. An additional system of managing snow is offered through a system of swathing at alternate heights. The concept is based on the fact that snow usually drifts until it is deposited on the leeward side of wind barriers, such as standing stubble. Under continental climatic conditions, winds often reach blizzard intensity, and modifications in wind barriers may be needed (Nicholaichuk and Read, 1981). The system was tested over a seven-year period at Swift Current, Saskatchewan, on plots 4 to 4.8 ha (10 to 12 ac) in size. The average depth of snow in the alternate-height stubble was 23 cm (9 in.) compared to 20 cm (7.9 in.) for the uniform stubble. This amounted to 1.7 cm (0.07 in.) of moisture or a 30% average increase. Approximately 92% of the trapped snow was stored as soil moisture, but the amount may depend on slope and runoff.

13. The Industrial Revolution that removed workers from the farm to an assembly-line job placed a strong demand on bread for lunch-box sandwiches, thereby securing the place of wheat in world agriculture for bread and sandwich production. Switching from wheat to other crops is not compatible with the world's food demands. Perhaps human food preferences can continue to be met through the development of a perennial form of wheat. Although perennial wheat appears as a radical concept, it is not drastically different from ratoon cropping in rice and sorghum. The problems of winter hardiness,

however, that now exist in winter annuals may be a limitation with a perennial wheat crop.

14. An underdrainage system to remove water percolating through the soil would prevent lateral water movement and hence saline-seep development. A system of drainage is expensive, however, and economics may render this infeasible.

It is evident that changes in land use have upset a delicate balance, and early warnings of a potential collapse of dryland agriculture are evident. Change into an unplotted course will be slow to develop, yet change must occur. Consideration must be given to possible long-term implications and chain reactions.

CONCLUSIONS

Winter hardiness, cold tolerance, and soil salinity are three production aspects that play an important part in world agriculture. These production aspects cannot be considered as separate entities or as topics of exclusive attention. They, together with other production aspects, plant breeding, management, and a number of interacting variables, must be considered in total and integrated into a system of successful crop production. Individual aspects may be studied and considered by individual scientists. It is the task of the producer, however, to integrate the many factors into a systems approach. This chapter helps emphasize the need for an integrated cropping system.

QUESTIONS

1. *a.* What grain yield could be expected by seeding winter annuals in the spring?
   b. Discuss the difference between reproductive development caused by vernalization and bolting in spring-seeded crops such as sugar beet and that in garden vegetables such as red beet and carrot.

2. *a.* Carefully consider why winter cereals can outyield spring-seeded cereals in terms of total photosynthate produced.
   b. Under what circumstances might spring-seeded cereals outyield winter cereals?

3. *a.* What temperature conditions prevail under a uniform and deep snow cover?
   b. Snow may be a protective barrier or winterkiller. Comment and explain.

4. *a.* Recommendations for seeding dates of winter cereals usually have been given as calendar dates. Determine the recommended seeding dates for various winter cereal crops for a specific location.
   b. Determine the interval between the above calendar dates and the end of the growing season for the location in question.

5. Consider why spring × winter cereal hybridization programs have not been used more extensively in the past.

6. Consider summer fallowing under the following topics:
   a. From the standpoint of machinery available at the turn of the century.
   b. As a technique to control weeds before the advent of herbicidal weed control.
   c. Under conditions of a vastly smaller world population.
   d. As a modern technique.

7. Develop a history of fallow in ancient as well as modern times. Refer to Buller (1919) and MacEwan (1969) for historical materials; use state or national statistics for up-to-date figures.

8. Can a parallel in soil-salinity problems be drawn between ancient and modern times?

9. Comment on the ability of humans to implement a systems approach that can reduce or stop the spread of saline soils.

10. What conclusions can be drawn from this chapter regarding the state of agriculture as a science or an art?

11. A shortage of water for irrigation purposes has resulted in a shift to more economical use of water by such methods as drip and sprinkler methods in contrast to flooded soils.

a. What impact might more efficient irrigation methods and intensive agriculture have on soil salinity?

b. How might salinity problems be reduced by irrigation methods?

c. Consider how summer fallow may increase saline-seep areas.

12. Consider the merits of cold tolerance in selected cereal crops in terms of:

a. World agriculture.

b. Agriculture in your region.

c. Vegetative growth.

d. Grain yield.

e. A breeding objective.

References

Anonymous. 1968. "Project to Combine the Best of Spring and Winter Wheats" *CIMMYT News* 3: 7–8.

———. 1978. *CIMMYT Report on Wheat Improvement.* 310 pp.

———. 1979. "Grass Barriers Greatly Reduce Erosion." *Crops and Soils* 31 (7): 27.

———. 1980. *CIMMYT Review 1980.* pp. 47–49.

Bauer, A., and R. A. Young. 1966. "Soil Fertility and Water Relations." *North Dakota Farm Research* 24 (6): 8–11.

Begg, J. E., and N. C. Turner. 1976. "Crop Water Deficits." *Advances in Agronomy* 28: 161–217.

Black, A. L., and F. H. Siddoway. 1976. "Dryland Cropping Sequences Within a Tall Wheatgrass Barrier System." *Journal of Soil and Water Conservation* 31: 101–105.

Bullen, E. R. 1977. "How Much Cultivation?" *Span* 20 (2): 53–54.

Buller, A. H. R. 1919. *Essays on Wheat.* New York: The Macmillan Company. 339 pp.

Cherry, M. 1980. "How Rising Salt Causes Land Loss." *Span* 23 (3): 139–140.

Clegg, M. D., J. D. Eastin, and L. A. Nelson. 1983. "Field Evaluation for Cool-Tolerance in Grain Sorghum." *Crop Science* 23: 23–26.

Cloutier, Y. 1982. "A New Test for Frost-Hardiness in Winter Cereals." *Canada Agriculture* 27 (3): 33–36.

Day, A. D., and S. Intalap. 1970. "Some Effects on Soil Moisture Stress on the Growth of Wheat (*Triticum aestivum* L. em. Thell). *Agronomy Journal* 62: 27–29.

Dexter, S. T. 1956. "The Evaluation of Crop Plants for Winter Hardiness." *Advances in Agronomy* 8: 203–237.

Eagles, H. A., and A. K. Hardacre. 1979. "Genetic Variation in Maize (*Zea Mays* L.) for Germination and Emergence at 10°C." *Euphytica* 28:287–295.

Eilers, R. G., et al. 1977. "Soils." *Principles and Practices of Commercial Farming.* Faculty of Agriculture, The University of Manitoba. 5th ed., pp. 17–59.

Epstein, E., et al. 1980. "Saline Culture of Crops: A Genetic Approach." *Science* 210: 339–404.

Everson, E. H., and C. R. Olien. 1975. "Breeding Wheat for Winter Hardiness." *Proceedings of the Second International Winter Wheat Conference.* Zagreb, Yugoslavia. pp. 74–78.

Fenster, C. R., and G. A. Wicks. 1982. "Fallow Systems for Winter Wheat in Western Nebraska." *Agronomy Journal* 74: 9–13.

Fowler, D. B. 1981. "Fall Growth and Cold Acclimation of Winter Wheat and Rye on Saline Soils." *Canadian Journal of Plant Science* 61: 225–230.

———. 1982. "Date of Seeding, Fall Growth, and Winter Survival of Winter Wheat and Rye." *Agronomy Journal* 74: 1060–1063.

———, and L. V. Gusta. 1982. "Fall Growth and Cold Acclimation of Winter Wheat and Rye Differentially Fertilized with Phosphorus." *Agronomy Journal* 74: 539–540.

George, R. P., and S. C. Bull. 1979. "The Causes of Salt Land—A Case Study at Espeerance." *Journal of Agriculture of Western Australia* 20: 3.

Greb, B. W., and R. L. Zimdahl. 1980. "Ecofallow Comes of Age in the Central Great Plains." *Journal of Soil and Water Conservation* 35: 230–233.

Gusta, L. V., and D. B. Fowler. 1977. "Factors Affecting the Cold Survival of Winter Cereals." *Canadian Journal of Plant Science* 57: 213–219.

Halvorson, A. D., and A. L. Black. 1974. "Saline-Seep

Development in Dryland soils of Northeastern Montana." *Journal of Soil and Water Conservation* 29 (29): 77–81.

Hoefer, R. H., G. A. Wicks, and O. C. Burnside. 1981. "Grain Yields, Soil Water Storage, and Weed Growth in a Winter Wheat-Corn-Fallow Rotation." *Agronomy Journal* 73: 1066–1071.

Hunt, L. A. 1980a. "Winter Cereals in the World Scene." *Notes on Agriculture* 16 (1): 4–6.

———. 1980b. "Winter Rye." *Notes on Agriculture* 16 (1): 13–14.

International Rice Research Institute. 1979. *Report on a Rice Cold Tolerance Workshop*. Los Banos, Philippines. 139 pp.

Jenkins, G. 1970. "Winter Barley in Britain." *National Agricultural Advisory Service Quarterly Review* 89: 74–83.

Keim, D. L., and W. E. Kronstad. 1981. "Drought Response of Winter Wheat Cultivars Grown under Field Stress Conditions." *Crop Science* 21: 11–15.

Kern, K. R. 1980. "Probing the Gene Pools. Spring × Winter Crosses in Bread Wheat." *CIMMYT Today*, No. 12. 12 pp.

Kirkland, K. J., and G. H. Keys. 1981. "The Effect of Snow Trapping and Crop Sequence on Moisture Conservation and Utilization in West-Central Saskatchewan." *Canadian Journal of Plant Science* 61: 241–246.

Kronstad, W., et al. 1975. "Spring × Winter Crosses for Winter and Spring Wheat Improvement." *Proceedings of the Second International Wheat Conference*. Zagreb, Yugoslavia. pp. 105–107.

Loomis, R. S. 1976. "Agricultural Systems." *Scientific American* 235 (3): 99–105.

Lyons, J. M. 1973. "Chilling Injury in Plants." *Annual Review of Plant Physiology* 24: 445–446.

MacEwan, G. 1969. *Harvest of Bread*. Saskatoon: Prairie Books. 180 pp.

Mahon, J. D. 1983. "Limitations to the Use of Physiological Variability in Plant Breeding." *Canadian Journal of Plant Science* 63: 11–21.

McKersie, B. D. 1981. "Strength in Adversity: Winterhardiness in Cereals." *Highlights of Agricultural Research in Ontario*. 4 (4): 12–15.

McLeod, J. G. 1980. "Winter Cereals in Western Canada." *Notes on Agriculture* 16 (1): 15–17.

Metcalf, E. L. et al. 1970. "Relationship Between Crown Moisture Content and Killing Temperature for Three Wheat and Three Barley Cultivars." *Crop Science* 10: 362–365.

Meyer, J. 1983. "Ecofallow: Its Impact on Dryland Farming." *Crops and Soils* 35 (5): 17–18.

Mock, J. J., and S. A. Eberhart. 1972. "Cold Tolerance in Adapted Maize Populations." *Crop Science* 12: 466–468.

Mock, J. J., and W. H. Skrdla. 1978. "Evaluation of Maize Plant Introductions for Cold Tolerance." *Euphytica* 27: 27–32.

Morrison, K. J., et al. 1981. "Interrelation in Common Winter Wheat of Cold Hardiness and Emergence in Dry Soils." *Agronomy Journal* 73: 952–956.

Nicholaichuk, W., and D. W. L. Read. 1981. "Snow Management by Swathing at Alternate Heights." *Agriculture Canada* 26: 25–26.

Nordquist, P. T. 1971. "Use of Earliness and Cold Tolerance in Breeding Sorghums." *Proceedings of the Annual Corn and Sorghum Research Conference* 26: 33–41.

Pelton, W. L. 1969. "Influence of Low Seeding Rates on Wheat Yields in Southwestern Saskatchewan." *Canadian Journal of Plant Science* 49: 607–614.

———. 1982. "Tillage for Cereal Crop Production." *Canada Agriculture* 27 (1): 22–26.

Poysa, V. W. 1980. "Winter Triticale: The Developing Potential." *Notes on Agriculture* 16 (1): 15–17.

Rennie, D. A. 1979. "Intensive Cultivation—The Long Term Effects." *Agrologist* 8 (4): 20–21.

———. 1982. "The Deteriorating Soils of the Canadian Prairie." *Span* 25 (3): 99–101.

Rhykerd, C. L., C. F. Gross, and E. F. Sullivan. 1960. "Sorghum Grows Best at Warm Temperatures." *Crops and Soils* 12: 24.

Rosenberg, N. J. 1977. "Sown Windbreaks: Saving Water and Increasing Crop Production." *Span* 20 (1): 12–14.

Salmon, S. C. 1917a. "The Relation of Winter Temperature to the Distribution of Winter and Spring Grains in the United States." *Journal of the American Society of Agronomy* 9: 21–24.

———. 1917b. "Why Cereals Winterkill." *Journal of the American Society of Agronomy* 9: 353–380.

Single, M. V. 1975. "Frost Injury." (In) A. Lazenby and E. M. Matheson (eds.), *Australian Field Crops*. Vol. I. *Wheat and Other Temperate Cereals*. Sydney: Angus and Robertson Publishers. pp. 364–383.

Spratt, E. D., and S. L. Chowdhury. 1978. "Improved Cropping Systems for Rainfed Agriculture in India." *Field Crops Research* 1: 103–126.

Steponkus, P. L. 1978. "Cold Hardiness and Freezing Injury of Agronomic Crops." *Advances in Agronomy* 30: 51–98.

Stickler, F. C., A. W. Pauli, and A. J. Casady. 1962. "Comparative Responses of Kaoliangs and Other Grain Sorghum Types to Temperature." *Crop Science* 2: 136–139.

Stoffer, R. V., and G. E. V. Riper. 1963. "Effect of Soil Temperature and Soil Moisture on the Physiology of Sorghum." *Agronomy Journal* 55: 447–466.

Stoskopf, N. C. 1972. "Winter Hardiness of Spring × Winter Wheat Crosses." *Proceedings of a Colloquium on the Winter Hardiness of Cereals.* Martonvasar, Hungary. pp. 143–150.

Trione, E. J., and R. J. Metzger. 1970. "Wheat and Barley Vernalization in a Precise Temperature Gradient." *Crop Science* 10: 390–392.

Warnes, D. D., and V. A. Johnson. 1972a. "Crown-Freezing and Natural Survival Comparisons of F_2-F_4 Bulk Populations from 25 Winter Barley Crosses." *Crop Science* 12: 403–405.

———. 1972b. "Crown-Freezing and Natural Survival Comparisons in Winter Wheat." *Agronomy Journal* 64: 285–288.

Yakar, K., and F. Altay. 1980. "Winter × Spring Wheat Breeding in Turkey." *Proceedings of the Third International Wheat Conference.* Madrid, Spain. pp. 137–151.

8

Cereals as Forages, Livestock Feeds, and in Mixtures

INTRODUCTION

Perennial grasses and legumes generally are considered as the basis of forage feeding programs for ruminant animals. The reasons are broadly recognized:

1. Perennial forages can produce a high yield of palatable roughage with up to 20% protein content.
2. They are adapted to several cuts a year, tolerate livestock grazing, or can be harvested repeatedly as green crop each season.
3. Legumes can convert nearly inert nitrogen in the atmosphere to a form usable by the plant or subsequent crops.
4. Perennial forages have persistence, so that they do not have to be seeded annually.
5. They form a protective cover against erosion year-round and contribute to soil structure and stability.
6. They are an effective break crop in a rotation to curb plant-disease cycles.

Despite these reasons and the historical record of perennial forages, annual cereal crops also play major roles as a source of forage and as high-energy rations for many classes of livestock. Cereals are finding increasing use as forage and livestock feeds and under certain circumstances are displacing perennial forages from their historical role.

CHARACTERISTICS AND IMPORTANCE OF CEREAL GRAINS AS FEED CROPS

Cereal crops are basic food crops and have been developed primarily for grain production. Any cereal crop, however, can be used both as an annual fodder crop and as seed grain. From a forage-production viewpoint, cereal plants exhibit distinct characteristics, such as:

1. Comparatively few, large tillers with elongated growing points, resulting in a plant that does not recover very well from cutting or grazing, compared to perennial forage grasses.
2. Ease of establishment because of their large seeds compared to forage grasses and legumes. Seed can be sown into a rough seedbed prepared with zero- or minimum-tillage techniques. Planting depth is less critical than in perennial forages.
3. Inexpensive seed because of the high grain yield compared to that of forage species bred for vegetative, indeterminate growth.

From a forage-production viewpoint, cereal crops have the following features:

1. A range in variability among cereal crops that provides livestock producers with considerable choice and flexibility. Spring-seeded cereals such as corn, sorghum, oat, or barley generally produce a high yield from a single cut. Winter annuals such as rye, triticale, and wheat can be pastured in late autumn or early spring and in favored areas during the entire winter period. Fall pasturing of cereal crops can extend the fall grazing period and shorten the time required for winter feeding. Millet and sorghum can be seeded under hot, dry conditions. Barley is suitable for late seeding. A specific crop and management program can be put together to accommodate almost any situation.
2. A capacity to produce an abundance of high-quality forage during the hot part of the season in mid to late summer, when production of cool-season perennial forages slumps off. Ruminant livestock operations can benefit from the relatively high protein yields and dry-matter digestibility that can be produced by cereal crops during this period.
3. Rapid establishment and production. Every year a significant amount of perennial forage crops are lost by winter damage or other reasons. Needs arising from unforeseen loss of regular hay or pasture fields or from changes in weather or livestock numbers can be met by supplementary cereal crops. Most cereal crops can be seeded throughout the growing season with minimal risk of failure and can be used as forage four to six weeks after emergence.
4. Flexibility in use. Cereals can be used for pasture, for green chop brought to the feedlot, for hay, for silage, or may be allowed to mature as a grain crop. Choices of crop and cultivar depend on soil type, soil-erosion problems, the time when feed is needed, the type of livestock, intended grazing management, and availability and price of seed.
5. The high total dry-matter yield and high energy value produced by cereals such as corn and sorghum means that they may be part of a planned program and not only a supplementary program. Year-round containment of animals fed from stored feed has resulted in a major shift to cereal crops, notably corn, as a forage crop.

Traditionally, corn, oat, and barley have been regarded as feed grains. They are produced extensively as such and can be fed with relative safety in volume to livestock to meet the energy requirements for rapid growth, high productivity of milk or other products, or animal maintenance.

An important criterion for a feed grain is the ability to produce high levels of energy per unit area. Food crops such as wheat have been developed carefully to preserve desirable food-quality aspects, such as strong gluten and high protein for bread making or soft gluten and low protein for pastry use. Modern civilization consumes wheat mainly as food rather than as animal feed. Economic principles dictate, however, that when wheat is available and comparable in cost to other sources of nutrition for livestock, it be used for feed. If quality is reduced by environmental conditions, wheat unsuited for bread can be a valuable livestock feed (Anonymous, 1970).

In the late 1960s, some traditional food grains such as wheat, rye, and triticale began to be regarded as potential feed grains. The historical view of cereal crops as food crops began to change for the following reasons:

1. Evidence that yields could be increased either by plant breeding or through better management practices if traditional food-quality factors could be disregarded. This fact was demonstrated by new levels of wheat and rice yields produced by cultivars associated with the green revolution of the 1960s. One wheat cultivar, Pitic 62, in 1969 was licensed for sale in Canada, a country noted for high-quality bread and durum wheats. Pitic 62 was licensed as a feed grain and was eligible only for a low grade, regardless of the actual quality of the sample. A low grade was associated with a low price, and this opened the door for wheat as a feed crop. The lower price was offset usually by higher yields. Breeders may produce a yield advantage of 10% to 20% by emphasizing yield without attention to quality aspects. More net profit per unit area of land may be realized from feed or utility wheat grades than from quality bread wheats. The market share of feed or utility wheat increased as the price spread attracted buyers who blended low-grade with high-quality, more expensive bread wheat. Feed wheat grades,

therefore, provided breeders with an opportunity to exploit more fully the yield potential of some crops.

2. The possibility that substantial yield increase could be achieved with F_1 hybrids (Chapter 9) that were produced by crossing unrelated parents but deviated from traditional quality aspects. F_1 hybrids represented one method of approaching the record yields reported for various crops. Record yields, in themselves, are not very important except to show the potentials within present resources. The goal is not simply to match or exceed these high yields occasionally, but to stabilize production much above the average. This may be achieved by breeding efficient, responsive cultivars that remain healthy and free from pests all season, and by developing soil moisture and fertility management to approach as closely as possible the optimum conditions required. These aspects may be achieved if crop breeders and producers are not constrained by quality factors.

3. The discovery that some cultivars of triticale have an extremely high nutritive value that stimulates rates of gain in growing animals. The nutritive value was higher than in any other cereal tested (Table 8-1).

4. Unusually strong international demands for feed grain, which have developed new and potential markets and opened the door for cultivars designed for special needs and conditions. In countries where surplus grain stocks exist and where producers have difficulty marketing wheat, four markets can be developed: domestic food, export, feed, and industrial. If producers are to be given an opportunity to express their capacity to produce, increasing amounts will have to be marketed as livestock and poultry feed.

Table 8–1
Comparisons of Protein Efficiency Ratios (PER), an Indicator of Nutritive Value, among Three Cereal Crop Species

	Range in PER	
Cultivars of	*Low*	*High*
Triticale	0.07	3.88
Wheat	0.76	2.00
Rye	2.07	2.99

Source: Data from F. C. Elliott, Michigan State University, East Lansing, and reported by R. C. McGinnis (1970).

5. The development and potential of multilines or blends. By this means, secondary spread of diseases can be restricted, and yield stability is increased. The development of several cultivars phenotypically similar but differing in disease reaction may be an important step. Quality aspects may be lacking if time and costs do not allow them to be fully considered.

6. The potential of perennial rye, which has opened new possibilities for this crop as feed. Perennial habit can be obtained by crossing common fall rye cultivars with wild rye species. McGinnis (1970) reported that for six consecutive years, perennial rye overwintered without any obvious loss in vigor and seed set. Perennial rye offers promise as a dual-purpose crop for both grazing and grain.

7. A demand for specific traits, such as resistance to deer grazing for crops to be sown near woodlots, or high-protein cereals from traditional low-protein cultivars for breakfast foods or specialty products. Dwarf stature has become important for lodging resistance and for more efficient distribution of photosynthate. The image of a bald wheat head in favor of one bearded to increase photosynthetic area and to support grain in the swath is a modern innovation.

8. An increase in livestock and poultry numbers. In some areas, animal numbers increased faster than feed-grain supplies. The use of traditional food grains such as wheat for livestock feed is widespread in countries where surplus production is a problem. The amount of wheat that is fed to livestock varies widely according to the food demands on the crop, and, although it is a relatively small portion of the total crop, the volume is considerable. The

countries where wheat as a livestock feed is widespread are Australia, Canada, the United States, and Western Europe.

9. Recognition of the value as grain as a feed. Corn is the highest-energy grain for livestock feeding, but the higher protein content of wheat makes the two grains about equal in value as livestock feed. Wheat is an excellent feed grain for all classes of livestock and poultry and can make up all the grain in a ration. Barley and oat have a lower protein than wheat and may contain a hull high in fiber that dilutes the high energy of the kernel, making them a safe feed. Oat grain has a higher fat content than barley.

CEREAL BY-PRODUCTS AS FEED

To assess the total impact of cereals grains as livestock feed, by-products of cereal processing must be taken into account. Milling of cereal grains involves the separation of the various parts of the kernel. In the milling of wheat for white flour, the most nutritious parts of the kernel became feed and include the bran, germ, middlings, and shorts. These products are produced by removal of the outer layers of the kernel to expose the starchy endosperm, which is ground into flour.

Feed by-products derived directly or indirectly from barley include barley pearlings or **bran,** malt culms, brewers' and distillers' spent grains, residues, and dust from cleaning.

In the production of **malt,** the sprouts and rootlets are removed from the kiln-dried malt and are used in annual feeds. These products are known as malt culms, coombs, and cummins. Many of these valuable feed products would go to waste if livestock were not available to consume them.

Malt and grain used to produce alcoholic beverages have a high proportion of the soluble sugars and starch removed, and the residues are high in fiber, protein, fat, and ash. Brewers' and distillers' spent grains may be fed as a wet mash (70% to 80% moisture) or dried (10% moisture). Wet spent grains must be fed rapidly to avoid deterioration. Spent grains, spent hops, and surplus yeast from breweries are valuable as feed for ruminants.

Feed materials are produced from the wet milling process of corn, in which corn starch and oil are extracted. The feed products are mainly protein and fiber and amount to about 33% of the original corn dry weight. Corn gluten meal contains about 60% protein and, if yellow corn is used, contains **xanthophyll,** a pigment source in feeds. Straw residues may be chemically treated and used as a ruminant feed.

HIGH-MOISTURE GRAIN FOR FEED

Grain for livestock feeding may be stored and fed dry after being ground to increase digestibility. Grain also may be stored wet under anaerobic conditions in a silo and fed wet as it is removed. The feed value of wet stored grain is maintained.

Crops that are produced exclusively for feed grain for on-farm use can be handled efficiently as **high-moisture grain** in a semiensiled state. Some of the advantages include:

1. Energy conservation, as the need for energy inputs for drying full-season crops such as corn and sorghum is eliminated.
2. An opportunity to harvest corn for high-moisture grain two to three weeks earlier, thereby reducing the risk of crop loss. Kernel moisture for storage in a silo as shelled grain corn or for kernels of other crops such as barley should be in the 25% to 30% range, a level too high for bin storage without supplementary drying.
3. A reduction in losses, as field shattering is eliminated, crop lodging may be reduced, and storage losses are minimized. Field losses from birds, insects, and weather are reduced.
4. A reduction in handling operations, as grain is ready to feed when it comes from storage.
5. An increase in combine harvesting efficiency, especially with small cereals, as **swathing** can be eliminated and only the heads need be harvested, which reduces the volume of material passing through the combine harvester (Figure 8-1).
6. The removal of early-developing crops, such as winter barley, as high-moisture grain may allow for a second crop to be seeded and a double-cropping system developed.

Figure 8-1. Cereal grains can be direct combine harvested as high-moisture grain for storage in a silo in a semiensiled state. (Courtesy of Ontario Ministry of Agriculture and Food)

As a feed for most classes of livestock, high-moisture grain has been found to be fully as satisfactory as, and in some cases better than, dry grain when expressed on a dry-matter basis. High-moisture grain sorghum is superior to dry sorghum as a livestock feed.

Whole grain fed to livestock is not masticated thoroughly, and as a result an undigested portion is passed intact into the feces. Both dry and wet grain should be ground before feeding. High-moisture grain can be processed with roller mills fitted with scrapers to prevent the moist particles from building up on the rollers.

If high-moisture grain is to be used in a feeding system, it must be recognized that rapid spoilage of unconsumed grain may occur and that it must be fed as removed from the silo. High-moisture grain usually is fed on the farm where it is produced; because of the potential for spoilage, it cannot be marketed easily, and then only as a feed product.

A variation to the concentrated high-energy feed value of cereal grains is the incorporation of some roughage for ruminant feed. In corn, this can be achieved by chopping the entire ear to produce corncob meal (Figure 8-2). For whole-plant corn silage, the amount of roughage can be influenced by regulating the height of cut when harvesting. Cutting height may vary from ground level by 40 cm (16 in.) in the case of short grain sorghum to 90 cm (36 in.) in corn but always including the ear. Sorghum cut midway on the stalk produces a 60% to 65% grain mixture that is excellent for beef feeding programs. Dead bottom leaves and much of the fibrous stalk are left in the field.

Feed-composition comparisons of various cereal crop products are shown in Appendix D.

The main function of grain feeds is to provide energy, but because of their high inclusion rates, cereal grains also provide considerable quantities of protein and essential amino acids. Lawrence (1978) reported that the total intake of protein from cereals amount to 59%, 36%, and 12% of the intake of pigs, poultry, and ruminants, respectively.

Considerable research has been directed to providing balanced rations to obtain highest rates of gain or production. Basically, however, any of the cereal grains can be fed safely and satisfactorily to all classes of livestock.

Figure 8-2. Corncob meal consists of the entire ear, which is chopped and ensiled to produce a livestock feed. (Courtesy of Ontario Ministry of Agriculture and Food)

MYCOTOXINS IN FEED GRAINS

When grain is stored under damp and warm conditions, mold may develop on its surface, making it unfit or undesirable for food purposes. Molds (fungi) produce substances known as *mycotoxins*, which are toxic to other organisms. Different toxins are produced by different fungi under different conditions. Toxins differ in their effects on animals, some appearing to have few or no effects, others appearing to have very serious ones. The *aflatoxin* group of mycotoxins is potentially harmful in reducing feed intake and may be a threat to animal health and efficient production. The mold *Aspergillus flavus* is capable of producing aflatoxins when grain has been stored too wet, but it may also develop in the field prior to harvest under certain environmental conditions.

If cereal grains show any evidence of mold, they should be checked for mycotoxins before being used. If aflatoxin is present, the grain should be fed accordingly; it may be diluted by mixing it with aflatoxin-free grain, or the contaminated grain may be detoxified.

In swine, a disease known as *estrogenic syndrome* can be found on grain corn produced by the fungus *Fusarium roseum* (also called *Giberella zeae* or *Fusarium graminearium*).

When toxin production on various substrates was examined, corn grain was found to be one of the most suitable media. The greatest amount of toxin production takes place when a two- to three-week period of warm weather (24°–27°C or 75°–80°F) is followed by a period of constantly or intermittently lower temperatures (13°–16°C or 55°–60°F) along with high relative humidities. Toxin production is encouraged at a grain moisture of 45% to 50%, often found in a hybrid that is too late for the area in which it is grown. Aflatoxin problems are common in the corn-growing regions of the northern United States and Canada, Europe, and the United Kingdom.

Aflatoxins do not develop in the absence of air, which means that they will not be produced on properly ensiled high-moisture grain corn. The extent to which aflatoxins may develop in cribbed corn is not clear, and the fungus is designated as a "field pathogen," as opposed to "storage mold" which grows at lower moisture levels.

A mycotoxin called *vomitoxin* resulted in the grading of considerable quantities of wheat as unfit for food purposes in the early 1980s. In a detailed profile of *Fusarium graminearium* in wheat and corn, Sutton (1982) traced outbreaks of this fungus disease to weather, soil, insects, birds, and the host at various stages of the life cycle of the pathogen.

ERGOT IN FEED GRAINS

Ergot is a fungus disease of the grasses in which the floret is replaced by an ergot body or sclerotia, which contains alkaloids that may cause toxic effects when consumed. Ergot is common in rye but is also found in other cereals (Figure 8-3), including wheat, barley, and triticale, and in other grasses. Ergot is common at northern latitudes.

If grain contains more than 0.1% ergot (one gram per thousand grams of grain), it should be considered dangerous for livestock feeding. The danger level is specified by weight, as ergot bodies vary in size. In wheat, the ergot body usually is the same size as the normal kernel; in rye and barley, often much larger. Expressed in other terms, 0.1% by weight is 42 ergot bodies in a liter of wheat (47 per quart) and 34 ergot bodies in a liter of barley (38 per quart). Feed grades of cereals may contain up to 0.24%.

When ergot-contaminated feed is fed to livestock, some of the following effects may be observed: behavioral changes such as convulsions, muscle incoordination, lameness, difficult breathing, salivation, and diarrhea; dry gangrene and loss of extremities such as hooves, tails, or ear tips; abortion; altered reproduction rate and reduced lactation; feeding effects such as unpalatability, reduced feed intake and weight gain, and increased feed/gain rates; and death (Young, 1980).

Ergot levels can be reduced by blending infected with uninfected grain, but ergot infected grain cannot be fed to pregnant animals and cannot be recommended for feed under any

Figure 8–3. Ergot in barley seed. Ergot is a fungus disease that replaces the kernel with a purple-to-black-colored sclerotium which contains alkaloids that are toxic if consumed. (Courtesy of Ontario Ministry of Agriculture and Food)

circumstances, blended or otherwise. Often ergot can be removed by screening, but ergot bodies that are similar in size and density to grain kernels cannot be removed easily.

RYE GRAIN PALATABILITY

In addition to rye's susceptibility to ergot, a palatability problem has been observed when rye grain comprises a large portion of the ration. Animals fed full rye rations are reported to have reduced rates of gain and a reduced feed intake (Friend and MacIntyre, 1969; Friend, 1970; Smith and MacIntyre, 1969).

In a thorough study of this problem, Wieringa (1967) identified *alkylresorcinol* as the cause of decreased food intake and reduced growth rates. The compound was found to be concentrated in the oil fraction, and if rye oil was fed, the resulting growth depression was comparable to that produced by an equivalent amount of rye. The growth of pigs on rations containing 50% rye oil was 11% to 12% lower than on a barley ration.

In general, pigs above 50 kg (110 lb) in weight could be fed a ration with 50% rye, and poultry 20%, without measurable effects. From nearly all feeding experiments showing the adverse effect of rye, it can be concluded that the diminished growth rate is related primarily to a decrease in the intake, for which the bitter taste of rye is held responsible. Rye is believed to also cause a sticky mass in the mouths of animals.

Evans, Dedio, and Hill (1973) screened 397 spring and winter rye cultivars and found that the alkylresorcinol content ranged from a low of 0.036% to a high of 0.322%. If alkylrescorcinol is the "growth-inhibiting factor" of rye, selection may be effective in decreasing its content.

TRITICALE AS A FEED GRAIN

Interest in triticale as a feed has been generated because of a high protein level and improved amino acid composition as compared to other cereal grains. With good management practices, triticale can be used satisfactorily as a feed grain. It may be considered competitive to wheat, corn, grain sorghum, and barley as a source of energy (Allee, 1974).

In some feeding trials, the overall intake of triticale by animals has been lower than with other grains. Sometimes a loss of appetite has been observed in swine fed large quantities. Less than optimum performance may be attributed to a low level of ergot.

Generally triticale can replace other grains without altering performance appreciably in poultry, swine, and ruminants. With chicks and broilers, the improved amino acid composition helped meet their total protein requirement.

Triticale has considerable potential as an animal feed, either as grain or as forage, in livestock-producing areas of the world.

A PERSPECTIVE ON CEREAL GRAINS AS FEED

Cereals are the staple foods of most of the world's population, and the rapid increase in the use of cereal grains for livestock feeding since 1950 has raised questions about the wider implications for world food supplies. Concern exists that the use of basic foods to produce animal products is too inefficient and detracts from the food supply in poor countries. In the United Kingdom, animals consume three times as much metabolizable energy as humans and provide only one-third of the consumer's energy intake (Lawrence, 1978).

If a moral issue exists about feeding animals on products that could be consumed by humans, the counterargument is economic. Some nations are too poor to purchase grain and, unless they are given food as aid, their needs may not be met. Others can afford expensive animal products and see livestock as a means of marketing their surplus grain as feed. Under such circumstances, the economic argument prevails but the conflict is not resolved.

A new dimension to the feed-versus-food debate has been developed through an awareness of energy inputs. In North America, about twice as much fossil fuel is expended to provide the consumer with a given quantity of food energy through animal products as through plant products. A large part of this difference is generated in the off-farm stages of the food system in terms of transportation, processing, and packaging. At the farm production stage, much obviously depends on the relative importance of grassland and energy-intensive feeding systems. Energy and associated costs have raised the debate about ruminant animals, which can convert waste and by-products to food, versus nonruminants, which compete with humans for the same food products. When viewed as harvesters, gleaners, and consumers of products that might otherwise go to waste, some classes of animals such as ruminants add to world food supplies.

A difficult question is the maximum global population that food and feed suppliers ultimately will be capable of nourishing adequately. Generally, agriculturists are optimistic about technological progress in meeting food and feed supplies. Nevertheless, it is possible to envision a future in which land resources, population numbers, and economic resources have been so mismanaged as to make it imperative to use every precious plant product in the most economical manner. Even under these circumstances, it is difficult to conceive of a world in which ruminants do not have a vital role.

The release of cereals from livestock feeding in developed countries was regarded by Abercrombie (1982) as unlikely to be of much help to undernourished people in developing countries. No direct link between under- and overnutrition appears to exist, except in the special circumstances of a short-term world food crisis. A more direct relationship may be found within an individual developing country.

To complete the picture, the amount of grain fed is an important consideration. Of the total world consumption of cereals in 1978, 36% was fed to livestock (FAO estimate reported by Abercrombie, 1982). This figure becomes 43% if rice, of which less than 1% is fed, is excluded. Corn is the largest feed crop, and in 1978, 61% of world consumption was fed to livestock. Barley is the second largest feed crop, and 68% of world consumption went into livestock feeds, with most of the remainder used in the brewing industry. As high as 17% of the wheat consumption is used for feed.

Feeding cereals to livestock occurs predominantly in developed countries and was estimated in 1978 to amount to 56% of cereal consumption, compared to 13% in developing countries (Abercrombie, 1982). Cereals fed to livestock originate almost entirely from domestic stocks or from other developed countries. In many developed countries, the volume of cereals destined for livestock feeding forms a reservoir about twice the size of conventional carry-over stocks. Part of this could be diverted to food used in an emergency and provides a source of security from human starvation in developed nations, should a major crop failure occur.

WINTER CEREALS AS FORAGE CROPS

The practice of grazing fall-sown winter cereals has been widely known and extensively conducted in many parts of the world. Late fall grazing is common where the risk of winterkill is low or when fall growth is excessive. In the

United Kingdom, grazing is common in the early spring months.

In the southern Great Plains region of the United States, when warm-season perennial forages are dormant in the cool winter months, the winters are sufficiently mild to permit growth of cool-loving cereals. Winter cereals produce a highly palatable and succulent growth and are grazed extensively in this region. Usually livestock are removed sufficiently early for the crop to develop and to be harvested for grain. Under conditions where moisture for maturing of the crop is limited, or where the value for forage exceeds that for grain, the crop may be pastured until it matures (Atkins et al., 1970).

Forage produced from winter cereals is of excellent quality, having a protein content of 20% to 30%, and cattle gains are usually quite high, often exceeding 1 kg/day (2.2 lb/day) (Mathers, Stewart, and Grunes, 1982). Young, leafy cereal green feed has a digestibility of 80% on a dry-matter basis, when dry matter is between 14% and 18%. As the crop becomes reproductive, quality declines as a result of fiber increase and stem elongation. Cereals infected with diseases such as rust, mildew, or yellow dwarf virus may reduce the quality of the green feed.

Under irrigation and in high-rainfall areas, the value of the crop for winter pasture may equal or exceed the value of the grain crop. If allowed to develop grain, the economic return for winter cereals includes both the value of the grain and the value of the animal products produced. If grain production is not desired, hybrid wheats that lack fertility restoration (Chapter 9) and that have an extended vigorous vegetative growth period because they cannot set such seed are preferred as grazing types. Male sterile wheat does not give a producer the flexibility, however, that fertile wheat offers.

Winter cereals grown for winter pasture or for pasture and grain production should grow four to six weeks before livestock are permitted to graze. Close or early grazing, overstocking, severe tramping when the soil is soft, or grazing in the spring after the spike has started to develop may reduce grain yields. Early-maturing cultivars are more frequently injured by grazing than late-maturing ones. In this respect, triticale cultivars often develop after wheat and are preferred for grazing.

Hazards to winter grazing also exist. Grass tetany and frothy bloat are observed frequently among cattle grazing succulent winter cereals. Grass tetany or hypomagnesemia is caused by low levels of magnesium in the animal's blood due to low absorption. Incidents of grass tetany are high for grazing ruminants when temperatures fluctuate. Losses of livestock grazing winter cereals may reach and exceed 5% in certain years (Mathers, Stewart, and Grunes, 1982).

When grain yields of various cereal crops have been measured following clipping or grazing, results have varied widely with location and year. Dunphy, McDaniel, and Holt (1982) observed that responses ranged from a decrease in grain yield, to no effect, or to a marked increase in grain yield. Holliday (1956), in a review of the effect of forage utilization on the grain yield of cereals, summarized the results of 35 different studies; 24 of these reported a yield decrease; 5 a yield increase attributed to forage removal; and the remainder no effect. The degree of yield reduction was associated with the intensity of grazing and ranged in magnitude from 3.2% to 87%. The experimental evidence reviewed by Holliday allowed Aldrich (1959) to formulate the following general conclusions:

1. Increases in grain yield normally occur only when soil fertility conditions are so good that lodging would take place if the crop was not grazed. Grazing results in less vegetative growth, which allows for better light penetration and associated cell differentiation of the lower internodes.
2. Under conditions of average or low soil fertility, losses in yield may be expected to result from grazing.
3. The later in the season grazing takes place, the more severe are the yield reductions from grazing. Grazing when the growing point has elongated sufficiently to become susceptible to damage can be extremely harmful to grain yield.

Removal of vegetative growth on two winter wheat cultivars in Texas indicated that forage utilization should be terminated by the early joint stage or earlier to obtain satisfactory grain yields (Dunphy, McDaniel, and Holt,

1982). Delaying the final forage harvest generally resulted in reduced tiller survival and fewer seeds per head but had little effect on the average weight per seed. Grain-yield reductions due to forage utilization ranged from 4% to 84% (Table 8-2).

Grazing of winter rye resulted in a less severe yield reduction than with winter oat (Table 8-3) under conditions in New Zealand (Scott, 1978) when several studies were averaged. The reasons for these differences were suggested by Scott (1978) to lie in the different growth habits of the two species. Oat has a capacity for early and rapid elongation of vegetative tillering. Rye has the capacity to produce more regrowth as dead tillers are replaced.

In the dryland region of southwestern Saskatchewan fall rye grain yields were reduced when grazed. Spring grazing reduced grain yield by 10%, fall grazing by 17%, and grazing in both fall and spring by 25% (Kilcher, 1982). It is not unreasonable to expect this level of reduction on very sandy soil, especially when precipitation is average or below. Nevertheless, the positive value of grazing may be of sufficient magnitude to more than offset the 10% to 25% loss in grain yield.

Winter cereals may be spring-seeded if production is exclusively for pasture. When planted in the spring, winter cereals remain vegetative without a vernalization treatment. Unvernalized winter wheat is not noted for its summer growth, as it suffers from attack by various leaf diseases. Preliminary evidence with winter triticale, however, indicates that the regrowth can be harvested up to three times for silage or green chop. Yields of tall winter triticale cultivars have produced very high forage yields of up to 5 tonnes/ha per cut (Anonymous, 1983) with crude protein levels near 25% under the cool conditions of 42° N latitude.

In areas where winter triticale can be grown, the crop can be planted early in the fall and used as a late fall pasture. Winter triticale can be handled as a silage crop and ensiled as whole-plant material in the dough stage.

Table 8–2
Effect of Forage Removal at Different Stages of Plant Development on the Forage and Grain Yield of Winter Wheat: Values Averaged over the Three-Year Period 1977 to 1979 for Two Cultivars Grown in Texas

	Forage Yield		Grain Yield	
Treatment[a]	kg/ha	lb/ac	kg/ha	lb/ac
Early joint	1778	1588	2097	1872
Mid joint	2078	1855	1596	1425
Late joint	2498	2230	1182	1055
Not cut	—	—	2557	2283

Source: Dunphy, McDaniel, and Holt (1982).
[a]Jointing is described as the stage when the growing point begins to elevate above ground level. Growing points were not removed by any of these cutting treatments.

Table 8–3
The Effect of Midwinter Grazing on the Yield of Winter Oat and Rye Crops in New Zealand Averaged over Several Studies

	Dry-Matter Yield			
	Ungrazed		Grazed	
Crop	kg/ha	lb/ac	kg/ha	lb/ac
Rye	2366	2112	2034	1816
Oat	3277	2926	1866	1666

Source: Scott (1978).

CORN FOR SILAGE

Corn is probably the most versatile crop available for supplementary feed production and is grown extensively as a forage crop. In its major use as whole-plant silage the entire plant is chopped and ensiled (Figure 8-4) as a stored feed for ruminant animals. Corn also can be planted early or late to produce silage green chop or pasture. Corn intended for silage or grain may be harvested for green chop but where possible should be left for silage and feed production as it reaches maturity. Corn for green chop should be regarded as a form of insurance in an emergency situation.

Figure 8–4. Corn is used extensively as a forage crop as a feed for ruminant animals. The entire plant is chopped and ensiled as a stored feed using a forage harvester and self-unloading wagons. (Courtesy of Ontario Ministry of Agriculture and Food)

Corn is without equal as a silage crop. As whole-plant silage, it can produce as much total dry matter in one cutting as can several cuts of the leading perennial forage crops. Yields of 13 to 20 tonnes of dry matter per hectare (6 to 9 tons/ac) are common.

Corn is readily ensiled because of an abundance of fermentable carbohydrate in the ears and stalks. Corn silage mass has a low buffering capacity, which means it has a low resistance to reduction in acidity. If properly ensiled, desirable lactic acid is produced; the acidity drops to a *p*H of 3.8 to 4.0 required for storage. Once preserved, silage can be stored for long periods.

At an average moisture level of 65% to 70% for the entire plant, corn can be ensiled readily, and at this moisture level maximum dry-matter yields are produced. A moisture level of 65% to 70% occurs two to three weeks prior to the attainment of maximum grain dry weight. If ensiled at a moisture level above 65% to 70%, the wet mass excludes oxygen so effectively that a temperature rise associated with aerobic fermentation and required for lactic acid production is limited. Acetic acid is produced, which ionizes too slowly, and undesirable butyric acid is produced. Butyric acid does not lower the acidity sufficiently, and undesirable bacteria that break down protein occur. A putrid-smelling silage is produced. When corn is ensiled too wet, losses occur from seepage. If ensiled too dry, too much oxygen is trapped, and excessive aerobic respiration results in excessively high temperatures, which cause caramelization and possibly combustion. Mold flourishes under such conditions, and the silage is unpalatable.

At 65% whole-plant moisture level corresponds to an ear moisture of approximately 48% and a grain moisture of 44% (Table 8-4). At this stage, kernels are well dented and are doughlike in texture. Allowing corn to be frosted before ensiling reduces the moisture content but causes a loss of green color and vitamin A content.

It is often assumed that the best grain-corn hybrids are usually the best for silage as well (Hunter, 1978). This is based on the recognition that 50% of the total plant weight is contained in the ear and that over 70% of the feed value comes from the ear. In animal feeding trials it has been concluded that corn silage should not be considered as a roughage, but rather as a mixture of a roughage and grain (Huffman and Duncan, 1954). Proponents of the theory that the sink (the ear) rather than the source (the photosynethic area) was the limiting factor argued the necessity of a large ear.

Hybrid recommendations for grain and silage in many geographical regions were based on one list of hybrids recommended for grain. A hybrid suitable for silage differs only in relative maturity. Hybrids that are about two weeks later in maturity than if grown for grain are considered satisfactory for silage production.

Based on evidence in the literature, Hunter (1978) suggested that corn breeders should select forage genotypes from corn that produces a reasonably high proportion of final dry matter as grain but fulfills three basic objectives for forage-corn production:

1. High dry-matter yield per unit area of land.
2. Sufficient dry-matter content to ensure proper fermentation and high animal dry-matter intake.
3. High feeding value.

Table 8–4
Relationships Between the Moisture Percentage and Dry-Matter Yield of Various Plant Components Averaged over Eight Corn Hybrids and Measured on Eight Dates in 1970 and 1971 at Guelph, Canada

	Moisture Percent			*Yield in Percent of Maximum*	
Date	*Grain*	*Ear*	*Whole Plant*	*Grain*	*Whole Plant*
9/4	60	62	77	59	91
9/10	55	58	75	68	94
9/14	50	54	71	77	97
9/18	45	49	66	86	99
9/26	40	44	60	93	100
10/4	35	40	54	98	100
10/10	30	36	48	100	97
10/18	25	31	42	99	94

Source: Daynard and Hunter (1975).

Establishment of silage-corn breeding programs must be accomplished by testing programs. Silage-corn trials may include measurements on whole-plant dry-matter yield; moisture at harvest; digestibility; and crude protein content.

For maximum silage yield and when using hybrids developed for grain, cultural practices are nearly identical to those for grain production except that plant populations may be moderately higher. High plant populations increase plant stress. An increase in plant population of 1.0% to 20% is possible for silage-corn production in areas where the resulting increases in evapotranspiration rates are not likely to cause drought stress.

If corn is to be used for pasture or green chop, it can be sown in rows 18 to 36 cm (7 to 14 in.) apart at rates of 56 to 112 kg/ha (50 to 100 lb/ac). For pasture or green-chop purposes, F_2 seed from the feed bin, rather than more expensive F_1 seed (see Chapter 9), can be used for seed, provided germination is satisfactory. High plant populations can be achieved with conventional corn-planting equipment by double or triple passes back and forth over the field. Fertility levels should be comparable to those for crops grown for silage or grain. Harvesting can begin when the plants are 90 cm (36 in.) tall. Only as much should be cut as livestock will consume in a couple of hours. If too much is cut and heating occurs, palatability decreases. Tramping waste is high if corn is pastured, as plants can be broken over. If cattle are allowed to pasture corn, the area should be divided into sections and rotationally grazed. There is no regrowth from corn once harvested. Corn probably is the highest producer of green chop or pasture among the annual forages where it is adapted

SORGHUM

Several cultivars in the *Sorghum* genus can be used for forage and include Sudan grass, Sudan grass hybrids, and sorghum × Sudan grass hybrids. These hybrids came into use in the mid 1960s.

Sorghum can be classed into four types:

1. Forage.
2. Grain.
3. Grass (Sudan grass).
4. Broomcorn.

Forage sorghum (Figure 8-5) is tall (3.6 m or 11.8 ft) and is capable of producing heavy yields of forage or silage but relatively low grain yields. The stalks are sweet and juicy and provide an abundance of fermentable sugars for the production of lactic acid in silage. Highest silage yields are obtained, however, when the grain is in the stiff-dough stage.

Grain sorghum has relatively short stalks (0.6 m or 2 ft) (Figure 8-5) that are not sweet and are low producers of forage and relatively high producers of grain.

Sudan grass (Figures 8-5 and 8-6) and sorghum-Sudan grass hybrids tiller abundantly, have finer stems than forage sorghum, and produce an abundance of grassy leaves. Sorghum-Sudan grass crosses are taller than Sudan grass, have thicker stems, and can yield more than Sudan grass but often produce more harmful **prussic acid.**

Sudan grass differs from Johnson grass, a serious weed in the southern United States, by the fact that Sudan grass has no rootstocks that cause it to spread. Seed of Sudan grass is larger than Johnson grass seed.

Sorghum cultivars such as Sudan grass, forage, and grain sorghum hybridize readily. Hybrids are vigorous, high-yielding, and exhibit considerable disease resistance. Commercial production of F_1 hybrids was made possible through the use of **cytoplasmic male sterility** (Chapter 9).

Figure 8–6. Sudan grass and sorghum-Sudan grass hybrids shown in this photo are an important forage species for livestock feed. Sorghum-Sudan grass hybrids tiller profusely, are tall in stature, and have an abundance of grassy leaves. (Courtesy of Ontario Ministry of Agriculture and Food)

Figure 8–5. Forage sorghum, characterized by tall stature, coarse stems, and a seed panicle, is shown on the left. Dwarf grain sorghum with sturdy, lodging-resistant stems is shown in the middle. On the right is grass sorghum, also known as Sudan grass and characterized by tall leafy growth. (Courtesy of Ontario Ministry of Agriculture and Food)

Sorghums are well adapted to green chop and can be used as an emergency pasture, but difficulties in mechanical handling and curing of the crop restricts its use as hay. Forage sorghums can produce high-quality forage in mid to late summer, when cool-season perennial forages have low production, and therefore can play an important role in a summer pasture program. Dairy and some beef operations can benefit from the relatively high protein levels and dry-matter digestibility of these crops, under careful management systems, compared with perennial forages in summer (Reich and Atkins, 1970).

Sorghums are characterized by slow, early growth, followed by a surprising growth rate after reaching a height of nearly a meter (36 in.). Sorghum is a warm-weather crop that cannot be planted until the soil is warm and is well adapted to late planting after a previous crop is removed. The crop is unique in that it can remain dormant during stress periods and renew growth when conditions improve. Under rainfall conditions that restrict corn growth, forage sorghum will produce more silage and total digestible nutrients than corn, because sorghums are drought-tolerant. Under conditions suitable for corn production, forage sorghum may outyield corn but may produce less total digestible nutrients per unit area. Sorghum responds as well or better than corn to favorable moisture conditions. Forage sorghum for silage in livestock-producing areas is an excellent form of insurance against dry weather.

Forage sorghums have a similar range of adaptation to grain sorghum. Cultivar selection according to the length of the growing season is essential. Full-season cultivars outproduce shorter-season cultivars, but if the cultivar is too late, it may not reach a 65% moisture level to produce good silage before a killing frost.

Forage sorghums remain green and retain a moisture level optimum for silage over a longer period than corn. Corn is in the proper stage for ensiling for a ten- to 14-day period; sorghum remains in the proper stage for 20 to 30 days but at high latitudes may become frozen if not ensiled as soon as it is ready.

Forage sorghums are considered by many as the standard annual forage. Yet they occupy an area less than one-tenth that of silage corn even in areas where both are fully adapted (Pedersen et al., 1982). Sorghum is often relegated to land that is marginally or poorly suited for growing corn because of limited rainfall or frequent drought.

The large area seeded to silage corn may be due to a number of factors but primarily to the lower feeding value of forage sorghum (Schmid et al., 1976). The reasons for a lower feeding value have not been clearly defined, but Pedersen et al. (1982) tested 49 F_1 forage-sorghum hybrids and found that in-vitro dry-matter disappearance was the only trait that varied among hybrids. Tannins, prussic or hydrocyanic acid content, relatively low animal intake, digestibility and dry-matter content, and a high cell-wall content may be associated factors for the lower feed value or usage.

Precautions must be taken in feeding Sudan grass and sorghum-Sudan grass hybrids to animals to prevent losses due to prussic acid or hydrocyanic acid poisoning. All known Sudan grasses and sorghums contain adhurrin, a cyanogenetic glucoside, that yields hydrocyanic acid when hydrolyzed enzymatically in disrupted plant tissues or in the rumen of consuming animals (Gorz, Haskins, and Vogel, 1982). Under low-temperature conditions, high-nitrogen fertility levels, and in the initial seedling or regrowth stage, sufficient prussic acid could be produced to poison ruminants that consume the green leaves. A giant step toward a reduced prussic acid level was taken about 1940 when cultivars low in prussic acid were found (Quinby and Martin, 1954). The crop can be fed safely, provided sound management is practiced to prevent prussic acid poisoning (Gorz, Haskins, and Vogel, 1982).

In the future, sorghum cultivars may be produced that can be safely utilized in all stages of growth. The development of a simple, rapid, nondestructive spectrophotometric procedure based on leaves from seven-day-old seedlings grown under controlled conditions may facilitate breeding programs to reduce hydrocyanic acid (Gorz et al., 1977). Using this technique, substantial progress was made in changing the levels of hydrocyanic acid in two cycles of selection (Gorz, Haskins, and Vogel, 1982).

A popular and widespread belief exists that all sorghum seed contains tannins that adversely affect nutritive value. In 1928, a cultivar with palatable white seed was found, known as sweet sorghum. The low level of tannin in the white-seeded cultivar was in sharp contrast to the high levels in the bitter-tasting, unpalatable, brown-seeded cultivars (Quinby and Martin, 1954). The palatability problem was not solved entirely, however, as the lower acceptance by animals appears in part to be associated with the presence of tannin in the plant itself. On the positive side, tannins in the leaves of vegetative plants may provide a possible protection against bloat in grazing animals (Sarkar, Howarth, and Goplen, 1976). In the seed, tannins are contained in the testa layer, and the testa, therefore, plays an important role in reducing digestion and assimilation of grain protein and amino acids. Exactly why tannins reduce livestock performance and digestibility has not been explained (Kofoid, Maranville, and Ross, 1982).

Depending on the nature and proposed use of sorghum crops, either an increased or decreased tannin content might be a reasonable breeding objective. Tannin in sorghum generally is associated with bird resistance. In areas where losses from bird damage is severe, sorghum cannot be grown successfully without the incorporation of bird resistance. Birdproof cultivars frequently show superior agronomic characteristics.

When birdproof sorghum cultivars are used for malt to be brewed with sorghum beer, neither enzyme production and activity nor taste is affected by the presence of tannins (Daiber, 1975). Tannin in sorghum also is associated with decreased preharvest seed germination (Harris and Burns, 1970), reduced seed molding (Harris and Burns, 1973), and seed dormancy (Clark, Collier, and Langston, 1968).

In testing forages for tannins content in the vegetative stages, Walton, Haskins, and Gorz (1983) cautioned against testing procedures that may lead to false positive results. The level of tannins indicated to be present may be inaccurate.

The fact remains that although sorghum and sorghum derivatives are used extensively as a forage, animal performance is limited unless adequate feed supplementation accompanies sorghum feeding programs (Wheeler, 1980). Until the feeding value of forage sorghum can be improved, it may continue to play a lesser role than corn silage.

The incorporation of daylength insensitivity and the development of relatively short-season cultivars has greatly expanded the area of sorghum production. The result has been an enormous increase in testing, breeding, and experimentation with sorghum.

Under climatic conditions in the tropics, the perennial habit of sorghum permits the production of successive seed crops from an initial planting. Cultural treatments involving tillage and fertilization to favor regrowth, termed ratooning, may serve to illustrate management techniques for forage production. Based on a study in Hawaii (Escalada and Plucknett, 1977), it was recommended that 200 to 250 kgN/ha (180 to 220 lbN/ac) be applied to ratoon sorghum harvested three times at cutting heights of 13 cm, 3 cm, and 8 cm (5, 1, and 3 in.).

Sudan grass, a Sudan grass hybrid, and six sorghum × Sudan grass hybrids cut at 46 cm (18 in.) in height resulted in five harvests; when cut at 92 cm (36 in.), three harvests; and at the hard-dough stage, one harvest under conditions at Ames, Iowa (Wedin, 1970). Percentages of crude protein and in-vitro digestible dry matter were highest for frequently harvested forage (Table 8-5), although total dry-matter yields were reduced by frequency of cutting. Other workers have reported that yields tend to be higher when the plants are cut only a few times in the season rather than when defoliated frequently (Beuerlein, Fribourg, and Bell, 1968; Burger and Hittle,

Table 8–5
Mean Dry-Matter Yields, Percentage Crude Protein, and Percentage In-vitro Digestible Dry Matter of a Sudan Grass, a Sudan Grass Hybrid, and Six Sorghum × Sudan Grass Hybrids in 1965 and 1966 at Ames, Iowa

Treatment	Dry-Matter Yield		Crude Protein	In-vitro Digestible Dry Matter
	kg/ha	lb/ac		
One cut at dough stage	13,460	12,018	5.8	57.0
Cut at 46, 92, and 137 cm (18, 36 and 54 in.)	8,120	7,250	15.1	67.7

Source: Wedin (1970).

1967; Edwards, Fribourg, and Montgomery, 1971).

Some management suggestions for forage sorghum include:

1. Choose an adapted cultivar with demonstrated yield performance and sow at about 12 to 24 kg/ha (11 to 22 lb/ac) when the soil and weather conditions are warm. Row widths can vary from 36 cm to 1 m (14 to 40 in.). The high seeding rate is used with narrow-row seeding
2. Do not apply excessive levels of nitrogen fertilizer because of the increased risk of prussic acid poisoning. A balanced level of soil fertility can be achieved by split applications of nitrogen. A rate of 225 kgN/ha (200 lbN/ac) was reported to be about optimum (Sumner, Martin, and Etchegaray, 1965) as a split application when pasturing or green-chopping. Higher applications do not appreciably increase yields but do account for relatively high rates of accumulated nitrate in the harvested forage.
3. Avoid pasturing succulent young shoots, as they are high in prussic acid. Sorghums can be pastured in about eight weeks but not before they are at least 60 cm (24 in.) high.
4. Do not pasture or feed sorghum and Sudan forages that have been subjected to stress conditions such as frost or prolonged drought. Do not utilize these forages for at least 72 hours after wilting or a killing frost.

PEARL MILLET

Pearl millet is an important forage crop in various regions of the world. Several features make millet an attractive forage crop (Gupta, 1975):

1. Its robust, quick growth means that a successful crop can be produced if planted late or after the removal of an early crop. Millet is efficient in its utilization of moisture and has a higher level of heat tolerance than does sorghum or corn.
2. It has a high fodder yield because of its large stems, leaves, and heads.
3. An important characteristic of millet for grazing is its comparatively low hydrocyanic acid content. This factor, along with disease resistance, has stimulated interest in millet. Oxalic acid may be present in millet and so must be managed accordingly. Millet has been reported to depress butterfat production in dairy herds (Clark, Hemken, and Vandersall, 1965).

Millet must be managed carefully to produce a palatable feedstuff. If sown thickly and harvested green before the seeds are formed, excellent quality fodder can be produced comparable to that of corn or sorghum. Unlike corn and sorghum, however, the straw does not remain green when grain is maturing. As a result, late-harvested millet has low feeding value.

OAT

The oat crop is well known for its contribution to forage production under an emergency situation and is used most commonly as an emergency pasture among the spring-seeded small grains. It makes as good an annual pasture as any crop. It can be sown at 95 kg/ha (85 lb/ac) in the spring for summer pasture or in the summer for fall pasture. Oat fields can be grazed in about six weeks when they are in the early heading stage. If grazed too early, oat has a laxative effect on cattle.

Summer production is low when seeded late in the spring, but for early fall pasture oat is much superior to fall rye, providing nearly three times the amount of feed.

The oat crop makes good-quality silage when harvested in the early to mid dough stage. Stage of maturity at harvest can greatly affect yield and quality of silage (Gardner and Wiggans, 1961). If oat harvest for silage is delayed from the boot or heading to the dough stage, dry-matter yield may increase as much as twofold (Table 8-6). Grain formation is desirable as a source of fermentable carbohydrate in silage production. At the dough stage, the moisture level is a desirable 65%. At this stage of development, the oat crop is green, leafy, and succulent, which makes silage or hay palatable.

Although the protein content is less at the dough stage than at earlier stages, the large increase in dry matter and the natural adjustment of moisture to a optimum level may offset the lower protein content. The time interval between

Table 8–6
Composition and Yield of Vegetative and Grain Parts of Four Oat Cultivars Cut as Silage in 1958

Stage of Maturity	Total Yield of Dry Matter kg/ha	Total Yield of Dry Matter lb/ac	Composition, Vegetative	Percent Grain
Boot	3,227	2881	100	0
Heading	4,799	4285	100	0
Milk	8,376	7479	78	22
Early dough	10,218	9123	71	27
Late dough	10,636	9496	67	33

Source: Gardner and Wiggans (1961).

early and late dough stages may vary from four to eight days, so timeliness of operation is important. The oat crop will not regrow after harvest.

Highest dry-matter yields are produced with a tall, late-maturing cultivar, although such cultivars may be adversely affected by drought, diseases, or lodging.

The oat crop is frequently used as a companion crop with forage seedlings. Removal of the companion crop as green chop or pasture when the oat crop is 38 to 40 cm (15 to 16 in.) in height reduces competition and favors the forage stand. Care in pasturing is essential to minimize damage to seedlings.

RYE

Fertile tetraploid forms of rye are available that make this crop attractive for forage. Rye is vigorous in growth, tolerates relatively poor soil conditions, and is tall.

To help meet the expectations of a forage rye, the cultivar Wheeler was developed (Anonymous, 1972). It is capable of high forage yields. Feeding trials measuring rates of gain in dairy heifers and milk production in lactating cows produced equal results with rye and alfalfa haylage.

Fall-seeded rye can be spring-grazed when the rye is 20 cm (8 in.) in height. Rye grazing occurs before perennial forage crops are sufficiently productive to be grazed. Wet spring conditions that make the soil soft and subject to tramping may be a problem.

Fall rye seeded four to six weeks earlier than that desired for grain production can produce a fall pasture. Grazing can begin in seven weeks or at a height of 23 to 25 cm (9 to 10 in.), whichever occurs first.

Rye can be seeded following another spring cereal crop and can be plowed under after spring grazing to produce another crop, such as corn.

GENERAL VIEW OF CEREAL CROPS AS FORAGE

The energy input required to seed cereals annually for forage is a major deterrent compared to perennial forage crops. Energy input-output relationships will determine the extent to which annual cereal crops are grown for forage. It is probable that high-output crops such as corn and sorghum will play a major and expanded role in livestock feeding programs. Small grains will be of major value as an emergency or insurance crop when an abrupt or unforeseen change occurs.

Cereals such as wheat, oat, and barley can be handled as a hay crop, but they are difficult to handle in this fashion because the coarse, hollow stems with a fleshy node are slow to dry. Mechanical conditioners or stem crimpers that crust and break the stem are an aid to drying.

When ensiled, the hollow stems of cereals create special problems for packing, and if too much air is trapped, excessive aerobic respiration results. Timeliness of handling is essential. Grain in the advanced dough stage is ideal. If ensiled too early, the material is wet and difficult to ensile. Barley produces the best silage when the plant begins to turn color in the ripening stage. When ensiled at the firm dough stage and fed to dairy cows, triticale has been found to be low in energy, digestibility, and milk production.

The oat crop is a popular choice for forage production, as it remains palatable. Of the spring-seeded cereals, barley can produce a high dry-matter yield and a high percentage of digestible dry matter in adapted areas (Table 8-7). Yield differences among the cereals are a minor consideration, however, when choosing a cereal as a forage crop.

Seeding rates for forage production are as follows:

Oat:	95 kg/ha (85 lb/ac)
Barley:	134 kg/ha (120 lb/ac)
Wheat:	168 kg/ha (150 lb/ac)
Rye:	156 kg/ha (140 lb/ac)
Triticale:	156 kg/ha (149 lb/ac)

A fertilizer application of perhaps 220 kg/ha (200 lb/ac) containing equal portions of nitrogen, phosphorus, and potash is suggested, depending on soil fertility.

CROP MIXTURES FOR FEED PURPOSES

The use of cereals for annual forage production and for high-energy grain feed has provided flexibility not only in livestock feeding systems but in crop mixtures in an attempt to increase yield and feeding value. Cereal mixtures may consist of mixtures of cultivars of different species in varying proportions. The concept of **mixed grain** follows logically from mixtures of perennial forages that may consist of one or more grasses grown with one or more legume species.

Cereal mixtures produced for feed grain often consist of an oat-barley mixture in varying proportions of barley to oat by volume ranging from 35 : 65 to 50 : 50 to 65 : 35. In an effort to increase the feeding value of the oat-barley mixture, wheat or pea may be added. Seed mixtures are blended with 15% spring wheat, 20% barley, and 65% oat or 15% pea, 35% barley, and 50% oat. These same mixtures may be used for forage production and may be seeded in alternate rows (Walton, 1975).

Table 8–7
Dry-Matter Yields and in-vitro Digestible-Dry-Matter Concentration of Four Small-Grain Crop Species Averaged over Six Maturity Stages and Two Locations in Minnesota over the Years 1978 and 1979

Spring Cereal	*Dry-Matter Yield*		*Digestible Dry Matter, Percent of Dry Weight*
	kg/ha	*lb/ac*	
Barley	6300	5625	71.4
Oat	5900	5268	65.3
Triticale	5875	5246	69.1
Wheat	5375	4799	66.1

Source: Cherney and Marten (1982).

Crop mixtures are grown for the following reasons:

1. The hope that an increase in yield will occur by the complementary habits of associated genotypes.
2. The hope for an increase in yield stability over locations and seasons due to the ability of at least one genotype in the mixture to yield well in adverse conditions. Taylor (1978) suggested that a mixture of two species will compensate each other in yield depending on the environment.
3. To restrict the secondary spread of diseases (Wolfe and Minchin, 1977; Clark, 1980; Browning and Frey, 1969).
4. To minimize the effects of soil variability associated with topography, drainage, or type.
5. To increase nutrient feed value by adding a component such as pea or wheat. The amount of each component varies with growing conditions, and this does not constitute a sound method of formulating a balanced ration.

In some countries, corn may be grown with other crops in an effort to increase yield and/or feeding quality (Enyi, 1973; Okigbo and Greenland, 1976; Willey, 1979). Warren (1980) tested mixtures of corn hybrids with sunflowers at Ottawa, Canada, because they appeared compatible for forage production under cool-temperature conditions. Highest dry-matter yields were found when corn was grown alone. A reduction in digestible dry matter of the corn-sunflower mixture meant that such a mixture should be avoided.

Ross (1965) compared the yields of five grain-sorghum hybrids with yields in all possible 1 : 1 blends derived from them. No consistent advantage in grain yield was observed for any of the blends.

In small cereals, there is little scientific evidence to substantiate a grain-yield or dry-matter-yield advantage of mixtures as evaluated in small plots (Fejer, Fedak, and Clark, 1982; Syme and Bremner, 1968). Yields are usually less than that of the highest-yielding component. Small yield increases over the mean of components in monoculture are common (Jensen, 1952; Klages, 1936) and detract producers away from growing the highest-yielding component of a mixture. Perhaps small plot yields grown on uniform land may be too restricted to express the advantages of versatility in mixtures, and any restriction on the spread of diseases may not be effective in small, experimental plots. This observation has merit, and for feed purposes on variable feeds, mixtures of cereal grains may be warranted.

The hope that selected mixtures compounded in various proportions or **blends** of phenotypically similar or dissimilar cultivars can outyield the best cultivar in monoculture production presents a formidable array of possible blends and mixtures. The performance of 28 biline blends grown in a 1 : 1 genotypic ratio with similar and dissimilar phenotypes of winter barley and compared to the right components grown in monoculture gave no yield advantage over the high-yield component (Palvakul, Finkner, and Davis, 1973). The possibility that blends or mixtures can express superiority over pure stands needs critical evaluation. If mixtures or blends are to be used successfully, some basic principles need to be understood.

YIELD PRINCIPLES OF MIXTURES AND BLENDS

Every effort is made in crop production to eliminate weeds, because producers are well aware of competitive stresses that exist between weeds and crop plants. Areas of the field missed by effective herbicides serve as striking examples of the adverse impact of weed competition.

Cereal producers are aware also of the small unproductive plants associated with an excessive seeding rate in areas of the field double-seeded because of the inadvertent overlap or turning on the headlands. It may be these observations that have led crop producers to the view that competition is undesirable and should be eliminated or avoided. Yet without realizing the consequence of their actions, crop producers seed in row widths and at rates that competitively push plants to the limit of tolerance. Competition cannot be avoided or eliminated if environmental variables such as light, carbon dioxide, fertilizer nutrients, and oxygen are to be used to maximum advantage in photosynthesis. Donald (1963) noted:

> It is a surprising thought that, in the production of a successful and "healthy" field crop, such intense **competition** is developed that in quantitative terms, the individual plants are markedly subnormal. It is the community of suppressed plants which gives the greatest yield. (p. 5)

Three characteristic plant responses were described by Ford (1975) and serve to clarify Donald's observation. The three competitive responses are:

1. Density-dependent mortality.
2. Plastic responses.
3. Domination or hierarchy of exploitation.

Density-dependent mortality in pure stands reduced the number of wheat plants per unit area by 7.6% and 35.6% at seeding rates of 62 and 366 kb/ha (55 and 327 lb/ac), respectively (Table 8-8) in the first 119 days of plant development. In the next 63-day period, competition further reduced plant number by 9.4% and 35.6%, respec-

Table 8–8
Competitive Effects Expressed as "Density-Dependent Mortality" on Wheat Plants per Square Meter at Two Seeding Rates and at Three Stages of Crop Development When Grown at Adelaide, Australia

	Density of Plants/m² (ft²)		
Seeding Rate	*At Emergence*	*At 119 Days*	*At 182 Days*
62 kg/ha	184	170	154
(55 lb/ac)	(16.6)	(15.3)	(13.9)
336 kg/ha	1078	694	447
(327 lb/ac)	(97.0)	(62.5)	(40.2)

Source: Data from Donald (1963).

tively. Density-dependent mortality did not reduce plant population to a level that gave maximum grain yield, however.

Plastic responses were observed in the same wheat plants when seeded at two extreme rates of 2.4 and 366 kg/ha (2.1 and 327 lb/ac) (Table 8-9). From the low to the high seeding rate, dry weight per plant decreased 43 times; tillers per plant decreased 25 times; plant height was increased owing to the elongation of cells under low light; and leaf area per plant decreased 53 times.

Plastic adjustments throughout the life cycle of the barley plant were measured at six plant populations ranging from 50 to 1600 plants/m² (4.6 and 148/ft²) (Kirby and Faris, 1970). Increasing density reduced leaf number from a mean of 10.2 and 8.0 and caused stem internode elogation to start earlier and at a lower internode. Final stem length was reduced, owing to earlier cessation of growth. The lamina and sheath lengths of lower leaves were increased with increasing density, and lamina width decreased. Development of the spike began six days earlier at the highest density than at the lowest density, and this difference persisted throughout the development of the spike. Initiation of new florets on the spike stopped abruptly at the lower densities, and then successively at the lower densities. Early cessation of growth at the higher densities led to a shorter ear.

Kirby and Faris (1970) suggested that early spike development led to earlier competition for photosynthate; plant leaves at a high population

Table 8–9
Expressions of Plant Plasticity in Wheat at the Beginning of Reproductive Development When Seeded at Two Rates at Adelaide, Australia

	Agronomic Feature						
	Plant Dry Matter			Height		Leaf Area per Plant	
Seeding Rate	g	oz	*Tillers per Plant*	cm	in.	cm²	in.²
2.4 kg/ha (2.1 lb/ac)	47.6	1.7	29.5	83	32.7	2660	426
366 kg/ha (327 lb/ac)	1.1	0.04	1.2	108	42.5	50	8

Source: Data from Donald (1963).

could not receive enough light to supply the necessary photosynthate. The result was starvation of the tip of the spike, and consequently fewer spikelets were developed. Cereal producers must aim at a plant population that will intercept sunlight to best advantage yet avoid severe competition for light and other factors.

A third form of competition occurs when two species or cultivars of the same species are grown in a mixture. One may yield less and one may yield more compared to monoculture stands. One is the aggressor and is able to exploit more than its "share" of the environment; the other is suppressed. This interaction is known as the *hierarchy of exploitation*.

Recognizing that no single cultivar has all the desirable or necessary features to obtain maximum production, producers often mix two or more crops of different species or blend two or more cultivars of the same species together. In livestock-producing areas, approximately 50% of the barley and oat feed produced is from mixtures (Fejer, Fedak, and Clark, 1982). Blends of crops are sold on the premise that they have superior yield performance. Blends may be a means of disposing of surplus seed of a specific cultivar, which may be blended with other cultivars in a ratio advertised to be superior but possibly reflecting the volume of seed to be disposed of.

In all mixtures and blends, four principles developed by Donald (1963) invariably apply:

1. Mixtures will usually yield less than the yield of the highest-yielding component grown in pure culture.
2. Mixtures will usually yield more than the lowest-yielding component grown in pure culture.
3. Mixtures may yield more or less than the average of the pure cultures, but generally less.
4. Although **cooperation** among plants is believed to exist, there is limited evidence to show that mixtures can exploit the environment better than a pure stand.

These are believed to be the basic principles of competition, and they can be demonstrated in most yield trials involving mixtures. A specific example using mixtures of sorghum, and based on total dry-matter yields, illustrates all the principles (Table 8-10). In addition the same principles apply to quality factors as measured by

Table 8–10

Total Dry-Matter Yield, Crude Protein of Leaves, and In-vitro Digestible-Dry-Matter (IVDDM) Comparisons of Three Tall (265 cm or 103 in.), Three Medium (171 cm or 67 in.), and Three Short (113 cm or 44 in.) Sorghum Cultivars in Pure Stands and in Mixtures Grown at Lincoln, Nebraska, and Averaged over 1977 and 1978

	Dry-matter yield			
Description	*tonnes, 1 ha*	*tons, 1 ac*	*Crude Protein, % (Leaves)*	*IVDDM % (Leaves)*
Tall (T)	19.5	8.8	11.2	55.2
Medium (M)	15.6	7.0	10.2	51.1
Short (S)	10.2	4.5	12.0	51.9
Mean of pure stands	15.1	6.8	11.1	52.8
TMS	13.8	6.2	10.4	52.2
TM	18.1	8.1	10.1	52.7
TS	14.5	6.5	11.1	53.7
MS	13.7	6.2	10.1	50.7
Mean of mixtures	15.0	6.8	10.4	52.3

Source: Data from Gorz and Haskins (1982).

crude protein and in-vitro dry-matter digestibility of the sorghum leaves.

Crop producers continue to grow crop mixtures, and attempts will continue to be made to find superior blends. Producers, however, should be under no illusion that mixtures will give superior performance in terms of dry matter, protein, or any other component. Mixtures for disease and yield stability may be warranted as a production item.

CONCLUSIONS

For many years cereal crops were considered exclusively as food crops. Their use as forage was reserved for emergency situations, and only low-quality or surplus grain was considered for livestock feeding purposes. Corn and sorghum, however, have reached dominant roles as producers of forage or grain for feed. It is possible that further increases in hybrid vigor and associated dry-matter yields will help promote these crops to even greater contributions to livestock feeding programs. Certainly the development of short-season hybrids of corn and sorghum will extend the range of production of these crops.

Increases in energy prices have greatly increased irrigation costs. This fact, coupled with a lowering of the water table in some areas of the world and the accompanying limits on irrigation, may promote shifts to crops such as millet, sorghum, and corn that have a low evapotranspiration rate compared to perennial forage crops.

Although the inefficiency of food production from livestock may be questioned, it is possible that livestock may play a growing role in meeting human food needs. Increases in livestock numbers will help increase the importance of cereal crops both as producers of forage and as feed grains.

QUESTIONS

1. Consider cereal grains as a livestock feed in terms of:
   a. Moral issues.
   b. Economics.
   c. By-products.
   d. Efficiency.
2. What are the alternative uses for cereal grains if they are not used as livestock feed?
3. Summarize precautions that must be taken when cereal crops or their products are used as a livestock feed.
4. How was the name "sweet sorghum" derived?
5. Compare regional statistics for your state or region to determine the extent to which crop mixtures are produced, or select a livestock region to determine the relative importance of crop mixtures.
6. What conclusion do you reach about the merits of producing a crop mixture, after you have reviewed the reason for producing mixtures and the yield principles, if:
   a. Yield is the prime objective?
   b. Factors other than yield are important?
7. Is plant competition desirable or undesirable?
8. Are cereal crop mixtures or blends recommended and sold in your area?
9. Apply the principles of competition to yield results of various mixtures to determine their merit. If regional yield data are unavailable, use the results reported by Fejer, Fedak, and Clark (1982).
10. Compare the advantages and disadvantages of forage production from perennial forages and annual cereals. Consider:
    a. Yield.
    b. Quality.
    c. Seasonal productively.
    d. Cultural practices.
    e. Agronomic and ecological aspects.
    f. Energy efficiency.

11. *a.* Consider the future of cereals as producers of forage and feed grains in terms of:

(1) Future population.

(2) Future yields of cereal crops.

(3) Competing for land areas used for food production.

(4) Water-deficient areas.

b. What general conclusions do you reach regarding cereal crops as producers of livestock feed?

12. Consider breeding programs of cereal crops for forage and livestock feed in terms of:

a. Past contributions.

b. Future opportunities.

References

Abercrombie, K. 1982. "Intensive Livestock Feeding: Reckless Folly or an Emergency Food Reserve?" *Ceres* 15 (1): 38–41.

Aldrich, D. T. A. 1959. "The Effects of Grazing Management on the Response of Winter Wheat to Spring Defoliation." *Empire Journal of Experimental Agriculture* 27: 10–16.

Allee, G. L. 1974. "Triticale as a Feed." (In) C. C. Tsen (ed)., *Triticale: First Man-Made Cereal.* St. Paul, Minn.: American Association of Cereal Chemistry. pp. 272–279.

Anonymous. 1970. *Wheat in Livestock and Poultry Feeds.* Proceedings of an International Symposium. Stillwater: Oklahoma State University. 200 pp.

———. 1972. *Wheeler Rye—for Forage and Green Manure.* East Lansing: Michigan State University Agricultural Experiment Station. Research Report 172. 4 pp.

———. 1983. *Research Newsletter.* Chatham, Ont.: King Grain. Spring. 4 pp.

Atkins, I. M., et al. 1970. *Wheat Production in Texas.* Texas Agricultural Experiment Station Bulletin 1095. 30 pp.

Beuerlein, J. E., H. A. Fribourg, and F. F. Bell. 1968. "Effects of Environment and Cutting on the Regrowth of a Sorghum-Sudangrass Hybrid." *Crop Science* 8: 152–155.

Browning, J. A., and K. J. Frey, 1969. "Multiline Cultivars as a Means of Disease Control." *Annual Review of Phytopathology.* 7: 355–382.

Burger, A. W., and C. N. Hittle. 1967. "Yield Protein, Nitrate, and Prussic Acid Content of Sudangrass, Sudangrass Hybrids and Pearl Millets Harvested at Two Cutting Frequencies and Two Stubble Heights." *Agronomy Journal* 59: 259–262.

Cherney, J. H., and G. C. Marten. 1982. "Small Grain Crop Forage Potential. I. Biological and Chemical Determinants of Quality and Yield." *Crop Science* 22: 227–231.

Clark, L. E., J. W. Collier, and R. Langston. 1968. "Dormancy in *Sorghum Bicolor* (L.) Moench. II. Effects of Pericarp and Testa." *Crop Science* 8: 155–158.

Clark, N. A., R. W. Hemken, and J. H. Vandersall. 1965. "A Comparison of Pearl Millet, Sudangrass and Sorghum Sudangrass Hybrid as Pasture for Lactating Dairy Cows." *Agronomy Journal* 57: 266–269.

Clark, R. V. 1980. "Comparison of Spot Blotch Severity in Barley Grown in Pure Stands in Mixtures with Oats." *Canadian Journal of Plant Pathology* 2: 37–38.

Daiber, K. H. 1975. "Enzyme Inhibition by Polyphenols of Sorghum Grain and Malt." *Journal of the Science of Food and Agriculture* 26: 1399–1411.

Daynard, T. B., and R. B. Hunter. 1975. "Relationships Among Whole-Plant Moisture, Grain Moisture, Dry Matter Yield, and Quality of Whole-Plant Corn Silages." *Canadian Journal of Plant Science* 55: 77–84.

Donald, C. M. 1963. "Competition Among Crops and Pasture Plants." *Advances in Agronomy* 15: 1–18.

Dunphy, D. J., M. E. McDaniel, and E. C. Holt. 1982. "Effect of Forage Utilization on Wheat Grain Yield." *Crop Science* 22: 106–109.

Edwards, N. C., Jr., H. A. Fribourg, and M. J. Montgomery. 1971. "Cutting Management Effects on Growth Rate and Dry Matter Digestibility of the Sorghum-Sudangrass Cultivar Sudax SX-11." *Agronomy Journal* 63: 267–271.

Enyi, B. A. C. 1973. "Effects of Intercropping Maize or Sorghum with Cowpeas, Pigeon Peas or Beans." *Experimental Agriculture* 9: 83–90.

Escalada, R. G., and D. L. Plucknett. 1977. "Ratoon Cropping of Sorghum. III. Effect of Nitrogen and

Cutting Height on Ratoon Performance." *Agronomy Journal* 69: 341-346.

Evans, L. E., W. Dedio, and R. D. Hill. 1973. "Variability in the Alkylresorcinol Content of Rye Grain." *Canadian Journal of Plant Science* 53: 485-488.

Fejer, S. O., G. Fedak, and R. V. Clark. 1982. "Experiments with a Barley-Oat Mixture and Its Components." *Canadian Journal of Plant Science* 62: 497-500.

Ford, E. D. 1975. "Competition and Stand Structure in Some Even-Aged Plant Monocultures." *Journal of Ecology* 63: 311-333.

Friend, D. W. 1970. "Comparisons of Some Milling Products of Barley and Rye When Fed in Diets of Rats." *Canadian Journal of Animal Science* 50: 345-348.

———, and T. M. MacIntyre. 1969. "Digestibility of Rye and Its Value in Pelleted Rations for Pigs." *Canadian Journal of Animal Science* 49: 375-381.

Gardner, F. P., and S. C. Wiggans, 1961. "Yield, Moisture and Protein Composition of Spring Oats Cut for Silage at Different Stage of Maturity." *Agronomy Journal* 53: 251-254.

Gorz, H. J., et al. 1977. "Assay of *p*-hydroxybenzaldehyde as a Measure of Hydrocyanic Acid Potential in Sorghums." *Crop Science* 17: 578-582.

Gorz, H. J., and F. A. Haskins. 1982. "Performance of Blends of Short, Medium and Tall Sorghum for Forage," *Crop Science* 22: 223-226.

Gorz, H. J., F. A. Haskins, and K. P. Vogel. 1982. "Divergent Selection for Hydrocyanic Acid Potential in Sudangrass." *Crop Science* 22: 322-325.

Gupta, V. P. 1975. "Fodder Improvement in Pennisetums." *Forage Research* 1: 54-60.

Harris, H. B., and R. E. Burns. 1970. "Influence of Tannin Content on Preharvest Seed Germination in Sorghum." *Agronomy Journal* 62: 835-836.

———. 1973. "Relationship Between Tannin Content of Sorghum Grain and Preharvest Seed Molding." *Agronomy Journal* 65: 957-959.

Holliday, R. 1956. "Fodder Production from Winter-Sown Cereals and Its Effect upon Grain Yield." *Field Crop Abstracts* 9: 129-135, 207-213.

Huffman, C. F., and C. W. Duncan. 1954. "The Nutritive Value of Corn Silage for Milking Cows." *Journal of Dairy Science* 37: 957-966.

Hunter, R. B. 1978. "Selection and Evaluation Procedures for Whole-Plant Corn Silage." *Canadian Journal of Plant Science* 58: 661-678.

Jensen, N. F. 1952. "Intra-Varietal Diversification in Oat Breeding." *Agronomy Journal* 44: 30-44.

Kilcher, M. R. 1982. "Effect of Cattle Grazing on Subsequent Grain Yield of Fall Rye (*Secale cereale* L.) in Southwestern Saskatchewan." *Canadian Journal of Plant Science* 62: 795-796.

Kirby, E. J. M., and D. G. Faris, 1970. "Plant Population Induced Growth Correlations in the Barley Plant Main Shoot and Possible Hormonal Mechanisms." *Journal of Experimental Botany* 21: 787-798.

Klages, K. H. 1936. "Changes in Proportions of the Components of Seeded and Harvested Cereal Mixtures in Abnormal Season." *Agronomy Journal* 28: 935-940.

Kofoid, K. D., J. W. Maranville, and W. M. Ross. 1982. "Relationship of the Testa to Agronomic and Nutritional Traits in Sorghum." *Crop Science* 22: 352-357.

Lawrence, T. L. J. 1978. "Cereals for Humans and Processed Cereals for Farm Livestock: Present and Future." (In) M. G. Hughes (ed.), *Are Today's Cereals Tomorrow's Crop?* Proceedings of the Twelfth Annual Conference of the Reading University Agricultural Club. Reading University. pp. 57-65.

Mathers, A. C., B. A. Stewart, and D. L. Grunes. 1982. "Effect of a Nitrification Inhibitor on the K, Ca, and Mg Composition of Winter Wheat Forage." *Agronomy Journal* 74: 569-573.

McGinnis, R. C. 1970. "Breeding Feed Grains in Canada." (In) *Feed Grains of Canada*. Proceedings of the General Council of the Canada Grains Council, Winnipeg, Manitoba. pp. 58-64.

Okigbo, B. N., and D. J. Greenland. 1976. "Intercropping Systems in Tropical Agriculture." (In) *Multiple Cropping*. Special Publication 27. Madison, Wis.: American Society of Agronomy. pp. 63-102.

Palvakul, M., V. C. Finkner, and D. L. Davis. 1973. "Blendability of Pheonotypically Similar and Dissimilar Winter Barley Cultivars." *Agronomy Journal* 65: 74-77.

Pedersen, J. F., et al. 1982. "Variability for Quality and Agronomic Traits in Forage Sorghum Hybrids." *Crop Science* 22: 853-856.

Quinby, J. R., and J. H. Martin. 1954. "Sorghum Improvement." *Advances in Agronomy* 6: 305-359.

Reich, V. H., and R. E. Atkins. 1970. "Yield Stability of Four Population Types of Grain Sorghum, *Sorghum Bicolor* (L.) Moench, in Different Environments." *Crop Science* 5: 593-594.

Ross, W. M. 1965. "Yield of Grain Sorghum (*Sorghum vulgare* Pers.) Hybrids Alone and in Blends." *Crop Science* 5: 593-594.

Sarkar, S. K., R. E. Howarth, and B. P. Goplen. 1976. "Condensed Tannins in Herbaceous Legumes." *Crop Science* 16: 543-546.

Schmid, A.R., et al. 1976. "Relationships Among Agronomic Characteristics of Corn and Sorghum Cultivars and Silage Quality." *Agronomy Journal* 68: 403–406.

Scott, W. R. 1978. "Winter Green Feed Crops: The New Zealand Experience." *Span* 21: 106–109.

Smith, R. E., and T. M. MacIntyre. 1970. "The Feeding of Rye to Growing Chickens." *Canadian Journal of Animal Science* 50: 107–113.

Sumner, D. C., W. E. Martin, and H. S. Etchegaray. 1965. "Dry Matter and Protein Yields and Nitrate Content of Piper Sudangrass (*Sorghum sudanense* (Piper) Stapf.) in Response to Nitrogen Fertilization." *Agronomy Journal* 57: 351–354.

Sutton, J. C. 1982. "Epidemiology of Wheat Head Blight and Maize Ear Rot Caused by *Fusarium graminarum*." *Canadian Journal of Plant Pathology* 4: 195–209.

Syme, J. R., and P. M. Bremner. 1968. "Growth and Yield of Pure and Mixed Crops of Oats and Barley." *Journal of Applied Ecology* 5: 659–674.

Taylor, B. R. 1978. "Studies on a Barley-Oats Mixture." *Journal of Agricultural Science, Cambridge* 91: 587–591.

Walton, M. F., F. A. Haskins, and H. J. Gorz. 1983. "False Positive Results in the Vanillin-HCl Assay of Tannins in Sorghum Forage." *Crop Science* 23: 197–200.

Walton, P. D. 1975. "Annual Forages Seeding Rates and Mixtures for Central Alberta." *Canadian Journal of Plant Science* 55: 987–993.

Warren, F. S. 1980. "Forage Production of Corn and Sunflower Mixtures." *Canadian Journal of Plant Science* 60: 1377–1382.

Wedin, W. F. 1970. "Digestible Dry Matter, Crude Protein, and Dry Matter Yields of Grazing-type Sorghum Cultivars as Affected by Harvest Frequency." *Agronomy Journal* 62: 359–363.

Wheeler, J. L. 1980. "Increasing Animal Production from Sorghum Forage." *World Animal Review* 35: 13–22.

Wieringa, G. W. 1967. *On the Occurrence of Growth Inhibiting Substances in Rye*. Waginingen, The Netherlands: Institute of Storage and Processing of Agricultural Produce. Publication No. 156. 68 pp.

Willey, R. W. 1979. *Intercropping: Its Importance and Research Needs*. "Competition and Yield Advantages Part 1. Agronomy and Research Approaches Part 2." Madison, Wis.: American Society of Agronomy. Publication 32. pp. 1–10 and 73–81.

Wolfe, M. S., and P. N. Minchin. 1977. "Powdery Mildew of Barley." *Annual Report of the Cambridge Plant Breeding Institute* 1976: 112–114.

Young, J. C. 1980. "Ergot Contamination of Feeds and Food." *Canada Agriculture* 25 (3): 25–28.

9

Hybrids

Hybrids, or the first-generation plants produced after a cross, often exhibit hybrid vigor or heterosis for yield, improved seed quality, drought resistance, vigor, or other desirable traits. The success of hybrid crops such as corn and sorghum suggests that some day all crops will be hybrid. This possibility must be considered in relation to how plants propagate themselves. Cereals, for example, reproduce sexually by means of seeds, but in two categories—seed set through self-pollination and seed set largely through cross-pollination. Another group of crops includes species that are asexually propagated through the planting of vegetative parts or grafting.

The cornerstones of plant-breeding programs include the conscious introduction of genetic diversity and the selection of superior plants from the genetic diversity achieved. The appropriateness of a breeding methodology is determined mainly by the sexual nature of the crop—inbreeding (self-pollinating) or outbreeding (cross-pollinating)—its genetic structure, its physical structure, and the breeding objectives.

HYBRID CORN DEVELOPMENT

Corn is a cross-pollinated plant that is highly heterozygous after centuries of natural crossing. The corn plant bears male and female flower organs separately on the same plant and is unique among the major cereals in this respect (Figure 9-1). The ear contains only female ovaries, each with a silk to receive the male pollen. The ear is enclosed in husks and is located in a leaf axil midway up the plant. The tassel located at the top of the plant bears a thousand or more male flowers, each with three anthers that each contain about 2500 pollen grains (Figure 9-2). A single tassel is capable of producing several million pollen grains that are easily carried by the wind to other plants, thereby causing cross-pollination. Random pollination may produce a genetically superior combination, but since the parentage is unknown and because corn plants are heterozygous, the superior combination cannot be repeated. Conventional corn-breeding techniques outlined in Chapter 5 were not outstandingly successful because of the highly heterozygous nature of the corn plant.

The production of hybrid corn offers a means of increasing yield, but, more important, provides control over matings so that repeatable results can be obtained. The system provides also a method for judicial selection of superior inbreeds to be used for crossing purposes.

The method involves inbreeding or selfing of corn plants for six to ten generations to produce true-breeding pure lines. Although in-

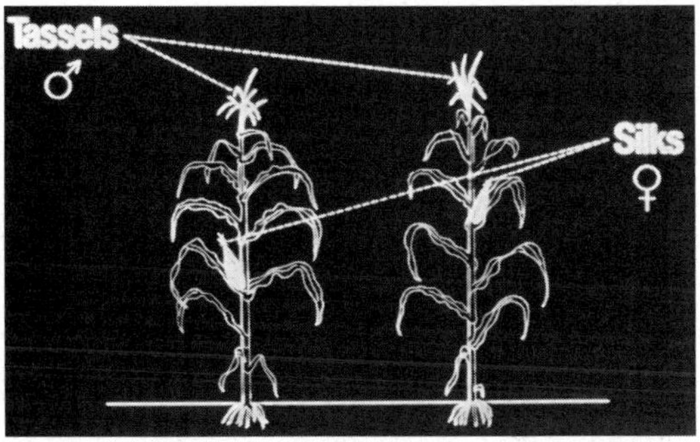

Figure 9–1. The corn plant bears male and female flower clusters separately on the same plant. At the top of the plant is the tassel or pollen-producing male inflorescence. The ear contains several hundred ovaries, each attached to a silk to receive the pollen. The entire ear is encased in husks, with the silks exposed as the tip of the ear. The height of the ear on the stalk may vary, as shown in the diagram. (Courtesy of Pioneer Hi-Bred International, Inc., Des Moines, Iowa, U.S.A.)

breeding is associated with a marked decline in plant vigor (Figure 9-3), inbreeding provides an opportunity to identify and discard undesirable inherited abnormalities that are controlled by recessive genes, which are masked in the heterozygous state. Inbreeding must be repeated until plants are homozygous and recessive genes can be identified. Abnormalities that can be eliminated by roguing include defective seeds, dwarfs, **albinos,** and chlorophyll deficiencies or striping.

When two unrelated inbreds or pure lines are crossed, the F_1 progeny usually are vigorous and high-yielding. Combining two inbreds produces a single-cross progeny (Figure 9-4). Superior F_1 hybrids for commercial production can be produced at will by combining selected hybrids.

HISTORY OF HYBRID CORN

Basic facts associated with a reduction in vigor by inbreeding and the recovery of hybrid vigor were reported first in 1876 in a publication by Charles Darwin, "The Vegetable Kingdom." Darwin's work may have led to work on hybrids by W. J. Beal at Michigan State College (Sprague, 1955). Beal reported a 10% to 50% yield advantage with F_1 hybrids over their parental open-pollinated cultivars in the first recorded yield trial.

Despite the yield advantages reported by Beal and others, active interest in hybrid corn did not develop until 1909, when G. H. Schull and E. East proved conclusively that although vigor was lost during inbreeding, crossing inbred lines resulted in an explosive recovery of vigor in the F_1 cross that was called hybrid vigor or heterosis. Seed for hybrid corn production produced from single crosses, however, was borne on a small inbred plant that produced a small ear

Figure 9–2. Photograph of a single branch of a corn tassel or pollen-bearing flower cluster. The tassel, located at the top of the plant, bears a thousand or more flowers, each containing three anthers. Each anther produces about 2500 pollen grains, which are carried by the wind to other plants to effect cross-pollination. The entire tassel is capable of producing several million pollen grains. (Courtesy of Pioneer Hi-Bred International, Inc., Des Moines, Iowa, U.S.A.)

Figure 9–3. Inbreeding of the normally cross-pollinated corn crop reduces plant vigor but provides an opportunity to identify and discard inherited abnormalities that are controlled by recessive genes and that become apparent as homozygosity is achieved. This photo shows a portion of an inbred test block and the small-stature inbred lines that evolve. (Courtesy of Pioneer Hi-Bred International, Inc., Des Moines, Iowa, U.S.A.)

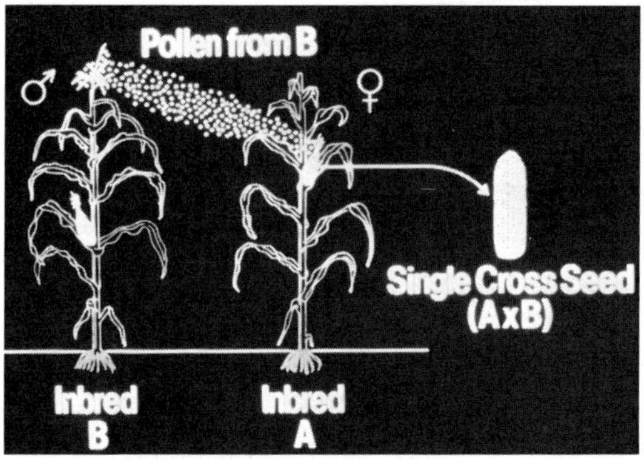

Figure 9–4. Single-cross seed is produced by combining two inbreds A and B. One of the inbreds is designated as the pollen parent (inbred B—the male, ♂) and the other the seed parent (inbred A—the female, ♀). Note that the tassel of the female seed parent has been removed to prevent self-pollination. The seed produced on inbred parent A is used for seed in the F_1 generation. The cross A × B is described as a single cross. (Courtesy of Pioneer Hi-Bred International, Inc., Des Moines, Iowa, U.S.A.)

(Figure 9-5) with small seeds. Low seed yields resulted in prohibitively high seed costs, and small seed did not attract the attention of farmers, who believed that large ears and large seed were essential to the production of a successful corn crop. The commercial exploitation of single-cross F_1 seed was a real problem. This difficulty was overcome by D. F. Jones in 1917, when he developed double-cross hybrids produced by mating two unrelated single-cross hybrids. In other words, single-cross F_1 plants produced from inbreds A and B were crossed with single-cross F_1 plants produced from inbreds C and D. Commercial seed was produced on an F_1 single-cross hybrid, reducing the cost of hybrid seed to the producer.

It is interesting to note that it took producers until about 1965 to accept single-cross seed commercially. By 1983, nearly 70% of the corn produced in North America was single-cross seed. Steady improvements in vigor toward a larger, more vigorous inbred helped this transition, however.

A feature of hybrid crops is that seed must be produced and purchased every year. Hybrid seed could not be used extensively until a system

Figure 9–5. An initial hindrance to the acceptance of hybrid seed was the fact that single-cross seed is produced on an inbred plant. The small ears (center) resulted in high seed costs. By mating two single-cross hybrids, seed is produced on an F_1 hybrid plant that produces larger ears, as portrayed in the two outer ears of the photo. (Courtesy of Pioneer Hi-Bred International, Inc., Des Moines, Iowa, U.S.A.)

of inbred line development, testing for superior combinations, producing, distributing, and marketing seed was developed. The establishment of Pioneer Inc., the first private hybrid seed company, by H. A. Wallace in the mid-1920s was an essential step (Figure 9-6). Additional commercial and private companies developed and established a competitive marketing system for hybrid corn. The use of hybrids was delayed by the disastrous economic depression of the 1930s, but by the early 1940s hybrid production began to expand. By the mid 1950s hybrid corn dominated North American corn production to the virtual exclusion of open-pollinated cultivars.

Since the commercial introduction of the first hybrids in the early 1940s, a steady improvement in hybrid corn superiority has taken place. This has led to corn as the number one crop in terms of area and yield per unit area where it is adopted.

The special needs of third-world countries—the economic circumstances of farmers and the lack of the infrastructure necessary for hybrid success—has meant that open-pollinated populations are used in most developing countries.

Figure 9–6. Seed of hybrid crops must be renewed yearly to avoid segregation and loss of vigor in subsequent generations. An enormous infrastructure exists in hybrid corn production. Included is inbred development, testing of inbreds for hybrid performance, testing and comparing of hybrids in yield trials as shown in this photo, producing hybrid corn seed, and distributing, advertising, and marketing of hybrid seed. Pioneer, Inc., was the first private hybrid seed company to develop in the mid-1920s. (Courtesy of Pioneer Hi-Bred International, Inc., Des Moines, Iowa, U.S.A.)

HYBRID SORGHUM

Mechanical manipulation of corn to promote selfing or crossing is facilitated by separate male and female organs and by the fact that cross-pollination normally occurs. The success of hybrid corn led researchers to consider the extension of hybridization to other crops.

Sorghum is normally a self-fertilized crop, but a low level of cross-pollination consistently occurs, and no barriers to cross-pollination exist. The panicle of sorghum is borne at the top of the plant and contains both male and female organs. Controlled self-pollination to avoid contamination by foreign pollen is accomplished readily by placing waterproofed paper bags over the panicles.

To effect cross-pollination, self-pollination must be prevented. This is accomplished by the use of cytoplasmic male sterility, a factor (inherited through the cytoplasm and not through the nucleus) that prevents viable pollen from being produced. Since no pollen is produced, plants are described as being male-sterile, are designated as the A line, and must always be used as the female parent. The A line is pollinated by a genetically similar maintainer B parent and gives rise again to a male sterile line which serves as one parent of a cross used to produce a hybrid. A second fertile restorer or R line parent is required to produce a commercially fertile hybrid. Plants carrying restorer genes can be identified by crossing them onto a male-sterile plant and raising the progeny to check that fertility has been restored (Figure 9-7). Tests should be conducted at various locations to assure that restoration is complete and reliable.

The discovery and development of cytoplasmic male sterility and an associated fertility-restoration mechanism made hybrid sorghum cultivars commercially possible. The introduction of hybrid sorghum in 1956–1957 had an impact on agriculture in the United States comparable with that of hybrid corn.

Figure 9–7. Sorghum crossing block. The sorghum panicle may be placed in a waterproofed paper bag to avoid contamination by foreign pollen and to assure self-pollination; male-sterile plants may be bagged so that pollen from selected parents may be placed on the stigmas to make specific crosses; or male-sterile plants may be bagged so that pollen from selected plants can be tested for genetic fertility restoration. (Courtesy of Pioneer Hi-Bred International, Inc., Des Moines, Iowa, U.S.A.)

Research on methods of producing hybrid sorghum began in the early 1930s. Initially genetic male sterility was reported in India and in the United States (Anonymous, 1980a).

Evidence of cytoplasmic male sterility was announced by Stephens, Kuykendall, and George (1952) and more clearly defined by Stephens and Holland (1954). Male sterility is caused by an interaction between two grain sorghums, milo and kafir. By replacing the entire milo genome with that of kafir by repeated backcrossing with milo as the female parent, male sterility is produced. Stephens and Holland (1954) reported that the proportion of male-sterile plants increased substantially with one backcross of kafir to the F_1 plant and to plants of milo × kafir, and more than 99% male sterility was obtained in the second backcross.

Two recessive genes msc_1 and msc_2 (Anonymous, 1980a) are responsible for male sterility. The dominant alleles MSc_1 and MSc_2 of these genes cause plants to be male-fertile. Fertility restoration in the F_1 may not be complete, owing to different alleles at the major loci or to modifier genes at the other loci.

HYBRID PEARL MILLET

Pearl millet is an annual, cross-pollinated crop that has both male and female reproductive organs contained in each flower of the inflorescence on the top of the plant. Pearl millet exhibits a mechanism known as *protogyny* that serves to reduce self-pollination. This is achieved by the development and maturation of ovaries prior to another development and pollen release (Figure 9-8). About 80% of the flowers are naturally cross-pollinated, but complete cross-pollination is rare, because stigmas of later tillers often receive pollen from flowers of earlier tillers, and occasionally anthers may pollinate later-emerging stigmas on the same head. Furthermore, variations in the degree of protogyny may influence the extent of natural cross-pollination.

For controlled selfing, heads are enclosed in bags prior to the emergence of the stigmas, when most of the head is in the boot. If protogyny results in an extended time lag between stigma emergence and pollen release, poor seed set on selfing results.

For controlled cross-pollination, heads of the female parent are enclosed, before emergence of the stigmas, in glassine bags. When most of the stigmas are *extruded*, pollen obtained from the male parent is dusted on the head.

A few pearl millet cultivars set seed equally as well when selfed as when crossed. Genetic male sterility may be responsible for a lack of seed set (Burton, 1958). In the United States, where pearl millet is used largely as a forage crop, fertility restoration is not needed. Sterility in the commercial hybrid would prevent a producer from using F_2 seed and experiencing the associated yield losses resulting when segregating generations are produced. Sterility may help also to maintain the hybrid longer in a desirable vegetative condition for forage production.

The mechanisms needed to produce commercial millet hybrids for grain were first

Figure 9–8. Inflorescences of pearl millet at three stages of development: at protogyny (left), anther emergence (center), and grain maturity (right). Protogyny is a mechanism that encourages cross-pollination, as the female reproductive structures mature before anthers develop and release pollen. The stigmas of the pistil are exerted first in the top third of the head, proceeding downward until stigma extrusion is complete in about three days. Emergence of anthers occurs after stigma emergence is complete and in the same order. Stigmas remain receptive for two or three days, but cross-pollination is the rule. [Courtesy of International Crops Research Institute for the Semi-Arid Tropics (ICRISAT), Hyderabad, India.]

observed in 1955–1956 at Tifton, Georgia (Burton, 1958). Included in the findings were cytoplasmic male sterility or the A line; a maintainer line B; and a source for fertility restoration, the R line. These were released freely for hybrid development and were known as Tift 23 A and B.

The A and B are genetically similar, and when the male-sterile A parent is pollinated with a genetically similar maintainer B parent, seed is set that again gives rise to male-sterile plants. Cultivars with male-sterile cytoplasm have a recessive gene (*ms*) for maintaining sterility. The sterility-maintainer parent has the normal cytoplasm and the same recessive gene (*ms*). When fertile hybrids are desired for grain production, the cytoplasmically male sterile A lines are crossed with restorer (R) parents.

In 1961 and 1962, additional cytoplasmic male-sterility sources were discovered at Ludhiana, India, in two different populations (Athwal, 1965). Additional and different sources were welcomed, as breakdown of cytoplasmic male sterility was observed to occur sometimes when male-sterile cultivars were grown in environments that differed widely from those in which they were originally selected. The two cytoplasmic male-sterile sources were reported to possess distinctly different sterile cytoplasm (Burton and Athwal, 1967).

Likewise, not all restorer mechanisms are reliable under diverse environments. Although sterility maintainers and partial and reasonably reliable fertility restorers occur abundantly (Burton, 1977), sources capable of complete, reliable fertility restoration in all sterile × fertile hybrid plants are scarce (Burton and Powell, 1968).

The procedure for incorporating cytoplasmic male sterility in new cultivars is achieved by using an existing male-sterile A line as the female which is crossed with an inbred plant to be sterilized. A backcross system is used with the F_1 as the female, because it is sterile, and the line to be sterilized as the recurrent male parent. Backcrossing is continued until the recurrent parent is reconstituted.

The first hybrid of pearl millet for grain was released in India in 1965 and was reported to be capable of yielding nearly twice as much grain as adapted open-pollinated cultivars (Burton and Powell, 1968). Despite initial problems with fertility restoration, superior grain-yielding hybrids, using the male-sterile cultivars released from Georgia, were developed with remarkable speed. The era of hybrid millet had arrived. In a 15-year period following the development of hybrid millet, eight additional hybrid cultivars were developed and released (Anonymous, 1980a).

HYBRID CROPS IN GENERAL

The success of hybrid corn, sorghum, and millet, both from a commercial and a grain- or forage-yield viewpoint, has raised the concept that some day all crops will be hybrid. This premise is based on Vavilov's Law of Homologous Series and the concept of parallel variation. Parallel mechanisms necessary for F_1 hybrid development have been found in many crops. Continued research will mean that parallel situations will be extended to an increasing array of crops (Rutger and Carnahan, 1981).

Hybrids have been developed using cytoplasmic male sterility in all pollination systems including self-pollinated crops, the most difficult in which to develop F_1 hydrids. By far the most common method of developing F_1 hybrids is the use of cytoplasmic male sterility. Other mechanisms can be used, however.

F_1 hybrids have been produced in several crops through the application of techniques involving apomixis, e.g., buffelgrass (Taliaferro and Bashaw, 1962). Hybrid vigor can be fixed and perpetuated through vegetative propagation, e.g., Bermuda grass (Burton and Hart, 1947). Burton and Hart (1947) used self-incompatible but cross-compatible Bermuda grass to produce hybrids. In Sudan grass Craigmiles (1961) used the conventional cytoplasmic sterility for hybrid development.

The technology required to produce commercial hybrid cotton was developed in the 1960s (Davis, 1978) using male-sterile and restorer stocks and bee-pollinators for the often-cross-pollinated species. The use of insecticides is therefore limited, making hybrid cotton seed production difficult. Hybrid cotton cultivars are believed to offer more opportunity to the cotton industry than any other single genetic possibility on the horizon because of the level of heterosis found (Bhatt and Rao, 1980).

Hand emasculation and pollination in tomato has been replaced by large-scale production of inexpensive hybrid seeds through the use of male sterility (Daskalov and Georgiev, 1976). Genetic male sterility in the tomato crop is controlled by a single gene maintained in the heterozygous condition, and fertile plants are rogued out. Hybrid seed is produced by hand pollination on the male-sterile plants.

Quality and yield of such vegetable crops as onions, carrots, cole crops (cauliflower and broccoli), brussel sprouts, muskmelon, squash, and cucumber have been improved by F_1 hybrid development (Dorsman, 1976; Jones and Clarke, 1947). Male sterility is used to produce hybrids, and, since vegetables are produced for their vegetative growth, fertility restoration is of no consequence.

The use of F_1 hybrid seeds of several of the herbaceous bedding plants—petunias, zinnias, pansies, and marigolds—has resulted in superior flower types and plant vigor. Flower breeding has been concerned with heterosis from the very beginning of F_1 hybrid production because of hand pollination and asexual techniques of propagation (Reimann-Philipp, 1976). Several genetic systems providing self-sterility have been introduced and used in practical breeding programs. F_1 hybrid production has been extended to a remarkably large number of flower species.

The potential use of hybrids in the production of perennial forage crops remains somewhat uncertain. Hybrids have been tested in bromegrass (Craigmiles, Crowder, and Newton, 1965; Dunn and Wright, 1970), crested wheatgrass (Knowles, 1970), orchard grass (Christie, 1973), alfalfa (Pedersen and Hill, 1972), and red clover (Anderson, Taylor, and Kirthavip, 1972). In general, moderate levels of heterosis have been reported, but Melton (1969) concluded that the breeder could make as much progress with *synthetic cultivars* as with hybrids.

In the self-pollinated crops of sesame and safflower, male sterility offers promise that F_1 hybrids can be developed (Heaton and Knowles, 1982; Osman and Yermanos, 1982).

Yams have traditionally been propagated by vegetative means. The development of techniques to make the dioecious and unisexual yam set viable seed has opened the way for selection of hybrid vigor and for perpetuation of this vigor by subsequent asexual propagation (Sadik and Rockwood, 1975).

The discovery of cytoplasmic male sterility in sunflower in 1969 and the subsequent identification of genes for fertility restoration in 1974 has resulted in the widespread production of sunflower hybrids (Beard, 1981; Dedio, 1979; Reddy and Thammirau, 1977; Whelan 1980).

In asexually propagated crops such as sugar cane, potato, cassava, poplar, and willow tree, carefully selected parents can be cross-pollinated and the hybrid perpetuated indefinitely through asexual cuttings.

Successful F_1 hybrid production means more than obtaining hybrid vigor expressed in grain or forage yield. Attention must also be directed to a multiplicity of qualitative and quantitative factors, such as disease and insect resistance, machine harvestability, plant height and architecture, lodging resistance, maturity, and quality factors. Increases in photosynthesis must be utilized effectively through satisfactory sink-source relationships.

The self-pollinated cereals of wheat, rice, barley, and oat offer possibilities for hybrid development with a challenge for commercial success. Many of the approaches to hybrid development are based on perceptions and techniques formed from hybrid corn development. The initial objective of hybrid corn was to develop a technique to control the genotype so that a specific hybrid cultivar could be reproduced as desired. Careful selection of parents led to hybrid vigor for grain yield, with the result that hybrid development has become synonomous with a yield increase. The extent of this increase, realized over many years of hybrid corn development, cannot be measured by comparing hybrid corn with open-pollinated cultivars, because concentrated work on hybrids to the exclusion of open-pollinated cultivar development does not make for an accurate comparison. Generally, however, the increased yield associated with hybrid development is considered to be about 25% (Mangelsdorf, 1951).

In addition to grain yield, hybrid corn offers the advantage of uniformity in ear height and maturity, improved disease and insect resistance, shorter and stronger stalks suitable for mechanical harvesting, increased tolerance to high plant populations and other stress conditions, and flexibility to modify specific gene resistance. Corn has set a standard of performance for hybrids.

Development of hybrids in other crops may require a change in the rationale, procedures, and outlook formed from hybrid corn production. To illustrate the challenges to be met to make hybrid cereals a success, the following sections examine hybrid possibilities and problems in these crops.

HYBRID WHEAT

A pioneering step toward hybrid wheat development was taken by the Japanese scientist (Kihara (1951), who was the first to report cytoplasmic male sterility in hexaploid wheat. Kihara's colleague, Fukasawa (1953 and 1955), reported the second requirement, that of fertility restoration in several genetic backgrounds.

The male-sterile cultivars developed by Kihara, although in a hexaploid background, had severe limitations for potential commercial use, as did the male-steriles of tetraploid background developed by Fukasawa (1953). Likewise, the fertility-restoration mechanism reported by Fukasawa (1955) was limited to specific genotypes.

The first truly promising stable male-sterile cultivars in a hexaploid background were developed by Wilson and Ross (1962) at the Kansas Agricultural Experiment Station at Fort Hayes. The source of cytoplasmic male sterility was *Triticum timopheevi*. In the same year, Wilson and Ross (1962) reported on the successful restoration of pollen fertility through the use of fertility-restoring factors transferred likewise from *Triticum timopheevi* to the hexaploid wheats.

Basic stocks for use in the development of hybrid wheat were made available to plant breeders, starting a wave of excitement in research stations around the world. Many research programs were initiated on hybrid wheat development, with visions of yield increases amounting to a quick and easy 20%. Captions such as "greatest find in 50 years" stimulated the interest of both wheat growers and seed producers.

But hybrid wheat was not developed quickly and easily. Many more complicated problems were involved than first anticipated by many researchers. After a major shift to hybrid wheat research, most public agencies and several private agencies have dropped or drastically curtailed their hybrid wheat research efforts. Reasons for discouragement were listed by Hayward (1975) as follows:

1. High cost of research.
2. A questionable level of heterosis in a self-pollinated polyploid.
3. The **nicking** and seed-set problems, the total amount of pollen available, and the efficiency of cross pollination when depending upon wind to transfer the heavy pollen grains from the anthers of one plant to the stigmas of another.
4. The questionable economic profitability of wheat hybrids when considering production costs, seeding requirements, the large low-yielding, dryland areas, and the competition of more profitable crops.
5. The belief that yield gains can be achieved as rapidly by conventional pure-line methods.
6. The decision by many public institutions to do more basic research, leaving more of the applied hybrid research and development to private agencies.
7. The failure to find adequate high-yielding pollen restorer systems of good agronomic types that perform well in hybrid combinations under varying environmental conditions.
8. The cost, personnel, and difficulty involved in finding isolation areas for A-line increases and hybrid seed production.

The number and extent of hybrid wheat development problems have discouraged many. A 12-year feasibility study by Canada Agriculture showed that biological components for producing hybrid wheat indeed are available. However, continued research for hybrid soft white winter wheat in southern Ontario requires more years and resources than the uncertain results warrant. As a result, in 1978, after years of effort, Canada Agriculture opted out of hybrid wheat research (Sampson, 1978).

Yet, despite the apparent problems, hybrid wheat research programs continue in Japan, Bulgaria, and the United States. Sufficient optimism prevails that a few public institutions and several U.S. commercial companies have continued or expanded their hybrid wheat programs in the hope that it will pay off in areas where environmental factors do not limit the yield potential.

After two decades of research and development, hybrid wheat remains more of a promise than a reality. The prospect for commercial development is one of degree, and continued research is directed to the challenges encountered.

HYBRID WHEAT DEVELOPMENT CHALLENGES

Multiplication Ratio

Hybrid wheat seed production has three separate and distinct steps in seed production: restorer increase, male-sterile increase, and hybrid seed production (Figure 9-9). A low seed-multiplication ratio hinders economic development in each of the three steps.

In wheat and other small-grain crops, high plant populations commonly are used (Table 9-1) that result in seeding rates several times that of corn, sorghum, millet, or sunflower. As well, in these latter crops, a single hand pollination produces enough seed to conduct yield trials at several locations; the same single hand pollination in wheat, oat, barley, or rice produces fewer than 30 seeds. The labor and costs of experimental hybrid testing are thereby increased, limiting the number and diversity of hybrids tested. To produce sufficient seed for experimentation, the female cereal parent must be converted to its male-sterile counterpart; both the male and female parents must be increased; and hybrid seed must be produced in isolated crossing blocks in which a single fertility-restorer male plant is used to pollinate several male-sterile female parents.

Consideration for plant population and seeds harvested resulted in a multiplication ratio of 25 for wheat verses 215 and 450 for corn and sunflower, respectively (Table 9-2). Using a 1 : 1 male-sterile : maintainer ratio, with 60% seed set in the male-sterile parents, an area 2.3 times as large is required in order to produce the same amount of seed of a conventional cultivar (Lucken, 1981).

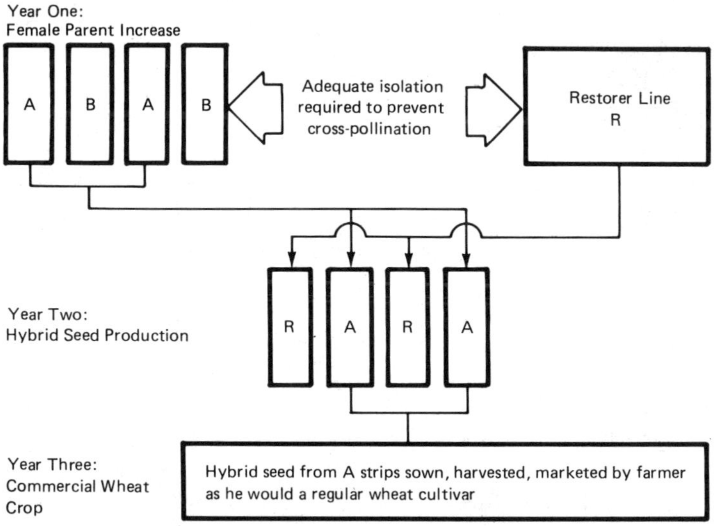

Figure 9–9. In contrast to the production of ordinary self-pollinated common wheat, in which a seed grower has only one type of seed to grow for each cultivar, hybrid wheat seed production involves the production of three types of seed. The three are called A, B, and R cultivars. The B and R lines are self-pollinated and produced like ordinary cultivars. Seed of A cultivars comes from the cross A×B, and the hybrid seed that the producer grows comes from the cross A×R.

The A or female parent is cytoplasmically male-sterile and will not set seed until it receives windborne pollen from normal wheat. To breed true, the A parent should receive pollen from a genetically similar B or maintainer parent but lacking male sterility. Maintainer parents are produced by backcrossing. The progeny of this cross are male-sterile.

Commercial hybrid seed is produced by crossing the male-sterile A parent with a parent carrying restorer genes.

Cross-pollination must be accomplished on a field scale by drilling alternate strips of parents to be crossed. At maturity the parental strips must be harvested separately.

Table 9–1
Comparative Seeding Rates of Selected Major Crops

	Seeding Rate		*Plant Population (in thousands)*	
Crop	*kg/ha*	*lb/ac*	*hectare*	*acre*
Corn	11–20	10–18	40–75	16–30
Sorghum	2–12	2–10	17–140	6–57
Millet	4–9	4–8	100–250	40–100
Sunflower	3–8	3–7	45–104	18–42
Wheat and other cereals	40–160	36–140	140–6000	58–2500

Table 9–2
Comparative Seed-Multiplication Ratios for Wheat, Corn, and Sunflower in Hybrid Seed Production Fields, with Values Adjusted for Female : Male Ratio and Seed Set

Crop	Planted Population (Seed)	Yield	Number of Hybrid Seeds Harvested	Multiplication Ratio
Wheat	1,000,000	1.817 kg/ha (1622 lb/ac)	25,000,000	25
Corn	20,000	3366 kg/ha (3005 lb/ac)	4,300,000	215
Sunflower	18,000	897 kg/ha (800 lb/ac)	8,100,000	450

Source: Lucken (1981).

Both seeding rates and seed multiplication ratios dictate a rather long lead time, about five years, for the increase of the male-sterile female parent. This is a difficult and expensive proposition, in that male-sterile increase fields must be isolated from other wheat fields to avoid contamination by out-crossing with undesirable pollen sources. Mixing of normal, fertile seed with male-sterile seed must be painstakingly avoided.

Producing sufficient hybrid wheat seed for the entire area seeded to wheat worldwide has considerable commercial appeal but does present a formidable challenge.

Adequate Cross-Pollination

Wheat and other small grains, except rye, normally are self-pollinated crops, and for successful hybrid development natural cross-pollination under field conditions must occur. The crux of commercial hybrid seed production economics is cross-pollinated seed set on the male-sterile parent. Both environment and genotype have a large effect on cross-pollination.

Studies in North Dakota during the early 1970s, using 0.81-ha (2-ac) crossing blocks with drill-strip widths in approximate ratios of 1 : 1, 2 : 1, and 3 : 1 of male-sterile to restorer, showed the magnitude of environmental impact. Highest seed set was achieved at a 1 : 1 ratio, but seed set differed sixfold (Table 9-3). Seed production from male-sterile plants is affected by the same environmental factors that effect the grain yield of the pollinators, as shown by the restorer parent yields in Table 9-3. Prolonged rainfall in more humid areas can reduce cross-pollination even when grain yields are high (Lucken, 1981).

The amount of cross-pollination is influenced also by genotype. Lucken (1981) found that some male-sterile parents consistently had higher seed set and seed yields: some restorer parents are more effective pollinators than others. Genotype can result in yield differences as large as twofold, or threefold, or more. It is essential that proper matching or "nicking" of male-sterile florets with the pollinator parent occurs.

Another extrusion varies substantially among wheat cultivars, influences cross-pollination, can be scored visually at anthesis, can be improved by selection, and also is an indicator of fertility-restoration potential (Bitzer and Patterson, 1967; Joppa, McNeal, and Berg, 1968; de Vries, 1973; Milohnic and Jost, 1976; Atashi-Rang and Lucken, 1978; Komaki and Tsunewake, 1981).

During male-sterile seed increase high anther extrusion for cross-pollination is desirable. Many semidwarf wheat cultivars that otherwise would be excellent female parents have low anther extrusion and cannot be increased readily.

Breeding to develop high anther extrusion in maintainer and restorer parents is possible and necessary. Determination of plants for

Table 9–3
Mean Seed Yields of Restorer and Male-Sterile Seed Yields (Hybrid Seed) Grown at Eight Environments in North Dakota, 1972 to 1974

Location and Year	*Restorer Parent*		*Male Sterile : Restorer Ratio* 1 : 1		*Male Sterile : Restorer Ratio* 2 : 1	
	kg/ha	lb/ac	kg/ha	lb/ac	kg/ha	lb/ac
Larimor, 1972	2320	2071	1750	1563	1010	902
Wyndmere, 1973	1520	1357	1320	1179	850	759
Erie, 1973	1810	1616	1060	946	1010	902
Minot, 1973	2310	2063	1360	1214	1140	1018
Casselton, 1974	1390	1241	360	321	270	241
Langdon, 1974	2400	2143	2150	1920	1520	1357
Langdon, 1974	2400	2143	1190	1063	610	545
Larimor, 1974	740	661	240	214	200	179
Wyndmere, 1974	1560	1393	590	527	540	482
Mean	1830	1634	1110	991	790	705

Source: Miller and Lucken (1977).

anther extrusion can be done rapidly and easily on either a single-plant or a plot basis (Ghiasi, Lucken and Hammond, 1982a).

The finding of male-sterile parents that have high seed-set potential is essential to economical hybrid seed production. In addition to anther extrusion, floral features on the female parent may be important. Lodicule size may be one example, but floral traits in the female inducive to cross-pollination have not been described.

Lucken (1981) observed that although wheat normally is self-pollinated, it does appear to have a latent ability to cross-pollinate, which can be enhanced by using the considerable variation in floral traits that exists. He reported on impressive levels of seed production attained by DeKalb Agresearch Inc. with the right parents in the right environments (Table 9-4). A 3 : 1 female : male ratio was used under irrigation. Female strips averaged 16 m (52 ft) wide with male strips 5 m (16 ft) wide. Positive results shown in Table 9-4 provide encouragement that cross-pollination can be successful, and such results are helpful in extending hybrid seed production to less favorable environments. As

Table 9–4
Hard Red Winter Wheat Hybrid Seed Production Averaged over 18 Crossing Fields over a Two-Year Period in Three Geographical Regions

Year and Location	*Grain Yield* *male*		*Grain Yield* *female*	
	kg/ha	lb/ac	kg/ha	lb/ac
Southwest Kansas 1979	4241	3787	4254	3798
Texas Panhandle 1980	3655	3263	2921	2608
Southwest Texas 1980	3419	3053	3460	3089

Source: Lucken (1981).

production stresses become known, the necessary alterations in floral characteristics of the parents can be recognized.

Reliable and Complete Fertility Restoration

A major challenge in hybrid wheat development has been complete fertility restoration as well as the development of fertility-restorer (R) parents that restore complete male fertility under a range of environmental conditions.

Results from inheritance studies have shown that at least two or three genes plus modifier genes are necessary for full restoration of male sterility induced by *Triticum timopheevi* cytoplasm (de Vries, 1971; Sage, 1972; Sage, 1976).

Selection procedures for restorer parent development are complicated by the need to determine the fertility restoration of each potential restorer parent by test crossing. The testcross system used involves an initial selection for self-fertility as expressed in the cytoplasmic male-sterile background. Desirable restorer genotypes are then further evaluated by their breeding behavior in test crosses with male-sterile parents. Test crossing is necessary to identify genotypes that carry the complement of fertility-restorer genes that are fully effective in the heterozygous F_1 hybrid. The test-crossing procedure is time-consuming and expensive; it restricts the effective population size and the selection pressures that can be applied for agronomic traits and reduces the rate and efficiency of restorer parent improvement.

Since Wilson and Ross's (1962) discovery of male sterility from *Triticum timopheevi* cytoplasm, over 15 different cytoplasms have been recognized, several of which induce male sterility in wheat and could be the basis for alternative systems of hybrid wheat production (Maan, 1975; Mukai and Tsunewake, 1980). Unless alternative systems have a clear advantage over the *timopheevi* system, the considerable time and expense invested in incorporating *timopheevi* cytoplasm and restorer genes into agronomically suitable wheat will limit changes to alternative systems. No advantage is apparent (Ghiasi and Lucken, 1982).

Some degree of F_1 sterility can be found in many hybrids under stress conditions and appears to be a more common problem in the soft red and white winter wheats and in the hard red spring wheats than in the hard red winter wheats (Lucken, 1981; Trupp, 1976). Environmental causes of variation in F_1 hybrid fertility as observed in common sets of hybrids grown in several countries are not evident (Lucken, 1981).

The research to find, incorporate, and refine a ctyoplasmic male-sterility fertility-restoration system has required ana enormous effort. It indicates that the species "gap" between *Triticum aestivum* and *Triticum timopheevi* is relatively large.

Satisfactory Hybrid Performance

Yield increases of hybrid wheat must be of sufficient magnitude to more than offset the increased seed costs. Superimposing heterosis or F_1 vigor upon the yield base of the present high-yielding conventional commercial production of F_1 hybrids using cytoplasmic male sterility is expensive, and so the question of a yield advantage must be answered if hybrid wheat is to be commercially successful.

After more than 20 years of hybrid wheat research, the yield question has not been answered satisfactorily. As yield data are accumulated, the complexity of the yield question increases. Perhaps the situation cannot be fully explored until a reliable system of F_1 hybrid production has been perfected.

It goes without saying that both A and R parents of a successful hybrid have to have a number of desirable features, including resistance to lodging, disease resistance, winter hardiness if winter types are produced, acceptable height, and milling and baking quality. In the initial stages of research, many crosses were made that lacked adequate agronomic features. After years of breeding and backcrossing the better restorers and A parents are comparable to conventional cultivars for most agronomic traits.

For heterosis to be maximized, genetically diverse parents must be crossed. No particular advantage or disadvantage appears to be associated with hybrids, and milling and baking quality values usually fall within the range of the parents. In an effort to increase genetic diversity, crossing a hard with a soft quality parent might reduce quality aspects. At high yield levels, low protein content may occur, a desirable situation for soft pastry wheats but undesirable for strong gluten bread wheats.

Lucken (1981) reviewed the extensive number of experiments conducted to evaluate hybrid wheat. Many involved hand-produced hybrid seed, often from **diallel crosses,** sometimes tested as single potted plants, sometimes as small unreplicated plots seeded at low seeding rates, sometimes as an F_2 generation, but well tended and carefully measured. Because the trials were conducted with limited seed, usually tests were conducted at one site only, so that the impact of environment could not be determined. The average yield increase measured in such plots over a four-year period, 1964 to 1967, for 36 wheat hybrids was 31% (Livers and Heyne, 1968). The results of such tests have been inconclusive, because there are several standards against which the performance of hybrids can be compared. By definition, comparisons with the mid-parent or high parent provide an indication of hybrid vigor. The deciding comparison for a producer considering hybrid production is that between the best available cultivar and the hybrid. This is called standard heterosis.

A wheat breeder may perceive hybrid performance in terms of genetic differences between parents of a cross, in terms of ability to combine the two parents, or in terms of relative rates of progress in improving yield from conventional breeding or hybrid programs.

Actual and potential hybrid performance must be compared. As deficiencies in existing parents and hybrids are eliminated, hybrid performance can be improved. Continued efforts with hybrid wheat will lead to continued improvements.

These considerations should be kept in mind when the hybrid test results are examined. For example, during the 1970s, disappointing yield results relative to commercial cultivars were obtained from field-scale experimental hybrid trials (Table 9-5). These early field results gave yield comparisons, but, more important, they tested the entire hybrid wheat system and prepared the way for pilot-scale and commercial production.

Increasingly extensive yield trials have produced yields 0%, 5%, 10%, and sometimes 15% more than those of the best commercial cultivars. Evidence of superior hybrids and superior parents for specific crosses boosted the yield advantage

Table 9–5
Grain-Yield Comparisons of Hard Red Winter Wheat F_1 Hybrid Field Trials from the Great Plains of the United States

	Grain Yield			
	1976[a]		*1977*[b]	
	kg/ha	lb/ac	kg/ha	lb/ac
Check cultivar	3231	2885	3231	2885
Hybrids[c]	2962	2645	3164	2825
Best hybrid (standard heterosis)	3164	2825	3366	3005

Source: Johnson (1977) as reported by Lucken (1981).
[a]Eleven sites in five states.
[b]Ten sites in four states.
[c]Fifteen hybrids in 1976; 16 hybrids in 1977.

from 12% to 21% based on 11 trials over a three-year period (Lucken, 1981).

From disappointing initial tests, progress to modest but promising results is evident. Continued progress is certain to occur. Lucken (1981) listed seven factors affecting hybrid performance:

1. Agronomic merit of male-sterile and restorer parents.
2. Combining ability of parents.
3. Genetic diversity of parents.
4. Genetic system used to produce hybrids.
5. Quality of seed produced on male-sterile plants.
6. Purity of hybrid seed.
7. Hybrid vigor inherent in wheat.

Hexaploid wheat is a polyploid with the amphiploid kind of chromosome pairing; i.e., it contains three sets of genomes and is in fact a "permanent heterozygote," in which positive and negative heterotic interactions are effectively fixed. Heterosis implies values outside the parental range, beyond that produced from genomic interactions, which cannot be separately evaluated. Heterotic effects and ploidy must be considered jointly.

F_1 hybrid vigor can be identified readily in wheat, but few crosses exhibit a substantial increase in vigor. Initially, few parents had the required combination of fertility-restoration, agronomic, and seed-production attributes, so that the number of hybrid combinations was restricted. As hybrid breeding programs progress, a greater number of potential parents will be produced and greater levels of hybrid vigor may be found.

The fact that inbred cultivars can be easily produced and tested offers wheat a significant advantage over cross-pollinated crops such as corn. Perhaps the emphasis in hybrid development should be directed to testing parents rather than testing their hybrid progenies. Effort in hybrid wheat development would result not only in higher-yielding hybrids but in superior conventional parental wheats as well. Bitzer, Patterson, and Nyquist (1982) suggested that successful hybrid wheat will most often come from high- × high-yielding crosses because of the preponderance of general combining-ability effects in wheat. Hybrid wheat must be more than simply a cross between conventional cultivars; instead it must be the product of a hybrid breeding program involving superior parents, carefully evaluated for genetic diversity and combining ability. Superior parental cultivars should produce better hybrids which may be used to develop improved conventional cultivars. Hybrid and conventional breeding programs, therefore, are at the same time complementary and competitive. What should emerge from hybrid wheat research is high-yielding wheat for both hybrids and parents. The promise of reward is very great, even though completion of the task may be difficult.

HYBRID RICE

The fact that more people rely on rice for their daily food supply than on any other crop makes the promise of increased yield through hybrid development an exciting prospect with worldwide implications.

As in wheat, hybrid development in rice is challenged by the fact that it is a self-pollinated crop. Several distinct advantages over other self-pollinating small grains are found in rice, however: (1) Under tropical conditions rice can behave as a perennial crop. (2) Although seeding rates may be as high as for other cereals (Table 9-1), in labor-intensive areas where transplanting of seedlings is practiced, seed requirements are reduced substantially (Figure 9-10). Expensive seed associated with F_1 hybrids may be feasible economically because of widespread transplanting of seedlings. (3) In rice-breeding programs, multigenerations can be obtained within a year in early-maturing rice cultivars.

Although Craigmiles, Stansel, and Flinchum (1968) concluded that hybrid rice of the world had potential in the transplanted rice areas of the world, and although they suggested a possible method of producing hybrid rice, a major problem has been out-crossing (Carnahan et al., 1972). It was not until the early 1980s that the

Figure 9–10. The rice seedlings in this photograph have been transplanted by hand, a practice that drastically reduces the seed requirements of this crop. Transplanting is a desirable practice for hybrid rice production, as it lends itself economically to the use of expensive F_1 hybrid seed. Tillering, a characteristic for which rice is noted, is associated with high yield performance under spaced conditions. The possibility for yield increases with hybrid rice has worldwide implications, because rice is the main source of nourishment for over half the people of the world. [Courtesy of the International Rice Research Institute (IRRI), Los Banos, Philippines]

Chinese produced hybrid rice commercially (Virmani and Edwards, 1983).

Research on hybrid rice was initiated at IRRI in 1970 by incorporating additional sources of cytoplasmic male sterility (Erickson, 1969; Athwal and Virmani, 1972) into indica rice (Anonymous, 1980b). About the same time, rice researchers in the People's Republic of China started exploring the possibilities of hybrid rice.

Despite seemingly insurmountable problems in seed production, production of about 5 million hectares (over 2 million acres) of F_1 hybrid rice occurred in 1980 in Hunan province of China (Lin and Yuan, 1980). Seed set in hybrid seed production fields was raised from an unacceptably low 15% in China by two labor-intensive practices (Azzini and Rutger, 1982). (1) The flag leaf on both the female and male parents was cut back to remove these obstacles to pollen dispersal. The flag leaf of the female was completely removed, while two-thirds of the male parent flag leaf was cut back. (2) Pollination was facilitated by pulling a rope over the panicles of the male rows at 20- to 30-minute intervals three to five times daily on calm days (Virmani and Edwards, 1983).

In the United States out-crossing on three short-stature male-steriles of rice averaged 5.4% in 1978 and 32.5% in 1979 when planted in alternate rows with pollinators constituted of a mixture of tall cultivars. These values include a 3.5% selfing. The higher seed set in 1979 apparently was due to better synchronization of flowering of male and female plants (Azzini and Rutger, 1982). This unacceptably low amount of outcrossing was achieved even with a low ratio of male-steriles to pollinators, but was higher than the 16% reported by Carnahan et al. (1972) with the same cultivar.

In addition to the cytoplasmic male sterility, maintainers, and restorers needed for F_1 hybrid development, Rutger and Carnahan (1981) proposed a fourth element in hybrid rice seed production—a recessive gene for tallness. Usually, tall mature plant height is dominant to shortness in cereals. This results in tall hybrid plants in short-by-tall crosses, which is an undesirable height in situations where short-stature plants are known to be more productive. Using a tall male parent carrying a recessive gene for elongated internode, an obvious advantage exists in pollen dispersal among semidwarf or dwarf female parents. Use of a recessive tall in pollinator parents results in the desired short F_1 hybrids for the commercial crop. In addition, since the pollinator is about 25 cm (10 in.) taller than the female parents, it might be possible to eliminate most of the pollinators by clipping off their panicles or by rope-wick application of herbicide soon after pollination. The tall pollinators must be removed soon after pollination, because they eventually bend down into the canopy as grain filling progresses. Removal of the pollinators facilitates harvest of the entire field for hybrid seed production. A small percentage of missed tall plants can be tolerated in the F_1 field if sufficient heterosis occurs to make hybrids feasible.

HYBRID BARLEY

Methods involving cytoplasmic male sterility exist in barley (Ahokas, 1979) but are less developed than in wheat. Incorporation of male sterility from wild barley into suitable cultivars remains to be done. The second requirement, fertility restoration, is as yet incomplete.

Major gene-controlled male sterility has been pursued as an avenue for F_1 hybrid barley development (Hayes and Foster, 1976). A recessive gene for male sterility has been utilized in two main ways: (a) by use of balanced tertiary **trisomic** plants, and (b) by a susceptibility/resistance reaction.

Use of Balanced Tertiary Trisomics

Male sterility in barley can be obtained by means of an extra chromosome obtained through a reciprocal translocation and called a *balanced tertiary trisomic (BTT)*. The extra chromosome of the BTT carries a dominant fertile allele of the male-sterile gene, and the male-sterile recessive alleles are carried on a pair of normal chromosomes (Ramage, 1965). Selfing a BTT plant in which the extra chromosome has been transmitted through the ovules and not through the pollen produces 30% male-fertile BTT plants identical to the parent and 70% diploid plants homozygous for male sterility. The mixture of seed can be separated by seed size. Generally the trisomic seed is smaller and produces less vigorous plants than the normal diploid seed. The BTT seed either can be used to produce pollen parents or male-sterile diploid plants or can be multiplied in isolation for multiplication.

The procedure to produce hybrid barley is to screen out the larger seed, which is sown as the male-sterile diploid female parent. This seed is sown thickly, so that remaining and less vigorous BTT plants are eliminated by competition. The smaller BTT seed is sown thinly to enable the morphologically distinct male-sterile diploid plants to be removed by hand.

A problem with this system is the lack of absolute homogeneity of F_1 seed, because the separation of the diploid from trisomic seed by weight is not complete. A more satisfactory separation might be achieved when the extra chromosome is incorporated into a fully adapted two-row genetic background rather than on six-row cultivars, which produce smaller grains on the laterals than on the medium spikelets of the ear. An additional problem has been the fragmentation of the translocated chromosome with associated genetic changes (Hayes and Foster, 1976).

One advantage of the trisomic breeding method is the elimination of seed losses due to incomplete fertility associated with cytoplasmic male sterility.

The first hybrid barley using the BTT method was produced in 1968 in Arizona and yielded 15% to 20% higher than the commercial cultivar.

Problems of cross-pollination, mechanics of seed planting and harvesting, and level of heterosis have caused economic considerations in barley similar to those outlined in wheat (Campbell and Kasha, 1981).

Susceptibility/Resistance System

A scheme for producing F_1 hybrid barley using diploid plants only was proposed by Wiebe (1960). A genetically controlled susceptibility/resistance reaction can be used to eliminate male-fertile plants in a population segregating for male sterility. The system requires very close linkage between the recessive genes for resistance to the chemical and male sterility. No such scheme has yet been put into practice.

HYBRID RYE

Hybrid development in rye began in the early 1970s, when cytoplasmically inherited male sterility and nuclear restorer genes were detected (Becker, Geiger, and Morgenstern, 1982; Geiger and Schnell, 1970). Because rye is a cross-pollinated crop, commercial development appears easier, and many hybrid breeding programs have been started in Europe, where rye is grown more extensively than in other areas of the world. Results in rye appear to be promising, and a refinement in methodology and breeding goals may result in commercial adaptation.

Studies in Germany (Becker, Geiger, and Morgenstern, 1982) that took into account yield performance, stability, and economy of seed production, revealed a double cross or an equivalent form with a two-line synthetic as pollinator parent as the most favorable hybrid type for rye at present.

HYBRID CEREALS AND DISEASE

Loose smut[1] and ergot[2] are possible problems in F_1 seed production because of the susceptibility of the open-flowering male sterile lines of normally self-pollinating crops. Systemic fungicides will continue to control loose smut, but a virulent new form resistant to the prevailing chemical is an ever-present threat. Loose smut is carried in the germ and unless checked could cause serious losses and contamination in the F_1 generation.

The ergot fungus is common on cereals and grasses throughout the temperate regions. Infection can originate from grasses in the fence rows. Florets infected with windborne ergot spore produce an ergot body that may be harvested with the grain or fall to the ground. Under conditions of cross-pollination a longer time is required for fertilization to take place than if self-pollination occurs and susceptibility is increased.

F_1 hybrids, however, present an efficient means of using dominant genes for resistance to diseases and insect pests other than those infecting the flowering organs. Combining two different dominant genes in the F_1 hybrid provides a method of bringing together heterozygosity for a high level of race nonspecific resistance. The spectrum of resistance could be changed by use of a different parent.

CHEMICAL HYBRIDIZING AGENTS

An attractive alternative to the development of male sterility and the associated requirement of fertility restoration is a system involving chemicals that selectively induce male sterility. Using this technique, conventional cultivars could be turned into female parents by treatment with the chemical prior to anthesis. Alternating drill strips of untreated male parents serve as the pollinator to produce hybrid seed. Research programs of several chemical companies are directed to the synthesis identification and the evaluation of the potential of these compounds, which appear sufficiently promising to warrant the issuing of patents for them and their use (Lucken, 1981; Virmani and Edwards, 1983).

Environmental stresses such as frost, drought, or high temperatures that cause male sterility but not female sterility suggest that an inherent differential sensitivity exists between the male and female reproductive organs, a situation imperative to the success of a chemical hybridizing agent (CHA). It is known also that separate nuclear genes or cytoplasmic mechanisms influence male sterility. Nevertheless, the requirements of CHAs are exacting, for not only must they selectively induce complete or near-complete male sterility, but they must not reduce female fertility, must be sufficiently persistent to sterilize both early and late tillers, and should have a minimal interaction with genotype and environment. Preferably, CHAs should be **systemic** in their action. Other effects should be minimal, with no residual effect in the F_1 generation.

The first chemicals tested as possible CHAs were deficient in one or more aspects, in that they reduced internode elongation, interfered with spike emergence from the boot, thereby reducing cross-pollination, upset normal spike development, caused leaf chlorosis, delayed heading date, and affected seed development (Rowell and Miller, 1971 and 1974; Jan, Qualset, and Vogt, 1974; Fairey and Stoskopf, 1975; Hughes, 1976; Stoskopf and Law, 1972; Dotlacil and Apltauerova, 1978; Miller and Lucken, 1977; Johnson and Brown, 1978; Lucken, 1981). Many of these factors could reduce yield, and, because male sterility is frequently only 90%, further progress is needed for its successful commercial use. Because of continued progress and success at minimizing these problems, hope for a CHA prevails.

[1]Loose smut—*Ustilago* spp.
[2]Ergot—*Claviceps purpurea*.

In addition to reducing breeding programs needed to manipulate cytoplasm and fertility-restorer genes, CHAs promise to reduce problems associated with the expensive and time-consuming male-sterile increases. Lead time for parental increase is reduced. Cultivars with poor anther extrusion, including semidwarfs and dwarfs, could be used as female parents without concern about anther extrusion in maintainer parents. Chemically induced male sterility may provide a rapid method for screening various parental combinations when seeking F_1 hybrids that show positive heterosis for yield.

CHAs are not a panacea for hybrid crop development. They will not eliminate the problems associated with cultivars that lack floral and spike characteristics associated with economical cross-pollination in the female. Timeliness of CHA application may be a problem under rainy conditions. Prolonged windy conditions may make it difficult to avoid spray drift to the female parent if seeding is in alternate strips.

Perhaps future advances in this area may involve female-sterilizing agents patterned after male CHAs. This could be useful if the compound could be applied to the pollen-producing male parent, so that a mixture of male and female plants could be grown.

Alternatively, a chemical capable of fertility restoration in genetic male-steriles might have greater impact than chemicals that induce male sterility (Hockett, Baenziger, and Steffens, 1978). In the absence of the chemical treatment, a genetic male-sterile plant could be completely male-sterile, thereby eliminating the need for a maintainer parent. In barley, an array of genetic male-sterile plants are available to test this theory.

Chemical use for hybrid crop development has a strong appeal, and it may prove to be the technique of greatest value in changing hybrids from a biological reality to a commercial reality.

HYBRID VIGOR OR HETEROSIS

Heterosis is of outstanding importance in practical plant breeding and has been given intense consideration in all fields of genetic research. There is not doubt that the causes of hybrid vigor are a result of genetic processes, yet the nature of heterosis has not been revealed completely. After years of research, neither genetic nor physiologic approaches satisfactorily explain it.

From a practical viewpoint, heterosis has become synonymous with an increase in yield. Without a yield increase to offset its cost, commercial hybrid seed production is not an attractive proposition. Yield increases must be of sufficient magnitude to compensate for the additional inherent expense in the production of hybrid seed.

From a genetic standpoint, two theories, those of "dominance" and "overdominance," are among the older explanations of heterosis (Powers, 1944; Shull, 1948).

The dominance theory suggests that deleterious recessive genes are masked by dominant genes, e.g., $aaBB \times AAbb = AaBb$. The plant *AaBb* has a dominant gene at each allele that results in heterosis.

The overdominance theory suggests that heterozygosity by itself may be desirable. The heterozygous condition *Aa* is believed to be superior to either homozygous condition *AA* or *aa* because of a "buffering" action. Overdominances at different loci are then added up to form heterosis (Srivastava, 1981).

From a plant-breeding viewpoint, Shull's concept of a luxuriant, unfixable superiority of the hybrid over the parents has merit. Heterosis may not, however, be associated with superior grain yield. In the situation where forage production is the economic yield, heterosis may be expressed in association with sterility, and no reproductive capacity exists.

From an evolutionary standpoint, heterosis is regarded as a tool for maintaining genetic flexibility, and from this aspect it appears logical to compare hybrid and parental mean. In contrast, a crop producer seeks progeny that are superior to both parents.

Sage and Hobson (1973) measured heterosis from a physiological aspect based on the extraction and assay of **mitochondria** from plant shoots. This technique might have possible value as a preliminary screening of potential parental lines to facilitate the rational choice of suitable material for producing hybrid wheat. The results obtained by determining the oxidative efficiency of mixtures of the mitochondria of potential parents were suggested by Lupton (1976) to give a measure of the yield potential of the crosses concerned, but this measure might be invalidated by the incidence of disease or by other agricultural limitations to crop performance. Lupton was not able to relate yield heterosis to differences in the photosynthetic activity of seedling plants. Perhaps sink-source relationships prevented higher photosynthetic rates from being utilized.

Part of the difficulty in understanding heterosis appears to be associated with the number of vague theories postulated. Perhaps heterosis cannot be separated into discrete aspects designed to meet individual perspectives.

Konarev (1976) approached heterosis from a comprehensive viewpoint. He described heterosis as a complex biological phenomenon manifested in the superiority of hybrids over parent forms according to the rate of development of one or more complex characters. Konarev observed that heterosis is expressed very clearly in many crosses between species and in a number of crosses between cultivars of one species. Heterosis occurs mainly in the F_1 progeny and falls off sharply in the following generations. When inbreds are used as parents, specific combinations produce the greatest level of heterosis. By crossing inbreds, heterosis can be controlled genetically in a number of crosses.

Six hypotheses for heterosis that take a comprehensive approach to the question were summarized by Konarev (1976). Heterosis may appear in a hybrid as a result of:

1. Suppression of negative action of some recessive genes by dominant ones.
2. Combined action of favorable dominant factors.
3. Compensation of a defective recessive gene (e.g., mutant) of one parent form by a normal gene of the other.
4. Superiority of heterozygotes (*Aa*) over homozygotes (*AA*, *aa*).
5. Forming of complete genetic complexes by including additional genes.
6. Changing of the genetic balance of a genome in the process of interaction of different genotypes.

Perhaps all six of these causes are involved in heterosis, but their appearance level of expression and degree of manifestation may vary. On this assumption, Konarev (1976) suggested that a general theory of heterosis be presented that binds together all available data. The hypothesis suggested is that of *genetic balance and genetic complex completeness.*

Genetic balance assumes that the performance of any character is determined by the relationship of many factors. According to this view, inbreeding results in the development of forms that are not balanced genetically. When two inbreds are crossed, genetic balance is improved.

The second component of the hypothesis assumes that the causes of heterosis are expressed through additive and modified gene action. Contemporary views consider that the plant is a system and that any plant character is determined not only by primary activity of associated genes but also by the condition of the whole genetic system. Frequently, however, heterosis appears to be expressed in increased development of discrete characters, such as plant height, root weight, leaf area, disease resistance, and cold tolerance. Heterosis seldom appears to influence the entire plant organism. It is concluded, therefore, that a series of separate genetic systems (the six listed) interact to produce heterosis.

MacKey (1976) considered heterosis from three aspects:

1. *Direction.* Heterosis may be positive or negative. This deviates from its definition as the "increase" of size, yield, vigor, and so on given by Shull (1948) in 1914 when he introduced the term heterosis. Negative heterosis was demonstrated in the number of locules per tomato fruit. F_1 progeny had fewer locules than either parent, a negative but desirable form of heterosis.

2. *Expression.* Heterosis may be luxuriant, as expressed in increased plant size, adaptive, selective, and/or reproductive in its advantage or disadvantage.

3. *Transmissibility to the next sexual generation.* Heterosis may be unfixable, or it may be fixable either in a balanced heterozygous or homozygous constitution.

Despite the complexity of the explanations of heterosis and despite the fact that there is no generally accepted method of proving its existence, an understanding of the mechanisms associated with heterosis is highly desirable. Heterosis is of more than academic interest, and may be useful in the measurement and understanding of **combining ability** of parental lines, a major component in the production of hybrid cultivars of any crop. Advance prediction of performance of any possible combination would be of enormous value, as commercial breeders are faced with an almost infinite choice of combinations of inbred cultivars of corn or other crops to form hybrids. Measurement of combining ability and related heterosis is achieved by producing sufficient quantities of hybrid seed to allow comparative yield trials to be conducted. Estimation of potential combining ability without the necessity for producing the actual hybrid combination would be an enormous asset.

THE FUTURE IN HYBRID CROPS

The suggestion that F_1 hybrids will be potentially available in all crops is a real possibility. Each passing year gives greater assurance that parallel variation will provide the mechanisms needed to produce hybrids. Biotechnology offers a promise that specific traits may be transmitted from one crop species into another (Chilton, 1983). Biologically, the success of hybrids is assured, but their commercial success is less certain.

With each crop in which hybrid development is attempted, new problems arise. These are among the most fascinating plant-improvement problems confronting agronomists. Attempts to solve them will lead to advances in plant improvement in both conventional cultivars and hybrids.

A widespread switch from conventional cultivars to F_1 hybrids may not be desirable, however. The example of corn illustrates the reason for concern.

Since the development of hybrid corn in 1909, hundreds of inbreds have been tested and evaluated, but only a few have been used extensively, because simple economics dictates that inbreds which produce only the highest-yielding F_1 hybrids be used. A survey of commercial seed companies in 1970 revealed that 71% of all the inbreds produced in the United States were developed from six inbreds, and three of the six originated from the same synthetic source (Batty, 1975). The resulting high degree of uniformity among plants of a specific hybrid cultivar and the lack of diversity among cultivars has created a favorable environment for pest attack.

A second concern is the potential reduction in the gene base from which new inbreds are derived as a result of the virtual elimination of open-pollinated corn cultivars. Open pollination assures genetic diversity. Any reduction in genetic variability might restrict improvement and limit new sources of resistance to pests and other agronomic traits (Mangelsdorf, 1951).

A third concern deals with the lack of yield stability under variable conditions, especially with single-cross hybrids. Hybrid corn may possess high yield ability but lack yield stability.

Scientists differ in their view about the genetic diversity present in elite inbred parents or hybrids. Modern hybrids are not delicate and unstable but have hybrid vigor which provides them with stress tolerance.

As well, a number of commercial seed companies compete in marketing hybrids. The existence of competitive firms provides assurance that diversity of genotypes will exist among the cultivars marketed. By purchasing seed of several hybrids from one seed company, or purchasing seed of hybrids from several companies, genetic diversity is ensured.

The essential nature of gene banks, which preserve older cultivars or open-pollinated cultivars of corn and provide a hedge against the loss of genetic diversity, becomes more apparent.

Hybrid cereals offer promise but should not be viewed as the only avenue for future development. Conventional plant breeding will continue to contribute to commercial production as well as to progress in hybrid development (Borlaug, 1983). Perhaps a combination of procedures will be needed to shatter the barriers to advances in plant breeding. This view is not incompatible with the statement that "some day, all crops will be hybrid."

Questions

1. Does a double-cross hybrid meet the definition of a hybrid and thereby qualify as a true hybrid?
2. Name several commercial private corn companies and consider their contribution to agriculture.
3. *a.* Review the reasons why hybrid seed must be purchased every year or why producers do not grow F_2 seed produced from a previous hybrid crop.
   b. What problems have arisen from this fact, especially in underdeveloped countries?
   c. Can these problems be overcome?
4. The production of hybrid seed is accomplished by cross-pollination. One inbred is selected as the male or pollen-producing plant and the other as the female or seed-bearing plant.
   a. On what criterion might an individual inbred be designated as male or female?
   b. Consider the choice of male or female parent in crops where cytoplasmic male sterility is involved.
5. The first hybrid of pearl millet in India was released in 1965. Since then, eight more have been released. Despite grain-yield and other advantages, only about 10% of the total millet-producing area was sown to hybrid seed in 1980.
   a. Consider possible technical and social reasons for this seemingly slow growth of hybrid millet in India.
   b. Consider the impact of hybrid corn or millet worldwide.
   c. What problems are associated with hybrid crops worldwide?
6. For hybrids to be successful, yield increases must be sufficient to more than offset the increased seed costs.
   a. Consider the factors influencing F_1 hybrid seed costs in corn and wheat.
   b. At current prices, can you determine the amount of heterosis needed to warrant commercial hybrid seed production?
7. In determining the degree of heterosis, comparisons may be made with the mean yield of the two parents, with the yield of the highest-yielding parent of the cross, or with the yield of the present highest-yielding conventional commercial cultivar. What yield comparison is most meaningful?
8. Heterosis may be expressed in various plant parameters such as winter hardiness, vigor, tillering capacity, seeds per spikelet in wheat, and cold tolerance. Is it possible that hybrid crops may be desirable for other reasons in addition to yield, or are all factors reflected in grain yield?
9. Consider the possibility of using in other areas of the world the labor-intensive practices associated with hybrid rice development in China.
10. Cytoplasmic male sterility, maintainer systems, and fertility-restorer systems have been found in corn, sorghum, wheat, barley, rice, and other plants.
    a. Comment on Vavilov's Law of Homologous Series and the concept of parallel variation (see Rutger and Carnahan, 1981).
    b. Commercial F_1 hybrids are successful in corn, sorghum, millet, and other crops. Does Vavilov's law suggest that F_1 hybrid systems will be successful in other crops?

11. *a.* On the basis of the problems, challenges, and evidence of success in commercial hybrid development, consider the statement that "some day, all crops will be hybrid."

b. On the basis of any new findings or developments that have been brought to light since this chapter was prepared, consider the statement in part *a* above.

12. Consider the time frame associated with selected crops in relation to hybrid development in conjunction with the speeding up of hybrid progress with each subsequent contribution to the development of knowledge.

13. Clearly determine the reasons for inbreeding of crops such as corn to produce hybrid seed.

14. List the advantages of hybrid seed.

15. Every day ingenious ideas and methods are born, bringing the possibility of hybrid crops closer to reality. Consider hybrid production possibilities in terms of the following:

a. A chemical capable of fertility restoration in genetic male-steriles.

b. A chemical capable of inducing female sterility.

c. Marker color genes so that self- or cross-pollinated seed could be identified.

d. A chemical hybridizing agent (male or female) that could be applied to the seed.

e. Whether or not is is possible to fix heterosis in pure breeding lines.

f. The use of apomixis to fix heterosis (see Reddy, Schertz, and Bashaw, 1980).

g. Hand emasculation and pollination.

References

Ahokas, H. 1979. "Cytoplasmic Male Sterility in Barley. III. Maintenance of Sterility and Restoration of Fertility in the *msml* Cytoplasm." *Euphytica* 28: 409–419.

Anderson, M. K., N. L. Taylor, and R. Kirthavip. 1972. "Development and Performance of Double-Cross Hybrid Red Clover." *Crop Science* 12: 240–242.

Anonymous. 1980a. *Improvement and Production of Maize, Sorghum and Millet.* Vol. 2. *Breeding, Agronomy and Seed Production.* FAO Plant Production and Protection Paper 24/2. Rome: Food and Agriculture Organization of the United Nations. p. 216.

———. 1980b. *Annual Report for 1979.* Los Banos, Philippines: International Rice Research Institute. pp. 157–158.

Atashi-Rang, G., and K. A. Lucken. 1978. "Variation, Combining Ability, and Interrelationships in Anther Length, Anther Extrusion, Glume Tenacity, and Shattering in Spring Wheat." *Crop Science* 18: 267–272.

Athwal, D. S. 1965. "Current Plant Breeding Research with Special Reference to *Pennisetum.*" *Indian Journal of Genetics and Plant Breeding.* 26A: 73–85.

———, and S. S. Virmani. 1972. "Cytoplasmic Male Sterility and Hybrid Breeding in Rice." *Rice Breeding.* Los Banos, Philippines: International Rice Research Institute. pp. 615-620.

Azzini, L. E., and J. N. Rutger. 1982. "Amount of Outcropping on Different Male Steriles in Rice." *Crop Science* 22: 905–907.

Batty, R. G. 1975. "Facing the Reality of a Possible Hybrid Corn Extinction." *Crops and Soils* 27 (7): 16–18.

Beard, B. H. 1981. "The Sunflower Crop." *Scientific American* 244: 150–161.

Becker, H. C., H. H. Geiger, and K. Morgenstern. 1982. "Performance and Phenotypic Stability of Different Hybrid Types in Winter Rye." *Crop Science* 22: 340–344.

Bhatt, J. G., and M. R. K. Rao. 1980. "Heterosis in Growth and Photosynthetic Rate in Hybrids of Cotton." *Euphytica* 30: 129–133.

Bitzer, M. J., and F. L. Patterson. 1967. "Pollen Dispersal and Cross-Pollination of Soft Red Winter Wheat (*Triticum aestivum* L.)." *Crop Science* 7: 482–484.

———, and W. E. Nyquist. 1982. "Hybrid Vigor and Combining Ability in a High-Low Yielding, Eight-

Parent Diallel Cross of Soft Red Winter Wheat." *Crop Science* 22: 1126–1129.

Borlaug, N. E. 1983. "Contribution of Conventional Plant Breeding to Food Production." *Science* 219: 689–693.

Burton, G. W. 1958. "Cytoplasmic Male Sterility in Pearl Millet (*Pennisetum glaucum* (L.) R. Br.)." *Agronomy Journal* 50: 230.

———. 1977. "Fertile Sterility Maintainer Mutants in Cytoplasmic Male Sterile Pearl Millet." *Crop Science* 17: 635–637.

Burton, G. W., and D. S. Athwal. 1967. "Two Additional Sources of Cyloptasmic Male Sterility in Pearl Millet and Their Relationship to Tift 23A." *Crop Science* 7: 209–211.

———. 1967. "Use of Self-Incompatibility to Produce Commercial Seed-Propagated F_1 Bermudagrass Hybrids." *Crop Science* 7: 524–527.

Burton, G. W., and R. H. Hart, 1947. "Breeding Bermudagrass for the Southeastern United States." *Journal American Society of Agronomy* 39: 551–569.

Burton, G. W., and J. B. Powell. 1968. "Pearl Millet Breeding and Cytogenetics." *Advances in Agronomy* 20: 49–89.

Campbell, K. W., and K. J. Kasha. 1981. "Selection for Cross-Pollination in Barley Using Segregating Bulk Populations." *Canadian Journal of Plant Science* 61: 17–24.

Carnahan, H. L., et al. 1972. "Outlook for Hybrid Rice in the USA." (In) *Rice Breeding.* Los Banos, Philippines: International Rice Research Institute. pp. 603–607.

Chilton, M. 1983. "A Vector for Introducing New Genes into Plants." *Scientific American* 248 (6): 50–59.

Christie, B. R. 1973. "Performance of Hybrids in Orchard Grass." *Canadian Journal of Plant Science* 53: 783–789.

Craigmiles, J. P. 1961. "The Development, Maintenance and Utilization of Cytoplasmic Male-Sterility of Hybrid Sudangrass Seed Production." *Crop Science* 1: 150–152.

———, L. V. Crowder, and J. P. Newton. 1965. "Methods of Breeding Smooth Bromegrass." *Crop Science* 5: 15–16.

Craigmiles, J. P., J. W. Stansel, and W. T. Flinchum. 1968. "Feasibility of Hybrid Rice." *Crop Science* 8: 720–722.

Daskalov, Chr., and Chr. Georgiev. 1976. "New Trends and Results of the Use of Heterosis in Tomato." (In) A. Janossy and F. G. H. Lupton (eds.), *Heterosis in Plant Breeding.* Proceedings of the Seventh Congress of Eucarpia. Amsterdam: Elsevier Scientific Publishing Company. pp. 303–305.

Davis, D. D. 1978. "Hybrid Cotton: Specific Problems and Potentials." *Advances in Agronomy* 30: 129–157.

Dedio, W. 1979. "Hybrid Sunflowers Are Here to Stay." *Scientific American* 244: 150–161.

de Vries, A. Ph. 1971. "Flowering Biology of Wheat, Particularly in View of Seed Production—A Review." *Euphytica* 20: 152–170.

———. 1973. "Some Aspects of Cross-Pollination in Wheat (*Triticum aestivum* L.). 2. Anther Extrusion and Ear and Plant Flowering Pattern and Duration." *Euphytica* 22: 445–456.

Dorsman, C. 1976. "F_1 Hybrids in Some Out-Pollinating Crops." (In) A. Janossy and F. G. H. Lupton (eds.), *Heterosis in Plant Breeding.* Proceedings of the Seventh Congress of Eucarpia. Amsterdam: Elsevier Scientific Publishing Company. pp. 197–208.

Dotlacil, L., and M. Apltauerova. 1978. "Pollen Sterility Induced by Ethrel and Its Utilization in Hybridization of Wheat." *Euphytica* 27: 353–360.

Dunn, G. M., and J. A. Wright. 1970. "Performance of Diallel Crosses in *Bromus inermis* for Forage Yield." *Crop Science* 10: 50–58.

Erickson, J. R. 1969. "Cytoplasmic Male Sterility in Rice (*Oryza satival L.*)." Agronomy Abstracts. p. 6.

Fairey, D. T., and N. C. Stoskopf. 1975. "Effects of Granular Ethephon on Male Sterility in Wheat." *Crop Science* 15: 29–32.

Fukasawa, H. 1953. "Studies on Restoration and Substitution of Nucleus of *Aegilotricum*. I. Appearances of Male-Sterile Durum in Substitution Crosses." *Cytologia* 18: 167–175.

———. 1955. "Studies on Restoration and Substitution of Nucleus (Genomes) in *Aegilotricum.* II. The Inter-relationships between *ovata* Cytoplasm and Fertility Restoration Factors." *Cytologia* 20: 211–217.

Geiger, H. H., and F. W. Schnell. 1970. "Cytoplasmic Male Sterility in Rye (*Secale cereale* L.)." *Crop Science* 10: 590–593.

Ghiasi, H., and K. A. Lucken. 1982. "Agronomic and Male Fertility Restoration Characteristics of Wheat Restorer Lines with *Aegilops speltoides* and *Triticum timopheevi* cytoplasms." *Crop Science* 22: 527–530.

———, and J. J. Hammond. 1982a. "Mass Selection for Anther Extrusion and Seed Set in Fertility Restorer Bulks of Wheat." *Crop Science* 22: 410–414.

———. 1982b. "Selection Procedures for Developing Fertility Restorer Line Wheat." *Crop Science* 22: 523–527.

Hayes, J. D., and C. A. Foster. 1976. "Heterosis in Self-Pollinating Crops, with Particular Reference to Barley." (In) A. Janossy and F. G. H. Lupton (eds.), *Heterosis in Plant Breeding.* Proceedings of the Seventh Congress of Eucarpia. Amsterdam: Elsevier Scientific Publishing Company. pp. 239–256.

Hayward, C. F. 1975. "The Status of and Prospects for Hybrid Winter Wheat." *Proceedings of the Second International Winter Wheat Conference.* Zagreb, Yugoslavia. pp. 84–104.

Heaton, T. C., and P. F. Knowles. 1982. "Inheritance of Male Sterility in Safflower." *Crop Science* 22: 520–522.

Hockett, E. A., P. S. Baenziger, and G. L. Steffens. 1978. "A Proposal for Increased Research on Chemical Induction of Fertility in Genetic Male-Sterile Barley." *Euphytica* 27: 109–111.

Hughes, W. G. 1976. "Ethrel Induced Male Sterility in Wheat (*Triticum aestivum* L.)." (In) A. Janossy and F. G. H. Lupton (eds.), *Heterosis in Plant Breeding.* Proceedings of the Seventh Congress of Eucarpia. Amsterdam: Elsevier Scientific Publishing Company. pp. 111–114.

Jan, C. C., C. O. Qualset, and H. E. Vogt. 1974. "Chemical Induction of Sterility in Wheat." *Euphytica* 23: 78–85.

Johnson, R. R., and C. M. Brown. 1978. "Use of DPX 3778 to Produce Hybrid Wheat Seed." *Crop Science* 18: 1026–1028.

Johnson, V. A. 1977. "Small Grains." *Twenty-Third Farm Seed Conference.* American Seed Trade Association. pp. 27–34.

Jones, H. A., and A. E. Clarke. 1947. "The Story of Hybrid Onions." *U.S.D.A. Yearbook* (1947). pp. 320–326.

Joppa, L. R., F. H. McNeal, and M. A. Berg. 1968. "Pollen Production and Pollen Shedding of Hard Red Spring (*Triticum aestivum* L. em. Thell) and durum (*Triticum durum* Desf.) Wheat." *Crop Science* 8: 487–490.

Kihara, H. 1951. "Substitution of Nucleus and Its Effects on Genome Manifestations." *Cytologia* 16: 177–193.

Knowles, R. P. 1970. "Performance of Synthetics and Sib-Lines in Fairway Crested Wheatgrass, *Agropyron cristatum* (L.) Gaertn." *Crop Science* 10: 642–645.

Komaki, M. K., and K. Tsunewake. 1981. "Genetical Studies on the Difference of Anther Length Among Common Wheat Cultivars." *Euphytica* 30: 45–53.

Konarev, V. C. 1976. "Physiological and Biomechanical Aspects of Heterosis." (In) A. Janossy and F. G. H. Lupton (eds.), *Heterosis in Plant Breeding.* Proceedings of the Seventh Congress of Eucarpia. Amsterdam: Elsevier Scientific Publishing Company. pp. 81–93.

Lin, Shi-Cheng, and Loring-Ping Yuan. 1980. "Hybrid Rice Breeding in China." *Innovative Approaches to Rice Breeding.* Los Banos, Philippines: International Rice Research Institute. pp. 35–51.

Livers, R. W., and E. G. Heyne. 1968. "Hybrid Vigor in Hard Red Winter Wheat." *Proceedings of the Third International Wheat Genetics Symposium.* Canberra: Australian Academy of Science. pp. 431–436.

Lucken, K. A. 1981. *The Breeding and Production of Hybrid Wheats.* Paper presented to the American Society of Agronomy. Atlanta, Georgia.

Lupton, F. G. H. 1976. "The Physiological Basis of Heterosis in Wheat." (In) A. Janossy and F. G. H. Lupton (eds.), *Heterosis in Plant Breeding.* Proceedings of the Seventh Congress of Eucarpia. Amsterdam: Elsevier Scientific Publishing Company. pp. 71–79.

Maan, S. S. 1975. "Cytoplasmic Variability and Speciation in *Triticinae.*" (In) M. K. Wali (ed.), *Prairie: A Multiple View.* Grand Forks, N.D.: University of North Dakota Press. pp. 225–281.

MacKey, J. 1976. "Genetic and Evolutionary Principles of Heterosis." (In) A. Janossy and F. G. H. Lupton (eds.), *Heterosis in Plant Breeding.* Proceedings of the Seventh Congress of Eucarpia. Amsterdam: Elsevier Scientific Publishing Company. pp. 17–33.

Manglesdorf, P. C. 1951. "Hybrid Corn." *Scientific American* 185 (2): 39–47.

Melton, B. 1969. "Forage Yields of Alfalfa Crosses and Synthetic Varieties." *New Mexico Agriculture Experiment Station Bulletin* 545. 12 pp.

Miller, J. F., and K. A. Lucken. 1976. "Hybrid Wheat Seed Production Methods for North Dakota." *Crop Science* 16: 217–221.

———. 1977. "Gametocidal Properties of RH-531, RH-532, RH-2956, and RH-4667." *Euphytica* 26: 103–112.

Milohnic, J., and M. Jost. 1976. "Pollen Production and Anther Extrusion of Wheat (*Triticum aestivum* L. em. Thell)." *Hybrid Wheat Research in Yugoslavia.* University of Zagreb, Yugoslavia. pp. 105–111.

Mukai, Y., and K. Tsunewake. 1980. "Comparison of

the Alloplasmic Wheats Produced in Kyoto University, Japan, and North Dakota State University, U.S.A." (In) K. Tsunewake (ed.), *Genetic Diversity of the Cytoplasm in Triticum and Aegilops*. Tokyo: Japanese Society for the Promotion of Science. pp. 131–159.

Osman, H. Elg., and D. M. Yermanos. 1982. "Genetic Male Sterility in Sesame: Reproductive Characteristics and Possible Use in Hybrid Seed Production." *Crop Science* 22: 492–498.

Pedersen, M. W., and R. R. Hill, Jr. 1972. "Combining Ability in Alfalfa Hybrids Made with Cytoplasmic Male Sterility." *Crop Science* 12: 500–502.

Powers, L. 1944. "An Expansion of Jones' Theory for the Explanation of Heterosis." *American Naturalist* 78: 275–280.

Ramage, R. T. 1965. "Balanced Tertiary Trisomics for Use in Hybrid Seed Production." *Crop Science* 5: 177–178.

Reddy, P. S., and B. Thammirau. 1977. "Inheritance of Fertility Restoration in Sunflower (*Helianthus Annuus* L.)." *Euphytica* 26: 409–412.

Reddy, C. S., K. F. Schertz, and E. C. Bashaw. 1980. "Apomictic Frequency in Sorghum R473." *Euphytica* 29: 223–226.

Reimann-Phillip, R. 1976. "Breeding of F_1-Hybrids in Flowers." (In) A. Janossy and F. G. H. Lupton (eds.), *Heterosis in Plant Breeding*. Proceedings of the Seventh Congress of Eucarpia. Amsterdam: Elsevier Scientific Publishing Company. pp. 135–146.

Rowell, P. L., and D. G. Miller. 1971. "Induction of Male Sterility in Wheat with 2-Chloroethylphosphonic Acid (Ethrel)." *Crop Science* 11: 629–631.

————. 1974. "Effect of 2-Chloroethylphosphonic Acid (Ethephon) on Female Fertility of Two Wheat Varieties." *Crop Science* 14: 31–34.

Rutger, J. N., and H. L. Carnahan. 1981. "A Fourth Genetic Element to Facilitate Hybrid Cereal Production—A Recessive Tall in Rice." *Crop Science* 21: 373–376.

Sadik, S., and W. G. Rockwood. 1975. "Introducing the Hybrid Yam." *Span* 18 (2): 67–68.

Sage, G. C. M. 1972. "The Inheritance of Fertility Restoration in Male-Sterile Wheat Carrying Cytoplasm Derived from *Triticum timopheevi*." *Theoretical and Applied Genetics* 42: 233–243.

————. 1976. "Nucleo-Cytoplasmic Relationships in Wheat." *Advances in Agronomy* 28: 265–298.

———— and G. E. Hobson. 1973. "The Possible Use of Mitochondrial Complementation as an Indicator of Yield Heterosis in Breeding Hybrid Wheat." *Euphytica* 22: 61–69.

Sampson, D. D. 1978. "Canada Opts Out of Hybrid Wheat." *Canada Agriculture* 23 (4): 19–21.

Shull, G. H. 1948. "What Is 'Heterosis'?" *Genetics* 33: 439–446.

Sprague, G. F. 1955. "Corn Breeding." *Corn and Corn Improvement*. New York: Academic Press. pp. 221–292.

Srivastava, H. K. 1981."Intergenomic Interaction, Heterosis, and Improvement of Crop Yield." *Advances in Agronomy* 34: 117–195.

Stephens, J. C., G. H. Kuykendall, and D. W. George. 1952. "Experimental Production of Hybrid Sorghum Seed with a Three-Way Cross." *Agronomy Journal* 44: 369–373.

Stephens, J. C., and R. F. Holland. 1954. "Cytoplasmic Male-Sterility for Hybrid Sorghum Seed Production." *Agronomy Journal* 46: 20–23.

Stoskopf, N. C., and J. Law. 1972. "Some Observations on Ethrel as a Tool for Developing Hybrid Cereals." *Canadian Journal of Plant Science* 52: 680–683.

Taliaferro, C. M., and E. C. Bashaw. 1962. "Inheritance and Control of Obligate Apomixis in Breeding Buffelgrass, *Pennisetum ciliave*." *Crop Science* 6: 473–476.

Trupp, C. R. 1976. "Fertility Restoration in Soft Winter Wheat Hybrids." *Crop Science* 16: 453–456.

Virmani, S. S., and I. B. Edwards. 1983. "Current Status and Future Prospects for Breeding Hybrid Rice and Wheat." *Advances in Agronomy* 36: 145–214.

Whelan, E. D. P. 1980. "A New Source of Cytoplasmic Male Sterility in Sunflower." *Euphytica* 29: 33–46.

Wiebe, G. A. 1960. "A Proposal for Hybrid Barley." *Agronomy Journal* 52: 181–182.

Wilson, J. A., and W. M. Ross. 1962. "Male Sterility Interaction of the *Triticum aestivum* nucleus and *Triticum timopheevi* cytoplasm." *Wheat Information Service* 14: 29–30.

10

Seed Production and Technology

Seed produced by sexual union of male and female gametes is the essential lifeline for each generation of crops. A seed provides the means by which embryonic life can be almost suspended and then revived to new development, even years after the parents are dead and gone. In homozygous self-pollinating plants, reproduction by seed provides an opportunity to reproduce the parent as precisely as by asexual reproduction. Cross-pollination of unrelated parents provides opportunity for genetic recombination and desirable variability. Seeds carry a promise, therefore, that life can be sustained over a period when conditions for growth are suspended; they carry a promise to replenish our food supplies to sustain human life; and they carry a promise either to reproduce the parent per se or to serve as vehicles for essential genetic diversity.

SEEDS AND GERMINATION

Most seeds consist of an embryo, an endosperm, and a pericarp or seed coat. In all seeds, the embryo is a tiny plant which possesses all the main features of the mature plant—root, stem, leaves, and buds. The embryo is attached to the nutrient reserve or endosperm in the cereal grains by a fragile structure through which dissolved foods pass. The nutrients from the endosperm sustain the embryo until it has emerged from the soil and can meet its food needs through photosynthesis. Should the connection between the embryo and the endosperm reserves be damaged, the food supply will be disrupted and germination may be prevented. Such damage is usually caused by excessive mechanical impact during harvesting, threshing, cleaning, transporting, or even sowing.

Cracked endosperms are commonly found in cereal seeds and may result in inadequate food supply to the germinating embryo.

The food-utilizing germination process may begin soon after seeds are planted in a moist, warm soil. Each kind of seed has its own requirement and preference for moisture, temperature, air, and light.

Moisture, the first requirement for seed germination, must be imbibed by the seed. For lowland rice to germinate, it must be submerged completely in water. When moisture is sufficient, the next requirement is a suitable temperature. Oxygen, a third requirement, is needed for respiration and release of food energy. Most seeds germinate best in the dark, but cereal seeds will germinate when exposed to sunlight if they are consistently moist. If any of the conditions needed for germination are lacking, viable seeds will not germinate.

Some seeds will germinate and grow as soon as conditions are inducive. Other seeds require an after-ripening process before germination will occur, because they enter a state of dormancy as they ripen. Such seeds will not germinate even when supplied with the conditions known to promote germination. Seed dormancy assures that seeds will not germinate out of season, an essential feature of wild plants that shatter their seed at maturity. Selection against any forms of dormancy in domesticated cereals has resulted in preharvest seed-sprouting problems under humid, warm conditions.

LIFE SPAN OF SEEDS

When the Nuremberg State Opera House in West Germany was demolished in 1955, the cornerstone, inscribed 1832, was found to contain 11 corked vials sealed with wax. Each vial contained seeds of a different species. The wax on the vials containing oat and barley seeds remained intact, and seed from these vials germinated after 124 years (Aufkammer and Simon, 1957).

The keeping quality of cereal grains is of interest, not only because good seeds are vital to good crops, but also because of the food value of the grain. Conditions that favor seed longevity also maintain good food value and taste.

Just how long seed viability can be maintained is uncertain. One article (Anonymous, 1843) reported the germination of one of 12 wheat seeds taken from an alabaster vase in a tomb not visited by humans for 3000 years. The resulting plant was reported to have grown to maturity and produced 27 grains, which, in turn, germinated and produced vigorous plants. Several observers on this matter have taken the view that the seeds could not possibly have been viable or that a hoax of some kind was involved (Barton, 1961).

The life span of cereal grains has been reported by various authors as follows: wheat, six to 32 years; oat, six to 29 years; barley, two to ten years; rye, nine to ten years (reviewed by Barton, 1961). With increasing age, both germination and vitality were found to decline. Both can be influenced by storage conditions, mainly humidity and temperature.

Preservation of seed germplasm is a critical priority that has received considerable emphasis. Optimum conditions must be provided to ensure maximum seed preservation for long-term maintenance of genetic resources in gene banks. In the tropics seed longevity is an acute problem, often because of inadequate storage facilities. An increasing awareness and desire for seed vigor have raised considerable interest in factors affecting seed storage. Differences in seed-storing characteristics among crops also are important.

Three environmental factors—seed moisture (relative humidity), temperature, and atmosphere environment (other than water vapor)—have been suggested as major factors influencing seed deterioration in storage (Bass, 1975; Bass and Stanwood, 1978). Generally, as seed moisture and storage temperature increase, deterioration in most species is more rapid.

Because species vary so widely in their response to environmental variables, individual species need selective attention. Some species observations are summarized below.

WHEAT, RYE, AND OAT. In the southern United States, seed germination of these crops has been a problem, especially when seed is held over for a second planting season (Morey, 1976). Rye seed is notorious for poor germination and low vitality after one or more years of warehouse storage. It is generally believed that rye seed stored in the southern United States region in burlap or cotton bags in warehouses, or in paper bags or envelopes at room temperature and humidity, will deteriorate in germinability and vitality within 15 to 18 months after harvest. Several reasons were mentioned, including damage from storage insects, seed-borne diseases, and vulnerability to mechanical damage. The pointed embryo of rye is especially vulnerable to becoming cracked or broken, and such damage may prevent normal germination. Desirable seed moisture levels of 12% are difficult to maintain under conditions of high temperature and humidity.

Rye, wheat, and oat seed stored at 11.5% moisture or lower were found to retain good germination under laboratory conditions for three years in paper containers (Morey, 1976). When stored in a refrigerator or deep freezer, the same seed in paper containers retained its viability for five years. The composition of the container did not influence germination percentages. No direct effect of storage organisms on seed germination was found.

Tests indicated that oat seed is least affected by storage conditions; wheat is intermediate; and rye is the most difficult to store. Rye seed stored in impermeable containers at 11.5% moisture averaged 40% germination after four years and only 12% after five years. At 9.5% moisture, rye seed averaged 79% germination after four years and 51% after five years of storage.

Low moisture content is a key to long-time storage of small-grain seed. If chemical treatment is applied to control insects and diseases, these cereals can be stored in paper containers for up to three years in warehouses in the southern United States without significant germination reduction. The same seed should give good results after five or more years in cool or cold storage conditions.

RICE. In humid regions of the world where rice is a common crop, seed is harvested at a grain moisture of about 16% and is subject to postharvest invasion by storage fungi, resulting in decreased germinability (Christensen, 1969), discoloration, and sometimes toxin production. In India, rough rice is more subject to deterioration than any other cereal crop because it is stored under humid and often primitive conditions (Mallick and Nandi, 1979), and a large percentage of the annual grain yield is lost in storage.

In studies in India (Mallick and Nandi, 1979) and in West Bengal (Mallick and Nandi, 1982) rough rice stored at 12% to 21% moisture at 70% and 90% relative humidity showed marked losses in germinability and increased invasion by moisture content from high humidity over the 135-day storage period.

Rough rice reached an equilibrium moisture content of 13.8% to 13.9% within ten days when exposed to a relative humidity of 80% at 22°–28°C (72°–82°F), a full percentage point below that of polished rice or of rice with the hulls removed. At moisture contents above 14.5%, invasion of rough rice by storage fungi and decrease in germination percentage of the seeds were proportional to increasing moisture content and increasing length of storage (Christensen and Lopez, 1965).

SORGHUM. Over a 16-year storage period, sorghum seed was best preserved at −12°C (5°F) regardless of seed moisture content, atmospheric environment, or whether seed was sealed in metal containers or paper envelopes (Bass and Stanwood, 1978). At 32°C (90°F) and 10% seed moisture, all seeds were dead in eight years.

At moisture contents of 14.5% and above, invasion by storage fungi and decrease in germinability were proportional to increasing moisture content and increasing time of storage (Christensen, 1970). Sorghum seed exposed to relative humidities of 75%, 80%, and 85% reached moisture equilibrium of 14.9%–15.8%, 15.9%–16.6%, and 17.0%–17.5%, respectively. Sorghum seed stored for planting could deteriorate to feed grade within a few months, probably as a result of invasion by storage fungi resulting from high temperature and humidity.

CORN. In samples of corn stored at 14.2% and 15.5% moisture content at 10°C (50°F), there were no changes in germination, numbers and kinds of storage fungi, or fatty acid values within one year, and only slight changes after nearly two years (Sauer and Christensen, 1968). In samples stored at the same moisture contents and at 25°C (77°F), germination decreased to zero within a year, invasion by storage fungi increased, and fatty acid value increased.

Specific requirements for safe storage of various kinds of seed must be based on numerous experiments. Governments of various countries are becoming increasingly aware of the possibilities of retaining both food and seeding value by proper storage.

Postharvest deterioration can be summarized as being caused by development of storage fungi, loss of germinability, discoloration, depletion of nutritional value, and production of toxic chemicals. These factors are influenced by moisture content of seeds; the effects can vary over a broad range but can be summarized as follows:

1. Above 40% to 60% moisture, seeds germinate.
2. Above 18%–20% heating may occur.
3. Above 12%–24% fungi grow on seed.
4. Above 8%–9% insects are active and multiply.
5. Below 8% seed is stored most safely.

Storage temperature is equally important to kernel moisture. At temperatures between 0° and 50°C (32° and 122°F) and with seed stored at moisture levels of 5% and 14%, seed longevity can be summarized as follows (Peterson, 1979):

1. For each 1% increase in seed moisture the life of the seed is halved.
2. For each 5°C (9°F) rise in temperature the life of the seed is halved.

Because of the relationship between relative humidity and seed moisture levels it is possible to use simple climatic data to assess the suitability of an area or a site for seed-storage purposes. Delouche et al. (1973) suggested a general guide as follows:

1. If the sum of the temperature in degrees Celsius and the relative humidity in percent is 80, the seed will begin to deteriorate after one to nine months, depending on the crop species and the initial quality of the seed.
2. If the sum of the two previous factors is 70, the safe storage period is extended to 18 months.
3. If the sum is reduced to a value of 30 to 45, most seeds will survive for three to five years.

For example, if the temperature of the seed is 20°C and the relative humidity 50%, the sum of the two is 70, and it can be taken that seed might be stored up to 18 months. On the other hand, if the temperature is 25°C and the relative humidity 60%, conditions are unsuited to seed storage, and an alternate site should be considered or controlled storage units erected.

SEED DRYING

Under a maritime climate, cereal grain crops may be harvested at a moisture content of between 17% and 22%. Corn may be harvested at moisture contents of up to 40%. Safe storage cannot be expected at such high moisture levels, and artificial drying may be required. Driers come in many styles, including low-temperature driers, continuous-flow forced-hot-air driers, and batch driers. Drying is ineffective, however, if storage conditions allow high-humidity air to penetrate the grain after it is removed from the drier.

PEDIGREED SEED PRODUCTION

High production goals, increased production costs, and a growing awareness that good seed is a sound investment have led farmers to demand and get seed of the highest quality. Demands for better seed quality have resulted from changes in farming practices, such as precision planting, high-density planting, earlier planting, or planting under stress conditions.

A seed takes on a special significance when it represents a cultivar of a particular crop. Considerable time and expense have been directed to shaping the genotype of such a seed. Much trust is placed in a seed to perform according to the genetic makeup, and it becomes the responsibility of seed growers to maintain the genetic constitution and the responsibility of producers to use the best seed available.

An awareness of the importance of good seed, coupled with the need, in the case of hybrids, to purchase seed for each crop, has resulted in a large and important seed industry. This industry is responsible for providing adequate quantities of genetically pure, high-quality, vigorous seed.

Seed may be multiplied by agricultural experiment stations, government agencies, crop improvement associations, cooperatives, private organizations, or individuals. To assure the consumer that seed has been produced according to accepted and supervised methods, standards of certification have been established by which

pedigreed status can be achieved. Many countries of the world have established a set of guidelines for pedigreed seed. The Canadian approach to handling seeds has evolved over a long period and is outlined here as an example of a highly successful system whereby seed is produced and marketed under careful control.

All seed grown or offered for sale for seed purposes in Canada comes under the jurisdiction of the Seeds Act and is administered by the federal government. The sale of seed of all cereal crops by cultivar name is restricted to those cultivars that are licensed according to the Canada Seeds Act. Licensing guarantees that, before a cultivar may be sold, it must undergo careful testing and be of value to crop production. Inspection assures that the seed matches the cultivar name and pedigree under which it is offered for sale. Grading verifies the quality as labeled.

The Canadian Seed Growers' Association, established in 1904, is the official pedigreeing agency responsible for prescribing standards and issuing crop certificates for Canadian-produced seed. The U.S. counterpart is the American Seed Trade Association. It is beyond the scope of this text to outline variations in pedigreeing systems among states, regions or countries. The Canadian situation is used here as an example only.

CLASSES OF CEREAL SEEDS

Seed of the self-pollinated, nonhybrid cereal crops may be divided into pedigreed classes based on breeding and genetic purity (see Chapter 5): breeder, select, foundation, registered, and certified.

Breeder seed usually is produced in limited quantities and maintained by the originating or sponsoring institution or organization. Breeder seed originates from the bulking of many lines, examined and kept pure each year by the plant breeder (see Chapter 5). Breeder seed is supplied to a farm producer recognized by the Seed Growers' Association and is increased as a select seed plot. In some countries, breeder seed may be increased by the originating institution and released for foundation seed production, so that four pedigreed classes exist.

Certified seed, the last pedigreed class, is the seed of commerce used for foods or feed production. Certified seed gives the farm producer a pure cultivar of quality seed at a moderate price. All classes of seed prior to the certified class are used by seed growers for multiplication.

Normally each class represents one generation, but variations are possible. Foundation seed may be promoted directly to certified status if it is not required as registered seed or if it has not met standards of purity or isolation to be labelled as registered. Every effort is made to introduce seed into commercial channels as quickly as possible for early farm evaluation on a widespread basis. Classes of seed and production systems vary around the world, but the fact that seed production is regulated is of importance.

Seed produced from certified seed becomes nonpedigreed or commercial seed. Commercial seed may be of any source that meets grade requirements. Commercial seed is sold in unsealed containers and labeled only as to kind (wheat, oat, barley, rye) and grade. Because of the uncertainty as to cultivar purity, nonpedigreed seed should be purchased only when the reputation and reliability of the producer are known.

Use of cultivar names is restricted to the pedigreed classes. Nonpedigreed or commercial seed may be sold legally but must not be advertised or labelled as to cultivar name. This is to assure buyers that seed is not misrepresented as to cultivar; only pedigreed seed can be so guaranteed.

Cereal seed mixtures can be labelled to show cultivar names only if the mixture was made from certified seed.

CEREAL SEED GRADES

Three criteria are used in Canada to grade cereal seeds: quality, purity, and germination. *Quality* is based on three criteria: (1) freedom from undesirable seeds, such as undersized, light, immature, or damaged seeds, (2) seed weight (weight per 1000 kernels, and (3) color.

Purity is based on the presence of weed seeds or seeds of other crops. Pedigreed seed does not mean 100% purity or freedom from weed seeds or other crop seeds. The number and type of seeds are controlled and limited, however.

Grade according to *germination* is based on a minimum percentage germination and is the same for all pedigreed classes at 85% to 75% for grades 1 and 2, respectively. These are the minimum levels, and germination values may be much higher.

All classes of seed are graded Canada No. 1 or Canada No. 2 seed, based on quality, purity, and germination. Seed lots are labelled with white, purple, and blue tags to indicate foundation, registered, and certified seed grades, respectively. A seed label may read "Canada Certified No. 1." Nonpedigreed seed is named Canada No. 1 or Canada No. 2, depending on grade. Minimum standards used in grading are shown in Table 10-1. Provision can be made for sale of bulk lots of certified seed.

Seed with pedigreed status provides assurance by law that established standards have been met. Genetic purity of pedigreed seed is maintained by strict isolation and by land requirements regarding previous crops produced. Frequent roguing removes off-types, noxious weeds, and diseased plants. All fields are inspected, and rogues are checked and recorded. After cleaning, the seed is again checked to ensure quality, purity, and germination and to provide assurance that the finished product has not been altered.

PEDIGREED STATUS FOR HYBRID CROPS

Specific regulations for hybrid crops have been established in Canada and vary for each crop. In the case of hybrid corn, breeder and certified are the only classes. Breeder seed includes all classes of parents, which are inspected for isolation, roguing, effectiveness of detasselling, and removal of male rows. Certified seed is again the seed of commerce.

Specific details for each crop in Canada are available from the Canadian Seed Growers' Association in a publication entitled "Regulations

Table 10–1

Standards Used in Grading a Nonhybrid Wheat Crop in Canada and Administered by the Food Production and Inspection Branch of Agriculture Canada: Values Indicate the Maximum Number of Seeds per 500 g Except Where Otherwise Indicated

		Noxious Weed Seeds				
	Grade Name	*Primary*	*Primary plus Secondary*	*Total Weed Seeds*	*Total Seeds of Other Crops*	*Minimum Percentage Germination*
Used for Seed Increase	Can. Foundation No. 1	0	0	3	0.25	85
	Can. Foundation No. 2	0	1 per 10 kg	10	1.0	75
	Can. Registered No. 1	0	0	3	0.25	85
	Can. Registered No. 2	0	1 per 10 kg	10	1.0	75
Used for Food or Feed Production	Can. Certified No. 1	0	0	5	1.0	85
	Can. Certified No. 2	0	2 per 5 kg	10	3.0	75
	Canada No. 1	0	1	15	10.0	85
	Canada No. 2	2	8	20	50.0	75

Source: Schedule 1, Table 1, The Canada Seeds Act.

and Procedures for Pedigreed Seed Crop Production." Comparable details are available from agencies in other countries.

Comparable regulations should be obtained for your own country, state, or province.

CULTIVAR VERIFICATION TESTING

Postcontrol cultivar verification testing provides a final check on the pedigree system. Samples of pedigreed seed are compared to breeder seed of the same cultivar. Any lot found to have an excess of off-types can have its pedigree withdrawn. This kind of control is important in monitoring the grower's effectiveness in keeping pedigreed seed lots distinct, correctly labelled, and free from contamination.

Growing seeds outdoors to maturity takes a full season and is expensive. Laboratory tests that readily identify a cultivar would be of value in shortening cultivar verification testing. Considerable success has been achieved in finding rapid, reliable, and accurate methods of identifying crop cultivars. Verification may be based on a combination of measurements and observations, including kernel morphology; the reaction of the seed to chemicals such as phenol, caustic soda, or hydrochloric acid; the fluorescence of seed under ultraviolet light; seedling characteristics in growth-chamber tests; accelerated flowering in the growth chamber; reaction to diseases; and biochemical characteristics detected by **chromatography** or **electrophoresis.**

Just as no two people have identical fingerprints, so, too, a system of "fingerprinting" can be developed to distinguish among the many cultivars of each crop. Fingerprinting may depend on several criteria, including extraction of proteins, enzymes, or other biochemical compounds using start-gel electrophoresis. Kahler and Allard (1970) used esterases in barley to fingerprint cultivars. Esterases in barley comprise different enzymes, of which many are present as different isoenzymes. Fedak and Rajhathy (1972) used the system to effectively characterize Canadian barleys.

Using morphological features such as the rachilla, the presence or absence of teeth on the lateral nerves of the lemma, and the edges of the ventral furrow, Andersen (1982) was able to classify 47 barley cultivars into five groups. Other morphological features are less reliable because they are influenced by environmental factors. Combining morphological traits with isoenzymes of amylases in sprouted barley grains, the 47 cultivars could be classified in 32 groups.

Fingerprinting techniques may be useful to:

1. Verify visual identification traits based on morphological traits.
2. Determine cultivar identity.
3. Check against contamination of one cultivar in another.
4. Provide evidence in legal cases as a result of legislation involving disputes over a proprietary cultivar.

HYBRID SEED PRODUCTION

Hybrid seed production represents a special challenge, as the success of hybrid crops depends on the use of F_1 seed. Seed produced from commercial hybrid seed should not be used to produce subsequent crops if high agronomic performance is desired. Second-generation seed may appear physically satisfactory, may germinate well, and may produce a normal-looking crop but is likely to yield from 10% to 40% less grain because of a loss of hybrid vigor and a lack of uniformity (see Chapter 5). Rickey, Springfield, and Sprague (1934) reported a 5% to 24% decrease in grain yield of F_2-generation corn produced from ten double-cross hybrids in Ohio. An average 15% reduction over the F_1 hybrid crop yield resulted in about the same grain yield or a slight reduction of an open-pollinated cultivar.

In single-cross hybrids, the F_2 generation showed a 30% yield reduction (Neal, 1935). In general, Neal concluded that F_2 generation yields will be reduced one-half, one-third, and one-fourth when two, three, or four inbred lines comprise the parents of the F_1 hybrid. Results similar to those suggested by Neal were obtained by Grogan and Zuber (1958) and Kiesselbach (1960).

To produce sufficient commercial seed annually for the entire hybrid area to be seeded is an enormous task. Included in this large enterprise are the production, processing, sale, and distribution of seed. The practical aspects of commercial production of hybrid seed can be illustrated with three crops: corn, a monoecious species that is naturally cross-pollinated by wind; sorghum, a self-pollinated species that has been successfully adapted to commercial hybrid seed production; and wheat, a self-pollinated species that has not been adapted yet as a commercial hybrid on a large scale. Cultural practices for the commercial seed production of hybrid crops are similar in many respects, and the above three crops can be used to illustrate the principles. An essential aspect of all hybrid seed production is a high degree of purity, achieved by careful selection and isolation of a field, roguing of off-types, and careful harvesting and seed conditioning.

AGRONOMIC PRACTICES FOR HYBRID SEED PRODUCTION

Cultural practices associated with seed production of any class or pedigree are similar to those for commercial crop production. Involved in any seed field is the need for a balanced fertilizer program that optimizes yield but will not cause lodging in susceptible crops such as wheat. Pest control can be best achieved by specific pesticides, and spray programs may be imperative on parental lines that are susceptible to diseases and insects.

Plant populations should be compatible with the genotype and the climatic conditions to produce satisfactory yields. Plant populations of the female parents may be similar to those of commercial fields. The plant population of male rows may exceed commercial levels if inbred male parents have limited pollen production.

HYBRID TYPES

Single-cross seed is produced by crossing two unrelated inbred lines; it provides the greatest opportunity for hybrid vigor and is highly uniform. From a commercial seed-production viewpoint, single-cross seed of corn offers the disadvantage that seed is produced on the inbred female parent, which usually produces lower yields than the female parents used in other types of crosses.

To overcome the problem of low seed quality and seed yield on inbred parents, a double-cross system of hybrid corn seed production was adapted. A double cross involves crossing two single-cross hybrids in the field [(A × B) × (C × D)]. During the 1940s and 1950s double-cross seed was the standard hybrid seed available. It meets the definition of an F_1 hybrid—the first-generation cross from two unrelated parents. In practice, double-cross seed has less vigor than crosses involving fewer parents.

Commercial seed of a three-way cross [(A × B) × C] can be produced on an F_1 hybrid female parent (A × B). Modified crosses are produced by crossing two closely related inbred lines (A^1 and A) and using the resulting single cross as the female parent. A modified single cross [(A^1 × A) × B] or a modified three-way cross [(A × B) × (C^1 × C)] can be produced with this method.

Generally it is accepted that good seed yields can be produced on inbred female parents of wheat and sorghum. Single crosses are used exclusively in these crops.

To reduce seed costs, however, it may be advantageous to produce seed on an F_1 female parent. A three-way cross can also offer a broader genetic base to provide yield stability over a range of climatic conditions, especially under variable rainfall conditions in India and where the full hybrid genetic potential may be limited by climate. Preliminary work on three-way sorghum crosses indicated that yields were comparable to single-cross yields (Walsh and Atkins, 1973; Patanothai and Atkins, 1974). Seed yields were greatly improved, and Bhale and Borikar (1982) reported seed yields from the two highest-yielding single-cross female parents exceeded those of the normal inbred parent by 59% and 45%. In addition to yield, male-sterile single-cross female parents may exhibit other advantages, such as stability of flowering over environments, which reduces synchronization problems; a longer period of blooming; and prolonged stigma receptivity.

PARENTAL MATERIALS

An initial step to commercial seed production is to obtain breeder seed and to increase these parental seed stocks. There are no limits on the number of generations an inbred can be grown to produce breeder seed stocks. Parental material must be multiplied to quantities required for production of commercial seed.

A strict schedule of procedures is required to produce hybrid seed. No matter how well a hybrid may perform, it will not become a commercial reality unless seed can be produced economically, reliably, and with reasonable ease. The male parent must produce adequate pollen that coincides with receptivity of the female organ to assure that the female plant produces a reasonable seed yield.

Interest in corn hybrids is expanding to third-world countries, where hybrids have not been used because the small-scale farmer too often is unable to afford hybrid seed or is unaware of hybrid advantages and because genotype may not be the primary factor limiting yield. In underdeveloped countries a system of hybrid seed production may not yet exist; seed growers must come to recognize the vulnerability of hybrid seed production if fields are not regularly inspected and kept at a safe distance from other corn fields.

The accepted procedure for the field production of corn hybrids, established by the first commercial seed company in the mid 1920s, has been to remove by hand the tassels from female parent to ensure cross-pollination by the male parent. In the early 1950s Rogers and Edwardson (1952) outlined procedures for utilizing cytoplasmic male-sterile plants that eliminated detasseling in the commercial production of corn hybrids. The cytoplasmic male-sterile system received a setback with the outbreak of southern corn leaf blight[1] in 1970 (Tatum, 1971; Ullstrup, 1972).

[1]Southern corn leaf blight—*Helminthosporium maydis* Nisikado and Miyake.

The system used to eliminate fertile pollen in the female parents determines the types of parents to be increased. The cytoplasmic-genetic male-sterility system of hybrid production requires three types of parents: the male-sterile A line, the male-fertile maintainer B line, and the restorer R parent with male-fertile pollen. Parental seed increase of the male-sterile A line resembles that of commercial F_1 hybrid seed production, with the A line being the female and the B line the male. To ensure adequate pollination, a low ratio of the female A line to the male B line is usually used, e.g., a 1 : 1 ratio in sorghum (Wright, 1980). Double crosses involve the multiplication of A and B lines of two inbreds (four sets) and male-fertile inbreds of the C and D parents (two sets) for a total of six parental seed stocks. If detasseling is used in corn to eliminate fertile pollen in the female parents, all parents are male-fertile, and parental seed of each is produced in isolated fields to prevent out-crossing of these open-pollinated parents.

CHOICE OF FIELD FOR PARENTAL OR COMMERCIAL SEED PRODUCTION

In full-season crops such as corn and sorghum, seed fields must be located in an area where adequate heat units are accumulated to mature the crop. To prevent stress at pollination time and to assure maximum grain yields, the area normally should receive adequate moisture.

Seed fields need not be chosen in terms of their adaptation for a particular hybrid cultivar. parental selection determines the genetic constitution, and the area of production has no impact on the subsequent genotype. Seed fields are chosen on the basis of maturity of the hybrid parents, spring and fall frost dates, and other environmental variables suitable for seed production. Most hybrid seed is produced under conditions of natural rainfall, but irrigation may be applied as needed to ensure seed production.

Seed for short-season corn hybrids is usually grown 300 to 600 km (200 to 400 miles) south of where they are best adapted in order to reduce risk of a damaging frost and to increase seed size and yield per plant.

Guidelines for isolation of seed fields for pedigreed status must be met to prevent pollen contamination from other corn fields or from weeds such as Johnson grass and Sudan grass that may be a contaminating pollen source for grain sorghums. For corn, sorghum, and wheat the minimum isolation distance is 200 m (650 ft) (Wright, 1980) from fields of the same crop species. This may be hard to achieve in areas where the same crop is grown commercially on an extensive scale. In such cases, a barrier of the male parent may be planted between the source of contamination and the seed-production area, so that pollen from the barrier plants has a greater chance of pollinating the female parent than does that from other sources. By means of a barrier, isolation distance may be reduced.

Seed growers are responsible for roguing out off-type plants in a seed field before they produce pollen. Off-types can result from undesirable cross-pollination, mixing, or a volunteer plant from a previous crop. Fields that produced the same crop the previous year may be unacceptable for pedigreed status, but this limitation varies with the crop and with state or national standards. In wheat seeded in conventional narrow rows, roguing for off-type plants is less practical, but roguing for male-fertile plants in a male-sterile parent is essential. Roguing is required both in fields of parental seed stock production and in seed fields for commercial hybrid use. Removal of off-type plants in both the male and female parental rows is desirable to maintain genetic purity. Roguing should involve removal of the entire plant, especially if it is done when the plants are small, to prevent secondary tillers from arising.

Figure 10–2. Cross-pollination for commercial production of hybrid seed is achieved by planting alternating strips of male and female parents. In this hybrid seed field of corn, the female seed parent rows have been detasseled to ensure cross-pollination by the male parent rows. The planting pattern of this field is such that the female seed parents are not more than two rows from the pollen parent and have a 1:2 plant ratio. A 1:3 ratio may be achieved by planting two male rows to six female rows or four male to 12 female rows. (Courtesy of Pioneer Hi-Bred International, Inc., Des Moines, Iowa, U.S.A.)

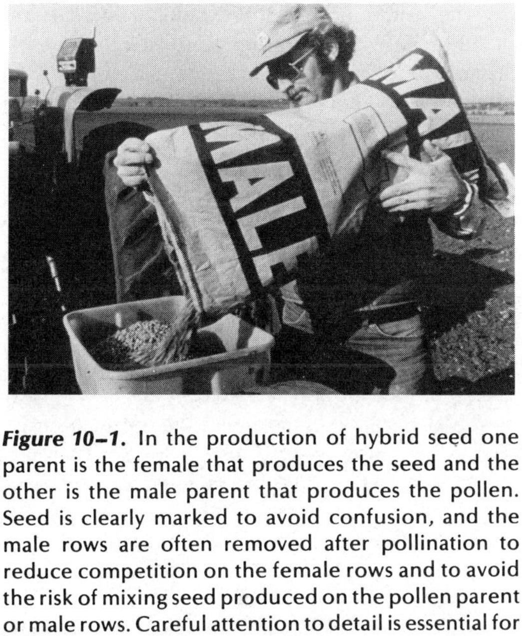

Figure 10–1. In the production of hybrid seed one parent is the female that produces the seed and the other is the male parent that produces the pollen. Seed is clearly marked to avoid confusion, and the male rows are often removed after pollination to reduce competition on the female rows and to avoid the risk of mixing seed produced on the pollen parent or male rows. Careful attention to detail is essential for successful commercial hybrid seed production. (Courtesy of Pioneer Hi-Bred International, Inc., Des Moines, Iowa, U.S.A.)

Careful attention to identifying and treating male and female rows accordingly is critical to hybrid seed production. To reduce error, seed is clearly marked (Figure 10-1). Seed of male parents is interplanted among the female rows to ensure an adequate amount of pollen when the female flowers are receptive. Ratios of male and female rows or strips vary from 1 : 3 to 1 : 6 in sorghum, 1 : 2 to 1 : 4 in corn, and 1 : 1 to 1 : 3 in wheat (Wright, 1980). The combination of male and female rows may vary among regions or commercial firms (Figure 10-2) but is primarily influenced by the adequacy of pollen production. In the production of double-cross seed a ratio of 1 : 4 may be used because of an abundance of pollen produced on the male parental rows, which are F_1 hybrids.

To ensure that pollen release coincides with female receptivity, various techniques may be employed to hasten or delay one parent. In corn and sorghum, delayed or split planting is the most effective and common procedure (Figure 10-3). By determining the number of heat units or growing degree-days required to reach flowering in each parent, approximate planting dates can be calculated. Delays in seeding are less effective in winter wheat, however, as a ten- to 14-day delay in fall seeding may result in only a one- to two-day delay in flowering date.

Additional practices to delay flowering in corn, sorghum, and wheat include early removal of vegetative growth, variable planting depths, variable amount and analysis of fertilizer applications, and timing of irrigation water. Treatments are directed to the male parent whenever possible so as not to reduce seed set on the female parent rows.

Figure 10–3. Differences in maturity between the male and female parents may require different planting dates to ensure synchronization of flowering parts. One parent may be hastened or delayed to obtain pollen release when the female silks are receptive. (Courtesy of Pioneer Hi-Bred International, Inc., Des Moines, Iowa, U.S.A.)

Figure 10–4. Under conditions of hand pollination, pollen is collected by bending the stalk and tapping the bag to dislodge the pollen. In commercial seed-producing fields, hand bagging and pollination is not economically feasible. (Courtesy of Pioneer Hi-Bred International, Inc., Des Moines, Iowa, U.S.A.)

CONTROLLING POLLINATION

In the dioecious plants such as wheat and sorghum cross-pollination is controlled by the use of male sterility. Plants that are male-sterile become the female parent. Hand emasculation of female plants is not possible on a commercial scale.

Cross-pollination in the monoecious corn plant is possible, but in breeding programs the bagging of both the male and female parts is conducted (Figures 10-4 and 10-5) to force selfing or crossing as desired.

On a field basis, female plants are emasculated by **detasseling** the plants by hand or machine before pollen is released. Detasseling of the seed-parent rows is one of the most expensive and painstaking operations in the production of corn hybrids. Cytoplasmic male sterility received a setback when the extensively used Texas or T male-sterile cytoplasm was found to be susceptible to a new mutant strain of corn leaf blight, which devastated 15% of the North American corn crop (Tatum, 1971). The blight was brought under control the following year by elimination of the T cytoplasm and a return to hand detassel-

Figure 10–5. Under conditions of hand pollination, pollen is transferred from the tassel to the silks of the female flowers. The ear shoot is covered until fertilization is complete. (Courtesy of Pioneer Hi-Bred International, Inc., Des Moines, Iowa, U.S.A.)

ing. Additional sources of male sterility have allowed some return to male sterility as a form of pollen control (Craig, 1976).

All tassels must be removed from the female rows before they have shed any pollen and as soon as they are free from the uppermost leaf sheath. The time period from tassel emergence to pollen release is limited. If tassels are pulled too soon, one or two leaves may be removed with the tassel, or the tassel may break and not be removed completely.

Careless detasseling may result in seed pollinated by the ear-producing parents, with the result that hybrid vigor is lost. Rigid standards for pedigreed status must be met. A crop will not be eligible for pedigreed status if on any one inspection more than 1% of the female seed parent plants possess tassels, or if the total for three inspections on different dates exceeds 2% when 5% or more of the female seed parent plants have receptive silks. Tassels are considered to be shedding pollen when 2.5 to 5 cm (1 to 2 in.) of tassels on tillers, portions of tassels, or tassels on the main plant have anthers extruded from their glumes.

Detasseling must be achieved with a minimum of plant damage so that seed parent plant yields are not reduced. Tassel removal alone causes a minimum of physiological damage to the plant and results in an increase in ear size and grain yield. A large portion of the yield response from detasseling was shown by Hunter et al. (1969) to result from the elimination of tassel light interception; an increased amount of light was available for leaf photosynthesis when tassels were removed. A second explanation may be that considerable photosynthetic energy is required to produce the 25 million pollen grains on an average corn tassel. Removal of this energy-requiring sink may allow more photosynthetic energy to be directed to grain yield.

Removal of one or two leaves with the detasseling process can cause a yield reduction of sufficient magnitude to warrant serious consideration in seed-production fields. Smaller inbreds with fewer leaves appear to be affected more by leaf removal than large inbreds (Table 10-2). Kiesselbach (1945) summarized that removal of one, two, or three leaves with the tassel resulted in grain reductions of 4.3%, 9.3%, and 16.4%, respectively.

The tassel on an inbred plant may be pulled during a period of six to 18 hours, but because plants differ in time of tasseling, the period involved in detasseling a field may be from five to ten days or up to two weeks to remove late tassels on tillers.

Detasseling crews must patrol each row four to seven times during the season and must go over the fields every 24 to 48 hours. Crews walk or ride through the field. Propelled platforms, high enough to pass over the corn plant, can be used to transport the detasselers between the rows. High-clearance platforms allow the detasselers to look down into the leaf whorl and to more effectively and easily remove the tassel (Figure 10-6). Detasseling on foot may be difficult if a tall single-cross parent is grown. Under wet soil conditions, however, personnel carriers may not be able to be used. The tassel is removed by a gentle upward pull and dropped to the ground. Some parents are easier to detassel than others.

Mechanical detasselers can be used to either cut or pull the tassels. Cutting action is achieved by a rotating blade; pulling is achieved by me-

Table 10–2
Average Grain-Yield Performance Following Several Detasseling Treatments of Five Short-Season and Five Long-Season Inbreds Grown at Two Sites in Ontario in 1971

	Early Inbreds		*Late Inbreds*	
Treatment	*kg/ha*	*lb/ac*	*kg/ha*	*lb/ac*
Check	3736	3360	3054	2727
Tassel removed	3908	3489	3331	2974
Tassel plus one leaf	3795	3388	3312	2957
Tassel plus two leaves	3525	3147	3324	2968
Tassel plus three leaves	3136	2800	3086	2755

Source: Hunter, Mortimore, and Kannenberg (1973).

chanically guiding the tassels into v-shaped rubber rollers. Mechanical detasselers remove the top leaves along with the tassel. High labor costs have encouraged the use of mechanical detasselers despite problems of yield reduction.

In a uniform seed field in which the tassel is well extruded ahead of pollen shedding, mechanical detasselers can remove 70% or more of the tassels in one or two passes over the field with a minimum of leaf damage (Craig, 1976). Tassels on late or short plants missed by mechanical detasselers must be removed by hand. Mechanical detasselers should be viewed as an aid to detasseling to be used in combination with hand detasseling. Cost savings with mechanical detasseling may be offset by seed parent yield reduction.

Figure 10–6. Detasseling crews using a powered platform ride through a seed-corn field to detassel female seed parent rows. The tassel is removed by a gentle upward pull and dropped to the ground. The driver is able to provide a final check on the thoroughness of detasseling. Not more than 1% of the tassels can be shedding pollen in any one day for pedigreed status to be provided. Tolerances can be met by repeat passes through the field. (Courtesy of Pioneer Hi-Bred International, Inc., Des Moines, Iowa, U.S.A.)

A selective chemical hybridizing agent that could control pollen release would be of tremendous value in helping produce corn hybrids. In corn, such a compound could reduce the need for tassel removal from seed-production fields. Some tassel removal may be required to eliminate any plants on which the chemical was not totally effective.

HARVESTING HYBRID SEED

Harvesting equipment must be designed and operated to harvest only the female seed-bearing rows. Harvesting equipment must be considered in terms of the ratio of male to female rows or strips (Figure 10-7).

Physiological maturity in corn occurs when the kernels contain 30% to 38% moisture and is the point at which harvesting for seed can occur. By harvesting shortly after physiological maturity is reached, losses from shattering are reduced, losses by birds, insects, or ear and stalk rot are

Figure 10–7. The ratio of male to female rows is determined by the amount of pollen produced from the male parents and the size of the harvest machinery. Corn seed is harvested as soon as the kernels reach physiological maturity, which occurs when kernel moisture is approximately 30% to 38%. Ears of corn are normally harvested rather than as shelled corn to minimize seed damage and to facilitate uniform drying. Ears that are diseased, damaged, or with off-type seed can be removed. (Courtesy of Pioneer Hi-Bred International, Inc., Des Moines, Iowa, U.S.A.)

minimized, and damage from freezing is eliminated.

Seed germination can be reduced by frost damage (Craig, 1976; Rossman, 1949). Rossman (1949) studied maturing ears containing 30%, 40%, and 50% moisture, which were subjected to controlled laboratory freezing conditions intended to simulate typical freezes that might occur in the field. Temperatures of −7°C (20°F) for eight hours, with husks on, reduced subsequent germination of 25 inbred parents. At 50% moisture, germination was reduced to an average of 51.6% compared to 88.6% for the unfrozen check; at 40% moisture, germination averaged 79.6% compared to the check at 93.5%. Differences in freezing tolerance of the seed were found among the inbreds and was related to seedling vigor and to physiological maturity. Less injury occurred to inbreds more physiologically mature and vigorous. Freezing damage was dependent on seven factors: (1) temperature, (2) duration of exposure, (3) seed moisture, (4) inbred, (5) husk protection, (6) stage of maturity, and (7) rate of drying after freezing.

Ears of corn are picked intact with some of the husks remaining, which are removed before ear sorting and drying at the processing plant. By harvesting the entire ear rather than using a picker-sheller, seed damage at 30% to 38% moisture level is minimized, undesirable ears can be discarded manually at sorting time, and seed on the cob can be dried more uniformly, which results in high-quality seed.

If seed is harvested as shelled corn, grain moisture may need to be at 20% or less to keep mechanical damage at a minimum. In short-season areas, only the earliest of the parental cultivars may dry down to this level in the field. Where a picker-sheller is used, field roguing must be very thorough.

Wheat and sorghum are allowed to dry in the field to a level at which they can be stored without artificial drying. Cylinder speed of the combine must be adjusted so as to avoid mechanical damage to the seed.

An essential step in harvesting is to carefully clean seed-handling machinery and equipment to avoid mechanical mixing of different genotypes. Because combines cannot be cleaned completely without dismantling, several hectoliters (bushels) of seed harvested of a new parent are discarded.

PROCESSING-PLANT OPERATIONS

Hybrid-seed processing plants perform a series of operations to the seed as it is delivered from the field. During harvest, seed producers are asked to deliver seed of specific cultivars at specific time in order to reduce the opportunity for mixing. Each grower is paid on the basis of total weight and moisture level; in corn, shelling percentage is an additional parameter. As the ears of corn are conveyed to the drier, inspection crews remove any husks and undesirable ears (Figure 10-8).

Drying should begin as soon as possible after harvest to reduce the risk of mold, deterioration of germination, and loss of seed vigor. Corn is

Figure 10–8. Hybrid seed corn delivered to the processing plant is conveyed by moving belts to the temperature-controlled hot-air driers, to be shelled, and to bin storage for grading and bagging. Inspection crews remove diseased, wet, and off-type ears and husks. In this photo, situated on the roof of the drying unit, a group of students examine the ears. (Photo courtesy of Doug Wagner)

dried on the cob at 40° to 44°C (105° to 111°F), measured at the point of contact with the seed, for 60 to 75 hours. Temperatures above these levels may reduce germination. At the conclusion of the drying period, kernel moisture is approximately 12%. Seed is then removed from the ears and moved directly to bulk storage to be sized, treated, and bagged.

Sizing involves separating corn seed into various sizes to improve the accuracy of planting and spacing, especially with plate-type planters. Dent corn may be separated into eight different flat sizes and eight different round sizes. Any remaining seed may be graded into three plateless-planter sizes. Kernel counts are commonly done on the various seeds, which are bagged with a minimum number of seeds per bag; the number varies from 60,000 to 100,000, with 80,000 kernels/bag being the most popular size. Sale of seed based on kernels per bag rather than weight per bag has become commonplace, because growers can determine readily the number of bags needed based on the intended plant population and area to be seeded.

Seed of wheat and sorghum is cleaned and graded but is not sized. After treatment with a fungicide and insecticide, these seeds are bagged in units of 22.7 kg (50 lb). Containers of hybrid seed are given a certified label to identify cultivar name and number, contamination, and viability as shown in Table 10-1. Seed is stored in an air-conditioned warehouse that maintains the air temperature below 15°C and commonly at 10°C (50° to 60°F) and 50% humidity.

RECORD-KEEPING AND QUALITY CONTROL

The production of hybrid seed that is genetically pure and of high quality demands attention to detail at all stages. Proper quality control involves accurate planning and records (Figure 10-9). Previous-crop isolation, planting procedures, male and female planting dates and ratios,

Figure 10–9. Field isolation, previous crop, ratio of male and female parents, planting dates, and roguing are important aspects of commercial seed production. Seed growers may produce hybrid seed under contract with a commercial firm, who provide seed, instructions and supervision. The farmer-cooperator must be concerned about soil fertility status, tillage, pest control, and harvest procedures. (Courtesy of Pioneer Hi-Bred International, Inc., Des Moines, Iowa, U.S.A.)

roguing, detasseling of corn, and harvesting are performed on the farm. At the processing plant, sorting, drying, cleaning, sizing of corn, treating, bagging, labelling, and storing are associated with quality control. Additional steps involve tests for germination and vigor.

Germination tests under standardized procedures are routinely conducted on samples from all seed lots. Germination tests have been widely accepted as an index of seed quality. A tremendous amount of time and effort has gone into their improvement and standardization, more than for any other quality measure.

Labelling for vigor is viewed by many as a desirable step, but problems and controversy have arisen. A major problem is the lack of an acceptable and workable definition of seed vigor (Sayers, 1976), owing in part to the lack of a clear objective as to what vigor tests are to achieve.

Germination vigor is regarded by some as being of greatest significance (Spain, 1976). Vigor is viewed as a transitory quality, and the questions of expiration date and tolerances need to be addressed. Seed vigor may be of great value in direct-seeded, space-planted crops such as corn. For drill-seeded crops, vigor may be of less value, especially where tillering provides a compensation mechanism for variation in plant populations (Spain, 1976).

Objectives for measuring seed vigor were listed in a survey (Grabe, 1976) as follows:

1. To indicate stand establishment ability.
2. To select seed lots for planting.
3. To provide a general ranking of seed lots for vigor.
4. To indicate relative storability of lots.
5. To adjust planting rate.
6. To detect mechanical damage.
7. To indicate potential yield ability.
8. To indicate potential uniformity of the crop.

SEED VIGOR AND PLANT PERFORMANCE

Some batches of seed do not grow as well in the field as others of the same cultivar, and this can lead to significant yield reductions (Anonymous, 1976). Performance testing among different cultivars may not reflect genetic potential if seed sources vary in vigor. Seed-vigor determinations are providing an added dimension to seed quality. Seed lacking in vigor can be alive and germinate but lack the ability to develop as fast as they should and under stress conditions can fail completely. Weak plants that contribute little or nothing to yield consume water and plant nutrients and behave like weeds.

Lots of wheat seed in England all showed germination percentages above 90%, but in the field, seedling emergence in one lot was 30% compared to 88% in another (Anonymous, 1976).

The need for additional seed-quality measures to supplement the standard germination test are documented, and the value of seed vigor and vigor testing is well recognized in the seed industry. Vigor tests have a definite and important place, but not necessarily in any regulatory capacity. Problems include standardization of procedures, interpretations of results, and utilization and application of test results. The worth and reliability of vigor tests must be clearly documented before official sanction or status can be given to them.

Vigor tests are often regarded as an in-house quality measure in an effort to benefit the consumer as well as the seedsman. By meeting his customers' demands for better-quality seed, the seedsman may improve his own business position (Sayers, 1976).

SEED-VIGOR DETERMINATIONS

Direct and indirect tests can be used to measure seed vigor. Direct tests are based on the interaction of the seed with the environment; indirect tests measure enzyme activity or other physiological characteristics without germinating the seed. Various tests were outlined by Grabe (1976) as follows:

THE COLD TEST. This was designed to predict the performance of corn seed under cool, wet soil conditions in the laboratory and is used routinely. Seed is planted in field soil, which is watered at 70% of saturation under 10°C (50°F) conditions to seven days followed by four days at 25°C (77°F). The number of normal seedlings is counted, and 95% germination indicates vigorous corn seed. For satisfactory field performance, 70% to 85% germination is suggested.

Presumably, vigorous seeds and seedlings are able to resist decay organisms to a greater extent than weaker seeds. Cold tests are often performed on seed following treatment with a fungicide, and under such circumstances the test primarily measures how adequately the seed is protected with a fungicide. Different soil types, microorganisms present in the nonsterilized soil, and variations in soil moisture make standardization of procedures difficult.

ACCELERATED AGING TEST. This test was developed (Delouche and Baskin, 1973) to predict longevity of seed in storage. Seed samples are placed in 41°C (106°F) and nearly 100% relative humidity conditions for four to five days, followed by a germination test.

TETRAZOLIUM TEST. Quick information on seed viability can be obtained with the chemical tetrazolium, which turns living tissue red while dead tissue remains colorless. With this test, it is possible to evaluate internal seed injury, insect damage, the destructiveness of frost, and the viability of dormant seed. The test is relatively easy to standardize, but readings are subjective. Both viability and vigor readings may be determined from the same seeds.

SEEDLING GROWTH RATE. An indication of vigor may be obtained by measuring length or weight of shoot, root, or entire seedling grown in an appropriate germination medium after four or more days.

In a four- or five-day test, however, seedling length may be influenced unduly by the length of time required for germination. Small wheat seed may germinate more rapidly than large seed, thereby giving the impression that small seed has greater vigor. Growth subsequent to germination may be a more meaningful indicator of seedling growth rate.

This test may vary from placing seed in a rolled moist towel to providing a six-week growth period in a greenhouse with soil.

SPEED OF GERMINATION. Based on the concept that more vigorous seeds germinate faster than less vigorous ones, speed-of-germination tests have been developed for vigor evaluation. Differences in seed size may confound this test. As well, this test may measure attributes similar to those measured by seedling-growth-rate tests.

CONDUCTIVITY TESTS. When seeds are soaked in water, the amount of leachate or electrolyte released can be measured by electric conductivity. The concept is that low-vigor seeds release more electrolytes, which under field conditions nourish soil microorganisms resulting in seed decay and poor stand establishment. Equipment is available for measuring the average conductivity of 100 individually compartmentalized seed very quickly and accurately under standardized conditions.

SEEDLING VIGOR CLASSIFICATION TEST. An added dimension of standard germination tests is to visually classify the seedlings as strong or weak. Despite its subjective nature, valuable information on relative vigor of seed lots can be obtained easily by this test.

BRICK-GRIT TEST. Sterilized brick grit, which is heavy and acid in nature, is used as a 3-cm (1.18-in.) barrier to seedling emergence. Weak or strong seedlings are determined by their ability to penetrate this layer after 14 days.

RESPIRATION TESTS. Seed respiration rate is believed to reflect seed vigor and can be measured with specialized equipment (Woodstock, 1973).

MECHANICAL DAMAGE TEST. Internal bruises, fractures, or cracks in the seed coat can reduce seedling vigor and can be evaluated by several methods. External damage such as breaks in the seed coat can be detected visually with the aid of a dye or under low-power magnification. Internal damage may be detected by x-rays.

ENZYME-ACTIVITY TESTS. The presence and relative activity of several enzymes have been demonstrated to be related to seed germination, seedling growth rate, storability, and yield. Enzymes are a very sensitive indicator of physiological deterioration in storage. Specialized equipment similar to that for measuring respiration is used to measure enzyme activity.

SEED-SIZE TEST. Although seed size is relative and is not directly related to yield, knowledge of seed-size distribution within the lot of seed may be of value. Seed lots with the larger proportion of seeds in the large-seed categories are believed to be superior to lots with the largest proportion of seeds in the small-seed categories.

SEED UNIFORMITY, SIZE, AND CROP PERFORMANCE

Association of seed size with seedling vigor has been reported in different crops by various workers. The effect of seed size on crop development and grain yield remains ambiguous because of the confounding effect of depth of seeding and environmental conditions. Once the plant has emerged from the soil, photosynthesis takes over as the source of food for the seedling. Large food reserves in the endosperm are more important when the seedling is slow to emerge from the soil, owing to deep seeding or other stress conditions, than when the seedling emerges quickly and photosynthesis proceeds. Failure of some seedlings to emerge from the seed may result in lower grain yield.

Uniformity in seed size may be as important as size itself. At high plant populations competition among plants is severe. Variation in seed size may allow a vigorous plant to dominate an adjacent weak seedling.

Large embryo size and food reserves that resulted in rapid seedling emergence and early photosynthesis activity resulted in an extension in the duration of photosynthesis in crops. This resulted in an increase in final grain yield (Wood, Longden, and Scott, 1977).

For many years, uniform and accurate planting was achieved with corn seed by means of graded seed for plateplanters. The development of plateless planters capable of uniformly and accurately planting ungraded seed, and the lower prices of ungraded or small-sized seed, have attracted interest to these classes of seed. As well, the shift to single-cross hybrid seed has resulted in a marked reduction in seed size.

Contrary to the report of Wood, Longden, and Scott (1977), tests with corn seed varying in size from 23 to 39 g/100 seeds from one single-cross hybrid and planted in the field on May 9, 19, and 29 in 1967 at three shallow seeding depths of 2.5, 5.0, and 7.5 cm (1, 2, and 3 in.) showed no seed-size influence (Hunter and Kannenberg, 1972). No effect of seed size was observed on number of days to 50% emergence, final percent emergence, or grain yield (Table 10-3) under the conditions in southern Ontario.

A similar test was conducted over a two-year period in Minnesota (Hicks et al., 1976) to determine yield differences among seed grades at an early and late May planting date. No yield differences were found among large round, large flat, small round, and small flat grades of corn. Yields from ungraded seed were slightly lower than yields of graded seed. In three of four location years, no yield differences among seed size were found, but again ungraded seed lots yielded less. Cool soil conditions had little effect on grain-yield response to seed size except at one location, where large seed produced higher yields in one hybrid.

In triticale, Dhillon and Kler (1978) and Srivastava and Nigram (1973) found that seed size did not influence grain yield, whereas Bishoi (1977) did find that heavier seeds of triticale produced higher germination, emergence, dry weight of 40-day-old plants, grains per ear, and grain yield.

Studies in wheat (Austenson and Walton, 1970; Dasgupta and Austenson, 1973) and in barley (Pinthus and Osher, 1966) found higher grain and straw yields from large seeds. On the average, Pinthus and Osher found that plants from large seed were taller and produced 24% more grain, owing almost equally to a greater number of spikes per plant and to a greater number of kernels per spike. McDaniel (1969) reported a

Table 10–3
Days to 50% Emergence, Final Percent Emergence, and Grain Yield for Large (39 g/100) Seeds and Small (29 g/100) Seeds for a Single-Cross Hybrid Averaged over Three Planting Dates (May 9, 19, and 29) and at Three Depths of Planting at Guelph, Ontario, in 1967

Measured Parameter	Planting Depth		Large Seed	Small Seed
	cm	in		
Days to 50% emergence	2.5	1.0	13.3	13.3
	5.0	2.0	14.9	14.8
	7.5	3.0	16.1	15.9
Mean			14.8	14.7
Final % seed emergence	2.5	1.0	94	95
	5.0	2.0	91	95
	7.5	3.0	90	93
Mean			92	95
Grain yield, kg/ha (lb/ac)	2.5	1.0	6905 (6165)	6717 (5997)
	5.0	2.0	7049 (6294)	6999 (6249)
	7.5	3.0	7012 (6261)	6767 (6042)
Mean			6989 (6240)	6828 (6096)

Source: Hunter and Kannenberg (1972).

positive relationship between seed weight and fresh seedling weight.

Under both nonstress and stress conditions achieved by irrigation of pearl millet in India, large seeds produced more vigorous seedlings and large ears, yield, and seed (Chhina and Phul, 1982).

In corn, Burris (1975) found a positive relationship between seed weight and fresh seedling weight, but differences dissipated with time. Seed size in corn was not related to grain yield, plant height, or tasseling date (Hicks et al., 1976).

Large seeds of sorghum had higher levels of germination than smaller seeds, but seed characteristics did not affect final emergence, stand, or grain yield (Maranville and Clegg, 1977).

SELECTING A VIGOR TEST

Hybrid vigor may be expressed in seed vigor as a genetic component. As a genetic trait, it may be possible to select for vigor in breeding programs through a number of traits. Bittenbender and Ries (1977) observed that the site of seed vigor in rice appeared to be localized in the embryo, while the endosperm is thought to be responsible for seedling vigor. The environment may determine how fully the genetic potential is realized.

In choosing a vigor test, laboratories and seed companies must determine what information is desired (Mathews, 1981). A single test may be inadequate. A four-test series including a growth-rate test, a stress test, a biochemical test and a mechanical damage test may be needed, and such a series is not uncommon among commercial companies.

Vigor tests, like a soil test, must be interpreted for the customer. The most challenging task of interpreting is to relate laboratory measurements to field performance, and this can be determined only by field trials. It is possible that under favorable conditions, wide differences in laboratory vigor ratings may not influence field performance.

The greatest value of vigor tests may be in monitoring seed-storage conditions. Periodic testing and charting against established standards can reveal variations of specific cultivars.

Vigor testing may be useful when selection of plants from segregating generations is aimed at improving a specific trait. But plant breeding should not be regarded as a solution to vigor problems. Proper cultural practices, including seeding depth, may solve poor emergence associated with a short coleoptile or low seed vigor.

PLANT BREEDERS' RIGHTS OR PLANT CULTIVAR PROTECTION

In many countries the seed industry has evolved along two distinct avenues, those of the public and private sectors. The public sector, financed through state funds, for many years has released new cultivars developed at public institutions for public consumption without concern about proprietary control over the cultivars released. Prior to 1960, the majority of plant breeding of small grain cereals was undertaken on a national basis almost exclusively at government expense (Ruttan, 1982).

With the advent of commercial hybrids produced by private firms, the associated inbred parents provided sufficient natural protection to enable the private sector to establish proprietary control over hybrid cultivars. By the mid 1970s in the United States, over 80% of the corn and sorghum cultivars used in commercial production were controlled by private interests (Hanway, 1978). In nonhybrid crops, where this natural protection did not exist, over 80% of the small cereal grains grown commercially were public cultivars.

In general, the combined efforts of the public and private interests were effective in generating new cultivars, but the escalating costs of plant breeding programs placed greater demands on the public purse and reduced the opportunity for profit in the private sector. Some form of protection was needed to allow an expansion in breeding programs into the private sector.

The opportunity to patent or provide equivalent protection to new crop cultivars could serve as an incentive to expand research and development programs. It does take ten to twenty years to produce a new crop cultivar in sexually produced plants, and substantial capital investments are required. Protective legislation would provide a much-needed economic incentive, result in additional cultivars to help diversify genotypes produced in an area, increase cropping flexibility for producers, promote resourcefulness to secure superior genotypes, and stimulate a healthy competitive situation within the private sector.

Legislation has been introduced in a number of countries, principally in the United States and Western European countries. The format of most legislation is based on the suggestions of the International Convention for the Protection of New Varieties of Plants held in Paris in 1961.

Legislation to provide protection is known as **Breeders' Rights.** The stated justification of such legislation is "to encourage the development of novel varieties of sexually produced plants and to make them available to the public, providing protection available to those who breed, develop, or discover them, and thereby promoting progress in agriculture in the public interest."[2] Although Breeders' Rights was introduced principally to stimulate plant breeding in the private sector, products of state stations can be eligible also for protection to provide unfair competition with private breeders and to provide a return to the taxpayer. In England, the National Seed Development Organization Ltd. (NSDO) was formed to act as the general commercial agent for the state-aided plant-breeding programs.

To be protected in Europe, a new cultivar must incorporate agronomic attributes found in previous cultivars and must contribute an additional feature of merit. In Europe, testing to determine merit required for registration is done by a government agency. An extensive testing program is needed to achieve this end. In the United States, the onus is on the owner to demonstrate the novelty of his or her cultivar. In addition to novelty, stability and uniformity criteria must be met. This means that a cultivar must be

---

[2]Preamble, United States Plant Variety Protection Act.

distinctive so it can be identified from other cultivars, that essential and distinct genetic characteristics must be retained when sexually reproduced, and that any genetic variants must be described and predictable (Anonymous, 1982). The objective is to reduce the use of different names for the same cultivar in different countries. F_1 hybrids are not eligible for protection, but parental cultivars used to develop hybrids are eligible.

An important aspect of Breeders' Rights is the opportunity for reciprocity agreements among countries, even though the protection in the countries concerned need not be identical. Plant-breeding products may be extensively cultivated in foreign countries. Two-row barley cultivars, for example, have been grown extensively in many different countries or continents. Other crops have comparatively narrow areas of adaptation and may be restricted by sensitivity to daylength, spring or winter habit, or other specific climatic requirements.

Breeders' Rights provides the plant breeder with a means of recovering the costly investment in such research. Without protection, developers of new cultivars will soon find others selling the same cultivars at a lower cost, since the others do not have the development cost to cover. But the advantages of Breeders' Rights are not solely to plant breeders. Every consumer benefits, because food security is increased. Seed costs may rise slightly to cover the royalty fee, but volume of seed sales could mean this increase is small. Seed costs are a small proportion of crop production costs, and royalty fees may not be reflected in the price of food to the consumer.

The countries adopting this important form of legislation are growing in number but are found almost exclusively within the industrialized parts of the world. Countries such as France, Germany, the Netherlands, Belgium, Italy, Denmark, Switzerland, and Great Britain signed the 1961 Convention of Paris shortly after it was established (Weibull, 1964). Refinements have taken place in many countries. In 1964, the Plant Varieties and Seeds Act was signed in the United Kingdom; it included a provision for plant breeders to obtain royalties.

In the United States, a Plant Patent Act was passed in 1930 that extended patenting rights to breeders of certain asexually reproduced plants. Sexually reproduced plants were not included for fear that such legislation would inhibit the free exchange of genetic materials and lead to excessive concentration of proprietary control in the seed industry. A Plant Variety Protection Act developed by the American Seed Trade Association and fashioned after the Paris Convention was passed in the United States in 1970; it included seeds of sexually reproduced plants but omitted seed of several vegetable species. Amendments in 1980 extended the 1970 bill to cover vegetables not included previously, and the period of protection was extended to 18 years to conform with international standards.

In Canada, although much of the concern that cultivar protection would reduce public breeding programs has disappeared, and despite encouragement by the Canadian Seed Trade Association and other agricultural organizations, finalization of a bill has languished. Passage of a bill was anticipated in 1984. Internal systems for collecting royalties exist in Canada, but without legislation there can be no reciprocity agreements. Cultivars developed through governmental support at public institutions can be marketed through SeCan (Seeds Canada).

Initial concerns that Breeders' Rights might threaten genetic resources, hinder the free exchange among public and private plant breeders, and produce a flood of worthless material have not materialized. The fact is that the tasks of plant breeding are unlimited in terms of developing new cultivars resistant to various diseases, improving winter hardiness, meeting quality needs, and producing cultivars to meet agronomic demands. The opportunities associated with genetic engineering provide an enormous additional dimension. Plant cultivar protection provides a desirable and positive incentive to develop these possibilities and to increase genetic diversity.

Questions

1. *a.* List the various types of corn hybrids.
   b. Why was the development of double-cross corn seed considered a giant step forward?
   c. Why has single-cross corn seed gained such prominence?
2. Consider the various types of hybrids in selected crops from the standpoint of:
   a. Yield ability.
   b. Yield stability.
   c. Seed yield for commercial seed production.
   d. Cost of seed production.
3. Commercial hybrid corn seed production and marketing represents a success story of enormous magnitude both in North American and in the world. Popularizing hybrid corn production, however, was a slow process that got underway in 1926 when Henry Wallace established the Hi-Bred Corn Company. Perhaps history offers some lessons that can be learned today in introducing a new concept.
   a. Briefly trace the spread of hybrid corn throughout the United States and the world [an account of the Garst and Thomas Hybrid Corn Company written by Rajki (1980) provides an interesting human-interest story on one of the pioneer seed-corn producers].
   b. What factors have prevented hybrid corn from being universally adopted?
   c. To what extent do these factors apply to other hybrid crops, and how might they be overcome?
   d. Could hybrid seed be produced in North America for use in other parts of the world?
4. What is the primary purpose of legislation to provide Plant Breeders' Rights?
5. List the advantages of Plant Breeders' Rights.
6. What has happened or may happen to public plant breeding when Breeders' Rights are introduced?
7. Do Plant Breeders' Rights provide an avenue for the market to become flooded with inferior cultivars?
8. What is the likely impact of Plant Breeders' Rights on the price of seed?
9. Is it conceivable or likely that Plant Breeders' Rights legislation may make it possible for larger companies to make huge profits at the farmer's expense?
10. Might Plant Breeders' Rights make it possible for large companies with fertilizer and chemical divisions to develop cultivars requiring specific pesticides and fertilizers?
11. *a.* Corn seed is carefully sized into several categories. Consider the value of seed sizing to crop production.
    b. What do corn producers look for in seeding machinery? Consider accuracy, damage, and speed of planting.
    c. What changes have taken place in corn production to affect the plate planter? (See Reynolds, 1968.)
12. Review seed germination and seedling development and consider the effect of depth of seeding and seed size on shoot emergence.
13. *a.* Consider the impact of mechanical damage on seed germination, seedling vigor, and plant yield.
    b. Will cracked or broken seeds germinate and produce a plant?
14. *a.* What factors are essential for germination to occur in cereal grains?
    b. Do cereal grains require dark or light conditions to germinate?
15. *a.* Compare seed longevity of crops stored under ambient environment conditions in the tropics and in the temperate zones.

b. What factors are mainly responsible for a reduction in germination and vigor?

c. What conclusions do you reach about the need for controlled storage units in the two regions?

16. Some cereal growers believe that after a cultivar has been grown for a period of years, it goes through some kind of change that causes it to yield less than when it was first released. Such seed is said to have "run out." The term is not usually applied to reduction of yield by disease, lodging, or other obvious cause. "Running out" is used when yield goes down but the grower cannot put his finger on the exact cause. Consider the concept of "running out":

a. From a genetic viewpoint.

b. As an explanation of how bearded plants appear in a beardless field, or two-row barley in a six-row field, or white chaff wheat in a red chaff cultivar, or chess plants in wheat fields.

c. From the viewpoint of purchasing certified seed of a nonhybrid, self-pollinated crop frequently or every year.

d. From the viewpoint of purchasing F_1 hybrid seed each season.

e. As fact or fiction.

References

Andersen, H. J. 1982. "Isoenzyme Characters of 47 Barley Cultivars and Their Application in Cultivar Identification." *Seed Science and Technology* 10: 405–413.

Anonymous. 1843. "Mummy Wheat." *Gardeners' Chronicle*. pp. 787–788.

———. 1976. "Cereals: Vigour Tests Could Increase Yields." *Span* 19: 3.

———. 1982. *A Chronicle of Plant Variety Protection*. Kalamazoo, Mich.: Asgrow Seed Company. 10 pp.

Aufkammer, G., and U. Simon. 1957. "Die Samen Landwirtschaftlicher Kulturpflanzen im Grundstein des Ehemalizer Nurnberger Stadttheaters und ihre Keimfahigkeit." *Zeitschrift für Acker and Pflanzenbau* 103: 454–472.

Austenson, H. M., and P. D. Walton. 1970. "Relationship Between Initial Seed Weight and Mature Plant Characters in Spring Wheat." *Canadian Journal of Plant Science* 59: 53–58.

Barton, L. V. 1961. *Seed Preservation and Longevity*. London: Leonard Hill (Books) Ltd. 216 pp.

Bass, L. 1975. "Seed Moisture and Storage." *Seed Science and Technology* 3: 743–746.

———, and P. C. Stanwood. 1978. "Long-Term Preservation of Sorghum Seed as Affected by Seed Moisture, Temperature, and Atmospheric Environment." *Crop Science* 18: 575–577.

Bhale, N. L., and S. T. Borikar. 1982. "Male-Sterile Single Crosses for Hybrid Seed Production in Grain Sorghum." *Seed Science and Technology* 10: 373–378.

Bishoi, U. R. 1977. "Effect of Increasing Test Weight on Germination and Seedling Performance in Triticale." *Cereal Research Communications* 5: 377–385.

Bittenbender, H. C., and S. K. Ries. 1977. "Germination and Growth of Rice Seedlings from High and Low Protein Seeds after Exposure to Various Storage Conditions." *Journal of Seed Technology* 2(1): 62–72.

Burris, J. S. 1975. "Seeding Vigour and Its Effect on Field Production in Corn." *Annual Corn and Sorghum Research Conference* 30: 185–193.

Chhina, B. S., and P. S. Phul. 1982. "Association of Seed Size and Seedling Vigour with Various Morphological Traits in Pearl Millet." *Seed Science and Technology* 10: 541–545.

Christensen, C. M. 1969. "Influence of Moisture Content, Temperature and Time of Storage Upon Invasion of Rough Rice by Storage Fungi." *Phytopathology* 59: 145–148.

———. 1970. "Moisture Content, Moisture Transfer and Invasion of Stored Sorghum Seeds by Fungi." *Phytopathology* 60: 280–283.

———, and L. C. Lopez. 1965. "Relation of Moisture Content and Length of Storage to Changes in the Microflora and Germination Percentage of Rough Rice." *Phytopathology* 55: 953–956.

Craig, W. F. 1976. "Production of Hybrid Corn Seed." (In) G. F. Sprague (ed.), *Corn and Corn Improvement*. Madison, Wis.: American Society of Agronomy. pp. 671–719.

Dasgupta, P. R., and H. M. Austenson. 1973. "Relations

Between Estimates of Seed Vigor and Field Performance in Wheat." *Canadian Journal of Plant Science* 53: 43–46.

Delouche, J. C., and C. C. Baskin. 1973. "Accelerated Aging Techniques for Predicting the Relative Storability of Seed Lots." *Seed Science and Technology* 1: 427–452.

Delouche, J. C., et al. 1973. "Storage of Seed in Sub-Tropical and Tropical Regions." *Seed Science and Technology* 1: 671–700.

Dhillon, G. S., and D. S. Kler. 1978. "Influence of Seed Size on Growth and Yield of Triticale." *Seed Research* 6: 110–117.

Fedak, G., and T. Rajhathy. 1972. "Esterase Isozymes in Canadian Barley Cultivars." *Canadian Journal of Plant Science* 52: 507–516.

Grabe, D. F. 1976. "Measurement of Seed Vigor." *Journal of Seed Technology* 1 (2): 18–32.

Grogan, C. O., and M. S. Zuber. 1958. "Comparative Performance of Various Combinations of F_1 and F_2 Generation Seed of Maize." *Agronomy Journal* 50: 88–89.

Hanway, D. G. 1978. "Agricultural Experiment Stations and the Variety Protection Act." *Crops and Soils* 30 (5 and 6), Parts 1 and 2: 5–6; 5–7.

Hicks, D. R., et al. 1976. "Seed Grade Effects of Corn Performance." *Agronomy Journal* 68: 819–820.

Hunter, R. B., et al. 1969. "Effect of Tassel Removal on Grain Yield of Corn, *Zea Mays* L." *Crop Science* 9: 405–406.

Hunter, R. B., and L. W. Kannenberg. 1972. "Effects of Seed Size on Emergence, Grain Yield, and Plant Height in Corn." *Canadian Journal of Plant Science* 52: 252–256.

Hunter, R. B., C. G. Mortimore, and L. W. Kannenberg. 1973. "Inbred Maize Performance Following Tassel and Leaf Removal." *Agronomy Journal* 65: 471–472.

Kahler, A. L., and R. W. Allard. 1970. "Genetics of Isozyme Variants in Barley. I. Esterases." *Crop Science* 10: 444–448.

Kiesselbach, T. A. 1945. "The Detasseling Hazard of Hybrid Seed Corn Production." *Journal of American Society of Agronomy* 37: 806–811.

————. 1960. "Performance of Advanced Generation Corn Hybrids." *Agronomy Journal* 52: 29–32.

Mallick, A. E., and B. Nandi. 1979. "Role of Moisture Content in Deterioration of Rough Rice in Storage." *Seed Science and Technology* 7: 423–429.

————. 1982. "Deterioration of Stored Rough Rice. 5. Grain in Private Storage." *Seed Science and Technology* 10: 527–533.

Maranville, J. W., and M. D. Clegg. 1977. "Influence of Seed Size and Density on Germination, Seedling Emergence and Yield of Grain Sorghum." *Agronomy Journal* 69: 329–330.

Mathews, S. 1981. "Evaluation of Techniques for Germination and Vigour Studies." *Seed Science and Technology* 9: 543–551.

McDaniel, R. G. 1969. "Relationship of Seed Weight, Seedling Vigour and Mitochondrial Metabolism in Barley." *Crop Science* 9: 923–927.

Morey, D. D. 1976. "The Effect of Storage Conditions and Containers on the Viability of Rye, Wheat and Oat Seed." *Seed Science and Technology* 4: 313–323.

Neal, N. P. 1935. "The Decrease in Yielding Capacity in Advanced Generations of Hybrid Corn." *Agronomy Journal* 27: 666–670.

Patanothai, A., and R. E. Atkins. 1974. "Yield Stability of Single Crosses and Three-Way Hybrids of Grain Sorghum." *Crop Science* 14: 287–290.

Peterson, J. R. 1979. "Seed Moisture—How It Can Work For You." *Agricultural Gazette* 90 (1): 12–14.

Pinthus, M. J., and R. Osher. 1966. "The Effect of Seed Size on Plant Growth and Grain Yield Components in Various Wheat and Barley Varieties." *The Israel Journal of Agricultural Research* 16 (2): 53–58.

Rajki, S. 1980. "50th Anniversary of the Garst and Thomas Hybrid Corn Company." *Acta Agronomica Academiae Scientiarum Hungaricae* 29: 417–424.

Reynolds, R. 1968. "Saga of the Seed Plate." *The Furrow* 73: July–August Editorial.

Rickey, F. D., G. H. Springfield, and G. F. Sprague. 1934. "The Loss in Yield That May Be Expected from Plating Second Generation Double Crossed Seed Corn." *Agronomy Journal* 26: 196–199.

Rogers, J. S., and J. R. Edwardson. 1952. "The Utilization of Cytoplasmic Male-Sterile Inbreds in the Production of Corn Hybrids." *Agronomy Journal* 44: 8–13.

Rossman, E. C. 1949. "Freezing Injury of Inbred and Hybrid Maize Seed." *Agronomy Journal* 41: 574–583.

Ruttan, V. W. 1982. "Changing Role of Public and Private Sectors in Agricultural Research." *Science* 216: 23–29.

Sauer, D. B., and C. M. Christensen. 1968. "Germination Percentage, Storage Fungi Isolated From, and Fat Acidity Values of Export Corn." *Phytopathology* 58: 1356–1359.

Sayers, R. L. 1976. "Seed Trade Viewpoint—Seed Vigour." *Journal of Seed Technology* 1(2): 7–12.

Spain, D. F. 1976. "Seed Vigor, A Seed Control Official's Viewpoint." *Journal of Seed Technology* 1(2): 13–17.

Srivastava, J. P., and S. N. Nigram. 1973. "Effect of Seed Size on Yield and Other Agronomic Characters in Wheat (*Triticum aestivum*)." *Seed Research* 1: 52–57.

Tatum, L. A. 1971. "The Southern Corn Leaf Blight Epidemic." *Science* 171: 1113–1116.

Ullstrup, A. J. 1972. "The Impacts of the Southern Corn Leaf Blight Epidemic of 1970–71." *Annual Review of Phytopathology* 10: 37–50

Walsh, E. J., and R. E. Atkins. 1973. "Performance and Within Hybrid Viability of Three-Way and Single Crosses of Grain Sorghum." *Crop Science* 13: 267–271.

Weibull, G. H. 1964. "Breeders' Rights Systems in Europe." *Plant Breeders' Rights*. Madison, Wis.: American Society of Agronomy. Special Publication No. 3. pp. 4–19.

Wood, D. W., P. C. Longden, and R. K. Scott. 1977. "Seed Size Variation, Its Extent, Source and Significance in Field Crops." *Seed Science and Technology* 5: 337–352.

Woodstock, L. W. 1973. "Physiological and Biochemical Tests for Seed Vigor." *Seed Science and Technology* 1 (1): 127–157.

Wright, H. 1980. "Commercial Hybrid Seed Production." (In) W. R. Fehr and H. H. Hadley (eds.), *Hybridization of Crop Plants*. Madison, Wis.: American Society of Agronomy—Crop Science Society of America. pp. 161–176.

11

Crop Residues

INTRODUCTION

Grain production is associated with the plant's photosynthetic machinery as well as with roots and stems that support its seed-bearing structures. It is not biologically possible for us to produce precisely what we require for human food. For example, animals cannot be produced without **offal**, vegetables without stems and leaves, or fruit without cores or pits. High-quality produce cannot be obtained without at least some low-quality produce.

In the slender-stemmed cereals, and in the coarse-stemmed corn, sorghum, and millet, any above-ground plants parts that remain after the seeds have been harvested are known as *straw* and *stover*, respectively. Leaves, stems, and roots comprise from 50% to 75% of the total cereal biomass produced. We are challenged to use this seemingly waste material in an economic manner.

This chapter is concerned primarily with cereal crop residues remaining after grain harvest. Crop residues occur also in other forms, however. They can include the by-products from livestock production that consist of undigested and unabsorbed components of the feed or forage roughage, collectively known as manure. Industrial processing of grain crops may produce other crop residues, such as spent brewers' grain, hulls, corncobs, and cleanings of meal from oil-extraction processes.

CROP RESIDUES—WASTE OR RESOURCE?

When grain is regarded as the economic yield in crop production, the associated dry structural material is often of secondary importance. Cereal crop residues consist of lignified cells, sometimes with an accumulation of silica in the cell walls, which provide rigidity needed to support the seed-bearing structures. Often they are durable, decay-resisting materials that can be used as a roofing, known as thatch. The durability and abundance of cereal residues creates a problem involving their economic and inoffensive disposal. Intensive cash-crop producers often regard crop residues as waste material as a nuisance to be disposed of quickly and economically, often by being burned in the field. Other producers may regard crop residues as a tremendous natural resource to be used for feed, livestock bedding, or industrial purposes, or as a protective mulch cover or an essential soil amendment.

Depending on one's perspective, crop residues can be viewed as desirable or undesirable. A story by Taiganides (1979) is useful to illustrate the point:

When hens were introduced into a country unfamiliar with chicken production, the people were confused when two items were excreted. An expert committee was instructed to solve the problem of which was the product and which the by-product of the feed they gave the chickens daily. Initially, the manure was regarded as the product, because it was produced in a quantity of about five times that of the egg. Further deliberation led to the concept that quality and not quantity was the criterion, and on a dry-weight basis the protein level of the manure was found to exceed that of the egg. Continued study led to the declaration that on the basis of odor, the manure was the waste and the egg the resource. By the time the study was nearing completion, the egg left in the sun had rotted and smelled as bad as the manure. No decisive conclusion was reached regarding which was residue or resource.

Just as the egg, a resource, became waste, so too crop residues can become a waste when neglected. Crop residues also can be a valuable resource. The conclusion is that there is no such thing as waste if handled properly.

A SYSTEMS APPROACH TO CROP RESIDUES

Cereal crop residues historically were regarded as a resource and were collected and used as livestock bedding or feed, thereby being converted into manure, which was regarded as essential for successful crop production. Changes in harvesting methods, a shift away from straw as bedding material, specialization of farming systems, and an increase in the volume of residues produced have resulted in changes in attitude toward crop residues. How crop residues are perceived can vary with labor availability and cost, degree of mechanization, soil condition, energy-cost relationships, alternative uses and demand, degree of crop intensification, market value, and other factors. What may be regarded as a sound approach to the management of crop residues in one area of the world under a specific system may be completely impractical for other systems. Sweeping generalizations as to the value of straw and how it should be handled cannot be made to suit all situations.

One common thread does exist among various farming systems, and it pertains to the long-term productivity of soils and the ability to meet the food needs of an ever-expanding population. Growing evidence of soil depletion and deterioration has raised serious concern about cropping methods and cultural practices involving rotations, green manure, and the management of crop residues.

Failure to address the soil-productivity question adequately has led to reduced organic-matter levels, which in turn are associated with reduced soil fertility, an increase in soil compaction, a reduction in ease of tillage, less favorable oxygen-temperature relationships, reduced water penetration, increased wind and water erosion, and yield reductions. Crop productivity is maintained by increased inputs. These problems are universally recognized. These concerns help underline the need for the wise use of crop residues and help place agricultural residues in perspective as a resource rather than a waste.

In the decade of the seventies and early eighties, several key issues related to the wise use of crop residues came to the forefront, forcing a reevaluation of production systems. Among these issues are energy supplies, availability, and cost; declining soil organic-matter levels and associated problems; and soil salinization, waterlogging, and alkalinization.

Six production models or systems, each requiring a different approach to crop-residue management, can be identified:

1. Land-livestock based models.
2. Livestock-based models without a land base.
3. Multicommodity models.
4. Specialized models (intensive farming systems adopted by choice or limited by climatic variables).
5. Conservation models.
6. Biologically correcting models.

Land-Livestock Based Models

Self-sufficiency in traditional mixed agricultural systems often involves the production of one or more classes of livestock and a series of feed crops grown in rotation. Under this system, crop residues are treated as an integral part of the farming system, and straw may be collected and valued for a number of purposes. Straw may serve as an inexpensive, clean, dry bedding material to provide animal comfort and to help hold and conserve animal excreta. When the manure is returned to the soil, it has value as a soil amendment and as a source of plant nutrients.

When manure is properly piled and allowed to partially decompose, the associated odors and pollution problems are minimized and are regarded as being compatible with the rural environment. Disposal problems are minor, as manure can be spread at several points throughout the system. Manure applications may be made prior to plowing, so it is completely incorporated; on perennial forage fields after the crop has been removed after grazing; or following the removal of an annual crop.

Traditional land-livestock systems were aided by the binder-reaper, which cut and bound small cereals (Figure 11-1) that were stooked (Figure 11-2) for field drying and later hauled (Figure 11-3) to a stationary thresher. The pertinent feature of this system of harvesting was that the straw was concentrated in a pile or stack at one central point.

Binder harvesting allowed cereals to be cut as soon as or even before physiological maturity was reached and when the nodes were green. Slightly immature grain could complete its ripening and drying in the stook. Early cutting gave no evidence of yield loss, and grain yields may have been higher because of lower grain shedding; early cutting also reduced the risk of loss through lodging.

Because the crop was cut when slightly immature, the straw was less lignified; leaves were intact and were included in the residue along with the chaff. Such straw could be fed as a low-energy maintenance feed for overwin-

Figure 11–1. For nearly a century the binder-reaper was used to cut and bind small cereal grains into sheaves. An essential feature of binder harvesting was that the crop could be cut before the grain had dried down and when the leaves and stalks were slightly immature. The result was a less lignified product with leaves intact and included in the residue along with the chaff. (From the R. R. Sallows Collection, Ontario Ministry of Agriculture and Food)

Figure 11–2. After cutting and binding into sheaves, the crop was stooked in the field with five to seven sheaves per stook to allow for drying. The stooks in this picture are those of the oat crop. (From the R. R. Sallows Collection, Ontario Ministry of Agriculture and Food)

Figure 11–3. After the stooks were adequately dried, the sheaves were hauled to a stationary thresher. Straw residues were concentrated in a pile or stack and served as a convenient source for bedding or feed purposes (From the R. R. Sallows Collection, Ontario Ministry of Agriculture and Food)

tering ruminants, such as heifers or dry cows, or could be used as bedding material.

In the mechanized areas of the world the combine harvester has replaced the binder-reaper (Figure 11-4). Harvesting is begun generally when the straw is fully mature and dry. Lignification is complete, residues are brittle, and nutritional value is lower than if cut when slightly immature. Recovery of residues following combine harvesting adds a second operation; it is less thorough because of a high cut to reduce the volume passing through the combine and because of loss of chaff, leaf tissue, and brittle fragments. Under wet weather conditions, recovery operations may be delayed, with the risk that the straw may be left in the field uncollected in windrows that add to the difficulty of subsequent field operations.

High-yielding hybrid cultivars of full-season corn and sorghum crops resulted in a major shift away from the lower-yielding, short-season cereal crops. Stover residues from grain corn and sorghum are coarse and pithy; they are considerably less easy to handle and less attractive as bedding or feed materials than the slender-stemmed cereal crops. The full-season nature of these crops and the introduction of dwarf sorghum cultivars added to the difficulty and reduced the desirability of collecting residues after grain harvest.

Stover residues on the field add to the complexity of subsequent tillage operations; if left on the surface they may reduce herbicide effectiveness and delay soil warming at high latitudes.

As intensity of livestock production increased, residue-handling problems grew because of greater volumes produced, and the times when manure could be spread were reduced by the full-season annual corn and sorghum crops. Corn and sorghum also replaced some of the perennial forage crops as a forage roughage because of their high yields and ease of ensiling. Annual corn and sorghum crops provided no protective winter cover and added to soil-erosion problems.

A lack of suitable and adequate bedding materials, or perhaps the labor and cost involved in obtaining bedding materials, led to alternative manure-management systems, such as liquid-manure tanks with livestock on slotted or sloped

Figure 11–4. The combine harvester replaced the binder as a labor-saving device. Combining occurs after the straw is fully mature and dry. Crop residues are dropped behind the combine, and attempts to recover these residues are less thorough and produce a lower-quality product for bedding or feed than produced with a binder. (Ranson Photographers Ltd., Edmonton)

floors. Residues, generally devoid of bulk, are pumped for rapid disposal on the fields. The anaerobic fermentation associated with a liquid-manure pit produces an environmentally offensive product with a strong odor. Application is limited to areas sufficiently distant from residential areas, where the land is dry enough to prevent compaction by large delivery tanks, and (at high latitudes) where conditions of snow depth and subsequent runoff are acceptable.

Land-based livestock systems offer an opportunity to utilize waste products from one enterprise to help increase the efficiency and productivity of another. Without an adequate system, residues can become a problem. Every effort must be made to accommodate crop residues in the system.

Livestock Production Without a Land Base

After World War II a trend toward greater specialization and intensification occurred in many areas of the world, and in keeping with this trend **feedlot** production systems on a small land base were established. In such operations, manure disposal becomes a problem to the point where residues are regarded as a nuisance. A partial solution was the development of a system of methane fermentation of animal wastes (Haga, Tanaka, and Higaki, 1979).

An alternative system is to use undiluted poultry droppings or cattle manure from stock fed a high-energy ration as feed supplements. Such manure contains nitrogenous compounds that could be a protein source for ruminants or undigested fragments of energy feed that could be utilized by swine. Although feeding manure may appear aesthetically offensive, there is no intrinsic reason why it should not be acceptable. Poultry droppings can be used as a feed supplement without affecting the flavor or nutritive value of the end product and apparently without introducing any health hazards to man or animals (Hudson, 1977). Passing manure by-products to a second class of livestock does not, however, solve residue-disposal problems in the absence of a land base.

Multicommodity or Crop-Rotation Model

Traditional farming systems usually involve the production of several crops grown in **rotation.** Rotation systems provide a stability of production, diversity of income sources, diversity of diet, and distribution of labor. Rotational systems usually are associated with livestock production.

Partly by choice and partly because of off-farm pressures, crop-rotation models became increasingly intense, and in some instances livestock production was discontinued. Without livestock, perennial nitrogen-fixing forage legumes were excluded from the rotational system. Use for bedding or feed of straw or stover residues disappeared on farms free of livestock. Further, it often was uneconomic to collect and transport crop residues to other farms or for industrial utilization. Residues left in the field were often in greater abundance than could be rotted without impeding tillage operations or tying up nitrogen because of the high carbon : nitrogen ratio common to cereal crop residues.

An awareness of the importance of rotational cropping systems to provide a protective cover crop overwinter at high latitudes, to provide a nitrogen-fixing legume, and to allow for a reasonable economic return has focused attention on rotational systems. Crop residues, however, may limit rotational systems. For example, Hughes (1979) noted the increased area seeded to fall-sown winter cereals in Britain. Because of the late harvest of spring-seeded cereals that allows for only a brief period between removal and fall seeding, straw from the spring crop became a nuisance and a hindrance to the successful seeding of the fall crop. In Britain straw is commonly burned to facilitate a quick turnaround.

Innovative approaches may be needed to make rotational systems viable. For example, in short-season areas a corn, soybean, winter wheat rotation is desirable but difficult to achieve, as the soybeans normally are not removed until well after the recommended seeding time for wheat. Overseeding the soybean crop with wheat seed prior to the soybean leaf drop, which later provides a protective mulch cover, not only can make the rotation work but takes advantage of the soybean leaf residues. Subsequent stem residues must be thoroughly spread over the established wheat seedlings at soybean harvest time to prevent smothering of the winter wheat.

Multiple cropping can increase crop-residue handling problems, as one crop is seeded into another before harvest. A triple multiple-cropping system, illustrated in Figure 11-5, involves winter wheat, corn, and soybeans. Although such a system does not lend itself to mechanization and thus entails hand labor, it is important to note that about 90% of one of the world's most important food crops, rice, is produced by hand labor (Figure 11-6). Multiple cropping offers a system capable of increasing world food supplies (Andrews and Kassam, 1976), as it is a form of intensification in time and space. Such a system intensifies crop-residue problems. In a study in the southern part of the United States involving a soybean-wheat double-cropping system, the greatest average net return on a short-run basis was the treatment in which burning of the wheat straw was followed by no-till soybean seeding (Sanford, 1982).

Figure 11–5. This photo illustrates a labor-intensive system of multiple cropping found in the People's Republic of China. The sheaves are those of winter wheat that have been cut with a sickle during the first half of June. Corn seedlings have been planted in the tramways between the strips of wheat earlier in the season. Soybeans will be interseeded between the corn as soon as the wheat sheaves can be removed and the wheat stubble tilled. Water management is a major consideration under such an intense cropping system and may require supplemental irrigation. (Photo by N. C. Stoskopf)

Figure 11–6. Much of the world's rice crop is produced by hand, including seeding, harvesting, and threshing. Systems of multiple cropping may not be inducive to mechanization, but where labor is abundant, greater flexibility may be possible. (Courtesy of International Rice Research Institute)

Specialized Models

Prior to 1974, inexpensive and abundant supplies of mineral fertilizers encouraged specialization of cropping programs. Inexpensive commercial fertilizer conveniently replaced fertility obtained from crop residues, manure, and nitrogen-fixing legumes. It was more convenient and economic to buy inorganic fertilizer elements than to transport manure any distance. Because it was believed that fertilizer needs could be met entirely from inorganic sources, crop residues were regarded as a nuisance and were deemed unnecessary for successful crop production.

Mechanization in many areas facilitated timely seeding of a single crop over a large area. The need for diversification to spread the workload was reduced. Monoculture cropping systems allowed producers to concentrate their managerial skills and to reduce the amount of expensive machinery needed for greater diversification. Genetically resistant crop cultivars, effective pesticides, precise fertilizer applications, and specialized machinery allowed for intensification and specialization of cropping programs. Crop residues were not held in high regard.

In the specialized cereal-growing areas of the world, there has been little demand for straw. Where cereals are combine-harvested, the crop may be uniformly dried in a swath or windrow consisting of grain cut from a strip 15 m (50 ft) wide. If the straw is not to be removed, the same straw and chaff ideally should be distributed over the same area from which it was gathered. This does not happen, since the straw is too light to be flung that far by the distributing fan of the combine. A concentration of crop residues results in uneven crop growth the following year, as moisture-retention differentials exist. A concentration of straw residue increases the complexity of tillage and, if incorporated into the soil, it causes denitrification or the temporary loss of available soil nitrogen, because bacteria that break down the carbon in the straw need nitrogen in order to grow and thrive. For many dryland farmers, the most economic and satisfactory method of disposing of the straw has been field burning (Figure 11-7). The effects of this action in reducing soil organic matter are of concern, and a better straw-disposal system is needed.

The validity of returning crop residues to the soil needs to be demonstrated economically, however. Although it seems logical to expect that incorporation of straw in the soil would benefit both the soil and the following crop, there is evidence to the contrary. In Britain, when straw was returned to land over an extended period, the soil organic carbon increased by less than 0.2% (Pierce, 1982). Where direct drilling or minimal cultivations are practiced and the soil is no longer inverted by plowing, roots appear to enhance the organic-matter status of the soil even when the straw is burned on the surface.

Hughes (1979) questioned the value of returning straw and other crop residues to the land. He noted that on the majority of soils in Britain, the roots of most crops, including cereals, can supply the necessary amount of organic matter to maintain stability of soil structure.

Figure 11–7. Despite decreasing soil productivity associated with declining organic matter and increased soil erosion, burning of straw is a common practice. In areas of intensive agriculture, a concentration of straw residue increases the complexity of tillage, and, if incorporated into the soil, it causes a temporary loss of available soil nitrogen. For many farmers the most economical and satisfactory method of disposing of the straw has been field burning. (Ranson Photographers Ltd., Edmonton)

It appears from these observations that straw presents a disposal problem rather than a source of fertility.

The presence of toxic substances produced from decaying residues under wet and cool conditions (Cochran, Elliott, and Papendick, 1982) is a strong reason why residues may not be incorporated into the soil.

Conservation Model

In dryland regions, water conservation is essential for successful cereal grain production. Conservation models use cropping (summer fallow) and management (mulch or snow retention) systems to conserve moisture, especially in areas that receive less than 500 mm (20 in.) annual rainfall, although this varies with rainfall effectiveness. On the central Great Plains of North America, Greb (1979) estimated that improved stored soil moisture levels associated with several factors, including stubble mulch, increased yields 45%. No single system of moisture conservation or crop residue management is advocated, however.

Under conditions at Pullman, Washington, Cochran, Elliott, and Papendick (1982) noted that direct drilling of winter wheat into surface cereal residues reduced soil erosion but frequently resulted in stunted plants and lower grain yields than conventional planting methods. Yields in the Pacific Northwest of the United States and in Britain may be reduced 20% to 30% when plantings are made in heavy surface cereal residues. Related factors include low fertility, high crown set, suboptimal soil temperatures, poor seed-soil contact, and phytotoxicity from decomposing surface residues. Growth of seedlings is retarded most severely in areas of the field where chaff or straw are concentrated but may be satisfactory where there is only standing stubble or where residues have been removed mechanically or burned.

Yield depressions appear to be greatest under cool, wet weather in the fall, because under these conditions decomposing crop residues may release short-chain fatty acids that retard seedling growth of wheat (Cochran, Elliott, and Papendick, 1982). Also, winter wheat seedlings growing in heavy surface residues often set a high crown that exposes adventitious roots to herbicides, frost damage, and phytotoxins produced during straw decomposition. Use of a disk drill to direct-seed cereals usually places the seed in proximity to surface residues and may press the straw down the soil slit, enclosing the seeds in a straw envelope. This results in poor seed-soil contact and enhances exposure of the seed to any phytotoxins produced. Positioning of residues may be useful.

In a study on managing crop residues, Cochran, Elliott, and Papendick (1982) used spring and winter wheat stubble to compare plant stands, grain yields, and water-use efficiency between direct drill-seeded and conventionally seeded winter wheat. Residue management treatments included standing stubble, complete residue removal, moving the crop residue from the seed row, and residue incorporation. During the three-year study, winter wheat yields from plots direct-drilled into surface cereal residues were equal to those from plots tilled and seeded conventionally.

Until more conclusive research demonstrates methods of crop residue managements, precise recommendations cannot be made.

Biologically Correcting Models

Productivity of agriculture in some areas of the world has been increased by replacing manual labor with machines and implements, by replacing crop residues with fertilizer nutrients, and by relying heavily upon pesticides and other chemicals to control insects, weeds, and diseases. Using conventional tillage programs, it was observed that more and more inputs are required to sustain high yield levels and that in some instances soil productivity declines owing to erosion, compaction, and reduced organic matter. In the corn belt of the United States, where intensive row cropping is practiced, the average annual soil loss from erosion exceeds 18 tonnes/ha (8 tons/ac) (Berg, 1979). Soil-productivity concerns have stimilated interest in **organic farming**[1] systems of food production. "Sustaining the soil productivity" was selected in 1981 as the foremost U.S. agricultural research priority by more than 100 prominent scientists and leaders at a multiagency national workshop on soil and water resources (Larson et al., 1981).

In a review of organic farming methods, Parr, Papendick, and Youngberg (1983) concluded (p. 198):

> In the near term, increased soil erosion and the rapidly rising costs of commercial fertilizer nitrogen will no doubt provide impetus for continued interest in organic farming technology. We believe that the time is not far away when many farmers will reduce their inputs accordingly because of these high costs. At that time they will have no choice but to adopt modified crop rotations with legume crops and/or organic

---

[1]Glossary definition based on report by USDA (1980).

recycling programs to reduce the deficit of nitrogen and other plant nutrients.

As the growing world population requires more food, perhaps more and more inputs will be required to meet this need, not fewer inputs as suggested by organic farming methods. But as finite sources of inputs are depleted or rise in price, organic inputs undoubtedly will be of greater and greater value. As priority for food production increases, the future role and scope of organic farming is unclear. It is inevitable that where mismanagement has caused soil productivity to deteriorate, biologically correcting models or regenerative agriculture systems will be adopted.

Organic producers advocate the return of all crop residues to the soil to the exclusion of nonorganic fertilizer sources. Although such a proposal has merit, it must be recognized that organic matter may not be an adequate, complete, or balanced nutrient source. As well, temperature and soil moisture variations may cause differences in the rate of decay. In a world that demands steady increases in crop yield, the rate of decomposition of organic matter may be too slow to supply adequate nutrients. Manufactured fertilizers are needed to overcome deficiencies and to promote optimum plant growth, even though abundant organic matter is present in the soil (Aldrich, 1977).

Enhancement of the activity of nitrogen-fixing bacteria to reduce nitrogen fertilizer inputs for cereals is an exciting concept. Bacteria in close symbiotic association with plant roots may provide the cereal plant with nitrogen compounds in exchange for carbohydrates produced by the plant via photosynthesis, as is

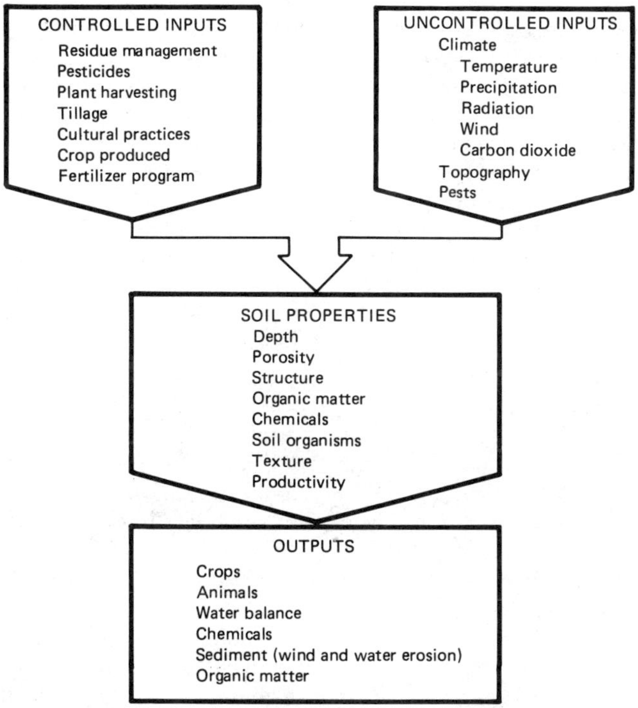

Figure 11–8. Factors to be considered in constructing a mathematical model to quantify the effects of crop-residue management on erosion and crop productivity. (After USDA, 1981)

common in legume plants. Although research continues on this possibility, and crop biologists remain convinced that symbiotic nitrogen fixation by grasses is theoretically possible, it appears that this achievement will not occur rapidly or easily.

Considerable advances have been made in the use of blue-green algae, which fix atmospheric nitrogen as a source of nitrogen in lowland rice (FAO, 1977b; 1981).

MAGNITUDE OF THE CROP-RESIDUE MANAGEMENT PROBLEM

The foregoing brief descriptions of six production models indicate that flexibility, imagination, and computer modeling may be required to integrate the many facets of a system related to crop-residue management. Mathematical modeling and field experiments to support the models need to be conducted and integrated.

A major factor in crop-residue management is the effect of erosion on soil productivity. Figure 11-8 shows the many interacting factors that must be considered in building a mathematical model to quantify the effects of erosion on crop production. Until relationships between crop-residue management, soil erosion, economically sound crop-productivity inputs, and numerous other factors are adequately developed and understood, the selection of strategies that involve crop-residue management and that maximize long-term productivity will be impossible. Poor decisions can result in underusing or damaging resources.

The specific models needed for each of the six systems outlined cannot be developed because of a lack of research information. Even empirical relationships are difficult to develop, because data are limited and because improved technology has masked the effect of crop-residue, mismanagement, erosion, or low soil organic-matter levels.

An apparently successful system of recycling organic wastes, but one that involves a high labor input, is illustrated in the People's Republic of China (Figure 11-9). Study of this flow diagram reveals the magnitude and complexity of the system. Prominent in the model is the use of green manure, legumes, and **azolla** to maintain soil productivity, fertility, and plant growth. The use of nitrogen sources such as crop residues, **night soil,** kitchen refuse, sewage, garbage, and manure is strongly evident. Regarding such items as input resources rather than as undesirable wastes is essential to the success of this system.

NATURE OF CEREAL STRAW AND STOVER

Volume Produced

Modern changes in agricultural systems have meant that cereal crops increasingly are produced for their grain yield, with straw and stover being of less importance and under some systems a nuisance. This shift in emphasis has encouraged plant breeders to develop higher-grain-yielding genotypes by partitioning more of the photosynthate into grain and less into straw.

In contrast to grain yield, which can be readily sampled and weighed, straw yields are reported with less accuracy because of variations in height of cutting under field conditions and loss of chaff and leaves when sampling is done following a combine harvester. Crop residues are seldom measured, because they have high bulk and volume, because roots (an integral part of crop residues) cannot be accurately or readily measured, and because of crop lodging. Larson, Holt, and Carlson (1978) suggested that 37.5% of corn residue remains on the soil after machine collection. For wheat, 560 kg/ha (500 lb/ac) of residue was believed to remain after collection. These workers found that residue yields generally vary directly with grain yields.

Researchers have sought to obtain reliable estimates of crop-residue yields by establishing a fixed grain-to-straw ratio (MacKenzie, 1926; Gately, 1975; Donald and Hamblin, 1976). Gallagher and Biscoe (1978) found that in wheat and barley, less than 10% variation occurred between grain and straw yields for individual

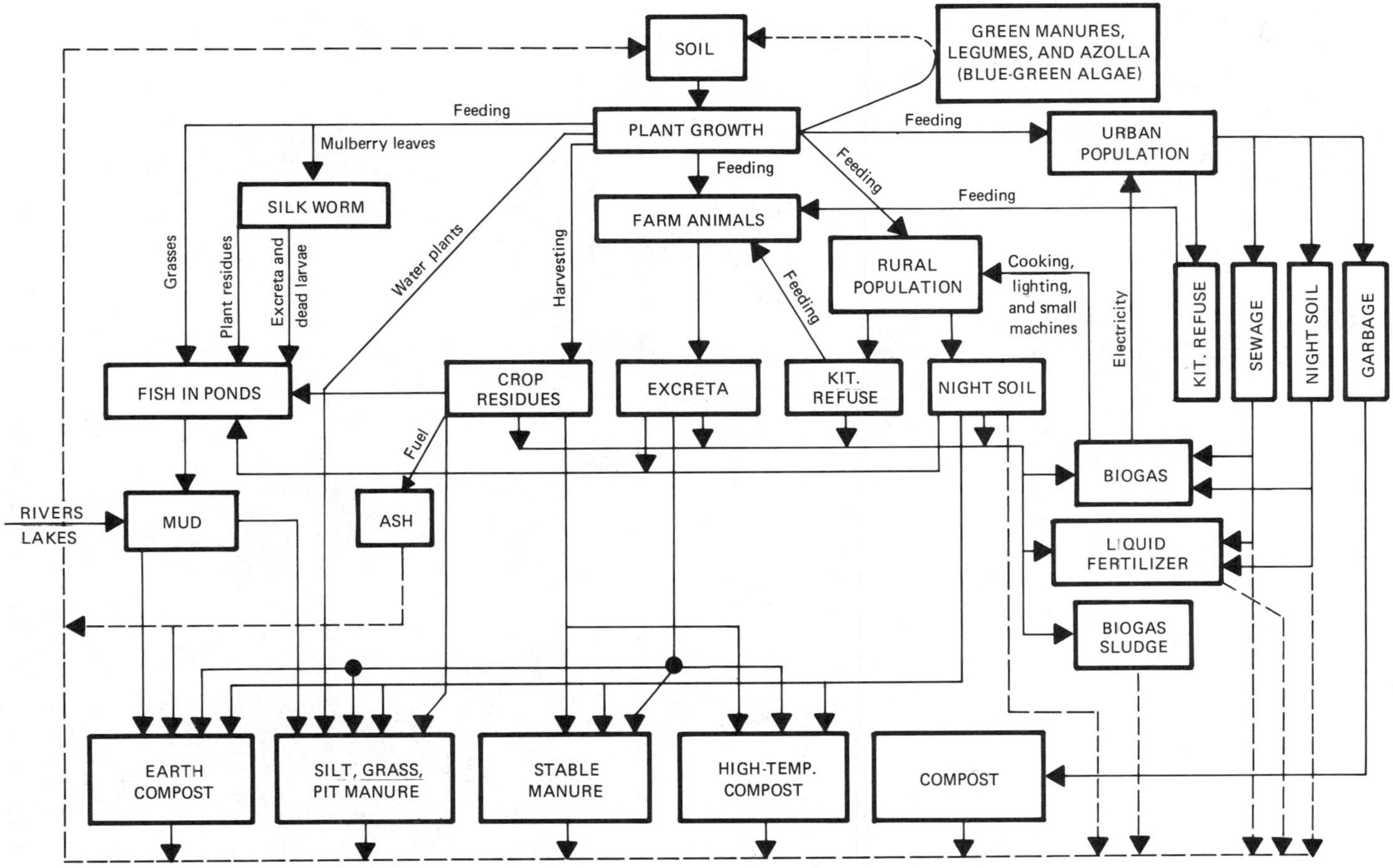

Figure 11–9. A system of recycling organic wastes in the People's Republic of China is illustrated in this model. Of paramount importance in the system is the use of green manure, legumes, and azolla along with night soil, sewage, and kitchen refuses to maintain soil productivity and plant growth. (Courtesy Food and Agriculture Organization of the United Nations. Reference FAO, 1977a)

cultivars, even when grain yields varied widely and over a large number of observations.

In the slender-stemmed cereals a general rule-of-thumb is that the straw-to-grain ratio is one-to-one. A summary compilation of straw-to-grain ratios by Staniforth (1979a) showed only minor deviations from this generalization (Table 11-1). For comparative purposes values for seven cereal crops grown in the United States are shown in Table 11-2. The superior ratios of corn and sorghum should be noted. The ratios in Table 11-2 were used to determine residues produced from grain yields. A known yield of 5000 kg/ha of barley will probably have 7500 kg/ha of straw associated with its production.

Table 11–2
Estimated Straw : Grain Ratios for Selected Cereals in the United States

Crop	*Straw : Grain Ratio*
Barley	1.5 : 1
Corn	1.0 : 1
Oat	2.0 : 1
Rice	1.5 : 1
Rye	1.5 : 1
Sorghum	1.0 : 1
Wheat, winter	1.7 : 1
Wheat, spring	1.3 : 1

Source: Larson, Holt, and Carlson (1978).

Chemical Composition

Cellulose, hemicellulose, and lignin are the three main organic compounds of cereal crop residues (Table 11-3). Rice has the highest content of silica, which helps reduce decomposition and makes it useful as thatch for roofing but also dulls cutting blades and inhibits combustion and digestion.

The chemical content of straw as well as the percentage of protein, waxes, and other components varies according to soil, cultivar, time of cutting, and environmental factors. For example, Jackson (1977) reported the silica content of 11

Table 11–1
Approximate Straw : Grain Ratios for Five Cereals Produced in Various Regions

Cereal	*Region*	*Straw : Grain Ratio*
Wheat, normal height	UK	0.8 : 1
Wheat, semidrawf	UK	0.4 : 1
Wheat	France	0.8 : 1
Wheat (total but not harvested)	USA (Kansas)	1.7 : 1
Wheat	India	1.7 : 1 to 1.1 : 1
Wheat (total but not harvested)	Australia (NSW)	1.0 : 1
Wheat	Asia and Africa	1 : 7 : 1
Barley (mean of six cultivars)	UK	0.6 : 1 to 0.5 : 1
Barley (mean of 126 trial sites from high to low nitrogen)	Ireland	0.8 : 1 to 0.7 : 1
Barley	France	0.8 : 1
Barley	Asia and Africa	1.7 : 1
Rice	Japan, Korea, Taiwan	0.8 : 1
Rice	Tropical Asia	1.4 : 1
Rice	USA (California)	1.3 : 1
Rice	Asia and Africa	1.4 : 1
Oats	France	0.9 : 1
Rye	France	1.1 : 1

Source: Literature reviewed by Staniforth (1979a, p. 39).

Table 11–3
Comparison of the Chemical Composition[a] of Straw from Four Cereal Crops, Expressed in Percent Dry Matter

Crop	Cellulose	Hemicellulose	Lignin	Silica
Barley	44	27	7	3
Oat	41	16	11	3
Rice	33	26	7	13
Wheat	39	36	10	6

Source: Jackson (1977).
[a]The remaining percentage consists of cell contents such as small amounts of protein, waxes, sugars, and salts.

dwarf paddy rice cultivars to vary from 13.5% to 16.5%, with reductions in silica content of both rice and wheat being related to irrigation practices.

Determinations of the chemical composition of different fractions are shown in Table 11-4. Protein generally is higher in the leaves, chaff, and rachis than in nodes and internodes. Fiber, which may be taken to reflect the cellulose in Table 11-3, is highest in the internodes. The leaf fraction contains the highest percentage of digestible materials.

Manley and Wood (1978) noted changes in the analysis of different fractions of winter wheat and spring barley grown in Britain and sampled over a three-week period immediately preceding harvest. A general tendency was found for the fiber content to increase and for the digestibility of the straw to decrease during this period. If cereal grain residues are intended for feed, harvesting should be done as soon as the grain reaches physiological maturity in order to increase the feeding value of the straw.

The composition of the straw varies from top to bottom and, unless fiber is desired, a high cut may improve chemical composition (Table 11-5). In corn cut for silage, this is an important consideration.

Table 11–4
Comparison of the Chemical Composition of Five Fractions of the Straw of Spring Barley (SB) and Winter Wheat (WW) at Harvest When Grown in the United Kingdom in 1977, Given as Percentages of Dry Matter and Averaged in Each Case from Two Cultivars

	Crude Protein		Fiber[a]		In-vitro Digestibility		Total Ash	
Fraction	WW	SB	WW	SB	WW	SB	WW	SB
Nodes	3.6	2.7	45.2	48.7	38	39	7.9	5.1
Internodes	2.5	1.5	52.5	57.9	39	30	5.6	2.7
Rachis	3.7	3.4	49.4	43.3	36	42	3.9	4.5
Leaves	5.5	4.2	47.4	47.3	41	53	14.7	6.8
Chaff and awns	4.9	2.8	45.2	43.4	33	45	8.3	13.3

Source: Manley and Wood (1978).
[a]Modified acid detergent method.

Table 11–5
Comparison of the Chemical Analysis of Barley Straw Divided into Top, Middle, and Bottom Cuts and Expressed in Percentages of Dry Matter When Grown in Britain

Cut	Crude Protein	Crude Fiber	Ash	Moisture
Top third	4.0	36.9	6.3	9.0
Middle third	3.0	48.0	4.3	8.8
Bottom third	2.9	50.2	4.3	8.7
Mean	3.5	45.0	5.0	8.8

Source: Wilson and Brigstocke (1977).

CROP RESIDUES AND ORGANIC MATTER

Crop residues are an important source of soil organic matter. Three kinds of organic matter can be found in soils:

1. The visible root system of plants.
2. The partly decomposed remains of plants, still retaining much of their original structure.
3. The well-decomposed brown or black organic matter, commonly called humus.

The fate of crop residues in the soil, as shown in Figure 11-10, illustrates the importance and problems of incorporating crop residues. The positive aspects of soil organic matter can be listed as follows:

1. Organic matter obtained from incorporation of crop residues or from deep-rooted crops reduces the risk of surface crusts, tillage hardpans, or soil compaction. Organic matter is essential to good soil structure.
2. Crop residues that are left to accumulate on the surface reduce water and wind erosion by a cushioning effect on rainfall and reduce wind speed. Organic matter increases water infiltration and reduces surface runoff. Humus in the soil cements soil particles together and reduces erosion. Instability of structure can develop in sand, very fine sand, or silt when organic matter levels fall below 3%.
3. Humus or stabilized organic matter has a high cation-exchange capacity and acts similarly to clay in holding available plant nutrients against leaching. Humus maintains the nutrients in a form available to higher plants and has a 3% to 6% nitrogen content.
4. Almost all the life in the soil is dependent on organic matter for energy and nutrients to nourish microorganisms that keep the soil "alive" and productive.
5. Organic matter makes soils friable and increases ease of cultivation.
6. The decomposition of crop residues releases carbon dioxide, an essential ingredient in photosynthesis. Crop producers have observed the "beneficial effect" of manure over comparable fertility levels achieved by manufactured fertilizers. The release of CO_2 during decomposition of crop residues is of tremendous value in "feeding" photosynthesis. The addition of commercial fertilizers to a soil low in organic matter adds nutrients but cannot produce carbon dioxide.
7. Soil organic matter can significantly affect soil water content, soil temperature, and soil aeration. These factors of the physical environment are associated with a favorable microclimate for good root growth.
8. Humus absorbs large quantities of water, resulting in swelling and shrinking that is important in forming soil structure.

There are also negative aspects. Hughes (1979) pointed out major penalties for leaving straw and other residues undecomposed at or near the soil surface. These include:

1. Under minimum-tillage systems, residues present a physical hindrance to the establishment of a new crop.

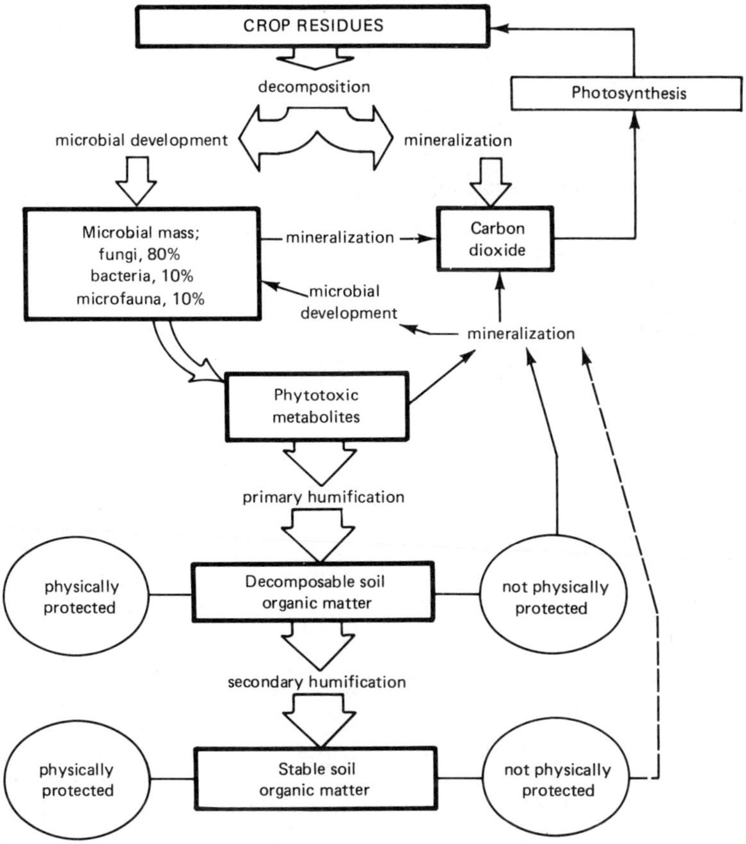

Figure 11–10. Fate of crop residues in soil. Based on this type of flow diagram, mathematical models have been formulated to describe the stages. Inadequate information on many of the stages means that such models are inadequate. (After Lynch, 1979)

2. Residues can produce phytotoxic metabolites that hinder the growth of cereal seedlings.
3. Buried residues provide an ideal breeding habitat for specific pests and a medium for the carryover of disease organisms.
4. Transferring straw from one field to another can spread weed seeds.
5. Wads of decomposing straw can lower the efficacy of soil-applied pesticides.
6. The existence of surface residues in various stages of decomposition reduces soil temperature in short-season areas, delaying seeding operations and subsequent germination. Early spring seeding or timeliness of establishing fall-seeded cereals is important in cereal production.
7. Dealing with crop residues either by physical removal, incorporation (Figure 11-11), or chopping increases energy inputs. The existence of straw residues can reduce the effectiveness of costly and energy-intensive inputs intended to control pests.

It is apparent that the advantages outweigh the disadvantages, yet organic-matter levels have been allowed to decline. Reductions in soil organic matter may be explained as follows:

Figure 11–11. Crop residues left to accumulate on the surface may delay soil warming, reduce the efficiency of soil-applied pesticides, require a high energy input to incorporate, and present a physical hindrance to the establishment of a new crop. Plowing, as achieved in this photo, turns up buried seeds, exposes desirable soil organic matter to the elements, and renders the soil prone to erosion by wind and water. Incorporation may tie up soil nitrogen during the period of decomposition. (Courtesy of Ontario Ministry of Agriculture and Food)

1. Economic reasons. If prices warrant, crop residues may be removed and sold. If prices are low, residues may be burned in the field to reduce subsequent tillage or seeding inputs and to avoid energy costs in their handling and disposal.

2. A widespread belief that the application of commercial fertilizers could replace nutrients obtained from decomposing crop residues or nitrogen obtained by biological fixation of legume crops. Under a regime of inexpensive commercial fertilizers, it was more convenient and less costly to supply the nutritional needs of the plant from the fertilizer bag. Recent years have seen sharp increases in costs of fertilizer, especially nitrogen [it requires 1300 m^3 (39,300 ft^3) of natural gas to produce a tonne (ton) of nitrogen fertilizer].

3. Tying up of available soil nutrients during decomposition of crop residues. Crop residues added to the soil must decompose to release the nutrients they contain and before stable organic matter or humus is produced. The largest component of cereal crop residues in carbon. Nitrogen is present at very low levels in comparison (Table 11-6). Carbon is utilized by microorganisms only if sufficient nitrogen is available. For every 30 units of carbon in the crop residue, at least one unit of nitrogen must be present, or decomposition will slow down. If a carbon : nitrogen ratio of 30 : 1 does not exist, the soil microorganisms will utilize nitrogen already present in the soil or use nitrogen added for plant use. As a result the crop may be deprived of nitrogen it needs for growth and development.

Depending on the nitrogen used in producing a crop, the resulting crop residue commonly has a C : N ratio of 85 : 1 or 90 : 1. Thus for each 1000 kg (2205 lb) of crop residue added to the soil, approximately 30 kg (66 lb) of nitrogen must be added. Barber (1979) reported an average annual return of 9050 kg/ha (8080 lb/ac) of corn stover in the corn belt of the United States, requiring 272 kg N/ha (243 lb N/ac) either present or applied to the soil to prevent nitrogen deficiency. Barber reported highest corn yields over a 12-year period when 200 kg N/ha (179 lb N/ac) was applied when all crop residues were returned. In the short term it was easier to apply nitrogen to meet the direct needs of the growing crop rather than to apply nitrogen to decompose the residues as well as meet the crop needs.

Increased volumes of crop residues associated with increased grain yields have increased the crop-residue handling problem.

Table 11–6
Analysis of Wheat Straw

Constituent	*% by weight*
Carbon	43.3
Oxygen	36.2
Moisture	10.0
Hydrogen	5.3
Ash	4.2
Nitrogen	0.9
Sulphur	0.1
Total	100.0

Source: Lester (1977).

4. A view that the loss of organic matter was the inevitable consequence of soil cultivation and crop production. It was observed that organic-matter levels would decline rapidly at first, then at a decreasing rate until they stabilized.

In a review of the depletion rate of organic matter in soils cultivated for small-grain production (Tiessen, Stewart, and Bettany, 1982), it appears that about 1% of organic carbon is lost per year during the first 20 to 30 years of cultivation. At individual sites, however, actual losses of carbon ranged from 0% to 2.5% per year. Initial losses are rapid and become slower as organic matter levels drop.

After an initial rapid decline in soil organic matter, it frequently is suggested that levels stabilize at a level possibly 40% to 60% of the original content, or at about 1.0% to 4.0%, depending on cultural practices. Rennie (1979) suggested that levels may equilibrate at 75% of the initial 4% or 5% level or at 3.0% to 3.8%. In Indiana, Barber (1979) found that organic matter decreased to 90% of the original level after ten years. In Iowa, Larson et al. (1972) found that organic matter decreased to 87% of the initial level after 12 years when all surface residues were removed.

Initially it was believed that degradation of organic matter effectively ceased after 25 years of cultivation; later this was revised to between 60 and 70 years, and more recently it was suggested that organic-matter losses may continue indefinitely, owing to accelerated erosion of cultivated fields by wind and water (Tiessen, Stewart, and Bettany, 1982; Voroney, Van Veen, and Paul, 1981).

5. A popular belief that the application of nitrogen fertilizer to stimulate plant growth would rebuild soil organic matter quickly and easily if all the crop residue was returned to the soil. It has been demonstrated, however, that under conventional tillage methods soil organic matter cannot be maintained at anything approaching its virgin levels. Soil cultivation that exposes the soil to the elements is like the opening of air vents on a furnace, in that breakdown of soil organic matter is speeded up (Meints et al., 1977; Parr and Papendick, 1978).

From 8% and 18% of crop residues are converted to humus (Barber, 1979; Larson et al., 1972). To increase the soil organic-matter level by 1%, 7250 kg/ha (6473 lb/ac) would have to be added annually for 24 years with no annual cultivation losses, assuming an average 13% conversion rate. A small increase of 0.2% over a long period of time after returning straw residues was reported by Harvey (1959).

The disappointingly low increase achieved by incorporating crop residues has discouraged attempts to increase soil organic-matter levels. Emphasis may be directed most effectively to maintaining soil organic-matter levels and guarding against losses. Crop producers must recognize the difficulty of increasing organic matter and take precaution to maintain existing levels.

ORIGINAL ORGANIC-MATTER LEVELS

Comparisons of present-day levels of organic matter and those when cropping began are difficult to make because original levels were not determined. It is clear that they were higher under grassland than under forest soils. Brady (1974) estimated organic-matter levels to vary from 2% to 8% for virgin soils in the north central Great Plains regions of North America. Evidence of sod with high organic matter representing centuries of accumulated grass roots and decayed material produced above ground was revealed by Hardeman (1981) in his descriptions of pioneer farms. The sod was so firm that settlers often made their first corn plantings by chopping holes in it with axes. House walls constructed of the firm sod substituted for wood or stone.

To break the heavy virgin sod, a professional group of plowhands emerged, known as "breakers." The nature of their equipment bears testimony to the difficulty of sod breaking: three to seven yoke of oxen were required on a single-furrow plow, which had a heavy 4.3-m (14-ft) wood beam and a 27- to 57-kg (60- to 125-lb) plowshare. The furrow turned by this large plow varied from 40 to 76 cm (16 to 30 in.) wide and from 5 to 13 cm (2 to 6 in.) deep.

HANDLING CROP RESIDUES

Problems associated with handling organic matter have prevented a precise recommendation. New theories and alternative plans need to be checked in long-term studies. Until well-defined management systems have been achieved, incorporation of crop residues must be regarded as a means of disposal of unwanted material, rather than as a method of conservation and a method of amending soils. Changing conditions tend to invalidate long-term studies.

Two well-known long-term studies related to soil productivity and handling of crop residues are the classical experiments at Rothamsted founded in 1843 and the Morrow Plots at the University of Illinois (Thorne, 1978; Welch, Melsted, and Oldham, 1976). In general, these studies have indicated that if soils have been depleted but have not deteriorated, productivity can be restored. In areas more subject to erosion, soil deterioration may be of greater consequence than revealed by the Morrow or Rothamsted plots.

Other short-term studies are summarized here to reveal the complexity associated with crop-residue management studies.

An experiment on the use of wheat straw, plowed in or composted at the rate of 6576 kg/ha (5863 lb/ac) every second year, with fertilizers applied as needed to maintain fertility, was conducted in England between 1933 and 1958 (Patterson, 1960). The treatments appear to have no effect on crop yields through improvements to the soil caused by better structure or increased organic-matter content. When the straw was incorporated, about 10 kg N/ha (9 lb N/ac) was needed for each ton of straw to overcome losses in yield due to nitrogen deficiencies.

Hughes (1979) suggested that in England and Wales there are few advantages and many disadvantages to incorporating straw. Under minimum-tillage conditions, cereal roots can supply the necessary amount of organic matter to maintain stability of soil structure on the majority of soils in Britain. It should be noted that cereal crop roots are fine and fibrous compared to corn roots, which are coarser and may be less inducive to good soil structure.

Over 18 years of observation on clay, chalk, and sand soils in which the straw was incorporated, burned, or removed, Hughes (1979) concluded that high cereal yields can be maintained by the use of adequate levels of fertilizer, especially nitrogen, regardless of the method of straw disposal and without deterioration in soil conditions. Organic-matter levels in this study were below 3.9%. Levels declined slightly where the straw was burned or removed.

In tests established in Kansas in 1969, various methods of managing irrigated winter wheat and grain sorghum residues on yield and soil chemical properties were measured. Methods of residue management included incorporation of twice the quantities of residue produced, and burning. In 1979, plots on which residues had been physically removed or burned were lower in organic matter and potassium and had a higher *p*H. Nitrogen was leached to a greater extent on these plots. During the early years of the experiments there were no differences in yield due to residue treatments. However, there was a trend toward lower yields on the burned and physical-removal plots.

BURNING STRAW

A controversial but common practice is the field burning of crop residues on farms where these residues cannot be, or as yet have not been, incorporated economically into the farming system. In Britain, farmers are anxious to maximize the area of winter cereals because of their greater profitability. The questionable need to plow for seeding winter cereals following the removal of spring cereals has increased field burning of crop residues (Ellis, 1979). Burning of straw is recognized as the most economical way of disposing of this unwanted commodity, as it does not lead to soil compaction by removal equipment, can destroy or can encourage germination and subsequent elimination of volunteer cereal and weed seeds, and may reduce the incidence of disease. Research over a short period supports the burning of straw (Table 11-7).

Burning is recognized as having the following disadvantages: it results in the loss of protection against soil erosion; the ash that contains plant nutrients can be lost easily by wind or water erosion; most nutrients in the ash are in a soluble form and are easily leached through the soil; the opportunity to increase desirable soil organic matter is reduced (Hooker, Herron, and Penas, 1982).

Advantages of specific farming systems may outweigh these disadvantages. Burning the straw may be justified if efficient alternative methods of straw management or disposal are lacking. Short-straw cultivars, for example, have helped reduce the magnitude of the problem but have not eliminated it.

Burning may be justified under the following circumstances: to facilitate harvesting, as when a flash fire of sugarcane fields eliminates excessive undesirable leaf residues; when an excess of residues interferes with the efficient operation of no-till seeders; to control disease; to remove an accumulation of residues to reduce the risk of accidental fire; to reduce residue phytotoxicity.

Field burning of straw is environmentally undesirable and, if undertaken, may have to conform to a "straw and stubble burning code" established for geographic areas. Care is demanded to prevent the undesired spread of fire and to protect wildlife and the environment.

Alternative methods to burning usually cannot be achieved without additional energy-demanding inputs. In some dryland areas, where straw volume is not so great, stubble-mulch farming is practiced to advantage. This method can reduce soil losses from wind and water erosion, protect the soil surface against structural deterioration, and increase water infiltration and conservation (Skidmore and Siddoway, 1978).

PHYTOTOXICITY FROM CROP RESIDUES

In areas of the world where conservation tillage systems are employed, phytotoxicity associated with crop residues is a potential problem (Figure 11-9). Evidence is lacking that phytotoxic substances from cereal crop residues occur in sufficient concentration to inhibit plant growth under field conditions. Evidence does not exist

Table 11–7
Comparisons of Winter Wheat Yields Established after Contrasting Methods of Cultivation and Residue-Management Treatments over a Three-Year Period in Britain, 1975–1977

Cultivation Treatment and Straw Management System	*Grain Yield*	
	kg/ha	*lb/ac*
Direct-drilled:		
Burned	5100	4554
Baled and removed	3100	2768
Chopped	2800	2500
Fine-cultivated:		
Burned	5000	4464
Baled and removed	4500	4018
Chopped	4300	4732
Plowed:		
Burned	5000	4464
Baled and removed	4600	4107
Chopped	4400	3929

Source: Ellis (1979).

that leaving residues on or near the surface may reduce subsequent cereal crop growth and yield as compared with removing residues or using conventional tillage systems (Elliott, McCalla, and Waiss, 1978). Allelopathic chemicals may be released from residues of crop plants and perennial and annual weeds (Bhowmik and Doll, 1982).

Evidence of phytotoxicity is associated with wet conditions that result in the production of acetic, proprionic, and butyric acid from wheat and barley straw (Lynch, 1979). These acids can reduce the germination and subsequent elongation of barley, corn, oat, and wheat seedling roots. The production of phytotoxic substances appears more likely when a certain sequence of cropping is adopted, such as wheat after wheat (Ellis, 1979; Hale and Moore, 1979). Grain sorghum, grown in the greenhouse, had depressed growth when aqueous sunflower-leaf extracts were added (Schon and Einhellig, 1982).

Toxic activity was most pronounced when seed was placed in direct contact with decomposing residue, under anaerobic conditions, and in the early stages of crop establishment. Toxins were produced seven to ten days after incorporation; phytotoxicity reached a maximum in three weeks and declined after six to seven weeks (Tousson et al., 1968).

Under anerobic conditions, fungal colonization on the seed can lower the oxygen availability to the germinating seed (Lynch and Pryn, 1977). Whether fungal growth is associated with the presence of crop residues or is due primarily to anaerobic conditions is unclear.

Until any possible harmful effects of straw are eliminated, producers will be reluctant to incorporate crop residues.

ALTERNATIVE USES OF STRAW

Straw as a Ruminant Feed

A ruminant feed is considered deficient if it does not contain 8% protein with a reasonably high biological value. Crude protein of corn plant parts is low (Table 11-8), as is biological value. If cereal residues are fed, they must be supplemented with a nitrogenous source. The presence of lignin in the cell wall of straw adversely affects digestion and makes it of low value as a ruminant feed.

The digestibility of straw, however, can be improved by either physical or mechanical processing (chopping into fine segments), by chemical treatments with sodium hydroxide (NaOH) to dissociate lignin from the cell wall, and by supplementation of straw feed with minerals, additional sources of nitrogen, and a source of readily available carbohydrate. These treatments can occur singly or in combination.

Crop residues such as cereal straw, corn stalks, leaves, husks, and cobs may be used as a low-grade wintering ration for beef cows and heifers. Feed materials may be gleaned by livestock after harvest. In heavy-snowfall areas, crop residues for feed may be stored as bales, as stalkage, or as huskage in a silo. Fine-chopping the feed as it goes into the silo enhances both digestibility and preservation and helps prevent

Table 11–8
Composition of Corn Plant Parts Following Grain Harvest and Compared to the Grain, Expressed in Percent and Averaged over Two Years

Item	*Leaf*	*Stalk*	*Husk*	*Cob*	*Grain*
Dry matter	73	33	65	56	76
Crude protein	6.2	2.7	3.3	3.6	9.6
In-vitro digestibility	54	50	63	45	90

Source: Perry (1975).

cattle from sorting the roughage in feed bunks. A source of nitrogen such as urea may be added to act as a protein source in the rumen.

Use of crop residues as a roughage and a substitute for forage crops will not alleviate the crop-residue problem as long as quality roughage can be produced economically. Although the treatment of cereal crop residues can increase their potential as an alternative source of feed energy, economic aspects may limit its success (Epstein, Alpert, and Calvert, 1978; Staniforth, 1979a).

Straw as an Energy Source

Renewable biological sources of energy sprang to prominence with the OPEC decision to unilaterally increase oil prices by 300% in 1973 (Lewis, 1981). Fossil fuels produced from photosynthetic plant products by heat and pressure millions of years ago are part of the same carbon cycle as the fuels that can be produced currently from crops and crop residues.

Energy can be produced from crop residues by direct combustion (Peill, 1981), by anaerobic digestion to produce biogas, by fermentation to produce ethanol or methanol, and by thermochemical conversions (pyrolysis) (Pankhurst, 1983). In some areas of the world, crop residues are already important energy sources, and continued high cost of fuels and the threat of diminished supplies may mean that fuel from crop residues could become of major importance worldwide.

Several critical issues need to be resolved regarding the use of plant biomass as an energy source. First and foremost is the impact on crop productivity of the widespread use of crop residues. Overuse of crop and animal wastes, including wood, can result in lowered agricultural productivity (Dunkerley and Ramsay, 1982). If wastes are used as fuel, energy-intensive fertilizer substitutes must be made available (de Montalembert, 1983).

Second, initial successes in energy production from crop residues may lead to the use of high-energy food grains for energy production (Shih, Gascho, and Raki, 1981; Hodgson, 1982). Competition between energy and food could impact adversely on world food supplies.

A third concern is the question of embodied energy and the efficiency of utilizing renewable energy sources produced from inherently low levels of photosynthetic efficiency. Input-output analyses that calculate the total direct plus indirect energy required to produce food are not favorable to systems that produce energy from crop residues. Research is needed toward fuel production as an integrated component of a system whereby the by-product from one enterprise can offset the cost of another enterprise (Stoskopf, 1983). A system involving livestock and crop production may offer a viable approach because crop residues can complement livestock production. In the production of fuel alcohol, the stillage could form an excellent high-protein feed. The animal manure could be used to produce biogas (CH_4), which removes only carbon and hydrogen, reducing the volume by 5% to 7% but without affecting the fertilizer nutrient value of the manure. Associated forages allow for a crop-rotation system and the conversion of inert atmospheric nitrogen into a biologically usable form to reduce energy-dependent nitrogen fertilizer needs for cereal production such as corn or sorghum.

Although energy input-output efficiency ratios of highly mechanized modern agriculture are not favorable, the larger considerations may be more important (Costanza, 1980; Daynard, 1983). The facts are that (1) a joule of energy in the form of food is far more valuable to humankind than a joule in the form of crop residues, petroleum, or electricity, (2) efficiencies in energy cannot produce the food for the billions of people on earth, and (3) energy efficiency may not be an economically sound approach.

Straw for Commercial Use

It is apparent, even to the casual observer, that huge quantities of straw are produced from the world's cereal crops. Ways of utilizing this annual renewed resource in a profitable, commercial manner long have attracted attention.

Staniforth (1979a and b; 1982) reviewed the commercial uses of straw fiber in Britain and listed industrial uses as follows:

Paper and board
Fuel and chemicals
Single-cell protein
Nutritionally improved straw for feed
Edible mushroom production
Cucumber production
Thatching
Building boards
Crafts

In the manufacture of paper from straw, the chopped material is digested by a strong alkali which separates the lignin and parts of the hemicellulose. A component of hemicellulose is xylose, a sugar with five carbon atoms, which is not readily metabolizable by monogastric stomachs. Reduction of xylose with hydrogen produces xylitol, an attractive alternative to glucose or common sugar and one that raises very few medical objections. Xylitol is acceptable to diabetics and does not contribute to dental caries (Rijkens, 1977).

Furfural can be produced from xylose, a proven chemical used for years in the purification of edible lubricating oils, in plastics, in pharmaceuticals, and as a solvent in the lubricating oil production processes (Rijkens, 1977). Furfural is difficult to produce from petroleum because it is readily **polmerized.** Furfural is produced from by-products rich in hemicellulose, such as corncobs or oat hulls (Stacey, 1976).

The use of straw and cereal residues for commercial use must take into consideration the reliability of supply. Although straw is an abundant and renewable commodity, it must be collected and transported to processing plants. The machinery exists to achieve this end, but the cost and bulky nature of straw must be considered. When grain corn was harvested on the cob, central hulling facilities produced a supply of cobs for furfural production. Picker-shellers that remove the grain from the cob at harvest (see Figure 1-1) and scatter the cob and other residue over the field have reduced the opportunity to use cobs for furfural production. As well, the use of corncobs as an abrasive material for cleaning purposes and in cosmetics has created a competitive situation for accumulated corncobs.

CONCLUSIONS

The historical view of crop residues as having value for a number of purposes has given way to a modern view of their being a nuisance and a waste. A further change may come, as residues increasingly are regarded as an energy source.

Residues from cereal crops are highly versatile materials. Under modern systems of intense crop production, however, although some producers are using residues to advantage, many cannot.

A growing concern about conservation of resources has aroused worldwide unease about how crop residues are often burned or wasted. Many agriculturists are reexamining crop-residue handling systems and are seeking new ways to use residues in a cropping system. They represent a valuable energy resource. Precise recommendations, however, on how to introduce crop residues into farming systems cannot be given because information is still insufficient.

Questions

1. Describe and discuss economic factors, shifts in emphasis, changes in agricultural practices, or other factors associated with cereal crop production that have changed the way in which crop residues are regarded in your area.

2. Consider the validity in the statement: "Crop residues are a valuable resource to be added to the soil." Does this concept apply to all systems?

3. *a.* Develop a workable system for residue management for an enterprise in your area similar to that shown in Figure 11-8. Review the section on crop residues as an energy source.

b. Compared to the Chinese model shown in Figure 11-8, what major shortcomings exist in most other models?

4. Consider rotations in terms of residue management and as part of a farming system.

a. What is implied by the term **rotation?** (See the glossary definition.)

b. Elaborate on the advantages and disadvantages of crop rotations.

c. Is a wheat summer fallow a rotation? Is monoculture production of corn a rotation?

d. Describe with examples an intense crop-rotation system.

e. Do systems involving livestock add to or reduce the intensity of crop rotation?

5. *a.* Determine the amount of crop residues produced from cereal grains grown in your area.

b. What cereal grains are the highest producers of crop residues in general?

6. Consider crop residues as a feed source obtained from the binder and from the combine harvester on the basis of chemical composition as shown in Table 11-4.

7. Consider crop residues as a source of carbon dioxide.

a. How much carbon dioxide is required to produce a given crop?

b. What are the sources of carbon dioxide (1) in a cereal crop field? (2) in the production of a successful traditional "hot bed" for starting vegetables?

8. What long-term effect has "slash and burn"-style agriculture on tropical soils had on crop production?

9. Speculate on the possible evolutionary significance of root exudates, allelopathic chemicals, or phytotoxicity produced from crop residues.

10. What circumstances will help precipitate a trend toward or against the recognition of cereal crop residues as a valuable resource?

11. *a.* Consider what constitutes a "long-term" research projection on factors affecting soil productivity

b. What changes may occur in a "long-term" study that invalidate results?

c. In view of factors in *a* and *b*, should "long-term" studies be undertaken?

References

Aldrich, S. R. 1977. "Organic Farming Can't Feed the World." *Farm Chemicals* 141: 17, 20, 22.

Andrews, D. J., and A. H. Kassam. 1976. "Importance of Multiple Cropping in Increasing World Food Supplies." (In) *Multiple Cropping*. Madison, Wis.: American Society of Agronomy. Publication 27. pp. 1–10.

Barber, S. A. 1979. "Corn Residue Management and Soil Organic Matter." *Agronomy Journal* 71: 625–627.

Berg, N. A. 1979. "Soil Conservation: The Physical Resource Setting." (In) *Soil Conservation Policies: An Assessment*. Proceedings of the Soil Conservation Society of America. Ankeny, Iowa. pp. 8–17.

Bhowmik, P. C., and J. D. Doll. 1982. "Corn and Soybean Response to Allelopathic Effects of Weed and Crop Residues." *Agronomy Journal* 74: 601-606.

Brady, N. C. 1974. *The Nature and Properties of Soils.* New York: The Macmillan Company. 8th ed. 637 pp.

Cochran, V. L., L. F. Elliott, and R. I. Papendick. 1982. "Effect of Crop Residue Management and Tillage on Water Use Efficiency and Yield of Winter Wheat." *Agronomy Journal* 74: 929-932.

Costanza, R. 1980. "Embodied Energy and Economic Valuation." *Science* 210: 1219-1224.

Daynard, T. B. 1983. "Energy Considerations in Crop Production" (In) *Energo '83 Conference Papers.* Ontario Ministry of Energy. Toronto, March 6-9. pp. 176-179.

de Montalembert, M. R. 1983. "Biomass Resources for Energy: The Critical Issues." *Ceres* 16 (1): 40-44.

Donald, C. M., and J. Hamblin. 1976. "The Biological Yield and Harvest Index of Cereals as Agronomic and Plant Breeding Criteria." *Advances in Agronomy* 28: 361-405.

Dunkerley, J., and W. Ramsay. 1982. "Energy and the Oil-Importing Developing Countries." *Science* 216: 590-595.

Elliott, L. F., T. M. McCalla, and A. Waiss, Jr. 1978. "Phytotoxicity Associated with Residue Management." (In) *Crop Residue Management Systems.* Madison, Wis.: American Society of Agronomy. Special Publication No. 31. pp. 131-146.

Ellis, F. B. 1979. "Agronomic Problems from Straw Residues with Particular Reference to Reduced Cultivation and Direct Drilling in Britain." (In) E. Grossbard (ed.), *Straw Decay and Its Effect on Disposal and Utilization.* Chichester: John Wiley & Sons Ltd. pp. 11-20.

Epstein, E., J. E. Alpert, and C. C. Calvert. 1978. "Alternative Uses of Excess Crop Residues." (In) *Crop Residue Management Systems.* Madison, Wis.: American Society of Agronomy. Special Publication No. 31. pp. 219-229.

FAO. 1977a. *China: Recycling of Organic Wastes in Agriculture.* Rome: Food and Agriculture Organization of the United Nations. Soils Bulletin 40. p. 3.

———. 1977b. "Bio-Fertilizers." (In) *China: Recycling of Organic Wastes in Agriculture.* Rome: Food and Agriculture Organization of the United Nations. Soils Bulletin 40. pp. 29-40.

———. 1981. *Blue-Green Algae for Rice Production.* Rome: Food and Agriculture Organization of the United Nations. Soils Bulletin 46. 102 pp.

Gallagher, J. N., and P. V. Biscoe. 1978. "A Physiological Analysis of Cereal Yield. II. Partitioning of Dry Matter." *Agriculture Progress* 53: 51-70.

Gately, T. F. 1975. "Effects of Fertiliser Nitrogen and Number of Years of Tillage on the Yield of Barley Straw (cv. Nessa)." *Irish Journal of Agricultural Research* 14: 237-244.

Greb, B. W. 1979. "Technology and Wheat Yields in the Central Great Plains—Commercial Advances." *Journal of Soil and Water Conservation* 34: 269-273.

Haga, K., H. Tanaka, and S. Higaki. 1979. "Methane Production from Animal Wastes and Its Prospects in Japan." *Agricultural Wastes.* An International Journal. London: Applied Science Publishers Ltd. Vol. 1: 45-55.

Hale, M. G., and L. D. Moore. 1979. "Factors Affecting Root Exudation II: 1970-1978." *Advances in Agronomy* 31: 93-124.

Hardeman, N. P. 1981. *Shucks, Shocks and Hominy Blocks. Corn as a Way of Life in Pioneer American.* Baton Rouge: Louisiana State University Press. 271 pp.

Harvey, P. N. 1959. "The Disposal of Cereal Straw." *Journal of the Royal Agricultural Society, England* 120: 55-63.

Hodgson, H. J. 1982. "Fuel from Crops: Part 1, Grain and Sugar Crops. Part 2, Lignocellulosics." *Crops and Soils* 34 (Nos. 7 and 8): 5-7; 5-6.

Hooker, M. L., G. M. Herron, and P. Penas. 1982. "Effects of Residue Burning, Removal, and Incorporation on Irrigated Cereal Crop Yields and Soil Chemical Properties." *Journal of the Soil Science Society of America* 46: 122-126.

Hudson, J. P. 1977. "Disposal of Straw and Slurry." *Span* 20 (1): 35.

Hughes, R. G. 1979. "Arable Farmer's Problems with Straw." (In) E. Grossbard (ed.), *Straw Decay and Its Effect on Disposal and Utilization.* Chichester: John Wiley & Sons Ltd., pp. 3-10.

Jackson, M. G. 1977. "The Alkali Treatment of Straws." Review Article. *Animal Feed Science Technology* 2: 105-130.

Larson, W. E., et al. 1972. "Effects of Increasing Amount of Organic Residues on Continuous Corn. II. Organic Carbon, Nitrogen, Phosphorus, and Sulfur." *Agronomy Journal* 64: 204-208.

Larson, W. E., et al. (eds.) 1981. *Soil and Water Resources: Research Priorities for the Nation* (Executive Seminar). Madison, Wis.: Soil Science Society of America. 45 pp.

Larson, W. E., R. F. Holt, and C. W. Carlson. 1978. "Residues for Soil Conservation" (In) *Crop Residue Management Systems.* Madison, Wis.: American Society of Agronomy, Special Publication No. 31. pp. 1-16.

Lester, T. W. 1977. "The Direct Firing of Wheat Straw in a Municipal Generating Station." Manhattan:

Kansas State University. Report CES.41. pp. 25–48.

Lewis, C. W. 1981. "Biomass Through the Ages" *Biomass* 1: 5–15.

Lynch, J. M. 1979. "Straw Residues as Substrates for Growth and Product Formation by Soil Micro-Organisms." (In) E. Grossbard (ed.) *Straw Decay and Its Effect on Disposal and Utilization.* Chichester: John Wiley & Sons Ltd. pp. 47–56.

———, and I. S. J. Pryn. 1977. "Interaction Between a Soil Fungus and Barley Seed." *Journal of General Microbiology* 103: 193–196.

MacKenzie, W. A. 1926. "Note on a Remarkable Correlation Between Grain and Straw, Obtained at Rothamsted." *Journal of Agricultural Science* 16: 275–279.

Manley, A. C., and R. S. Wood. 1978. "The Analysis of Fractions of Cereal Crops on Three Dates Prior to Harvest." *Agriculture Progress* 53: 71–76.

Meints, V. W., et al. 1977. "Long-Term Trends in Total Soil N as Influenced by Certain Management Practices." *Soil Science* 124: 110–116.

Pankhurst, E. S. 1983. "The Prospects for Biogas—A European Point of View." *Biomass* 3: 1–42.

Parr, J. F., and R. I. Papendick. 1978. "Factors Affecting the Decomposition of Crop Residues by Micro-Organisms." (In) *Crop Residue Management Systems.* Madison, Wis.: American Society of Agronomy. Special Publication No. 31. pp. 101–139.

———, and I. G. Youngberg. 1983. "Organic Farming in the United States: Principles and Perspectives. *Agro-Ecosystems* 8: 183–201.

Patterson, H. D. 1960. "An Experiment on the Effects of Straw Ploughed in or Composted on a Three Course Rotation of Crops." *Journal of Agricultural Science* 54: 222–229.

Peill, J. E. 1981. "Straw as Energy." *Agrologist* 10 (2): 12–13.

Perry, L. J., Jr. 1975. "Stover for Your Stock." *Crops and Soils* 27 (4): 9–11.

Pierce, R. T. R. 1982. "Why Farmers Burn Straw." *Span* 25 (2): 65.

Rennie, D. A. 1979. "Intensive Cultivation. The Long-Term Effects." *Agrologist* 8 (4): 20–22.

Rijkens, B. A. 1977. "Some Possibilities for Multiple Use of Straw." *Agricultural Progress* 52: 69–75.

Sanford, J. O. 1982. "Straw and Tillage Management Practices in Soybean-Wheat Double Cropping." *Agronomy Journal* 74: 1032–1035.

Schon, M. K., and F. A. Einhellig. 1982. "Allelopathic Effects of Cultivated Sunflower on Grain Sorghum." *Botanical Gazette* 143: 505–510.

Shih, S. F., G. J. Gascho, and G. S. Raki. 1981. "Modeling Biomass Production of Sweet Sorghum." *Agronomy Journal* 73: 1027–1032.

Skidmore, E. L., and F. H. Siddoway. 1978. "Crop Residue Requirements to Control Wind Erosion." (In) *Crop Residue Management Systems.* Madison, Wis.: American Society of Agronomy. Special Publication No. 31: 17–38.

Stacy, M. 1976. "Straw as a Potential Raw Material for Chemicals." *Agricultural Progress* 51: 69–75.

Staniforth, A. R. 1979a. *Cereal Straw.* Oxford: Oxford University Press. 175 pp.

———. 1979b. "Industrial Uses for Straw and the Relevance of Decay to Such Uses." (In) E. Grossbard (ed.), *Straw Decay and Its Effect on Disposal and Utilization.* Chichester: John Wiley & Sons Ltd. pp. 209–216.

———. 1982. "Commercial Uses for Straw Fibre." *Outlook on Agriculture* 11: 21–26.

Stoskopf, N. C. 1983. "Biomass—A Crop Scientist's View." (In) *Energo '83 Conference Papers.* Ontario Ministry of Energy. Toronto, March 6–9. pp. 188–193.

Taiganides, E. P. 1979. "Wastes Are—Resources Out of Place." *Agricultural Wastes.* An International Journal. London: Applied Science Publishers Ltd. Vol. 1: 1–9.

Thorne, M. D. 1978. "A Symbol of Hope for Today's Society." *Agronomy Journal* 70: 1–5.

Tiessen, H., J. W. B. Stewart, and J. R. Bettany. 1982. "Cultivation Effects on the Amounts and Concentration of Carbon, Nitrogen and Phosphorus in Grassland Sorts." *Agronomy Journal* 74: 831–835.

Tousson, T. A., et al. 1968. "Nature of Phytotoxic Substances Produced During Plant Residue Decomposition in Soil." *Phytopathology* 58: 41–45.

United States Department of Agriculture. 1980. *Report and Recommendations on Organic Farming.* A special report prepared for the Secretary of Agriculture. Washington, D.C.: U.S. Government Printing Office. 164 pp.

Voroney, R. P., J. A. Van Veen, and E. A. Paul. 1981. "Organic C Dynamics in Grassland Soils. 2. Model Validation and Simulation of the Long-Term Effects of Cultivation and Erosion and Rainfall Erosion." *Canadian Journal of Soil Science* 61: 211–224.

Welch, L. F., S. W. Melsted, and M. G. Oldham. 1976. "Lessons from the Morrow Plots." *Illinois Research.* Winter: 3–4.

Wilson, P. N., and T. Brigstocke. 1977. "Re-Cycling Farm Waste for Animal Feed." *Agriculture Progress* 52: 49–58.

12

Cereal Crop Quality

THE CONCEPT OF QUALITY

Quality refers to the desirability of the product and may include various physical and chemical aspects, depending on the intended purpose. Physical aspects of cereal grain quality include weight per unit volume, kernel weight, kernel size and shape, kernel texture, vitriousness, color, damaged kernels, impurities, or malting and milling properties. Chemical aspects of quality include moisture content, protein quantity, protein quality, fat acidity, and crude fiber and ash content. Other quality factors may be imposed, such as taste, color appeal, starchiness, and texture when prepared for food. If considered as livestock feed, quality aspects may include digestibility, percent hull, palatability, or value as whole-plant forage.

Quality factors are important to a manufacturer attempting to produce a consistent product. Since manufacturers cannot improve upon the quality of a cereal grain, they purchase products that meet specified standards according to grade. Grading may be determined by one or more quality factors recognized as being desirable to meet the wants of the consumer.

A difficult concept to grasp is the interpretation of "quality" in cereal products based on subjective evaluation. Food preferences are based on subjective factors such as eye appeal, custom, ancestry, social status, and geographic location. Quality is not an absolute value, may vary from area to area, is influenced strongly by the consumer, and may have little scientific basis. To provide an insight into the complexity, variability, and importance of quality, this chapter deals with selected aspects of cereal quality.

PROTEIN CONTENT AND QUALITY

Cereals, as the primary food source for humankind, are regarded primarily as a source of carbohydrate and energy because of their large, starchy endosperms. Rice kernels, for example, consist of 80% starch and 6% to 8% protein. The heavy dependence on cereal grains in many areas of the world means that they also serve as a primary protein source. Worldwide, 70% of the protein consumed comes from cereal grain.

The protein content of many cereal grains has been neglected because of the view that cereal crops are primarily sources of energy food. Attention to protein quantity and quality was directed for many years primarily to major sectors of cereal technology such as brewing and bread making. Work on bread wheat was dominated by one form of protein called gluten; in brewing, a combination of high starch and low nitrogen content was desired. Attention was directed to protein quantity and quality when protein deficiencies were observed in areas in which cereal grains dominated the diet. Protein-deficiency symptoms, known as **kwashiorkor,** could be overcome by the inclusion of animal products or several vegetable protein sources in the human diet. It came to be recognized that both the quantity and the quality of the protein are important. Energy needs of the body must be met, however, before protein needs can be satisfied. Consumption of protein with inadequate energy will likely result in protein's being converted to energy.

The quantity of protein in a food can be measured using the **Kjeldahl** method of analysis, which measures nitrogen content only. Based on nitrogen content, protein levels are calculated by using an arbitrary factor such as 5.70 in wheat and triticale, 6.25 in pearl millet, and 5.95 in rice. Total protein levels do not, however, reflect biological value or quality aspects when consumed. Proteins are polymers of amino acids linked by peptide bonds; the quality or biological value of proteins is determined by the content of these amino acids. Of the eighteen amino acids commonly found in natural edible protein, eight are not synthesized in the human body but are essential for normal metabolic functions in adults. The eight essential amino acids are:

Lysine
Tryptophan
Phenylalanine
Methionine
Leucine
Isoleucine
Threonine
Valine

The quality of protein supplied, as well as the quantity, varies with plant species, and the needs of different animal species vary also. The amino acid composition of cereal grains is usually unbalanced. Lysine content is often inadequate, and it is the amino acid most deficient in all cereals. The second next most deficient are threonine in oat, barley, rye, and wheat and tryptophan in corn. Methionine and valine are commonly low in cereal grains. Humans or nonruminant animals receiving inadequate quantities of the essential amino acids will exhibit protein-deficiency symptoms characterized by unsatisfactory growth and development.

Efforts aimed at improving cereal grain quality in terms of total protein have been unsuccessful because the protein fraction that was increased was the alcohol-soluble part containing the amino acid, glutamic acid. The alcohol-soluble portion of protein is low in lysine and other essential amino acids, but glutamic acid is in excess. Selection for total protein content in corn increased the level to 19.5% (Woodworth, Leng, and Jugenheimer, 1952), but as total protein increased, the percentage of essential amino acids decreased. A breakthrough was needed to reveal the possibility of increasing amino acid composition in cereal grains.

Such a breakthrough occurred with the discovery of mutant genes known as "opaque" and "floury" (Mertz, Bates, and Nelson, 1964). When genes governing these characteristics are present in the homozygous recessive state, the usual hard, vitreous corn kernels become soft in texture and with a faded, dull color. The opaque mutants were so named because, in contrast to the transluscence of kernels in normal corn, their endosperm does not transmit light; the floury mutants were named for their soft, fine-textured kernels, also present in the opaque mutant.

A second mutant in the opaque group, called opaque-2, was able to increase the lysine content of an ordinary strain of corn by as much as 69%. Increases in tryptophan and several other amino acids accompanied the lysine increase, but some amimo acids decreased. Table 12-1 lists the amino acids present in corn and the changes associated with opaque-2 genes compared to normal corn. A gene called floury-2 was discovered to induce properties similar to those of opaque-2 (Nelson, Mertz, and Bates, 1965).

These genes and associated visual markers have proved to be of great importance in breeding programs, as they provide a reliable

Table 12–1
Amino Acid Content Comparisons of Opaque or High-Lysine Corn with Normal Corn Expressed in Percentage

Amino Acid	Opaque-2	Normal
Lysine	3.39	2.00
Tryptophan[a]		
Histidine	3.35	2.82
Amide ammonia	3.41	3.28
Arginine	5.10	3.76
Aspartic acid	8.45	6.17
Glutamic acid	19.13	21.30
Threonine	3.91	3.48
Serine	4.99	5.17
Proline	9.36	9.67
Glycine	4.02	3.24
Alanine	6.99	8.13
Valine	4.98	4.68
Cystine	2.35	1.79
Methionine	2.00	2.83
Isoleucine	3.91	3.82
Leucine	11.63	14.29
Tyrosine	4.71	5.26
Phenylalanine	4.96	5.29

Source: Mertz, Bates, and Nelson (1964).
[a]Tryptophan values are not given because it is destroyed by the hydrolyzing agents used to break protein down into its amino acid units.

and distinct marker to help trace the presence of the genes associated with improved amino acid composition. Time-consuming, expensive, and demanding analysis or biological assays normally associated with amino acid determinations are unnecessary.

The ability to develop corn with an improved amino acid composition was a giant step forward in protein nutrition in areas where corn comprises the bulk of the diet. If grain from corn cultivars containing opaque-2 or floury-2 genes were used in the diet, the essential amino acids for adequate growth could be available (Mertz and Bates, 1965). It is possible that such corn cultivars could be used almost exclusively as food or feed for monogastric animals (humans, hogs, chickens) without supplementation and meet their needs for essential amino acids, proteins, fats, and carbohydrates. Production problems have prevented the widespread adaptation of high-lysine corn cultivars.

To demonstrate the improved biological value of high-lysine corn hybrids, Harpstead (1971) conducted feeding trials with pigs. In the early growth stage, pigs grew 3.5 times faster on high-lysine corn than on normal corn as the sole protein source. The feed ratio for the high-lysine corn was 3.3 to 1, compared to 7.0 to 1 for normal corn—i.e., 3.3 or 7.0 units of feed were required per unit of weight gain. In comparison with normal corn rations balanced with soybean meal, growth with high-lysine corn was equal. In the finishing stage of development of the pigs, a 50% faster growth rate was observed than on normal corn without a supplement. High-lysine corn is of no benefit to ruminant animals, which can produce protein in their digestive system from nitrogenous sources.

When high-lysine corn was given to children showing protein malnutrition, rapid recovery from protein-deficiency symptoms was observed (Harpstead, 1971).

PROTEIN QUALITY IN CEREALS OTHER THAN CORN

On the basis of the Law of Homologous Series, high-lysine cultivars parallel to the improved amino acid composition found in corn should exist. On this basis, considerable effort has been directed toward genetic improvement of protein quality in cereals. In cereals in which poor amino acid quality existed, improvements generally have been made and are reviewed here.

IN WHEAT. Investigations into a broadened amino acid composition were begun in 1966. The United States Department of Agriculture world wheat collection was screened for protein and lysine differences using a chemical analyzer (Johnson, Mattern, and Schmidt, 1972).

Success has been somewhat limited compared to the corn crop. Analysis of more than 20,000 hexaploid, tetraploid, and diploid accessions indicated that genetic variation for total grain protein existed for at least 5 percentage points in hexaploid wheat. Genetic variation for lysine, however, was limited, amounting to no more than 0.5 percentage point (Johnson and Mattern, 1979). Likewise Villegas, McDonald, and Gilles (1968) found little variation in lysine content and reported the highest lysine level in durum to be 3.4% and in bread wheat 3.03%. Generally, lysine in wheat ranges from 2.55% to 3.74%.

IN RYE. Total protein content of rye varies from 6.5% to 15.0%. Rye protein has a higher biological value than wheat protein, owing primarily to a higher lysine content. Lysine content in rye ranges from 2.42% to 4.26%.

IN TRITICALE. In some instances triticale, produced from a wheat-rye combination, combines the high total protein content of wheat and the high lysine content of rye. For 191 cultivars of triticale grown in Mexico in 1969–1970, Zillinsky and Borlaug (1971) reported total protein contents of 12.8% to 17.9%.

IN BARLEY. A barley mutant was found at Svalof, Sweden, with high lysine content called Hiproly, a two-row barley from Ethiopia and selected from the world barley collection (Munck et al., 1970). The discovery of additional cultivars from this region suggests a mutational origin of the high-lysine gene (Walker, 1973). Numerous analyses have confirmed the increase in lysine in the cultivar Hiproly as 20% to 30% and with high biological value. Actual values are 4.1% for lysine and 19.6% for total protein. In addition to lysine, threonine, valine, methionine, isoleucine, alanine, glycine, and aspartic acid were higher in Hiproly barley, whereas cystine, glutamic acid, proline, and ammonia were lower when compared to other cultivars. Hiproly barley, however, needs agronomic improvement in order to be acceptable in modern agriculture.

IN OAT. Examination of oat seed has revealed that this cereal already possesses protein with a lysine level of about 3.8% (Kent, 1957). Despite a favorable amino acid composition, the high husk content and consequently low energy value of oat seed has produced relatively little enthusiasm for the crop, other than as a feed crop.

IN SORGHUM. Seeds from over 9000 accessions in the world collection were examined phenotypically to identify floury endosperm for subsequent chemical analysis for lysine content (Singh and Axtell, 1973). Two floury cultivars of Ethiopian origin had improved lysine levels of 3.34% and 3.13% at an overall protein level of 15.7% and 17.7%, respectively. The high-lysine gene altered the amino acid pattern relative to normal endosperm by increased lysine, arginine, aspartic acid, glycine, and tryptophan concentrations and decreased amounts of glutamic acid, proline, alanine, and leucine.

The biological value of the high-lysine cultivars was much higher than that of average sorghum cultivars as determined by feed trials with rats.

The importance of nutritional improvement of grain sorghum can be emphasized by the fact that sorghum is the fourth most important cereal crop in the world (Singh and Axtell, 1973), consumed often as the primary food in the diet in areas of Asia and Africa. Lysine is recognized as the first limiting amino acid in normal sorghum, which is characterized generally by low nutritional quality. Several human diseases have been associated with the poor nutritional quality of sorghum. Populations subsisting principally on sorghum may exhibit **pellagra** symptoms, associated with an imbalance in the leucine : isoleucine ratio.

Wide differences in the digestibility of total protein in sorghum cultivars have been reported that appear to be independent of amino acid composition. The differences may be due to phenolic compounds which, if not actually toxic, block nutrition. Existing genetic variability offers opportunity and considerable scope to improve the quality aspects of this crop.

IN PEARL MILLET. Like sorghum, pearl millet grain is low in nutritional quality. Its nutritive value has not received extensive attention, so the data available are limited (Kumar, Gupta, and Andrews, 1983).

Pearl millet is grown extensively as a grain crop for human consumption in the semiarid regions of Africa and Asia as well as for use as a fodder crop, and it ranks among the six leading cereals in providing calories and protein for humans (Evans, 1975). The crop generally is grown on poor, depleted soils, has low protein levels, and is deficient in lysine, valine, and histidine (Vasi and Chintalapudi, 1979). The range in variation of four essential amino acids was reported to be: lysine 1.8% to 4.2%; methionine, 1.4% to 2.6%; tryptophan, 1.1% to 2.4%; and threonine, 2.5% to 6.9% as reported from several sources by Rachie and Majmudar (1980). The genetic variation for amino acid composition appears to exist, and the importance of improving amino acid quantity and quality in areas where millet is the major food source cannot be overemphasized.

IN RICE. As much as 80% of the total calorie intake for millions of people in Asia consists of rice. Brown rice, with outer bran layers intact, contains about 8% protein; milled rice, with bran layers removed, contains about 7% (Juliano, 1972). Whole brown rice is nutritionally best among cereal proteins; the problems are that rice is inherently low in total protein and that a large percentage of the amino acids are removed in milling to produce a white rice product.

Starting in 1966, the IRRI world rice collection of rice cultivars was screened to identify genetically high-protein cultivars to be used in breeding programs (Beachell, Khush, and Juliano, 1972). Of the 7760 cultivars examined, total protein values ranged from 5% to 17%, with an overall mean of 10.6%. Selection of progeny for total protein from high-protein cultivars resulted in a decrease in lysine and tryptophan values. Increases in protein caused by either genetic or environmental factors had comparable effects. High-protein cultivars of rice that ranged from 11.4% to 11.8% total protein contained 3.3% to 4.0% lysine; 3.4% to 4.1% threonine; and 0.9% to 1.2% tryptophan. Progress in protein quality is possible but not as dramatic as with marker genes, such as in corn.

GENERAL ASPECTS OF IMPROVING PROTEIN CONTENT

Three major problems exist in breeding plants high in protein (Burrows, 1977):

1. As selection pressure for high total protein is increased, grain yield has been observed to decrease. This observation has been accepted by many as inevitable, but Johnson and Mattern (1979, pp. 40–41) found strong evidence that true genetic differences in protein exist in wheat and that increases in protein by breeding can be accomplished without grain-yield loss. The strong environmental influence on both yield and protein necessitates simultaneous monitoring of both traits for breeding progress. Selection for one trait, without attention to the other, will most certainly perpetuate the belief that increased grain protein can be achieved only at the expense of yield.
2. As the nutritionally inferior alcohol-soluble protein fraction that is low in essential amino acids is increased by efforts to increase total protein, a decline in the nutritional quality of the protein occurs. For crops with a high content of alcohol-soluble protein, it may be better to breed for high-yield potential than for increased protein content. An exception to this rule may be the oat crop.
3. Environmental variables that influence seed protein content make selection for genetic increases in protein difficult.

In general, genetic improvement of protein quality has been successful. The use of high-lysine cultivars of corn and barley has produced some problems, however. Softness and floury texture associated with high-lysine cultivars is associated with vulnerability to insect attack both in the field and in storage. The soft kernels are easily broken, and when milled, the flour does not have the functional quality of standard cultivars. Grain corn yields averaged 6% to 11% lower than normal corn because of lower kernel density (Lambert, Alexander, and Dudley, 1969; Sreeramulu and Bauman, 1970; Wessel-Beaver and Lambert, 1982). Generally, kernels with opaque-2 genotype weigh 10% to 15% less than normal kernels. To gain public acceptance, improvements in yield and harder seed are required, or new milling and/or processing techniques must be developed (Martin et al., 1982). Since the discovery of high-lysine corn cultivars, advances in texture and yield have been slow.

Protein quantity in cereals is influenced by climate, soil, and management practices. Of the climate variables, rainfall and temperature are most significant. Protein content tends to be higher in crops grown in hot dry conditions than in cool moist climates. Low soil water availability may restrict growth and allow more photosynthate to be directed and differentiated into protein. If drought coincides with grain-filling period, differentiation into protein can raise protein level.

The main soil factor influencing protein content is nitrogen. When nitrogen is abundant, the protein content of the seed increases. Management practices include nitrogen fertility levels. If applied to the soil at planting time, nitrogen will promote rapid and abundant growth, which intercepts sunlight energy and produces photosynthate used for increased grain yield. Nitrogen levels over and above that used for growth may contribute to increased protein content. Too much nitrogen can stimulate an excess of vegetative growth, and lodging will occur, thereby reducing crop quality.

It is possible to increase total grain protein by a foliar spray on the leaves at flowering time. Any yield increase by such an application will be modest and likely related to an increase in seed size, because tillering already has taken place and head size has been established. In one test, total protein of wheat grain increased from 14.0% to 18.8% when 50 kg N/ha (45 lb N/ac) was applied (Dubetz and Gardiner, 1980). Excessive foliar nitrogen applications may cause leaf burn damage. Surface vehicles will cause tramping of the grain unless **tramlines** are used.

Total protein of irrigated cereal crops can be increased by management. If water is withheld and stress conditions result, protein may be increased by differentiation of photosynthate. If water stress is too severe, shrunken kernels may reduce yield, but if fertilizer rates and irrigation are managed properly, both yield and protein content can be maximized.

The cost of producing high-protein cereal grains is high. If high-protein crops are desired, producers may be encouraged to produce them by an economic incentive based on protein content. For milling purposes, it is unnecessary and perhaps even undesirable to use wheat that contains more than 18% protein. Protein levels above 18% may be desired for other food and feed purposes.

In areas where adequate alternate protein sources are available or where supplementation with lysine is possible, cereals should be produced under management practices inducive to high grain yield. Where cereals form the main source of dietary protein, efforts aimed at increased protein quality and quantity are justified.

TEST WEIGHT OF CEREALS AS A QUALITY FACTOR

The weight of a measured volume of grain, known as test weight, is a traditional quality factor in grain grading. On the metric scale, test weight is expressed as kilograms per hectoliter; on the imperial scale, in pounds per bushel.

Traditionally, low test weight in wheat has been associated with low flour yield (Yamazaki and Briggle, 1969). Such a relationship appears to hold true when low test weight is a result of stress factors that cause kernels to shrivel. Some soft wheat cultivars, however, may exhibit a genetically controlled low test weight together with acceptable flour yield, provided kernels are otherwise normally developed. This has caused some excellent wheat samples to be downgraded.

To resolve the controversial association between test weight and flour yield, Ghaderi, Everson, and Yamazaki (1971) conducted test weight-kernel density relationship studies. They found that test weight is the product of kernel density, and the absolute volume of grain occupying the container. The volume of the grain, when expressed as percentage of the volume of the container, is referred to as packing efficiency.

Packing efficiency was found to be closely related to test weight. Grain shape and surface characteristics such as humping, dorsal or lateral depressions, and wrinkling can result in variations in the void-space ratio and thus random packing efficiency. These surface and shape characteristics are controlled genetically among cultivars and may be modified by the environment. The length-width ratio of the kernel had no relationship with kernel volume.

Kernel density is not associated with the genotype of soft wheats but is affected by the rate and extent of kernel filling and drying. Kernel filling in turn is influenced by environmental factors including alternate wetting and drying.

Variations in density are due, in part, to air spaces within the kernel. In the endosperm cells, granular starch particles of diverse size have a characteristic packing factor and porosity. Kernel weight can be increased when protein fills the interstices between the granules. In soft wheat, which characteristically has a low protein content of about 7.5% to 9.5%, there may be insufficient protein to fill the void spaces or interstices. Low protein content in soft wheats tends to be associated with air spaces in the kernel, which lowers grain density. In hard wheat cultivars with a high protein level this problem does not exist.

Another test of quality is the weight of a given number of kernels, usually 1000. This kernel weight is closely related to test weight, but there appear to be no genes in common controlling test weight and kernel weight in the soft wheat cultivars studied. The conclusion is that the environmental factors that affect kernel weight also affect test weight.

Industrial processing firms and grading institutions must recognize that some low-protein wheat cultivars may have a low test weight without an invariably low flour yield.

PREHARVEST SPROUTING DAMAGE

A widespread problem in any of the cereal crops is preharvest germination, especially under humid, warm, moist conditions. The net result of this unwanted sprouting is a yield loss during threshing and a reduction in grade due to reduced test weight—a direct loss to the grower. For the processor, sprouting is associated with the formation of alpha-amylase, which along with other enzymes reduces the suitability of the grain for malt, bread, or pastry purposes. Preharvest sprouting that adversely affects marketability, grade, net return to the producer, processing properties, and quality of consumer products is therefore a major consideration in cereal production in many geographic regions.

Preharvest sprout damage commonly occurs where rain and humid conditions prevail near harvest time. Some crops and crop cultivars are more vulnerable than others. Wheat cultivars, for example, grown in northern Europe, in some regions of Chile, Argentina, Brazil, eastern New Zealand, South Africa, Zimbabwe, Kenya, and the Great Plains region of North America periodically suffer sprouting damage. In the Pacific Northwest and eastern wheat regions of the United States and Canada, and in the northern wheat belt of Australia, an extremely high risk of preharvest sprouting damage prevails among vulnerable white wheats because of environmental conditions inducive to sprouting (Derera, 1980).

When water is imbibed into the cereal grain, the germination process is initiated immediately. Germination involves a series of steps normally occurring prior to the emergence of the radicle from the germ. The steps involve water uptake, activation of enzymes such as amylases and hormones such as **gibberellins,** and the subsequent release of energy to initiate growth (Mayer and Shain, 1974).

The amount of germination activity and sprouting may vary from that which doesn't show at all to severe damage that is readily apparent (Figure 12-1). Purchase of grain at elevator centers may involve an assessment of the degree of sprouting either visually or by the use of the Hagberg falling number test (Hagberg, 1960). This is a relatively simple and quick test that is widely accepted both by the processing industry and grain elevator centers. It involves the detection of low levels of alpha-amylase with an apparatus that compares the brightness of light passing through a suspension (Campbell, 1980; Kruger and Tipples, 1981). Little relationship was found between visual evaluations of sprouting and measured levels of alpha-amylase (Kruger and Tipples, 1982). Even a slight rain shower can cause bleaching of the

Figure 12–1. Unwanted preharvest germination is a widespread problem among cereal crops produced in humid, warm, and moist regions of the world. In addition to harvest losses during threshing and a reduction in grade, activation of enzymes and hormones occurs. Enzyme activity initiates the conversion of the starchy endosperm to sugar, rendering the grain unsuitable for bread, pastry, or malting purposes. Sprouting damage may range from the extreme level in this photograph to one that does not show any visual damage. (Courtesy of Ontario Ministry of Agriculture and Food)

grain, and the vitreous texture of wheat may become more mealy. Small amounts of moisture intake will initiate hormonal activity, and the amount of amylases will increase rapidly. Cereals with too high an amylase content for use in industry are sold as feed, where the increased amylase activity is not regarded as detrimental.

Sprouting reduces wheat flour qualities, resulting in low-grade breads and pastries regardless of the baking method used. Rain damage on durum wheat reduced appearance, but the quality of **paste products** was unaffected.

In barley and sorghum used for malt production, preharvest sprouting reduces malt quality. In the production of malt from these crops, rapid and uniform germination is desirable to produce maximum alpha-amylase. Sprouting in the field may mean that these crops cannot be used to produce malt, as it will germinate and malt unevenly, be susceptible to storage damage, and may not have the required longevity after malting is completed (Derera, 1980).

Triticale is very sensitive to preharvest sprouting and may sprout prior to harvest even without rain.

PREVENTION OF UNDESIRABLE PREHARVEST SPROUTING

Attempts to inhibit sprouting are complicated by the fact that germination may be controlled in many different ways (Taylorson and Hendricks, 1977). Breeding for sprouting-resistant cultivars is the most obvious solution. A misconception commonly exists, however, that preharvest sprouting resistance is due predominantly to **dormancy** of the caryopsis (Harrington, 1949; Belderok, 1976; Piech, Ruszkowski, and Jaworska, 1970) and that cultivars possessing red pericarp may exhibit dormancy (Derera, 1973; Everson and Hart, 1961; Gfeller and Svejda, 1960; Moss, Derera, and Balaam, 1972). Such a view appears to be based on incomplete evidence. More complete evidence suggests that preharvest sprout damage resistance cannot be related solely to dormancy. Dormancy may be related to the sensitivity of the seed to gibberellin, differences among cultivars in producing alpha-amylase, germination inhibitors in the glumes, and genotypic differences in embryo response to the inhibitors that determine the degree and extent of preharvest sprout resistance. Germination inhibition, therefore, may be related to a number of internal factors in the seed as well as external factors.

Instead of referring to dormancy as preharvest sprouting resistance, Derera (personal communication) suggested it be denoted as *robur contra pluvium* (RCP) or resistance to rain.

WHEAT QUALITY ASPECTS

Wheat can be categorized according to three distinguishing features:

1. Growth habit—spring or winter.

2. Kernel color—red, white, or amber.
3. Kernel texture—hard or soft.

Kernel color and texture are criteria of quality, and five general classes of wheat can be described:

1. Hard red winter.
2. Hard red spring.
3. Durum (spring).
4. Soft white winter.
5. Soft red winter.

Depending on the genotype as well as the environment and resulting level of gluten-forming protein, three types of flour are produced: hard, medium hard, and soft. These may be milled separately or blended to produce **all purpose wheat flour** and **pastry flour.** A fourth type, **whole wheat flour,** may be blended to produce a satisfactory loaf volume and still retain the natural goodness of the whole grain.

Gluten strength determines flour use. Protein and related gluten level cannot be genetically fixed in wheat any more than grain yield can be fixed. Protein content and yield are biological variables, and the environment in which the wheat is grown is the major influence on both traits. The genetic potential for high protein, like high yield, can be genetically incorporated into cultivars. High-protein cultivars have more efficient and complete translocation of nitrogen from the plant to the grain. They do not have a greater uptake of nitrogen by roots (Johnson, Mattern, and Schmidt, 1967).

Physically hard wheat yields a strong flour, high in gluten and other forms of protein. The desirability of gluten and other proteins arises from its favorable effects on loaf volume and texture. High-quality hard wheats are produced in areas having a warm and fairly dry growing season, so that gluten and protein are differentiated, but probably with a sacrifice to total grain yield. Canada and the United States are the main producers of hard and medium-hard wheat. Milling wheat into flour reduces the protein content below that of the whole grain. Protein reductions range from 0.5 to 1.5 percentage points or roughly 10%.

Soft wheat is produced under conditions of adequate rainfall or irrigation water that promotes high grain yield and limits differentiation. Most wheat grown in Western Europe and in the high-rainfall areas of North America is of the soft type. Variation in quality is expected, however, and a common practice to standardize dough strength is to add a high-quality hard wheat.

Wheat growers select cultivars and management treatments to minimize variation in quality. Additional nitrogen fertilizer applications on hard red spring wheat in North Dakota increased grain protein 1.6 to 2.1 percentage points, depending on initial soil fertility levels (Syltie, Dahnke, and Harrold, 1982).

Hard wheat is preferred for bread flour largely because of its relatively high protein content and desirable gluten quality and because it yields flour of a desired granular character. Bread flour and water form a glutenous elastic mass of dough which captures carbon dioxide produced from yeast and causes the bread to rise to produce a **leavened** product. Bread flour can also be used in the production of unleavened or flat bread.

Quality bread flour has 11% to 14% total protein. Pure white flour is preferred because it is associated with purity, but this also means that the more nutritious bran and coarser fractions of the wheat kernel are removed. About 72% of a wheat kernel can be used as flour; the remaining 28%, consisting of bran and shorts, it used as livestock feed. Possibly white bread, once served only to royalty and the very rich, represents a certain status symbol, now attainable by all. A high dependence on bread with its inadequate nutritional properties can lead to nutritional disorders.

Cake, cookie, and pastry flour is milled from soft red or soft white wheats lower in protein than those used for bread flour. A gentle action is required in milling to produce flour having smaller average particle size, lower ash, lower protein, and lower starch damage than flour milled from hard wheats (Figure 12-

Figure 12–2. Pastry flour is produced from soft wheat and has a protein content of 8% to 11%. The weak gluten flour is used to produce cookies and cakes, ice cream cones, pretzels, crackers, and breakfast foods. (Courtesy of Ontario Ministry of Agriculture and Food)

2). Soft wheat flour has 8% to 11% protein. A common product made from pastry flour is the traditional pie, consisting of a pastry shell filled with fruit or custard (Figure 12-3).

Semolina is produced from durum wheat used in the manufacture of paste or pasta products. A distinction is made between semolina and flour, both of which can be milled from durum wheat. Semolina has larger particles than flour and will pass through a 30-mesh sieve but not a 100-mesh sieve.

Pasta products are produced from a basic mixture of semolina and water and formed into specific shapes, then either immediately cooked and eaten or dried for consumption at a later time. Typical products (Figure 12-4) include:

1. Macaroni, produced as hollow tubes of various shapes and diameters.
2. Spaghetti, produced as solid rods, generally of small but variable diameter.
3. Noodles, either produced as flat strips (rolled and cut) or extruded as oval rods.
4. Shapes of stars, letters of the alphabet, Greek symbols, shells, and other forms, stamped from sheets of dough.

An important quality factor of durum wheat is color. Semolina color was found to be a highly heritable trait controlled primarily by additive gene effects (Johnston, Quick, and Hammond, 1983). Improvement of semolina color was believed to be possible by early-generation selection, although the intensity of expression

Figure 12–3. As the name implies, pastry flour is used to produce pastry products such as pie crusts. Pie crust can be filled with a variety of ingredients and is eaten either as a sweet dessert, as pictured, or as a main course. (Courtesy of Ontario Ministry of Agriculture and Food)

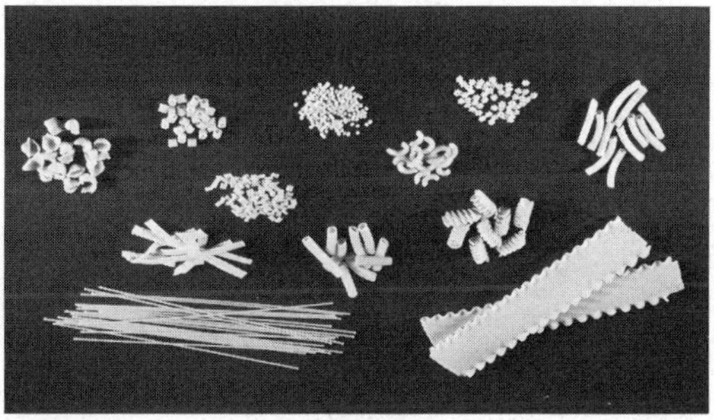

Figure 12–4. Semolina, or coarse-textured durum wheat flour, is used to produce pasta products known as macaroni, spaghetti, or noodles. These products are produced in a variety of shapes and sizes for use in the preparation of many specialty dishes. Pasta products can be stored because they are in a dry, stable form. (Courtesy of Ontario Ministry of Agriculture and Food)

was influenced by environment. Durum kernels may range in color from white to orange-yellow to red. Manufacturers want macaroni products that are deep amber or golden and may blend different durums to get the color they want. Durums that are white, grey, or mottled are rejected. The desirable yellow color results from a high level of the enzyme lipoxidase. Quality characteristics of specific cultivars can be determined by a high carotene and a low lipoxidase content.

Durum wheats with a high percentage of strong gluten, and hence a high protein content of about 13.5% to 14.0%, are desired so that the product does not disintegrate if overcooked.

Flour yields from durum wheat average 58% compared to 72% for bread wheat flour, based on cleaned wheat. Large durum wheat kernels are desired, because large kernels have a high proportion of endosperm to bran and hence a higher yield of semolina.

Yellow berry is a condition in which the normally hard vitreous starch is present as floury starch and gives the kernel a pale or white discolored appearance. The more yellow berry a kernel has, the higher the proportion of flour and the lower the semolina yield. Yellow berry occurs in durum, bread wheats, and triticale and adversely affects milling and baking quality.

Total protein levels in kernels with yellow berry are 1 to 1.5 percentage points lower than in normal kernels. Sharma, Paul, and Bietz (1983) reviewed the literature on yellow berry and reported the cause to be nitrogen deficiency and excess soil moisture. Waines et al. (1978) reported 9% to 28% less protein, higher lysine and threonine, and lower glutamic acid levels in kernels with yellow berry compared to normal wheat kernels. Increased nitrogen fertilizer in triticale reduced the incidence of yellow berry (Sharma, Paul, and Bietz, 1983). Some triticale cultivars produce more yellow berry kernels, implicating genetic factors in yellow berry expression.

BREAD

The factor that places wheat in an outstanding position among the crops of the world is its bread-making quality. Milling and baking quality refers primarily to the suitability of wheat cultivars for a highly mechanized food-processing industry and has little to do with nutritional quality.

Improving the amino acid composition of wheat protein may not be entirely compatible with accepted standards of bread wheat processing quality of the western world. High-grade white flour is composed largely of kernel endosperm. Endosperm protein levels are low and relatively poor in lysine (2%) compared with the nonendosperm protein (over 4%). Since the nonendosperm proteins are mostly eliminated from wheat flour during milling, increasing their lysine content would not change much the lysine content of wheat flour.

Whole wheat flour, cracked wheat, flaked wheat, and **farina** which have bran and germ included and a superior nutritional value, are sold to a rather limited extent because the stability of these products is limited. Bran and germ oils will become rancid unless the product is specially processed.

Wheat, and to a lesser degree rye and triticale, are the only cereal grains yielding flour from the endosperm capable of trapping carbon dioxide liberated in a mass of fermenting dough and resulting in the porous structure associated with leavened bakery products (Figure 12-5). Wheat breads can be described as having a light crumb and golden-brown crust. The crumb structure is fine, silky, and soft-textured, with a mild, yet aromatic, delectable taste. White flour breads can be eaten with zesty as well as sweet foods.

Rye flour is darker, heavier, and coarser than wheat flour. Rye bread is less digestible and less attractive in appearance than wheat bread because of it coarse texture and darker crust. Rye bread has a pronounced aromatic-sour taste, which stems from the multistage **sour dough** fermentation or from the ready-to-use dry sours method of production.

Triticale is intermediate between rye and wheat in its bread-making properties. Triticale has yet to establish itself as a bread-making crop because it is relatively new.

Figure 12–5. Wheat, and to a lesser extent rye and triticale, are the only cereal grains yielding flour from the endosperm capable of trapping carbon dioxide liberated in a mass of fermenting dough. The result is a porous structure associated with leavened bakery products. White pan bread (upper left) is characterized as having large pores and soft texture with a mild yet aromatic, delectable taste. The French loaf in the center is made from a flour lower in gluten, resulting in a thick chewy crust and full-bodied flavor. The whole wheat on the upper right is light brown in color. The slices of pumpernickel are produced from rye meal and dark rye flour with sour dough. Pumpernickel has a typically deep brown color produced by low baking temperatures that produce numerous browning reactions. The rye bread on the lower left has a very dark, black crust, heavy texture, and a pronounced aromatic sour taste. (Courtesy of Ontario Ministry of Agriculture and Food)

Special-grain breads may have flour or meal of grains such as barley and oat added to wheat flour. By blending wheat and rye, and perhaps wheat and triticale, triticale and rye, or possibly wheat-rye-triticale, numerous composite flours can be made. Composite flours may be used to boost nutritional quality, to reduce ingredient costs, to extend a limited wheat supply, to improve use of nonwheat flour plants operating below capacity, to use a nation's nonwheat crop surplus, and to save foreign exchange by reducing wheat importation.

Composite flours can result in a variety of products. In West Germany, Pomeranz (1983) reported on approximately 200 truly different types of bread. The 200 types can be divided into four classes based on flour composition:

1. Wheat breads with a minimum 90% wheat components.
2. Mixed breads with 51%–89% wheat components.
3. Mixed-rye breads with 51%–89% rye components.
4. Rye breads with a minimum of 90% rye components.

Specialty breads are distinguished by the addition of special, eye-catching, taste-governing and enhancing additives such as raisins, poppyseed, soft cheese, sesame seeds, peel, or nuts.

Pumpernickel bread is produced from rye meal and dark rye flour with sour dough and is baked slowly at low temperatures that produce numerous browning reactions. The color is typically deep brown.

French bread is characterized by a crisp, eggshell crust, 3 to 4 mm thick, and an overall, thick chewy crust (Tweed, 1983). The loaf has an open grain, a full-bodied flavor, and characteristically is produced with seven lateral cuts on the upper surface.

Unleavened or flat bread, produced without benefit of yeast and fermentation, is the typical bread consumed in the Middle East. In some areas where unleavened bread is consumed, zinc deficiencies have been observed among humans, despite intakes that equal or exceed those of populations where deficiency symptoms are unknown (Reinhold, 1972). In its fully developed form, human zinc deficiency is characterized by failure to undergo the sexual development and growth spurt associated with puberty. Reinhold (1972) concluded that leavening and fermentation reduce the concentration of phytate (phytic acid), which can render zinc unavailable. In areas where unleavened bread is consumed extensively, phytic acid is an important quality characteristic.

BREAD MAKING

Gluten gives flour the capacity to absorb water and, when yeast is added, causes the dough to rise. Bakers wishing to produce more loaves per unit of wheat seek flours with a high protein and associated high gluten content. As well, bread made from high-protein flour, described in the trade as hard, strong flour, tends to stay fresh longer. French bread, which is made from soft, weak flour, low in protein and gluten, is a delight when eaten fresh but has the disadvantage of staling quickly and should be purchased and consumed daily. White pan bread is produced in North America primarily from a high-protein, hard, strong flour. In other countries, such as Holland, the United Kingdom, and Germany, bread is produced from blends of soft, locally grown wheat flour with hard, high-protein wheat (often imported) flour, which helps produce a satisfactory loaf that keeps longer.

During fermentation, gluten undergoes certain physical and chemical changes whereby it takes on a desirable elastic or extensible nature. In bread making, several processes have been developed as follows:

1. The *straight dough method* refers to any process where all the ingredients are combined and mixed into a dough in one step. The dough is then allowed to rest and ferment under favorably warm conditions for a period of time before it is divided and baked. The dough may be punched or knocked back once or twice to remove excess carbon dioxide produced during fermentation. The straight dough method is used widely by small bakeries around the world. Fermented doughs are used in over 80% of all bakery foods produced in the United States (Dubois, 1981).
2. The *continuous mix process*, developed in the United States, works the dough mechanically, using high-speed mixers to produce acceptable loaves of bread while eliminating the bulk fermentation stage characteristic of the straight dough method.
3. The *Chorleywood method* developed in the United Kingdom involves intense mechanical mixing of the dough. The high energy input associated with the high-speed mixing action produces a high loaf even from flour with low protein and gluten quality because of the inclusion of extra yeast, fat, and water by a more complete blending.

The Chorleywood method has had a substantial impact on the marketing of wheat, because it can produce a satisfactory leavened product by blending high-protein premium wheats with low-protein, lower-quality wheat. Blends based on very high-protein premium wheat allow for the inclusion of a greater quantity of low-protein domestic wheat. This has put pressure on exporting countries to market wheat on the basis of guaranteed minimum protein content. Canada adopted this method of marketing in 1971.

The Chorleywood method has caused a minor revolution in wheat production. A case in point is that of a cultivar of hard red spring wheat named Glenlea, developed at the University of Manitoba and licensed for production in Canada in 1972 (Evans et al., 1972). Because Glenlea was assessed as being inferior in quality to the standard of comparison, Marquis, it could not qualify for the Canada Western (CW) red spring class of wheat, and a utility class of wheat was established as a result. Glenlea is noted for its large kernels and superior yield and low price as a utility wheat. Large kernels give a high flour yield. Although Glenlea has inferior bread-making quality when used in conventional processes, it appears satisfactory in blends (Bushuk, 1980).

Wheat producers are faced with the difficult question of whether they should grow quality or quantity. Canada has historically produced high-quality wheat and demanded a premium price, but such cultivars may be less capable of producing top yields. In contrast, utility wheats, which are low in price, lacking in quality, but capable of a high grain yield in the field, offer an interesting alternative. New baking processes suggest that low-quality wheat may be of increasing acceptability.

4. *Chemical dough processes* achieve development of gluten by the combined action of a fast chemical reducing agent and a slower oxidizing agent in conjunction with standard dough mixing. This process, developed in Australia, also enables the use of weaker flours.

MALT PRODUCTION OF CEREAL CROPS

The production of malt from cereal grains involves the conversion of starch in the endosperm to sugars through enzyme actions associated with germination. In the process the hard and starchy kernel is transformed, by a series of chemical and physical processes, to malt, which is highly enzymatic, friable, and flavorsome. The purpose of producing malt is to release starch for brewing into alcoholic drinks and for other food uses such as sweeteners and flavorings. An additional objective is to produce enzymes that are not present in the ungerminated grain and that are necessary for the breakdown of malt and adjuncts during brewing.

Any cereal grain can be converted to malt, and over the centuries many cereals have been used to make alcoholic beverages. In temperate countries, where production of barley, wheat, and rye is abundant, these three crops are malted commercially. In hot, semiarid regions such as Africa, sorghum is the grain of preference for malting. Sorghum produces a more traditional brew, often lacking in clarity but which could be improved with research (Glennie, Harris, and Liebenberg, 1983).

Triticale can be malted successfully and produces a malt with high extract of good quality, but the amount of nitrogenous material in the brew is higher than from barley malt (Pomeranz, 1974). Beer produced from triticale is darker and has a higher *p*H value than barley beer.

The malting process can be outlined briefly in very simple terms. Cleaned and graded grain is immersed in water until the grain has reached a specified moisture level. After the water is drained off, the moisture retained in the grain initiates germination. Associated with germination is the production of a coleoptile or leaf initial, known in the malt trade as the acrospire, arising from the top of the embryo. At the base of the embryo, rootlets develop. As germination proceeds, the endosperm undergoes "modification" by enzymes (alpha and beta amylase), which hydrolyze the starch. When modification is sufficient, the mass is dried in hot, dry air to stop germination. After the dry and brittle rootlets and sprouts have been removed, the product is malt, a stable, storable commodity.

When crushed malt is mixed with warm water, the enzymes catalyze materials in the malt or in materials mixed with it to sugars and other simple soluble substances called *wort*. If yeast is added to the wort, fermentation converts the sugars into alcohol. The malt confers color, aroma, and flavor to the product, to which hops may be added for additional flavoring and preservation. The resulting brew is called beer.

Barley is the preferred grain for malt production for several reasons:

1. Only barley, wheat, and rye produce two enzymes, alpha and beta amylase, when germinated. The conversion of starch to dextrins and fermentable sugars is more efficient when combinations of the two enzymes are present.
2. Only barley has a caryopsis with a tightly adhering lemma and palea (hull). During the malting process, the acrospire or coleoptile grows beneath the lemma and pericarp, which prevents damage to the tender shoot. This is important, because at intervals the grain is mixed and turned to provide more uniform growth and to prevent the roots from matting together. If the acrospire is broken or damaged during the germination process, enzyme action ceases and starch conversion may be incomplete.
3. The hulls form a natural filter and aid in separating soluble materials from undissolved particles in the production of wort.
4. Steeped barley has a firmer texture than other grains and can be handled with less damage at high moisture levels.

Malt quality is determined by the quality of the raw grain and by the malting techniques used during processing. Economic considerations have forced the maltster to favor cultivars that will malt quickly, have a high germination percentage, and produce high extracts. On the basis of a number of physical and chemical criteria, various cultivars have been designated as having or not having acceptable malting quality.

QUALITY IN RICE

A popular innovation of the twentieth century is the milling of wheat and rice to remove the outer layers of the kernel to produce a white product. In rice this milling process is known as polishing or whitening.

Nearly all rice consumed as food is prepared by some type of milling operation which leaves part of the kernel intact rather than producing a flour or meal. After threshing, each rice grain is encased in an easily removed protective hull and is referred to as rough or paddy rice. The kernel with the hull removed has dark brown pericarp or bran layers covering the white endosperm and is referred to as brown rice. Under the bran layer is the aleurone layer, which is rich in protein.

Milling of rice removes the germ, bran, and aleurone layer. Rice bran consists mainly of the coarse outer bran coatings and the germ. The white inner bran and the aleurone layer are removed in the final stages of milling and together comprise the fraction known as rice polish. Milled or white rice is the finished product.

Rice is stored dried as paddy or rough rice and milled only when it is needed for immediate use or sale. When stored as paddy rice it can be protected by pesticide sprays or powders. Storage of up to three to four months helps improve cooking quality, in that the rice expands more and absorbs more water than freshly harvested rice, becoming more flaky. In Japan, where sticky rice is preferred, aging of rice is regarded as being inducive to deterioration.

Although, among cereals, rice protein has the best nutritional values, the protein value in milled rice is low. Removal of the outer layers of the kernel reduces the nutritional value of the product. The proportion of proteins, fats, and vitamins decreases, while the proportion of carbohydrate increases (Table 12-2).

Threshed or paddy rice consists of approximately 20% hulls, 8% bran, 2% polish, and 70% milled product. Milled rice may be sold directly for food or may be parboiled or converted into quick-cooking forms as a convenience item.

Rice flour is lacking in gluten and cannot be used for bread-making. It is used as a thickening agent because it produces a smooth-textured paste with a bland flavor.

Parboiled rice is produced by soaking rough or paddy rice in water and then pressure-cooking it to gelatinize the starch. Following this treatment the kernels are slowly dried and milled to remove the outer layers. During the steeping operation, water-soluble vitamin B compounds and minerals are infused into the endosperm from the bran, germ, and hull. Higher nutritional values may exist also because such rice is undermilled compared with rough rice, since it is more resistant to milling. Parboiled rice has a rubbery texture and can be milled without breaking the cooked kernels.

Changes associated with parboiling are mainly physical, and the process results in harder and more translucent grain. Protein bodies are disrupted, and the starch granules are completely gelatinized. Rice is parboiled in many Asian countries.

Table 12–2
Composition of Rice and Rice Products

	Brown Rice	*Rice Bran*	*Rice Polish*	*Milled Rice*
Carbohydrates, %	87.2	46.6	66.8	91.5
Protein, %	8.3	14.6	13.2	7.6
Fat, %	2.0	13.4	10.7	0.3
B vitamins, µg/g				
Thiamine	4.2	27.9	23.9	0.80
Niacin	47.2	408.6	384.7	18.1
Pyridoxine	10.3	32.1	30.8	4.5
Pantothenic acid	17.0	71.3	92.5	6.4
Riboflavin	0.53	2.22	1.34	0.26

Source: Matz (1970).

The advantages of parboiled rice over raw rice are that it is more nutritious, is more resistant to insect infestations, and does not disintegrate when used in soups or puddings. Parboiled rice is less likely to become mush when cooked than raw milled rice. Although parboiled rice is darker than raw milled rice and has a slightly different flavor, it is widely accepted and often preferred to white milled rice, except in areas where extreme whiteness is a measure of quality.

Although parboiling results in complete cooking, the rice is not a quick-cooking rice because of the density of the grains, which are slow to take up water. Parboiled rice takes nearly twice as long to cook as white rice.

Quick-cooking rice is made as a convenience food to be cooked in minutes. It is produced in a number of ways by precooking in water and drying under closely controlled conditions or with dry heat.

Eating of the nutritionally superior brown rice is rare, in part because of the high fuel consumption needed to cook it for 35 to 45 minutes to make it soft enough to eat. Boiling of brown rice takes two or three times as long as for white rice, and in fuel-deficient areas this is of major importance. Precooked rice can be prepared in as little as two to fifteen minutes.

A report of a light polishing process (Tainsh and Bursey, 1982) resulting in a pale golden rice of excellent taste and texture, which cooks as quickly as white rice and which loses only 1% instead of the usual 17% in weight in polishing, offers promise of better nutrition. The advantages of lightly milled golden rice include:

1. The milling machines use one-third the amount of energy required to mill a white rice because of their slow speed and short operation time.
2. The milling away of the outer waxy epicarp on the rice grains removes 65% of the total wax on the bran layers. The pale golden layer so exposed has very little wax and a hard surface. Removal of the oil and wax reduces the liability of rancidity compared to brown rice. The golden rice stores well, as the light polishing process does not cut into the remaining bran layers, leaving only the hard, low-wax layer exposed to the atmosphere.
3. Fewer kernels are broken in the gentle process, so that the product can be sold as "whole" grain rice directly without additional equipment for separation of brokens.
4. More rice remains for food consumption.

ENRICHMENT

To overcome severe deficiencies in B-complex vitamins that occur when highly polished rice is the main food source (80% of the caloric intake), enrichment with thiamine, niacin, and iron is common. These vitamins and minerals reduce the incidence of and mortality from **beriberi.** About 80% of the thiamine, 56% of the riboflavin, 65% of the niacin, 60% of the pantothenic acid, and 55% of the pyridoxine are removed when brown rice is milled. Milled rice grains can be enriched by coating some of the rice kernels with the enriching ingredients, adding a powder to the rice, or supplying an enriched wafer to be added when the rice is cooked.

Diets could be improved and the need for enrichment eliminated if consumers could be educated to consume brown rice. This is unlikely, as consumers adhere to their food habits with an irrational rigidity. Perhaps partial milling to produce the golden rice described by Tainsh and Bursey (1982) may prove of value. Parboiling can be considered a form of enrichment, as some of the vitamins and minerals are retained in the finished product.

SPECIFIC QUALITY ASPECTS OF RICE

Rice grain consists mainly of two types of starch: amylose, which makes cooked rice fluffy, and amylopectin, which makes cooked rice sticky. In predicting the cooking, eating, and processing characteristics of rice, amylose content is considered the single most important criterion. Cultivars with a high amylose content cook to a dry and fluffy consistency and have grains that tend to remain separate. In processing, they lose a relatively low percentage of solids and show minimal splitting and fraying of kernel edges and ends.

Waxy rice differs from normal rice in that its starch consists of nearly 100% amylopectin, which stains rust color with iodine solutions, in contrast to amylose starch which stains blue. Waxy rices are moist and chewy upon cooking, and grains tend to clump together. Waxy rice flour, however, produces an ideal thickening agent for white sauces, gravies, and puddings, because it does not separate when frozen, stored, and subsequently thawed.

Nonwaxy rices are translucent; waxy rices are opaque. Despite the unique cooking qualities, steamed waxy rice is a staple food in some areas of the world.

The alkali spreading score, which provides an estimate of gelatinization temperature, is an additional starch characteristic considered to be of prime importance to rice quality (McKenzie and Rutger, 1983).

The quality of rice must meet the needs of the consumer, which may vary with personal preferences, eating habits, and procedures for cooking. Rice is used:

1. As a boiled rice for direct consumption. This accounts for by far the majority of rice consumption throughout the world.
2. As canned rice included in soups, rice pudding, and Spanish rice.
3. As an "adjunct" in beer to serve as a source of carbohydrate in the brewing process.
4. As dry breakfast foods, such as puffed rice or rice flakes.
5. In baby foods.
6. For industrial uses, including rice starch and rice paper.
7. As rice flour to make an ideal thickener in sauces, gravies, and puddings.
8. In alcoholic beverages, such as rice wine (known as sake).

The concept of quality for steamed or boiled rice varies tremendously with consumer preferences. In Indonesia and the Philippines, rice that remains soft or moist on standing after cooking (as a result of 17% to 22% amylose in the starch) is preferred. A large grain size and a high degree of flakiness are desired in North Vietnam, Sri Lanka, Malaysia, Bangladesh, and parts of Thailand. In Laos and parts of Thailand, waxy rice is preferred. North Americans want rice to have a firm texture when cooked and for as long as possible while being kept warm for serving. In general, rice should not be sticky or clump together, have mushy or elastic kernels, or have a slimy fluid associated with the kernels.

Gelatinization temperature, a physical property of starch, is the range of temperatures within which the starch granules start to swell irreversibly in hot water. Final gelatinization temperatures range from 55° to 79°C (131° to 174°F). Medium-and short-grain rice cultivars grown in the southern part of the United States are low in amylose and have a lower gelatinization temperature (61° to 66°C or 142° to 150°F) than long-grain cultivars. Milled rice with a high protein content or a high gelatinization temperature requires more water and a longer time to cook than rice with a lower gelatinization temperature (Juliano, 1972).

Although rice is grown primarily for human consumption, by-products such as rice bran and polish provide excellent feed for livestock. Rice hull is of low nutritive value but can be fed to ruminant animals if adequately supplemented with protein.

QUALITY IN SORGHUM

Sorghum is often regarded as nutritionally inferior. This view has resulted from the presence of tannin in sorghums that are resistant to bird damage. Most sorghums in North America, however, do not contain condensed tannins. The nutritional value of North American sorghum for livestock feed is considered to be 95% or more the value of corn (Rooney, 1979).

Sorghum has many potential uses as a food, and it can be dry-processed into grits and other milled products for use in numerous items including beer, syrup, alcohol, baked foods, snacks, prepared breakfast foods, and tortillas. In milling, the pericarp and the germ are removed to produce an endosperm product that is light in color and has a bland taste. Sorghum with a white pericarp and testa is preferred.

Prototype products made with sorghum have been developed that look and taste good. Sorghum can be puffed, popped, extruded, shredded, and flaked to produce delightful snacks and breakfast food products. Until there is a change of sorghum's image as a crop "fit for feed only," and until adequate products of improved sorghums exist, such products will not be marketed (Rooney, 1979) in the industrial world.

CORN QUALITY

Corn is primarily a livestock feed crop, with about 85% to 90% of the total North American crop being fed on the farms where it is grown. The remaining corn is used to manufacture hundreds of edible (Figure 12-6) and nonedible products. Possibly more industrial products are produced from corn than any other grain.

Starch, the most prominent and important constituent of the corn kernel, normally makes up 72% of its dry weight (Table 12-3). The high starch content and the sugars present serve as a readily available and economic source of carbohydrates for industrial fermentation processes.

Figure 12–6. This display of commercial corn products resting on a bed of corn kernels, and with intact ears in the foreground, symbolizes the many products made from corn. Products shown include syrup, oil, starch, alcoholic beverages, snacks, breakfast foods, and citric acid. Additional products with a corn oil base are shown. (Courtesy of Ontario Ministry of Agriculture and Food)

Table 12–3
Average Dry-Matter Composition of the Corn Kernel

Component	*Percent*
Starch	72.0
Protein	10.0
Oil	4.8
Fiber	8.5
Sugars	3.0
Ash	1.7

Source: Neukom and Buchi (1979).

Alcohol, organic acids (e.g., citric acid), antibiotics, enzymes, and vitamins are produced by the action of yeasts, molds, or bacteria on the carbohydrate of corn. Bourbon whiskey in the United States and rye whiskey in Canada as well as many other distillation products are made principally from corn. A typical mash may contain 75% corn and 15% rye with 10% malt to provide enzymes for the conversion of starch to fermentable sugars.

Wet milling, in which the kernel is steeped for 30 to 40 hours in a mixture of water and sulfur dioxide, allows the endosperm to be separated from the hull (6%) and the germ (12%), and under a high-speed centrifuge the differential weights of starch and protein are separated. Pure starch is dried for dry-starch products or for conversion into syrup and sugar such as glucose, and high-fructose corn syrup to provide sweetness in many food products. High-fructose corn syrup has become a major success (Figure 12-7) since it was commercially introduced in the late 1960s and can replace completely sucrose from sugar beet or sugar cane (Anonymous, 1982). In addition to many food uses in confectionary dairy products, jams and jellies, brewing, baking, and soft drinks, these sugars are used in nonfood products, such as premixed concrete, gypsum wallboard, adhesives, and paper coatings (Craig, 1977).

The starch of waxy corn, like that of waxy rice, consists of branched-chain starch or amylopectin. Both starches can be used as

Figure 12–7. Corn starch can be converted into sweeteners such as glucose and dextrose for use in many food products. Corn sweeteners have proven to be a major success and can replace sucrose from sugar beet and sugar cane. (Courtesy of Ontario Ministry of Agriculture and Food)

thickeners separately or in combinations to obtain specific properties. Amylopectin is used in the glue on postage stamps and envelopes and helps to give the glossy finish on paper for magazines and books (Crookston, 1979). Starch from waxy corn has properties similar to **tapioca** starch. Waxy corn is grown under contract for specialty products for food or industrial uses.

Corn oil is extracted from the corn germ, which contains 4.5% oil. It is used principally as a salad or frying oil. Corn oil is valued because of its pleasant taste, resistance to the development of off-flavors, and excellent frying characteristics. A high content of polyunsaturated fat, consisting mainly of linoleic acid, and the possible desirability of polyunsaturated fat in the diet have reinforced the value of corn oil (Reiners, 1982).

Corn endosperm is used to manufacture a variety of corn products such as grits, flour, breakfast cereals and tortillas.

QUALITY IN TRITICALE

Triticale has yet to establish its role as a food crop in society. Preliminary observations using milling and baking techniques adopted from wheat have revealed that triticale has considerable promise in the production of "standard" products. With time, techniques and procedures may be adapted to suit the quality characteristics, and acceptable new products may be developed.

In dry milling, using procedures suited to wheat milling, flour yields from triticale averaged 60% to 64% as compared with 69% to 74% for spring and durum wheat (Bushuk and Larter, 1980). Modifications in milling techniques, however, did not change flour yield, which may be associated with the shrunken and wrinkled seeds of triticale.

In the milling process, a greater percentage of the protein of triticale is removed in the bran and shorts, and the protein advantage of triticale over wheat was generally lost when triticale was milled into flour used to produce white bread.

Because of the different properties associated with triticale flour compared to wheat flour, a satisfactory bread can be produced from triticale flour if modifications in procedures are made, such as a reduction in mixing speed. Batches must be prepared by the straight dough procedure, using a paddle instead of a hook for mixing to prevent sticking to the edge of the mixing bowl. Because of the inherently high alpha-amylase activity of triticale flour, optimal fermentation times are considerably less than for wheat doughs. Although bread of quality equal to that of bread baked with good-quality wheat flour was produced from a specific triticale cultivar (Lorenz, 1974a), generally the quality of pan bread from triticale is lower than that from bread wheat. Some consumers prefer the flavor of triticale bread, which has the characteristics of a very mild rye bread.

By blending triticale and wheat flours, acceptable bread can be produced from appropriate blends. Triticale bread was found to stale about twice as fast as wheat bread (Tsen, Hoover, and Farrell, 1973).

The use of 100% whole triticale meal to produce bread with the objective of maintaining the high-protein advantage of triticale grain over wheat was unsatisfactory (Lorenz, 1974b).

Breads baked with up to 30% triticale meal were acceptable, however, and had a higher protein content and a better amino acid balance than breads made from whole wheat flour.

Pancakes and waffles were of excellent quality and were preferred for flavor and taste. Cakes and cookies were acceptable only if modifications were made in milling and if emulsifiers and other ingredients were added (Bushuk and Larter, 1980). Breakfast cereals produced from triticale were satisfactory.

In the manufacture of tortillas and chapatti, triticale flour produced satisfactory results except for a darker color in the chapattis (Villegas, 1973).

The potential of triticale as a food product continues to be developed, and selected products can be purchased.

Questions

1. *a.* Select a typical day or meal in your life and consider what cereal products you consumed.
   b. How important are quality aspects in the cereal food products you consume regularly?
   c. Name foods that contain cereal flour that are not obvious to the casual observer.
2. Name ways in which pasta products can be served.
3. *a.* Contracts for the production of malting barley may call for controlled nitrogen fertilizer applications. Explain.
   b. Consider how high-nitrogen fertilizer applications would affect the quality of the following crops:
   (1) Bread wheat.
   (2) Corn for starch production.
   (3) Pastry wheat.
   (4) Durum wheat.
   (5) Rice.
4. Comment on the value of high-lysine corn:
   a. In a dairy or beef feeding system.
   b. In areas where 8000 to 10,000 food items can be found on the shelves in a grocery store.
5. "A popular innovation of the twentieth century is the milling of wheat and rice to remove the outer layers of the kernel to produce a white product." Comment on this statement in terms of:
   a. Factors causing change in this outlook.
   b. The impact of education.
   c. Enrichment.
   d. Prospects for the future.
6. List products manufactured from a specific cereal crop of your choice.
7. Consider triticale as a new food source from the following perspectives:
   a. Potential new products.
   b. Marketing of new products.
8. Discuss the statement that people stick to their food habits with an "irrational rigidity." Consider your own background in terms of traditional food habits maintained in your household.
9. In your view, are quality aspects exaggerated or realistic?
10. *a.* Should quality aspects be the responsibility of the plant breeder, the crop producer, or the processor?
    b. Consider the extent to which each can influence quality aspects.

References

Anonymous, 1982. "Cane, Beet Sugar Markets Cut by Corn Sweetners." *Crops and Soils* 34 (7): 12–13.

Beachell, H. M., G. S. Khush, and B. O. Juliano. 1972. "Breeding for High Protein Content in Rice" (In) *Rice Breeding.* Los Banos, Philippines: International Rice Research Institute. pp. 419–428.

Belderok, B. 1976. "Changes in the Seed Coat of Wheat Kernels During Dormancy and After-Ripening." *Cereal Research Communications* 4: 135–137.

Burrows, V. D. 1977. "The Future of Cereal Protein." *Seed Scoop* 25 (1): 1–2.

Bushuk, W. 1980. "The Baking Potential of Glenlea Wheat." *Canadian Journal of Plant Science* 60: 737–739.

———, and E. N. Larter. 1980. "Triticale: Production, Chemistry, and Technology." (In) Y. Pomeranz (ed.), *Advances in Cereal Science and Technology.* St. Paul, Minn.: American Association of Cereal Chemists, Inc. pp. 115–157.

Campbell, J. A. 1980. "A New Method for Detection of Sprout Damaged Wheat Using a Nephelometric Determination of Alpha-Amylase Activity." *Cereal Research Communications* 8: 107–113.

Craig, W. I., 1977. *Corn Sweetners.* Canadex 111.80. Agriculture Canada. 2 pp.

Crookston, R. K. 1979. "The Story of Waxy Corn." *Crops and Soils* 31 (9): 11–13.

Derera, N. F. 1973. "Should Red Wheat Be Introduced into the Northern Wheatbelt?" *Journal of the Australian Institute of Agriculture Science* 39: 48–50.

———. 1980. "The Audit of Sprouting." (In) M. D. Gale and V. Story (eds.), *Proceedings of the Second International Sprouting Symposium.* England, 1979. *Cereal Research Communications* 8: 15–22.

Dubetz, S., and E. E. Gardiner. 1980. "High-Protein Wheat-Production and Use." *Canada Agriculture* 25 (2): 10–12.

Dubois, D. K. 1981. "Fermented Doughs." *Cereal Foods World* 26: 617–622.

Evans, L. E., et al. 1972. "Glenlea Red Spring Wheat." *Canadian Journal of Plant Science* 52: 1081–1082.

Evans, L. T. 1975. "Crops and World Food Supply, Crop Evolution and the Origin of Crop Physiology." (In) L. T. Evans (ed.), *Crop Physiology.* Cambridge University Press. pp. 1–22.

Everson, E. H., and R. B. Hart. 1961. "Varietal Variation for Dormancy in Mature Wheat." *Quarterly Bulletin of Michigan State University Agriculture Experiment Station* 43 (4): 820–829.

Gfeller, F., and F. Svejda. 1960. "Inheritance of Post Harvest Seed Dormancy and Kernel Color in Spring Wheat Lines." *Canadian Journal of Plant Science* 40: 1–6.

Ghaderi, A., and E. H. Everson. 1971. "Genotype-Environment Studies of Test Weight and Its Components in Soft Winter Wheat." *Crop Science* 11: 617–620.

———, and W. T. Yamazaki. 1971. "Test Weight in Relation to the Physical and Quality Characteristics of Soft Winter Wheat (*Triticum aestivum* L. em. Thell)." *Crop Science* 11: 512–518.

Glennie, C. W., J. Harris, and N. V. D. W. Liebenberg. 1983. "Endosperm Modification in Germinating Sorghum Grain." *Cereal Chemistry* 60 (1): 27–31.

Hagberg, S. 1960. "A Rapid Method for Determining Alpha-Amylase Activity." *Cereal Chemistry* 37: 218–222.

Harpstead, D. D. 1971. "High-Lysine Corn." *Scientific American* 225 (2): 34–42.

Harrington, J. B. 1949. "Testing Cereal Varieties for Dormancy." *Scientific Agriculture* 29: 538–550.

Johnson, V. A., P. J. Mattern, and J. W. Schmidt. 1967. "Nitrogen Relations During Spring Growth in Varieties of *Triticum aestivum* L. Differing in Grain Protein Content." *Crop Science* 7: 664–667.

———. 1972. "Wheat Protein Improvement." (In) *Rice Breeding.* Los Banos, Philippines: International Rice Research Institute. pp. 407–418.

Johnson, V. A., and P. J. Mattern (principal investigators). 1979. *Genetic Improvement of Productivity and Nutritional Quality of Wheat.* Final Report of Research Findings. Agronomy Department, University of Nebraska, Lincoln, Nebraska, and Agency for International Development, Washington, D.C. 110 pp.

Johnston, R. A., J. S. Quick, and J. J. Hammond. 1983. "Inheritance of Semolina Color in Six Durum Wheat Crosses." *Crop Science* 23: 607–610.

Juliano, B. O. 1972. "Physiochemical Properties of Starch and Protein in Relation to Grain Quality and Nutritional Value of Rice" (In) *Rice Breeding.* Los Banos, Philippines: International Rice Research Institute. pp. 389–405.

Kent, N. L. 1957. "Recent Research in Oatmeal." *Cereal Science Today* 2: 83–91.

Kruger, J. E., and K. H. Tipples. 1981. "Modified Procedure for Use of the Perkin Elmer Model 191 Grain Amylase Analyzer in Determining Low Levels of Alpha-Amylase in Wheats and Flours." *Cereal Chemistry* 58: 271–278.

———. 1982. "Comparison of the Hagberg Falling Number and Modified Grain Amylase Analyzer Method for Estimating Sprout Damage in Rye." *Canadian Journal of Plant Science* 62: 839–844.

Kumar, K. A., S. C. Gupta, and D. J. Andrews. 1983. "Relationship Between Nutritional Quality Characters and Grain Yield in Pearl Millet." *Crop Science* 23: 232–235.

Lambert, R. J., D. E. Alexander, and J. W. Dudley. 1969. "Relative Performance of Normal and Modified Protein (Opaque-2) Maize Hybrids." *Crop Science* 9: 242–243.

Lorenz, K. 1974a. "The History, Development, and Utilization of Triticale." *Critical Review in Food Technology* 5: 175–280.

———. 1974b. "Triticale—A Promising New Cereal Grain for the Baking Industry?" *Baker's Digest* 48 (3): 24–26, 30, 32.

Martin, S. K., et al. 1982. "Selection Indices for the Improvement of Opaque-2 Maize." *Crop Science* 22: 478–485.

Matz, S. 1970. *Cereal Technology*. Westport, Ct.: The Avi Publishing Company, Inc. p. 349.

Mayer, A. M., and Y. Shain. 1974. "Control of Seed Germination." *Annual Review of Plant Physiology* 25: 167–193.

McKenzie, K. S., and J. N. Rutger. 1983. "Genetic Analysis of Amylose Content, Alkali Spreading Score, and Grain Dimensions in Rice." *Crop Science* 23: 306–313.

Mertz, E. T., and L. S. Bates. 1965. "Growth of Rats Fed Opaque-2 Maize." *Science* 148: 1741–1742.

———, and O. E. Nelson. 1964. "Mutant Gene That Changes Protein Composition and Increases Lysine Content of Maize Endosperm." *Science* 145: 985–987.

Moss, H. J., N. F. Derera, and L. N. Balaam. 1972. "Effects of Pre-Harvest Rain on Germination in the Ear and Alpha-Amylase Activity of Australian Wheat." *Australian Journal of Agriculture Research* 23: 769–777.

Munck, L., et al. 1970. "Gene for Improved Nutritional Value in Barley Seed Protein." *Science* 168: 985–987.

Nelson, O. E., E. T. Mertz, and L. S. Bates. 1965. "Second Mutant Gene Affecting the Amino Acid Pattern of Maize Endosperm Proteins." *Science* 150: 1469–1470.

Neukom, H., and W. Buchi. 1979. "Industrial Utilization of Maize." (In) *Maize*. Basle, Switzerland: CIBA-GEIGY Ltd., pp. 95–96.

Piech, J., M. Ruszkowski, and K. Jaworska. 1970. "Inheritance of the Seed Dormancy Stage Duration in Winter Wheat (*Triticum aestivum* L.)." *Genetica Polonica* 11: 227–240.

Pomeranz, Y. 1974. "Malting of Triticale." (In) C. C. Tsen (ed.), *Triticale: First Man-Made Cereal*. St. Paul, Minn.: American Association of Cereal Chemists. pp. 261–270.

———. 1983. "From the Land of 200 Breads and 1200 Small Baked Goods." *Cereal Foods World* 28: 387–390.

Rachie, K. O., and J. V. Majmudar. 1980. *Pearl Millet*. University Park: The Pennsylvania State University Press. p. 199.

Reiners, R. A. 1982. "Corn Oil." *Notes on Agriculture*. University of Guelph. Vol. 18 (1): 22–23.

Reinhold, J. G. 1972. "Phytate Concentrations of Leavened and Unleavened Iranian Breads." *Ecology of Food and Nutrition* 1: 187–192.

Rooney, L. W. 1979. "Sorghum for Food: Opportunities, Problems and Potential." *Proceedings of the Thirty-Fourth Annual Corn and Sorghum Industry Research Conference*. pp. 180–187.

Sharma, G. C., A. D. Paul, and J. A. Bietz. 1983. "Nitrogen Fertilization Effects and Anatomical, Protein, and Amino Acid Characteristics of Yellow Berry in Triticale." *Crop Science* 23: 699–703.

Singh, R., and J. D. Axtell. 1973. "High Lysine Mutant Gene (*hl*) That Improves Protein Quality and Biological Value of Grain Sorghum." *Crop Science* 13: 535–539.

Sreeramulu, C., and L. F. Bauman. 1970. "Yield Components and Protein Quality of Opaque-2 and Normal Diallels of Maize." *Crop Science* 10: 262–265.

Syltie, P. W., W. C. Dahnke, and R. L. Harrold. 1982. "Nutritional Value of Hard Red Spring Wheat Grain Protein as Influenced by Fertilization and Cultivar." *Agronomy Journal* 74: 366–371.

Taylorson, R. B., and S. B. Hendricks. 1977. "Dormancy in Seeds." *Annual Review of Plant Physiology* 28: 331–354.

Tainsh, J. A. R., and E. C. Bursey. 1982. "Golden Rice." *World Crops* 34: 68–69.

Tsen, C. C., W. J. Hoover, and E. P. Farrell. 1973. "Baking Quality of Triticale Flours." *Cereal Chemistry* 50: 16–26.

Tweed, A. R. 1983. "A Look at French Bread." *Cereal Foods World* 28: 397–399.

Vasi, I. G., and R. S. Chintalapudi. 1979. "Locational Variation in the Essential Amino Acids Present in Pearl Millet (*Pennisetum typhoides* S. and H.) Grown in Ahmedabad District." *Composite Physiology and Ecology* 4: 31–34.

Villegas, E. 1973. "Improving Nutritional Quality of Triticale." (In) *Triticale Breeding and Research at CIMMYT*. International Maize and Wheat Improvement Centre. Research Bulletin No. 24. pp. 55–63.

———, C. E. McDonald, and K. A. Gilles. 1968. *Variability in the Lysine Content of Wheat, Rye, and Triticale Proteins*. International Maize and Wheat Improvement Center. Research Bulletin No. 10. 32 pp.

Waines. J., et al. 1978. "Protein and Amino Acid Profiles of Normal and Yellow Berry Bread Wheat." *Crop Science* 18: 590–592.

Walker, J. T. 1973. "Breeding Cereals for Protein." *Span* 16 (1): 7–9.

Wessel-Beaver, L., and R. J. Lambert. 1982. "Genetic Control of Modified Endosperm Texture in Opaque-2 Maize." *Crop Science* 22: 1095–1098.

Woodworth, C. M., E. R. Leng, and R. W. Jugenheimer. 1952. "Fifty Generations of Selection for Protein and Oil in Corn." *Agronomy Journal* 44: 60–65.

Yamazaki, W. T., and L. W. Briggle. 1969. "Components of Test Weight in Soft Wheat." *Crop Science* 9: 457–459.

Zillinsky, F. J., and N. E. Borlaug. 1971. *Progress in Developing Triticale as an Economic Crop*. International Wheat and Corn Improvement Centre. Research Bulletin No. 17. pp. 1–27.

13

Cereal Crop Pests

INTRODUCTION

The struggle between man and crop pests began with the domestication of the first crops perhaps 12,000 to 18,000 years ago. As agriculture and associated crop production expanded, pest losses increased to the point that down through the ages, pests were an ever-present limitation to increases in crop yield. Without effective means of managing crop pests, cyclical outbreaks of a specific pest often resulted in crop failure. Widespread pest damage of epidemic proportions resulted in famine. The trend toward a few key cereal crops increased the intensity of crop pests, because most pest organisms were favored by large areas of a single crop.

The pest problem is characterized by tremendous diversity, continuous change, and few final answers. Despite the most sophisticated efforts at pest management of modern times, which includes plant breeding and genetic resistance, powerful pesticides, machinery to remove or incorporate infected crop residues, an arsenal of other control measures, and the ability to utilize several techniques simultaneously, pest problems have not been eliminated or brought under complete control. New pest problems continually arise as a result of evolutionary developments in many pest organisms, introduction of new pests to an area, changing cultural practices, or crop intensification.

Pests include a number of major organisms: insects, diseases, weeds, **nematodes,** mites, spiders, rodents, birds, worms, and domestic or wild animals. Pests generally can be defined as unwanted and undesirable organisms that limit crop productivity or damage or destroy food.

Within each major class of pest there are numerous variations. Diseases, for example, can be categorized according to the causal agent, such as **bacteria, fungi, viruses, mycoplasmas,** or nematodes. As well, each disease can be considered according to cause, distribution, symptoms, and control.

A weed may be classed as a perennial, **biennial,** or annual, as a grass or broadleaf, and according to root structure, reproduction, and seed-distribution mechanisms. Weeds, like other pest organisms, exhibit tremendous variation.

The number of insect species discovered and described may some day total two million. The **metamorphosis** of insects may vary from primitive forms, in which young are hatched from an egg and develop gradually into an adult; to more advanced forms involving stages in which nymphs progress to adulthood; to the most advanced forms involving egg, larva, pupa, and adult stages. Mouthparts of insects can vary from chewing to piercing-sucking, lapping, siphoning, or sponging, which may have variable effects on plants.

It is evident that a comprehensive coverage of the pests attacking the major cereal crops cannot be included in one chapter. Even a cursory coverage of major pests is beyond the scope of this book. Numerous books are available on each major pest that provide details. Identification of the common diseases of cereals was provided by Zillinsky (1983). To those who wish to delve deeper into the study of plant diseases, books such as the *Compendium of Wheat Diseases* (1977) and the *Compendium of Barley Diseases* (1982), published by the American Phytopathological Society, are available. This chapter provides a perspective on cereal pests through a generalized treatment of the subject.

EXTENT AND NATURE OF PEST PROBLEMS IN CEREAL CROPS

Effectively integrated systems of production, storage and handling, processing, marketing, and distribution are required to supply food to the billions of people on earth. Pests of many kinds attack cereal grain crops at all stages of their development in the field as well as cereal crop products at any stage in the food system. Pest damage can occur to the seed, at any stage of vegetative or reproductive growth, before and after harvest, during transport to market, in storage, in warehouses, elevators, ships, after processing, in the supermarket, or in the home after purchase. Precautions must be taken to safeguard cereal grains at any and every step. If precautions are inadequate or unsuccessful, and if major portions of cereal grains are lost, the very basis of our food supply is threatened. Pest management, therefore, is of major importance in the food system.

The number of organisms, worldwide, capable of destroying food supplies is staggering. In the United States alone, crop yields are reduced by a host of pests including some 160 species of bacteria, 250 kinds of viruses, 8000 species of insects, 8000 species of **pathogenic** fungi, and 2000 weed species (Batra, 1982; Ennis, Dowler, and Klassen, 1975).

Losses from pests were estimated by Pimentel et al. (1978) to amount to 33% of the potential crop production in the United States despite the widespread use of advanced pest-management techniques.

None of the major cereal crops is exempt from pest damage (Table 13-1). In rice, the major food crop for the majority of the world's population, worldwide losses from insect, disease, and weed pests are a phenomenal 47%. In rye, grown on the fringe of cereal production, in relatively small amounts, and where seasonal extremes in climate help keep pests in check, losses averaged 15% (Table 13-1). The potential for pest losses in the tropics is much greater than in temperate zones because of a greater number of pests and more favorable climatic conditions.

Table 13–1
Estimated Losses (in Percent) in World Cereal Crops from Insect, Disease, and Weed Pests

Crop	*Insect*	*Disease*	*Weed*	*Total*
Wheat	5	9	10	24
Oat	8	9	10	27
Barley	4	8	9	21
Rye	2	3	10	15
Rice	27	9	11	47
Millet and sorghum	10	11	18	39
Corn	12	9	13	34

Source: Estimates obtained from values quoted by McEwen (1978).

The number of major pests of rice grown in the tropics was suggested by Haskell (1982) to be about 28, compared to 12 in the United States. Rice in temperate zones was suggested to be subject to 54 pathogenic infections, and in the tropics to over 500. For corn, the figures for pathogenic infections were 85 and 125 for the temperate and tropical zones, respectively.

Losses from pests are universal but highest in Africa and Asia (Table 13-2). Worldwide, pest losses amount to 30% to 40% (Haskell, 1982). Even in technically advanced areas of the developed world, and in major cereal-producing regions of northern Europe and North America that have extremes in seasonal variations where pest cycles are broken, crop losses are high and average 25% to 28% (Table 13-2). Air pollution in industrialized areas, although not caused directly by pests, can cause crop damage and has caused yield reductions of up to 40% in affected areas (Hofstra and Ormrod, 1979). In the less developed areas of the world, losses from all pests range from 21% to 51% (Gooding, 1983). Worldwide, losses due to weeds alone amount to a quantity of food sufficient to feed more than 250 million people.

Losses credited to pests dwarf single advances made by man through improved genotypes and production practices, which may account for yield increases of 2% to 10% and are achieved at great effort after years of development and research.

Disease and insect losses generally are unpredictable, but the potential for crop losses is always present. In cereal crops, losses due to disease are an annual occurrence, but generally not of sufficient magnitude to warrant additional inputs to lessen crop losses. Insect losses in cereal crops often are cyclical in nature, and, except in instances of an outbreak of epidemic proportions, no action may be taken. In most years, yield reductions from insects or diseases are minimal per unit area, but because of the vast area involved in cereal crop production, total losses are staggering. Small but consistent widespread losses help explain why losses average 30% to 40% when pest losses are considered collectively, even in developed nations where the technology for more complete pest management is available. Weed losses tend to be more consistent and predictable than losses from insects or diseases. Many broadleaf weeds can be removed from cereal crops effectively with herbicides. In instances where management of crop losses may not be practical, economical, or possible, pests can cause losses well in excess of 30% to 40%, and losses of 100% are possible.

Quality is an important aspect in cereal crops produced for human food. Losses due to pests may be associated with a reduction in quality due to reduction in cosmetic standards. Shrunken kernels have a low volume weight and an associated low flour yield in wheat may result in downgrading to the status of feed. Grain may be destroyed as human food by insects (Figure 13-1) feeding on the endosperm, thereby reducing the weight, with an accompanying loss in food value and germination. Adult insects feeding on grain may devour their own weight of food each week and many times

Table 13–2
Estimated Regional Losses (in Percent) in Crop Production from Insect, Disease, and Weed Pests

Region	*Insect*	*Disease*	*Weed*	*Total*
Africa	13	13	16	42
Asia	21	11	11	43
Europe	5	13	7	25
North America and Central America	9	11	8	28
South America	10	15	8	33

Source: Estimates obtained from values quoted by McEwen (1978).

Figure 13–1. Direct losses of cereal grains may occur by insects. A reduction of cosmetic standards due to damage or contamination with insects or insect parts, rodent hairs, or excrement may render the product unacceptable for human food. (Courtesy of Ontario Ministry of Agriculture and Food)

their own weight during larval development. Grain contaminated with pests or pest parts generally is unacceptable for human food.

To combat crop pests totalling over 18,000 species (Batra, 1982), capable of various forms of damage at numerous places in the food system, no simple or single approach exists. The development of new forms of **virulence** in most pests adds to pest-management problems. As well, the widely observed phenomenon of pesticide-resistant forms, as part of the endless process of evolutionary adaptation to a changing environment, means that no single approach is favored. Such evolutionary changes assure us that the struggle against pests is an endless problem.

Combating a single pest in itself is a challenge, but in any field-crop situation several pest organisms may be present. Pests may move or be carried into a field at repeated intervals; protective sprays may be washed away with rain or rendered ineffective with age. Because of associated pest problems, the value of a specific measure to combat a pest increases exponentially. A treatment giving 99.7% control is 70% better than one giving 96.7% control because of population dynamics (Grainger, 1982).

From a crop-production standpoint, infestation by a pest results in a yield reduction. Pests that reduce the size and/or effectiveness of the photosynthetic area, or that compete with the plant sink (the grain) for photosynthate, directly reduce crop yield. Production practices aimed at utilizing environmental factors such as increased plant population, fertilizer, or irrigation water favor pest development and spread. Tender, succulent leaves are ideal sites for pests to attack; restricted air movement or light penetration encourages the growth and development of many pests. As yield increases are attempted, pest management becomes more intense. This is an axiom that must be recognized by opponents of pesticides (Weir and Schapiro, 1981) and by those encouraging greater food production by the use of pesticides (Barrons, 1981). Pesticides include **insecticides, fungicides, rodenticides, herbicides, nematicides,** and **aracicides.**

Despite the problems and limitations of pesticides, there can be no doubt that they are essential for the maintenance and expansion of world food production, especially in areas favorable for pest development and survival (Gooding, 1983). It must never be overlooked, however, that all pesticides are toxic; they must be handled safely and with extreme caution so as to avoid or reduce excessive and costly waste, environmental concerns, crop damage (Figure 13-2), damage to adjacent crops by spray drift, injury to the applicator, excessive contamination and residues, and injury to beneficial organisms such as honeybees. It is not the use but the misuse of pesticides that has resulted in serious problems. A basic rule to follow is to read the label and to follow directions precisely.

With the number and population of insects, spores of pathogens, weeds, or other pest organisms potentially present at any time, it is inconceivable that a crop could be produced outdoors free of pests without Herculean efforts. It essentially is impossible to control or eliminate the presence of pests. A generally accepted view is that attempts should be directed, not to the elimination or eradication of a pest in a given area, but only to the reduction of its numbers so that it does not cause unacceptable economic

Figure 13–2. All pesticides are toxic and must be handled safely. An excessive or untimely herbicide application resulted in damage to the barley heads in the illustration with an accompanying yield reduction. (Courtesy of Ontario Ministry of Agriculture and Food)

damage. A possible objective may be to reduce losses from insects, diseases, and weeds from a worldwide level of 30% to 40% to perhaps 25% to 30%. To this end, a combination of pest-management practices may be required, including compromises in production techniques such as lower plant populations, which reduce stress but may not maximize yield.

An essential component of pest management is the economic aspect. In the production of cereal crops, characterized by a relatively low return per unit area, any input must be balanced against economic return, and expensive inputs cannot be justified. Generally, most pest-management systems are expensive, and for economic reasons most cereal crops of the world receive no additional pesticide applications other than that required as a seed treatment or weed-control measure. As suggested in Chapter 1, cereal crops are ecologically safe crops. Good management is used to replace more expensive pest-management methods with no special economic input above normal production costs. Approximately 30% of potential pest losses in the west of Scotland, for example, were considered to be controllable by "costless" management methods (Grainger, 1982).

Some pests occur in every crop produced and must be addressed on an ongoing basis to achieve a profitable yield. Such organisms are called *primary* or *key pests.* For example, corn borer[1] is one of the most serious of all pests of corn (Figure 13-3) and can be expected to occur wherever corn is produced. Likewise, mildew on barley[2] can be expected in moist, warm environments when susceptible cultivars are produced.

Secondary pests are ones that attain crop-damaging levels infrequently or on a cyclical basis. Secondary pests arise when natural enemies are decimated or unusual environmental conditions prevail. Many insect pests are secondary pests because of their cyclical nature.

Among the various pests that attack crops, insects can exert a direct effect by the destruction or injury of photosynthetic area through chewing or sucking. Indirect effects result when insects serve as the vector of plant disease. Insects are capable of transmitting such disease-

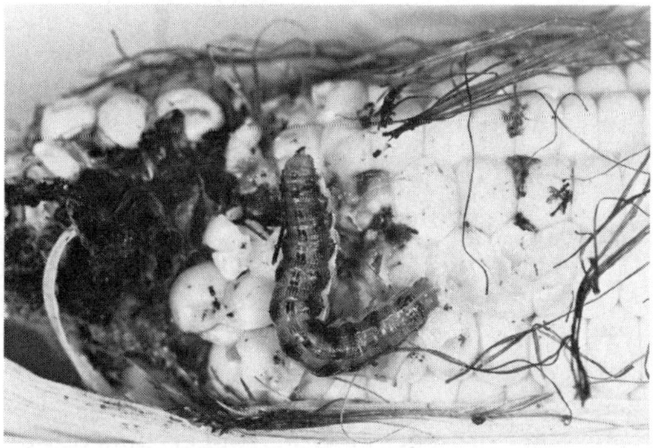

Figure 13–3. The corn borer or corn earworm is one of the most serious of all pests of corn. The illustration shows the naked, brownish, pinkish, greenish, or almost black larva in a typical pose. The unsightly appearance of the ear is one that commonly greets the consumer when the husks are removed from sweet corn. (Courtesy of Ontario Ministry of Agriculture and Food)

[1]Corn earworm or borer—*Heliothis armigera.*

[2]Powdery mildew of barley—*Erysiphe graminis hordei.*

Figure 13–4. Barley seed infected with loose smut gives rise to smutted heads having brown-to-dark-brown spore masses enclosed in a fragile, gray membrane that ruptures and releases the spores to be wind-borne to healthy flowers. Loose smut is found wherever barley is grown and is a highly visible disease when heading occurs. The obvious presence and grain loss is alarming, but yield losses may be less severe than those to more insidious diseases whose presence is less obvious. (Photo by N. C. Stoskopf)

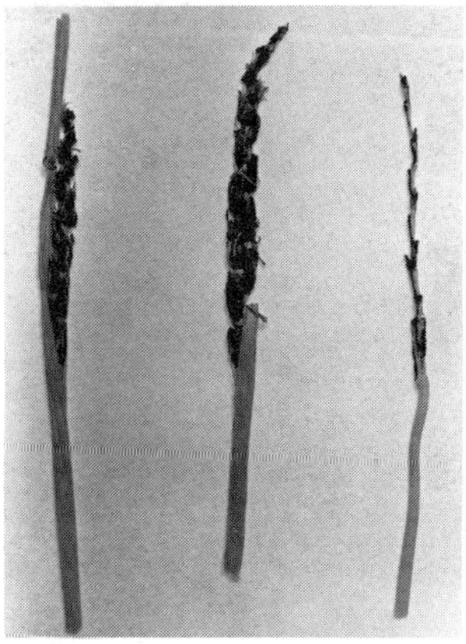

Figure 13–5. Loose smut is spread by spores infecting the flower of adjacent healthy plants, where they survive as dormant mycelia within the embryo of seed. At this stage the smut is not visible, and infected seed appears healthy and normal. When infected seed germinates, the mycelia grow systemically within the plant and eventually produce the smutted heads. The smutted heads on the left and center are as they appear when they emerge from the boot. Only a naked rachis remains after spores are dispersed (right). (Courtesy of Ontario Ministry of Agriculture and Food)

producing entities as bacteria, fungi, and viruses. Fortunately, most insect vectors are specific in the disease they transmit, and only a few insect species transmit more than one disease.

Some pests are more visible than others, and their presence may alarm cereal producers. Loose smut in wheat or barley[3] (Figure 13-4) is highly visible, and the presence of smut may cause serious concern about yield losses. After the smut spores are washed away, the naked rachis becomes less conspicuous (Figure 13-5). The smut **mycelium** is carried in the embryo of seed. Infected and healthy seeds cannot be distinguished visually, and unless a systemic fungicide is applied, the mycelium will grow systemically within the plant to produce smutted heads (Figure 13-6).

Other pests on cereals may be less visible but capable of causing severe losses. Ear rot on corn and vomitoxin[4] on small-cereal grains may render the entire infected lot unfit for human consumption and undesirable as a livestock feed, although the grain may be excellent in other respects. As well, control of insect-spread virus diseases may require management of insect populations long before any symptoms of

[3]Loose smut of wheat—*Ustilago nuda tritici;* loose smut of barley—*Ustilago nuda hordei.*

[4]Ear rot and vomitoxin—*Fusarium* species.

Figure 13–6. Loose smut can be controlled by treatment with a systemic fungicide seed protectant. The untreated portion on the left of the illustration has numerous smutted heads, which emerge a couple of days earlier than healthy heads and protrude slightly above the healthy plants. The treated plants on the right are free of smut. Although cultivars genetically resistant to loose smut are available, seed treatment is advised to control virulent races of the disease. (Courtesy of Ontario Ministry of Agriculture and Food)

the virus appear. Knowledge of the characteristics of each pest is essential for effective management. A producer cannot respond only to visible symptoms.

The most widely recognized and perhaps the most destructive of cereal crop diseases are the rusts.[5] At least five rust species attack cereal crops, and one or more species occur in almost all areas of the world where cereals are grown. Rusts are capable of attacking any vegetative portion of the plant at any stage of development. A rust that infects the oat crop but not wheat or barley is identified as a species.[6] Within each of these species there are physiologic forms or races. Cereal rusts are a constant threat wherever cereals are grown (Figure 13-7).

Rusts are a constant threat because of their flexibility. Wheat stem rust, for example, can spread by following either of two distinct life cycles. One cycle infects and reproduces only on growing wheat plants. Wind-borne **urediospores** are produced every seven to ten days as long as growing wheat plants are available. These **spores** spread the disease north or south. In North America, the rust overwinters mainly on winter wheat in northern Mexico or southern Texas. During spring and summer, wind carries the spores successively farther north along a route described by Knott (1972) as the "Puccinia Path," where they infect first winter and then spring wheats. The rust does not survive the hot summer in the southern zone where it overwinters. Thus in the fall, spores must be blown back from the north into the overwintering area. Each year rust goes through a cycle over thousands of kilometers or miles, starting in the south, moving north, and returning to the south.

---

[5]Cereal rusts—*Puccinia* species.

[6]Oat stem rust—*Puccinia graminis avenae;* wheat stem rust—*Puccinia graminis tritici* (infects wheat, barley, and triticale).

Figure 13–7. Pustules of stem rust of wheat. These pustules may occur on the stems, leaf sheaths, leaf blades, glumes, and beards of infected plants. The epidermis of the leaves and culms is ruptured, becomes pushed back, and releases urediospores. Prior to plant maturation telia are formed, which are oblong-to-linear and dark-brown-to-black in color. Stem rust reduces cereal yield and quality by destroying the photosynthetic area of the plant and competing for photosynthate. (Courtesy of Ontario Ministry of Agriculture and Food)

The disease in this cycle persists from cycle to cycle primarily as urediospores or as mycelium established in the cereal host tissue and is dependent on growing wheat plants.

The second, more complete life cycle involves the development of overwintering **teliospores,** which survive on wheat residue and which germinate in the spring to form **basidiospores** that are wind-borne and infect young leaves of the alternate host, the barberry,[7] where they form **pycnia** on the upper leaf surface (Figure 13-8). Basidiospores cannot infect cereals or grasses. Exchange of genetic material contained in the nuclei of rust spores occurs in pycnia. To assist in cross-fertilization a thick, sticky, sweet liquid, attractive to and spread by insects, helps spread the **pycniospores** from one pycnia to another to assure genetic recombinations. Following fertilization, an **aecium** develops on the lower leaf surface of the barberry, directly below the pycnium. Aecia release **aeciospores** that are wind-borne to wheat plants. Infected wheat plants release urediospores that spread the rust epidemic. The development of new genetic combinations may be able to attack formerly resistant wheat cultivars, and the use of genetically resistant wheat cultivars may therefore provide only temporary resistance. The elimination of the barberry as the alternate host prevents genetic recombinations and a source of inoculum early in the season. Failure to manage wheat stem rust could mean an epidemic over continental North America.

[7]Barberry—*Berberis* species.

METHODS OF REDUCING PEST LOSSES IN CEREAL CROPS

Repeated reference has been made to the need for multidisciplinary or systems approach to food production. In the case of pest management, observations on the number of pests, the

Figure 13–8. In the complete life cycle of wheat stem rust, wind-borne basidiospores developed from overwintering teliospores from pycnia on the upper leaf surface of the barberry plant, which acts as the alternate host. Recombination of genetic materials occurs in the pycnia and results in cross-fertilization which produces new virulent physiologic rust races. Aeciospores, released from aecia that develop on the lower leaf surface beneath the pycnia, are capable of infecting adjacent wheat plants. (Courtesy of Ontario Ministry of Agriculture and Food)

extent of damage, variation in survival, and mode of action suggest that a systems approach is imperative. Twelve major techniques of pest management are reviewed below.

Genetic Resistance

Genetically or cytoplasmically governed plant features offer important aspects for pest management and may be achieved by a physical trait (pubescence or thick waxy cuticle) or physiological trait. Such traits prevent a step or steps from being achieved in insect, disease, or nematode attack or development. Selecting cultivars that are resistant to prevalent diseases or other pests, or to prevalent races of the organism, offers a useful and efficient method of pest management. Genetically resistant cultivars especially are useful when pests remain below levels at which additional action to control them is required. Genetic resistance may avert an outbreak of epidemic proportions such as those reported throughout history.

Where pest-resistant cultivars are available, they offer the most economical, carefree and nonpolluting method of pest management. The cost of seed for a resistant cultivar may be no more than for a nonresistant cultivar, and often no additional control measures are required. In some instances where an inexpensive treatment is available, such as in loose smut of barley, seed of genetically resistant cultivars should be treated, because such cultivars may be resistant only to the prevalent races of smut fungus and may not resist other races. Seed treatment, in this case, not only offers an inexpensive form of assurance but may deter the development and spread of new smut races.

Genetic resistance is one of the oldest means of reducing crop losses from pests and can continue to be of major importance. The two major problems directly affecting the plant breeder are those of finding and developing the desired resistance and of maintaining it to provide an element of yield stability. Genetic resistance to a specific pest organism may not be known to exist, or, if it does exist, may not be capable of providing an enduring form of resistance. Evolutionary changes in the pest organism resulting from mutational changes or genetic recombination, as outlined for wheat stem rust fungus, may result in a loss of resistance. Rapid breakdown of genetic resistance could result in the depletion of readily available genotypes. A continual breeding effort to incorporate new genes for resistance may result in only maintaining the yield performance, as too little breeding effort remains to select for yield ability.

Agronomic or Management Practices

There is little doubt that a complete and sound system of agronomic practices can be effective in pest management. Modern, intensive cropping systems that short-circuit sound techniques complicate and add to pest-management problems. Nine management, cultural, or agronomic practices are described below.

CROP ROTATION. Rotations are of benefit in reducing the build-up of species-specific pests, because a change in crop species will break the life cycle of a pest organism. Many plant pests survive or overwinter in infected plant residue. If the same crop is grown in that soil the following growing season, the pest is present to infect the new crop. Substituting a nonhost plant in a rotation prevents the pest from increasing its numbers through growth on crop plants. Crop rotation is unlikely to eliminate a plant pest.

An excellent example of a pest organism that is reduced to manageable levels by crop rotation is corn rootworm,[8] which thrives only on corn roots. Including a "break-crop" in a continuous corn production program will reduce the pest to tolerable levels. If crop rotation is not practical, an application of a soil insecticide may be required. Crop rotation can be viewed as providing a direct effect on pest management.

The use of crop rotation also can provide indirect effects on crop pests. Cereal crops produced on a legume-grass sod with good drainage, aeration, and organic matter may result in healthy, vigorous plants that are more

---

[8]Corn rootworm—*Buphonella murina*.

tolerant to plant pests, thereby indirectly helping reduce crop losses. Corn grain in a rotation that included a legume sod always outyielded monoculture corn in a measured experiment, regardless of how much nitrogen was applied (Rohweder, Shrader, and Templeton, 1977).

Continuous barley production is extensively practiced in some cereal-growing areas and generally is regarded as acceptable but not as a preferred system. The inclusion of a forage crop for two successive years in the rotation resulted in striking yield increases (Sheard, 1977). Continuous barley may encourage an insidious build-up of the fungus disease Helminthosporium root rot or blight.[9] The incidence of the disease may increase steadily and without causing a major failure accept under conditions very favorable to the disease.

Fungus diseases such as take-all[10] and eyespot,[11] also called Cercosporella footrot, footrot, strawbreaker, and culm rot, can be controlled by a rotation involving a nongrass crop, since these diseases do not survive in the soil in the absence of cereal plants. When wheat follows wheat, as much as one-third of the second and third crops may be destroyed. In the third or fourth crop year of a continuous wheat cropping program, the incidence of the disease reaches a peak that may result in a total crop failure. However, the build-up of organisms antagonistic to these fungi may allow for wheat production in subsequent years with a reduced and acceptable incidence of the disease. A crop producer wishing to grow continuous wheat under humid environmental conditions that favors these fungi may not be able to tolerate yield losses prior to the establishment of an equilibrium between the fungi and the antagonistic organisms.

Under intensive cropping programs, a rotation of more than three years may not be economically desirable. Such short rotations are generally helpful in pest management, but where pests can remain in the soil for several years, longer rotations may be required to be effective.

SANITATION. Pests that survive or develop on previous crop tissue can be reduced in number by removal, incorporation, or destruction of these residues. Incorporation through deep plowing hastens decomposition of plant residue that may be used as a substrate for pathogens or other pests. As residues decompose, competition with other soil microorganisms takes place and a suppression of pest organisms may occur. Incorporating crop residues prevents pests from being in a position on the soil surface to be easily carried to nearby plant hosts. A trend toward minimum or zero-tillage, which eliminates plowing and promotes surface residues to reduce wind and water erosion, is counter to sanitation techniques once commonly practiced.

Pests can be transported from place to place by seed, on cereal residues used as packing material, feed, or bedding, on soil carried by machinery, on transplants, or by a general lack of hygiene. Sanitation is especially important in seed storage units to prevent the carryover of insect pests from one season to another.

FLEXIBILITY IN CROP USE. Crop loss may be minimized and the spread of pests reduced by flexibility that allows alternative crop uses in the event of a pest outbreak. Corn seeded as grain and attacked by one or more of stalk rot, ear rot, corn rootworms, or corn borer may be used advantageously as silage to minimize grain-yield losses. Almost any cereal crop may best be managed as forage in the event of a major pest infestation.

CROP MIXTURES. Despite competitive forces associated with mixtures of oat and barley crops, a widespread custom is to grow a mixture of two or more crops together. Although yields may not be maximized under favorable conditions, under conditions of a pest infestation a mixture may provide a greater degree of yield stability by reducing the severity of the pest attack and its spread from plant to plant.

---

[9]Helminthosporium root rot—*Helminthosporium sativum.*

[10]Take-all—*Ophiobolus graminis.*

[11]Eyespot—*Pseudocercosporella herpotrichoides.*

A refinement of crop mixtures is the development and production of multiline cultivars, involving a blend or mixture of cereal cultivars of one species that are genetically very similar but differ in specificity to races of a disease. The concept was first proposed by Jensen (1952). Components of a multiline closely resemble each other in height, maturity, yield performance, and leaf characteristics, so as to minimize competition among them. When a new race of a pest appears, only a specific component of the multiline may be susceptible. The typical epidemic that spreads from plant to plant is thus checked.

USE OF CLEAN SEED. Many pathogens survive in or on the seed (such as smut fungi of cereals). If infected seed is used, a source of inoculum is introduced into the field. Seed contaminated with weed seeds may introduce new species to a given field or add to an existing weed population (Figure 13-9). The use of pedigree seed, certifying a high germination and vigor, freedom from disease, and a minimum of weed seeds, is a primary step in pest management. Depending on the nature of a disease, the presence of infected plants in a seed field may result in a seed lot's not being certified.

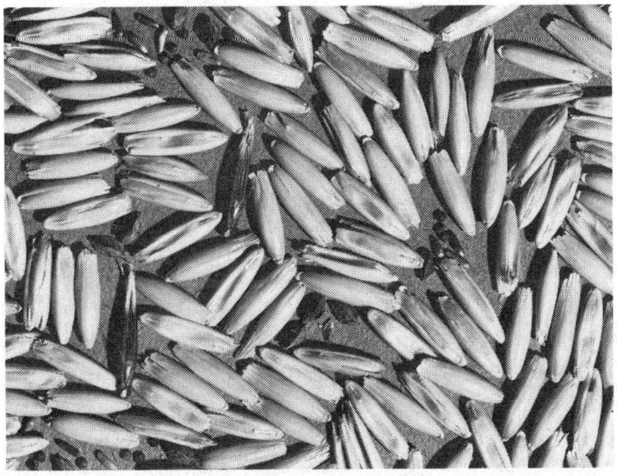

Figure 13–9. Planting of clean seed is perhaps the most important weed-management technique to control pest losses. The oat seed shown here is heavily contaminated with a variety of weed species that could reduce the ability of the crop to utilize sunlight, carbon dioxide, water, and plant nutrients needed for photosynthesis. (Courtesy of Ontario Ministry of Agriculture and Food)

In addition to the use of pedigree seed, seed treatment should be applied to kill surface inoculum and in some instances inoculum in the seed.

WEED CONTROL. Besides competing with crop plants for light, water, carbon dioxide, and plant nutrients, weeds may interfere by secreting allelopathic substances that suppress crop growth. Weeds may act also as a host for plant pests, providing an adjacent source of inoculum and allowing pests to survive or proliferate even in the absence of crop plants. Finally, weeds may serve as alternative hosts to allow for sexual recombination of specific diseases or as a means of surviving a stress period. Weed management, including hand removal of weeds, will help reduce weed seeds in the soil and in crop seed.

ROGUING OF INFECTED PLANTS. A source of inoculum and subsequent spread of a disease may be reduced by roguing infected plants before spores are released, especially in seed fields.

ESCAPING OF PESTS. Early seeding or the use of fall-seeded winter cereals may allow for early maturity that helps plants escape pest development.

USE OF INTERROW CULTIVATION WITH DISCRETION. In row crops such as corn and sorghum, interrow cultivation should be avoided when foliage is wet, as some bacteria require a film of water to penetrate and infect a plant. Equipment that causes minor wounds or carries the organism from plant to plant during conditions inducive to spread of the disease should be avoided. Under moisture stress conditions, interrow cultivation may further stress the plant by causing wilting.

Chemical Pest Control

Modern science has provided a number of pesticides that are specific in their action and provide the overwhelming amount of crop protection available for pest management. Pesticides are highly efficient and provide an economical means of pest management—so much so, that the modern agricultural scene would be inconceivable without the application of chemical crop-protection measures. Pesticides not only provide a means of stabilizing crop production but also serve as a means of increasing food output and preserving what is produced. Pesticides have come into worldwide use and have provided increased scope for pest management.

Chemical pest control is not without problems and critics. Many examples of misuse in the past have resulted in a blanket condemnation of pesticides. During the relatively short history of pesticides, a rapid progression from an exploratory to an exploitive to a more intelligent reflective stage has occurred (McEwen, 1978).

In the exploratory and exploitive stage, the ability to provide nearly 100% pest control in a given situation raised speculation that individual pests could be eliminated as a production problem. A belief existed that if a little was good, more would be better. Pesticides were applied when pest levels were at an economically low level. Often pesticides were applied without regard to their toxicity to humans or nontarget organisms. During the exploratory stage, the view was held that pesticides could be used as the primary and perhaps sole approach to a specific pest problem.

A number of considerations—failure of pesticides to eradicate a pest, destruction of nontarget organisms, environmental concerns, and fear of food contamination—resulted in a reappraisal of pesticides and their use. This reflective stage led to a more mature view of the place and roll of pesticides in a pest-management program. The new rationale recognizes that pesticides are indispensable tools in crop pest management but that they must be used with discretion. A reflective evaluation of specific compounds led to the elimination of undesirable products and uses and resulted in improved pesticides to overcome many past objections (Pimentel and Edwards, 1982). Improved application techniques have reduced the quantities needed; pesticides have been designed to be highly selective with regard to nontarget organisms; and their longevity in the environment has been reduced.

The major advantages of pesticides were reviewed by Ennis, Dowler, and Klassen (1975):

1. An array of pesticides is available, capable of suppressing most pests at practical costs.
2. Pesticides are fast-acting and are effective against high pest populations that could destroy a crop rapidly and before other management techniques become effective.
3. Equipment to apply pesticides is widely available, and large areas can be covered rapidly and thoroughly (Figure 13-10).
4. Pesticides can provide a recovery treatment when a component in the pest-management system has failed.
5. Many pesticides, including herbicides, can be used selectively to destroy a pest without damage to the crop. Wild oats can be removed from a barley or wheat crop during vegetative growth stages, without damage to the cereal crop.

Reliance on pesticides as a single line of defense is not advised, because it has created problems of pesticide-resistant pest forms, has suppressed natural pest enemies, and has allowed for secondary pest outbreaks that were not suppressed by natural competitive forces. An examination of specific pesticides will help increase the understanding of pesticides as a form of pest management.

HERBICIDES. A mixed population of annual and perennial broadleaf weeds and grasses are found in most cereal crops. Use of herbicides can manage such a weed problem best in corn and sorghum and has reduced or eliminated the need for interrow cultivation or, within the row, hand weeding. A consequence has been that herbicide residues toxic to other cereal crops reduced the opportunity for rotational cropping and encouraged continuous corn production. In time, a shift in weed species and the emergence of weeds resistant or tolerant to specific herbicides developed, begging a reevaluation of weed-management systems heavily dependent

Figure 13–10. Equipment is widely available that can apply pesticides accurately and rapidly to large land areas. (Courtesy of Ontario Ministry of Agriculture and Food)

on herbicides. Mahn and Helmecke (1979) suggested that reliance on a single herbicide could result in quantitative changes in the structure of the weed population in as few as five years. As weed population stresses are shifted by herbicides, weeds formerly of secondary importance emerge as primary weed problems. Such problems may be avoided by an integrated system of weed management, possibly involving rotation of chemicals as well as rotation of crops.

FUNGICIDES. Chemical control of diseases is achieved by direct toxicity to the pathogen or by the development of toxic derivatives by the pathogen or host plant tissues. Protectant sprays, with their associated problems of being washed off, interactivated, or ineffective on new unsprayed growth (Rowell, 1972), generally are less desirable than systemic fungicides. The development of systemic fungicides that are highly efficient and are selective in their toxicity has helped establish new standards of pest management (Geissbuhler, Brenneisen, and Fischer, 1982).

The term *systemic fungicides* implies movement through the plant system, which is misleading, as systemics move in the plant with the transpiration stream. Leaves are the primary transpiring organs, and movement within plants is from the treated seed or soil to the expanded leaves (Edgington et al., 1980). Very young leaves, reproductive organs, and grain do not transpire significant amounts of water. Seed residues from systemics, therefore, are minute, and may be no more than expected with conventional protectants. So-called systemic compounds are really partially systemic. Because of their effectiveness and safety, the number of systemic fungicides has increased in one decade to one-third of the total fungicides in use (Edgington et al., 1980).

INSECTICIDES. An unchecked outbreak of an insect pest can ravage a crop. Insecticides provide a strategic resource to minimize crop losses from an attack against which otherwise there may be no defense. Insecticides can be applied as a seed treatment, as a preplant broadcast application, as a band application, as a granular foliage application to be caught in the **whorl** of corn, or as a foliar spray application.

Physical and Morphological Resistance

A thick leaf culticle, or cork layer, or leaf pubescence are among the morphological or physical factors that prevent or deter parasites from attacking plants. Leaf pubescence, called hairs or trichomes (Figure 13-11), offers a form of resistance to the cereal leaf beetle,[12] as it interferes with both egg-laying and the survival of larvae (Webster, 1972; Ringlund and Everson, 1968). Comparison of pubescent and nonpubescent cultivars showed yield losses ranging from 17.6% to 25.6% (Webster, Smith, and Hoxie, 1982) in the susceptible nonpubescent cultivars.

Trichomes found in some germplasm lines of sorghum confer resistance to the shootfly,[13] a serious pest of sorghum in South Asia and several parts of Africa (Gibson and Maiti, 1983; Maiti and Gibson, 1983).

[12]Cereal leaf beetle—*Oulema melanopus* (L).
[13]Shoot fly—*Atherigona soccata* Rond.

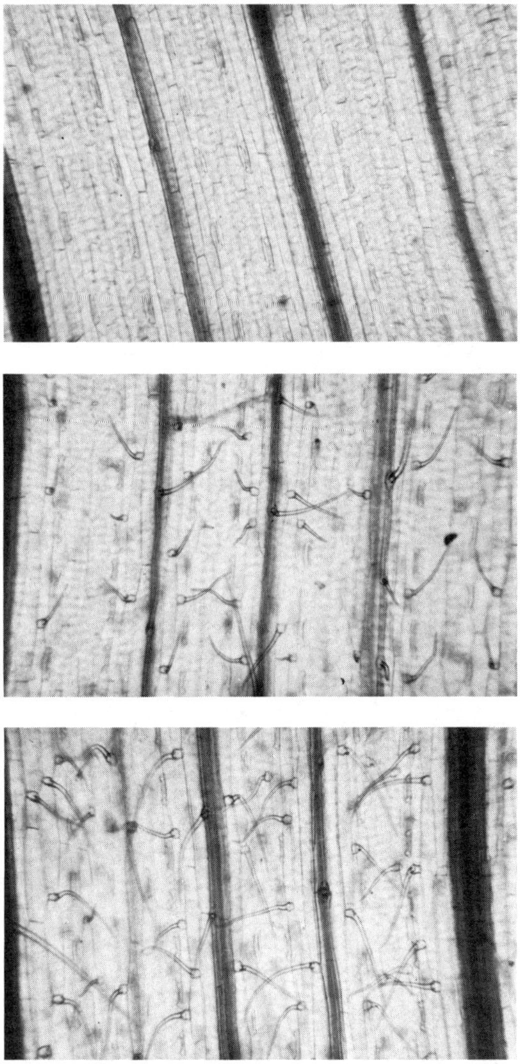

Figure 13–11. Photomicrographs of leaf samples from three wheat cultivars showing differences in pubescence density. Cultivar A (top) has 1 hair, cultivar B (center) 23 hairs, and cultivar C (bottom) 54 hairs per 10.8 mm̀ area. Resistance to cereal leaf beetle is associated with pubescence density. (Courtesy of Michigan Agricultural Experiment Station)

Pheromones

Insects give off and are programmed to respond to chemical compounds called *pheromones*. The pheromone exuded by a female insect, for example, attracts males to the same species over a large area. Other pheromones identify members of a colony, trigger fight or flight reactions, or are used to mark a path toward food sources.

Natural and synthetic chemical pheromones can be used as attractants, repellants, or disruptants to affect insect behavior. Such compounds may serve as a component of pest management, and their possible uses were reviewed by Silverstein (1981) and Tette (1974) as follows:

1. To attract and trap insects for purposes of surveying and monitoring population number or of detecting new areas of infestation at an early stage. These surveys allow for insecticides to be applied only as needed or when populations exceed a predetermined level.
2. To lure insects into specific areas where they can be treated or infected with substances that they spread to the rest of the population.
3. To mass-trap insects so as to suppress the population.
4. To suppress insect population growth by disrupting mating procedures.

Pheromones cannot be regarded as solving the problems of insect control or the associated spread of insect-borne diseases, but they can play an important role in a pest-management system. Pheromones can help reduce the use of insecticides to lessen environmental concerns, and they can serve as supplements to insecticides that by themselves have not provided acceptable insect management.

Biological Control

Biological control may be defined as the balance between a species and biological stress factors that limit the potential population of a pest organism such as an insect species (after Dicke, 1972). Three major divisions of biological control in insects are recognized:

1. The use of predators and parasites.
2. Insect pathology.
3. Radiation and genetics.

Biological control of insects is complex, because many factors in the environment are involved in keeping a population balance of desirable species. When a control agent is introduced and becomes successfully established, it becomes a permanent part of the **agroecosystem,** and there is a risk that it may become a pest to another desirable species.

Introducing parasites and predators to combat introduced insect pests is not new and was actively practiced in the early twentieth century. Since 1863, Batra (1982) reported that 294 organisms were deliberately introduced into new areas of the world. Environmental concerns about chemical usage and the evolution of forms resistant to pesticides have resulted in renewed enthusiasm for biological control despite its complexity and risks. A modern example of biological control is the introduction of a bacterium, *Bacillus thuringiensis,* that attacks the European corn borer (footnote 7).

When a pest is introduced into an area, the natural enemies which normally keep them in check are absent. For example, the cereal leaf beetle (footnote 12), native to Eurasia, inadvertently was introduced into North America and was identified in 1962 in the Chicago area of the United States. Without natural enemies, the insect spread unchecked, and by 1966 insecticide had to be applied to over 65,000 ha (over 1,660,000 ac) to prevent severe yield losses, reported to be as high as 55% in untreated spring wheat fields (Wellso, 1982). Despite widespread use of insecticide the insect moved rapidly eastward, owing perhaps to prevailing westerly winds, damaging oat, barley, wheat, rye, and corn crops.

Although effective insecticides were available against the pest, a search for alternative management measures was begun as early as 1963. Extensive spraying was viewed with concern, because resistant strains of the pest might develop, economic and environmental concerns existed, and spraying led to tramping of grain. Biological control was undertaken in the form of five species of wasps introduced from France. Four of the wasp species became established and were effective in reducing the cereal leaf beetle to economically acceptable levels in most areas by the mid 1970s (Wellso, 1982).

Biological control in insects is in principle applicable to native as well as introduced pest species (Carl, 1982). Purely biological control measures provide excellent protection against many pests, but their contribution in the 1970s and early 1980s to crop protection as a whole has remained relatively small.

Figure 13–12. The European buckthorn is the alternate host of oat crown rust or oat leaf rust. The complete life cycle of this fungus disease is parallel to that of wheat stem rust. Basidiospores produced from overwintering teliospores of the rust infect only buckthorn. Oat producers in cold climates where species of buckthorn are present should learn to identify the dark green leaves. The plants should be destroyed to reduce infection of oat plants with crown rust and to reduce sexual recombinations of the rust spores that lead to new physiologic races. Plants can be identified readily by the dark green leaves with veins that curve into the midrib. (Courtesy of Ontario Ministry of Agriculture and Food)

Destruction of Alternate Host

European buckthorn,[14] a shrub or small tree with round black berries and dark green leaves with veins that curve into the midrib (Figure 13-12), is the alternate host for oat leaf rust, also called crown rust.[15] Although crown rust is a significant disease of the oat crop and is prevalent wherever the crop is grown, it does not affect other cereal crops.

The complete life cycle of oat crown rust is similar to that of wheat stem rust on barberry plants (footnote 7). Local residents should learn to identify these alternate hosts of rust fungi and

[14]European buckthorn—*Rhamnus carthartica.*
[15]Oat crown or leaf rust—*Puccinia coronata.*

Figure 13–13. Ergot infection in heads of mature rye. Some of the kernels are replaced by sclerotia or ergot bodies that protrude from the florets. The dark ergot sclerotia can be detected in the threshed grain as well and are toxic when consumed. (Courtesy of Ontario Ministry of Agriculture and Food)

destroy them. The common barberry is a shrub ¾ to 1 meter (2 to 3 ft) in height with three-pronged spines on the stems, saw-toothed leaf edges (Figure 13-8), and oblong red berries which remain on the plant through fall and winter. Rust losses commonly are downwind from the alternate hosts, as rust spores are wind-borne over a wide area. Destruction of the alternate host is effective only if eradication is conducted over the entire crop-producing area.

Elimination of the alternative hosts of cereal rusts prevents the disease from sexually recombining to produce new and virulent races. Removal of the alternate host will not prevent the incidence of disease, however, as spores overwinter in warm climates, and repeated rust cycles move with the developing crop as the seasons progress. Destruction of the alternate host will add stability to genetically resistant cultivars and help extend their commercial life.

Throughout the world, a fungus known as ergot[16] commonly occurs on rye but is also found on wheat, triticale, barley, oat, millet, and many grasses. Ergot is more prevalent in cool, temperate climates and on crops in which the florets remain open for extended periods during the flowering stage. The fungus attacks only the seed-producing organs and results in purplish-black ergot bodies or **sclerotia** that replace the kernels (Figure 13-13). Direct yield reductions are not as serious as indirect losses from discounted grain quality. Ergot bodies contain a mixture of chemicals, including alkaloids that may cause toxic effects when consumed by humans or animals.

The degree of infection is highly variable and may be influenced by a number of factors. The prevalence of the disease may be related to the susceptibility of native grasses in these areas. These grasses produce ergot bodies that overwinter on the soil surface or in plant debris. In the spring and early summer, sclerotia produce spores that infect the stigmas and ovaries of small grains and grasses. Removal of the grass host plants can reduce the incidence of the disease.

Use of Pest Deterrents

Pest losses of stored grain can be avoided or reduced if measures are taken to prevent or reduce infestation and damage. Rodent-proof steel or concrete bins are a form of deterrent, just as chemical protectants or physical deterrents prevent pest attack.

Systemic chemicals are effective deterrents in the case of aphid-transmitted viruses. Most viruses require some time in the aphid vector before they can be transmitted, and an effective systemic chemical may destroy the vector before incubation is complete.

Quarantine Regulations

To reduce the possibility of carelessly or inadvertently introducing plant pests from one area to another, restrictions or quarantines governing the importation of certain plants or plant products may be enforced. Plant inspectors enforce quarantine regulations on seed or plants in many countries. Straw or hay possibly carrying pests may be prohibited from export for use as feed or bedding or as packing material. Plants con-

[16]Ergot—*Claviceps purpurea.*

tained in soil may be prohibited free access from one area to another because of the risk of transfer of nematodes or other pests.

Pest Monitoring

In temperate zones, many pests become dormant during cold weather and many are killed by harsh winter conditions. As seasonal cycles of pests rise and fall, pest monitoring serves a useful purpose to trace or record pest numbers and to inform area producers of the build-up of specific pests and of threshold levels that warrant action. Associated knowledge of the life cycle has resulted in a reduction of pesticide applications, making pest management effective with less economic and environmental stress (McEwen and Sutton, 1977).

Passive Crop Protection

Management of some cereal-damaging birds that feed on the maturing grain, especially in lodged sections or at the edge of a field, may be achieved by passive action. A bird-scaring kite, tethered by a cord above a cereal crop field and kept buoyant by helium-filled balloons or air currents, has the effect of a bird of prey hovering over birds that are feeding on grain. The scaring effect is completed by any swooping or diving action resulting from variable air currents.

Variations include the attachment of a flashing strobe light to scare birds away, or the use of amplified sound recordings of distress calls or the attacking sounds of birds of prey such as falcons. Calls delivered at random intervals during daylight hours may be more effective than systematic signals or acetylene bangers that go off at regular intervals.

The effectiveness of a passive system using a single device is limited to a small field area.

Integrated Pest Management

A crop producer frequently can determine a precise method of managing a single pest, but in many instances a number of pests attack over the entire growing season. Because reliance on one method of pest management may be inadequate in terms of cost, timeliness (Figure 13-14), and effectiveness, the concept has developed of jointly using a number of appropriate forms of pest management and integrating them into a comprehensive system.

Figure 13–14. Timeliness is one aspect of an integrated approach to pest management. Final crop yield is influenced by the stage of growth of crop and weed and the time that control measures are applied. Hand hoeing and/or interrow cultivation to remove weed interference in this photo will not eliminate the adverse effects that weeds have already exerted on the crop. (N. C. Stoskopf)

The introduction of systemic chemicals that provide more lasting and more selective control than protectants may have stimulated a comprehensive approach to pest management. Perhaps the development of pesticide-resistant forms of pests or the evolution of physiological races of a pathogen revealed the need for the integration of a number of pest-management inputs into an effective system.

A successfully integrated system requires careful management. Integrated pest management is complex because it requires assembling a great deal of information on pest populations and related factors of weather, stage of plant development and vulnerability, acceptable pest population levels, probable price of crop commodity, other means of pest management and the conditions under which each is most effective, life cycles of pests, and how various management systems may interact. Then all this information must be integrated into a workable program. To do this may require access to computerized information sources.

The trend toward greater cropping intensification, intercropping, and sequential cropping has placed greater and greater demands on pest management (Litsinger and Moody, 1977). These intensive systems have burdened an already stressed program of pest management that relies on one or more methods. The need for a comprehensive or integrated system of pest management is abundantly evident.

Despite the apparent complexity of an integrated pest-management program, success has been achieved with a limited number of important crops. Full development of the system offers considerable promise for more crops.

Unlike exploratory or exploitive initial approaches, whereby a new and powerful force such as a chemical is aimed at eliminating a pest, integrated pest management is designed to maintain pests at economically acceptable numbers, at the least cost, and with minimum disruption to the environment. These objectives are met, in part, by using pesticides and other inputs more selectively, in the smallest possible doses, and only when a proven need exists.

Integrated pest management may be illustrated with the ergot fungus (footnote 15), for which there is no single direct control method. A combination of five methods may be considered:

1. Remove unwanted grass species in fence rows or adjacent to cereal fields that serve as host sources of inoculum.
2. Use cultivars that are less subject to ergot infection or have some resistance to the disease (Platford and Bernier, 1970). Resistance may be associated with florets that remained closed and inhibit entry of the fungus spores.
3. Incorporate ergot sclerotia into the soil at sufficient depth so they will not develop spores to be released into the air to cause infection. Soil microorganisms will cause buried ergot sclerotia to decompose in a year, so that subsequent cultivation will not bring them up to the surface.
4. Use ergot-free seed. Frequently ergot bodies are longer and larger than normal seed and can be removed by seed cleaning.
5. Harvest as soon as possible after the grain is physiologically mature to prevent sclerotia from falling onto the soil. Subsequent screening will help remove remaining sclerotia from grain intended for seed purposes.

Treating seed for loose smut (footnote 3), including genetically resistant cultivars, provides a form of insurance against infestation by a virulent race and helps reduce the development of virulent races. Using more than one approach to pest management and not depending on a single form of pest management is a form of integrated pest management. In the case of cereal leaf beetles (footnote 12), Webster, Smith, and Hoxie (1982) recommended that resistant cultivars be produced in localities where the beetle has been a problem, despite the establishment of biological control measures in the form of parasitoids. An outbreak of the pest beyond a threshold level may prompt the need for a pesticide. Greater security of pest management is provided, therefore, through an integrated approach.

CEREAL YIELD STABILITY AND CEREAL PESTS

Two aspects of yield stability are essential in both extensive and intensive systems of agriculture. One is stability of production year after year, because humans depend on cereal crops as their main food source. A second involves extending the commercial life span of cereal cultivars, relieving plant breeders of the ongoing struggle to incorporate genetic resistance to pests and freeing them to concentrate their efforts on yield, quality, hardiness, lodging resistance, or other agronomic traits.

Yield stability can be increased by a number of means. We will consider it here on a macro rather than on a micro or individual field basis. Factors associated with yield stability are as follows:

Genetic Resistance to Cereal Pathogens

For some time it was believed that yield stability could be achieved through incorporation of major genes for resistance. Major genes are available for such diseases as rice blast,[17] mildew, smut, and rust (footnotes 2, 3, and 5, respectively). The development of new **races** of the pathogen capable of overcoming the defense mechanisms of a host has destroyed the concept of lasting resistance, especially that achieved through major genes. This phenomenon is inevitable and warrants further explanation.

When a cultivar with genetic resistance to a specific pest or races of a disease is produced, the plants may remain healthy and vigorous as long as no virulent races of the pest are present. In such an event, competition among races is reduced substantially, thereby allowing a previously unrecognized or minor race to become dominant. Genetically new races may be developed on the alternate host or arise through mutation. Because of the rapid evolution of virulent races of oat crown rust (footnote 15), prior to the 1960s a major concern in most oat-breeding programs was to incorporate crown rust resistance. The single major-effect genes relied on for resistance were effective only three to seven years before being overcome by new sources of virulence (Murphy, 1965). Some potential cultivars become susceptible to virulent physiologic races even before regional testing was completed and before licensing. In response, oat breeders were having to add a new gene for resistance with each cycle of cultivar development, which greatly restricted opportunity for improvement of traits other than crown rust resistance. Plant breeders were devoting 50% to 90% of their efforts to developing disease resistance and having to ignore other agronomic aspects. The result was that genetic gains for grain yield in the oat crop were substantially below that of other cereal crops (Rodgers, Murphy, and Frey, 1983). Even more alarming was the fact that readily available sources of resistant genes were being depleted.

Likewise, in rice, resistance to the fungus disease called blast (footnote 17) is one of the most important breeding objectives in many rice-growing areas. In Japan, the use of resistant cultivars was begun in 1944, and repeated incorporation of new genes to give resistance has not resulted in any lasting resistance. Kiyosawa (1972) reported that the breakdown of resistance occurred two to six years after release of the cultivar, thereby reducing yield stability and increasing the risk that the crop could be diminished in yield or destroyed by virulent races.

More durable resistance exists against mildew and smut (footnotes 2 and 3, respectively), but in general it is apparent that techniques must be developed to give a cultivar a longer commercial life and greater stability. This may be achieved by greater longevity of genetic resistance.

The development of new virulent races was studied by Flor (1940, 1946, 1947, 1955, 1956, 1971) in the rust attacking flax.[18] Flor studied the genetics of both members of a host-parasite system and concluded that for each gene determining resistance in the flax plant, there is a specific and related gene determining virulence in the rust. Flor's classic piece of work led to the concept of a "gene-for-gene" relationship that was found to be applicable to most host-parasite relationships in which resistance is conditioned by major genes. Person (1959) concluded that gene-for-gene relationships are to be expected as a general rule and not as an isolated event. Virulence in the parasite occurs in a stepwise fashion. In essence the adaptive forces in the pathogen population are proportional to the magnitude of the resistance in the host population, and vice versa.

The gene-for-gene concept extends beyond that of disease resistance and includes specific races of Hessian fly[19] able to attack resistant wheat cultivars (Hatchett and Gallun, 1970). Eleven genes for resistance to the Hessian fly have been identified (Oellermann, Patterson, and Gallun, 1983; Stebbins, Patterson, and Gallun, 1983).

The implications to plant breeding of a gene-for-gene relationship were summarized by Kiyosawa (1972) as follows:

---

[17]Rice blast—*Piricularia oryzae.*

[18]Flax rust—*Melampsora lini.*

[19]Hessian fly—*Mayetioloa destructor.*

1. There are many gene pairs for resistance and avirulence.
2. Highly specific relationships exist between a resistant and an avirulent gene.
3. The number of resistance genes depends upon the number of avirulence genes contained in the fungus strains used and vice versa. This point can be used to explain why a susceptible cultivar is often not susceptible in another country.
4. Resistance usually is dominant over susceptibility.
5. Races can be identified readily when single genes for susceptibility and resistance are found. Differential cultivars can be used to identify the specific physiologic race present.

The interactions between host and parasite based on the gene-for-gene relationship may be explained as follows. A "universal suscept" is a cultivar that possesses no gene for resistance and is attacked by every race of the pathogen, regardless of the number of additional genes for virulence they possess. Race R_1 limited to attacking cultivar C_1, can attack no cultivar with any additional genes for resistance and possesses no additional genes for virulence. C_2 is susceptible to only one race, R_2. Other races with one or more genes for virulence can attack all cultivars except those whose genes for resistance are not overcome by related genes for virulence in the race. Only those genes for resistance that are not overcome by related genes for virulence present in the parasite population are of value to the plant breeder to incorporate into a cultivar.

The commercial life of a cultivar was reported by Kiyosawa (1972) to be influenced by:

1. The amount of pathogen around the field, which is related to the extent to which susceptible cultivars are produced.
2. Mutation frequency of an allele for avirulence to virulence.
3. The amount of virulent physiologic races present at the time of release of the cultivar.
4. Multiplication rate of virulent races between seasons.

Work on gene-for-gene relationships has made it abundantly clear that additional pest-management aspects in addition to major-gene resistance are needed to provide stability to cereal cultivars.

Multilines

The idea that a mixture may reduce the incidence of foliar diseases has led to the development of a multiline. A multiline is defined as a blend or mixture of cultivars that are genetically very similar because they are derived from a series of crosses with disease-resistant cultivars that are backcrossed to a common parent. Multilines are formulated to include a number of nearly identical cultivars but differing in genes having specific resistance. Multilines are distinct from blends, which are formulated without specific guidelines.

A multiline may exhibit stability for the following reasons;

1. Resistant components of the multiline mixture check the spread of an outbreak of a virulent race in a specific field.
2. By checking the spread of a virulent race, a multiline cultivar may have an extended commercial life span.
3. A multiline will check the spread over a geographical area, thereby providing wide geographical stability.

There is evidence that multilines do provide insurance against failure due to virulent pest strains (Anonymous, 1976), but stability is difficult to evaluate using standard experimental plot techniques. In the absence of a pest attack, basic principles of crop competition dictate that grain yield of multiline cultivars generally will be lower than a pure-line cultivar or of the multiline components grown separately. In tests to this effect, only composite No. 3 outyielded the **check cultivar** and then by only a small margin (Table 13-3). It is under conditions of a pest outbreak that the real value of multilines can be demonstrated and then mainly under field conditions rather than small-plot tests. In 1976 at Toluca, Mexico, disease destroyed the check cultivar but several multilines withstood the attack (Anonymous, 1976).

Table 13–3
Grain Yield of Twelve Multiline Composites of Spring Wheat in Comparison to a Check Cultivar, Siete Cerros, and to the Average Yield of Their Components When Grown in Pure Culture: Test Conducted in 1975–1976 in the Yaqui Valley of Mexico

Composite Number with Number of Components in Brackets	Mean Yield of Composite in Percent of Check	Mean Yield of Components in Percent of Check
No. 1 (5)	90	97
No. 2 (4)	95	91
No. 3 (6)	102	94
No. 4 (9)	96	92
No. 5 (3)	95	92
No. 6 (5)	88	88
No. 7 (4)	79	87
No. 8 (4)	91	89
No. 9 (5)	86	89
No. 10 (3)	87	93
No. 11 (8)	87	87
No. 12 (11)	90	93
Check (Siete Cerros)	100	100
Actual yield (kg/ha)	7583	7626
(lb/ac)	6771	6809
Check (kg/ha)	8384	—
(lb/ac)	7486	—

Source: CIMMYT report on wheat improvement (1976); Anonymous (1976).

The multiline approach is regarded as a sound method for use of resistance genes (Browning, 1974; Browning and Frey, 1969). Within multiline cultivars, resistance genes may vary in time by replacement of component cultivars. Among cultivars, the resistance genes may vary by the use of different genes in the component cultivars. Together the two create a dynamic diversity, which can be controlled by humans (Parlevliet and Zadoks, 1977).

Several problems exist with multilines. First, they are expensive to produce and may slow crop improvement by taking most of the available resources. Second, it may be difficult to maintain sufficient uniformity in a multiline, particularly for characteristics such as baking quality in wheat. Thus, seed growers may have difficulty maintaining the original composition of the mixture if roguing is allowed. Production of the component cultivars of a multiline will add to seed production costs.

Crop Diversity

Trends toward monoculture production on specific fields as well as on large geographical regions reduce yield stability (Wade, 1972). In North America in 1953, rust destroyed 75% of durum and 35% of the bread wheat crops because large geographical areas seeded to common cultivars produced no restrictions to the rapid spread of the disease. Southern corn leaf blight reached epidemic proportions in North America in 1970. Most of the corn grown over a large geographic region contained genes for susceptibility (Tatum, 1971).

The alternative to uniformity is crop diversity (Adams, Ellingboe, and Rossman, 1971). To promote greater yield stability the uniform host must be replaced by a variable one, so that a pest capable of attacking a specific genotype will not prevail over a wide area. A highly specialized pest will be unsuited to a diversity of genotypes. Knott (1972) suggested that the path of rust in North America should be controlled by a patchwork of cultivars differing in resistance to rust races to serve as a selection sieve to achieve long-lasting control of the disease.

Generalized Resistance

Disease resistance in plants may be conditioned by many genes (polygenic resistance) or by a few genes (oligenic resistance). There is no dividing line between the two forms of resistance. Polygenic resistance acts by retarding the rate of disease increase rather than by delaying its onset, as is the case in specific resistance. With polygenic resistance, reduced amounts of disease will occur, but crops are protected from serious losses under normal field conditions. Generally, polygenic resistance is not rendered useless by changes in the pathogen and can provide lasting protection. The concept of polygenic resistance was first proposed by Van der Plank (1963).

Polygenic resistance is also referred to as race-nonspecific, generalized, or horizontal resistance, in contrast to vertical resistance which involves an interaction between cultivars of the host and races of the pathogen. One cultivar will be resistant to certain races of a pathogen but susceptible to others, whereas a second cultivar may show the opposite reactions. Simons (1972) reviewed polygenic resistance and suggested that resistance is horizontal when it is spread evenly against all races of the pathogen and there is no interaction. With horizontal resistance, a horizontal line results when the reaction of a cultivar is plotted against races of the pathogen—hence the term horizontal.

Rust reactions of vertical and horizontal resistance may be obscured by maturity of the crop, and initial infectibility early in the growing season is important in determining subsequent disease damage. In the case of wheat leaf rust,[20] exclusion of the rust by stomata of various plant parts was suggested by Romig and Caldwell (1964) to illustrate the diversity of horizontal resistance. They observed that leaf rust developed abundantly on the leaf blades of some cultivars but that the sheaths and culms were relatively free of infection. Exclusion of the rust spores by stomata was probably effective against all races of the fungus and therefore could be regarded as general or horizontal resistance.

In the case of loose smut of barley and ergot (footnotes 3 and 16, respectively), resistance is achieved by the fact that the glumes remain closed during and after pollination, so that spores of these fungi cannot enter the floret. This is a form of horizontal or general resistance, as all races are excluded.

Multilines, consisting of several cultivars containing different genes for vertical resistance that restrict the infection site and process of parasites, provide a form of stability. Multilines exhibit horizontal resistance by curbing disease increase by reducing the amount of effective inoculum of the right races at the right places (Nelson, 1972).

The polygenic approach was regarded by Parlevliet and Zadoks (1977) as a sound method of achieving yield stability and cultivar longevity. They suggested that because breeders cannot recognize the different minor genes, there is a high probability that the various cultivars produced contain sets of polygenes differing from one another. The replacement of cultivars over time for agronomic reasons results in the pathogen population's being confronted with different sets of polygenes in time as well as in space. A form of dynamic diversity is achieved, similar to that produced through the multiline approach.

The multiline and polygene approach can provide effective yield stability, but reliance on these measures alone is not advised. Multiline and polygene cultivars should be backstopped by other disease-management measures to assure that resistance will be adequate and that pest resistance will last for the commercial life of the cultivar.

Breeding for Stability

Polygenically inherited resistance to disease to provide yield stability and to extend the com-

[20] Wheat leaf rust—*Puccinia recondita*.

mercial life of a cultivar may not require any different breeding programs than for any other quantitative character. Breeding programs designed to develop horizontal resistance may, however, require recognition of the objectives and use of techniques to identify the desired plants. Examples of techniques used may illustrate specific approaches.

To develop horizontal resistance to northern corn leaf blight,[21] nine corn populations were grown in the field and inoculated with the disease (Jenkins and Robert, 1959; Jenkins, Robert, and Findley, Jr., 1954). At pollination, plants with the fewest diseased leaves were selected for interpollination. Pollen was collected and mixed and then used to pollinate the plants from which it had been collected. Seed harvested from these plants was mixed to produce the next cycle. After three generations of such recurrent selection, resistance was increased, suggesting that a form of general resistance had been achieved.

For self-pollinated crops, Jensen (1970 and 1978) proposed a diallel selective mating system to provide breeders of self-pollinated crops ways to incorporate broader germplasm into breeding projects through a two-stage crossing scheme, involving single crosses as parents in diallel double-crossing arrangements. The resulting double-cross seeds are blended for the next generation of selection. The important features of the diallel selective mating system are a widening of the germplasm base of the breeding project, economy of effort required, and helpful assistance by competition.

In an effort to increase cultivar stability in barley, a plant population of a composite cross was grown over a long time period in three environments in California and Montana (Hockett et al., 1983). When grown in sixteen diverse environments, yield stability was found to be inconsistent, although some populations were superior in yield stability. If composite-cross population breeding methods are combined with other breeding methods and if such populations are continually upgraded, a means of increasing cultivar stability may be developed. Composite crosses in conjunction with male sterility allow for natural cross-pollination and exchange of germplasm.

Diversity in crops grown and cultivars used and the use of stable cultivars achieved by multilines or horizontal resistance may reduce the degree of acquired resistance in pests.

[21] Northern corn leaf blight—*Helminthosporium turcicum*.

Acquired Resistance to Pesticides

A commonly recognized phenomenon is the gradual loss of pesticide effectiveness on a given pest population after repeated use. This loss of effectiveness results from development of a resistant form or **biotype** in the pest. Resistance to pesticides can be found among many organisms, including plant pathogens (Johnson, 1972; Wade, 1982); weeds (Bandeen and McLaren, 1976; Gasquez and Compoint, 1981; Holliday and Putwain, 1977; Ryan, 1970; Thompson, Schumacher, and Rieck, 1974; Warwick and Black, 1980); insects (Perrin, 1983; Sawicki, 1979); and nematodes (Johnson, 1972).

Acquired resistance or tolerance of pests to pesticides has been known since the early 1900s. By the mid-fifties, however, only about 25 pest species were known to have developed resistance to pesticides (Anonymous, 1982). By the early 1980s, Perrin (1983) reported over 400 instances of species resistant to one or more types of chemicals, 260 of these being agricultural pests. Worldwide, Sawicki (1979) suggested at least 350 species of insect pests were resistant to insecticides to which they were originally susceptible. Resistance to individual pesticides as well as to whole classes of pesticides has occurred. Day (1983) reported 61 cases of pathogen resistance to 45 chemicals in one fungicidal class. At least 30 common weed species are known to have developed resistance to one important class of herbicides.

Development of resistance is a result of genetic selection through the mechanism of survival of the fittest and happens in all forms of pests. Once resistance to a chemical occurs, the use of that chemical must be abandoned or modified.

Pesticide resistance has presented a worldwide challenge to pest control using chemical means. The significance of this phenomenon at the field level can be evaluated on the experience of a chemical used to control rice blast (footnote 17). Within ten years after a new chemical control measure was introduced, resistant strains of rice blast were identified in a district of Japan where the pesticide had been used intensively and exclusively. Within a year, 97% of the blast pathogen in that district appeared to be resistant. Within three years of stopping application of the chemical, the level of resistance had dropped below 20% (Anonymous, 1982). This permitted renewed use of the chemical, but at levels that did not place high selection pressure upon the pathogen.

The development of new pesticides is extended over a long test period and is expensive, so that obsolete pesticides cannot be replaced fast enough to combat pesticide resistance. Despite this, tenuous management programs dependent largely on chemicals still exist. Perrin (1983) pointed out, however, that resistance usually occurs in isolated pockets, notably where pesticide usage has been most intense, and that pest resistance is not serious on a macro scale. It is apparent, however, that pesticides do not offer a panacea to pest management.

The development of resistance to pesticides can be delayed by several methods suggested by Day (1983):

1. Use pesticides that break down quickly when applied.
2. Use pesticides to which no known forms of resistance exist.
3. Avoid continuous-release formulations.
4. Apply pesticides when the pest is most vulnerable and under the best environmental conditions.
5. Use pesticides intermittently at a minimum dosage rate but sufficient to destroy all the pests contacted.
6. Avoid sweeping, areawide campaigns, and use pesticides locally only when needed.
7. Use band applications or side-dressing instead of broadcast applications to minimize the amount applied.
8. Use unrelated pesticides in rotation.
9. Combine control measures where possible.

IMPACT OF PEST MANAGEMENT ON CROP PRODUCTION

Effective modern pest management has had a major impact on cropping practices and has meant considerably more than increasing crop yield or freeing workers from specific tasks that traditional pest-management practices demanded. One aspect of pest management, weed control, has allowed for major changes in cropping systems. In row crops, such as sorghum and corn, weed control has been spectacular (Figure 13-15) and in these crops has had a major impact. The practices of after-harvest cultivation or summer fallowing to encourage the germination of weed seeds, and the removal of seedlings by frequent cultivation, upon which weed control once depended, are no longer necessary.

Figure 13–15. Effective pest management such as weed control, as demonstrated in this corn plot, has meant much more than a reduction in interrow cultivation or freedom from hand hoeing. Additional changes that have become possible include changes in plant architecture and morphology, changes in rotational systems, opportunities for minimum or zero-tillage, and an associated contained weed-seed population. (Courtesy of Ontario Ministry of Agriculture and Food)

Chemical control of vegetation has introduced new concepts in soil preparation and tillage. Tillage serves only as a temporary means of weed control, as the soil contains many ungerminated weed seeds. Plowing may bury weed seeds at a depth that prevents germination but may also expose other, once deeply buried weed seeds to conditions inducive to germination. A deeply buried or contained weed-seed population will not be spread by wind, birds, or surface erosion and will result in an extended dormancy period for such seeds. Surface seeds that germinate can be removed by a herbicide.

The ability to control vegetation without primary tillage has made minimum or zero-tillage a possibility (Cussons, 1980). Elimination of plowing and secondary tillage has overcome some of the problems associated with intensive row-cropping and erosion.

Successful pest management has resulted in more intensive forms of agriculture, such as monoculture production and a shift to corn or sorghum. In turn, greater stress has been placed on pest-management systems. Effective weed control in row crops such as corn and sorghum has allowed for narrow rows and higher plant populations.

Management of weeds means that crop plants no longer have to be able to compete for environmental factors. This fact has allowed plant breeders to effectively develop plants with short stature, upright leaves, or uniculm characteristics.

Effective modern pest-management systems are essential to the success of new innovations in cultural practices, architectural and morphological changes, and rotational systems. Additional food needs will increase the stress on pest-management systems, and a continued struggle will be required to assure success.

Questions

1. *a.* Name major pests of cereal crops grown in your area.
   b. Name pests that occur every year as well as ones that are cyclical in nature.
   c. Describe the extent to which specific pest-management inputs are employed in the production of cereal crops in your area and compare these with inputs in high-value crops in which cosmetic appearances are important, such as in fruit and vegetable production.
   d. Consider the control measures that are routinely adopted to control pests, both direct and indirect, in your region.
2. Review the reasons why cereal crop losses from pests are of the magnitude of 33% even in developed countries, where the technology for pest management is available.
3. Consider reasons why pest losses in rice reported at 47% are much higher than in other cereals, especially rye.
4. Outline precautions that should be taken when pesticides are used in relation to:
   a. Storage.
   b. Disposal of pesticide containers and contaminated material.
   c. Protective clothing.
   d. Sprayer calibration.
   e. Pesticide drift.
   f. Honeybee protection.
   g. Seed treatment.
5. Suggest examples or describe forms of integrated pest management.
6. *a.* Standard plot-testing techniques have developed whereby yield performance, lodging resistance, plant height, or disease reaction can be evaluated. How can the stability element of a multiline cultivar be evaluated?
   b. How effective are standard plot-testing techniques to determine horizontal or general resistance to disease?
   c. What is the best criterion for measuring the stability of multilines or horizontal resistance in cultivars?

7. Review the impact of effective modern pest management on crop production.
8. Suggest, with reasons, what you consider to be the weakest link in pest management of cereal crops in your region.
9. Consider the problems associated with pest control in the sense of pest elimination or eradication in comparison with pest management.
10. With the aid of modern science, the alkaloids contained in ergot have been identified and their effects on animals or humans determined. One of the most powerful of the alkaloids present is an hallucinogenic drug.
    a. Review the effects of ergot consumption on human health before the problem of eating ergot-infected food was realized (see Caporael, 1976).
    b. Speculate on the impact of ergot infection on rye bread consumption and the related production of the rye crop.
11. Review and evaluate problems associated with multilines.
12. Knott (1972) suggested a patchwork of cultivars differing in resistance to rust races be planted across the length and breadth of the wheat-growing regions of North America. Such a patchwork of cultivars could serve as a sieve to restrict the "Puccinia Path" across the continent. Consider the problems and merits of this proposal.
13. Compare major gene resistance with that of polygene or multiline stability in terms of:
    a. Presence or absence of disease symptoms early in the season.
    b. Presence or absence of disease symptoms late in the season.
    c. Impact of disease on yield in the short and long term.
    d. The length of commercial life.
14. Review the impact of effective pest management on crop production in your area in terms of:
    a. Changes in cultural practices.
    b. Any apparent beneficial impact.
    c. Any apparent undesirable impact.

References

Adams, M. W., A. H. Ellingboe, and E. C. Rossman. 1971. "Biological Uniformity and Disease Epidemics." *BioScience* 21: 1067–1070.

Anonymous. 1976. *CIMMYT Report on Wheat Improvement 1976*. El Batan, Mexico. pp. 5–6.

———. 1982. "Pest Resistance Poses Challenge to Chemical Control." *Ceres* 15 (4): 4–5.

Bandeen, J. D., and R. D. McLaren. 1976. "Resistance of *Chenopodium album* L. to triazine herbicides." *Canadian Journal of Plant Science* 56: 41–42.

Barrons, K. C. 1981. *Are Pesticides Really Necessary?* Chicago: Regnery Gateway. 246 pp.

Batra, S. W. T. 1982. "Biological Control in Agroecosystems." *Science* 215: 134–139.

Browning, J. A. 1974. "Relevance of Knowledge About Natural Ecosystems to Development of Pest Programs for Agro-Ecosystems." *Proceedings of the American Phytopathological Society* 1: 190–199.

———, and K. J. Frey. 1969. "Multiline Cultivars as a Means of Disease Control." *Annual Review of Phytopathology* 7: 355–382.

Caporael, L. R. 1976. "Ergotism: The Satan Loosed in Salem?" *Science* 192: 21–26.

Carl, K. P. 1982. "Biological Control of Native Pests by Introduced Natural Enemies." Review Article. *Biocontrol News and Information* 3: 191–200.

Cussons, G. W. 1980. "Weeds in Cereals and Their Control." *Span* 23 (1): 30–32.

Day, B. E. (chairman). 1983. "Pesticides and the 'Superpest Problem'." *Science of Food and Agriculture 1 (2): 22–27.*

Dicke, F. F. 1972. "Philosophy on the Biological Control of Insect Pests." *Journal of Environmental Quality* 1: 249–253.

Edgington, L. V., et al. 1980. "Systemic Fungicides: A Perspective After 10 Years." *Plant Disease* 64: 19–23.

Ennis, W. B. Jr., W. M. Dowler, and W. K. Klassen. 1975. "Crop Protection to Increase Food Supplies." *Science* 188: 593–598.

Flor, H. H. 1940. "New Physiologic Races of Flax Rust." *Journal of Agricultural Research* 60: 575–591.

———. 1946. "Genetics of Pathogenicity in *Melampsora lini*." *Journal of Agricultural Research* 73: 335–357.

———. 1947. "Inheritance of Reaction to Rust in Flax." *Journal of Agricultural Research* 74: 241–262.

———. 1955. "Host-Parasite Interaction in Flax-Rust—Its Genetics and Other Implications." *Phytopathology* 45: 680–685.

———. 1956. "The Complementary Genetic Systems in Flax and Flax Rust." *Advances in Genetics* 8: 29–54.

———. 1971. "Current Status of the Gene-for-Gene Concept." *Annual Review of Phytopathology* 9: 275–296.

Gasquez, J., and J. P. Compoint. 1981. "Isoenzymatic Variations in Populations of *Chenopodium album* L. Resistant and Susceptible to Triazines." *Agro-Ecosystems* 7: 1–10.

Geissbuhler, H., P. Brenneisen, and H. P. Fischer. 1982. "Frontiers in Crop Production: Chemical Research Objectives." *Science* 217: 505–510.

Gibson, P. T., and R. K. Maiti. 1983. "Trichomes in Segregating Generation of Sorghum Matings. I. Inheritance of Presence and Density." *Crop Science* 23: 73–75.

Gooding, E. G. B. 1983. "Pesticides in the Third World. A Comment from the Tropics." *World Crops* 35: 90–92.

Grainger, J. 1982. "Maximum World Crop Production. Part 10: Controlling Crop Losses." *World Crops* 34 (5), 180–182.

Haskell, P. T. 1982. "Can Tropical Crops Be Effectively Protected?" *World Crops* 34: 108–109.

Hatchett, J. H., and R. L. Gallun. 1970. "Genetics of the Ability of the Hessian Fly, *Mayetiola destructor*, to Survive on Wheats Having Different Genes for Resistance." *Annals of the Entomological Society of America* 63: 1400–1407.

Hockett, E. A., et al. 1983. "Effects of Natural Selection in Advanced Generations of Barley Composite Cross II." *Crop Science* 23: 752–756.

Hofstra, G., and D. P. Ormrod. 1979. "Air Pollution and Crop Damage." *Highlights of Agricultural Research in Ontario* 2 (2): 7–9.

Holliday, R. J., and P. D. Putwain. 1977. "Evolution of Resistance to Simazine in *Senecio vulgaris* L." *Weed Research* 17: 291–296.

Jenkins, M. T., and A. L. Robert. 1959. "Evaluating the Breeding Potential of Inbred Lines of Corn Resistant to the Leaf Blight Caused by *Helminthosporium turcicum*." *Agronomy Journal* 51: 93–96.

———, and W. R. Findlay, Jr. 1954. "Recurrent Selection as a Method for Concentrating Genes for Resistance to *Helminthosporium turcicum* Leaf Blight in Corn." *Agronomy Journal* 46: 89–94.

Jensen, N. F. 1952. "Intra-Varietal Diversification in Oat Breeding." *Agronomy Journal* 44: 30–31.

———. 1970. "A Diallel Selective Mating System for Cereal Breeding." *Crop Science* 10: 629–635.

———. 1978. "Composite Breeding Methods and the DSM System in Cereals." *Crop Science* 10: 629–635.

Johnson, H. W. 1972. "Development of Crop Resistance to Diseases and Nematodes." *Journal of Environmental Quality* 1: 23–27.

Kiyosawa, S. 1972. "Genetics of Blast Resistance." (In) *Rice Breeding*. Los Banos, Philippines: International Rice Research Institute. pp. 203–225.

Knott, D. R. 1972. "Using Race-Specific Resistance to Manage the Evolution of Plant Pathogens." *Journal of Environmental Quality* 1: 227–231.

Litsinger, J. A., and K. Moody. 1977. "Integrated Pest Management in Multiple Cropping Systems." (In) *Multiple Cropping*. Madison, Wis.: American Society of Agronomy. Special Publication No. 27. pp. 293–316.

Mahn, E. G., and K. Helmecke. 1979. "Effects of Herbicide Treatment on the Structure and Functioning of Agro-Ecosystems. II. Structural Changes in the Plant Community after the Application of Herbicides over Several Years." *Agro-Ecosystems* 5: 159–179.

Maiti, R. K., and P. T. Gibson. 1983. "Trichomes in Segregating Generations of Sorghum Matings. II. Association with Shootfly Resistance." *Crop Science* 23: 76–79.

McEwen, F. L. 1978. "Food Production—The Challenge for Pesticides." *BioScience* 28: 773–777.

———, and J. C. Sutton. 1977. "Pest Monitoring to Reduce Pesticide Use." *Notes on Agriculture* 8 (1): 20–23.

Murphy, H. C. 1965. "Protection of Oats and Other Cereal Crops During Production." (In) G. W. Irving, Jr., and S. R. Hoover (eds.), *Food Quality: Effects of Production Practices and Processing*. American Association for the Advancement of Science. Publication 77. pp. 99–113.

Nelson, R. R. 1972. "Stabilizing Racial Populations of Plant Pathogens by Use of Resistance Genes." *Journal of Environmental Quality* 1: 220–227.

Oellermann, C. M., F. L. Patterson, and R. L. Gallun. 1983. "Inheritance of Resistance in 'Luso' Wheat to Hessian Fly." *Crop Science* 23: 221–224.

Parlevliet, J. E., and J. C. Zadoks. 1977. "The Integrated Concept of Disease Resistance; A New View Including Horizontal and Vertical Resistance in Plants." *Euphytica* 26: 5–21.

Perrin, R. M. 1983. "Pesticide Resistance in Aphids." *World Crops* 35: 93–94.

Person, C. 1959. "Gene-for-Gene Relationships in Host : Parasite Systems." *Canadian Journal of Botany* 37: 1101–1130.

Pimentel, D., and C. A. Edwards. 1982. "Pesticides and Ecosystems." *BioScience* 32: 595–600.

Pimentel, D., et al. 1978. "Benefits and Costs of Pesticides Use in U.S. Food Production." *BioScience* 28: 772–784.

Platford, R. G., and C. C. Bernier. 1970. "Resistance to *Claviceps purpurea* in Spring and Durum Wheat." *Nature* 226: 770.

Ringlund, K., and E. H. Everson. 1968. "Leaf Pubescence in Common Wheat, *Triticum aestivum* L., and Resistance to the Cereal Leaf Beetle, *Oulema melanopus* (L.)." *Crop Science* 8: 705–710.

Rodgers, D. M., J. P. Murphy, and K. J. Frey. 1983. "Impact of Plant Breeding on the Grain Yield and Genetic Diversity of Spring Oats." *Crop Science* 23: 737–740.

Rohweder, D. A., W. D. Shrader, and W. C. Templeton, Jr. 1977. "Legumes. What Is Their Place in Today's Agriculture?" *Crops and Soils* 29 (6): 11–14.

Romig, R. W., and R. M. Caldwell. 1964. "Stomatal Exclusion of *Puccinia recondita* by Wheat Peduncles and Sheaths." *Phytopathology* 54: 214–218.

Rowell, J. B. 1972. "Fungicidal Management of Pathogen Populations." *Journal of Environmental Quality* 1: 216–220.

Ryan, G. E. 1970. "Resistance of Common Groundsel to Simazine and Atrazine." *Weed Science* 18: 614–616.

Sawicki, R. M. 1979. "Resistance to Pesticides. 1. Resistance of Insects to Insecticides." *Span* 22 (2): 50–52, 87.

Sheard, R. W. 1977. "Forages in a Rotation." University of Guelph, *Notes on Agriculture* 8 (3): 11–13.

Silverstein, R. M. 1981. "Pheromones: Background and Potential for Use in Insect Pest Control." *Science* 213: 1326–1332.

Simons, M. D. 1972. "Polygenic Resistance to Plant Disease and Its Use in Breeding Resistant Cultivars." *Journal of Environmental Quality* 1: 232–240.

Stebbins, N. B., F. L. Patterson, and R. L. Gallun. 1983. "Inheritance of Resistance of PI 94587 Wheat to Biotypes B and D of Hessian Fly." *Crop Science* 23: 251–253.

Tatum, L. A. 1971. "The Southern Corn Leaf Blight Epidemic." *Science* 171: 1113–1116.

Tette, J. P. 1974. "Pheromones in Insect Population Management." (In) M. C. Birch (ed.), *Pheromones*. New York: American Elsevier Publishing Co., Inc. pp. 399–407.

Thompson, L. J., R. W. Schumacher, and C. E. Rieck. 1974. "An Atrazine Resistant Strain of Redroot Pigweed." *Weed Science Society of America Abstracts*, p. 196.

Van der Plank, J. E. 1963. *Plant Disease: Epidemics and Control*. New York and London: Academic Press. 349 pp.

Wade, M. 1982. "Resistance to Fungicides." *Span* 25 (1): 8–10.

Wade, N. 1972. "A Message from Corn Blight: The Dangers of Uniformity." *Science* 177: 678–679.

Warwick, S. I., and L. Black. 1980. "Uniparent Inheritance of Atrazine Resistance in *Chenopodium album*." *Canadian Journal of Plant Science* 59: 269–270.

Webster, J. A. 1972. "Developing Resistance to Cereal Leaf Beetle: Morphological Basis, Problems and Progress." *Proceedings North Central Branch, Entomological Society of America* 27: 98–102.

———, D. H. Smith, Jr., and R. P. Hoxie. 1982. "Effect of Cereal Leaf Beetle on the Yields of Resistant and Susceptible Winter Wheat." *Crop Science* 22: 836–840.

Weir, D., and M. Shapiro. 1981. *Circle of Poison: Pesticides and People in a Hungry World*. San Francisco: Institute for Food Development Policy. 96 pp.

Wellso, S. G. 1982. "Cereal Leaf Beetle Under Control." *Crops and Soils* 39 (9): 7–9.

Zillinsky, F. J. 1983. *Common Diseases of Small Grain Cereals: A Guide to Identification*. CIMMYT. Mexico. 141 pp.

14

International Links to Meet World Food Needs

THE WORLD FOOD ECONOMY

World population is predicted to increase to 6.35 billion in 2000, a rise of over 50% from the 1975 level of 4 billion (Barney, 1980). By the year 2030 world population may be possibly 10 billion. Expressed otherwise, the annual increase of 1.8% may decline marginally to 1.7%, which means that 90 to 100 million people are added each year. Ninety percent of population growth will occur in the poorest countries, so that by 2000, 79% of the world population will be concentrated in the less-developed regions of the world (Table 14-1).

World food production over the period 1970 to 2000 is projected to increase 2.2% annually. Since the bulk of this increase will occur in countries that already have relatively high food production, food supplies in major populated areas of the world scarcely will improve or actually will decline below adequate levels.

Arable land is projected to increase only 4% by 2000, which means that most of the increased food production will have to come from higher yields per unit area. In the early 1970s, one hectare (2.47 ac) of arable land supported an average of 2.6 persons; by 2000 one hectare will have to support four persons. The area planted to cereal crops will increase, but not at a rate sufficient to maintain the area of land required to support each person (Table 14-2). Unfortunately most of the common inputs that contribute to higher yields, such as fertilizer, pesticides, machinery manufacture and operation, and irrigation, all depend to a large degree on expensive and depleting petroleum fuels.

Trend analysis into the not too distant future clearly indicates regional food gaps (Barr, 1981). The prospects are that as the maximum carrying capacity of these areas is reached and exceeded, two problems become paramount—preserving the carrying capacity and sustaining a reasonable quality of life. An acceleration of the rate of deterioration and loss of essential agricultural resources might occur through soil erosion, loss of plant nutrients, soil compaction, salinization, alkalinization and waterlogging, loss of cropland to urban development, crop damage and yield reduction due to increasing air and water pollution, the extinction of wild crop strains, and more frequent and severe regional shortages of quality water. Even optimists must recognize that the world food economy faces serious problems that are close upon us. The undermining of food production must be approached immediately with vigor and determination.

Table 14–1
Population Projections for the Last Quarter of the Twentieth Century for the World and Major Regions

Area	Population (millions) 1975	Population (millions) 2000	Percent Increase by 2000	Average Annual Percent Increase	Percent of World Population in 2000
World	4090	6351	55	1.8	100
More-developed regions	1131	1323	17	0.6	21
Less-developed regions	2959	5028	70	2.1	79
Major countries or regions:					
Africa	399	814	104	2.9	13
Asia and Oceania	2274	3630	60	1.9	57
Latin America	325	637	96	2.7	10
USSR and Eastern Europe	384	460	20	0.7	7
North America, Western Europe, Japan, Australia, and New Zealand	708	809	14	0.5	13
People's Republic of China	935	1329	42	1.4	21
India	618	1021	65	2.0	16

Source: Barney (1980).

FOOD-PRODUCTION CHALLENGES

Recognition of a few key cereal grains as the world's chief food source makes these crops the focal point for action of an international nature. The Great Plains of North America and of other exporting nations cannot be expected to balance world food deficiencies. The growing demand for grain is illustrated in Table 14-3. Prior to World War II, all geographic regions listed, except Western Europe, were net exporters. North America, one of several exporting countries but not the leading one, emerged from 1948 to 1980 as the major grain supplier in the export trade. North American surpluses can supplement world production but cannot begin to meet world needs. As well, North American grain is produced by enormous inputs of expensive fossil fuels for mechanization. As petroleum prices rise, so do grain prices—to the point where poor nations cannot afford to purchase.

There are no quick and easy solutions to grain-production problems. Mere increases of yield-enhancing, energy-intensive inputs may by themselves not be successful. National food economies are highly integrated and interde-

Table 14–2
World Population and Area Planted to Cereals, 1950 and 1980, with Projections to 2000

Year	Population (billions)	Area in Cereals (millions) ha	Area in Cereals (millions) ac	Area per Person ha	Area per Person ac
1950	2.51	601	1484	0.24	0.59
1980	4.42	758	1872	0.17	0.42
2000	6.20	828	2045	0.13	0.32

Source: Brown (1981).

Table 14-3
The Changing Pattern of World Grain Trade from 1934 to 1980, Expressed in Millions of Metric Tonnes of Net Exports and Net Imports[a]

Region	1934–1938	1948–1952	1960	1970	1976	1980
North America	+5	+23	+39	+56	+94	+131
Latin America	+9	+1	0	+4	−3	−10
Western Europe	−24	−22	−25	−30	−17	−16
Eastern Europe and the USSR	+5	—	0	0	−27	−46
Africa	+1	0	−2	−5	−10	−15
Asia	+2	−6	−17	−37	−47	−63
Australia and New Zealand	+3	+3	+6	+12	+8	+19

Source: Data from Brown (1975 to 1981).
[a]A plus sign indicates net exports, a minus sign net imports.

pendent. Decisions made within a given country may send shock waves through the entire food system. For example, the formation of a cartel by the Organization of Petroleum Exporting Countries (**OPEC**) in 1972 to hike the price of petroleum increased the cost of food production substantially; Latin America's failure to curtail population growth during the sixties and seventies resulted in this area's becoming a net grain importer rather than an exporter (Table 14-3) and a drain on the world food economy; shortfalls in grain production in the Soviet Union and the Soviet decision to import grain to maintain their livestock industry, rather than reduce consumption, triggered price increases around the world in 1973.

FOOD PRODUCTION IN THE TROPICS AND SUBTROPICS

It is in the tropical regions of the world, where most of the world's undernourished people live, that the outcome of the race between world food production and population largely will be determined. A climatically favored zone lies between the Tropic of Cancer and the Tropic of Capricorn at 23° 27′ north and south latitude, respectively. This area, in which the sun's rays are vertical between June 21 in the north and December 21 in the south, delineates the tropics. From an agricultural standpoint, this area is important for the following reasons:

1. About two-thirds of the world's people live in Asia, Africa, the northern part of South America, and the Caribbean (Table 14-1).
2. Although acid, infertile soils dominate tropical regions, favorable light and temperature conditions frequently exist for year-round crop production.
3. Tropical and subtropical regions are the homelands from which all our modern grain crops originated. The importation of tropically adapted plants and the need to find cultivars adapted to temperate-zone conditions may explain why plant breeding and production research became concentrated in temperate zones. To obtain new breeding material it is necessary to return to the genetic homeland to find new genes for resistance to be bred back into modern cultivars. Interest must be sparked to preserve wild plants growing along roadsides, waste areas, or uncultivated fields for future crop improvement. This interest can be developed through greater emphasis on plant breeding and research in tropical areas.
4. The tropics and subtropics account for 16.9% of the potentially arable land areas of the world, compared to 7.3% in the temperate zone (Table 14-4).
5. During the colonial period, little effort was devoted in the developing countries toward developing domestically used food crops, including cereal grains. Most crop research was directed to indigenous tropical cash crops for export to temperate areas where such crops could not be grown. At the conclusion of the colonial period, newly independent nations lacked research organizations directed at ce-

Table 14-4
Potentially Arable Land in Different Climatic Zones

Climatic Zone	Potentially Arable Land Hectares (billions)	Acres (billions)	Percent
Cold-temperate boreal	0.05	0.12	0.4
Cool-temperate	0.91	2.24	6.9
Warm-temperate			
Subtropical	0.55	1.37	4.2
Tropical	1.67	4.13	12.7
Total	3.18	7.86	24.2

Source: Data from Thomas, Curl, and Bennett, Sr. (1982).

real grains. The result was a basic and urgent need to produce cereal grains in the tropical areas and to develop research programs and related infrastructure.

INTERNATIONAL COOPERATION IN RESEARCH AND DEVELOPMENT FOR CEREAL GRAINS

Successes with Wheat

The first international effort at cooperation in cereal research, production, and training began in 1943, when the Government of Mexico and the private American Rockefeller Foundation entered into a cooperative program. The development of high-yielding cultivars of wheat and corn was a primary concern right from the start.

To speed up the wheat-breeding process and to produce a cultivar in as short a time as possible, two generations were grown annually at two sites with contrasting photoperiods and elevations. The main breeding nursery was grown in the winter at about 28° N latitude at only a few meters elevation above sea level; the second and summer breeding site was at 2200–2600 m (7200–8500 ft) elevation near Mexico City. Testing at two distant and distinct environments not only allowed for increased opportunity to screen for disease and yield performance, but resulted in the development of cultivars with wide adaptability. Among other factors, wide adaptability meant insensitivity to daylength or photoperiod. The degree of flexibility resulting from daylength insensitivity proved to be an asset, essential for international testing and cooperation.

An understanding of the value of daylength insensitivity and its action was clearly demonstrated in the first Inter-American Yield Test in 1959, in which wheat cultivars were grown in many countries and under a wide range of conditions. The cultivars separated themselves according to their adaptation to specific environments (Borlaug, 1972). The Canadian cultivars were triggered into reproductive development by the long days around June 21 in the northern hemisphere. When seeded in the fall, or in the subtropics where daylengths seldom reach the 14 hours needed to trigger reproductive growth, reproductive development did not occur.

Canadian wheat cultivars were unable to function economically below 39° N latitude. This prevented them from being used commercially or for breeding stock below this latitude. By selecting for daylength insensitivity, it was possible to modify the "biological clocks" of wheat so that they could be grown over a wide geographic range and without regard to planting dates, which made production and crossing much more flexible.

A second major advance was the incorporation of semidwarf genes found in Japanese wheats to reduce plant height and lodging.

Height reductions of 40% to 60% occurred. The Japanese semidwarf cultivars, introduced into the United States in 1947, were not adapted directly to North American growing conditions. The cultivar Gaines, selected from a cross with the semidwarf cultivar, produced world-record yields in the irrigated regions of the Pacific Northwest.

By incorporating the semidwarf genes into Mexican wheats, cultivars were produced that exhibited the semidwarf feature with wide adaptation.

The semidwarf feature was important not only because it allowed for increased grain production by a better partitioning of photosynthate into grain and straw, but also because of an increase in tillers with large ears and many grains per spikelet. Under conditions of good lodging resistance, production could be promoted by increased nitrogen fertility.

Producing cultivars with genetic disease resistance, semidwarfness, and photoperiodic insensitivity by themselves will not result in a yield breakthrough. A production package was developed that manipulated simultaneously and harmoniously fertilizer practices, pest and weed control, and moisture management. The production package, properly conducted, resulted in a substantial increase in grain yield. On-farm comparisons were an essential feature so that production increases could be readily observed. Farmer suspicions were dispelled, and in turn area farmers became enthusiastic advocates for new techniques.

Wide adaptability allowed for the launching of an international effort to extend the new and superior genotypes. Superior Mexican wheat cultivars were tested in 115 countries.

Performance under conditions of irrigation, pest control, and adequate fertilization resulted in spectacular yield increases over traditional cultivars in many countries. The impact of the new wheats is illustrated by wheat production in India. Wheat output rose from a 1967 output of 12.3 to 23.2 million tonnes in 1972, a feat unmatched by any major country in history.

The new wheat was widely accepted. By 1970, the semidwarf wheat cultivars were seeded on 10 million hectares (25 million acres) and in 1977 rose to 30 million hectares (74 million ac) or 48% of the developing countries' wheat land (Havener, 1980).

Successes with Corn

Traditionally, farmers in developing countries select superior open-pollinated corn cultivars and perpetuate them. An initial step aimed at corn-yield increases in Mexico was to collect thousands of cultivars developed over many centuries. From these, genotypes with disease resistance and high yield were selected, or produced by crossing, and sent out for regional testing and local use or for crossing with the best local cultivars. Cross-pollinated cultivars were maintained for a rapid initial advance and to avoid the high degree of technology involved in the development and maintenance of inbreds, the orderly crossing required to develop an F_1 hybrid, and the annual procurement of seed. F_1 hybrids are not practical without a well-developed infrastructure to produce and distribute adequate seed annually. Traditional small farmers in developing countries may realize the necessity of obtaining hybrid seed for each crop but may lack the funds to spend on commercial seed. Under conditions of yield-limiting inputs, hybrid corn is not likely to show any advantage over open-pollinated cultivars.

The major breakthrough in corn breeding came through the development of cultivars with wide adaptation, multiple disease and pest resistance, and improved protein quality.

Despite vigorous and productive research efforts on corn, the results have been less spectacular than with wheat and rice. Perhaps this can be explained by the fact that yield stability rather than yield ability was the primary motivation. Perhaps the vigor and yield associated with F_1 hybrids is required rather than open-pollinated cultivars.

Successes with Rice

The international success of the wheat program in Mexico led to the establishment in 1960 of the International Rice Research Institute (IRRI) in the Philippines. Following the example with wheat, a breeding objective was to develop high-yielding, short-strawed, lodging-resistant, nitrogen-responsive rice cultivars. Photoperiod insensitivity was an early breeding objective to meet the international scope of the institute. Under certain conditions, photoperiod sensitivity is desired to ensure that reproductive development coincides with cyclical environmental conditions.

A cross between a tall but vigorous rice cultivar from Indonesia and a semidwarf rice from Taiwan produced a superior cultivar named IR-8. This outstanding cultivar, if properly managed under the appropriate growing conditions, was capable of producing double the yield of most local cultivars. For the five crop years of 1967–1968 through 1971–1972, rice production in the Philippines increased an average of 12.4% annually (Jennings, 1974). The improved rice cultivars were widely adapted throughout the rice-growing regions of the world.

In both the high-yielding wheat and rice cultivars, plants with erect leaves were developed. Erect leaves permit light to penetrate more readily into the plant canopy, causing lower internodes to become well differentiated and lodging resistant rather than being elongated and weak. A better distribution of light more evenly over all the leaves produces a high rate of photosynthesis, with more photosynthate being directed into increased grain production and less into stalk development. Increased light penetration and photosynthesis promotes more productive and synchronous tiller development.

THE GREEN REVOLUTION

The improved yield potential of the new wheat and rice cultivars resulted in dramatic increases in output in developing countries in the tropics and subtropics and came to be described in glowing journalistic terms as the green revolution.

The term green revolution implied a simple, clever, convenient, effective, and rapid method of increasing crop production. Because of the apparent potency of the green revolution, a dangerous feeling of complacency developed. In reality, the green revolution was developed through a very complex mixture of technologies and strategies. A major increase in food production occurred, but formidable difficulties developed. The green revolution did not provide a once-and-for-all solution but rather helped mold agricultural strategies. It bought time to put these strategies in place (Borlaug, 1972).

At first, instead of welcoming the increased production, critics of the new and high-yielding cultivars assailed the high inputs needed to produce record yields. The high cost resulted in an unbalanced distribution of benefits to growers who could afford the inputs. Those who benefitted purchased land and machines, displacing former tenants, who became a landless group of unemployed. Rapid increases in production resulted in storage and transportation problems. Traditional importing countries became self-sufficient or exported surpluses and upset traditional markets. Inability to handle surplus production resulted in depressed prices or no markets at all.

A severe blow was dealt the green revolution in 1972 when fossil fuels, needed for nitrogen fertilizer manufacture, transportation of agricultural inputs and produce, machinery manufacture and operation, and petroleum-based pesticides, started to rise dramatically in price. The inputs needed to sustain the impetus of the green revolution became either too expensive or in short supply. Even with adequate fossil fuels, food production might not have kept pace. Although by 1972–1973 the additional yields were enough to feed about 160 million more people at the average consumption level of a developing country, the relentless growth in population made such advances inadequate (Chang, 1979).

The success of the green revolution must continue to be recognized as a model of what can be accomplished when international cooperation and research efforts are established. The green-revolution scientists demonstrated the opportunity for achievement, pride, and dignity associated with agricultural work. Perhaps few other areas of endeavor can offer greater personal reward than that of providing the means for increased food production.

The green revolution demonstrated only a beginning of the impact that scientific progress can achieve when conducted on a cooperative basis at the international level with sustained support and encouragement.

THE DEVELOPMENT OF A NETWORK OF INTERNATIONAL AGRICULTURAL RESEARCH CENTERS

As a parallel to the 1960 establishment of the International Rice Research Institute (IRRI), in 1966 the International Maize and Wheat Improvement Center (CIMMYT)[1] was formally established as an outgrowth of the cooperative program begun in 1943. Although work at CIMMYT was initially directed at wheat and corn, other grain crops such as barley were included, and later triticale was added.

The successful research and its widespread application that originated with the IRRI and CIMMYT programs led to the development of a series of international agricultural research centers[2] with five objectives:

1. The increase of production of the major foods in each region, involving not only crops, but also livestock and their diseases.
2. The training of people to increase food production in each region.
3. The strengthening of research and development capabilities of developing countries by the training of scientists from many developing countries.
4. The meeting of specific needs of agricultural producers, specifically that of improving the lot of the many, small, poor farmers in developing countries. The development of low-input soil-management technology has been included. The technique aims at making the most efficient use of scarce, purchased inputs by using cultivars that are more tolerant to existing soil constraints, thus decreasing the rate of fertilizer application while attaining reasonable but not necessarily maximum yields (Sanchez and Salinas, 1981).
5. The conservation and utilization of plant genetic resources.

Success of the IRRI and CIMMYT programs appeared to be fostered by a number of factors, and these were incorporated into each new center:

1. Carefully chosen scientists with outstanding ability.
2. The establishment of a team of scientists with interests in many areas, such as extension, pathology, plant breeding, physiology, and crop production.
3. Excellent research facilities.
4. A network of stations for testing and screening large plant populations for superior plants.
5. On-station specialist training programs. (See Figure 14-1.)

The establishment of a network of international centers is a conscious effort to bring the

Figure 14-1. Trainees from developing countries with ICRISAT principal training officer in an experimental sorghum field. Part of the success of the international agricultural research centers has been the development of extensive training programs. (Courtesy of the International Crops Research Institute for the Semi-Arid Tropics, ICRISAT)

[1]The acronym CIMMYT is derived from the Spanish terminology—Centro Internacional de Mejoramiento de Maiz y Trigo.

[2]Like IRRI and CIMMYT, most came to be identified by their acronyms.

world's major food commodities (numbering only fifteen) and the climatic zones of the developing world under study. Although the centers are strategically located to serve specific areas, many aspects of the research programs are of worldwide interest.

In 1971, after years of operation under the private American Rockefeller and Ford Foundations, and after several additional research centers had been added, the international center concept became the responsibility of the Consultative Group on International Agricultural Research (CGIAR), often shortened to the Consultative Group. CGIAR is an informal association of countries, international and regional organizations, and private foundations. The World Bank, the Food and Agricultural Organization (FAO), and the United Nations Development Program (UNDP) are cosponsors. Financial resources are obtained from contributing members (Bernstein, 1975; Bunting, 1981; Plucknett and Smith, 1982; Wade, 1975).

Crops receiving attention are not exclusively cereal crops. A listing of the international centers serves to emphasize the scope, location, and individual emphasis.

The International Rice Research Institute (IRRI)

IRRI is a one-crop center that within ten years left an indelible mark on rice production in the developing nations and set a pattern for institutional development. IRRI's impact has been expressed through development of outstanding rice cultivars, international testing programs, training programs, the assembling and distributing of germplasm, and the adopting of a systems approach to cropping programs.

The scope of IRRI's work on rice is illustrated by the International Rice Testing Program, involving a worldwide network for monitoring and testing rice. The program's purposes are:

1. To make the world's best rice cultivars available to rice scientists for their direct use in production or as parental material in breeding programs.
2. To provide rice scientists with an opportunity to test their cultivars over a wide range of environments.
3. To identify rice cultivars with broad resistance to major diseases, insects, and other stresses.
4. To monitor and assess the genetic variation of pathogens and insects.
5. To serve as a center for the exchange of information on the interaction between cultivars and the diverse rice-growing environments in the world.
6. To promote interaction among the world's rice scientists.

The magnitude of international rice testing is illustrated by the fact that as early as 1975, 82,000 seed packets were distributed in 457 nursery sets—mostly to Asian countries. By 1980, worldwide distribution meant a tripling of the program (Anonymous, 1981).

The result of testing hundreds of thousands of rice cultivars at hundreds of sites has provided an insight as to how new cultivars respond to biological, physical, and chemical stresses. Constraints to production can be identified and possibly overcome. A sense of community and a strong base of cooperation fostered by the testing network has helped in finding solutions to high and sustained rice production.

The International Wheat and Corn Improvement Center (CIMMYT)

CIMMYT is located in Mexico. Improving the quantity, quality, and stability of wheat, corn, barley, and triticale are the center's primary objectives, achieved through on-farm demonstrations and international plot testing. Each year CIMMYT prepares and sends out about 3000 cultivars for tests in 130 countries (Havener, 1980).

In its training program, over 1000 young agricultural scientists from over 70 countries have spent one crop season at the center, obtaining progeny of crosses for further selection and development in their homeland, and acquiring research skills through practical experience.

The International Center for Tropical Agriculture (CIAT)

CIAT, located in Colombia, South America, and established in 1967, is responsible for helping to develop the agricultural capability of the lowland tropics of the world. Emphasis is directed toward corn, rice, cassava, field beans, beef cattle, and pigs (Bower, 1976).

Close coordination between the rice programs at IRRI and CIAT has resulted in rapid yield advances through the development of improved cultivars adapted to the area. Open-pollinated corn cultivars with reduced height and improved amino acid composition adapted to the region have been developed through cooperation with CIMMYT.

An extensive outreach program is aimed at dealing with the soil problems of Latin America, which may include infertility, acidity, or poor physical structure common to **lateritic soil.** Latin America has a high rate of population and food production increases are urgently needed.

The International Institute of Tropical Agriculture (IITA)

IITA, established in 1967 in Nigeria, has as its primary objective to improve the quality and quantity of major food crops in the humid lowland tropics. IITA's mandate involves more than developing new cultivars and introducing new agronomic practices; it includes finding solutions to the problems associated with replacing shifting cultivation with more productive land-use systems.

The main emphasis of the IITA cereal improvement program is on corn and rice in cooperation with CIMMYT and IRRI. Research is also directed to cassava, yams, and grain legumes (cowpeas, lima beans, soybeans, and pigeon peas).

The International Potato Center (CIP)

CIP was established in 1971 in Peru as a one-crop potato center. Less than 5% of the genetic material available for improving the crop was used in the past for adapting the crop to European conditions (Sleeth, 1977). Many largely unexploited genetic resources suggest the potato as a prime subject for genetic improvement.

The International Crops Research Institute for the Semi-Arid Tropics (ICRISAT)

ICRISAT, established in 1972 in India, was the first unit to be developed by the Consultative Group. The task of ICRISAT is to improve crops and farming methods in the harsh region of limited and erratic rainfall and poor soils, found in parts of 50 countries and five continents. Nonirrigated crops of sorghum, pearl millet, chickpea, pigeon pea, and peanut (groundnut) were chosen for dryland research.

Research at ICRISAT is aimed primarily toward assisting small farmers of limited means, but large landowners would benefit as well. Progress has been made in the development of cultivars better suited to minimal fertilizer applications that maximize the use of water and in the development of cropping systems that employ simple but effective tools, including bullock-drawn implements.

Advanced breeding programs with pearl millet and sorghum have shown considerable promise in improving cultivars of these crops.

The International Laboratory for Research on Animal Diseases (ILRAD)

ILRAD was established in Kenya in 1973 to assist in the development of effective control measures for internationally important animal diseases through research and training programs. Diseases are one of the major constraints to the production, utilization, and food productivity of livestock in developing countries. Healthy livestock that act as gleaners, harvesters, and converters of crop products and land areas that might otherwise go to waste can contribute to the human food supply.

The International Livestock Center for Africa (ILCA)

ILCA was established in Ethiopia in 1974 with the aim of supporting systems of production and marketing of livestock and their products in tropical Africa in order to improve the standard of living of the region's people. The three principal activities of ILCA include research, training, and documentation (Anonymous, 1978).

The West Africa Rice Development Association (WARDA)

WARDA, located in Liberia, became part of the network in 1974 with the task of developing the tremendous potential in physical and ecological resources available for rice production in West Africa and making the region self-sufficient in rice. WARDA is a cooperative, intergovernmental, regional organization involving fourteen West African countries. To increase the quantity and quality of rice production in West Africa, emphasis is placed on appropriate research, development, and training programs, involved with rice.

The International Board for Plant Genetic Resources (IBPGR)

IBPGR was established at FAO headquarters in Rome, Italy, in 1974 to assist and encourage the collection, preservation, and exchange of plant genetic material. IBPGR represents an important step in coordinating gene banks, since the centers of diversity of many of the cereal crops are in the developing world. As outlined in Chapter 3, the genetic base of most of the important food crops has been rapidly narrowing, and in developing countries many traditional cultivars have been abandoned (Plucknett et al., 1983). For example, 95% of Greek wheat cultivars have been abandoned and replaced with superior genotypes since World War II. Virtually all the sorghum cultivars of South Africa have been replaced with high-yielding hybrids from Texas (Plucknett et al., 1983). Many of these traditional cultivars represent 10,000 years of evolution toward domesticated types and, along with their wild ancestors, comprise an irreplaceable reservoir of genetic material for continued improvement. The value of gene banks was evidenced when genotypes with superior amino acid quality were sought (Chapter 12).

The International Center for Agricultural Research in the Dry Areas (ICARDA)

ICARDA, located in Aleppo, Syria, was established in 1976 with the purpose of helping to increase and stabilize food production in developing countries of the temperate zone having an arid or semiarid climate (Darling, 1979). ICARDA's mandate encompasses an ecological zone made up primarily of areas with 300 to 800 mm (11.8 to 31.5 in.) of winter rainfall and includes North Africa, Iran, Afghanistan, Pakistan, part of the Sudan, Turkey, and parts of northern India and Tibet.

Research and training is directed to the improvement of farming productivity and quality of the principal crop of these dry, temperate areas and to the increasing of goat and sheep output. Barley, durum wheat, lentils, and broad beans are included in the research program, and corn and sorghum cultivars from CIMMYT and ICRISAT are tested for adaptation.

The Network and Additional Centers

The eleven international centers just described form a network of strategically located sites that serve developing countries in the tropics and subtropics, where research programs have historically been deficient. The locations of these eleven centers are shown in Figure 14-2. Most major ecological areas are served. Collectively the major populated land area is serviced, and hopefully higher yields will occur from associated research.

Additional centers have been developed to supplement or serve specific aspects:

The International Food Policy Research Institute (IFPRI), located in Washington, D.C., entered the system in 1975. The Institute serves to analyze strategies, monitor food production,

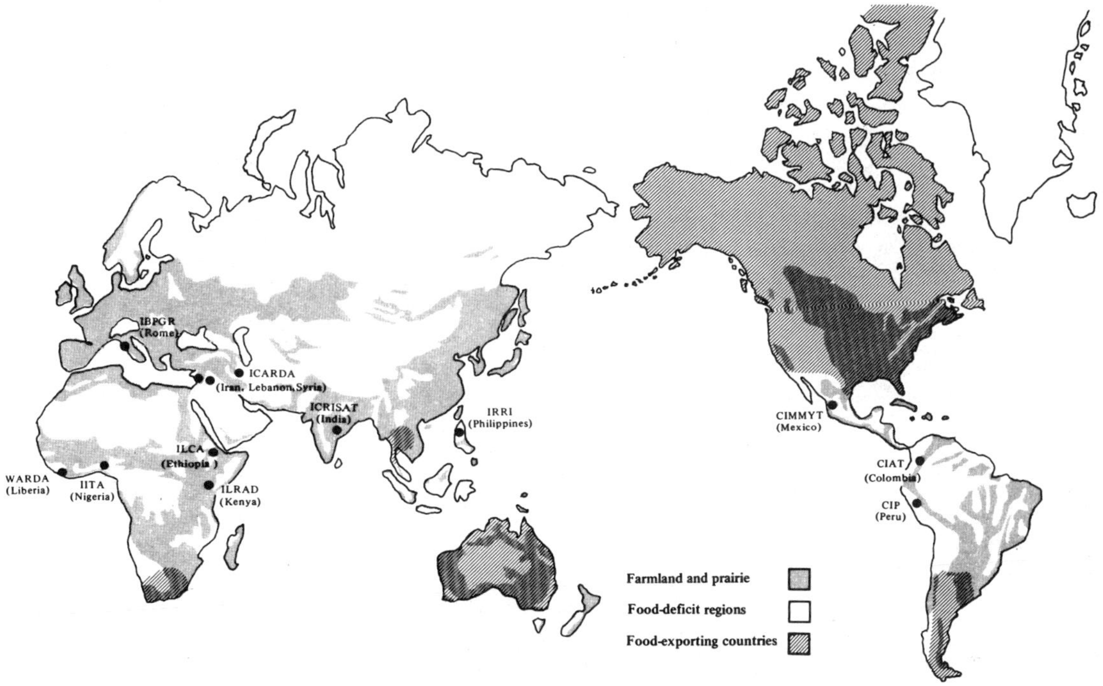

Figure 14–2. Geographic location of twelve International Agricultural Research Centers established by the Consultative Group on International Agricultural Research (CGIAR). (From the Consultative Group on International Research, United Nations Development Program, New York, 1976, pp. 6 and 7)

and recommend action and priorities to aid in meeting world food needs. In short, research is aimed at developing a global food policy.

The International Service for National Agricultural Research (ISNAR), located in Holland and established in 1979, was developed to help strengthen national agricultural programs so that research results benefit the inhabitants of developing countries. The objective is to strengthen the capacity of national agricultural research programs. The international research institutes described in this chapter are not intended to replace research at the national level. Many developing countries have effective national research programs that play a vital role in the transfer of technology and therefore are an essential link in the system. In other developing countries, national research programs are not effective or virtually lacking; ISNAR was developed to help strengthen national research in such countries.

The Asian Vegetable Research and Development Center (AVRDC), located in Taiwan, reports to and is funded by some Consultative Group members but is not included in the CGIAR group (Cowell and Luh, 1978).

The International Fertilizer Development Center (IFDC) has a similar arrangement to that of AVRDC and is located in Alabama in the United States. It was established in 1975 to develop new and improved fertilizer materials and processes.

THE INTERNATIONAL CENTERS IN PERSPECTIVE

Perhaps there are no shortcuts to overcoming the lack of agricultural technology among developing world farmers. It takes years of dedicated work before agricultural research findings from the laboratory or the fields are implemented on the fields of world farmers. The international system of centers is a positive move to speed up agricultural progress and production in the developing countries. The present system of international research was established, not as a permanent structure, but rather as an immediately needed input that is flexible to change as needs change. The success of such a system can be judged only over an extended time span.

International centers provide a means of concentrating strong scientific teams where they are needed without government constraints. In short, the network of centers was designed to extend the green revolution from wheat and rice to all the major crops in the developing world.

Although the centers are devoted to improving crops and cropping programs in the developing areas of the world, it is readily apparent that the developed world might benefit as well. A classic example where work in one area contributed to another occurred in British India. Indian scientists patiently developed a wheat that could ripen quickly between **monsoon** seasons. This short-season cultivar was used subsequently in adapting Marquis wheat, which become Canada's premium grain. The short, dry, northern growing season of the Canadian prairies proved well suited to this cultivar.

CIMMYT work on triticale and spring × winter wheat crosses has led to genotypes of value in North America.

Positive performance of the international centers can be identified as follows:

Table 14–5
Global Cereal Crop Production Changes over the Decade 1969–1971 to 1979–1981

	Production (in millions)				
	1969–1971[a]		1979–1981[a]		
Crop and Area	tonnes	tons	tonnes	tons	Percent Change
Developing world:					
Wheat	96.7	106.4	156.2	171.8	+61.5
Rice	286.2	314.8	370.8	407.9	+29.6
Barley	18.9	20.8	21.0	23.1	+11.1
Corn	102.3	112.5	148.5	163.5	+45.2
Rye	2.4	2.6	2.2	2.4	−8.3
Oat	2.1	2.3	2.1	2.3	0
Millet	30.6	33.7	26.8	29.5	−12.4
Sorghum	34.6	38.1	42.7	47.0	+23.4
Total	577.7	635.5	775.2	852.7	+34.2
Developed world:					
Wheat	231.2	254.3	287.7	316.5	+24.4
Rice	23.6	26.0	25.4	27.9	+7.6
Barley	106.4	117.0	137.5	151.3	+29.2
Corn	180.7	198.8	272.8	300.1	+50.1
Rye	28.4	31.2	22.7	61.3	−20.1
Oat	52.4	57.6	41.4	45.5	−26.6
Millet	2.6	2.9	1.7	1.9	−52.9
Sorghum	21.0	23.1	21.6	23.8	+2.9
Total	654.1	719.5	817.6	899.4	+25.0
World total	1231.8	1355.0	1592.8	1752.1	+29.3

Source: FAO Production Yearbook, 1981 (FAO, 1982).
[a]A three-year mean is taken to help level out variation due to environment.

Table 14-6
Cereal Crop Yields Changes over the Decade 1969–1971 to 1979–1981

	Yield per Unit Area				
	1969–1971		*1979–1981*		*Percent*
Crop and Area	kg/ha	lb/ac	kg/ha	lb/ac	*Change*
Developing world:					
Wheat	1149	1026	1622	1448	+41.2
Rice	2230	1991	2680	2393	+20.2
Barley	1063	949	1297	1158	+22.0
Corn	1466	1309	1903	1699	+29.1
Rye	997	890	1240	1107	+24.3
Oat	995	888	1275	1138	+2.8
Millet	697	622	666	595	−4.4
Sorghum	839	749	1071	956	+27.7
Average change	1463	1306	1859	1660	+27.1
Developed world:					
Wheat	1796	1604	2072	1850	+15.4
Rice	5202	4645	5243	4681	+0.8
Barley	2169	1937	2134	1905	−1.6
Corn	4041	3608	5293	4726	+31.0
Rye	1608	1436	1657	1479	+3.0
Oat	1856	1657	1710	1527	−7.9
Millet	886	791	596	532	−32.7
Sorghum	3109	2776	3321	2965	+6.8
Average change	2278	2034	2629	2347	+15.4
Overall change	1806	1613	2188	1954	+21.2

Source: FAO Production Yearbook, 1981 (FAO, 1982)

1. Food grain production of eight major cereal crops in the decade 1969–1971 to 1979–1981 increased when expensive fossil fuels and related nitrogen fertilizer inputs could have restricted output (Table 14-5). Part of this increase could have resulted from a shift away from other crops, but an examination of Table 14-6 reveals that yields per unit area did increase. The tables reveal that total productivity and yield per unit area in the developing world both show a greater increase than in the developed world.
2. CGIAR has brought together, and is building upon, a large body of expertise and knowledge related to the basic food commodities of the developing world. The full impact of this mobilization may not be realized for some years.
3. CGIAR has undertaken a long-term project with continued support to conduct productive long-term programs. Technologies leading to increased productivity cannot be developed by short-term and ad hoc projects. Support for international agricultural research has been faced with budgetary constraints and inflation (Lewin, 1982).
4. The international system has developed an intense, effective, and nonpolitical international network of communication to exchange information and genetic material. Past obstacles created by shifting, uncertain political and bureaucratic government bodies have been removed. Resources available for international agricultural research are used more effectively.
5. The system has elevated agricultural research to a level of prestige and respect it has been denied in the past in both developed and developing countries. Such recognition may help agriculture attract capable minds as well as other resources.

Agricultural research by itself cannot be expected to solve the food problems of the developing world. Economic, ecological, and social issues must be tackled. Specific problems include harvest crop losses and an effective agricultural extension service.

INTERNATIONAL MARKETING

The homogeneous and durable nature of cereal grains provides a product that is readily standardized, graded, stored, and shipped—features essential to successful year-round marketing on an international basis.

Quantities of cereal grains traded on world markets must be recognized as reaching substantial proportions, but quantities are small compared to world production or total demand. In 1980/81, 93.5 million tonnes of wheat and 104.2 million tonnes (102.9 million and 114.6 million tons) entered world trade (Simkins, 1981). Wheat was the largest single commodity traded in 1980/81 and amounted to 44% of the world grain exports. Milled rice accounted for about 6%, and coarse grains, of which the chief ones traded are corn, sorghum, and barley, accounted for 50%. The major suppliers for world export markets in 1980/81 included the United States, Canada, Australia, Argentina, and France, with the United States contributing nearly half of the total commodities traded.

A marketing system that endeavors to supply consumers worldwide involves considerable risk. It is impossible for any marketing system to exert any control over the countless grain growers, over the vast hordes of consumers, over climate and weather, pests, or any other factor that influences the supply and demand for grain. Marketing of one commodity is influenced by the availability of other commodities and by livestock consumption. Markets can change rapidly (Burley, 1983).

The result of so many uncontrollable factors is that marketing is a very complex and intricate system that must be highly flexible and sensitive to economic, political, and international developments that might stimulate or disrupt the flow of grain. A crop failure in a distant area of the world may be accompanied by purchases to avoid a deficiency that in turn may force grain prices to rise throughout the world. The resulting price increases have a vital bearing on the quantity and hence the price of virtually all necessities as well as luxuries for the world's consumers. No nation, region, or person is immune to the impact of the grain marketing system on the daily supply and price of food.

Basic Features of the Marketing System

Any marketing system is based on the premise that sellers and buyers must be brought together. A successful system will provide producers with an opportunity to obtain a maximum return on their produce yet provide consumers with the hope of purchasing their needs at a reasonable price. These two aspects are not in opposition to each other. A shortage in one region may cause a price rise in that area, but simultaneously a surplus in another region may cause a price reduction. If a marketing system brings the surplus into the deficient region, the consumers there will benefit; additional sales will benefit the producers in the surplus region. The fact that only 10% of all food produced by all nations reaches the export-import market means that market manipulation cannot be practiced to any great extent. Price stability is to the benefit of all concerned over the long-term.

The grain marketing system therefore serves to provide an orderly flow of grain on a worldwide basis. The marketing system allows for sales of new crop grain well in advance of harvest time. This allows regions with a late harvest to participate fully in supplying market requirements under competition with other countries. Advanced or "short" selling makes for a market situation with continuity and stability.

At the national or local level, an organized marketing system helps to avoid a glut of grain following the harvest. A glutted market results in depressed prices. As the glut is used up, supplies become short and prices rise. A miller producing flour for bread on a year-round basis may have more wheat offered at harvest time than handling facilities will permit. Toward the end of the crop year, this same miller may have to pay premium prices to stay in business. A steady and reliable market is obviously desirable.

FUTURES MARKETING

Cyclical variations in supply are solved through a **futures contract** that provide an incentive for traders, millers, merchants, and exporters to buy and sell grain all through the year. Futures contracts provide a guaranteed price at the time of delivery for some future date, thereby providing an incentive to hold grain in storage.

Futures markets do not exist for the exclusive use of the large exporters or dealers. Farm producers can use the futures market effectively in three major ways:

1. Contracting for the production of a crop before planting, or fixing the price of a crop before harvest. If grain futures appear to offer a price at which a grower should sell all or part of his prospective production, he sells futures contracts in advance. At harvest, if the price is higher than the contract calls for, he is obliged to sell at the agreed price, but the reverse is true also.
2. Removing the speculation associated with holding a crop in storage on the farm. Futures contracts provide a means of using on-farm storage space without the risk of a price decline. If the futures price for an advance date is higher than the price at harvest time, the grower sells futures in an amount equal to his stored crop.
3. Retaining possession of grain that has been produced. The method involves sale of the crop at harvest and the purchase of an equal amount of futures contracts. If the futures price does not meet on-farm storage costs, such action might be warranted.

The primary purpose of futures contracts is to provide insurance against the uncertainties and risks of grain marketing brought about by price fluctuations. Actual deliveries of futures contracts may not occur, as futures contracts are completed or liquidated before the specified delivery date. Futures contracts are a marketing tool that is used effectively in meeting a basic requirement of a marketing system—to provide a suitable return to the producer yet provide the consumer with a reasonably priced commodity. The following points need to be emphasized:

1. A futures contract is not a physical purchase or sale, only an agreement between two parties to buy and sell a commodity.

2. Most futures contracts are offset by opposite transactions—a practice known as **hedging.** The hedge avoids risk against unexpected and adverse price movements. Hedging consists of selling for future delivery quantities of grain equal to the cash purchases of the trader, or in buying for future delivery quantities equal to those sold. Downward prices would give the grain buyer a loss when he came to sell at a later date. Since the futures contract, which was sold as a hedge, has also declined in price, the futures contract can be bought back at a profit. If the loss of one transaction is offset by the gain on the other, it is a perfect hedge. The perfect hedge seldom occurs, and some risk to a buyer or seller is involved. This risk must be borne by someone, and the one who knowingly assumes this risk is known as a speculator. This individual is in the same category as farm producers who hold grain on their farms in anticipation of a price increase.

Another feature of futures trading in cereals is cross trading. By buying futures in more than one commodity or by buying on more than one exchange, traders are able to spread their risks. Loss on one commodity can be offset by a gain on another.

3. Only members of commodity exchanges may trade futures contracts. Farm growers employ an agent of a commission house to act on their behalf. Money is deposited to guarantee performance. More money must be deposited as losses occur or may be withdrawn if profits develop. Grain buyers, sellers, and processors also can use futures trading to hedge their transactions against risk from price changes. Hedging is an indispensable operation in a marketing system.

4. Grain can be and is sold for cash or "spot" grain for immediate delivery. Such transactions are conducted by cash grain brokers who work directly with the buyer and the seller. Spot grain sales, however, invariably are hedged and are accompanied by the transfer of futures contracts. The spot grain prices are based on prices recorded in the futures market.

5. Futures contracts must be recognized as tools to be used in managing risks, generating equity, and providing flexibility in selecting the time to price a crop. The existence of the futures market and its use for hedging purposes provides producers with a higher price and maintains the market for spot grain on a more stable basis.

The ability of grain traders to buy and sell contracts for the delivery of grain in the future lends stability to grain prices by discounting those changing factors in the supply of and the demand for grain which otherwise would cause violent and disturbing fluctuations in current prices. Traders endeavor to anticipate the cumulative long-term effects on grain prices after the effects of the influences affecting the market have ceased to be potential and have become actual. Major fluctuations are smoothed out.

Critics of the futures markets must realize that grain price fluctuations are not caused by the futures market or by speculators. The marketing system endeavors only to foresee the course of the market, not to determine its course. It is supply and demand that govern price.

The complexities of getting grain crops from the farm gate to the final consumer should be viewed as a major service function of a marketing system. The complex and multiple activities affecting marketing functions are performed more and more by specialists and less and less by grain producers. An effective marketing system in India (Chopra, 1982) has helped that nation distribute food internally. India has developed one of the largest food buying-and-selling organizations in the world.

To facilitate national and international marketing of grain, the Chicago Board of Trade was organized in the United States in 1848 and the system of futures contracts was established in 1865. Throughout the world there are about 60 individual futures markets in agricultural commodities located in 20 countries (Rossides and Simkins, 1981). The Chicago Board of Trade remains the largest and most active grain commodity market.

The international grain commodity marketing system is dependent on effective national grain-handling systems. The price of a grain commodity depends on grade and quality, possibly protein content, location, the availability of transport trucks, rail cars, barges, and ships to transport grain, and the availability of grain workers and handlers to get the grain to market. The 60 individual futures markets plot these variations throughout the world and place bids. The system has been described as "having an intricacy equivalent to that of playing a hundred simultaneous chess games" (Morgan, 1979, p. 283).

Figure 14–3. **Primary elevators, also known as line or country elevators, are a familiar part of the landscape in Western Canada. These elevators accumulate grain produced in the area for shipment by rail to terminal elevators. (Ranson Photographers Ltd., Edmonton)**

Each nation determines the export grades for each commodity to be marketed and undertakes the marketing of its grain. The Canadian marketing system can serve as an example, because it is considered one of the most powerful and prestigious systems.

THE CANADIAN MARKETING SYSTEM

Marketing services involved in the Canadian grain marketing system include primary elevators,[3] (Figure 14-3), railroads, terminal elevators,

[3] Also known as line or country elevators. They are a familiar part of the landscape in Western Canada.

exporters, and shippers along with agencies to control and organize the system. These include the Canadian Wheat Board, the Canadian Grain Commission, and the Winnipeg Commodity Exchange.

The Canadian Wheat Board

In Canada a single agency, the Wheat Board, handles all transactions in wheat, oat, and barley produced in Western Canada except sales of domestic feed grains. The Board developed from cooperative marketing systems called "pools," based on a network of elevators that were constructed in the 1920s and 1930s. All wheat delivered to the primary elevators comes under the Board, allowing for orderly marketing under a single agency. In Australia a similar single agency exists, and in the United States a number of private marketing companies exist.

The Canadian Grain Commission

Buyers and sellers use grades as a measure of quality in relation to price. Grades may vary from area to area and season to season according to environmental conditions.

The Canada Grain Act, which regulates the sampling, grading, handling, weighing, storage, and transportation of grain in Canada, is administered by the Commission. Grades are based on a specific cultivar of red spring wheat (Marquis), durum wheat, and barley. All new Canadian cultivars must meet or surpass the established standard of quality for the two top grades. Canadian wheat is sold on the basis of three grades (Anonymous, 1971): Nos. 1, 2, and 3 Canada Western Red Spring Wheat. The No. 1 grade is segregated for protein levels of 12%, 13%, 14%, and 15%.[4] Generally, about 75% to 80% of the wheat produced in Western Canada and qualifying for this grade falls into the 13% and 14% protein categories. In addition to these criteria, the degree of soundness, hectoliter weight (weight per bushel), and purity are used to determine grades (Anonymous, 1971).

The Canadian Grain Commission operates a research laboratory concerned with the milling and baking quality of Canadian wheat and compiles and publishes basic statistics on grain handling and storage.

The Winnipeg Grain Exchange

Practically all firms and agencies involved in marketing Western Canadian grain are voluntary members of the exchange. Membership includes representatives from primary elevators, railways, ship owners, millers, and foreign grain companies. Members receive information gathered from other exchanges around the world. Data are provided on weather, crop conditions, and grain movement and storage.

The exchange is a focal point for an efficient grain marketing system in Canada, where both spot purchases and futures contracts are handled in international trade. In comparison, the London Grain Futures Market in England concentrates on home-grown cereals and less on the international markets. The development of the European Common Market and the price controls associated with the Common Agricultural Policy have reduced the relevance of grain futures trading within Western Europe (Simkins, 1981).

Components of the Marketing System

In Canada, farmer producers deliver their grain and depend on the primary elevator to merchandise it. If space is available, elevators will store grain at an agreed price. If required, the elevator may clean the grain. Additional services include sampling, moisture testing, weighing, truck dumping, elevating, and railcar loading.

Grain delivered to the primary elevator is bought on behalf on the Wheat Board, either at an established initial payment price or at open-market prices for the elevators' own account. If the grain is to be sold privately, some elevator companies will act as commission agents on behalf of the farmer. Producers may receive a cash advance from the elevator company on grain which has not been sold.

Movement of grain from the primary elevators to terminal elevators is the responsibility

[4]As determined by Kjeldahl nitrogen × 5.7 and expressed on a 13.5% moisture level.

Figure 14–4. Terminal elevators are an integral part of the movement of grain into domestic or world markets. Lots of grain large enough to service ocean or inland waterway ships and railroads economically are handled by terminal elevators. (Courtesy of Ontario Ministry of Agriculture and Food)

of the railroad. Shipments for export or to national sites for domestic use originate from the terminal elevators (Figure 14-4). Grain is graded, cleaned, and dried in lots large enough to service ocean or inland waterway ships and railroads economically.

In addition to hard red wheat, Canada also produces soft white wheat. To prevent mixing, soft white wheat is marketed through the Ontario Wheat Producers' Marketing Board. This agency operates independently of the Canadian Wheat Board. Soft wheat is produced in less volume than hard red wheat and does not appear in the futures commodity marketing system. A system of storage units (Figure 14-5) provides producers with an orderly procedure of marketing.

WEIGHTS AND MEASURES

For a successful international marketing system, a uniform set of weights and measures is required. The metric version of weights and measures that began in Paris, France, in 1791 and achieved international status in 1875, when 17 nations signed the Treaty of the Meter, has grown to include approximately 98% of the world's population. An enhanced version of the metric system announced in 1960 is known as the International System of Units (SI) and is described in Appendix B. Widespread adoption of the SI has helped unify international trade.

CONCLUSIONS ON INTERNATIONAL GRAIN MARKETING

The system of futures trading has been developed over the long term. Its adaptability, widespread adaptation, resilience, and resulting stability have allowed it to survive all kinds of political and financial upheavals, widespread inflation, energy crises, and fluctuating currency. It may not be perfect, but it is the best system yet devised. A more sophisticated system, or one able to react more rapidly to the needs of so vast a trade as that in cereals, is not likely to be developed easily. The futures trading system is likely to endure.

INTERNATIONAL FOOD RESERVES

Successful marketing systems are dependent upon the ability of the purchaser to pay. Despite

Figure 14–5. A network of elevators is used to operate a pool marketing system for soft wheat whereby grain is assembled, sold, and moved by the Ontario Wheat Producers' Marketing Board. (Courtesy of Ontario Ministry of Agriculture and Food)

adequate food supplies, food shortages are common because of inadequate distribution related to inadequate ability to purchase. Food shortages resulting from climatic variables can be expected somewhere in the world. The simultaneous occurrence of crop failure in two or more areas can increase demand and result in an increase in world prices beyond the reach of a poor, needy nation. Marketing systems cannot solve the resulting crisis.

Since cereal crops are the world's main food source, and because they can be stored with reasonable precautions, grain reserves have been proposed as a means of stabilizing price and food supply (Josling, 1973; Sanderson, 1975; Richter, 1980). The idea of a buffer stock in times of crisis is by no means a modern concept, but never in history has an international storage system been developed.

Consideration of an international grain reserve is based on the premise that periodic crop failure is inevitable somewhere in the world, that such natural disasters do not occur at the same time and to the same extent, that surplus production will occur simultaneously in another area of the world, and that some degree of international risk-sharing is appropriate.

An international grain reserve might be built upon the following principles:

1. That reserve stocks would be over and above normal, privately held carryover stocks and would be of sufficient size to meet the contingencies of crop failure(s) somewhere in the world.
2. That grain reserves would be built up when world grain prices were in line with production costs and would be released either for famine relief in developing countries or when world prices were abnormally high. Although such action would interfere with the functioning of the world grain markets, it would influence the market in a predictable fashion. Buffer stocks would tend to support farm incomes when prices were low and to avoid excessively high prices in times of shortages, and so they could serve the interest of farmers as well as consumers in both exporting and importing countries.
3. That grain reserves would be made available rapidly to those areas where an unexpected shortfall occurred due to natural disaster.
4. That the cost of international grain reserves would be shared by as many countries as wished to participate but that the costs borne by a recipient would not coincide with a disbursement period.
5. That an international reserve would have no other function, such as price support, although it might affect prices indirectly.
6. That the political sensibilities of countries regarding their food supplies would be respected.
7. That nations can be grouped into three categories;
    a. Surplus producers.
    b. Developing countries having financial and technical resources to increase their productivity or to purchase supplies from the world market.
    c. Developing countries having low income, amounting to nearly half of the world's population, who lack a production base to produce more food and who cannot afford to purchase grain, especially in times of high prices during tight supply periods.

Agreement generally exists that reserves are desirable to provide greater price stability, but positive action toward implementation has not occurred. Grain reserves remain controversial, especially in terms of the following issues:

1. Under conditions of abundant grain production, livestock numbers will be increased, which in part constitute a form of food reserve. The danger lies in the fact that dependence on livestock becomes a way of life, so that under shortfall conditions grains may be used to maintain livestock numbers.
2. The reserve could increase to unacceptable levels, thereby weakening both market demands and the reserve's effectiveness.
3. Fears exist that reserves might be used for other than emergency food aid if such action enhanced world prices. As a result, poor countries might not be able to afford to purchase grain.
4. Food aid is an unsuitable solution for extended food shortages. Giving food aid does not change the fundamental and underlying reasons for long-term inadequate food production.

5. Many nations lack the infrastructure of ports, railways, highways or other storage, processing, and distribution facilities to effectively distribute grain to the needy.
6. In the long-term view, it may be that fewer and fewer nations will be in a surplus-production situation to contribute to international reserves (refer to Table 14-3).

The result of such controversy is that no official grain reserves have ever been established. Unofficial farmer-held grain in countries with surplus grain production has been cited as a contingency against shortages. Such reserves are subject to embargoes and restraints of sale that may be imposed to exert political pressure on a nation. Such action could result in lasting damage.

In the past, under conditions of continued surplus production and growing carryover stocks of grain, some nations took land out of production. Idled cropland was a form of reserve that provided security to the entire world, as it could be returned quickly to production.

Finally, there is a moral reason for reserves. International grain reserves assure a reserve supply of food for the needy. As well, a sharing of the financial burden of providing grain disaster relief would be possible, rather than falling heavily on the countries having grain surpluses.

UNRESOLVED ISSUES RELATED TO GRAIN RESERVES

Unanswered questions that have been raised (Josling, 1973) are reviewed here in order to paint a complete picture and to illustrate the immensity of the problem.

What Should Be the Size of the Reserve?

The answer to this question may depend on the variability of grain production from major surplus producers, the degree of price instability to be tolerated, the cost of carryover stocks, export controls, and the degree to which expected shortfalls would be met.

What Would Be the Composition of the Reserve?

Wheat is the product most commonly in surplus, but at least a third of the world's people depend on rice for more than half their food.

How Should Allocations from the Reserve Be Made?

Release would be subject to agreed international guidelines, perhaps administered by FAO.

Where Should Reserves Be Held?

The world's exportable grain reserves historically have been stored in North America. Perhaps reserves could continue to be stored there under national control. A corollary question arises: Is the reserve an addition to domestically held stocks or simply a scheme to coordinate them?

Should Withdrawals Be Purchased by the Recipient, or Should Grain Be Given Free or on Soft Terms?

The cost of providing grain for a reserve cannot be borne by the producer if production is to continue. If a needy people cannot pay, the morality of providing surplus food in a civilized world becomes a consideration. Human nature is such, however, that free food comes to be considered as a right, when in reality there is no free lunch.

In the absence of government intervention, the price of reserve grain stocks would be influenced by the amount of grain held by private traders, processors, and farm producers, along with the demand for such private stocks.

How Should Storage and Transportation Costs Be Financed?

Such costs are of minor consideration when survival is in question. Costs could be minimized if grain could be stored in existing facilities in both exporting and importing countries. Costs of implementing a grain reserve system would depend on an annual storage cost, which would reflect the amortization costs of storage, construction, and interest. Other factors include inflation and the length of time grain is stored.

CONCLUSIONS RELATIVE TO INTERNATIONAL RESERVES

Because of the unresolved issues reviewed here, and perhaps others, a system of international grain reserves is not in place or even planned. Inability to resolve the issues in question might be interpreted as leaving the world without any guarantee of a stable agricultural order.

An international system of food reserves will not solve the basic problems—slowing population growth and speeding food production. Perhaps other international orders outlined in this chapter provide some security against famine.

QUESTIONS

1. Alarmists such as Robert Thomas Malthus as early as 1798 predicted population growth would outstrip food supply. Nearly 200 years later, there is less hunger and poverty, percentagewise, than in the time of Malthus.

   Consider the world food economy from your own perspective. Collect evidence to support or discredit the likelihood of widespread food shortages resulting in nutritional upset or starvation (see Burton, 1983).
2. *a.* Why have the Great Plains of North America emerged as the dominant global food supplier (Table 14-3)?

   b. Evaluate the accuracy of the statement that the Great Plains of North America are the breadbasket of the world.
3. *a.* Elaborate on the observation that yield advances in corn in Mexico have been less spectacular than in wheat because yield stability rather than yield ability was emphasized.

   b. Compare this observation to the data for wheat and corn in the developing and developed worlds in Tables 14-5 and 14-6.
4. *a.* Consider the problems and possibilities of F_1 hybrid corn in developing countries.

   b. Consider whether commercial F_1 hybrids of corn cultivars are desirable in developing countries or not, and, if they are desirable, how steps to this end may be achieved.
5. The green revolution resulted in "formidable difficulties" that have been elaborated upon extensively in the literature (see Chang, 1979; Dalrymple, 1975; Wilson, 1978).

   a. Consider whether the green revolution was a success or a failure and outline reasons.

   b. Determine the "problems" that were associated with the green revolution and suggest how they might be avoided in a future situation of rapidly increasing food production.

   c. How might inputs into the green revolution be modified in hindsight?
6. Consider and discuss the various aspects and importance of developing national and global food policies. (Review the objectives of the International Food Policy Research Institute.)

   a. Can surplus food producers such as the United States and Canada develop effective national food policies in the absence of a global food policy?

   b. What should be the objective(s) of national and global food policies?
7. Additional international agricultural research centers may be established in the future.

   a. Suggest agricultural endeavors in which world-class centers might be developed.

   b. Suggest locations for such centers and give reasons for these choices.

8. Consider Tables 14-5 and 14-6 in terms of:
   a. What they illustrate.
   b. Their validity.
   c. Factors affecting the values reported.
   d. Explanation of millet yields.
9. Consider technical aid and research development inputs as outlined in this chapter versus food aid.
10. Consider "nonbiological" constraints associated with food production. That is, in addition to suitable cultivars, soil, equipment, and fertilizers, what additional factors are required to meet the food needs of a nation or geographic region?
11. Review the concept of international grain reserves in light of other international activities outlined in this chapter.

References

Anonymous. 1971. *Canada's New Grades of Red Spring Wheat*. Canada Department of Agriculture. 16 pp.

———. 1978. "International Research in Agriculture. 3. ILCA: International Livestock Centre for Africa." *Span* 21 (2): 85–87.

———. 1981. "Monitoring Work Essential Element in International Rice Testing Program Success." *The IRRI Reporter* No. 4: 1–6.

Barney, G. O. (study director). 1980. *The Global 2000 Report to the President of the U.S.* Vol. I. New York: Pergamon Press. pp. 105–110.

Barr, T. N. 1981. "The World Food Situation and Global Grain Prospects." *Science* 214: 1087–1095.

Bernstein, J. 1975. "International Agricultural Research Centres." *Span* 18 (3): 98–100.

Borlaug, N. E. 1972. "Breeding Wheat for High Yield, Wide Adaptation, and Disease Resistance." (In) *Rice Breeding*. Los Banos, Philippines: International Rice Research Institute. pp. 581–592.

Bower, C. E. 1976. "International Research in Agriculture. 1. CIAT: Centro Internacional de Agricultura Tropical." *Span* 19 (3): 110–112.

Brown, L. R. 1975. *The Politics and Responsibility of the North American Breadbasket*. Washington D.C.: Worldwatch. 43 pp.

———. 1981. "World Population Growth, Soil Erosion and Food Security." *Science* 214: 995–1002.

Bunting, A. H. 1981. "CGIAR: The First Ten Years." *Span* 24 (1): 3–4, 15.

Burley, T. M. 1983. "Agricultural Commodities: The Changing Pattern of International Trade." *Span* 26 (1): 26–27.

Burton, G. W. 1983. "Was Malthus Right?" *Science of Food and Agriculture* 1 (2): 10–13.

Chang, T. T. 1979. "The Green Revolution's Second Decade." *Span* 22 (1): 2–3.

Chopra, P. 1982. "Feeding India: The State Role in Marketing." *Ceres* 15 (5): 39–42.

Cowell, R. L., and C. L. Luh. 1978. "International Research in Agriculture. 4. AVRDC: Asian Vegetable Research and Development Centre, Taiwan." *Span* 21 (3): 104–106.

Dalrymple, D. G. 1975. *Measuring the Green Revolution: The Impact of Research on Wheat and Rice Production*. Foreign Development Division, Economic Research Service, U.S. Department of Agriculture in Cooperation with U.S. Agency for International Development. Washington, D.C. 40 pp.

Darling, H. S. 1979. "International Research in Agriculture. 5. ICARDA: International Centre for Agricultural Research in the Dry Areas." *Span* 22 (2): 55–57.

FAO, 1982. *FAO Production Yearbook, 1981*. Rome, Italy: Food and Agriculture Organization of the United Nations. Vol 35. 306 pp.

Havener, R. D. 1980. "International Research in Agriculture. 7. CIMMYT: International Maize and Wheat Improvement Centre." *Span* 23 (1): 37–39.

Jennings, P. R. 1974. "Rice Breeding and World Food Production." *Science* 186: 1085–1088.

Josling, J. 1973. *An International Grain Reserve Policy*. Published by the British-North American Committee. USA. 22 pp.

Lewin, R. 1982. "Funds Squeezed for International Agriculture." *Science* 218: 866–867.

Morgan, D. 1979. *Merchants of Grain*. New York: Penguin Books. 519 pp.

Plucknett, D. L., et al. 1983. "Crop Germplasm Conservation and Developing Countries." *Science* 220: 163–169.

Plucknett, D. L., and N. J. H. Smith. 1982. "Agricultural Research and Third World Food Production." *Science* 217: 215–220.

Richter, J. 1980. "Towards an International Grain Reserve." *Span* 23 (1): 10.

Rossides, S. C., and J. Simkins. 1981. "Commodity Futures Markets in Agricultural Produce." *Span* 24 (1): 11, 15.

Sanchez, P. A., and J. G. Salinas. 1981. "Low-Input Technology for Managing Oxisols and Utisols in Tropical America." *Advances in Agronomy* 34: 279–406.

Sanderson, F. H. 1975. "The Great Food Fumble." *Science* 188: 503–509.

Simkins, J. 1981. "World Futures Trade in Grains." *Span* 24 (3): 102–103.

Sleeth, S. H. 1977. "International Research in Agriculture. 2. CIP: Centro Internacional de la Papa." *Span* 20 (1): 15–17.

Thomas, G. W., S. E. Curl, and W. F. Bennett, Sr. 1982. *Food and Fiber for a Changing World.* Danville, Ill.: The Interstate Printers and Publishers, Inc. 2d ed., p. 82. 258 pp.

Wade, N. 1975. "International Agricultural Research." *Science* 188: 585–589.

Wilson, F. 1978. "The Failure Was Not Genetic." *Ceres* 11 (5): 49–50.

PART TWO

DETAILS OF SPECIFIC CEREAL CROPS

15
Rice

INTRODUCTION

Although rice is an ancient grain and the staple food for over half the people of the world, for many years the majority of the world's scientists paid little attention to this crop. The fact is, rice is not an important food or crop in the developed countries. As late as 1971, Ishizuka (1971) reported that only seven books on rice production had been published in English. Within a 20-year period, however, progress in rice research had led to impressive gains in production, as illustrated by data on production, area, and yield in the Philippines from 1960 to 1980 (Table 15-1). During this period rice production increased 108%, even though the area seeded increased only 8%. Yield per unit area increased 92%. This remarkable increase meant that about 3.2 million fewer hectares (7.9 million acres) were needed to achieve the 1980 production level than at the 1960 yield level. Such advances characterized much of the rice-growing areas of the world (Swaminathan, 1982).

The establishment of the International Rice Research Institute (IRRI) in 1960 and rapid advances in rice breeding and production transformed rice to a respectable crop (Figure 15-1). Success with rice research resulted in a global trend toward higher crop productivity. For Asian rice producers it meant more and better rice for the consumer and greater income and employment for the rice producer. Rice and IRRI set a pattern from which the network of International Agricultural Research Centers was established (Chapter 14). High-yielding rice and wheat cultivars were the crops that led to the green revolution.

Dramatic and widespread increases in rice yields were achieved through a number of accomplishments fitted into a systems approach to rice production. A major success was in cultivar improvement. The breeding program, outlined in Chapter 14, was directed toward short-stature, stiff-strawed, fertilizer-responsive, and photoperiod-insensitive cultivars. Such cultivars were early-maturing and tolerant to attacks of major insect and disease pests. Associated with IRRI's success was the establishment of a large and diverse germplasm collection for continued and long-term breeding possibilities. By 1982, the collection contained nearly 55,000 cultivars collected from around the world. A third effort was directed toward research related to inheritance and problems associated with shortness, seed dormancy, breeding methods, photoperiodism, pest tolerance, agronomic production, and maximizing of yield. An international training program was established and directed at the development of human resources. Trainees participated for periods of up to six

Table 15–1
Rice Production Area, and Yield in the Philippines, 1960, 1970, and 1980

	Rough Rice Production		Area (×1000)		Yield	
Year	tonnes	tons	ha	ac	kg/ha	lb/ac
1960	3705	4076	3198	7899	1160	1036
1970	5343	5877	3112	7687	1720	1536
1980	7723	8495	3459	8544	2230	1991

Source: Swaminathan (1982).

months, performing every operation necessary to produce a rice crop and to become aware of pest and nutrition problems associated with rice production.

An extensive international testing program was established involving four programs:

Figure 15–1. The International Rice Research Institute (IRRI), seen in the background, was established at Los Banos in the Philippines in 1960. Outstanding success in rice breeding and production at IRRI has focused on the rice crop the attention of scientists around the world. The image of rice has been transformed from that of a third-world crop, grown by peasant farmers using ancient practices such as cutting rice with a sickle (foreground), into that of a crop that has become a model for yield advances in other crops around the world. A former director of IRRI, Robert F. Chandler, Jr., has described progress in rice as an "adventure in applied science." (Courtesy of the International Rice Research Institute)

1. Breeding and evaluation nurseries that in 1981 totaled 1215 tests in more than 50 countries in Asia, Africa, and Latin America (Anonymous, 1982a; Beachell, Khush, and Aquino, 1972).
2. Tests to evaluate cultivar performance under different agroclimatic conditions (Anonymous, 1982b).
3. Trials to test fertilizer nitrogen sources, methods, and times of fertilizer application (Anonymous, 1982c).
4. A systems approach toward food production that includes seed-bearing legume crops, fertilizer efficiency, and long-term intensive cropping and tillage systems (Anonymous, 1982d).

The results of research and testing were made available throughout the world to both rice and nonrice scientists through an extensive outreach program involving conferences and publication of proceedings, research results, news bulletins, and books. The goal of other international research centers is to parallel these advances in other cereal crops. Rice has become a model for meeting the world's food needs.

A more detailed account of IRRI and progress in rice improvement is available in two books by Robert F. Chandler, Jr. (1979 and 1982), a former director of IRRI.

A PERSPECTIVE ON THE RICE CROP

Although rice characteristics are becoming more familiar to people around the globe, the image of rice varies. Rice, like most cereal crops, varies considerably in genetic traits, cultural practices, growing conditions, and crop adaptation. A brief description is included here specific to the rice crop, but reference should be made to Chapter 2 for more details on the botanical and physiological traits of cereal grasses.

GENERAL DESCRIPTION

Rice, being a member of the grass family, has many traits in common with other cereals. The root system is fibrous with one seminal root, but adventitious roots can arise from nodes above the basal node. Under suitable conditions of spacing and environmental factors many rice cultivars tiller profusely. The rice culm is hollow with 10 to 20 internodes. An open leaf sheath arises from each node, and at the juncture of the leaf blade a long ligule arises. Leaf blades are long and narrow with a characteristic long, hairy auricle. The rice leaf can be distinguished from those of other grasses by the presence of both the ligule and auricle.

Grain is produced on a loose panicle (Figure 15-2) carried on the uppermost node of the culm that is most commonly completely free from the sheath. Seed is carried in 75 to 150 spikelets per panicle, each spikelet usually with a single floret. Florets contain six stamens, two feathery stigmas in two styles and two lodiculae. Panicles are 10 to 25 cm long (4 to 10 in.).

The rice grain, enclosed in a hull consisting of the lemma and palea, is called paddy or rough rice and may vary in size, shape, color, and length of awn on the lemma. The hull of rice can be removed readily by an abrasive action to produce husked or brown rice. Grain variations fit into four categories: slender, long, medium, and short.

Plants vary in height from 40 to 700 cm (16 to 275 in.).

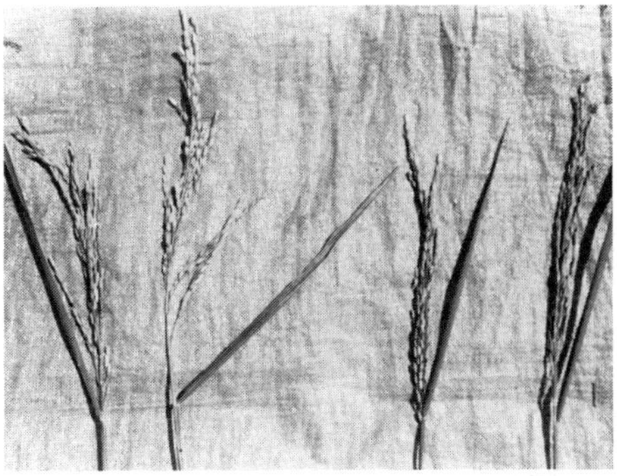

Figure 15–2. Grains of rice are produced on lax panicles carried on the uppermost node of the culm. Seed is produced in spikelets, each usually containing a single floret. The seed is enclosed by the lemma and palea. Leaves are long and slender and arise from an open leaf sheath at each node. Rice plants can be distinguished from other grasses and cereal crops by the presence of a large ligule and a long, hairy auricle distinctive to the rice plant. (Photo courtesy of Soon Jai Park)

Figure 15–3. Lowland rice is well adapted to submerged-soil culture and is the only major food crop that can be grown in standing water. The ability of rice to grow under semiaquatic conditions is an important attribute, as many poorly drained areas of the world can be used advantageously for food production. Although rice commonly is characterized as a semiaquatic plant, upland cultivars of rice also exist that can be grown under dryland conditions on hilly or sloping fields like any other cereal. (Courtesy of the International Rice Research Institute)

ADAPTATION

Rice commonly is characterized as a semiaquatic crop, well adapted to submerged anaerobic soil culture (Figure 15-3). Cultivars of rice also are adapted to dryland culture on aerobic soils and can be grown on hilly fields like any other cereal. Such cultivars are known as **upland rice** in contrast to **lowland rice.** The ability of rice to grow under semiaquatic conditions is an important attribute, as it allows the crop to be grown in the many great river basins and deltas of Asia, areas that might not otherwise be used advantageously for food production. By far the greatest proportion of the world rice crop, about 80%, is not irrigated and is dependent on natural rainfall (Moomaw and Vergara, 1965).

CLIMATE

Rice often is regarded as a tropical crop because it is so important in the tropical and subtropical regions of the world, where it often is the only small-grain crop grown. Rice is a C_3 crop and under high-temperature conditions has a high rate of photorespiration. Highest yields of rice are obtained in temperate areas. Rice was reported by Kratochvil (1956) to be grown as far north as 49° and as far south as 36° latitude. Rutger and Brandon (1981) suggested that rice is grown on all continents except Antartica and production extends from 53° N to 40° S latitude. Rice was established as a commercial crop in the United States in 1690 (Rutger and Brandon, 1981). Rice is not grown commercially in Canada but was produced successfully in test plots at 42° N latitude (Figure 15-4). Some cold-tolerant cultivars can be grown at elevations of 2400 m (7900 ft).

Figure 15–4. Although rice is the most important food crop of tropical and subtropical regions of the world, the zone of adaptation of this crop extends well into the temperate zone in Europe, Asia, and North America. The upland rice cultivars in this photograph are under test at 42° N latitude in Canada. (Courtesy of Soon Jai Park)

SOIL

Rice can be grown successfully on an enormous variety of soils and over a wide range of soil acidity and alkalinity. The ability of rice to tolerate wide variations in soil *p*H is related to its ability to grow in submerged soil. Under water, the *p*H of acid soils increases and of alkaline soils decreases. Some cultivars of rice are quite tolerant to soil salinity.

Because C_3 crops such as rice are not very water-efficient (refer to Table 2-5), supply and control of water is perhaps the single most important factor in rice production.

LOWLAND OR PADDY RICE

Rice that is grown under almost continuously flooded conditions is known as lowland rice and is essentially a **hydrophyte.** Most lowland rice cultivars are adapted to shallow-water conditions, but no quantitative classification of rice lands by water requirement or plant adaptation exists. In broad terms, lowland rice cultivars are adapted to shallow-water conditions, regions with intermediate water depth, and deep-water conditions. Water depth may be achieved by irrigation and may involve systems of double or single cropping, or it may depend on rainfall under **bunded** conditions to control water depth or for deep-water production. Lowland rice cultivars adapted to shallow-water conditions can be damaged by deep water, especially during panicle initiation (Moomaw and Vergara, 1965). Highest lowland rice yields are obtained with water depths controlled to about 15 cm (6 in.) or less. Rice is the only major food crop that can be grown in standing water.

Under flooded conditions, oxygen remains as a requirement for seed germination, root establishment, root-hair development, root absorption, and normal root metabolism. Without oxygen associated with wet or flooded conditions, most cereal crop plants turn yellow, wilt, and eventually die. Oxygen-deprived cereal plants show restricted shoot and root growth. Under normal, aerobic soil conditions, water and air fill the porous structure from which roots obtain the necessary oxygen to function. Lowland rice does not rely on soil-borne oxygen for survival and is able to grow under submerged-soil conditions because of three adaptive mechanisms (Naphade, 1971):

1. A highly developed system of anaerobic respiration, especially during early growth stages, allowing the plant to tolerate oxygen-deprived conditions.
2. The ability to transport oxygen from leaves to roots through tubelike structures known as **aerenchyma** cells in leaves, stems, and roots. Aerenchyma cells permit air to move from the leaves to the root surfaces, thus supplying the submerged roots with sufficient oxygen for normal respiration, subsequent growth, and nutrient absorption.
3. Development of very fine, abundantly branched, negatively geotropic roots on the soil surface at the time of panicle initiation. When stem elongation temporarily disrupts the internal downward flow of oxygen, oxygen is supplied to the roots by this special system.

Irrigated lowland rice is produced under flooded conditions achieved by regulating water depth from wells or rivers and containing the water by a dyke or bund (Figure 15-5). Rain-fed paddy rice is produced in fields that are bunded to contain the water from rains.

Figure 15–5. Lowland rice may be grown in bunded paddy fields to contain and control water depth. The bunds or dykes are maintained from season to season and are installed according to the land contour needed to maintain water depth. Rice may be transplanted by hand, generally before the rice seedlings are 21 days old. Controlled year-round irrigation facilities, provided by the well in the center background, permit the growing of two or more rice crops a year. (Photo by N. C. Stoskopf from the area of Beijing, China)

DEEP-WATER OR FLOATING RICE

In areas of Bangladesh, Thailand, India, Vietnam, Indonesia, and West Africa that experience uncontrollable flood conditions from monsoon rains every year, **deep-water** or **floating rice** cultivars are produced. Such cultivars can tolerate water depths of 1 to 6 m (3 to 19 ft) (Jackson et al., 1972). The internodes of deep-water rice cultivars can elongate as much as 4 to 30 cm (1.5 to 12 in.) per day as flood waters rise. As flood waters recede, the floating stems and leaves collapse in a tangled mass. A desirable characteristic of deep-water rice is its ability to become erect after the upper stems and leaves have been in a horizontal position on the surface of rising flood waters. This is a geotropic response and is known as **kneeing ability** (Figure 15-6).

Flood conditions associated with strong winds and heavy currents may uproot the plants. Uprooted plants are able to obtain nutrients from the flood waters from adventitious roots produced at upper nodes. The plants may become reestablished in the soil and continue growth and tiller production when the flood subsides. This unique set of circumstances gives deep-water or floating rice its name.

Deep-water rice is planted usually at the beginning of the first monsoon rains so that the plants can become well established before

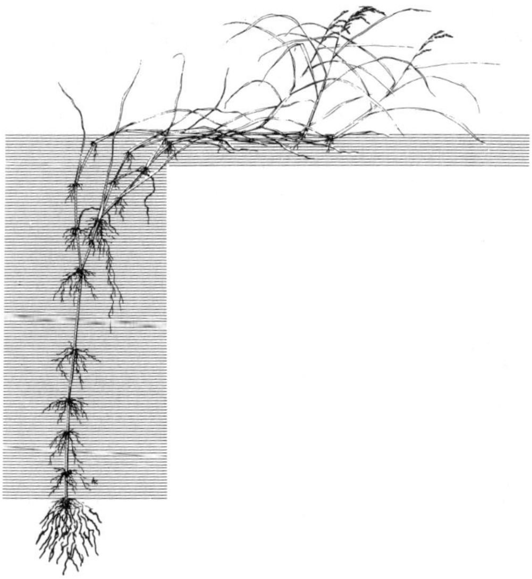

Figure 15–6. Deep-water or floating rice cultivars are adapted to deep-water conditions. As the water level rises, the internodes elongate, thereby keeping leaves above the water surface. Adventitious roots at nodes can absorb nutrients from silty water. Tiller branches emerge from nodes just below the water level. As flood waters recede, plants have the ability to become erect after the upper stems and leaves have been in a horizontal position on the water surface. The ability to bend upright is a response known as kneeing ability. (The *IRRI Reporter*, 1975)

the full monsoon onslaught. Seeds are usually broadcast on dry or moist soil. During the early period of irregular rainfall, the young plants grow as an upland crop, often under moisture stress. As the monsoon rains raise the water level, the rice plants adjust to the increasing water depths.

Photoperiod sensitivity provides floating-rice cultivars with a form of adaptation to deep-water conditions. The farmer can sow seeds whenever soil moisture conditions are favorable. Sensitivity to daylength stimulates flower development after the flood waters begin to recede, but before the monsoon rains end.

As the flood waters recede, the plants are left lying in all directions, making harvest difficult. Harvesting involves separation of the panicles from the long, tangled stems and curving shoots. To facilitate harvesting, cutting may begin before the waters recede, using boats. The harvest may be completed when the water completely recedes and the soil becomes dry and hard.

Deep-water or floating rice is seldom fertilized, because crop response is poor, application is difficult, and the fields are replenished with fertile silt every year. A starter fertilizer applied to the seedlings before the floods begin may increase seedling vigor and accelerate seedling growth.

Although deep-water rice yields are not high, no other food crop can be grown under such flood conditions. Deep-water rice yields seldom exceed 1.5 t/ha (1340 lb/ac) (Chattopadhyay, 1979) compared to double such yields in lowland rice (Swaminathan, 1982). On about 30% of the world's rice land, water depth cannot be controlled. Although deep-water rice yields are low, deep-water rice production is important in terms of world food supplies, and any yield increase through breeding or improved cultural methods would have a significant impact on world food supplies. Unfortunately, few crops are more difficult to work on than deep-water rice. Increasing the rice yields of deep-water and medium-water-depth cultivars represents a great challenge, but progress is evident (Chattopadhyay, 1979; Jackson et al., 1972; Anonymous, 1975).

UPLAND RICE

Upland rice is grown under a number of conditions, ranging from natural rain-fed areas in elevated dry-soil conditions to situations in which flooding may be provided for a considerable part of the plants' life cycle (Chang, Loresto, and Tagumpay, 1972). Upland rice may be distinguished by production that involves no bunding to contain the water.

Normally upland rice fields are cultivated prior to the rainy season. Rice seed is broadcast onto the worked soil, which may be hilly or have a sandiness that makes bunding impractical or ineffective.

Yields of upland rice usually are low because of erratic rainfall patterns, poor weed control under nonflooded conditions, accompanying low fertilizer use, and a high incidence of disease. In areas suited only to upland rice production, other cereal crops such as corn, sorghum, or millet that have more drought tolerance may be more advantageously grown.

TYPES OF RICE

At a symposium on rice genetics and cytogenetics held at the International Rice Research Institute in 1963, a committee on taxonomy recommended that 19 distinct and valid species of the rice genus *Oryza* be recognized (Chang, 1964). Chromosome number (2*n*) was either 24 or 48 for all 19 species.

Throughout most of Asia, Africa, Australia, Europe, and the Americas the 24-chromosome species, *Oryza sativa* L., is the most commonly and extensively cultivated. A second species, *Oryza glaberrima* Steud, found solely in West Africa, is the only other rice species grown for food. This species is considered to be inferior to *Oryza sativa* L. in yielding ability and disease resistance. Since the majority of research has been directed to *O. sativa* L., increasingly superior cultivars of this species are replacing *O. glaberrima* Steud, and its importance is declining.

A collection of over 1000 *O. glaberrima* cultivars has assured preservation of this gene pool and has allowed for a collective appraisal of its genetic variability. A more comprehensive observation and comparison has renewed interest in the genetic potential of this species. Crosses between the two commercial rice species results in sterility problems in the early-generation progeny.

BIOTYPES OF RICE

Three well-known biotypes, which were once considered separate species but which belong to *O. sativa* L., characterize most of the variability found in the crop and in part help explain the wide distribution and broad adaptation of rice. These biotypes, known as indica, japonica (also sinica or keng), and javanica (bulu), represent evolutionary adaptation to environmental variables.

The javanica rices are morphologically similar to the japonicas, except for wider and more pubescent leaves and a beard (the meaning of bulu) or hairlike awn. The indica and japonica biotypes are of interest because of contrasting characters. The combining of the most desirable features from the two types led to the yield advances associated with the successful high-yielding cultivars produced by IRRI. Earlier attempts to combine the contrasting morphological features had not been successful.

Comparisons between the two biotypes (Table 15-2) illustrates the range of variability found in them. Success in combining desirable features was achieved only under a suitable management system. An unprecedented grain yield of 8.23 t/ha (7347 lb/ac) was achieved (Chandler, 1979 and 1982).

A component of the system was the utilization of the abundant sunlight during the dry season instead of the reduced light intensity during the cloudy rainy season. The additional inputs needed to make this transition possible mean a more capital-intensive system of rice production.

Basic morphological changes and production principles associated with green-revolution or capital-intensive rice production were listed by Chandler, Jr. (1982) and are summarized here:

1. Short, thick culms are the key factors reducing lodging and nitrogen responsiveness. Plant height for lowland rice cultivars is considered to be ideally about 100 cm (39 in.).

2. Short and tall cultivars have a similar number of nodes and hence leaves, but the short cultivars have short internodes. Internode length is influenced by both environmental and genetic factors, but internode elongation is less in genetically dwarf than in tall cultivars. Internodes of genetically tall cultivars elongate the most under high levels of nitrogen fertility and

Table 15–2
Comparisons between Indica and Japonica Biotypes of Rice

Indica	*Japonica*
Tall stature	Short to intermediate stature
Heavy tillering with many unproductive tillers	Medium tillering but tillers remain productive
Adapted to tropics	Adapted to temperate zone
Drought tolerant	Drought intolerant
Disease and insect resistant	Susceptible to disease and insects
Lodging resistant	Susceptible to lodging
Varying photoperiod sensitivity	Low to no photoperiod sensitivity
Not responsive to nitrogen fertility	Responsive to nitrogen fertility
Broad to narrow light green leaves that flag over	Narrow, dark green upright leaves
Low grain-to-straw ratio	High grain-to-straw ratio
Low or intermediate gelatinization temp. (23% to 31% amylose)	Low gelatinization temp. (10% to 24% amylose)
Seed dormancy	No seed dormancy
Lacks cold tolerance	Cold tolerant

Source: Adopted from Chandler (1979 and 1982).

become more susceptible to lodging when fertilized with nitrogen than do short-stature cultivars.

3. High yield and lodging resistance can be achieved with short, upright-leaved cultivars that can intercept adequate sunlight to penetrate to lower leaves and cause lower internodes to remain short, sturdy, and lodging resistant.

4. Tillering is influenced by sunlight penetration to lower nodes, by temperature, and by nitrogen fertility. Upright leaves are essential in high-tillering cultivars, especially under high light intensities in the tropics and semitropics.

5. Cultivars with leaves that remain photosynthetically active from anthesis to harvest and that therefore have a long duration of grain filling are able to carry on photosynthesis for a period of time sufficient to fill the grain sink. Yield and components of yield of two rice cultivars nearly alike genetically but differing in growth duration because of differing sensitivity to photoperiod revealed the superiority of the cultivars with an 18-day-shorter vegetative period (Dat and Peterson, 1983a). A desirable balance in assimilate supply (source) and storage capacity (sink) produced superior grain yield in rice in a two-cultivar comparison (Dat and Peterson, 1983b).

6. Rice cultivars with short stature have a higher harvest index, which means more photosynthate is directed into grain and less into straw.

7. Under irrigation, a high grain yield can be produced during the dry season, when sunlight energy is more abundant and photosynthetic rates are higher than during the cloudy, rainy season.

8. A higher leaf-area index (LAI) is possible with cultivars having short, upright leaves than with cultivars having long, droopy or wide leaves (Figure 15-7).

9. Response to photoperiod should be modified to suit management conditions. Rice is basically a short-day plant, which means that the vegetative growth period of the crop is decreased under short-day conditions.

Photoperiod insensitivity has resulted in rice cultivars that are better adapted to the local environments and production requirements of many countries. Associated with photoperiod is temperature, which acts either alone or through an interaction with daylength and other factors in the period from sowing to floral initiation to influence yield potential. Temperature may set limitations on rice production. A maximum temperature of 22°C (72°F) may be the upper

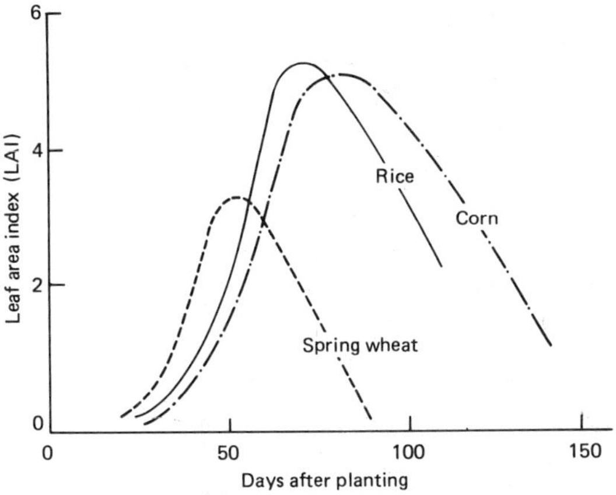

Figure 15–7. **Leaf-area index comparisons of successive growth stages of rice, corn, and spring wheat grown in the northern part of Japan (Hokkaido in 1975). Leaf-area index increased with growth of the plant, reached a maximum at a certain growth stage, and then decreased. The maximum leaf-area index achieved was above 5.5 in rice, 5.0 in corn, and about 3.5 in spring wheat. The decrease in the leaf-area index during later growth stages was slowest in corn, followed by rice and spring wheat. (After Tanaka and Osaki, 1983)**

limit of some rice cultivars, which rules out some areas and seasons of the tropics (Owen, 1971). Cool night temperatures may cause sterility and incomplete grain filling. Photoperiod insensitivity may be useful to help avoid conditions that are not inducive to rice production.

Historically, photoperiodically sensitive cultivars have been important to fix the date of maturity to coincide with the end of the rainy season and the beginning of weather favorable for sun-drying. The development of photoperiodically insensitive cultivars has resulted in cultivars with a shorter growth duration, which are important in areas where intensive farming methods are used and where machinery and drying equipment allow for harvesting during unfavorable weather. Photoperiodically insensitive cultivars are useful in multiple-cropping systems because they can mature in 100 days from seeding, allowing for three or more crops to be produced in one year.

Consideration of these nine factors and the 13 contrasting features of the two biotypes listed in Table 15-2 gives some indication of the complexity in selecting the most desirable morphological features. This feat of selection appears to have been achieved in the early cultivars produced by IRRI.

The rice plant clearly demonstrates the association of plant type with yield and of nitrogen responsiveness and yield with morphological features. Failure to recognize these relationships or to effectively integrate them into a production system was a reason for a lack of success in previous attempts at indica × japonica hybridization programs.

CHARACTERISTICS OF LOWLAND OR PADDY RICE SOILS

Lowland rice is grown in level paddies that are either rain-fed or irrigated. Lowland soils are those on which standing water can be maintained because they have poor internal drainage and remain flooded. Numerous soil types are suitable for lowland rice production.

To facilitate and control flooding, soils found in four physiographic landforms are used to produce lowland rice:

1. Bottomlands of present-day streams that are level and that may occasionally flood as the stream overflows its banks. These soils generally require low levels of phosphorus fertilizer for rice production.
2. Broad, nearly level alluvial plains or terraces that are not flooded by present-day streams. Such soils have a moderate to high need for phosphorus. Rice grown on alluvial plains covered with **volcanic tuff** responds only to nitrogen.
3. Depressed areas in uplands. Rice grown on such soils will respond to phosphate; where limestone is present, rice will respond to potassium as well as phosphate.
4. Terraced paddies constructed on sloped areas. These "upland" soils require phosphorus for rice production.

Water in rice paddies is controlled by bunds 60 to 90 cm (24 to 36 in.) high. The distance between the bunds depends on the slope of the field. Bunds follow the ground contours needed to achieve the desired water control.

EFFECTS OF FLOODING ON SOILS

Seedbed preparation may be achieved following irrigation to soften the soil. Immediately after plowing, a lowland field should be completely flooded. If flooded first, plowing should follow immediately.

The objectives of wet tillage, known as **puddling**, are to reduce the power requirements for seedbed preparation, to control weeds, to incorporate crop residues, to produce a soil layer that reduces percolation, and to prepare a soil suitable for transplanting.

In much of Asia, seedbed preparation of lowland soils occurs when the soil is flooded so that puddling occurs for transplanting the rice. For most of the growing period, flooding to a depth of several centimeters (inches) is maintained.

In the rice-growing regions of Europe, the United States, Australia, and some Asian countries rice land is prepared dry and flooded later. Such soils are neither puddled nor transplanted. Puddling therefore is not essential to a wetland rice soil.

After the flooding of a paddy field, the submerged soil undergoes physical, microbiological, and chemical changes entirely different from those of dry or upland soils. More efficient production of lowland rice may be achieved if the characteristics of flooded soils are understood. The complex changes associated with flooding were reviewed by Savant and De Datta (1982) and are summarized here briefly.

Most soils used for lowland rice production show a high need for nitrogen fertilizer. Efficiency of nitrogen fertilizer use by rice is as low as 50%. High costs of nitrogen fertilizer have raised the need for greater efficiency in its use. Factors affecting nitrogen fertilizer use are as follows:

Physical Changes

When lowland soils are flooded, the pore spaces become saturated with water, the soil swells, and hard clods soften and break into small aggregates. Puddling results in the soil's forming a slurry as aggregates are broken down. The result is a reduction in pore space, an increase in bulk density, restricted internal drainage, and increased water-holding capacity of the soil. The exchange of air between the atmosphere and the puddled soil is impeded.

The flooded puddled soil develops two zones. A thin layer, 1 to 10 mm thick, absorbs oxygen from the water, turns brown in color, and reacts to nitrogen like an unflooded upland soil. A second layer is formed in the remaining portion of the puddled soil that is deprived of oxygen. As iron compounds in the soil are reduced, the soil turns a dark gray or blue gray. Efficient nitrogen fertilization of lowland rice must take into consideration these two zones in flooded soils.

Microbial Changes

The absence of oxygen in a flooded, puddled soil drastically changes the kinds and numbers of microbes that decompose crop residues. Decomposition occurs at a much slower rate than in aerated upland soils, and toxic end products may result, including carbon dioxide (CO_2), methane (CH_4), hydrogen (H_2), nitrogen gas (N_2), and soluble iron and manganese. To avoid toxic effects from these end products, flooding of paddy rice fields should occur two or three weeks before transplanting. Incorporating crop residues before transplanting also reduces toxic compounds.

Under flooded conditions, the nitrogen content of crop residues is changed to ammonia ($NH^+{}_4$), which is stable under flooded conditions and can be used by the growing rice plants. If the soil is not flooded immediately after cultivation, nitrate nitrogen ($NO^-{}_3$) is formed as in upland soils. When the soil is flooded, nitrate nitrogen is changed to gaseous forms (N_2 and N_2O), which escape into the air (Figure 15-8).

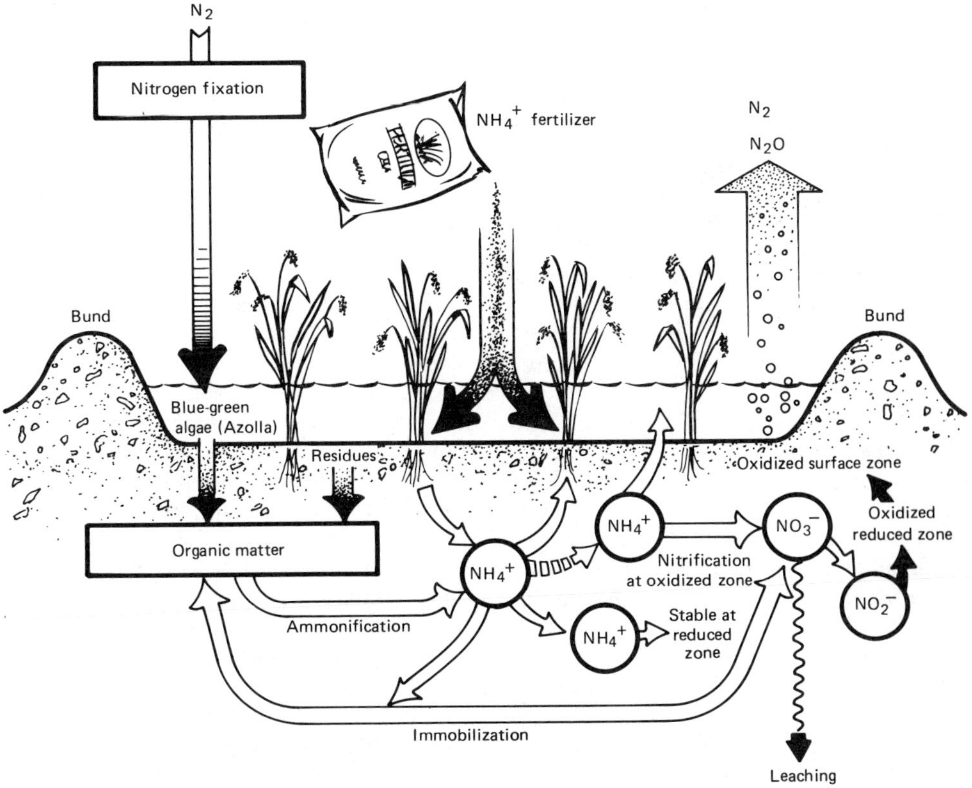

Figure 15–8. Submerged paddy rice soil undergoes physical, microbiological, and chemical changes entirely different from those of dry or upland soils. The fate of nitrogen in flooded soils is illustrated here. High nitrogen-fertilizer costs have raised the need for greater efficiency of fertilizer, and this may be achieved if the characteristics of flooded soils are understood. Most soils used for lowland rice production show a high need for nitrogen fertilizer. Rice is one of the few crops that can be grown on the same land year after year without serious problems developing. (*The Philippines Recommends—1969,* published at the College of Agriculture, University of the Philippines College, Lagune, p. 17)

Applications of ammonium forms of nitrogen fertilizer should be incorporated to prevent gaseous losses of nitrogen. Ammonium nitrate is transformed to nitrate by microbial action in the thin oxidized soil layer. Nitrate nitrogen is transformed to nitrogen gas and lost to the atmosphere. Starter applications of ammonium fertilizers should be incorporated directly and immediately into the reduced soil zone. This explains why placing nitrogen fertilizer at a depth of 10 to 12 cm (4 to 5 in.) and at a rate of 60 kg N/ha (54 lb N/ac) can give a yield increase comparable to that achieved by 100 kg N/ha (90 lb N/ac) that is not incorporated at sufficient depth to prevent gaseous losses. Losses can also be reduced by avoiding ammonium-forming nitrogen fertilizers such as ammonium sulfate, urea, or ammonium phosphate.

Chemical Changes

The availability of both native soil phosphorus and that added as fertilizers is increased by flooding, which is one of the advantages of growing rice under flooded conditions. Under flooded conditions iron, aluminum, manganese, and calcium become soluble and chemically release the phosphorus that under aerated conditions is chemically bound with these elements. The rise in *p*H of acid soils and the lowering in *p*H of alkaline soils under flooded conditions also increases phosphorus availability. Acid soils become less acid when iron (Fe^{++}) and manganese (Mn^{++}) precipitate and equilibrate with CO_2 and $HCO^-{}_3$ in the soil solution. In alkaline soils, the acidifying effect of CO_2 lowers the soil *p*H, which is buffered by calcium and magnesium carbonates.

Soils that are very acid or have high iron levels may require liming or application of sulfur-bearing materials to permit rice production.

The beneficial effects of flooding on rice production allow rice to be grown over a wide range of soil characteristics. Rice is one of the few crops that can be grown on the same land year after year without serious problems developing.

RESPONSE OF RICE TO NITROGEN

Modern, nitrogen-responsive rice cultivars are most productive when substantial amounts of nitrogen fertilizer are applied. Yield increases in rice can be demonstrated at nitrogen levels of 120 to 150 kg N/ha (107 to 134 lb N/ac). In contrast, the tall, lodging-susceptible traditional cultivars seldom respond to more than 30 or 40 kg N/ha (27 to 36 lb N/ha) and on fertile soils may show no response to additional nitrogen fertilizer applications. A yield loss may occur under high-nitrogen fertility conditions, where excessively tall and leafy growth is accompanied by early lodging, yield reductions associated with poor sunlight interception under lodged conditions, restricted carbon dioxide movement, disruptions in nutrient and assimilate translocation, and associated harvesting losses.

BIOFERTILIZERS

Legume crops possess the ability to convert the nearly inert atmospheric nitrogen, N_2, comprising nearly 80% of the atmosphere, into a biologically useful form. This transformation is achieved through a symbiotic relationship between the legume plant and Rhizobia bacteria. The importance of biological nitrogen fixation lies in the fact that such plants are at least partially self-sufficient in nitrogen. Nitrogen is required in large amounts by all forms of life and is an essential constituent of protein.

Nitrogen fertilizer is produced industrially through a process whereby nearly inert N_2 gas and hydrogen are combined under high temperature and pressure, normally with nonrenewable natural gas as the energy source. Energy equivalent to 82 liters of gasoline is required to make 100 kg of inorganic nitrogen fertilizer (20 gallons per 100 lb of N).

High fossil-fuel prices, the widening gap between supply and demand for nitrogen fertilizer, and the prevailing low purchasing power in the densely populated developing rice-growing regions of the world have caused serious concern about rice production with restricted nitrogen supplies.

Biological processes for fixing atmospheric nitrogen through symbiosis and by free-living nitrogen-fixing organisms offers a means of reducing the requirement for inorganic fertilizers. In tropical, flooded paddy rice conditions, two biological means of producting nitrogen are available:

1. Blue green algae.[1]
2. Azolla species.[2]

BLUE-GREEN ALGAE

An important technology that holds promise for lowland rice production is the utilization of

[1]Nitrogen-fixing blue-green algae—*Anabaena azollae*.

[2]Azolla species—e.g., *Azolla pinnata*.

blue-green algae capable of fixing 25 to 30 kg N/ha (22 to 27 lb/ac) every season (Venkataraman, 1981). Blue-green algae grows upon a free water surface and is ideally suited for propagation in the rice fields.

In Chapter 5, the transfer of symbiotic relationships into the cereal crops was suggested as a possible but giant step to help solve the nitrogen-fertilizer problem. However, nitrogen fixation is an energy-intensive process, requiring an input in the plant of about 61.5 kJ (147 kcal) to reduce one molecule of nitrogen (Venkataraman, 1981). This drain on assimilate could reduce grain yield. Symbiotic nitrogen fixation within a crop may not be desirable, as the yield reduction may not be affordable.

Blue-green algae live independently of the rice plant, contain chlorophyll, and present a self-supporting system of nitrogen fixation, as the energy required to reduce N_2 comes from the sun through photosynthesis. Blue-green algae have the advantages that they are self-regenerating and represent minimal environmental hazards associated with nitrogen-fertilizer manufacture, transport, and application.

Use of blue-green algae does not bring about the spectacular visual changes in crop growth and production observed with chemical fertilizer applications. Although some of the nitrogen fixed by blue-green algae is available to the rice plants, the bulk of it is contained in the organic matter produced by algae growth. As the organic matter residues remaining in the soil decay, nitrogen is released for use by subsequent crops. This delay presents some difficulty in the promotion of blue-green algae, and rice producers must realize the yield association with subsequent crops. Yield trials have shown a 33.6% yield increase on algae-grown plots for two months prior to transplanting rice and in the absence of any added fertilizers. An added problem is that herbicide applications for weed control may hinder blue-green algae establishment and reduce nitrogen fixation.

A number of blue-green algae strains can be applied to the soil. Their ability to survive, become established, and quickly colonize the flooded soil varies. Fast-growing strains adapted to a region should be selected. Dry algae material can be broadcast over the standing water in the rice field at a rate of 10 to 15 kg/ha (9 to 13 lb/ac) one week after transplanting the seedlings. Fresh algae can be established in a nursery plot and raised as inoculum in a newly cultivated and flooded rice paddy at the rate of 750 kg/ha (670 lb/ac). Addition of excess algae material is not harmful and may be beneficial, as it may accelerate establishment and multiplication in the field. Application may be made well in advance of transplanting rice.

Application of algae for at least three consecutive cropping seasons may be required to build up a suitable population. Once blue-green algae become established in the soil, their biological activity continues. Activity of blue-green algae is of short duration, extending from three to six months, after which they either perish or enter a dormant state. In subsequent seasons the algae will recover and flourish without any further inoculation, unless some unfavorable ecological conditions interact.

As noted earlier in this chapter, it is difficult to add nitrogen fertilizers to deep-water or floating rice under flood conditions. Such rice plants depend on nitrogen they can get naturally. Suitable blue-green algae that release much of their fixed nitrogen to deep-water rice plants could be of value.

AZOLLA

When nitrogen-fixing blue-green algae are grown in the presence of the water fern or azolla, the amount of atmospheric nitrogen fixed may increase by a factor of from 5 and 10, as compared with the same concentration of blue-green algae grown alone. As much as 312 kg/ha/yr of nitrogen (279 lb/ac) reportedly has been fixed by this symbiotic relationship (Anonymous, 1977).

Blue-green algae grow as a symbiont in the cavities of leaves of azolla. The azolla plant consists of a floating branched rhizome with small leaves and undeveloped roots which suspend into the water. It reproduces itself vegetatively.

Azolla strains grow in a temperature range between 20° to 35°C (68° to 95°F) and must be protected against temperatures outside this range. Protection from the cold can be achieved by storing azolla in pits or trenches covered with straw. Protection from heat is achieved by storing it in shaded areas with circulating water. Azolla does not survive temperatures lower than 0°C (32°F) or higher than 35°C (95°F). It grows well in standing water a few centimeters (inches) deep and is favored by phosphorus fertilizer.

Azolla can be established by inoculating rice fields at a rate of 7.5 t/ha (7000 lb/ac). Under suitable conditions azolla is capable of doubling its weight every five days, so that in a three- to four-month growing period it can produce 160 t/ha (71 tons/ac) when harvested up to ten times and contains about 425 kg (437 lb) of nitrogen. At this rate of production, azolla is a source of green manure or a livestock feed if harvested every seven to ten days.

Azolla can be plowed under before the planting of rice, or, if produced among rice seedlings, it is buried by hand to increase soil organic matter.

RATOON CROPPING OF RICE

Ratoon cropping is the practice of obtaining a second crop from tillers originating from the stubble of a harvested crop. Ratooning of rice offers a unique opportunity to maximize the use of land during a growing season that is too short to grow two planted crops. Ratooning may be suited to the southern United States (Evatt, 1966). Ratoon cropping also allows for the utilization of residual soil moisture in both irrigated or dryland tropical areas (Bahar and De Datta, 1977). It is a means to carry a crop over from one season to the next without replanting.

Rice ratooning has considerable appeal because it eliminates the need to prepare a seedbed, to reseed, and to begin a new crop from seed. Ratooning represents a step toward perennial production. Although rice ratooning has been tried in many areas of the world, its acceptance, particularly under intensive cultural systems, has been limited (Plucknett, Evenson, and Sanford, 1970). The reasons given include lower yields than in the plant crop, build-up of injurious insects, weed or disease problems, greater cost per unit produced, and problems with soil compaction.

Ratoon cropping may be improved by sound management. In a study with nitrogen fertilization and reflood timing on the ratoon-crop yield of two adopted cultivars, Mengel and Wilson (1981) demonstrated that ratooning has the potential to significantly increase rice production.

CONSIDERATIONS IN RICE PRACTICES

Transplanting of rice seedlings into puddled paddy fields is a common practice. Hand transplanting provides a method of uniform spacing, making more efficient use of sunlight energy and allowing for plant population control. Rice transplanted in straight rows allows the use of a rotary weeder for weed control.

Where labor for rice transplanting is too costly, direct seeding of rice is practiced, either by drilling the seed into dry soil or by broadcasting onto flooded sod. After direct drilling, the paddy is flooded. Direct seeding is less desirable than a transplanting because of poor stands, inability to cultivate between rows, and vulnerability to attack by rats, snails, and birds.

Rice seed usually will not emerge through both soil and flood water, and poor stands may result. To avoid this problem, seeds may be broadcast on rough seedbeds or covered shallowly with soil. If broadcast onto flooded soil, seeds tend to float and drift, resulting in uneven stands. To avoid this problem, seeds are presoaked to ensure that they will sink promptly. Presoaking initiates germination and results in rapid development of seedlings, making them better users of sunlight energy, more competitive with weeds, and more pest resistant.

Maximum yields cannot be achieved unless seedling density is satisfactory. Seedlings developed from seeds must emerge through the flood waters, usually 5 to 10 cm (2 to 4 in.) deep, before sunlight can be captured for photosynthesis. Shallow water, 2 to 5 cm (¾ to 2 in.), will facilitate rapid emergence, but deeper water will suppress weeds better. Flood water may be drained off during early seedling development to facilitate hand or rotary weeding.

Toxic effects from large amounts of plant residues can be reduced by draining intermittently until some cracks form on the surface and then reflooding. Better root activity is promoted as well.

Seedlings must become independent of the seed reserves before these nutrient sources are exhausted. Oxygen needed by the roots is obtained through the stomata of the leaves, once they emerge from the water. A lack of oxygen may result in the failure of rice to emerge from the water (Turner, Chen, and McCauley, 1981).

COMPARING SYSTEMS OF RICE CULTURE

Management systems for rice production vary from very large farms over 2000 ha (4940 ac) in size in developed countries to those of about 1 hectare (2.47 ac) in Southeast Asia. In developing areas the average farm holding is becoming smaller and smaller. Small units are labor-intensive; large units are highly mechanized.

Rutger and Brandon (1981, p. 42) described rice production in California as follows:

> The production of rice in California has evolved into one of the most highly mechanized agricultural operations in the world, including the leveling of fields by laser beam, the sowing of seed by airplane[1] and the harvesting of the crop by special combines that do not bog down in the mud of the rice paddy. The 1979 California yield of 6,450 pounds of rice per acre[2] was 50 percent higher than the average in the other rice-producing states of the U.S. and nearly three times higher than the world average of 2,360 pounds per acre.[3] Although the acreage devoted to rice in the U.S. is only 0.9 percent of the world total, the U.S. produces 1.7 percent of the world crop and in many years is the world's largest rice exporter.

Using the process method of energy analysis, cultural energy efficiency was determined by Freedman (1980) for various rice-producing systems. He compared traditional, transitional (labor-intensive), green-revolution (capital-intensive), and experimental (intermediate) systems. Cultural energy efficiency is determined by calculating the ratio of food energy output to cultural energy input.

For all methods of rice production, Freedman found a greater energy output of food production compared to cultural energy used for production. Values do not include processing, however.

For each unit of energy required in the production of rice, using traditional methods, 13.8 units of energy were made available as output. The traditional system of rice production gave the lowest yield. The labor-intensive transitional method produced an output/input ratio of 6.6; and for the green-revolution method the ratio was 2.9. The green-revolution system was the lowest of all the methods in cultural energy efficiency but produced a yield 3.52 times greater. The experimental method produced yields similar to that of the green-revolution method and had an output/input ratio of 4.6. This was more than 25% greater than the green-revolution value.

In a society conscious of energy, the experimental method must be regarded as superior. If, however, the goal is to increase yield per unit land area, the green-revolution system must be regarded as optimal. Since little additional crop land is available for increased food production in the major rice-growing areas of the world, it follows that capital-intensive agriculture must be adopted.

Despite the small size of farms in Southeast Asia, the success of the green revolution has demonstrated that small farm size in itself is not a handicap to the adoption of new technology. The small farm can facilitate intensive agriculture.

---

[1]Presoaked seed is used at a rate of 170 kg/ha (150 lb/ac).

[2]7224 kg/ha.

[3]2643 kg/ha.

CONCLUSION

A quote from M. S. Swaminathan (1982) may best sum up the story of rice (p. 15):

> We are running a race against time in our war against hunger and unemployment. We have the tools and the know-how to win this war. We should not wait while we look for a perfect variety or a perfect technology, but should speedily transfer the best available technology to farmers. Our individual strengths are variable, but our collective strength is considerable. The IRRI family of rice scientists located all over the world, to which I now have the good fortune to belong, has the capacity to provide the tools for making the problem of hunger in rice-eating countries a problem of the past. We should accept this challenge as the primary goal of our work.

QUESTIONS

1. Review the importance of photoperiodism of rice in terms of:
   a. Local adaptation.
   b. International adaptation.
   c. Crop production systems.
2. The energy-efficiency analysis conducted by Freedman (1980) is useful in determining the highest energy inputs into various systems of rice production. Based on the calculations by Freedman, which is the single largest energy input into green-revolution systems of rice production?
3. Review all the inputs associated with high-yielding rice production associated with the green-revolution system and explain why record high yields result.
4. Consider the merits of international rice-testing programs conducted by IRRI.
5. Review the key morphological changes associated with high-yielding rice cultivars.
6. The use of blue-green algae and azolla is a new and rapidly developing field. Review current successes and trends in the use of these plants as sources of nitrogen.
7. It was suggested by Chang, Adair, and Johnston (1982) that the rapid spread of the high-yielding semidwarf rice cultivars since 1966–1967 has greatly narrowed the genetic base of the rice crop.
   a. Explain why this view may or may not be correct.
   b. Consider the merits of a gene bank for rice (review Chang, Adair, and Johnston, 1982).
8. Chang, Adair, and Johnston (1982) reported that plant physiologists have estimated the maximum yield potential of rice in the tropics to be as high as 15.9 t/ha (14,194 lb/ac) for a single crop in the dry season and 9.5 t/ha (8481 lb/ac) in the wet season.
   a. Explain why the dry-season yields are estimated to be 60% higher than in the wet season.
   b. Explain why rice is able to produce such high yields. Note that the record yield reported by De Datta and Malabuyoc (1976) obtained at IRRI was 11.7 t/ha (10,445 lb/ac).
9. Attempts to combine indica and japonica rice were unsuccessful prior to the initiatives of IRRI. Lack of success may be explained on the basis that the many components of the production system were not properly integrated. Explain.

References

Anonymous. 1975. "Improved Rice Varieties Needed for World's Deep Water Regions." *The IRRI Reporter.* Los Banos, Philippines: The International Rice Research Institute. No. 3. 4 pp.

———. 1977. "Bio-Fertilizers." *China: Recycling of Organic Wastes in Agriculture.* Rome, Italy: Food and Agriculture Organization of the United Nations. pp. 29–40.

———. 1982a. "International Rice Testing Program. (IRTP)." (In) *Rice Research in the 1980's: Summary Reports from the 1982 International Rice Research Conference.* pp. 17–19.

———. 1982b. "Genetic Evaluation and Utilization (GEU) Projects." (In) *Rice Research in the 1980's: Summary Reports from the 1982 International Rice Research Conference.* pp. 21–23.

———. 1982c. "International Network on Soil Fertility and Fertilizer Evaluation for Rice (INSFFER)." (In) *Rice Research in the 1980's: Summary Reports from the 1982 International Rice Research Conference.* pp. 25–27.

———. 1982d. "Asian Cropping Systems Network (ACSN)." (In) *Rice Research in the 1980's: Summary Reports from the 1982 International Rice Research Conference.* pp. 29–33.

Bahar, F. A., and S. K. De Datta. 1977. "Prospects of Increasing Tropical Rice Production Through Ratooning." *Agronomy Journal* 69: 536–540.

Beachell, H. M., G. S. Khush, and R. C. Aquino. 1972. "IRRI's International Breeding Program." (In) *Rice Breeding.* Los Banos, Philippines: International Rice Research Institute. pp. 89–106.

Chandler, R. F., Jr. 1979. *Rice in the Tropics: A Guide to the Development of National Programs.* Boulder, Col.: Westview Press. 256 pp.

———. 1982. *An Adventure in Applied Science: A History of the International Rice Research Institute.* Los Banos, Philippines: International Rice Research Institute. 233 pp.

Chang, T. T. 1964. *Present Knowledge of Rice Genetics and Cytogenetics.* International Rice Committee Technical Bulletin No. 1.

———, C. R. Adair, and T. H. Johnston. 1982. "The Conservation and Use of Rice Genetic Resources." *Advances in Agronomy* 35: 37–91.

———, G. C. Loresto, and O. Tagumpay. 1972. "Agronomic and Growth Characteristics of Upland and Lowland Rice Varieties." (In) *Rice Breeding.* Los Banos, Philippines: International Rice Research Institute. pp. 645–661.

Chattopadhyay, S. B. 1979. "Towards Higher Yielding Deep Water Rice." *Span* 22 (3): 111–113.

Dat, T. V., and M. L. Peterson. 1983a. "Performance of Near-Isogenic Genotypes of Rice Differing in Growth Duration. I. Yields and Yield Components." *Crop Science* 23: 239–242.

———. 1983b. "Performance of Near-Isogenic Genotypes of Rice Differing in Growth Duration. II. Carbohydrate Partitioning During Grain Filling." *Crop Science* 23: 243–246.

De Datta, S. K., and J. Malabuyoc. 1976. *Climate and Rice.* Los Banos, Philippines: International Rice Research Institute. pp. 509–539.

Evatt, N. S. 1966. "High Annual Rice Yields in Texas Through Ratoon or Double-Cropping." *Rice Journal* 69: 10–12, 32.

Freedman, S. M. 1980. "Modifications of Traditional Rice Production Practices in the Developing World: An Energy Efficiency Analysis." *Agro-Ecosystems* 6: 129–146.

Ishizuka, Y. 1971. "Physiology of the Rice Plant." *Advances in Agronomy* 23: 241–315.

Jackson, B. R., et al. 1972. "Breeding Rice for Deep Water Areas" (In) *Rice Breeding.* Los Banos, Philippines: International Rice Research Institute. pp. 517–528.

Kratochvil, F. 1956. "Rice Moves Northward." *World Crops* 8: 146–147.

Mengel, D. B., and F. E. Wilson. 1981. "Water Management and Nitrogen Fertilization of Ratoon Crop Rice." *Agronomy Journal* 73: 1008–1010.

Moomaw, J. C., and B. S. Vergara. 1965. "The Environment of Tropical Rice Production." (In) *The Mineral Nutrition of the Rice Plant.* Baltimore: Johns Hopkins University Press. pp. 3–13.

Naphade, J. D. 1971. "Oxygen Relations in the Rice Plant." *Riso* 20: 157–163.

Owen, P. C. 1971. "The Effects of Temperature on the Growth and Development of Rice." Review article. *Field Crop Abstracts* 24: 1–8.

Plucknett, D. L., J. P. Evenson, and W. G. Sanford. 1970. "Ratoon Cropping." *Advances in Agronomy* 22: 285–330.

Rutger, J. N., and D. M. Brandon. 1981. "California Rice Culture." *Scientific American* 244 (2): 42–51.

Savant, N. K., and S. K. De Datta. 1982. "Nitrogen Transformations in Wetland Rice Soils." *Advances in Agronomy* 35: 241–302.

Swaminathan, M. S. 1982. "Improving the Productivity of Rice-Based Farming Systems: The Challenges Ahead." (In) *Rice Research in the 1980's: Summary Reports from the 1982 International Rice Research Conference.* Los Banos, Philippines: International Rice Research Institute. pp. 1–15.

Turner, F. T., C. C. Chen, and G. N. McCauley. 1981. "Morphological Development of Rice Seedlings in Water at Controlled Oxygen Levels." *Agronomy Journal* 73: 566–570.

Venkataraman, G. S. 1981. *Blue-Green Algae for Rice Production—A Manual for its Promotion.* Food and Agriculture Organization of the United Nations Soils Bulletin. pp. 1–45.

16

Corn

INTRODUCTION

William Cobbett, a prominent author, agricultural scientist, and farmer in New York State in the first quarter of the nineteenth century, described the corn plant not only as "the greatest blessing of the country" but also as "the greatest blessing God ever gave to man" (Hardeman, 1981, p. 2). Hardeman (1981) described corn as a "quiet monarch that has reigned supreme in America for thousands of years" (p. 2). Indeed, there are a number of reasons why corn is classed as a monarch among the grain crops. These reasons, some of which have been mentioned in earlier chapters, are summarized as follows:

1. Corn has the C_4 photosynthetic pathway, which is considerably more efficient than the C_3 pathway under high temperature and dryland conditions. The greater efficiency of C_4 plants, as outlined in Chapter 2, is believed to be associated with the lack of photorespiration of energy losses due to respiration during photosynthesis. C_4 plants are known as tropical plants, because they have a higher optimum temperature for photosynthesis than do C_3 plants, which are known as temperate species. Because corn is a C_4 plant, it is less adversely affected by high temperatures than C_3 cereal plants (Moss and Musgrave, 1971).

Although photosynthesis is the basis of all crop production, a direct cause-and-effect relationship between photosynthesis and yield has yet to be established. Yield represents a summation of numerous factors of physiological processes and morphological components. An enhanced capacity in corn to fix carbon dioxide provides this and other C_4 crops with greater scope for yield capacity under a broader range of conditions than C_3 crops. Crop producers are able to manage C_4 crops in terms of plant population and related row widths, planting patterns, and double cropping to better advantage than C_3 species. The photosynthetic light response curve is not linear for any species, and production practices must take this into consideration.

The favorable C_4 pathway present in corn, an appreciation and understanding of which developed after 1960, can account for some of corn's desirable attributes outlined in this section. In short, C_4 plants are the most productive in terms of food nutrients produced per unit land area, per unit of water transpired, and per unit of time (Figure 16-1) under the conditions for C_4 plants.

2. The corn crop has a wide adaptation and is

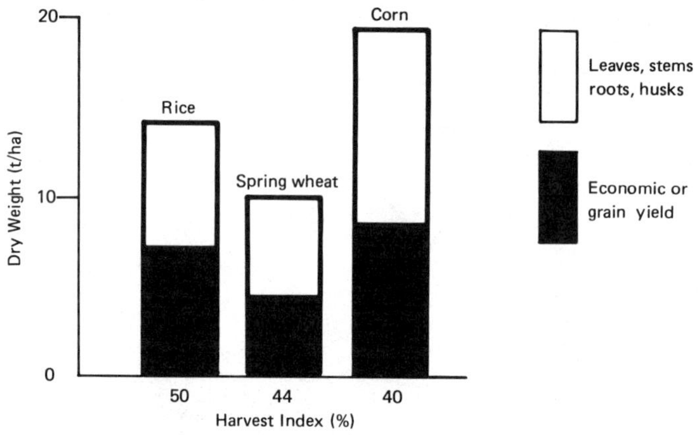

Figure 16–1. Biological or total biomass yields and harvest index or economic yield of three cereal crops averaged over the years 1975, 1976, and 1978 when grown in the northern part of Japan (Hokkaido). Biological yield is a function of growth duration and crop growth rate at successive growth stages. Economic yield of a crop is calculated from the biological yield by the harvest index value. Harvest index represents the partitioning of photosynthesis between harvested and nonharvested organs. Corn is a productive crop and has a high biological yield due to a high net assimilation rate owing to a high photosynthetic rate per unit leaf area. This is an advantage of the C_4 plant. (After Tanaka and Osaki, 1983)

able to grow in regions ranging from the semiarid, with an annual rainfall of 20 to 25 cm (8 to 10 in.), to those where annual rainfall may exceed 400 cm (157 in.). The crop can be grown as far north as the fiftieth parallel in Canada and over most of the United States, throughout Mexico and Central America, to as far south as central Argentina and Chile (about 35° S latitude) in South America. Corn is adapted to Africa, Central Europe, and Asia. Such a worldwide range means the crop is universal in its adaptation.

Corn can be grown over a wide range of altitude as well as latitude and is produced from elevations ranging from sea level to terraced plots 3960 m (13,000 ft) high in the Andean mountain range of South America. It cannot, however, be grown above the frost line.

The adaptation of specific open-pollinated or hybrid cultivars, however, may be rather limited because of genotypic differences in maturity, response to photoperiod, and resistance to pests.

3. Morphologically, corn exhibits a greater diversity of phenotypes than perhaps any other grain crop. Quantitative variations in the vegetative characters of corn were reported by Kuleshov (1933) as follows:

a. Plant height varied from 60 to 70 to 650 to 700 cm (2 to 22 ft).
b. Leaf number on the principal stalk varied from 8 to 48.
c. Leaf length extended from 30 to 152 cm (12 to 60 in.).
d. Leaf width measurements ranged from 4 to 15 cm (1½ to 6 in.).
e. Number of stalks, also known as suckers, tillers, or stools, arising from a single seed and hence one plant, varied from 1 to 12.

This range of variability provides plant breeders an opportunity to combine diverse genotypes and to select morphologically desirable phenotypes.

4. From the standpoint of plant architecture, corn exhibits many desirable features. Each leaf arises from a node and is separated by an internode, resulting in a leaf arrangement that is architecturally capable of maximizing sunlight interception with a minimum of mutual shading and that allows for free air circulation to replenish carbon dioxde. The upper leaf surface is pubescent and adapted for solar energy absorption; the lower leaf surface is glabrous and has numerous stomata, features that favor carbon dioxide absorption. The result is a photosynthetically efficient plant capable of high total dry-matter production.

Leaves and ears are borne on a stalk with a diameter of up to 5 cm (2 in.), which not only serves as a storage sink for surplus photosynthate but resists lodging associated with wind and rain. Although the ear may weigh 450 to 500 g (1 lb and over), prop or brace roots (Figure 16–2) help support the plant. Prop roots are adventitious roots arising from the basal nodes. Corn has no taproot, but roots can penetrate outward and downward into the soil sometimes as much as 2 m (6 to 7 ft) or more from the basal internode.

Reproductive organs arise from separate areas on the plant. The pollen-producing tassel located at the top of the plant facilitates cross-pollination by wind, yet allows for easy removal for controlled pollination in the production of hybrid corn (see Chapter 10). Both self- and cross-pollination can be manipulated. Corn is the only major cereal crop to exhibit the monoecious characteristic.

The ear or female reproductive organ arises between the leaf sheath and the stalk at an internode ½ to 1 meter (2 to 3 ft) below the tassel or about midway on the stalk. In contrast, the grain-bearing organ in other cereals is located at the top of the plant. The heavy ear of corn can be supported more readily midway on the stalk than at the top, thereby reducing lodging by stalk breakage.

A single corn stalk with a number of internodes is capable of initiating possibly up to six or seven rudimentary ears, of which one or two may develop, depending on the capacity of the source to fill the sink. Usually the ear below the primary one may have reduced grain set, as photosynthate is channeled into the primary ear. In prolific corn cultivars grown in southern regions, where the duration of photosynthesis is extended, several ears may develop. A properly fertilized and fully developed ear may bear 400 to 1000 grains.

Figure 16–2. Corn plants vary in height from 60 to 700 cm (2 to 22 ft) and support an ear weighing 450 g (1 lb) and over, but generally they are lodging resistant because of their sturdy 5-cm (2-in.) diameter stalks and brace or prop roots arising from lower internodes, as shown in an extreme case in this photo. Stalks also serve as a storage sink for surplus photosynthate. Large stalks with a high storage capacity serve to fill the ear if photosynthetic leaf area is destroyed by a fall frost. Sturdy, tall stalks with a large ear also make corn an ideal crop for forage (silage) production. (Courtesy of Ontario Ministry of Agriculture and Food)

5. The high multiplication ratio, whereby one seed is capable of producing a plant which in turn bears 400 to 1000 seeds (Figure 16–3), and where only 60,000 seeds are required to seed a hectare (24,000 seeds per acre), makes corn an excellent candidate for hybrid production. Single-cross seed is economically feasible.

Figure 16–3. In most of the developed areas of the world hybrid corn is grown to the exclusion of open-pollinated cultivars. Hybrid corn production is facilitated by separate male and female reproductive organs but also by a high multiplication ratio. One seed is capable of producing an ear bearing 400 to 1000 kernels. At a seeding rate of 60,000 seeds per hectare (24,000 seeds per acre) only about 80 plants are required to supply sufficient seed for a hectare or 32 plants for an acre. The yellow dent corn ear in this photo has over 800 kernels carried on 16 rows arranged in parallel ranks. The number of kernel rows is always even, because only the uppermost of the two florets in each spikelet is functional, resulting in paired rows of kernels. (Courtesy of Ontario Ministry of Agriculture and Food)

6. Each ear is protected by glumes or husks which form a tight, protective cover for the developing kernels. As the ear matures and the sheath of husks loosens, the ear bends downward (Figure 16-4) and protects the drying grain from rain and bird damage. Corn can remain in the field unharvested, should weather or other conditions cause an excessive delay in harvesting operations.

7. Genetically, corn is a diploid with ten pairs of chromosomes. The relatively low chromosome number, along with the fact that corn can be easily grown and manipulated, has led to extensive genetic studies. Chromosome maps have been prepared showing the location and description of one or more characteristic alleles at each locus (Coe and Neuffer, 1977).

Genetic variations in aleurone and scutellum color have served as readily identifiable marker genes. Corn color varies: red, blue, black, yellow, white, and variegated. Twenty classifiable shades varying from purple to a yellowish brown, from a pale yellow to dark brown, and from copper-red to orange have been produced by crossing.

The number of open-pollinated and hybrid corn cultivars is extensive. A listing of known cultivars, each differing genetically, would fill pages. The number of cultivars in existence is a reflection of the extensive breeding programs that have been directed to this crop.

8. From a physiological viewpoint corn has many superior features, chief among which is transpiration efficiency. For every 1000 units of

Figure 16–4. Ears of corn are protected by glumes or husks which form a tight covering during kernel development. As the ear matures, the husks loosen and drying is facilitated. A major feature of the corn crop is that the ears bend downward in advanced stages of maturity. Unharvested, mature ears in a suspended downward position are protected from rain and bird damage. (Courtesy of Ontario Ministry of Agriculture and Food)

water consumed, corn produces an average of 2.87 units of dry matter. The transpiration ratio is therefore about 350. Almost twice as much water is required to produce a unit of dry matter in rice, barley, wheat, oat, and rye.

9. Crop producers appreciate the response of corn to cultural practices. Area for area, corn outproduces wheat by nearly 2.5 times (Table 16-1). Higher record yields have been obtained with corn than with wheat.

Corn can be produced with a minimum of soil preparation. This important feature allowed the North American native Indians to produce corn in forested areas. Indians planted corn in **hills** to avoid cultivating the entire area. Settlers clearing land could plant corn before all the stumps and trees were removed. Variations of hill planting have been used to facilitated weed control by cultivation. Small grains other than corn were seeded broadcast, and the entire area had to be cultivated.

Modern corn production may employ various forms of conservation tillage. Corn may be planted without any tillage or in narrowly tilled strips with little or no cultivation between the rows.

Excessive vegetative growth or lodging is seldom a problem in corn, even under very high fertility levels. Early corn producers recognized this fact. In Peru, bird guano was used to boost grain yields; in New Mexico, fertilizer from bat caves was utilized by Hopi Indians. In the absence of a readily available fertilizer, early settlers copied Indian methods and placed a fish in each hill. Modern recommendations for nitrogen fertilizer range from 95 to 110 kg N/ha (85 to 100 lb N/ac) for a target yield of 6000 kg/ha (5400 lb/ac), depending on inherent soil fertility, previous crop, and residues incorporated.

Table 16–1
Corn and Wheat Yield Comparisons in the United States in 1974

	Average Yield		*Record Yield*	
Crop	*kg/ha*	*lb/ac*	*kg/ha*	*lb/ac*
Corn	4516	4032	19,255	17,192
Wheat	1882	1680	13,548	12,096

Source: After Wittwer (1975).

10. Corn is an outstanding forage crop as well as a grain crop (Chapter 8). It is ideal for the production of whole-plant corn silage because of the large yield of biomass and the ease with which a quality silage can be produced. Corn silage can be produced from corn when the plants are green and when the kernels of dent corn are well dented at an overall moisture level of 65% to 70%. Anaerobic fermentation converts the readily available carbohydrate in the grain and stems to a pH level of 3.8 to 4.0, whereby it be preserved and stored.

11. Corn is a highly versatile crop. Tiny, immature ears, 5 to 7 cm (2 to 3 in.) long, can be either pickled or boiled and eaten as tender vegetables. Ears of sweet corn can be eaten when they are full-grained but with soft milky kernels. Dry mature kernels can be ground into meal, flaked, or shredded to produce various breakfast foods. Popcorn can be exploded under heat as a confectionary food.

As a feed crop corn can be used as whole-plant silage, as center-cut silage in which the ear and the center portion of the stalk and leaves are harvested for feed, as high-moisture grain, as corncob meal, or as dry ground grain.

Corn stover can be used as a feed, as bedding, or as residue material to be incorporated into the soil or as a surface mulch.

12. In areas with a well-defined growing season, corn cultivars can be selected that utilize the entire growing season or frost-free period. By maximizing the duration of photosynthesis, sunlight energy conversion is used to advantage and stored energy is maximized. In contrast to full-season corn cultivars, some of the cereal grains use only a portion of the growing season because of their short-season growth habit.

13. Corn is adapted to production ranging from minimum to maximum levels of mechanization. Unlike small-grain cereal crops, in which primitive production practices involved throwing seed on the soil and accepting a meager yield because of weed interference, early corn producers placed a given number of seeds into a

hand-worked hill. The small area around the hill could be kept free of weeds. The large ear of corn could be hand picked, the husks removed (shucked), and hand shelled. The use of a sickle was not required. In mechanization of production, small-grain cereals were nearly a hundred years ahead of corn, largely because small grains could not be harvested as easily by hand and could not be left on the stalk very long without loss or rain damage. Corn kernels are embedded firmly on a cob and can be stored and dried in a crib (Figure 16-5), which facilitates free air movement to reduce spoilage.

In areas of the world that depend on mechanization for handling large-scale production rapidly, all aspects of corn production and handling can be fully mechanized. Mechanical units are available that achieve precision planting, application of a starter fertilizer, application of pesticides, and minimum tillage in a single pass over the field. Grain or silage can be harvested mechanically and placed in a silo or drier without any hand labor. Silos and grain storage units can be mechanically filled and unloaded, mixed into a ration, and conveyed to livestock on a conveyer or loaded for transport for market.

Figure 16–5. Corn is a full-season crop, and in many areas of the world kernel moisture levels exceed the level for safe storage in an enclosed bin. High fuel costs mean that high-temperature drying is expensive. A traditional method of drying corn involves the use of corn cribs, as shown in this photo. Corn kernels are embedded firmly on a cob, and if whole ears are placed in a crib, free air movement provides an effective means of reducing spoilage and results in a slow reduction in moisture levels. (Courtesy of Ontario Ministry of Agriculture and Food)

A review of the virtues of corn make it clear that corn is indeed a quiet monarch, the queen of the cereal crops. No crop is perfect, however, and corn has problems associated with it. These problems are perceived as follows:

1. Modern corn production, involving high-energy fossil-fuel inputs associated with mechanization, nitrogen fertilizer, and pesticides, has resulted in large production units capable of a high yield per unit area. It appears that under conditions of intensive production, an increasing amount of energy inputs are required to maintain yields, and several concerns are associated with high energy dependency. The first is that under conditions of high energy costs relative to a unit of corn, economic corn production may not be feasible, or under conditions of fossil-fuel scarcity, productivity may decline. A second concern is the efficiency of inputs relative to output. Despite the fact that corn is a photosynthetically efficient crop, capable of a high yield, the input/output efficiency in mechanized areas of the world is not high. The two factors raise the question of whether energy-intensive agricultural systems such as mechanized corn production are sustainable.
2. Corn is an annual row crop that offers little protection against soil erosion on sloping fields and little protective winter cover, especially if traditional methods of fall plowing are utilized. Corn roots are coarse and not inducive to good soil structure. Intensive monoculture production of corn has focused attention on declining soil productivity.

Whether the discovery in 1977 of a perennial form of teosinte with ten chromosome pairs[1] (Barney, 1980) will be useful in changing the growth habit of corn to a perennial form remains to be seen. Without the perennial habit,

[1]Perennial teosinte—*Zea diploperennis*.

corn lacks the protective cover offered by winter cereals, and unless alternate protective action is taken, corn production on rolling land or unstable soils may be restricted.

Studies on root-pulling resistance suggest that a large, profuse root system could be developed through breeding and selection (Peters, Shanks, and Nyquist, 1982). An increase in root size might consume more photosynthate, reducing the amount available for grain yield. No relationship was found, however, between resistance to root pulling and grain yield.

3. The full-season duration of corn growth and the desire among producers to select cultivars that utilize the full season in their area means that many cultivars are needed to serve geographic regions that have minor climatic differences. The need for numerous cultivars adds to the cost and complexity of seed production and marketing. This situation is complicated further by the fact that, in developed areas of the world, hybrid cultivars are used extensively. Scores of hybrid cultivars are available for a specific area, each requiring annual seed production.

4. Adding to the complex network of hybrids required to serve specific geographic regions is the fact that many hybrids are sensitive to photoperiod. Photoperiod influences the time period from germination to floral development. Corn is classified as a short-day crop, although photoperiod sensitivity varies enormously within the corn species.

Unlike the wide adaptation developed among rice and wheat cultivars, which facilitates worldwide testing and exchange, corn has narrowly adapted genotypes that must be tested in areas where they are adapted. A lack of freedom to exchange genotypes for plant-breeding purposes may result in the use of a narrow gene base to produce inbreds at a specific site.

The efficient breeding and selection of corn genotypes for different climatic regions requires a quantitative understanding of the plant's developmental responses to environmental factors such as temperature and photoperiod. For example, corn has determinate growth habit, and leaf initiation continues until floral initiation occurs at the apex, after which reproductive rather than vegetative development begins. Anthesis marks the start of the grain-filling phase. Photoperiod and associated temperature conditions influence both the rate and duration of many developmental events. Precise information as to the effects of photoperiod and temperature is essential if reliable and meaningful crop-production models are to be developed.

In general, an increase in photoperiod lengthened both the time between sowing and tassel initiation and the time between tassel initiation and anthesis (Warrington and Kanemasu, 1983). Both leaf-initiation and leaf-appearance rates increased with an increase in photoperiod. As photoperiod increased, leaf number increased. Temperature effects were expressed by rate rather than degree of development.

The use of winter nurseries to advance breeding material a generation is complicated by corn's sensitivity to photoperiod and temperature. Corn is a short-day crop. Cultivars adapted to high latitudes, where long-day conditions prevail through much of the growing season, will be stimulated by short-day conditions if grown near the equator. The result is that reproductive development will be stimulated prematurely relative to plant development; plants will not reach full stature and will have a small ear and low yield.

In contrast, when grown at a high latitude, corn cultivars adapted to short-day conditions near the equator will not be stimulated into reproductive development until a specific short day is reached late in the season. At this point the plants cannot be expected to mature, and crosses with locally adapted cultivars cannot be made.

It is to be noted that the response of photoperiodically sensitive crops is influenced by the dark period. So-called short-day plants respond to long nights; long-day plants respond to short nights. Long-day plants such as the barley and oat crops can be simulated into reproductive development when grown under short-day conditions by interruption of the long night period with artificial lights on a small plot area. This is not possible, however, with corn grown at a high latitude. The result is that free exchange of genetic stocks is hampered in corn.

5. Corn is believed to have evolved in the tropical areas of Central and South America. After hundreds of years of conscious or unconscious selection by native peoples of North America, corn adapted to favored areas throughout the continent was achieved. When grown in areas with long days, anthesis is delayed until days become sufficiently short to trigger reproductive development. By this time, however, the period of high sunlight that is photosynthetically most desirable has passed. The grain-filling period, the most critical time relative to maximum yield, is at a disadvantage (refer to Figure 4-7).

6. Corn can neither tolerate temperatures below freezing (refer to Figures 6-4 and 6-5) nor thrive under continued low but nonfreezing temperatures. Corn can be damaged by frost in both the seedling and adult stages. Attempts at early seeding, to promote a better coincidence of leaf area with the period of high light energy, may subject seedlings to late spring frost damage or cool conditions. Plants subjected to unfavorable conditions do not grow well and may suffer from seedling diseases, discoloration, and a general lack of vigor.

An improvement in the association between physiological behavior of corn to low temperature and its field performance could be important in selecting hybrids most suitable for production at high latitudes. Tolerance to low spring temperatures together with vigorous seedling growth and canopy expansion was suggested by Lee and Estes (1982) to be advantageous to plant survival, maturity, and yield.

Under controlled low-temperature conditions, Cal and Obendorf (1972) observed differences in percent emergence among corn hybrid cultivars. Superior seedling growth of early-maturing compared to later-maturing corn hybrids was noted under controlled low-soil-temperature conditions at Ottawa (MacLean and Donovan, 1973).

Improved tolerance to chilling temperatures and superior seedling vigor in an early-maturing corn hybrid were suggested by Lee and Estes (1982) to be associated with more stable chloroplasts. Precise relationships between seedling cold tolerance and final grain yield have not been clearly established. It may be that the associated smaller plant structure of an early, cold-tolerant hybrid can result in high grain yield only if plant population is increased. All factors of the production system must be integrated successfully.

7. In its natural state, corn is a highly heterozygous crop mainly because of cross-pollination and to a lesser extent gene mutations. Heterozygosity has resulted in a rapid response to selection pressures from both man and nature. The result is an extreme level of **polymorphism**.

The development and widespread use of hybrid corn produced from controlled crosses of a restricted number limits the extent of heterozygosity. Only a few selected inbred parents are crossed, with the result that natural evolution is restricted. In the long term, controlled breeding associated with hybrid production could limit the development of naturally adaped new genetic combinations. The possibility exists that the wide genetic natural diversity of corn may be reduced in areas where hybrid cultivars are grown exclusively.

8. A thick sturdy stalk and prop roots reduce lodging from stalk breakage or bending unless stalks become weakened by disease or insects. Despite prop roots, root lodging can be a problem and is influenced by soil fertility and other environmental factors. Efforts to improve lodging resistance in corn by selecting large root systems showed limited promise (Arihara and Crosbie, 1982). Lodging resistance may be reduced by root-rot organisms and corn rootworms.[2]

9. Barrenness, the failure of a plant to produce a normal ear, is a major limitation to efficient conversion of light energy into grain by corn grown at high plant densities. If genetic improvement in the ability of a corn plant to resist barrenness could be achieved, improved sunlight interception by higher plant populations might result in higher grain yield.

---

[2]Corn rootworms—*Diabrotic* spp.

Selection for photosynthetic rate under high plant populations was undertaken for five cycles of recurrent phenotypic selection for high carbon dioxide exchange in two populations of corn (Crosbie and Pearce, 1982). Few researchers have reported this approach to corn improvement, possibly because of negative results. Although no detrimental effects were reported by Crosbie and Pearce, the long-term effects are not clear.

Barrenness is a physiological response of the corn plant to inter and intraplant competition for environmental factors. The development of density-tolerant cultivars was examined by Smith, Mock, and Crosbie (1982). Although some progress may be possible, barrenness is a complex trait that will not be easily overcome. The problem cannot be solved simply by upright leaf orientation that allows for better sunlight penetration to the developing ear. Plant size may help. Early-maturing corn cultivars usually are smaller in stature than late-maturing hybrid cultivars. No interactions among plant densities could be found among early corn cultivars, however, in two studies to this effect (Cross and Hammond, 1982; Troyer and Hallauer, 1968).

10. A problem with corn and angiosperms in general is the rapid loss of germinability and vigor during long exposure to warm, humid conditions. In areas where hybrid corn cultivars are extensively grown, germinability and longevity are prime considerations. Hybrid seeds require major inputs in their development, multiplication, storage, and distribution. Seed must be obtained for each crop, and leftover seed must be capable of being carried over from one season to the next. To accomplish this, extensive controlled-temperature and -humidity storage conditions may be required. Without such storage facilities, hybrid corn development may not be feasible.

Improvement in seed longevity without adequate storage may assist a geographic extension of hybrid corn. As outlined in Chapter 9, temperature and humidity are key factors in seed longevity.

BOTANICAL ASPECTS OF CORN REPRODUCTION

Isolation of the reproductive organs on the corn plant means that barren plants can result when ear shoot development and pollen dispersal do not coincide. Pollen produced in the male flowers of the tassel at the top of the plant may be dispersed over a few days or a few minutes, depending on environmental conditions including temperature and humidity. Once pollen is dispersed into the air, it will remain viable only for a few minutes because of rapid desiccation.

The female organs are borne at the sixth or seventh node, below the top tassel in most North American corn hybrids. The stigma, known as a silk, is attached to an ovary on the cob. The silks must emerge from the tip of the ear husk to receive the pollen. There is one silk for each potential kernel, and the first silks to emerge are from the basal part of the cob.

The rate of silk emergence is a function of environmental factors including temperature, moisture, and photosynthate availability. Stress conditions may restrict silk development and emergence and hence ear formation. Silks may develop over a two-or three-day period and may remain receptive seven to ten days.

At each node below the top ear node, axillary buds are found that may develop into an ear. If they do, and if silks are pollinated and ovaries fertilized, more than one ear may develop per stalk. Multiple-eared stalks are known as prolific corn. Prolific genotypes exist that regularly develop two or more ears. In single-eared corn a relationship appears to exist between tassel and silk development. In prolific corn this relationship may not exist, and top ear shoot and tassel development may not coincide. Under conditions where the sink is the limiting factor to grain yield, a prolific corn may be desirable.

Cross-pollination is aided by wind currents. Because as many as 25,000 pollen grains may be produced for each silk on an ordinary ear, seed set is usually satisfactory under normal conditions.

VARIATION IN CORN

On the basis of physical appearance, seven classifications of corn can be made: flint, dent, flour, pop, sweet, waxy, and pod. Physical appearance is based on the quality, quantity, and pattern of endosperm composition in the kernel.

1. Flint corn kernels are characterized by hard, corneous or horny endosperm encased in a thick skin or pericarp, which remains smooth and rounded at maturity because the kernels do not shrink when mature and dry (Figure 16-6). Flinty kernels require more energy to grind than corn with soft endosperm, but they have the advantage of good storage qualities, as seeds will remain viable for a long time.

   It was flint corn that European settlers obtained from the North American native Indians when they arrived in New England in the seventeenth century. Flint corn was grown by the Indians as far north as the mouth of the St. Lawrence River in North America. A flint corn obtained from this area and known today as Gaspe flint matures in 60 days from planting and grows to a height of less than 1 m (1 yd). It is a source of genes for early maturity.

Figure 16–6. Flint corn has smooth and rounded crowns, because the hard endosperm does not shrink when mature and dry. The kernels on the two cobs in this photo show a minimum of denting compared with the softer kernels in Figure 16-3. The vitreous endosperm encased in a thick, tough pericarp is evident. Flint corn is often associated with early maturity, and flint-dent crosses are commonly made to combine the desired features of both kinds of corn. (Courtesy of Pioneer Hi-Bred International, Inc., Des Moines, Iowa, U.S.A.)

2. In contrast to flint corn, floury corn has soft starch throughout the kernel encased in a thin pericarp. The softness of these kernels makes them easy to grind, and they are preferred over flint corn where primitive grinding conditions exist. Floury corn has the disadvantage that in wet areas the kernels may mold on the cob, and so it is grown where the harvest season is dry. Floury corn kernels are similar in texture to the high-lysine mutant opaque-2.
3. Dent corn contains both hard and soft endosperm. The sides of dent kernels are flinty and the central core is floury. During drying, the soft or floury starch shrinks more than the hard sides, and the mature kernel has a dented crown (refer to Figure 16-3).
4. Popcorns are small-kerneled, flint-type corns. Popcorns with a sharp-pointed kernel are known as the rice type, those with a smooth rounded crown as the pearl type (Figure 16-7).
5. Sweet corn has a wrinkled and glassy appearance, resulting from the fact that a sugary gene prevents the normal conversion of sugar into starch during endosperm development.
6. Waxy corn carries a gene that causes 100% of the starch in the kernel to consist of a branched molecular form called amylopectin. The "normal" ratio for corn is 72% amylopectin and 28% amylose. Starch from waxy corn is used for food processing.
7. Pod or tunicate corn is so named because each kernel on the cob is enclosed in a glume similar to that in other cereals. The kernels of pod corn may be flinty, dent, or pod-popcorn. Pod corn is more of an ornamental than a commercially produced type of corn.

DOUBLE FERTILIZATION

In cereal crops, including corn and other angiosperms, seeds arise from developing ovules and their surrounding tissues. An essential step in sexual reproduction is the union of male pollen with the female egg. This process was described in Chapter 2 but is reviewed here to illustrate its importance in cross-pollinated plants.

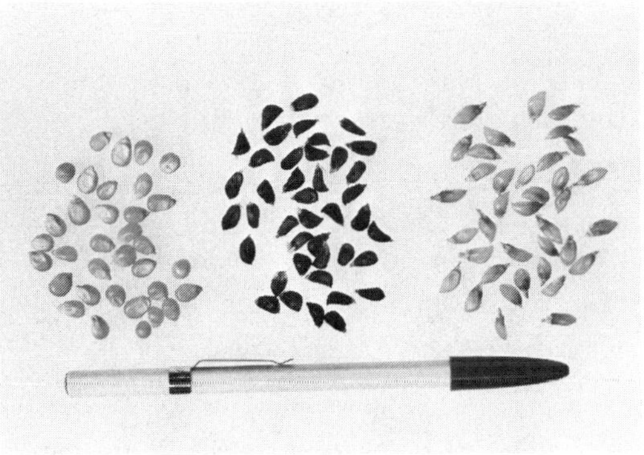

Figure 16–7. Kernels of popcorn with the pearl or round kernels on the left and rice or pointed kernels in the center and on the right. Popcorns are essentially a small-kerneled flint type with a thick pericarp that may be yellow or purple (center) or white. (Courtesy of Ontario Ministry of Agriculture and Food)

A common concern of popcorn and sweet-corn growers is the immediate effect on quality of out-crossing with a different type of corn. For example, will sweet corn cross-pollinated with dent corn have unsatisfactory eating quality; will popcorn cross-pollinated with dent corn be suitable for popping purposes?

During the process of fertilization, the pollen tube penetrates the embryo sac, and one male gamete unites with the egg to form the **zygote.** The zygote undergoes mitosis and differentiates into the embryo. New genetic combinations carried in the embryo will be expressed in the subsequent crop when the embryo carried in the seed forms a new plant. There is no immediate visible effect of fertilization on the embryo, even if widely different crosses occur, because it is meristematic tissue. The embryo serves to continue the hereditary stream.

The endosperm of a seed is also produced by fertilization with a second pollen nuclei and two polar nuclei formed in the egg. The complete process is referred to as **double fertilization.** The endosperm is triploid in its genetic constitution. As the endosperm develops, the effect of pollen becomes apparent and is known as xenia.

Classical examples of xenia occur in corn because of its heterozygous nature. When a cultivar with white endosperm is pollinated by one with deep yellow endosperm, the triploid endosperm is pale yellow. The genetic composition of the endosperm is modified only moderately, because it contains two-thirds the genetic composition of the mother plant and one-third that of the pollen parent.

POPCORN

Popcorn is a peculiarly American crop. It cannot be regarded as a staple food but is consumed directly, freshly popped, as a confection. When subjected to heat, popcorn pops with an explosive force and expands 20 to 30 times its original volume (Figure 16-8).

Exactly why popcorn pops is not well understood. Popping ability appears to be conditioned by the relative proportion of flinty or horny endosperm and the encasing of the starch grains in a tough, elastic, colloidal material. Steam pressure generated within the kernel

Figure 16–8. When popcorn seed is subjected to high temperature, it expands with explosive force into a dry, edible product as shown here. Popcorn expands 20 to 30 times its original volume and is consumed as a confection rather than a staple food. Popping corn is not an industrial process and can be achieved readily in the home. Popcorn kernels can be stored indefinitely and freshly popped as desired. (Courtesy of Ontario Ministry of Agriculture and Food)

upon heating is confined and resisted within the starch granule until it reaches explosive force (Weatherwax, 1922). Popping may be aided, not only by the embedding of the starch granules in a proteinacious matrix, but also by the fact that the inner soft starch is encased in hard endosperm as well as a tough pericarp that surrounds the kernel. Steam pressure can build up to considerable proportions (Richardson, 1959) until the kernel explodes.

Ability to pop is not limited to popcorn. Many flint types of corn pop fairly well, and even some of the dent corns with a large volume of horny endosperm pop a little. Some cultivars of sorghum with hard, flinty seed pop well. Breakfast food manufacturers use mechanical methods of confining steam pressure with the grains and releasing it suddenly. Puffed wheat and rice expand to several times their original volume.

Quality of popped corn is measured by high popping expansion, tenderness, flavor, kernel shape and color, and freedom from hulls. As the corn kernel is not composed of endosperm alone, popcorn quality cannot be measured by a single test. The germ contributes significantly to the texture of the popped corn. The harsh and unpalatable tip cap of the germ detracts from popcorn quality.

Popping expansion may be influenced by popping methods, differences in maturity, kernel moisture level, genotype, and measuring methods. In addition to the official volume test, a weight-volume test has been established. Popping expansion is generally recognized as being correlated with high quality. In the weight-volume test, the raw corn is weighed and the volume of the popped corn measured. The results are expressed as volume per unit of weight.

Tenderness includes the main portion of the popped kernel and the absence of a hard core at the base. The tough pericarp or hull is fragmented during popping, and these fragments become an important factor in popcorn quality. The thinner the pericarp, the more desirable the product. A thin, tender hull is desired that disintegrates in the mouth. The inheritance of pericarp thickness was reported by Richardson (1960) to depend on a single major dominant gene for thin pericarp and on a modifier complex which induces thick pericarp.

A sweet, pleasant flavor and not a strong, dent corn flavor is desired in popcorn. Brunson and Smith (1945) reported surprising variation in popcorn flavor when eaten without seasoning. Flavor was found to vary from none to pleasant and distinctly sweet, to strong and disagreeable. Selection among inbred lines can improve flavor in the resulting hybrids. Personal preference may determine the type selected. In general, white popcorn is more tender, but yellow cultivars have more flavor.

Shape of popped corn varies from a mushroom type, which fluffs up to a distinctly globular form, to the butterfly type, which turns inside out when it explodes (Figure 16-9). The globular type is characteristic of South American cultivars.

Color of popped corn varies from chalky white, to cream color, to a mottling of a deep yellow on the outside of the popped kernel. This

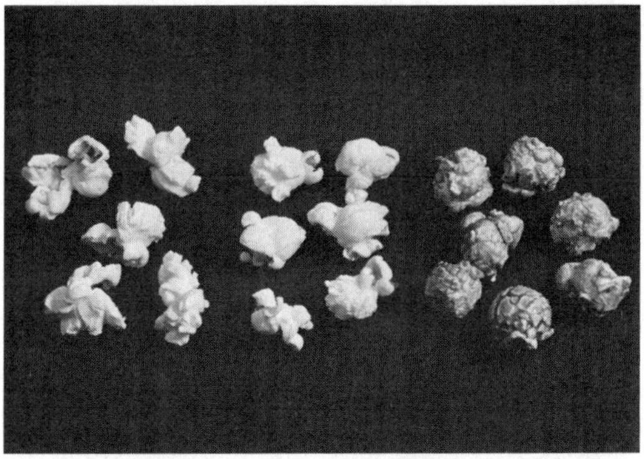

Figure 16–9. The shape of popped corn varies from a butterfly to a mushroom. The butterfly type shown on the left turns inside out when it explodes. The mushroom type, shown center and right, fluffs up to a distinctly globular form. The globular type on the right has been coated with caramel flavoring, which gives it a sweet taste and crisp texture. Caramel corn can be boxed, stored, and sold as a confection without becoming stale. (Courtesy of Ontario Ministry of Agriculture and Food)

latter is most pronounced in strains that have a mushroom shape when popped.

In North America, the development of hybrid popcorns has offered advantages in yield, uniformity, and standability similar to those found in dent corn hybrids. Improved hybrids of popcorn give an appreciably higher popping expansion than previously available (Brunson and Smith, 1945).

Extensive geographic production of dent corn has caused serious problems in providing adequate isolation for hybrid development in seed fields of popcorn. Under favorable condition, grains of pollen can be carried as far as 3.2 km (2 m) by wind (Brunson, 1955). Out-crossing of popcorn seed fields with dent corn will produce progeny intermediate between popcorn and dent corn, unsuitable for popping.

To overcome this problem, a "cross-sterile" gene can be incorporated into the popcorn parental inbreds by backcrossing. This will prevent out-crossing with "normal" dent corns (Brunson, 1955). Cultivars carrying the cross-sterile gene cannot be fertilized by pollen lacking the cross-sterile gene.

In fields grown for popping purposes rather than seed, xenia appeared to have a minor effect on popping expansion (Brunson, 1955). In a detailed study on the effect that 14 different endosperm genes had on popping, Richardson (1959) considered xenia to be almost always deleterious.

Popping quality was recovered in popcorn × dent corn crosses aimed at increasing genetic variability and popcorn field performance (Johnson and Eldridge, 1953).

If popcorn is stored properly, deterioration with age is minimal. Quality is maintained longer if stored on the cob than as shelled corn. Proper storage consists of maintaining moisture levels by storing it in tight containers, preventing pericarp damage, and preventing damage from storage insects. Based on several reported studies, Brunson (1955) concluded that popping expansion of good corn does not decline over a period of up to 14 years if properly stored.

Best popping results are obtained at about 14% moisture. Above or below this level, popping declines. Popcorn that is too dry for good popping can be corrected by exposing it to a moist atmosphere or adding a small quantity of water and mixing thoroughly.

SWEET CORN

The presence of a sugary gene makes sweet corn a desirable human food when the kernels are immature and succulent. Only immature corn with a moisture content of 72% to 75% is desirable.

Sweet corn may be boiled to heat-set the milky endosperm and served as whole ears, with the kernels to be eaten off the cob. It may also be served as kernels, canned or frozen. The kernels are removed from the cob by knives which cut them just above the glumes. Quality determines the market value and is measured by color, freedom from foreign material, cut, maturity, and flavor. Maturity is a major factor, as it influences tenderness or succulence (Smith, 1955).

Traditionally, sweet corn has been grown in the northern half of the United States and southern Canada, because high temperatures in the south are inducive to converting sugar in the kernels into starch with an accompanying loss of quality. A second reason has been the prevalence of insects and disease associated with the milder southern winters.

The development of hybrids with increased pest resistance has encouraged sweet corn production in southern areas of North America. The major reason for continentwide sweetcorn production, however, was the discovery of the gene known as shrunken. When the recessive shrunken gene is present in sweet corn, supersweet corn results (Cochran, 1962), having sweeter, plumper, crisper, and tastier kernels.

Normal sweet corn carries a double recessive for sugary and a dominant gene for shrunken. Supersweet corn has a dominant gene for sugary and a double recessive for shrunken. The genetic combinations resulting from a cross of sugary with shrunken are shown in Table 16-2. Mature supersweet kernels are slightly shrunken but not wrinkled like normal sweet corn.

Table 16–2
The Punnett Square Illustrates the Four Types of Kernels Produced in the F_2 Generation after a Cross of Sugary Sweet Corn (*su su Sh Sh*) and Shrunken Sweet Corn (*Su Su sh sh*)

		Gametes from Male Parent			
		Su Sh	*su Sh*	*Su sh*	*su sh*
Gametes from Female Parent	*Su Sh*	*Su Su Sh Sh*	*Su su Sh Sh*	*Su Su Sh sh*	*Su su Sh sh*
	su Sh	*Su Su Sh Sh*	*su su Sh Sh*	*Su su Sh sh*	*su su Sh sh*
	Su sh	*Su Su Sh sh*	*Su su Sh sh*	*Su Su sh sh*	*Su su sh sh*
	su sh	*Su su Sh sh*	*su su Sh sh*	*Su su sh sh*	*su su sh sh*

Summary:

9 normal (*Su-Sh-*)
3 sugary (*su su Sh-*)
3 shrunken (*Su-shsh*)
1 sugary shruken (*susu shsh*)
(this double recessive has brittle seed at maturity and cannot be used)

Field corn, sugary sweet corn, and supersweet differ in the manner in which simple carbohydrates manufactured in photosynthesis are converted to starch and stored in the kernel. In field corn, the products of photosynthesis are carried as sucrose sugar into the developing kernel, where it is converted rapidly to dextrin and immediately into starch. Starch does not taste sweet. Field corn carries the dominant counterpart for sugary and shrunken genes.

In sugary sweet corn, the presence of the sugary gene delays the conversion of dextrin into starch. The content of the sweet-tasting sucrose and the nonsweet dextrins is higher than in field corn. In the supersweet corn the recessive shrunken gene prevents the conversion of sucrose to nonsweet dextrin. The blocking of starch synthesis at an early stage increases the amount of sucrose sugar in the kernels. Supersweet kernels have five to eight times the sugar content of field corn and twice the sugar content of normal sweet corn at picking time.

A second major advantage of supersweet corn is that the sucrose content remains high after harvesting. When corn is picked, the flow of sucrose produced by photosynthesis is cut off. In field corn, any existing sucrose is converted to dextrin and starch, and an immediate taste loss occurs. With sugary sweet corn, the taste loss is less rapid but occurs within 48 hours, as the sucrose slowly converts to dextrin. Under high-temperature conditions, this conversion is speeded up. With supersweet, conversion of sucrose to nonsweet dextrin is blocked, and taste loss is not rapid. Some taste loss will occur as the sucrose is used in respiration. Sugary sweet corn loses two-thirds of its sucrose in the first 48-hour period after harvesting; supersweet loses about one-fifth.

In the production of hybrid sweet corn

cultivars, a double-cross method of hybrid production is possible with supersweet corn. When sugary sweet corn is used in hybrid production, single-cross seed is preferred. The reason is that conventional sugary cultivars pass very rapidly through the best stage for harvesting. Only single-cross offspring are sufficiently uniform to assure growers uniform ripening for high quality when the entire field is harvested for processing. With supersweet, high quality is maintained longer, so less uniformity of ripening is required. Double-cross seed is more economical to purchase.

TEOSINTE

Despite considerable research and speculation (Mangelsdorf, MacNeish, and Galinat, 1964; Mangelsdorf, 1974; Galinat, 1979), the origin of corn and the process of domestication remains a mystery. It is possible that teosinte,[3] a close wild relative of corn, is the ancestor. Teosinte is a tall grass native to certain sections of southern Mexico and Central America. Teosinte resembles corn in many respects and crosses readily with corn. It is possible that corn developed from teosinte or from some of the ancestors of teosinte. Perhaps corn descended from some unknown wild plant or arose through a cross of teosinte with some other plant such as *Tripsacum*.

Regardless of its ancestry, teosinte holds interest for plant breeders today. In 1910, a perennial cultivar of teosinte[4] was discovered (Barney, 1980). It was hoped that finding a perennial form of a possible ancestor of corn would open new possibilities for hybridization. The perennial form of teosinte had twice as many chromosomes as corn and could not be crossed. This discovery confirmed the hypothesis that a perennial form of teosinte had in fact existed. In 1977 a diploid, 20-chromosome form of perennial teosinte[5] was discovered that could be crossed with corn. The possibilities of perennial forms of corn, as discussed earlier, offer tremendous potential (Nault and Findley, 1982).

SOME PRODUCTION ASPECTS

It is impossible to give corn production recommendations on a world scale. Some major considerations are considered here in general terms.

Plant population is one of the most discussed issues in recent years, because success with new morphological types has resulted in dramatic yield increases. Every generation of corn producers, too, is anxious to increase sunlight interception and associated yield by an increase in plant population. Abrupt changes in plant population without full consideration of other aspects in the production system are seldom successful.

Herbicidal weed control has resulted in more dramatic changes in corn and sorghum production systems than in other cereal crops. A second factor that has allowed for plant population changes is the development of hybrids that do not tiller. Other factors influencing plant population include soil type, fertility, drainage, planting date, geographic location, and whether the crop is to be used for grain or silage. The ideal plant population is not a constant figure; for grain production, a final plant population of 47,000 to 60,000 plants per hectare (19,000 to 25,000 ppa) is recommended. When corn is sown for whole-plant corn silage, a 25% increase in final plant population is recommended.

It must be emphasized that as final stand densities are increased, the plants are under greater stress, since there are more plants competing for available moisture, nutrients, light, and carbon dioxide. With denser plant stands, additional stress such as drought, low or unavailable plant nutrients, and pests will have a detrimental effect on grain yield, grain quality, and lodging, and barrenness in plants could result.

Mechanization initially led to a series of operations associated with corn planting, includ-

[3]Teosinte—*Zea mexicana*.

[4]Perennial teosinte—*Zea perennis* (40 chromosomes).

[5]Perennial teosinte—*Zea diploperennis* (20 chromosomes).

ing plowing, disking, and/or cultivating and harrowing. Traditionally, tillage was required for three main purposes: weed control, incorporation of previous crop residues, and seedbed preparation. In an effort to reduce energy inputs, soil organic-matter losses, and erosion, chemical herbicides have replaced some of the tillage operations. For seedbeds, strips may be prepared into which rows of corn can be planted. Since a 3- to 5-cm (1¼- to 2-in.) seeding depth is satisfactory, shallow cultivation of 5 to 10 cm (2 to 4 in.) is sufficient. In areas that have cool temperatures at seeding time, cool seedbeds associated with zero-tillage have led to yields about 20% lower than those from traditional tillage practices.

Corn fits well into a system of crop rotation, but a trend toward intensive cropping and specialization has resulted in extensive production of monoculture corn. When corn is grown consecutively on the same land for a number of years, problems with pests and soil productivity commonly arise. Problems associated with soil productivity include poor soil structure, which has led to compaction, poor drainage, greater energy inputs to till the land, crusting, erosion, and ultimately reduced yield.

Corn can be grown effectively as the major crop in a rotation system. Corn grown in a rotation system will yield more than corn grown continuously. Inclusion of other cereal grains with their fibrous root system and greater ground cover, as well as a protective winter cover with winter cereals, helps to offset some of the detrimental aspects associated with row-crop and intensive monoculture production.

CONCLUSION

If corn is to continue to reign as the monarch among grain crops in North America, some changes in cultural practices may be required. Corn's many desirable morphological and physiological attributes may count for little value when offset by intensive, monoculture production in unsuitable soils or climates. Corn will continue to reign if grown in systems that play to its strengths and not its weaknesses.

Both in North America and in other areas of the world corn production is likely to expand as the superior attributes of corn are recognized.

QUESTIONS

1. Review the response of the corn crop to photoperiod.
   a. How is corn classified relative to photoperiod?
   b. When does corn reach anthesis relative to seasonally changing daylength at high latitudes?
   c. How might the date of anthesis be adjusted to more closely correspond with the photosynthetically more desirable long days early in the growing season?
2. How might corn adapted to the equator be stimulated into reproductive development when grown under long-day conditions at a high latitude?
3. What is the impact of altitude on corn production?
4. Review the C_4 photosynthetic mechanism in corn and consider its impact from the viewpoint of:
   a. Evolution.
   b. Crop production.
   c. Crop adaptation.
5. Corn is sensitive to frost damage in both the seedling and adult stages. In which stage would frost damage be most detrimental to corn production? Explain why.
6. Review general corn-production recommendations in terms of the following factors:

a. Seeding rate.
b. Planting depth.
c. Plant population.
d. Row width.
e. Cultivar selection.
f. Fertilization.
g. Pest control.

7. From among the positive and negative attributes listed in this chapter select a single item in each category that in your view is the most important item, and explain your choices.
8. Consider the merits and problems associated with perennial corn production if such a genotype were developed.

References

Arihara, J. and T. M. Crosbie. 1982. "Relationships Among Seedling and Mature Root System Traits of Maize." *Crop Science* 22: 1197–1202.

Barney, G. O. (study director). 1980. *The Global 2000 Report to the President of the U.S. Entering the 21st Century*. Vol. 1. *The Summary Report*. New York: Pergamon Press. p. 112.

Brunson, A. M. 1955. "Popcorn." (In) G. F. Sprague (ed.), *Corn and Corn Improvement*. New York: Academic Press, Inc. pp. 423–440.

———, and G. M. Smith. 1945. "Hybrid Popcorn." *Journal of the American Society of Agronomy* 37: 176–183.

Cal, J. P., and R. L. Obendorf. 1972. "Differential Growth of Corn (*Zea mays* L.) Hybrids Seeded at Cold Root Zone Temperatures." *Crop Science* 12: 572–575.

Cochran, D. E. 1962. "Sweet Corn Gets Sweeter." *Crops and Soils* 15 (3): 6–7.

Coe, E. H., Jr., and M. G. Neuffer. 1977. "The Genetics of Maize." (In) G. F. Sprague (ed.), *Corn and Corn Improvement*. Madison, Wis.: American Society of Agronomy. pp. 111–223.

Crosbie, T. M., and R. B. Pearce. 1982. "Effects of Recurrent Phenotypic Selection for High and Low Photosynthesis on Agronomic Traits in Two Maize Population." *Crop Science* 22: 809–813.

Cross, H. Z., and J. J. Hammond. 1982. "Plant Density Effects on Combining Abilities of Early Maize Synthetics." *Crop Science* 22: 814–817.

Galinat, W. C. 1979. "Botany and Origin of Maize." (In) *Maize*. CIBA-Geigy Agrochemicals Technical Monograph. pp. 6–12.

Hardeman, N. P. 1981. *Shucks, Shocks and Hominy Blocks: Corn as a Way of Life in Pioneer America*. Baton Rouge: Louisiana State University Press. 271 pp.

Johnson, I. J., and J. C. Eldridge. 1953. "Performance of Recovered Popcorn Inbred Lines Derived from Outcrosses to Dent Corn." *Agronomy Journal* 45: 105–110.

Kuleshov, N. N. 1933. "World's Diversity of Phenotypes of Maize." *American Society of Agronomy Journal* 25: 688–700.

Lee, S. S., and G. O. Estes. 1982. "Corn Physiology in Short Season and Low Temperature Environments." *Agronomy Journal* 74: 325–331.

MacLean, A. J., and L. S. Donovan. 1973. "Effects of Soil Temperature on Early Growth of Six Single-Cross Corn Hybrids." *Canadian Journal of Soil Science* 53: 128–129.

Mangelsdorf, P. C. 1974. *Corn: Its Origin, Evolution, and Improvement*. Cambridge, Mass.: Harvard University Press. 262 pp.

———, R. S. MacNeish, and W. C. Galinat. 1964. "Domestication of Corn." *Science* 143: 538–545.

Moss, D. N., and R. B. Musgrave, 1971. "Photosynthesis and Crop Production." *Advances in Agronomy* 23: 317–336.

Nault, L. R., and W. R. Findley. 1982. "Update on Perennial Corn Discovery: Rare Find Has Interesting Potential." *Crops and Soils* 34 (6): 10-13.

Peters, D. W., D. B. Shanks, and W. E. Nyquist. 1982. "Root-Pulling Resistance and Its Relationship to Grain Yield in F_1 Hybrids of Maize." *Crop Science* 22: 1112–1114.

Richardson, D. L. 1959. "Effects of Certain Endosperm Genes on Popping Ability in Popcorns." *Agronomy Journal* 51: 631.

———. 1960. "Pericarp Thickness in Popcorn." *Agronomy Journal* 52: 77–80.

Smith, C. S., J. J. Mock, and T. M. Crosbie. 1982. "Variability for Morphological and Physiological Traits Associated with Barrenness and Grain Yield in the Maize Population, Iowa Upright Leaf Synthetic #1." *Crop Science* 22: 828–832.

Smith, G. M. 1955. "Sweet Corn." (In) G. F. Sprague (ed.), *Corn and Corn Improvement*. New York: Academic Press, Inc. pp. 441–463.

Troyer, A. F., and A. R. Hallauer. 1968. "Analysis of a Diallel Set of Early Flint Varieties of Maize." *Crop Science* 8: 581–584.

Warrington, I. J., and E. T. Kanemasu. 1983. "Corn Growth Response to Temperature and Photoperiod. I. Seedling Emergence, Tassel Initiation and Anthesis. II. Leaf-Initiation and Leaf-Appearance Rates. III. Leaf Number." *Agronomy Journal* 75: 749–766.

Weatherwax, P. 1922. *The Popping of Corn.* Indiana Academy of Science Proceedings 1921: 149–153.

Wittwer, S. H. 1975. "Food Production: Technology and the Resource Base." *Science* 188: 579–584.

17
Sorghum

INTRODUCTION

The sorghum crop has historically been cultivated for grain to a varying degree in almost all tropical and subtropical countries of the world. In the past 50 years remarkable success in plant breeding has allowed sorghum production to extend into the temperate regions of the world. By 1980, sorghum production had spread throughout most of the United States including the northern areas (Posner, 1975) and into southern parts of Canada with economically competitive yield performance (Hume and Kebede, 1981). As a world food crop, sorghum ranks sixth in importance. As a forage crop, sorghum is almost as widespread as grain sorghum.

Sorghum is one of the most important food crops in the semiarid tropics, as it can be grown successfully under conditions too dry or too hot for corn and certainly unsatisfactory for cereals other than millet. Together, sorghum and millet supply 90% of the food energy requirements of the rural people in the hot, dry Sahelian region of Africa (Hibbler and Stanley, 1981).

Sorghum has undergone a substantial modification in adaptation, morphology, and performance since it was first introduced as a crop into the United States in the mid-nineteeth century (Martin, 1936). Exotic germplasms obtained from the world sorghum collection were transformed from tall, late-maturing, photoperiodically sensitive plants to short, early-maturing, photoperiodically insensitive counterparts. Morphological changes opened the way for machine harvesting and eliminated the slow and arduous traditional hand-harvesting methods.

The full extent of sorghum's adaptation may not as yet be fully realized. In the fringe areas of production, because of lack of experience with the crop and associated uncertainties, and because early-maturing corn hybrids have taken hold so well, there is less incentive to try sorghum. In dryland areas of western Canada, hybrid grain sorghum has outyielded wheat or barley in southern Alberta yield trials. Continued improvement in sorghum may elevate its status as a food and feed in regions currently considered on the fringe of sorghum production.

Much of the past research has been directed, understandably, to crop adaptation and general productivity. Breeding and selection programs aimed at cultivar development for specific environments will help the spread of sorghum. Phenological development studies specific to a local area are needed, as sorghum appears to respond differently to heat units or growing degree days than does corn. Associated with this may be different responses to day and night temperatures. The result is that the degree-day concept developed for corn may not be useful in predicting flowering time or maturity in sorghum (Quinby, Hesketh, and Voight, 1973). Sorghum may have minimum temperature requirements for development that differ from those of corn. In Canada, for example, sorghum yields may approach those of corn, but only in years with average or above-average temperatures (Hume and Kebede, 1981).

Little research has been directed to nutrition and cooking quality. Compared to wheat and rice, the amount of research in sorghum has been small. Two main achievements stand out, however: (1) the development of dwarf, combine sorghums, and (2) the development of hybrid sorghum.

A CHANGING PERSPECTIVE ON SORGHUM

Sorghum is often regarded as a poor cousin of corn that is grown in areas too hot or dry for corn. As well, indifferent animal performance on forage sorghum, related to tannins or hydrocyanic acid, has relegated forage sorghum to a secondary role as a feed crop.

The ability of sorghum to produce in areas unsuited to other cereal crops except millet is a definite asset that may become even more important under projected environmental changes. Increasing levels of carbon dioxide associated with industrial burning of fossil fuels could lead to a worldwide warming trend (Kerr, 1983; Kimball, 1983; Ramanathan, 1981). An increase in drought intensity in semiarid regions of the world is projected. The adaptation of sorghum means that the crop might become increasingly important under projected hot, dry conditions. To transport or pump water any distance or to pump it from increasing depth entails extensive and rising fossil-fuel costs and the ever-present risk of depletion and nonavailability.

Four outstanding features must be recognized in sorghum:

1. Its efficient C_4 photosynthetic pathway.
2. Its tolerance to heat.
3. Its productivity under drought conditions.
4. Its high biomass production.

The importance of these features is reviewed briefly:

1. C_4 crops are capable of a greater photosynthetic response than C_3 crops under conditions of increased carbon dioxide. Carbon dioxide concentration has increased over the past 50 years (Kieling et al., 1976) and may eventually be double the 300 parts per million found at the start of the industrial revolution (Gribbin, 1981).

Carbon dioxide is a major component in photosynthesis, and an atmospheric increase could have a beneficial impact on yield, especially among C_4 crops. Based on a view of more than 430 observations of yield of 37 species grown under conditions of carbon dioxide enrichment, Kimball (1983) concluded that a doubling of carbon dioxide may result in a yield increase of 33%. Based on 34 observations of C_3 grain crops, including barley, rice, and wheat, the mean relative yield increases were 1.32 compared to sorghum at 2.98 (Kimball, 1983).

2. Associated with an increase in carbon dioxide level is a rise in global temperatures from the **greenhouse effect** created. Based on several theoretical models, the mean global air temperature is predicted to increase 3° to 4°C (5.4° to 7.2°F) with a doubling of CO_2 concentration (Gribbin, 1981). Preliminary observations suggest that temperature increases will be greater near the poles than in the tropics. Heat tolerance coupled with C_4 photosynthesis makes sorghum suited to projected environmental conditions over a greater geographical range than the crop now occupies.

3. In general, increased heat will reduce precipitation effectiveness, and climatic shifts may reduce total precipitation. It is possible that the desert Sahel area of Africa will get more rain, but the grain-producing area of Canada, the United States, the Soviet Union, and China will get drier. The possibility exists that C_3 crops such as wheat, barley, and oat, produced extensively in the temperate zones, may suffer a yield reduction or be unsuited to the hot, dry climate. Research has shown that plant temperature is

strongly influenced by plant water status (Bartholic, Namken, and Wiegand, 1972; Diaz, Mathias, and Hanks, 1983). Grain yields of C_3 plants may be adversely effected under increased hot, dry conditions.

Under conditions of adequate moisture at heading and flowering, sorghum is capable of high productivity under otherwise drought conditions. Sorghum has the ability to slow its growth during a dry period and resume growth and development when moisture conditions improve. Under the very long days of summer at high latitudes, under good soil and fertility conditions in association with increased air temperature and dry soil conditions, sorghum may produce spectacular yields.

A second concern in the dry Great Plains region of North America has been the "mining" of groundwater—that is, the pumping of more water out of the ground than nature is putting back in—and its effect on the future productivity of crops with a high irrigation water requirements. In the central and southeastern parts of the United States, water shortages are increasingly evident (Sheridan, 1981). Irrigation agriculture in these areas is entering a period of declining water-resource availability and rapidly changing economic conditions. Garrity et al. (1982a) have suggested that the escalation of energy costs for irrigation pumping and the concurrent depletion of groundwater in some areas have forged a clear connection between increased efficiency in the use of a limited water supply and continued economic viability of irrigated farming.

Unless crops suited to dryland conditions are available, crop production may have to be abandoned in arid regions of the world as irrigation water becomes increasingly scarce. Sorghum has considerable promise as a low-moisture-requirement crop (Gardner et al., 1981).

A sorghum composite grown under marginal environmental conditions revealed traits that may be desired in future sorghum cultivars (Keim et al., 1983), but considerably more research and improvement is required and expected.

4. High biomass production has made sweet sorghum a leading contender in biomass production-energy systems. In addition to high biomass yield, sorghum produces a high percentage of readily fermentable sugars and combustible fiber (Shih, Gascho, and Rahi, 1981; McBee and Miller, 1982) as well as having the desirable physiological attributes just outlined. Sweet sorghum, therefore, holds an enormous potential as a field crop for sugar and ethanol production at diverse latitudes because it is adaptable to a wide array of growing conditions. In contrast, sugar cane, a crop with a high biomass yield potential, can be grown only in tropical and subtropical conditions. Additional research on sorghum biomass production is needed to understand how to manage the many factors associated with maximizing total plant production.

Forage sorghums are notable for their high yields of dry matter, their capacity for regrowth, and their efficient use of water. As C_4 plants, they are efficient in terms of nitrogen fertilizer compared to less efficient C_3 grasses on soils of low nitrogen content (Brown, 1978).

Indifferent animal performance was reviewed by Wheeler (1980). Potential and means of overcoming the hazards of sorghum appears to depend primarily on plant breeders' giving greater emphasis to digestibility and low hydrocyanic acid in cultivar development.

TAXONOMY AND CLASSIFICATION

An enormous range in morphological diversity exists in the sorghum crop. From a practical viewpoint, sorghum can be classed according to use:

1. Grain sorghums are used for food and feed.
2. Forage sorghums, also called sorgos, have sweet, juicy stalks, grow tall, and produce high forage yields but not high grain yields. Sweet sorghums can be used to produce syrup or molasses.
3. Grass sorghums, also known as Sudan grass, are grown for pasture or hay.
4. Broomcorn is grown for its long, fibrous panicle branches, used to make brooms.

A more comprehensive, taxonomically complete and definitive classification was presented by Snowden (1936). He divided the cultivated sorghums into 31 "species," despite the fact that all readily intercross. Snowden's classification included 712 **taxa,** and, although precise, it proved to be too cumbersome and confusing. A simplified classification was presented by Harlan and de Wet (1972). According to this system, all sorghums can be classified correctly as *Sorghum bicolor* L. Moench, including all morphologically distinct cultivars.

This simplified classification generally has been accepted, which means that the sorghum literature must be interpreted with discretion, as older reports may treat sorghum as consisting of several distinct species. For example, Sudan grass may be classed as *Sorghum sudanense.* Sudan grass can be crossed with other sorghum plants and is considered to be a cultivar of *Sorghum bicolor.* In addition to cultivated sorghum, a wild species called Johnson grass exists. A relative of *Sorghum biocolor,* it is a perennial with double the chromosome number. Tetraploid Johnson grass, classed as *Sorghum halepense* (L.) Pens, is an aggressive plant with spreading rhizomes. In many areas where it is adapted it is regarded as a noxious weed.

The classification of Harlan and de Wet (1972) is based largely on the inflorescence of sorghum, called a panicle. Sorghum panicles vary from very open and loose to very compact. Panicle variation was conveniently partitioned into nine classes. On the basis of spikelet morphology within the panicle types, five basic races were identified:

Bicolor
Guinea
Caudatum
Kafir
Durra

Sorghum production around the world utilizes the genetic variability represented by these five races, although local names are often applied. Genetic combinations among these five races have resulted in ten additional intermediate races:

Guinea-bicolor
Caudatum-bicolor
Kafir-bicolor
Durra-bicolor
Guinea-caudatum
Guinea-kafir
Guinea-durra
Kafir-caudatum
Durra-caudatum
Kafir-durra

Practically all the variation in cultivated sorghum can be accounted for by these five basic races and their intermediate combinations. The races, generally speaking, are easily identifiable by spikelet morphology alone, so that the chief advantage of the Harlan and de Wet system is its simplicity. The five basic races are described briefly as follows:

THE BICOLOR RACE. The race is complex, heterogeneous, and consists of Sudan grass, sorgo, broomcorn, and bicolor of the early classification. The long, clasping glumes, elongate seed, and open panicles suggest this race to be related to the primitive characteristics of wild sorghums.

The sweet-stalked cultivars in this race are used as forage for livestock and, although grown widely, occur on a small scale. Intermediate races involving bicolor are of minor importance, but durra-bicolor is grown extensively in the highlands of Ethiopia.

THE GUINEA RACE. Basically this is a West African race. The long glumes and open panicles are highly evolved. Seeds vary in size from small to large, are hard and corneous, and are resistant to insect damage. The open, pendulous panicles and gaping glumes probably help reduce mold damage under wet conditions.

In West Africa, cultivars of this race yield well, tolerate flooding, and are of tremendous importance. Some intermediate races involving guinea are also important. Guinea-caudatum is a major race in Nigeria, Chad, and Sudan. Guinea-kafir occurs in East Africa and is extensively grown in India.

THE CAUDATUM RACE. A characteristic of this race is the turtle-backed grains. It is dominant in parts of Sudan, Chad, Nigeria, and most of Uganda. The caudatum race is one of the most important from an agronomic standpoint. The intermediate races involved are important also. Cultivars with a more compact panicle are confined to the drier zones, and the open or umbellate head types are most common in high-rainfall areas. In Sudan, the name feterita is common for cultivars of this race. Other names are hegari and dobbs.

Kafir-caudatum is the most important of the intermediate races, and almost all of the modern North American hybrid grain cultivars are of this race. Yellow endosperm and large seed size used in American breeding programs are of the guinea-caudatum race.

Caudatum confers yielding ability, bright seed colors, and good seed quality upon its hybrid derivatives. This race has served as a most important source of germplasm for modern plant-breeding programs around the world.

THE KAFIR RACE. Kafir is a major race in East Africa from Tanzania southward. It is not well represented north of the equator. Kafir panicles are usually semicompact to compact. The race is important agronomically, and its intermediate derivations even more so.

The guinea-kafirs of India and the kafir-caudatums of North America have already been mentioned. The kafir-durras have provided valuable germplasm. The cytoplasmic male-sterile system used in hybrid grain sorghum and described in Chapter 9 involves durra cytoplasm and kafir derivatives as maintainers.

THE DURRA RACE. The durras are dominant in Ethiopia, Turkey, Syria, Iraq, Iran, Arabia, North Africa, along the fringes of the southern Sahara, and in India.

Cultivars of durra range in maturity from long- to short-season. Short-season cultivars of durra are either drought resistant or drought evasive and can be grown in the driest regions where sorghum is adapted. Durras were once important in the United States, where they were called milo, but they have been replaced largely by kafir-caudatum (Sorrells and Myers, 1982). The durras are the most specialized and highly evolved of all the sorghums, and many useful genes may be found in this race. The Ethiopian Center of Crop Diversity contains a wealth of genetic variability in the durra race (Gebrekidan, 1973; Harlan, 1969).

BOTANY OF SORGHUM

Sorghum is a warm-season, annual crop favored by high day and night temperatures. For seed germination, a minimum temperature of 7° to 10°C (45° to 50°F) is required, with an optimum germination temperature of about 15°C (59°F) for cool-temperature-tolerant cultivars. For heat-loving cultivars an optimum temperature of 23°C (73°F) (Kanemasu, Bark, and Choy, 1975) is preferred, as determined in ten cultivars in Kansas and when measured at seed depth. A temperature of 21°C (70°F) is considered as minimum for plant growth.

Flowering progresses normally at temperatures between 21° and 35°C (70° and 95°F) and is delayed by temperatures outside this range. When night temperatures drop below 12°C (54°F), flowering is limited the following day, even at optimum day temperatures (Adams, 1962). Sorghum seedlings are characterized by comparatively slow growth rates, especially under cool conditions.

Sorghum plants range in height from 61 cm (2 ft) to 6 m (20 ft). Cultivated grain sorghums in North America seldom exceed 2 m (6.5 ft). The solid stalks are coarse, vary in diameter from 1 cm to 5 cm (0.33 to 2 in.), and are juicy or dry, pithy, sweet or nonsweet in the forage grass cultivars.

Each plant has seven to 20 nodes and internodes. A leaf sheath arises from each node, and in dwarf cultivars sheaths overlap. Leaf blades arise from each sheath, arranged on opposite sides of the stalk, resulting in an alternate leaf arrangement. Leaves are similar to corn leaves in shape, but greater variation exists in length and width. Sorghum leaves are generally narrower than those of corn. Stalk for stalk, the leaf area of sorghum is about half that of corn.

The product of maximum length, maximum width, and a factor 0.747 was found to accurately estimate leaf area in six grain sorghum cultivars (Stickler, Wearden, and Pauli, 1961). Leaf number was influenced by temperature and photoperiod (Hesketh, Chase, and Nanda, 1969). Leaf number increased as temperature rose and as the days became longer.

Numerous bulliform cells are located near the midrib on the upper side of the leaf. Under drought stress, these cells result in a longitudinal rolling of the leaf that reduces transpiration and stress associated with wilting.

The leaf epidermis is commonly coated with a layer of white wax exudate known as bloom. This waxy surface layer, along with extensive cutinization, further limits water loss. Peterson, Suksayretrup, and Weibel (1982) determined the inheritance of three degrees of waxiness-bloom, sparse-bloom, and bloomlessness. Bloomlessness is influenced by recessive genes at two different loci and in a homozygous state. Bloomlessness has potential for use in combination with tolerance resistance genes to develop greenbug[1]-resistant hybrid sorghum.

Sorghum has a more extensive, finer, and more fibrous root system than does corn. An extensive root system that is able to extract available moisture from the soil may explain the ability of sorghum to tolerance drought stress. Some grain sorghum cultivars are suited to as little as 25 cm (10 in.) of seasonal rainfall. Sorghum is adapted also to 250-cm (100-in.) rainfall conditions.

The seminal root extends deeply into the soil at germination and has extensive lateral root development along its length. Adventitious roots arise from the basal node, with additional prop roots arising at higher nodes. Root density in two sorghum hybrids was found to increase until grain-filling time, followed by a decline toward maturity. A rapid decline in root density was associated with lower grain yield and possibly with senescence (Zartman and Woyewodzic, 1979). Sorghum roots have been observed to penetrate soil to a depth of 182 cm (6 ft).

A dense root system in sorghum-Sudan grass proved to be effective in preventing nutrient leaching from the soil when high nitrogen fertilizer rates of 250 to 500 kg/ha (223 to 446 lb/ac) were applied (Long, 1981).

Little is known about the root system in relation to ratoon cropping. Whether the old root system persists after the removal of the seed stalks and what part it plays to initiate regrowth of tillers is unclear.

Sorghum is a self-pollinated crop with no known barriers to cross-fertilization. Cross-pollination occurs at an average of 5% to 6% but ranges from 2% to 35%. Cross-pollination is achieved by wind and air currents.

Grain is borne on a branched, terminal panicle that ranges from compact (Figure 17-1) to loose or open (refer to Figures 8-5 and 8-6). Seed branches arise in whorls and terminate in spikelets containing paired florets. The fertile **sessile** spikelet has two outer glumes, a sterile (lower) floret, and a fertile (upper) floret. The sterile floret is able to produce pollen.

Fertile florets contain three stamens and an ovary with two long styles and feathery stigmas enclosed in a membranous lemma and palea. An awn, if present, arises from the lemma of the fertile floret. At the base of the floret are two lodiculae.

The glumes enclosing the spikelets may be black, red, brown, or straw-colored.

Flowering proceeds from the top of the panicle downward over a four- to seven-day period. Stigmas remain receptive for up to two days and from five to 16 days if unpollinated, depending on environmental conditions. Cool, wet weather delays flowering.

Between 800 and 3000 kernels are carried on a single panicle. Grain is small in size and varies from 2.36 to 4.29 mm in diameter. Seeds from a single panicle vary up to 10% in weight according to their position on the top, middle, or bottom portions of the panicle. For some hybrids, the top kernels are larger; for others, the bottom kernels are larger (Weibel et al., 1982).

[1]Greenbug (insect)—*Shizaphis graminum* (Rondani).

Figure 17–1. The inflorescence of sorghum is a branched, terminal panicle that varies from compact, as in this photo, to open or loose. Between 800 and 3000 kernels are carried on a single panicle. The solid stalks are coarse, juicy or pithy, sweet or nonsweet. Sorghum varies in height from 61 cm (2 ft) to 6 m (20 ft). The peduncle is the longest internode and usually extends the inflorescence free of the uppermost leaf sheath. Sorghum leaves are similar in shape to those of corn but show greater variation in leaf length and width. The leaf midrib varies in color from white or yellow in pithy-stemmed cultivars to green in the juicy-stemmed cultivars. The plants in this photo are those of grain sorghum. (Courtesy of Ontario Ministry of Agriculture and Food)

Physiological maturity is reached at a moisture content of approximately 30% (Bovey and McCarty, 1965) and occurs from 25 to 55 days after flowering in temperate-zone areas and from 34 to 70 days in the tropics. About two-thirds of the growing period in sorghum is spent from seeding to anthesis (Pauli, Stickler, and Lawless, 1964), depending on sensitivity to photoperiod.

When threshed, the floral bracts are removed, leaving a naked caryopsis that varies in color from white, chalky white, pink, yellow, red, buff, or brownish yellow to brown or reddish brown. Some white-seeded cultivars have spots of red, purple, or brown water-soluble pigmentation from the glumes.

Dehulling of sorghum is necessary before food use to remove the bran layer and the nutritionally undesirable pigments. Tannins usually are present in brown-seeded cultivars.

The seed consists of 84% endosperm, 10% germ, and 6% pericarp. The starch in the endosperm varies from 76% to 78% amylopectin to 100% amylopectin in waxy cultivars. Like corn, cultivars exist with sugary endosperms.

Sorghum seed is difficult to dehull because of variations in seed size and the amount of broken kernels that results (Wills and Ali, 1983).

PHYSIOLOGICAL ASPECTS

Tillering

Variation in the amount of tillering in sorghum is substantial. Tillering is influenced by the degree of apical dominance that is regulated by plant hormones. Apical dominance is a heritable characteristic that is modified by environmental factors such as temperature, photoperiod, and soil moisture as well as ecological factors such as plant population. These factors interact with the naturally occurring, internal hormone systems of the plant to determine the degree of tillering.

Tillering occurs when lateral, basal buds on internodes develop into side shoots. From an economic standpoint, tillering may contribute to grain yield, allow plants to compensate for low population densities, and contribute to yield of some forage sorghums. Development of tillers is essential when sorghum is ratooned for a second crop.

Under certain circumstances, tillering may be viewed as undesirable. Tillers may prevent the control of plant population by the planting of a controlled number of seeds; may cause delays in the harvesting of early-maturing main shoots; and, if excessive, may result in a leaf-area index that exceeds the optimum, so that lower, shaded leaves and plant parts become parasitic or senesce prematurely.

In contrast to tillering in small grains, which is synchronous, sorghum tillering is typically asynchronous. Synchronous tillering exists when the lead shoot reaches anthesis, followed closely by secondary tillers, with the cessation of subsequent tillering. In sorghum, the normal pattern of tiller bud outgrowth in the field is the production of some tillers during the vegetative growth period, followed by more extensive tiller production during and after anthesis of the main shoot (Isbell and Morgan, 1982). In an effort to make tillering more synchronous Isbell and Morgan studied tillering control mechanisms. They demonstrated the ability to manipulate apical dominance by the application of plant hormones and related plant growth regulators. Studies remain to be conducted on how tillering can be manipulated to produce highest grain, forage, or ratoon-crop yield.

Senescence and Nonsenescence

Under normal field conditions, leaves of many typical sorghum cultivars normally senesce after grain maturity. Older leaves at the base of the stalk normally turn brown and die. Under drought stress conditions, a pattern of leaf senescence and death involves a systematic progression up the stem, culminating in overall plant death, usually after the seed has reached physiological maturity. In nonstress situations, sorghum plants may senesce most of their leaves, but the basal and axillary buds are initiated, and tillers will grow until frost or a subsequent severe stress condition terminates them.

Sorghum cultivars that remain green and vigorous in the field after the grain reaches physiological maturity, when other cultivars have senesced, have been observed (Duncan, Bockholt, and Miller, 1981; Zartman and Woyewodzic, 1979). Desirable qualities of the nonsenescent genotype include increased pest resistance, tolerance to drought stress, improved capability for ratoon cropping, lodging resistance and increased yield potential. Nonsenescent cultivars maintained a higher root density than senescent cultivars (Zartman and Woyewodzic, 1979).

Photoperiodic Responses

Sorghum is a short-day tropical species that flowers readily when exposed to relatively short days (10 hr). Under long-day conditions (14 hr), the preflowering, vegetative growth period is extended, resulting in delayed flowering. Sorghum cultivars sensitive to short-day conditions and grown in the temperate zones grow vegetatively throughout the summer and flower too late in the fall for grain production. Under 14.5-hour daylengths, a spread in flowering from 40 to 100 days was observed (Miller, Quinby, and Cruzado, 1968). Under 11-hour daylengths, the spread in flowering was 42 to 64 days, suggesting a strong and specific sensitivity to photoperiod.

Successful introduction into long-day regions of the world was accomplished by breeding and selecting for cultivars that flower during early or mid summer (Quinby, 1973). As sorghum adaptation progressed into higher latitudes, a series of genetically defined sorghum cultivars were developed which differed in the time of flowering when grown at the same temperate-zone location. Quinby (1973) identified four genes as controlling the photoperiodic requirement for floral initiation in an otherwise uniform genetic background. These genes provided an excellent biological tool to study the mechanism(s) of photoperiodism.

Flowering is said to be controlled by a plant substance called "florigen" (Chailakhyan, 1975). Several plant hormones, including auxins, abscisins, and gibberellins, in an appropriate balance, were defined by Zeevaart (1976) and Dunlap and Morgan (1981) as being part of the "florigen" complex. These compounds are produced in the leaf, and night-interruption studies have demonstrated that the inductive process for flowering takes place at night (Hamner, 1940).

Its successful adaptation to environmental conditions outside the tropics and subtropics has allowed sorghum to become adapted to and to be widely grown in temperate climates under conditions of a long photoperiod.

Plant Height

The trend toward short-stature cereals began with sorghum. Dwarfing genes transformed traditionally tall grain sorghums from a dual-purpose role of supplying both grain and forage into a mass-production, industrial cash crop. Short or combine types of grain sorghum can be machine-harvested.

Traditional tall grain sorghums, called kafir, milo, or hegari, were not adapted to mechanical harvesting. Often they were hand-harvested from the standing stalks, a panicle at a time, still a custom in some developing countries. In more advanced areas, a row-binder was used to harvest the tall crop, and the panicles were threshed out of the bundle.

The need to direct-combine sorghums was evident, and the discovery of dwarf genes transformed the southern Great Plains of the United States into a fully mechanized sorghum-producing area in the early to mid 1940s. The changeover to dwarf sorghum cultivars occurred in a brief five-year period (Karper, 1949). For the first time in history sorghum reached the status of a large-scale cash crop in the United States, similar to that of corn and wheat.

Sorghum fields dried to near 14% grain moisture can be direct-combine harvested. The panicle threshes readily however, at a seed moisture content of 20% to 25%, followed by artificial drying to a safe storage level of 12% to 14%.

Initially, dwarf sorghum grain cultivars yielded less than or nearly equal to the older tall cultivars. Dwarf cultivars have a high ratio of grain to stover, and, under proper management that increases plant population in conjunction with reduced row width to intercept sunlight effectively, suitably high yields can be achieved. The subsequent development of hybrid vigor coupled with dwarf sorghums improved yield performance.

Plant height in sorghum is conditioned by four major genes (Quinby and Karper, 1954). These researchers used the terms 1-dwarf, 2-dwarf, 3-dwarf, and 4-dwarf to refer to genotypes recessive for one, two, three, and four dwarfing genes, respectively. Generally, the taller 2-dwarf cultivars outyielded the 3-dwarf cultivars, possibly because of better sunlight interception and less mutual shading related to a better spatial arrangement of leaves. Cultivars with three dwarfing genes are used in commercial production.

SORGHUM PRODUCTION

Planting Date

High temperatures of 23°C (73°F) at planting depth favor rapid germination, so sorghums must be planted after the soil is sufficiently warm. Suitable soil temperatures for sorghum generally occur ten to 20 days after corn-planting time in a particular locality. In short-season areas planting must proceed before the soil is ideal to avoid delays beyond a target ideal seeding date. Delays beyond an ideal seeding date decreased yield by 10% or more per week because of a shortened duration of panicle development and reduced tillering. In a two-year study in Southern Ontario, the target seeding date appeared to be mid-May (Hume and Kebede, 1981). In general, sorghum plantings may extend from midspring or early March to about July 1, in areas with an extended growing season (Stickler and Pauli, 1961). Late seeding may subject plants to insect damage.

Plant Depth

Because the seed of sorghum are small [with up to 10,000 seeds per kilogram (4540/lb)] and are unable to emerge from the soil if planted too deeply, or if the soil surface is hard or crusted, a shallow planting depth is recommended. In moist soil, sorghum seed is best planted about 2.5 cm (1 in.) deep, and in dry warm soil 5 cm (2 in.) deep. Generally, favorably high soil temperatures are found near the surface—a major reason for shallow seeding, especially in cool regions.

A study in Missouri with six planting depths (1.3 to 7.6 cm in 1.27-cm increments) (0.5 to 3 in. in 0.5-in. increments) showed that emergence was as low as 36.9% at a 27° and 17°C day and night temperature (81° and 63°F) regime and at 7.6 cm (3 in.) depth (Horrocks and Cloninger, 1974). Early-season planting should be less than 4 cm (1.5 in.). With the progression of the season and increase in soil temperature, planting depth can be safely increased. Even under warm soil-temperature conditions, however, emergence was reduced at the 7.6-cm (3-in.) depth. Deep planting should be considered only when the soil in the upper 5.5 to 6.0 cm (2 to 2.5 in.) is too dry to support successful germination and the probability of precipitation in the near future is low.

Seeding Rate

To seed a hectare 7.5 to 12.5 kg (6 to 10 lb/ac) of seed are required. With 10,000 seeds per kilogram (4540/lb) this amounts to 75,000 to 125,000 seeds and potential plants per hectare (30,000 to 50,000 plants/ac). Seeding rates of over 640,000 pph (260,000 ppa) may be used under favorable conditions, depending on row width. Seeding rates vary around the world, and this variation can best be understood by examining plants per unit area or planting density.

Planting Density

Most of the carbohydrate in the grain in cereals is produced after ear emergence, and so high grain yields depend upon a high rate or long duration of accumulation of dry matter after flowering. Evidence exists that the yield of improved grain sorghums is limited by the supply of assimilates during grain filling (Goldsworthy and Tayler, 1970; Muchow and Wilson, 1976).

Plant density is important during early stages of crop growth, as it determines the amount of leaf area available for maximum interception of sunlight energy. At a high plant density, changes in leaf length and width produce a more vertical orientation of the leaves. At too high a density, harvest index value may decline as total dry matter increases, because at high population densities some plants or tillers are not reproductive and constitute a loss in economic yield. Further loss in efficiency occurs when competition is so severe that some tillers and plants die.

Generally, under conditions of adequate moisture and fertility, maximum sorghum yields are obtained with rows 25 to 50 cm (10 to 20 in.) apart with seeds 5 to 10 cm (2 to 4 in.) apart within the row. Under less favorable conditions, wider rows and wider plant spacing within the row are recommended. Plant population also varies according to the tillering habit of the cultivars used.

Sunlight interception is a key to grain yield, but at extreme and at high latitudes poor penetration of photosynthetically active radiation into high-density stands may be inadequate, owing in part to low solar angles. Response to plant density may vary with latitude. Fischer and Wilson (1975) and Muchow and Wilson (1976) found highest yields at densities above 400,000 plants/ha (162,000 plants/ac). At 16° S latitude, Muchow et al. (1982) found grain yield did respond to row spacings below 75 cm (30 in.) and densities above 267,000 plants/ha (108,000 plants/ac).

In Queensland, Australia, Fischer and Wilson (1975) tested one hybrid cultivar for grain yield at low, medium, and high population densities of 14,352, 143,520, and 645,836 plants/ha (5,810, 58,105, and 261,472 plants/ac). The highest grain yield of 14,250 kg/ha (12,723 lb/ac) was obtained from the high-density planting. Plants were spaced precisely so that interplant competition was uniform. Such a high plant density would not be recommended under farm conditions in conventional row spacings.

This exceptionally high yield was measured at 12% moisture and achieved with a crop duration of 91 days. The superior yield of the high-density plants resulted from their greater production of total dry matter, while the harvest index of 45% was almost as high as that of the other densities.

A full-season, Nigerian sorghum cultivar and a short-season American hybrid were tested at three plant densities in Nigeria (Goldsworthy and Tayler, 1970). Nigerian cultivars may require 110 to 160 days for the heads to emerge, have between 25 and 35 leaves, grow 3.6 to 5.5 m (12 to 18 ft) tall, and have a low harvest index. Tests were conducted at 24,700, 123,500, and 247,000 plants/ha (10,000, 55,000 and 100,000 plants/ac). At each population, the grain yield of the short American hybrid was larger than that of the Nigerian cultivar. Yield per unit area was not reported.

Under optimum irrigated conditions, populations ranging from 135,850 to 553,280 plants/ha (56,000 to 224,000 plants/ac) revealed that 247,000 plants/ha (120,000 plants/ac) produced highest yields in all seven years of testing in Kansas (Grimes and Musick, 1960). Planting rates that gave average plant populations of 192,660, 266,760, and 375,400 plants/ha (61,000, 108,000, and 152,000 plants/ac) had little influence on grain yield of irrigated grain sorghum in Texas (Brown and Shrader, 1959).

Under conditions of a limited water supply in Nebraska, three grain sorghum hybrid cultivars that require 55, 60, and 70 days to heading were tested at 12 plant densities ranging from 29,000 to 279,000 plants/ha (11,740 to 112,955 plants/ac) (Blum, 1970). Grain yield of the late-maturing hybrid was highest under the low plant density and that of an early-maturing hybrid was highest under the high plant density. The highest yield in the two-year experiment was obtained with the earliest-maturing hybrid at the highest plant density.

With early-maturing grain sorghum hybrids tested in Southern Ontario (Hume and Kebede, 1981), grain yields at the highest plant populations approached those of grain corn. In 1975, increasing plant density from 75,000 to 300,000 plants/ha (30,364 to 121,457 plants/ac) increased grain yield from 4.3 to 6.0 tonnes/ha (3840 to 5359 lb/ac). In 1976, raising population from 150,000 to 450,000 plants/ha (60,730 to 182,186 plants/ac) increased average grain yields from 3.2 to 3.7 tonnes/ha (2858 to 3305 lb/ac).

Row Width

Grain sorghum traditionally has been grown in rows 76 to 102 cm (30 to 42 in.) apart, as determined by seeding, cultivating, and harvesting equipment. Flexibility in machinery design, the development of effective herbicide that reduced the need for interrow cultivation, and the use of short-stature, morphologically small plants with narrow, upright leaves have generated interest in narrow rows (40 to 76 cm) (16 to 30 in.) (Stickler and Anderson, 1964; Stickler and Wearden, 1965). Under drought stress conditions, however, narrow rows can increase plant stress, and row width under such circumstances should increase (Brown and Shrader, 1959).

Work in Kansas and Texas under conditions of favorable rainfall or irrigation generally indicated that narrow row spacings (25 to 50 cm or 10 to 20 in. apart) produce more grain yield than 102-cm (40-in. rows) (Brown and Shrader, 1959; Grimes and Musick, 1960; Porter, Jensen, and Sletten, 1960; Stickler et al., 1961). If the crop is to be direct-harvested with a combine, and herbicidal weed control is used, the closer row spacing is preferred. In the humid north, Robinson et al. (1964) reported that grain yield increased linearly as rows were narrowed from 102 to 25 cm (40 to 10 in.).

As row width decreased from 102 to 61, 51, 41, and 25 cm (40, 24, 20, 16, and 10 in.), soil temperature, light intensity at the soil surface, and water loss from the soil decreased (Robinson et al., 1964).

In 34 trials, grain yield of sorghum in 51-cm (20-in.) rows exceeded those in 102-cm (40-in.) rows by 10% in eastern Kansas and 7% in central Kansas (Stickler and Wearden, 1965).

The conclusion from these observations is that no single row width or planting density is best, and recommendations for sorghum production must consider cultivar, fertility, moisture level, and length of the growing season.

YIELD STABILITY

Stability of yield, defined by Heinrich, Francis, and Eastin (1983) as the ability of a genotype to avoid substantial fluctuations in yield over a range of environments, is a difficult breeding objective to achieve. The reasons for yield stability in a crop often are unclear, and a number of morphological and physiological mechanisms may be associated.

Four general mechanisms of yield stability were defined (Heinrich, Francis, and Eastin (1983):

1. Genetic heterogeneity.
2. Yield component compensation.
3. Stress tolerance.
4. Capacity to recover rapidly from stress.

Heinrich, Francis, and Eastin (1983) tested six cultivars in 14 environments in Nebraska and Kansas in 1978 and 1979. Mean yields of the six hybrids ranged from 5410 to 9840 kg/ha (4830 to 8786 lb/ac), and although all six had comparable yields in good environments, the three stable cultivars had higher yields in poor environments that amounted to 70% yield superiority. The stable cultivars had 40% more seeds per unit area of soil surface. Stable cultivars were able to maintain the size of their components of yield, such as heads per unit area, seeds per head, and seed weight, in poor environments. Tolerance of stable cultivars to stress conditions and the maintenance of all components of yield at relatively high levels appeared more important than compensation among the components.

Under drought stress conditions in Israel, Blum (1973) observed a decrease in grain yield associated with a reduction in tillering, but with an accompanying increase in panicle size. In contrast to drought tolerance suggested by Heinrich, Francis, and Eastin (1983), Blum considered that a compensatory effect existed whereby one or more yield components increased when tillering declined from drought stress. Drought-resistant hybrids performed better than drought-susceptible hybrids by producing a relatively high number of panicles per unit area and grains per panicle branch.

In a study on yield stability involving 54 grain sorghum cultivars tested in 48 environments in Nebraska and Kansas, it was found that early- and medium-maturing genotypes appeared to be more stable than late-maturing cultivars (Saeed and Francis, 1983), although exceptions exist.

It is interesting to note that yield stability is a realistic objective and that high-yield potential and stability are not mutually exclusive. The two most yield-stable cultivars maintained a yield advantage across nearly all the environments.

SORGHUM AND IRRIGATION

There is little doubt that irrigation agriculture in many areas, including the United States, is entering a period of management in which crop water deficits may not be totally avoided. Under such circumstances management of crop water deficits may be required to minimize yield reductions. Development of the capability to quantitatively predict the effects of crop water deficits will be essential to rational water management and irrigation scheduling. This capability is essential if limited irrigation water in different locations under a range of management conditions is to be optimized. As competition for limited water supplies increases in arid and semiarid regions, cost of irrigating are certain to rise. Where water is a limiting resource, the objectives of irrigation management may shift to that of obtaining maximum economic production per unit of water applied.

Blum (1974) observed differences of 20% among cultivars in their ability to extract stored soil moisture. The challenge of improving drought tolerance through plant breeding has special significance for drought-stricken regions that lack sufficient water for irrigation. Considerable plant-breeding effort has been conducted on the premise that a high-yielding cultivar under optimum conditions also will perform well under stress conditions. It may be that yield and drought are controlled by separate and individual genetic entities. If so, breeding procedures will involve the identification of definite drought-resistance attributes and their transfer to high-yielding and agronomically desirable cultivars.

Knowledge of evapotranspiration-yield relationships is fundamental in evaluating strategies for managing limited irrigation water. Work of Garrity et al. (1982b) indicated that water stress, regardless of timing, tended to reduce water-use efficiency. Reductions were smallest, however, when stress increased gradually throughout the growing season.

Under extreme drought stress, yield reductions of only 41% to 51% resulted from the gradual intensification of stress over the entire season (Garrity et al., 1982a). To maximize the efficiency of irrigation water use, it was suggested that an optimum limited irrigation strategy might be achieved by applying an irrigation amount equal to a fraction of the soil water deficit each week, allowing stress to gradually and progressively intensify throughout the season. The fraction chosen would depend on the degree to which the irrigation needs or desires required a cutback on water application.

This procedure was considered by Garrity et al. (1982a) to be better than a system that used the same water quantity and that scheduled a combination of full irrigation water needs in some periods and less than adequate in others. Yield reductions will occur, but the evidence over two seasons and three cultivars suggests that the reductions will tend to be lower per unit of irrigation water saved than under any alternative limited-irrigation strategy.

In a study with sorghum in Kansas with a soil moisture gradient, it was found that as water supply increased, dry-matter production, leaf area, and plant height increased (Chaudhuri and Kanemasu, 1982). Increasing the water levels increased water-use efficiency for total dry matter and grain yield.

LODGING

Despite the sturdy, solid culms, lodging in sorghum can be a problem. Lodging may be associated with diseases such as stalk rot, with movement of reserves out of the culms into the grain, or with morphologically thin rind or stalk walls, internode length, and whether the pith remains alive and strong. Lodging may occur before or after maturity.

Studies on the inheritance of lodging resistance in sorghum (Prest, Cantrell, and Axtell, 1983) suggest that breeding and selection for lodging resistance, including tall cultivars, is possible.

FERTILIZER REQUIREMENTS

In many areas where sorghum is grown, moisture is the limiting factor to sorghum production, not soil fertility. Fertilizer applications under dryland conditions are often uneconomical. This has led to the incorrect conclusion that sorghum does not respond to fertilizer and that the crop can be grown under low-fertility conditions.

Under irrigation of adequate moisture conditions, sorghum responds well to fertilizers, particularly to nitrogen. The degree of response does depend on the inherent soil-fertility level and other limiting factors of the environment.

Economic levels of nitrogen vary from 0 up to 180 kg/ha (0 to 160 lb/ac), depending on environmental factors (Blunt and Fisher, 1973), with the high level being applied to forage sorghums. Based on a 10% protein level, grain contains 1.5% nitrogen, so that a yield of 8000 kg/ha (7143 lb/ac) would require 120 kg/ha (107 lb/ac) of nitrogen for grain yield alone.

CONCLUSIONS

Sorghum is a major world crop and will remain important. Climatic changes that favor C_4 dryland crops such as sorghum may increase its importance. The prospects for an expansion of sorghum in North America and other water-deficient areas of the world where sorghum is not now grown look good.

The development of genotypes with an ability to tolerate cool conditions and short seasons and to resist disease will allow sorghum production to expand. The gene pool on which breeding programs can be based is enormous. A major factor favoring sorghum is the high level of heterosis found in F_1 hybrids.

Additional research on many aspects of sorghum is needed. Research on quality, processing, and new food products could increase the acceptability of this crop as human food and raise its status as a world crop.

Progress in sorghum has been made. Additional advances are needed and appear possible. In all respects, the future of sorghum is promising.

Questions

1. Sorghum is a tropical crop that has spread into the temperate zone of the world.
   a. What conditions or factors are prerequisite to the spread of a crop outside its normal area of application?
   b. What advances or changes in sorghum were key to its adaptation and acceptance in North America?
   c. Are there advantages or disadvantages worldwide in extending a crop out of its region of primary adaptation? If so, list them.
2. Sorghum is a crop with specific environmental requirements. When taken out of its area of adaptation, it may not reproduce because of environmental conditions. Suggest methods and problems of transferring desirable genes from alien sorghums into cultivars adapted to a specific environment. Refer to Stephens, Miller, and Rosenow (1967).
3. In a detailed study on the limiting factors to yield in sorghum, Muchow and Wilson (1976) concluded that yield was completely source-limited.
   a. Explain what is meant by "source-limited."
   b. Consider production practices that might help reduce this problem.
4. Consider the merits of tillering in sorghum. Compare its desirability with that of tillering in other cereals.
5. Suggest reasons why the highest average yield produced over a two-year test period was obtained with the earliest-maturing hybrid planted at the highest plant density of 279,000 plants/ha (112,955 plants/ac) under conditions of limited water supply in Nebraska (Blum, 1970).
6. Sorghum has been tested at plant densities ranging from 14,352 to 645,836 plants/ha (5810 to 261,472 plants/ac). It is apparent that seeding rates can vary widely for production of sorghum. Explain why such a wide range of plant densities for sorghum production exists.
7. Plant populations for corn seldom exceed 65,000 plants/ha (26,000 plants/ac) for grain production. Explain why sorghum for grain production may be seeded to advantage at populations twice as high as corn.
8. Review the progress that has been made in sorghum since 1900 in North America and consider the impact advances have had on sorghum production.
9. What are considered to be the weaknesses of sorghum?
10. Consider the future prospects of sorghum in your area.
11. Review the potential and limitations of ratoon cropping of sorghum. What physiological or morphological features might help make ratoon cropping more successful?
12. Speculate on the ability to effectively apply artificial compounds similar to "florigen" that induce tillering and flowering as desired. Consider such applications in terms of sink-source relationships.

References

Adams, J. E. 1962. "Effect of Soil Temperatures on Grain Sorghum Growth and Yield." *Agronomy Journal* 54: 257–261.

Bartholic, J. F., L. N. Namken, and C. L. Wiegand. 1972. "Aerial Thermal Scanner to Determine Temperatures of Soils and Crop Canopies Differing in Water Stress." *Agronomy Journal* 64: 603–608.

Blum, A. 1970. "Effect of Plant Density and Growth Duration on Grain Sorghum Yield under Limited Water Supply." *Agronomy Journal* 62: 333–336.

———. 1973. "Components Analysis of Yield Responses to Drought of Sorghum Hybrids." *Experimental Agriculture* 9: 159–167.

———. 1974. "Genotypic Responses in Sorghum to

Drought Stress. I. Response to Soil Moisture Stress." *Crop Science* 14: 361–364.

Blunt, C. G., and M. J. Fisher. 1973. "Production and Utilization of Fodder and Grain Sorghum as Forage for Cattle in the Ord River Valley, Western Australia." *Australian Journal of Experimental Agriculture and Animal Husbandry* 13: 234–237.

Bovey, R. W., and M. K. McCarty. 1965. "Effect of Preharvest Desiccation on Grain Sorghum." *Crop Science* 5: 523–526.

Brown, P. L., and W. D. Shrader. 1959. "Grain Yields, Evapotranspiration, and Water Use Efficiency of Grain Sorghum under Different Cultural Practices." *Agronomy Journal* 51: 339–343.

Brown, R. H. 1978. "A Difference in N Use Efficiency in C_3 and C_4 Plants and Its Implications in Adaptation and Evolution." *Crop Science* 18: 93–98.

Chailakhyan, M. Kh. 1975. "Substances of Plant Flowering." *Physiologia Plantarum* 17: 1–11.

Chaudhuri, U. N., and E. T. Kanemasu. 1982. "Effect of Water Gradient on Sorghum Growth, Water Relations and Yield." *Canadian Journal of Plant Science* 62: 599–607.

Diaz, R. A., A. D. Matthias, and R. J. Hanks. 1983. "Evapotranspiration and Yield Estimation of Spring Wheat from Canopy Temperature." *Agronomy Journal* 75: 805–810.

Duncan, R. R., A. J. Bockholt, and F. R. Miller. 1981. "Descriptive Comparison of Senescent and Nonsenescent Sorghum Genotypes." *Agronomy Journal* 73: 849–853.

Dunlap, J. R., and P. W. Morgan. 1981. "Preflowering Levels of Phytohormones in Sorghum. II. Quantitation of Preflowering Internal Levels." *Crop Science* 21: 818–822.

Fischer, K. S., and G. L. Wilson. 1975. "Studies of Grain Production in *Sorghum bicolor* (L. Moench). V. Effect of Planting Density on Growth and Yield." *Australian Journal of Agricultural Research* 26: 31–41.

Gardner, B. R., et al. 1981. "Relationships Between Crop Temperature, Grain Yield, Evapotranspiration, and Phenological Development in Two Hybrids of Moisture Stressed Sorghum." *Irrigation Science* 2: 213–224.

Garrity, D. P., et al. 1982a. "Moisture Deficits and Grain Sorghum Performance: Effect of Genotype and Limited Irrigation Strategy." *Agronomy Journal* 74: 808–814.

———. 1982b. "Moisture Deficits and Grain Sorghum Performance: Evapotranspiration-Yield Relationships." *Agronomy Journal* 74: 815–820.

Gebrekidan, B. 1973. "The Importance of the Ethiopian Sorghum Germplasm in the World Sorghum Collection." *Economic Botany* 27: 442–445.

Goldsworthy, P. R., and R. S. Tayler. 1970. "The Effect of Plant Spacing on Grain Yield of Tall and Short Sorghums in Nigeria." *Journal of Agricultural Science (Cambridge)* 74: 1–10.

Gribbin, J. 1981. "The Politics of Carbon Dioxide." *New Scientist* 90: 82–84.

Grimes, D. W., and W. T. Musick. 1960. "Effect of Plant Spacing, Fertility and Irrigation Management on Grain and Sorghum Production." *Agronomy Journal* 52: 647–650.

Hamner, K. C. 1940. "Interrelation of Light and Darkness in Photoperiodic Induction." *Botanical Gazette* 101: 658–687.

Harlan, J. R. 1969. "Ethiopia: A Center of Diversity." *Economic Botany* 23: 309–313.

———, and J. M. J. de Wet. 1972. "A Simplified Classification of Cultivated Sorghum." *Crop Science* 12: 172–176.

Heinrich, G. M., C. A. Francis, and J. D. Eastin. 1983. "Stability of Grain Sorghum Yield Components across Diverse Environments." *Crop Science* 23: 209–212.

Hesketh, J. D., S. S. Chase, and D. K. Nanda. 1969. "Environmental and Genetic Modification of Leaf Number in Maize, Sorghum and Hungarian Millet." *Crop Science* 9: 460–463.

Hibbler, M., and R. Stanley. 1981. "Prospecting for Research Gold." *IDRC Reports* 10 (1): 18–19.

Horrocks, R. D., and F. D. Cloninger. 1974. "Model for Predicting Emergence of Grain Sorghum." *Crop Science* 14: 365–367.

Hume, D. J., and Y. Kebede. 1981. "Response to Planting Date and Population Density by Early-Maturing Sorghum Hybrids in Ontario." *Canadian Journal of Plant Science* 61: 265–273.

Isbell, V. R., and P. W. Morgan. 1982. "Manipulation of Apical Dominance in Sorghum with Growth Regulators." *Crop Science* 22: 30–35.

Kanemasu, E. T., D. L. Bark, and E. C. Choy. 1975. "Effect of Soil Temperature on Sorghum Emergence." *Plant and Soil* 43:411–417.

Karper, R. E. 1949. "The Rise of Combine Sorghums." *Crops and Soils* 1 (5): 14–16.

Keim, K. R., et al. 1983. "Differential Survival of Genes in a Composite Cross of Sorghum." *Crop Science* 23: 781–784.

Kerr, R. A. 1983. "Carbon Dioxide and a Changing Climate." *Science* 222: 491.

Kieling, C. D., et al. 1976. "Atmospheric Carbon Dioxide Variations at Mauna Loa Observatory, Hawaii." *Tellus* 28: 538–551.

Kimball, B. A. 1983. "Carbon Dioxide and Agricultural Yield: An Assemblage and Analysis of 430 Prior Observations." *Agronomy Journal* 75: 779–788.

Long, F. L. 1981. "The Influence of Sorghum-Sudangrass Roots on Nutrient Leaching." *Agronomy Journal* 73: 537–546.

Martin, J. H. 1936. "Sorghum Improvement." (In) *Yearbook of Agriculture.* Washington, D.C.: United States Department of Agriculture. pp. 523–560.

McBee, G. G., and F. R. Miller. 1982. "Carbohydrates in Sorghum Culms as Influenced by Cultivars, Spacing and Maturity over a Diurnal Period." *Crop Science* 22: 381–385.

Miller, F. R., J. R. Quinby, and H. J. Cruzado. 1968. "Expression of Known Maturity Genes of Sorghum in Temperate and Tropical Environments." *Crop Science* 8: 675–677.

Muchow, R. C., et al. 1982. "Growth and Productivity of Irrigated *Sorghum bicolor* (L. Moench) in Northern Australia. I. Plant Density and Arrangement Effects on Light Interception and Distribution, and Grain Yield, in the Hybrid Texas 610 SR in Low and Medium Latitudes." *Australian Journal of Agricultural Research* 33: 773–784.

Muchow, R. C., and G. L. Wilson. 1976. "Photosynthetic and Storage Limitations to Yield in *Sorghum bicolor* (L. Moench)." *Australian Journal of Agricultural Research* 27: 489–500.

Pauli, A. W., F. C. Stickler, and J. R. Lawless. 1964. "Developmental Phases of Grain Sorghum (*Sorghum vulgare* Pers) as Influenced by Variety, Location and Planting Date." *Crop Science* 4: 10–13.

Peterson, G. C., K. Suksayretrup, and D. E. Weibel. 1982. "Inheritance of Some Bloomless and Sparse-bloom Mutants in Sorghum." *Crop Science* 22: 63–67.

Porter, K. B., M. E. Jensen, and W. H. Sletten. 1960. "The Effect of Row Spacing, Fertilizer and Planting Rate on the Yield and Water Use of Irrigated Grain Sorghum." *Agronomy Journal* 52: 431–434.

Posner, J. L. 1975. *Cultural Studies with* Sorghum bicolor *(L. Moench) in New York: Row Spacing, Date of Planting and Varieties.* M.Sc. Thesis, Cornell University.

Prest, T. J., R. P. Cantrell, and J. P. Axtell. 1983. "Heritability of Lodging Resistance and Its Association with Other Agronomic Traits in a Diverse Sorghum Population." *Crop Science* 23: 217–221.

Quinby, J. R. 1973. "Genetic Control of Flowering and Growth in Sorghum." *Advances in Agronomy* 25: 125–162.

————, and R. E. Karper. 1954. "Influence of Height in Sorghum." *Agronomy Journal* 46: 211–216.

Quinby, J. R., J. D. Hesketh, and R. L. Voight. 1973. "Influence of Temperature and Photoperiod on Floral Initiation and Leaf Number in Sorghum." *Crop Science* 13: 243–246.

Ramanathan, V. 1981. "The Role of Ocean-Atmosphere Interactions in the CO_2 Climate Problem." *Journal of Atmospheric Science* 38: 918–930.

Robinson, R. C., et al. 1964. "Row Spacing and Plant Population for Grain Sorghum in the Humid North." *Agronomy Journal* 56: 189–191.

Saeed, M., and C. A. Francis. 1983. "Yield Stability in Relation to Maturity in Grain Sorghum." *Crop Science* 23: 683–687.

Sheridan, D. 1981. "The Underwatered West: Overdrawn at the Well." *Environment* 23 (2): 7–13, 30–33.

Shih, S. F., G. J. Gascho, and G. S. Rahi. 1981. "Modeling Biomass Production of Sweet Sorghum." *Agronomy Journal* 73: 1027–1032.

Snowden, J. D. 1936. *The Cultivated Races of Sorghum.* London: Adland and Son. 274 pp.

Sorrells, M. E., and O. Myers, Jr. 1982. "Duration of Development Stages of 10 Milo Maturity Genotypes." *Crop Science* 22: 310–314.

Stephens, J. C., F. R. Miller, and D. T. Rosenow. 1967. "Conversion of Alien Sorghums to Early Combine Genotypes." *Crop Science* 7: 396.

Stickler, F. C., et al. 1961. "Row Width, Plant Population Studies with Grain Sorghum at Manhattan, Kansas." *Crop Science* 1: 297–300.

Stickler, F. C., and L. E. Anderson. 1964. "Comparative Response to Herbicides of Grain Sorghum Grown at Different Row Spacing." *Crop Science* 4: 497–500.

Stickler, F. C., and A. W. Pauli. 1961. "Influence of Date of Planting on Yield and Yield Components in Grain Sorghum." *Agronomy Journal* 53: 20–22.

Stickler, F. C., and S. Wearden. 1965. "Yield and Yield Components of Grain Sorghum as Affected by Row Width and Stand Density." *Agronomy Journal* 57: 564–567.

————, and A. W. Pauli. 1961. "Leaf Area Determination in Grain Sorghum." *Agronomy Journal* 53: 187–188.

Weibel, D. E., et al. 1982. "Relationships of Black Layer to Sorghum Kernel Moisture Content and Maximum Kernel Weight in the Tropics." *Crop Science* 22: 219–233.

Wheeler, J. L. 1980. "Increasing Animal Production from Sorghum Forage." *World Animal Review* No. 35: 13–22.

Wills, R. B. H., and M. R. Ali. 1983. "Effect of Grain Size on Dehulling of Sorghum." *Cereal Chemistry* 60: 12–14.

Zartman, R. E., and R. T. Woyewodzic. 1979. "Root Distribution Patterns of Two Hybrid Grain Sorghums under Field Conditions." *Agronomy Journal* 71: 325–328.

Zeevaart, J. A. D. 1976. "Physiology of Flower Formation." *Annual Review of Plant Physiology* 27: 321–348.

18
Wheat

INTRODUCTION

Wheat may have been the first crop to be domesticated and cultivated by humans. The process began possibly between 18,000 and 12,000 B.C. and probably with the collection of seed from the wild ancestors of present wheat plants. Wild ancestors of wheat had a brittle rachis, so that seeds were scattered effectively when mature. This was a necessary evolutionary feature for the annual reseeding and perpetuation of the species.

The gathering of seed resulted in a shift in direction of evolutionary processes. Variation for the nonshattering habit was favored, as only those seeds that remained on the rachis were collected. The eventual planting of seed and annual harvest from these plants favored the nonshattering habit. It may have taken centuries for the transformation to a domesticated form to occur, but its progression encouraged a stable agriculture based on crop production.

Domestication of wild wheat plants was not undertaken with a clear purpose and objective. But perhaps the variation that existed among wild plants was striking and could not be missed even by the untrained eye of early agriculturists. Consciously or unconsciously, domestication favored free-threshing forms from which the lemma and palea were removed readily; large seeds; and reasonably productive, high-tillering plants. Annual plants, rather than perennial forms, were favored by virtue of the fact that production occurred on the river flats that were flooded annually; flooding obliterated perennial forms, left a fertile silt deposit, and provided a seedbed nearly free of competing vegetation.

The process of domestication is obscure. We only can surmise the steps involved, the initial stage of crop production, the yields obtained, and how the grains were eaten. It may have been that domestication was not a single, isolated occurrence, but rather that controlled agriculture was an idea whose time had come. Perhaps many families independently selected seed from wild plants they considered desirable from among the wealth of diversity that existed in a center of origin. Perhaps the opportunity for selection, even to the untrained eye, was beyond the fondest expectations of a modern plant breeder. Perhaps numerous domesticated cultivars emerged, from which a limited number emerged as being preferable among the village producers. Once a satisfactory cultivar emerged, food production increased, civilization flourished, and segments of society, free of the task of seeking their daily food supply, developed art and culture.

If these views are reasonable, initial wheat production probably was associated with a center of origin somewhere in the Middle or Near East, possibly the rich soil area in the drainage basin of the Tigris-Euphrates rivers, an area known as the "fertile crescent."

Wheat was a logical crop to be domesticated. Early society could not have missed the fact that wheat seed provided a concentrated, high-quality foodstuff that could be stored, transported easily, and either eaten directly or transformed readily into a variety of food products. Wheat complemented food obtained in the hunt. Production provided greater assurance of a food supply and reduced anxiety when the hunt was unsuccessful. The straw residues may have attracted attention as a fuel, a building material, a bedding material for warmth and comfort, or a craft material for uses such as weaving.

The relative ease with which a wheat crop can be produced must have been instrumental in the transition from a hunting to an agricultural society. Under conditions of an inherent fertility system and minimal requirements for seedbed preparation, seed could be scattered and covered with a minimum of effort. Flood irrigation could be coaxed from the adjacent, slow-flowing river. Harvesting could be accomplished by stripping the spikelets from the rachis by hand, gathering the entire spike, or using a cutting device. Possibly the desire to obtain the straw residues led to the development of the first sickle to facilitate harvesting of the plant near the base. Perhaps the brittle rachis of wild plants was too abrasive for hand harvesting, and the sickle was a very early agricultural tool (Curwen and Hatt, 1953).

The cultivation of wheat spread to the Nile, where production was increased to a level sufficient to establish and sustain the Egyptian civilization, one of the most advanced and progressive of the ancient world. It was the Egyptians who recognized and utilized the unique baking properties of ground wheat that subsequently developed into bread making.

The development of agriculture marked a turning point in human progress. Technical advances in wheat and other crops based on selective development have been progressing ever since, but only in the past 100 years has progress been placed on a scientific basis through the knowledge of genetics and crop physiology. The foundation of our modern world to a large extent has been based on the progress and development of wheat.

Wheat production has flourished ever since, so that wheat ranks with rice in importance as a world crop. Its unique gluten content and associated bread-making properties assures its continued role in society. As outlined in previous chapters, wheat improvement has become synonymous with CIMMYT, the International Wheat and Corn Improvement Center in Mexico.

The importance of wheat worldwide has made a stylized wheat spike a symbol of food. For example, central to the FAO symbol is a bearded wheat spike with the Latin motto *Fiat Panis*, "Let there be bread." The United Nations Development Program (UNDP) shows the globe surrounded by two wheat spikes. CIMMYT, an organization that has contributed to wheat improvement worldwide, has a wheat spike as part of its international symbol (Figure 18-1). Bread

Figure 18–1. **Wheat, the world's most widely cultivated crop, is regarded by vast numbers of people as the "staff of life." For these reasons wheat has become a symbol of food worldwide, as exemplified by three international symbols or trademarks. A bearded wheat spike, for example, is central on the symbol of the Food and Agriculture Organization (FAO) of the United Nations. The Latin on the crest, *Fiat Panis,* translates as "Let there be bread." In the center, the United Nations Development Program shows two heraldic wheat spikes symbolically encircling the world. The International Wheat and Corn Improvement Center (CIMMYT) shows both a spike of wheat and an ear of corn superimposed on the globe.**

from wheat has been equated with the concept of sustenance.

A form of symbiosis has developed between wheat and humans. Just as many people worldwide are dependent on wheat for their daily bread, so too is wheat dependent on humans for its survival. Modern wheat cultivars lack the brittle rachis that was essential to independent survival, as it readily scattered and spread separate seeds in several directions. A tenacious awned glume helped bury the seeds that fell to the ground encased in the glumes. In modern wheat cultivars, the spike hangs tenaciously to the culm; if it is broken off, the seed may germinate the following season, but the competition arising from 40 or more seedlings emerging from an intact spike may prevent any from reaching reproduction. The efforts of humans to thresh and plant single seeds are essential to the survival of wheat.

ADAPTATION OF WHEAT

Wheat has become the world's most widely cultivated crop, occupying over 22% of the area devoted to cereal grain production compared to almost 13% for rice. The two crops share almost equally in fulfilling the energy needs of humans. Every month of the year a crop of wheat is harvested somewhere in the world, as far south as Argentina and as far north as Finland. Wheat is best suited to areas between 30° and 50° N latitude and 25° and 40° S latitude. Wheat is the major crop of the United States and Canada and is grown over a substantial area in almost every country of Latin America, Europe, and Asia. Wheat is of major importance in Oceania. Wheat is a C_3 crop that is not well adapted to tropical or semitropical conditions, although daylength-insensitive wheat cultivars developed by CIMMYT can thrive within 20° of the equator.

Wheat is the cereal crop that accounts for the greatest volume of international trade. Wheat production, perhaps to a greater extent than any other cereal crop, is subject to forces in addition to its environment adaptation. Actual production within the major wheat-producing areas is governed by political and economic considerations. Producers are guided in their decision-making process by the net returns to be obtained from wheat as compared to other farm products, in which animal product prices may be a deciding factor. Net returns for wheat are influenced greatly by prevailing national policies, with the result that world production patterns are strongly influenced by a number of factors other than agronomic considerations.

Well-defined environmental conditions must be met with respect to temperature, precipitation, frost-free period, and soil, if wheat production is to be successful. Winter wheat usually outyields spring wheat and is more extensively grown. Two major factors, however, limit winter wheat production: the ability to overwinter at high latitudes, and a vernalization or cold temperature-photoperiod interaction at lower latitudes.

Wheat grows well on most well-drained soils from sea level in the Netherlands to over 3000 m (9843 ft) in China.

The C_3 photorespiration of wheat means the crop is favored by cool temperatures. Minimum temperature for growth is about 3° to 4°C (37° to 39°F), optimum is 25°C (77°F), and maximum about 30° to 32°C (86° to 90°F). Daytime temperatures of 30° (86°F) encourage rapid growth and development with minimal differentiation, which is not inducive to high crop yield because of a low net assimilation rate.

The most important temperature effects were found by Warrington, Dunstone, and Green (1977) to occur during the ear-development phase. As expected of a C_3 crop, plants grown at low temperature during ear-development phase had long culms, large flag leaves, and potentially more fertile florets in each spikelet. The number of florets that produced harvestable grains, and the weight of these grains at maturity, were affected by temperature during the grain-growth stage (Table 18-1). Temperature prior to floral initiation was not of major importance to final ear weight, but it did have an effect on the number of mature ears present at harvest. Similar temperature results were reported by Johnson and Kanemasu (1983).

Early spring seeding of spring cultivars and the early and rapid growth characteristic of

Table 18–1
Total Grain Number per Main Stem Spike as Influenced at Each Growth Stage under Controlled Environmental Conditions

	Number of Grains per Spike at Day Temperature °C (°F):		
Growth Stage	15 (59)	20 (68)	25 (77)
Vegetative	57.8	59.8	56.8
Ear development	70.5	59.8	19.5
Grain filling	57.2	59.8	48.2

Source: Warrington, Dunstone, and Green (1977).

winter wheat, if combined with early maturity, may help avoid the vulnerable stage of high temperature that often occurs in midseason. Early seeding and early maturity are both recommended to help escape late-season drought as well.

Wheat is grown where annual precipitation averages 25 to 175 cm (10 to 70 in.), but about three-quarters of the land area in wheat averages from 37.5 to 87.5 cm (14 to 35 in.). For optimum yields, an annual rainfall of about 25 to 100 cm (10 to 39 in.) is required, but seasonal rainfall distribution is very important.

High rainfall, especially if accompanied by moderate or high temperature, is generally unfavorable to bread wheat production, because it intensifies disease and insect attacks, causes harvesting difficulties, induces undesirable seed sprouting in the field, and is usually associated with a lower protein content and reduced bread-making quality. For soft wheat production, adequate rainfall during growth and development promotes high yield and low protein—desirable factors for pastry production.

Using the spring wheat cultivar, Gabo, studies were conducted in pots at varying levels of water stress at the time of floral initiation (Angus and Moncur, 1977). After mild stress, plants showed hastened development; after severe stress, plants showed delayed development. The delay appeared to be due to the cessation of development of the shoot apex, and possibly cessation of all cell division during severe stress. Hastened development may be related to increased leaf temperature. Water stress can be an important factor affecting phasic development in wheat.

Transpiration rates tend to decrease as the soil water content decreases. Wheat leaves lack bulliform cells and associated longitudinal leaf roll to restrict transpiration. Stomatal closure, however, provides an adaptive mechanism which is related to changes in the osmotic potential of the leaves (Seaton, Landsberg, and Sedgley, 1977). Increased resistance to drought appears to confer significant advantages in terms of water conservation and adaptation to drought.

About 100 frost-free days are essential for the production of a spring wheat crop. Both spring and winter wheat seedlings will withstand reasonably low temperatures without adverse effects. A frost at anthesis can damage severely the reproductive organs and result in sterile florets. A late-season frost may cause kernel discoloration, hinder development, and lower grade and quality.

CLASSIFICATION OF WHEAT

Wheat belongs to the genus *Triticum*. A grouping within this genus can be made on the basis of chromosome sets. Wheat has four sets of chromosomes or four genomes designated A, B, D, and G. The genomic formulae of the three common allopolyploid groups are AA, AABB, and AABBDD. These three groups are diploid, tetraploid, and hexaploid with a $2n$ chromosome number of 14, 28, and 42, respectively.

Problems of wheat classification have been debated since the initial work of Linnaeus in 1753. One of the problems has been with a relative of wheat known as *Aegilops*. Bowden (1959 and 1966) discovered that two of the three genomes of hexaploid wheat were derived from *Aegilops*, and, because an intergeneric hybrid cannot be included legitimately in the genus of one of its parents, a classification was proposed that combines *Aegilops* and *Triticum* (Morris and Sears, 1967). Proper designation of common wheat under this system would be *Triticum aestivum* with ABD genomes.

Only three *Triticum* species are of any commercial significance: *T. aestivum*, *T. durum*, and *T. compactum*. Most of the wheat grown throughout the world is hexaploid (42-chromosome) *T. aestivum* or common bread wheat. *T. durum*, used for paste products, is a tetraploid (28 chromosomes) and is called durum wheat and has the AB genomes. Club wheat or *T. compactum* is a hexaploid with the ABD genomes. All three of these wheats developed from natural hybridization among ancestral types, none of which are grown commercially. A suggested, but not fully proven, evolutionary development of *Triticum* species was presented by Briggle (1980). Wheat species developed when sets of seven chromosomes, entire genomes, were contributed from a related, often wild, species through hybridization. Chromosome doubling may have occurred to produce the final chromosome number.

Minor production of several wheat species has and may still occur. If grown, the wheat may be used primarily for livestock feed. The species include *T. spelta* (spelt) (Lafever and Campbell, 1977), *T. dicoccum* (emmer), *T. polonicum* (Polish), and *T. turgidum* (rivet wheat). Production of these wheat species virtually has disappeared, though they may be grown for

Table 18–2
Wheat Species with Some of Their Characteristics

Scientific Name and Growth Habit	*Common Name*	*Morphological Features*	*2n Number*	*Genome Composition*
T. monococcum (winter and spring)	Einkorn	Single floret at each brittle rachis node; pale red kernels	14 (diploid)	AA
T. dicoccum (winter and spring)	Émmer	2 florets at each rachis node; kernels red or white	28 (tetraploid)	AABB
T. durum (mostly spring)	Durum	3–6 florets at each rachis node; kernels red or amber	28 (tetraploid)	AABB
T. turgidum (winter and spring)	Rivet or poulard	3–6 florets at each rachis node; kernels red or white	28 (tetraploid)	AABB
T. polonicum (spring)	Polish	3–4 florets at each rachis node; kernels white, amber, or red	28 (tetraploid)	AABB
T. spelta (winter and spring)	Spelt	3–5 florets at each rachis node; kernels pale red	42 (hexaploid)	AABBDD
T. aestivum (winter and spring)	Bread or common wheat	3–6 florets at each rachis node; kernels red or white	42 (hexaploid)	AABBDD
T. compactum (winter and spring)	Club	5–6 florets at each rachis node; kernels red or white	42 (hexaploid)	AABBDD
T. sphaerococcum (spring)	Shot wheat	6–7 florets at each rachis node; kernels red or white	42 (hexaploid)	AABBDD

scientific purposes. The species of *Triticum* and some characteristics are shown in Table 18-2.

A number of derived species have been produced using *Triticum*. A cross of *Triticum* with rye (*Secale*) has produced *Triticale;* with *Agropyron* has produced *Agrotricum* (Morris and Sears, 1967); and with *Haynaldia* has produced *Haynaltricum*. Some of this material has provided valuable genetic information or has contributed disease and insect resistance. The successful development of such wide crosses may be useful to the development of a perennial wheat and to gene diversity of cultivated wheat, but many are not useful at present.

Triticum timopheevi wheat, with 28 chromosomes, is a hexaploid with the AAGG genome composition. Whether this species varies enough to warrant the establishment of a group remains to be determined (Morris and Sears, 1967). *Timopheevi* was identified only in the twentieth century in Russia. *Timopheevi* has served as a source of cytoplasmic male sterility and is resistant to many diseases attacking other cultivated wheats. Table 18-2 lists nine *Triticum* species. N. I. Vavilov recognized 14 species of wheat. A listing of the 14 species is given by Mangelsdorf (1953).

BOTANY OF WHEAT

The inflorescence of wheat, known as a spike, is made up of separate clusters of flowers (spikelets) subtended by two glumes (Figure 18-2). Each spikelet of common bread wheat has two to five florets (four in Figure 18–2). Review Chapter 2 and Figures 2-3 and 2-5.

Florets within spikelets are enclosed by the lemma and palea. The tip of the lemma may bear an awn or be entirely awnless. If an awn is present, it may be short, medium, or long. Tip-awning occurs when the lemmas at the top third of the spike are short-awned.

Within the floret or between the lemma and palea are the sexual organs: three stamens, a single pistil consisting of an ovary bearing two short styles, both with a feathery stigma, and two small lodiculae. The stamen is composed of a filament and an anther, which contains the pollen.

Figure 18–2. A close-up of a wheat spikelet prior to flowering or anthesis. A spikelet consists of two to five florets and is contained between two empty glumes or chaff. Each floret within the spikelet is enclosed by a lemma and palea, which is removed readily from the kernel at threshing time. Awns, if present, occur at the tip of the lemma, and in this photo a short awn is present. Awns vary in length from short, medium, to long. Within the floret are the sexual organs, including three stamens, a pistil made up of the stigma, style, and ovary, and two lodiculae. Wheat is normally self-pollinated, and upon fertilization the ovary of the pistil develops in the grain or wheat kernel. (Courtesy of Ontario Ministry of Agriculture and Food)

During the flowering process, the lodiculae located at the base of the ovary swell and push the lemma and palea apart, a process that is completed within five minutes. During this same time period the filaments of the stamens elongate, suddenly and strongly, and force the anthers out of the floret. Simultaneously the stigmas quickly curve away from each other and spread out. Pollen released from the anthers during these processes falls within the same floret and results in pollination of the stigma and fertilization of the ovary to result in self-pollination. Part of the pollen may be shed outside of the floret and be effective in cross-pollination (De Vries, 1971), a step essential for hybrid wheat production. When the lodiculae lose their turgidity, the floret closes, and the flowering process is complete.

The number of florets and productive kernels developed per spikelet can influence yield. Primary and secondary kernels are largest, and generally the size of each order of kernel decreases. Although all yield is the product of photosynthesis, kernels per spikelet must be viewed as a valuable sink to store the carbohydrate produced by the source. The seeking of yield advances by selecting for photosynthetic roles is complex and usually ineffective, because yield is a complex character with components that are conditioned by many genetic and environmental factors. Selection for sink size has been found to be more efficient then selection for yield per se. Sink size is comprised of the number of spikes per plant, kernel number per spike (which is the product of number of spikelets per spike and kernel number per spikelet), and kernel weight. Most of these components are under simple genetic control. The lowest one to four spikelets of an ear are often entirely sterile (Figure 18-3).

Some cultivars have a higher number of kernels per spikelet than others. Kernel number per spikelet was determined by Ibrahim et al. (1983) to be controlled by four or eight genetic factors, depending on the cross. All of these genetic factors were found to be influenced greatly by the environment. Improvement by crossing and selection will be relatively slow, and large segregating populations will be necessary to recover desired cultivars.

Wheat is normally self-pollinated, with cross-pollination under natural conditions occurring at a level of 1% to 4%. The middle third of the ear is the first section to begin flowering, which progresses rapidly upward. The primary floret of a spikelet flowers first, followed by the second, third, and so on. The duration of time that a floret remains open varies according to genotype and weather. DeVries (1971) reviewed the duration of floret opening and reported it varied from 8 to 48 minutes. Once pollination has taken place, the lodiculae collapse. When male sterility is present, florets remain open much longer—a matter of days rather than minutes.

Figure 18–3. **The inflorescence of wheat is known as a spike, made up of spikelets carried on a zigzag, flattened, jointed central axis known as the rachis. About nine spikelets are found on each side of the rachis in the three wheat spikes shown. The lowest one to four spikelets of the spike are often entirely sterile. If each spikelet has three fertile florets, the average number of kernels carried in each spike would be 54. Spikes may also be produced on tillers. (Courtesy of Ontario Ministry of Agriculture and Food)**

Pollen remains viable for only a very short time, possibly only 15 to 20 minutes under field conditions (DeVries, 1971).

The grain or caryopsis of wheat consists of a germ or embryo which constitutes 2% to 3% of the entire grain by weight. The remainder consists of the endosperm. The wheat seed has a characteristically deep crease, which adds to milling difficulties. Variability for crease depth is not known to exist.

Physiological maturity of the seed is reached

at a moisture content of 25% to 35%. If harvested at this stage, the plants must be swathed to allow the seed to dry to a storable level before threshing. Wheat can be direct combine-harvested at a moisture content in the grain of 13% to 14% for safe storage and without artificial drying.

The stalk or culm usually is a hollow, jointed cylinder with about six nodes and internodes. The upper internode bearing the spike is called the **peduncle.** The length of the internodes increases from the lowest one. The stems are solid at the nodes. Solid-stemmed cultivars have been developed (Ekroth and McNeal, 1953) and provide resistance to the wheat stem sawfly[1] (Kemp, 1934), but have lower yield than hollow-stemmed wheat cultivars (McNeal et al., 1965; McNeal and Berg, 1979). The development of a solid pith uses food energy that might have otherwise gone into grain development.

A leaf sheath arises from each node that encloses a portion of the internode and from which a leaf blade arises. The leaf sheath is split over its entire length. The uppermost leaf on the plant is called the flag leaf. At the juncture of the leaf sheath and blade is a ligule and a pair of sharply curved auricles, small in size and fringed with minute hairs.

The wheat plant has two sets of roots, the seminal or seedling roots and the coronal, nodal, adventitious, or secondary roots. The seminal root emerges from the germ and helps the seedling become established. Seminal roots may remain functional for the life of the plant but comprise only a small portion of the total root system. Deep seeding encourages deep rooting only of the seminal or seedling roots, which may not be functional after the adventitious roots develop. Adventitious roots arise from the basal node, which forms near the soil surface. Spring wheat that was broadcast on frozen soil and remained on the surface produced yields higher than seed sown later (Stoskopf et al., 1968). In most instances, plants existed throughout the season on the seminal root, because wind action prevented adventitious roots from penetrating the soil.

Under favorable moisture conditions in Nebraska, winter wheat roots penetrated to a depth of 10 m (13 ft). Nitrogen fertilization increased root weights and moisture utilization in these tests (Kmock et al., 1957).

Tillers or lateral branches arise from axillary buds at the lower internodes. Tillers normally develop coronal roots and may function quite independently of the main shoot in advanced stages of development (Quinlan and Sagar, 1962).

A general trend in wheat breeding for northwestern Europe had been toward decreased tillering (MacKey, 1966). Likewise, the ideotype proposed by Donald (1968a and b) was that of a single culm along with other desirable agronomic features. The reason for favoring reduced tillering was the hope for an improved harvest index, because a greater proportion of the dry matter in the mature plant would be concentrated in the grain. The development of semidwarf wheat cultivars has led to an increase in tillering, possibly because of reduction in apical dominance. The resulting asynchronous tiller development of maturity in dwarf winter wheat was suggested as a limiting factor in productivity (Stoskopf and Fairey, 1975).

A uniculm or near-uniculm cultivar in wheat was described by Atsmon and Jacobs (1977). Spikes were large, with 30 spikelets per ear and up to 12 florets per spikelet and with seed set in up to nine florets. Vegetative plant parts were robust and vigorous and were described as exhibiting "gigas" feature. These gigas features were expressed, however, only under short days and low temperatures. Uniculm barley mutants, available for many years, have served only as theoretical models. Associated root deficiencies and low yield have restricted its commercial exploitation.

PRODUCTION OF WHEAT

Seeding Rate and Plant Density

Plastic responses of the wheat plant mean that no precise seeding rate can be recommended for

[1] Wheat-stem sawfly—*Cephus cinctus*.

the crop. At seeding rates that ranged from 0.4 to 366 plants/m^2 (0.4 to 327 plants/yd^2) dry weight per plant decreased 42 times; tillers per plant decreased 34 times; weight per tiller decreased 1.17 times; height increased 1.44 times; and leaf area per plant decreased 51 times (Donald, 1963). Wheat plants clearly are very plastic in their response to plant competition during vegetative development (Figure 18-4).

Numerous tests (Holliday, 1960) have shown that seeding rates can vary greatly without modifying final grain yield. Plastic responses of plants, however, will not compensate for improper seeding rates. Recommended rates are available for specific locations but should not be regarded as a fixed value. Fertility and moisture levels, date of seeding, and genotype must all be taken into consideration. As well, modern cultivars are generally shorter-strawed than older cultivars, may be more lodging-resistant, and have a higher harvest index value. New cultivars need to be examined under production systems and seeding rates best suited to each cultivar (Briggs and Aytenfisu, 1979; Ciha, 1983).

Laloux, Falisse, and Poelaert (1980) suggested for winter wheat cultivars in Belgium that 200 plants/m^2 (167 plants/yd^2) should be found at the end of winter. This is achieved by a seeding rate of 220 grains/m^2 (183 plants/yd^2) or about 100 kg/ha (90 lb/ac). The rate can be increased slightly where environmental conditions or late seeding necessitate but should never exceed 300 to 350 plants/m^2 (250 to 292 plants/yd^2) (Table 18-3). At a high plant density each plant competes for carbon dioxide, water, light, and nutrients, and supplies of these commodities may not be adequate for satisfactory production. Excessive interplant competition can result in lodging and increased incidence of certain diseases.

Soil Fertilization and Plant Nutrition

NITROGEN. Nitrogen is essential to all living organisms, and nitrogen is the most important applied fertilizer element for increasing both grain yield and grain protein (percentage and total). Nitrogen nutrition is involved in every stage of cereal growth and development. For example, Gallagher et al. (1983) applied increasing rates of nitrogen to two spring wheat cultivars and found a positive increase in many

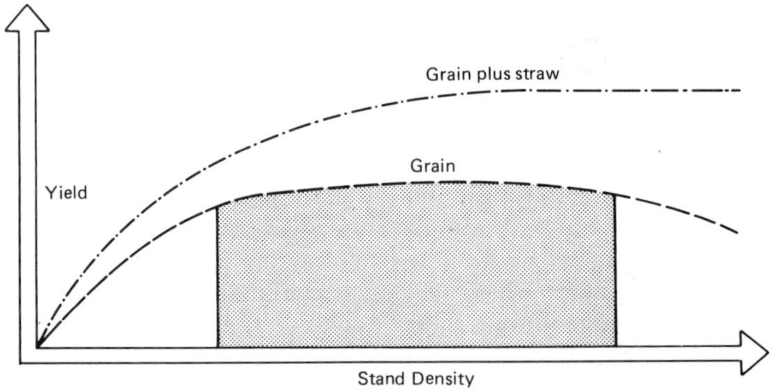

Figure 18–4. Seeding rate in wheat can vary greatly without modifying grain yield because of tillering and plant plasticity, or the adaptability of the plant to seeding rate. Seeding rates within the crosshatched section do not alter grain yield under normal growing conditions. Beyond a certain plant density, however, grain yield decreases because of plant competition, increased lodging, and the occurrence of disease. Producers can select a seeding rate within the range shown according to climate, moisture, soil, and fertility levels. (Adopted from Laloux, Falisse, and Poelaert, 1980)

Table 18–3
Winter Wheat Yields and Number of Grains Seeded per Unit Area Obtained at Four Densities over a Four-Year Period in Belgium

Seeding Rate		Seeds		Grain Yield	
kg/ha	lb/ac	/m²	/yd²	kg/ha	lb/ac
69	62	150–160	125–133	6949	6204
96	86	200–220	167–183	6953	6208
124	111	250–280	208–233	7030	6277
153	146	340–350	283–292	6970	6223

Source: Laloux, Falisse, and Poelaert (1980).

key factors influencing yield. Increased rates of nitrogen increased tissue nitrate concentrates, total aerial nitrogen accumulation, percent grain protein, percent straw protein, total grain protein, rate of leaf appearance, leaf weight and size, specific leaf laminae area, and light interception. Increases in the rate and duration of photosynthesis were associated with increases in grain yield.

Both the quantity of nitrogen applied and the date of application are important considerations. In winter wheat, an application of nitrogen in early spring can stimulate leaf and root growth. Excessive nitrogen can result in abundant vegetative growth, increased incidence of disease, and crop lodging. A moderate nitrogen application at the beginning of or during tillering will stimulate tillering, but if too much is applied, rapid and abundant leaf growth may shade lower internodes and restrict tillering. A second application coinciding with stem elongation stimulates the leaf growth and photosynthetic area needed for the period of high energy demands at anthesis and grain filling.

Nitrogen stimulates leaf growth, photosynthetic area, and leaf persistence. Maximum leaf area should coincide with the period when spike development and grain filling occur. The greatest portion of carbohydrates normally are produced after anthesis. Proper rate and timing influences harvest index.

In areas where moisture is not limiting, a split nitrogen application may be most effective to reduce excessive vegetative growth yet promote high grain yield. For example, on soils with a high clay content and low organic-matter level (1.8% to 2.0%), from which a previous nonleguminous crop was entirely removed, with wheat seeded at a plant density of 200 plants/m² (167 plants/yd²), and under a 75-cm (30-in.) rainfall, a total of 135 to 140 units of nitrogen were suggested per unit area (Laloux, Falisse, and Poelaert, 1980). This is a high application rate by world standards and, if applied to the plant at seeding in one application, surely would be uneconomic or result in excessive vegetative growth. Instead, only 20 to 30 of these units should be applied with the seed in order to prevent seed damage and to stimulate tillering and adequate but not excessive early vegetative growth. On completion of the tillering phase and at the onset of stem elongation, an application of 75 to 80 units would reduce the possibility of nitrogen starvation. A final application of 30 units at the time of **flag leaf** appearance will increase leaf persistence, photosynthesis in the important grain-filling flag leaf, and grain weight.

The importance of a split nitrogen fertilizer application was emphasized and the value of an adequate postanthesis sink was illustrated in a three-year study with a dwarf spring wheat under irrigated conditions in northwest Mexico (Fischer, Aquilar, and Laing, 1977). An increase in source can increase grain yield, but grain yield can be limited simultaneously by both source and sink. Split fertilizer applications provides greater opportunity for a sink-source balance to exist. A single nitrogen application of planting time or at the seedling stage may limit either the sink or source at a subsequent period critical to grain yield.

Wheat is a major food crop, and much of its nutritional value and its quality for bread making is determined by the grain protein. As grain yield increases, grain protein content often declines. Proper nitrogen fertilization can reduce this inverse relationship.

PHYSIOLOGY OF NITROGEN UTILIZATION. Nitrate (NO_3) is the most common form of nitrogen available to plants, since the reduced forms of nitrogen commonly applied rapidly undergo nitrification in well-aerated soils above 5°C (41°F). Once nitrate is absorbed by the plant, it must be reduced to the ammoniacal form prior to incorporation into amino acids for protein synthesis. The first enzyme, and one that is of central importance to the nitrate assimilatory pathway and plant metabolism is **nitrate reductase.** The amount of nitrogen reduced influences vegetative and reproductive features already considered, as well as grain protein. This latter component is dependent on nitrogen assimilation in the vegetative portion of the plant and on the translocation of the reduced nitrogen from the vegetative parts to the grain. Nitrate reductase may be the rate-limiting step in this reduction sequence (Beevers and Hageman, 1969).

The level of activity of nitrate reductase should be an index of the reduced nitrogen available to the plant for protein synthesis in tissues and grain. Studies on nitrate reductase with field-grown wheat at Urbana, Illinois, in the United States led to the following conclusions (Croy and Hageman, 1970):

1. A major influence on the level of nitrate reductase was the nitrate content of the tissue. The level of nitrate nitrogen is dependent on the ability of the plant to absorb adequate nitrate from the soil.
2. Leaf protein content was an indicator of nitrate reductase activity.
3. Supplemental nitrogen applications to the soil were related to the induction of nitrate reductase.
4. An increase in nitrate reductase activity was associated with increases in the percentage of total protein in the grain.
5. Total nitrate reductase activity was related to the amount of nitrate taken up by the plant over the growing season.
6. Some cultivars have a higher nitrate reductase activity than others.

A three-year test from 1968–1970 with two hard winter wheat cultivars at seven nitrogen fertility levels in Nebraska gave typical results (Table 18-4).

In a rate and date of nitrogen application study, Eilrich and Hageman (1973) found that split applications of nitrogen to winter wheat in

Table 18–4

Average Yield, Test Weight, and Protein Content of Two Hard Winter Wheat Cultivars in 12 Tests in Nebraska in 1968, 1969, and 1970 at Seven Nitrogen Levels Applied at Seeding Time

Nitrogen Rate		*Grain Yield*		*Test Weight, kg/hl.*	*Protein Content, %*
kg/ha	*lb/ac*	*kg/ha*	*lb/ac*		
0	0	2583	2306	78.8	11.7
22.5	20	2862	2555	78.7	12.3
45	40	3029	2704	78.3	12.9
67.5	60	3084	2754	78.0	13.8
90	80	3074	2745	77.0	14.3
112.5	100	3060	2732	76.8	14.7
135	120	3064	2736	76.8	13.7

Source: Johnson, Drier, and Grabouski (1973).

the spring increased nitrate reductase activity and increased grain protein content. Although grain yield reached a plateau, additional application of nitrogen to lodging-resistant cultivars maintained or increased grain protein and prevented the depression of percent grain protein frequently observed when grain yields are high.

Protein-yield relationships to nitrate reductase activity have suggested that selection for nitrate reduction capabilities may serve as a useful physiological criterion to select for increased grain yield (Croy and Hageman, 1970; Duffield, Croy, and Smith, 1972; Johnson, Schmidt, and Mattern, 1968). No identifiable physiological factor was found, however, that could serve as a single criterion in selecting wheat genotypes for efficient nitrogen utilization (Rao et al., 1977). In addition to nitrate reductase activity, other factors markedly influence the amount of grain protein. Two of these factors are the ability of the plant to absorb adequate nitrate from the soil and the translocation of reduced nitrogen from the vegetative parts to the developing grain.

The results from the tests shown in Table 18-4 suggest that nitrate reductase is a genetically controlled factor and that selection for high protein might effectively increase grain protein content. Gallagher et al. (1980) found that in two wheat cultivars, a single dominant gene accounted for most of the variability observed for nitrate reductase activity. This observation is in contrast to polygenic control of nitrate reductase activity suggested by Duffield, Croy, and Smith (1972). Deckard and Busch (1978) found that measuring nitrate reductase activity in parental cultivars was useful to predict those crosses that produced a higher frequency of high-yielding cultivars. However, the highest-yielding cultivars within a cross could not be differentiated from the lowest-yielding cultivars on the basis of nitrate reductase activity under the conditions of this test. Sampling occurred in the early seedling stage.

Nitrate reductase activity was found to be a measure of the level of grain nitrogen or protein content (Dalling and Loyn, 1977) in 24 wheat cultivars (Table 18-5). Assays were conducted on the first vegetative leaf of 21-day-old plants under controlled environmental conditions. Ability to identify and predict potentially useful cultivars in the seedling stage and to advance those worth keeping for further agronomic evaluation is of value.

Wheat cultivars differ in the efficiency with which they redistribute nitrogen, and this can be measured by the harvest index for nitrogen (ratio of grain nitrogen to total plant nitrogen). Variation in this value is shown in Table 18-5.

Like any biological phenomenon, increased grain yield or grain protein content is not a simple matter. Identifying plants with high nitrate reductase activity may serve as an additional tool to improve yield and protein content. Complete consideration of the system is required, including rate and date of nitrogen applications.

Much of the work on nitrate reductase has been directed toward the wheat crop, perhaps because of grain protein and quality relationships. Some work on nitrate reductase activity and total plant reduced nitrogen has been reported for corn (Deckard, Lambert, and Hageman, 1973), and for sorghum (Eeck, Wilson and Martinez, 1975).

OTHER FERTILIZER ELEMENTS. In addition to nitrogen, other **major elements** include phosphorus, potassium, calcium, magnesium, and sulphur. The latter three rarely present a problem in cereal production because they are present in the parent rock material and generally are abundant in the soil. These same elements may be used to correct soil *p*H.

Phosphorus and potassium levels vary among soils, and application should be based on analysis. A wheat grain yield of 8000 kg (3632 lb) removes approximately 80 kg (36 lb) of phosphorus (K_2O) from the soil. On many soils, a fertilizer maintenance program could be directed to replacing removed levels of these elements.

The effect on spikelet number of adding phosphorus at a rate of 40 kg/ha (36 lb/ac) at sowing was measured in Australia (Rahman and Wilson, 1977). The seven spring wheat cultivars responded similarly with increased spikelet number per spike, rate of spikelet initiation, and

Table 18–5

Nitrate Reductase Activity at the Seedling Stage, Grain Dry-Matter Yield, Grain Nitrogen Yield, and Harvest Index for Nitrogen of 24 Wheat Cultivars Grown in the Field at the University of Melbourne, Australia, in 1975

Cultivar	*Nitrate Reductase, Micromoles Nitrate Reduced per Hour per Gram Fresh Weight*	*Grain Dry-Matter Yield, Grams per Plant*	*Grain Nitrogen Yield, Milligrams of Nitrogen per Plant*	*Harvest Index for Nitrogen*
Celebration	12.0	5.55	102.9	60.6
Insignia	11.8	4.14	80.0	56.2
Javelin	11.6	5.72	102.4	75.0
Olympic	11.5	6.15	142.7	69.2
Atlas 66	11.1	5.19	109.5	46.6
Baldmin	10.0	4.94	109.2	52.6
Curlew	9.6	4.75	99.3	45.3
Gabo	9.5	3.52	64.2	57.4
Quadrat	9.4	4.56	86.1	74.5
Bencubbin	9.4	3.44	76.1	74.2
Fortunato	9.2	2.88	49.2	50.7
Florelle	8.6	7.31	133.0	63.9
Diadem	8.2	4.99	89.1	69.3
Coker 47-27	8.1	2.94	71.8	58.5
Thatcher	8.1	1.81	34.1	45.8
Pannacle	8.0	4.98	80.1	56.9
Sinvalocho	7.9	4.28	63.9	48.7
Piesa 4	7.3	4.12	69.9	55.3
Comeback	6.5	4.63	94.1	72.1
Koala	6.3	3.95	74.3	53.7
Petit Rojo	5.8	3.38	96.0	68.4
Kenya B	4.7	3.99	69.6	47.0
Argentine IX	4.1	2.76	74.5	46.8
Helgendorf	3.1	2.43	53.9	56.0

Source: Dalling and Loyn (1977).

apex length. Phosphorus had no effect on the duration of the vegetative phase, spikelet formation phase, or elongation phase or on leaf number. The increase in spikelet number may be related to the increase in the rate of cell division associated with phosphorus.

Minor or **trace elements** play an essential role in cereal crop physiology. Since most soils contain the essential trace elements, they often are applied as a foliar spray in solution when deficiency symptoms appear. Deficiencies may result from a change in soil *p*H which may render the elements unavailable to the plant.

A PERSPECTIVE ON WHEAT

Because wheat is regarded as the "staff of life," it has played a feature role throughout this book. A review of items covered in previous chapters will help place wheat in perspective:

1. Wheat has a wider geographical distribution than any other crop.
2. It plays a key role in international marketing (Chapter 14).
3. Wheat flour dominates the bread and pastry industry (Chapter 12).
4. Although normally self-pollinated, wheat has served for extensive and intensive research on hybrid development (Chapter 9).
5. Wheat doubles as a fall and winter pasture crop on an extensive area in the southern United States (Chapter 8).
6. Wheat is historically the crop associated with the development of agriculture, may have been the first crop to be domesticated, and as such has a prominent place in the annals of history.

7. Wheat and CIMMYT are synonymous with the basis for establishing the international network of agricultural research centers (Chapter 14).

The prominent role that wheat has played in the past makes it a prime candidate for a major role in the future. By no means, however, have the problems or challenges of wheat been met. Rajki (1983) outlined the colossal task facing the world in the future: that of doubling the food supply as the population of the world doubles in the next 50 to 75 years. Rajki suggested that over this short time period, the increase in the world's food production must be at least as great as that experienced during the 12,000 years it took for world population to reach the 4.5 billion people on earth in about 1975. Elfring (1983) emphasized this view and suggested that by 2015, the possibility exists that 8 billion people will live on earth. Food production must be more than doubled to accommodate rising dietary expectations. Agriculture must become more intensive, more selective, and more cost-effective if it is to feed these additional people.

To meet this demand, short-term improvements are needed, but a long-term plan is essential if the increase is to be met with minimal environmental risks. A major issue in the long term is the recognition and support of agriculture, a key element in survival.

As the "staff of life" throughout the world, improvement, production, and marketing of wheat is a matter of concern for producers, consumers, and governments. Aspects of food must become a central theme throughout the world.

In the developing world, the countries of India, Pakistan, and China have experienced rapid expansion in wheat production and an increased importance of wheat in the average diet. Production increases have come from both yield and area expansion, with yield increases being precipitated by cultivars and research developed by CIMMYT and the green revolution. Production increases have increased per capita supplies in the 1970s and 1980s and sharply reduced dependence on imports. The average yield increase over the two decades was 4.6% per year. Continuation of yield at this level is unlikely, because the price of nitrogen fertilizer is high relative to wheat.

At the same time, wheat consumption per capita has increased, and it may increase more slowly in the future or taper off. Owing to population growth, however, total demand will expand by 2% per year in developing countries.

In the developed countries, food use of wheat is unlikely to increase, as population increases are balanced by decreasing per capita consumption. Use of wheat for feed depends on its price relative to other feed grains.

Because wheat is the major food grain in international trade, world wheat trade is likely to continue to expand, especially if food deficits due to crop failure are experienced in any country or region. Technological advances to reduce crop variation and failure may mean that world wheat trade may increase less rapidly than in past years.

Wheat has undergone and will continue to undergo changes in physiology, morphology, and genetics. Further changes can be expected and are imperative, as the world strives to combat hunger and provide an adequate diet for all.

Although the evolution of wheat has been going on for thousands of years, the final or ideal product has not been reached. Breeders and producers must work constantly to keep the crop resistant to changing pest organisms, to increase the range of productivity, and to cope with unfavorable climatic circumstances or changes. Greater yield stability rather than yield ability may be required.

Wheat is genetically and morphologically very variable, and it is impossible to foresee genetic changes that may be achieved by genetic engineering and biotechnology. Future possibilities may include perennial wheat development, greater hormonal growth control, widespread use of F_1 hybrids, improved lodging resistance, increased winter hardiness, and the ability to fix nitrogen symbiotically. Greater crop diversity may in turn influence present standards of quality and systems of marketing.

Questions

1. *a.* Why is wheat a leading crop in international trade?
   b. Why has rice not assumed this position?
   c. Is corn or sorghum likely to replace wheat on the international market?
2. Wheat, like any other crop, has a well-defined area of climatic adaptation. Wheat, however, is adapted to a wide geographical range, so that the area of wheat produced is not restricted climatically. In other words, many forces in addition to agronomic and adaptation factors determine wheat production. Explain this view.
3. Based on your understanding of nitrogen uptake and utilization within the plant and on recommended application procedures, explain the results shown in Table 18-4.
4. *a.* Explain and define the vernalization requirements of winter wheat.
   b. What are the consequences of seeding winter wheat in the spring?
5. List some of the major pests of wheat and explain how they affect or harm the plant.
6. *a.* From the information in Table 18-5, select the top two or three cultivars in terms of yield performance and efficiency of nitrogen use.
   b. How do the top selections for grain yield or grain protein content compare in terms of nitrate reductase activity?
   c. In a similar manner consider cultivars with the lowest grain performance and relate this to nitrate reductase activity.
7. Describe some of the morphological features of wheat in terms of their desirability or undesirability for grain yield. Consider:
   a. The degree of awning.
   b. Florets that remain open because of swollen lodiculae.
   c. Degree of tillering.
   d. Spikelets with many fertile florets.
   e. Large droopy leaves or smaller upright leaves.
   f. Gigas features.

   Consider changes in production that may better utilize specific morphological traits.
8. What determines whether a creaseless kernel could be developed?
9. Explain why seeding rate in wheat is not a precise value (refer to Figure 18-4).
10. Consider the merits or disadvantages of deep seeding of wheat.
11. What factors determine the best time to seed winter wheat?
12. Consider plant-breeding objectives from the viewpoint of:
    a. The short term.
    b. The long term.

References

Angus, J. F., and M. W. Moncur. 1977. "Water Stress and Phenology in Wheat." *Australian Journal of Agricultural Research* 28: 177–181.

Atsmon, D., and E. Jacobs. 1977. "A Newly Bred 'Gigas' Form of Bread Wheat (*Triticum aestivum* L.): Morphological Features and Thermo-photoperiodic Responses." *Crop Science* 17: 31–35.

Beevers, L., and R. H. Hageman. 1969. "Nitrate Reduction in Higher Plants." *Annual Review of Plant Physiology* 20: 495–522.

Bowden, W. M. 1959. "The Taxonomy and Nomenclature of the Wheats, Barleys, and Ryes and Their Wild Relatives." *Canadian Journal of Botany*. 37: 657–684.

———. 1966. "Chromosome Numbers in Seven Genera of the Tribe *Triticale*." *Canadian Journal of Genetics and Cytology* 8: 130–136.

Briggle, L. W. 1980. "Origin and Botany of Wheat." (In) *Wheat*. Technical Monograph. CIBA-Geigy Ltd., Switzerland. pp. 6–13.

Briggs, K. C., and A. Aytenfisu. 1979. "The Effects of Seeding Rate, Seeding Date, and Location on Grain Yield, Maturity, Protein Percentage and Protein Yield of Some Spring Wheats in Central Alberta." *Canadian Journal of Plant Science* 59: 1139–1145.

Ciha, A. J. 1983. "Seeding Rate Seeding Date Effects on Spring Seeded Small Grain Cultivars." *Agronomy Journal* 75: 795–799.

Croy, L. I., and R. H. Hageman. 1970. "Relationship of Nitrate Reductase Activity to Grain Protein Production in Wheat." *Crop Science* 10: 280–285.

Curwen, E. C., and G. Hatt. 1953. *Plough and Pasture: The Early History of Farming*. New York: Henry Schuman. 329 pp.

Dalling, M. J., and R.H. Loyn. 1977. "Level of Activity of Nitrate Reductase at the Seedling State as a Predictor of Grain Nitrogen Yield in Wheat (*Triticum Aestivum* L.)." *Australian Journal of Agricultural Research*. 28: 1-4.

Deckard, E. L., and R. H. Busch. 1978. "Nitrate Reductase Assays as a Prediction Test for Crosses and Lines in Spring Wheat." *Crop Science* 18: 289–293.

Deckard, E. L., R. J. Lambert, and R. H. Hageman. 1973. "Nitrate Reductase in Corn Leaves as Related to Yields of Grain and Grain Protein." *Crop Science* 13: 343–350.

DeVries, A. Ph. 1971. "Flowering Biology of Wheat, Particularly in View of Hybrid Seed Production—Review." *Euphytica* 20: 152–170.

Donald, C. M. 1968a. "The Design of a Wheat Ideotype." (In) K. W. Finlay and K. W. Shepherd (eds.), *Third International Wheat Genetics Symposium*. Canberra: Australian Academy of Science. pp. 377–387.

————. 1968b. "The Breeding of Crop Ideotypes." *Euphytica* 17: 385–403.

————. 1963. "Competition Among Crop and Pasture Plants." *Advances in Agronomy* 15: 1–18.

Duffield, R. D., L. I. Croy, and E. L. Smith. 1972. "Inheritance of Nitrate Reductase Activity, Grain Protein, and Straw Protein in a Hard Red Winter Wheat Cross." *Agronomy Journal* 64: 249–251.

Eeck, H. V., G. C. Wilson, and I. Martinez. 1975. "Nitrate Reductase Activity of Grain Sorghum Leaves as Related to Yields of Grain, Dry Matter, and Nitrogen." *Crop Science* 15: 557–580.

Eilrich, G. L., and R. H. Hageman. 1973. "Nitrate Reductase Activity and its Relationship to Accumulation of Vegetative and Grain Nitrogen in Wheat (*Triticum aestivum* L.)" *Crop Science* 13: 59–66.

Ekroth, E. G., and F. H. McNeal. 1953. "Association of Plant Characters in Spring Wheat with Resistance to the Wheat-stem Sawfly." *Agronomy Journal* 45: 400–404.

Elfring, C. 1983. "Beyond the Green Revolution." *BioScience* 33: 617–618.

Fischer, R. A., I. Aquilar, and D. R. Laing. 1977. "Post-Anthesis Sink Size in a High-yielding Dwarf Wheat: Yield Response to Grain Number." *Australian Journal of Agricultural Research* 28: 165–175.

Gallagher, L.W., et al. 1980. "Major Gene Control of Nitrate Reductase Activity in Common Wheat." *Crop Science* 20: 717–721.

————. 1983. "Nitrogen Assimilation in Common Wheats Differing in Potential Nitrate Reductase Activity and Tissue Nitrate Concentrations." *Crop Science* 23: 913–919.

Holliday, R. 1960. "Plant Population and Crop Yield." *Field Crop Abstracts* 13: 159–167, 247–253.

Ibrahim, O. E., et al. 1983. "Inheritance of Kernel Number per Spikelet and Its Association with Kernel Weight in Two Winter Wheat Crosses." *Crop Science* 23: 927–931.

Johnson, R. C., and E. T. Kanemasu. 1983. "Yield and Development of Winter Wheat at Elevated Temperatures." *Agronomy Journal* 75: 561–565.

Johnson, V. A., J. W. Schmidt, and P. J. Mattern. 1968. "Cereal Breeding for Better Protein Impact." *Economic Botany* 22: 16–25.

Johnson, V. U., A. F. Dreier, and P. H. Grabouski. 1973. "Yield and Protein Responses to Nitrogen Fertilizer of Two Winter Wheat Varieties Differing in Inherent Protein Content of Their Grain." *Agronomy Journal* 65: 259–263.

Kemp, H. J. 1934. "Studies of Solid Stem Wheat Varieties in Relation to Wheat Stem Sawfly Control." *Scientific Agriculture* 15: 30–38.

Kmock, H. G. et al. 1957. "Root Development of Winter Wheat as Influenced by Soil Moisture and Nitrogen Fertilization." *Agronomy Journal* 49: 20–25.

Lafever, H. N., and L. G. Campbell. 1977. "Spelt—A New Look at an Old Crop." *Crops and Soils* 29 (9): 15–16.

Laloux, R., A. Falisse, and J. Poelaert. 1980. "Nutrition and Fertilization of Wheat." (In) *Wheat*. Technical Monograph. CIBA-Geigy Ltd., Switzerland. pp. 19–24.

MacKey, J. 1966. "The Wheat Plant as a Model in Adaptation to High Productivity Under Different Environments." Fifth Yugoslav Symposium on Research in Wheat. *Contemporary Agriculture* 14: 37–48.

Mangelsdorf, P. C. 1953. "Wheat." *Scientific American* 189 (1): 2–11.

McNeal, F. H., et al. 1965. "Relationship of Stem Solidness to Yield and Lignin Content of Wheat Selections." *Agronomy Journal* 57: 20–21.

McNeal, F. H., and M. A. Berg. 1979. "Stem Solidness and Its Relationship to Grain Yield in 17 Spring Wheat Crosses." *Euphytica* 28: 89–91.

Morris, R., and E. R. Sears. 1967. "The Cytogenetics of Wheat and Its Relatives." (In) K. S. Quisenberry and L. P. Reitz, (eds.), *Wheat and Wheat Improvement*. Madison, Wis.: American Society of Agronomy. pp. 19–87.

Quinlan, J. D., and G. R. Sagar. 1962. "An Autoradiographic Study of the Movement of ^{14}C-Labelled Assimilates in the Developing Wheat Plant." *Weed Research* 2: 264–273.

Rahman, M. S., and J. H. Wilson. 1977. "Effect of Phosphorus Applied as Superphosphate on Rate of Development and Spikelet Number per Ear in Different Cultivars of Wheat." *Australian Journal of Agriculture Research* 28: 183–186.

Rajki, S. 1983. "Grain Production and the Future of the Human Race." *Acta Agronomica Academiae Scientarum Hungaricae* 32: 233–252.

Rao, K. P., et al. 1977. "Nitrogen Nutrition and Grain Protein in Two Spring Wheat Genotypes Differing in Nitrate Reductase Activity." *Crop Science* 17: 283–286.

Seaton, K. A., J. J. Landsberg, and R. H. Sedgley. 1977. "Transpiration and Leaf Water Potentials of Wheat in Relation to Changing Soil Water Potential." *Australian Journal of Agricultural Research* 28: 355–367.

Stoskopf, N. C., et al. 1968. "Increased Yields from Cereals Sown on Frozen Soil." *Canadian Journal of Plant Science* 48: 428–430.

Stoskopf, N. C., and D. T. Fairey. 1975. "Asynchronous Tiller Maturity—A Potential Problem in the Development of Dwarf Winter Wheat." *Plant Breeding Abstracts* 45: 467–470.

Warrington, I. J., R. L. Dunstone, and L. M. Green. 1977. "Temperature Effects at Three Development Stages on the Yield of the Wheat Ear." *Australian Journal of Agricultural Research* 28: 11–27.

19

Rye

INTRODUCTION

Every crop that humans produce is plagued by one or more agronomic, genetic, physiologic, or morphologic deficiencies. The perfect crop or crop cultivar does not exist. The value of any crop is measured by the degree to which undesirable aspects are offset by virtues. Rye is a crop that is not held in high esteem because its virtues have been allowed to become overshadowed by negative attributes.

Rye is listed among the major cereal crops of the world. It may have achieved that status by default rather than design. Initially, rye was viewed as a weed in fields of wheat and barley, and it became accepted only when these crops proved to be unadapted to climatic extremes in northern Europe. Although rye is the only other major cereal capable of producing a leavened flour product, rye bread is dark, heavy, and a less prestigious food product than wheat bread and products.

Rye ranks as a world crop because of the area in production, which includes Argentina, South Africa, Canada, the United States, and—with the greatest area seeded—the USSR, Poland, and Germany.

In the short term, rye is not likely to increase in importance, but the distinctive nutty flavor of bread products manufactured from rye assures the crop a continued place among the cereals. In the long term, it is possible that as the geographic limits to increased food production are pressed, rye may assume a more commanding role as a source of genetic material and for food and feed production. In view of the mounting environmental concerns associated with intense production on land subject to erosion and lacking a suitable protective cover crop over winter, rye may have an ever-increasing role.

This chapter addresses the rye crop from the standpoint of recognized desirable features, considers its weaknesses, and attempts to view it in a specific context to place it in perspective as a world cereal grain.

VIRTUES OF THE RYE CROP

1. Compared with other winter cereals, rye is the most cold-tolerant cereal crop available. Its inherent winter hardiness means that it can serve as a reliable cereal crop in areas considered too severe for other cereals. Generally speaking, rye cultivars are hardier than the most hardy winter wheat cultivars.

Extensive and detailed temperature-rye relationships using phenological development were conducted by Nuttonson (1958). He reported that during a winter of limited, protective snowfall cover, winter rye was observed to withstand as low an air temperature as −35°C (−31°F). Overwintering, as in other cereals, is dependent on the survival of the nodal tiller to regenerate leaf and root growth in the spring. The northern limits of rye production are determined largely by winter temperatures. Rye has a deep root, which is protected by the soil, and which is able to capture and hold a protective snow cover needed to survive adverse conditions.

2. Of all the cereals, rye is the least exacting in its soil requirements. It is capable of satisfactory production under a wide range of soil moisture, soil acidity, and soil fertility. Rye can outyield other cereals on droughty, sandy soils of low fertility, yet it does respond to fertile and moist conditions. An extensive root system partially accounts for its drought resistance, but the crop also has an inherent adaptive ability, whereby maturity varies according to soil moisture conditions.

In areas considered too acid for wheat production, rye may be produced successfully, because it is able to tolerate a range of soil acidity. Although the optimum *p*H range for rye is 5.0 to 7.0, the minimum and maximum limits are 4.5 and 8.0, respectively.

3. Rye is most commonly recognized as a winter annual crop, but spring-seeded rye exists also. Collectively, the two forms of rye are capable of a wider potential geographical range than wheat, both as to latitude and altitude. Cold hardiness and minimal environmental demands help account for rye as the most widely adapted of all cereal crops.

At high latitudes, winter rye cultivars can be produced successfully through much of the spring wheat belts of North America, northern Europe, and the Soviet Union. Winter cereals are favored over spring cereals because of their potentially higher yields and protective soil cover. Spring rye cultivars are able to ripen their grain in regions lying a considerable distance to the north of the spring wheat belts.

Rye production is not restricted to high-latitude extremes but can be produced throughout the temperate and subtropical zones of the world. Rye can be grown from the semiarid regions near deserts to within the Arctic Circle in Scandinavia in the north to the southerly latitudes of Chile in the south.

In the Himalayan mountains in Asia, rye production is possible at an altitude of 4300 m (14,400 ft) (Nuttonson, 1958).

At such extremes, the season is too short for other cereals, but rye can be expected to reach maturity.

4. Rye has a high nutritive value. The biological value of rye protein is high because of a desirable balance of amino acids, especially the essential amino acid lysine.

5. Rye is an excellent rotation crop and may have a role in an integrated pest-management program. Rapid and vigorous fall and spring growth of winter rye allows the crop to effectively compete with weeds. With the exceptions of leaf rust[1] and ergot,[2] rye generally has good disease and pest resistance and helps break pest cycles if included in a rotation under a system of intense cereal production.

6. Winter rye provides an excellent cover crop because it can be seeded late in the fall, has an extensive root system that holds soil against erosion losses, and begins growth early in the spring. On farms with ruminant animals, these features may allow for an extension of the grazing season, either as fall or spring pasture.

On farms with a fixed ruminant population, rye provides a degree of much-needed flexibility in forage production. Rye either may be fall-grazed, used as a winter cover crop, and allowed to mature for grain, or may be spring-grazed after providing a protective winter cover and then plowed down as a green-manure crop to be followed by a warm-season corn, sorghum, or oil-seed crop.

The extent and timing of the grazing of the rye crop requires considerable care if the crop is to be used for grain production. This management can result in a serious grain-yield reduction (Kilcher, 1982), although the economic value of the forage must be balanced against any reduction in grain yield.

---

[1]Leaf rust of rye—*Puccinia rubigo-vera (secalis)*.
[2]Ergot—*Claviceps purpurea*.

7. Rye has been described as a pioneer crop to improve wasteland or unproductive soils by providing protective cover, abundant green manure, and an extensive and prolific root system. It has been used successfully on mine-tailings to provide a protective cover against leaching and wind erosion.

8. For biomass production to be used for renewable energy, rye is capable of high total dry-matter production compared to small grains. Vigorous rapid growth, the potential for relay cropping, and the tall stature of rye are valuable assets in biomass production.

Rye has a low grain-to-straw ratio, and harvesting rye for grain provides an abundance of straw residues. Rye straw is fibrous and tough, and, besides its use as bedding material, it may have potential in industrial uses such as strawboard manufacture.

High biomass production makes rye a possible crop for whole-plant sillage production in areas where corn or sorghum cannot be produced.

9. The potential of spring rye has remained largely untapped as a source of genetic variation that allows the frontiers of cereal production to be extended. Spring rye can be produced in climatic areas where late spring frosts may damage the reproductive organs of winter rye, or where the growing season is too short for reliable production of other spring cereals.

10. Rye possesses the potential to contribute individual genes or segments of chromosomes of value to wheat or other cereals (Driscoll and Anderson, 1967). Although transfers such as that of pubescent peduncle or "hairy neck" into wheat (Driscoll and Sears, 1963; O'Mara, 1946) have not proven of commercial value, the potential exists through genetic engineering to effectively transfer desirable genes or chromosome segments into other cereals.

Rye contributed an entire genome to triticale, and, although problems developed as a result of interaction with the wheat genome, some of the problem is in the inadequacy of the rye genome itself. Progress and improvement in rye through research and breeding may be needed for rye to reach its full contribution as a genetic donor to the gene pool as well as a crop in its own right.

Despite its valuable attributes, few grain crops have experienced such a sharp decline in importance as rye. During the 1970s the area of rye grown in the world declined 23.7% (Table 19-1). In the same period, total cereal area increased 6.7%. Rye was once the most common bread grain in northern and central Europe but has been replaced by bread wheat. Rye warrants a greater role in world food production than that of a feed grain or forage crop.

An examination of the problems with rye may provide an insight into the reasons for its decline. The purpose is not to further discredit this crop but to provide a complete and constructive evaluation. Recognition of weaknesses may be fundamental to avoiding them or improving the crop.

PROBLEMS OF THE RYE CROP

1. Although the protein of rye has a greater biological value than that of other cereal grains, protein availability is hampered severely by factors restricting animal intake and digestibility. Rye has been publicized as having a palatability problem in animals that results in low rates of gain. Animals on a full rye ration show a reduced feed intake as well (Friend, 1970; Friend and MacIntyre, 1969; Smith and MacIntyre, 1970).

The cause of reduced intake and low palatability was identified as 5*n*-alkyl-resorcinol, a growth-inhibitory substance concentrated in the oil fraction of rye grain (Wieringa, 1967). The growth of swine on rations containing rye oil resulted in a growth depression comparable to that of an equivalent amount of rye. The growth of swine on a 50% rye ration was 11% to 12% lower than on a ration in which barley replaced the rye component. Generally, it has been accepted that no problem exists when the amount of rye fed is limited by mixing it with other grains. Also, contrary to the view that greater feeding difficulties are encountered when freshly harvested rye is used, Wieringa (1967) observed no differences between fresh and that stored for up to 12 months.

Table 19–1
Decline in Area of Rye Harvested for the Decade of the Seventies from the Three-Year Average of 1969–1971 Compared with the Value for 1979–1981

	1969–1971 (in thousands)		1979–1981 (in thousands)	
Description	ha	ac	ha	ac
N. America	480	1,186	333	823
S. America	479	1,183	253	650
Asia	1,897	4,686	1,382	3,414
Europe	7,060	17,438	5,399	13,335
USSR	9,568	23,633	7,557	18,666
Oceania	39	96	24	59
Other	523	1,292	454	1,121
World total (rye)	20,046	49,514	15,302	37,796
World total (all cereals)	682,090	1,684,762	727,834	1,797,750

Source: 1981 Production Yearbook (FAO, 1982).

Age of swine and the level of rye in the ration may be important in determining how much rye can be included. Rations for pigs above 50 kg (110 lb) may contain 50% rye; for poultry, 20% rye can be included without measurable adverse effects.

Wieringa (1967) concluded that in nearly all feeding experiments showing the growth-inhibitory effect of rye, the diminished growth rate is caused primarily by a decrease in the feed intake. The bitter taste of rye associated with the presence of alkyl-resorcinol was held responsible. A second possibility is that rye meal forms a sticky mass in the mouths of animals that may result in lower feed intake.

The implication of alkyl-resorcinol as the growth-inhibiting factor and reason for low palatability led Evans, Dedio, and Hill (1970) to screen 397 spring and winter rye cultivars for level of alkyl-resorcinol. They found a range from a low of 0.036% to a high of 0.322%, suggesting that selection for a low level could be effective in reducing the problem. Spring rye cultivars have higher alkyl-resorcinol levels than winter rye cultivars.

2. Although foliar diseases generally are less severe in rye than in other cereals, ergot is a major concern in rye (refer to Figure 13-13). Ergot is best known as a disease of rye, although this fungus grows parasitically on a wide number of grasses and cereal crops. The alkaloids present in ergot bodies may cause toxic effects when consumed by humans or animals. Ergot poisoning has engendered a fear of rye as a feed or food crop.

3. In addition to ergot and alkyl-resorcinol, another aspect of rye flour that makes it less acceptable than wheat flour for human consumption is its dark color and the darkness and heaviness of rye bread products. White bread is held in higher esteem than dark bread. The possible social stigma associated with a reliance on rye bread may be a reason for a trend away from rye products.

4. The cross-pollinated nature of rye means that wind results in outbreeding, so that, when two or more cultivars are grown in close proximity, they are difficult to keep pure for seed increase or testing purposes. Perhaps this is a reason why breeding and selecting for improved rye cultivars has been limited. There are fewer rye cultivars than in wheat or other cereal crops.

5. Many rye cultivars are tall, and an abundance of straw with a low harvest index indicates a poor distribution of photosynthate. An abundance of straw generally is undesirable, as it adds to harvesting problems and is inducive to lodging. Genes for reduced plant height have

been found but generally need to be incorporated into an agronomically satisfactory genotype (Fedorov et al., 1970).

Rye test comparisons reported by Langille et al. (1981) and obtained from 17 tests over a six-year period gave an average height of 147 cm (58 in.) and ranged from 139 to 154 cm (55 to 61 in.).

6. Volunteer rye is a problem following a rye crop, adding an unsightly, tall weed in subsequent cereal crop fields.
7. Rye development is favored by cool spring temperatures. This can be an advantage to facilitate the early development of grain, escape summer heat or drought, or allow for grain removal in a system of relay cropping. A frost, however, during anthesis, when pollen is subject to damage, can reduce seed set.

For pasture purposes, early heading is undesirable, as the rye is a determinate crop, and vegetative growth stops when reproductive growth begins.

BOTANY OF RYE

The only cultivated species of rye is *Secale cereale*. The genus *Secale* consists of four to 12 species, depending upon the criteria used for species definition (Stutz, 1972). Both spring and winter forms of *S. cereale* exist. Spring rye is low-yielding and is not extensively grown, except under conditions where winter rye is unsuited, mainly in Siberia. The area of spring rye is very small and is of relatively little economic importance.

Spring rye is grown under conditions too severe for winter rye; as a replacement for winterkilled winter rye; in areas where autumn drought restricts fall-seeded winter rye; or where a late spring frost causes sterility in early-headed winter rye. A light frost of −2.2° to −1.1°C (28° to 30°F) may kill the pollen and partly or completely prevent fertilization of the flowers. Rye in the milk stage is not damaged by a light frost.

Spring rye cultivars have less desirable agronomic and lower end-use qualities than winter cultivars.

Purvis (1939) reported spring habit to be dominant and simply inherited.

Morphologically, rye most closely resembles wheat. Compared to wheat, its leaves are similar in shape, though the flag leaf is typically much smaller and less important as the photosynthetic source for grain sink development. Rye leaves exhibit a typical bluish color. The plants are typically taller than wheat and tiller less profusely.

Auricles on rye are very short or even absent. The sheath has a fine pubescence that helps distinguish rye seedlings from other cereal seedlings (refer to Figure 3-12). The ligule is short and rounded.

Roots of rye plants are deep, abundant, and highly branched.

The inflorescence of rye is a slender spike 10 to 15 cm (4 to 6 in.) long. Spikelets at each node of the rachis usually contain three florets; the two outer florets are fertile and the central one is sterile and so small it may be overlooked. Each floret contains three stamens that become fully extruded at anthesis, facilitating cross-pollination. The two feathery stigmas of the pistil extend out of the floret to receive pollen. Pollination usually occurs during cool-temperature conditions in the early morning.

As in wheat, the lemma and palea that enclose the floret are free-threshing. The lemma often bears barbs on the keel and often has an awn of intermediate length. The kernel is longer and more slender than wheat kernels.

Anthocyanin pigmentation in the aleurone or pericarp layers is associated with kernel color ranging from bluish-green to greenish-brown to brownish-olive. Yellow seed rye is free of anthocyanin pigmentation.

Rye differs from other small grains in that the crop is largely cross-pollinated. Most rye plants are self-sterile, and characteristically some florets fail to set seed. Leith (1925) reported that, in rye fields in Wisconsin, as many as 33% of the florets failed to set seed over a three-year observation period. Seed set from 60% to 80% was reported by Leith and Shands (1938); the percentage appears to be influenced by seasonal variations in weather and heritable differences. Ovule or zygote abortion under open-pollinated field conditions was suggested as a reason for floret sterility by Morey and Barnett (1980). Wells and Mahdawi (1953) assumed that the large differences in seed set between self- and cross-fertility of bagged plants were due to genetic differences.

In the development of hybrid rye, selfing is a desirable step in developing homozygosity and in selecting desirable inbreds. Leith and Shands (1938) inbred rye for 11 years and noted a decrease in vigor of vegetative growth. An increase in fertility occurred by selfing. Initially fertility averaged 6.5% and increased to 50% in the final years of selection.

Disagreement exists on the precise temperature conditions associated with seed germination. Starzycki (1976) gave the lowest temperature at which rye germinates as 3° to 5°C (37° to 41°F) and gave the optimum as 25° to 31°C (77° to 88°F). Nuttonson (1958) suggested that rye will begin to germinate and grow at a temperature as low as 0.6°C (33°F). Leonard and Martin (1963) stated that rye germination can occur at one or two degrees above the freezing point and gave 13° to 18°C (55° to 65°F) as optimum for germination.

Vegetative growth continues in the fall and begins in the spring at about 4°C (40°F), with minimal growth below this temperature.

Rye is a long-day plant; flowering is induced by 14 hours of daylight when accompanied by cool temperatures of 5° to 10°C (41° to 50°F). Under shorter daylengths, a delay in plant development occurs. At nine hours of daylength reproductive development may fail to occur. Rye plants subjected to short photoperiod persisted in a grasslike state for up to seven years (Starzycki, 1976).

ORIGIN OF RYE

The ancestry of rye remains obscure. From a commercial-use standpoint rye may have been domesticated recently relative to other cereals. Rye and oat were considered as weeds of cultivated wheat and barley fields and were spread inadvertently with other domesticated cereals as they were introduced into new areas. In northern parts of Europe, the rye weed plants were cultivated when they were observed to grow better than wheat and barley. Rye reached its greatest importance in northern Europe and Asia, where it is used as a bread grain.

One suggestion of the origin of rye is that a perennial form of *Secale montanum* Guss was one parent of *Secale cereale* (Stutz, 1972; Vavilov, 1917). This is an enticing concept in view of the desire for a perennial cereal. Hybrids between *Secale montanum* and *Secale cereale* were found to be superior for forage production (Pfahler, 1968a). For this interspecific hybrid to be successful, however, seed set must be adequate to supply an annual source of seed. Hybrids were generally intermediate to their parents in terms of seed set, with *S. montanum* being the lowest-seed-setting parent (Pfahler, 1968b). Under the conditions of this test, *S. montanum* acted as an annual.

PRODUCTION ASPECTS OF RYE

Rye should be seeded at about the same time as winter wheat, but because rye grows better than wheat at low temperatures, seeding date is less critical. In general, rye can be seeded safely as much as two weeks after wheat. If seeded for fall pasture or a cover crop, rye can be seeded two to four weeks earlier than for grain production.

Ideally, rye should be seeded early enough for the plants to become established before winter begins. Winter rye tillers mainly in the fall, which explains why it is an ideal fall pasture and cover crop. If seeded after an ideal date, tillering is restricted and grain yields may be reduced. Rye often follows row crops to serve as a protective winter cover crop, and since most row crops in the northern areas are harvested late, rye is often seeded too late for optimum yields.

Under conditions in south-central Alaska, when tests were made at three dates of planting—August 12 and 22 and September 8—the highest level of winter survival was for the August 22 planting date (Klebesadel, 1969).

In the north-central part of the agricultural area of Saskatchewan between 52° and 53° N latitude, the optimum dates for planting both winter rye and wheat determined over a three-year period were between August 15 and September 1 (Fowler, 1983). Seeding earlier than this usually resulted in reduced yield and lower 1000-kernel weight. Later seeding usually resulted in a yield reduction, delayed heading, later maturity, and reduced hectoliter weight. At an early seeding date and under conditions of these tests, rye cultivars were higher yielding, earlier heading and maturing, taller, and had a lower 1000-kernel and hectoliter weight (Table 19-2) than if seeded later.

In a parallel trial, involving two rye and four wheat cultivars seeded at two-week intervals between August 1 and October 15 over a three-year period (Fowler, 1982), it was found that rye accumulated shoot dry matter at twice the rate of wheat (Table 19-3). As expected, a decrease in shoot dry weight occurred with each succeeding delay in seeding date. Average survival was highest for the August 15 and September 1 seeding dates, which was four to six weeks prior to the onset of fall conditions favorable for cold acclimation.

Rye normally is seeded at 112 kg/ha (100 lb/ac) (Briggle, 1959), but when seeded late and if intended for a winter cover crop, seeding rate should be adjusted upward to provide effective cover and to reduce erosion losses. For example, Grant et al. (1983) found that a leaf-area index of 1.0 may be needed to minimize erosion. By adjusting seeding rate upward, the desired leaf-area index may be reached even under conditions of late seeding. Over a 25-day delay in seeding, the rate of seeding was progressively increased from 112 kg/ha to a high of 336 kg/ha (100 to 300 lb/ac) to achieve the desired level of protection against erosion.

Late seeding is not encouraged in rye unless removal of a previous crop or other circumstances restrict proper seeding date.

Seedbed preparations, seeding practice, and harvest methods are similar to those of wheat.

PHYSIOLOGICAL ASPECTS OF RYE

A low-temperature treatment to induce a shorter vegetative period and to hasten reproductive development is known as a vernalization treatment. Temperature treatments required for vernalization of winter cereals range from −0.2° to +10°C (30° to 50°F).

The time periods required for the vernalization phase of different cultivars of rye vary greatly. A period of 40 to 55 days was the maximum suggested by Nuttonson (1958) at a temperature of 1.1° to 3.8°C (34° to 39°F), and 15 to 25 days was the shortest.

Table 19–2
Influence of Seeding Date on Heading Date, Maturity, Height, and 1000-Kernel and Hectoliter Weight of Winter Rye over a Three-Year Period—1974, 1975, and 1976—between 52° and 53°N Latitude in Saskatchewan

Seeding Date	*Heading Date, day/mo*	*Maturity day/mo*	*Height*		*1000-Kernel Weight, g*	*Hectoliter Weight kg/hl*
			cm	*in.*		
Aug. 1	7/6	28/7	126	50	25.0	78.3
Aug. 15	7/6	29/7	128	50	27.3	78.2
Sept. 1	8/6	29/7	123	48	27.8	78.1
Sept. 15	11/6	1/8	119	47	27.6	77.0
Oct. 1	18/6	6/8	114	45	27.3	75.0
Oct. 15	19/6	8/8	111	44	27.2	75.2

Source: Fowler (1983).

Table 19–3
Comparisons between Frontier Winter Rye and an Average of Four Winter Wheat Cultivars, When Seeded at Two Dates of August 21 and September 15 and Sampled on October 24, in 1972 and 1973 at Saskatoon, Saskatchewan

	Winter Rye		Winter Wheat	
Component	Aug. 21	Sept. 15	Aug. 21	Sept. 15
Crown fresh weight (g/10 plots)	12.23	0.98	9.08	1.00
Crown water content (%)	76	78	74.3	78
Crown dry weight (g/10 plots)	2.90	0.22	2.31	0.22
Plant height (cm)	15.0	8.5	14.75	10.1
Crown depth (cm)	1.2	2.1	1.8	3.0
Leaves/plant	44	3.5	29	3.3
Tillers/plant	16	—	10.8	—
No. of crown roots/plant	9.2	—	6.9	—
Crown root length (cm)	2.2	—	2.3	—
Days to head	55	89	83.5	91.8
LT_{50}[a]	−23	−21	−18	−13

Source: Fowler and Gusta (1977).
[a]The temperature up to which 50% of crowns survive.

During vernalization, plants are hardened or acclimatized to winter conditions. Vincent (1972) reported that hardening ceases after the vernalization process has ended. Date of seeding, therefore, can influence winter hardiness.

Seeding-date studies (Fowler and Gusta, 1977) did show that early and late seeding results in plants that are less cold-tolerant than those from a recommended, intermediate seeding date, thereby confirming observations in the literature. Seeding after September 15 in the Saskatoon area, where winter stress conditions can be severe, corresponds to a late date of seeding.

Vernalization temperatures and daylength together may influence the impact of each separately. A short daylength at a low light intensity can substitute for low temperature. Proper date of fall seeding, whereby a shortening daylength, declining temperatures, and falling light intensity trigger acclimatization of plants to winter conditions is important.

Under controlled environmental conditions each environmental factor can be manipulated to result in vernalization of seedlings.

A commonly recognized phenomenon in cereals is that of postharvest dormancy, whereby newly harvested seeds become resistant to germination. In short-season areas, where an overlap between harvest and an optimum planting date can occur, postharvest seed dormancy presents a potential barrier to production. Producers who use freshly threshed seed may experience poor stands or delayed germination if after-ripening dormancy occurs.

Over a two-year observation period in which rye cultivars were cut and then seeded 0, 8, 15, 26, 36, and 98 days later without any dormancy-breaking treatment, no limits to production were observed (De Pauw, 1976). In all tests, some seed dormancy was observed, but this period was short. It was concluded that under the conditions of these tests the interval between cutting and threshing would be sufficient to meet all or part of the after-ripening requirement.

The length of the dormancy period in rye and other cereals depends on climatic factors during grain development and maturation. Under conditions of abundant rainfall and relatively high air temperatures during grain development and maturity, dormancy was not stabilized and preharvest sprouting may occur (Starzycki, 1976). Under dry, warm conditions, dormancy is increased, and several days of after-harvest maturation are needed to restore germination. The length of the dormancy period may vary.

Rye is a long-day plant, and shortening the daylength results in a delay in plant development. Starzycki (1976) reported that under short photoperiod, rye plants remained in a grasslike state for up to seven years. Concentration of rye in long-day regions of the world means that rye cultivars cannot be grown in areas with short-day conditions.

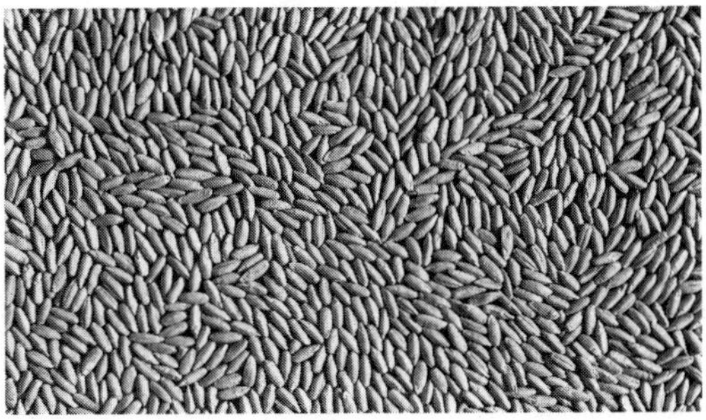

Figure 19–1. Doubling the diploid chromosome number of rye by means of the chemical colchicine results in an autotetraploid. Tetraploid rye exhibits gigas characteristics, as demonstrated by the large kernels in this photo. Despite an average increase in seed size of 53% over the diploid form, tetraploid rye has limited commercial value because of reduced tillering, increased sterility, and other undesirable features associated with tetraploidy. Rye seed ranges in color from bluish-green, to greenish brown, to brownish olive, to yellow depending on anthocyanin pigmentation in the aleurone or pericarp layers. (Courtesy of Ontario Ministry of Agriculture and Food)

TETRAPLOID RYE

Subsequent to the report of Dorsey (1936) on the production of tetraploid rye, a number of tetraploid cultivars were produced. The 14 ($2n$) chromosome number of diploid rye can be doubled using colchicine, resulting in an autotetraploid with 28 ($4n$) chromosomes.

Vegetative and reproductive organs of tetraploid plants commonly exhibit gigas features, which attracted the attention of producers and raised the hope for successful commercial development. One cultivar, Tetra Petkus, was grown extensively. Despite its large seeds (Figure 19-1), undesirable morphological and agronomic features resulted in limitations on production. Little or no economic advantage was found for producing tetraploid over diploid rye. With spaced-plant tests, an 18% to 19% yield increase was found (Muntzing, 1954) over that of the diploid parents used to produce three tetraploid cultivars. This advantage may not hold at normal plant populations.

The intensive effort directed at producing autotetraploid rye was reviewed by Muntzing (1951), based on observations of seven tetraploid cultivars. His conclusions were as follows:

1. Cell size and plant height exhibited a 15% increase over diploid counterparts. No difference in straw stiffness could be measured, but the increase in plant height was not viewed as desirable.
2. A 12% reduction in tillering was found in tetraploids.
3. The number of flowers per spike was lower in tetraploid, and a tendency to shed the basal spikelets before maturity was observed.
4. Kernel weight averaged 53% larger than in diploid rye. A greater weight of seed is required to achieve the recommended plant population.
5. Hectoliter weight was 5% to 6% lower in tetraploids, but this was a consequence of the large size of the kernels and the degree of compaction rather than an indication of shrunken or low-quality seed.
6. Seed set did not exceed 62% among the tetraploid cultivars observed, compared to 80% in the diploids under comparable conditions. Seed set was particularly low in secondary tillers. Pollen fertility of tetraploids appeared normal.
7. Seed set in tetraploid female parents was reduced when pollinated with diploid rye. When a tetraploid cultivar was tested among diploid yield

trials in Minnesota, it yielded only about one-half to two-thirds the yield when grown in isolation plots (Koo, 1958), an observation consistent with European results.

The reason for such a marked reduction in seed set is related to the formation of triploid zygotes, which fail to develop into mature seed because of abortion or shrivelled seed development. To avoid triploid development, tetraploid and diploid rye cultivars must be grown in isolation.

8. Tetraploid seed was superior to diploid seed in sprouting ability, as measured by the percentage of seedlings that develop from a given number of seeds sown.
9. Diploids and tetraploids were considered equal in degree of winter hardiness.
10. Data obtained from 18 different trials failed to reveal any differences in yield between diploid and tetraploid rye. Apparently large kernel size and superior sprouting ability can compensate for the reduced degree of seed setting, tillering, number of seeds per spike, and the tendency to shed basal spikelets.
11. The best tetraploid cultivars were considered equal to diploids with regard to yield, straw stiffness, earliness, and frost and drought resistance.
12. Tetraploid rye cultivars were more sensitive to drought than diploids.

AN OVERVIEW OF RYE

Of the cereal grains, rye is a species to which neither plant nor crop physiologists have paid much attention. The result is that many observations, such as that rye uses 20% to 30% less water per unit dry-matter formation than wheat, are unsubstantiated.

In more than one aspect, rye can be regarded as a peripheral crop. Rye has been given no research priority among the network of international research centers (Chapter 14). Additional research will be of indirect benefit only to developing countries in the tropics and semitropics. On the other hand, rye offers the best opportunity among the cereal crops to extend agriculture on what is regarded as a peripheral area of production. If rye continues to be ignored as a crop, a valuable tool and germplasm source may be overlooked, and without an objective approach, germplasm preservation may be sacrificed. A germplasm source population was established by Qualset and Hoskinson (1966) that utilized a limited number of North American cultivars, but international leadership is required to develop a more comprehensive germplasm bank.

Rye research has the additional problem that much of it extends across the political boundaries of centrally planned countries. Among scientists, however, such boundaries are not barriers to exchange of materials or information.

Perhaps the most compelling reason for research interest in rye is the formation of amphidiploids using the entire genomes. This has resulted in a new cereal species, known as triticale, covered in the next chapter.

Rye whiskey is a distinctive distilled alcoholic beverage produced in Canada. Canadian whiskey is produced from fermentable carbohydrate, mainly corn, with rye flavoring.

Bushuk (1976) concluded that "in spite of its many deficiencies, rye will continue to be an important crop because it has a number of advantages over the other crops. Its many uses and advantages far outweigh its deficiencies" (pp. 6–7).

The future of rye is by no means assured, however. To reverse the tide of public opinion against it, a breakthrough in some aspect of rye may be required. Rye may have the potential to provide such a breakthroough, provided sufficient attention is paid to the crop to recognize the breakthrough if and when it occurs.

Rye warrants continued research and improvement because it is a parent of triticale.

Questions

1. Compared to other cereal grains, relatively little effort is directed to breeding, selection, or physiological aspects of yield of rye.
   a. What has been the main preoccupation with work in rye?
   b. Compare grain-yield advances for rye with those of cereal crops over a selected time period.
2. a. Examine Table 19-1 and review the reasons for the decline in area of rye produced.
   b. Have any countries shown an increase in the area of rye produced?
3. Review the rye crop, its strengths and weaknesses, and consider which area(s) may offer opportunity for advancement.
4. What regions or countries are the major rye producers? Suggest reasons why.
5. a. Compare current world production of rye with those of other cereal crops.
   b. What is the percentage of total world cereal production represented by rye in terms of area harvested and total production?
6. How important is the rye crop, relative to other cereals, for purposes other than as a food crop?
7. Under what conditions might rye be an important crop?
8. Suggest reasons why rye is relatively free of many pest problems commonly found in other cereal grains.

References

Briggle, L. W. 1959. *Growing Rye*. Farmers' Bulletin 2145. Washington D.C.: United States Department of Agriculture. 16 pp.

Bushuk, W. 1976. "History, World Distribution, Production, and Marketing." (In) W. Bushuk, (ed.), *Rye: Production, Chemistry, and Technology*. St. Paul, Minn. American Association of Cereal Chemists. pp. 1-11.

De Pauw, R. M., 1976. "After-ripening in Fall Rye." *Canadian Journal of Plant Science* 56: 395–396.

Dorsey, E. 1936. "Induced Polyploidy in Wheat and Rye." *Journal of Heredity* 27: 155–160.

Driscoll, C. J., and L. M. Anderson. 1967. "Cytogenic Studies of Transec—A Wheat-rye Translocation Line." *Canadian Journal of Genetics and Cytology* 9: 375–380.

Driscoll, C. J., and E. R. Sears. 1963. "The Nature of a Spontaneous Transfer of Hairy Neck from Rye to Wheat." *Proceedings of the XI International Congress of Genetics*. p. 123.

Evans, L. E., W. Dedio, and R. D. Hill. 1973. "Variability in the Alkylresorcinol Content of Rye Grain." *Canadian Journal of Plant Science* 53: 485–488.

FAO. 1982. *1981 Production Yearbook*. FAO Statistics Series No. 40. Rome, Italy: Food and Agriculture Organization of the United Nations. Vol. 35. p. 104.

Fedorov, U.S., et al. 1970. The Genetics of Rye (*Secale cereale* L.) X. The Inheritance of Dwarfness. *Genetika* Moskva 6(3): 5-17. (In) Plant Breeding Abstracts 1970. p. 947. No. 7295.

Fowler, D. B. 1982. "Date of Seeding, Fall Growth, and Winter Survival of Winter Wheat and Rye." *Agronomy Journal* 74: 1060–1063.

———. 1983. "Influence of Date of Seeding on Yield and Other Agronomic Characters of Winter Wheat and Rye Grown in Saskatchewan." *Canadian Journal of Plant Science* 63: 109–113.

———, and L. V. Gusta. 1977. "Influence of Fall Growth and Development on Cold Tolerance of Rye and Wheat." *Canadian Journal of Plant Science* 57: 751–755.

Friend, D. W. 1970. "Comparisons of Some Milling Products of Barley and Rye When Fed in Diets of Rats." *Canadian Journal of Animal Science* 50: 345-348.

———, and T. M. MacIntyre. 1969. "Digestibility of

Rye and Its Value in Pelleted Rations in Pigs." *Canadian Journal of Animal Science* 49: 375–381.

Grant W. J., et al. 1983. "Seeding Rate Recommendations for a Winter Rye Cover Crop Based on Expected Leaf Area Index." *Journal of Soil and Water Conservation* 38: 361–363.

Kilcher, M. R. 1982. "Effect of Cattle Grazing on Subsequent Grain Yield of Fall Rye (*Secale cereale* L.) in Southwestern Saskatchewan." *Canadian Journal of Plant Science* 62: 795–796.

Klebesadel, L. J. 1969. "Winter Survival and Spring Forage Yield of Winter Rye Varieties in Subarctic Alaska as Influenced by Date of Planting." *Agronomy Journal* 61: 708–712.

Koo, F. K. S. 1958. "Deleterious Effects from Interpollination of Diploid and Autotetraploid Winter Rye Varieties." *Agronomy Journal* 50: 171–172.

Langille, J. E. et al. 1981. "Amino Fall Rye." *Canadian Journal of Plant Science* 61: 161–162.

Leith, B. D. 1925. "Sterility in Rye." *Journal of the American Society of Agronomy* 17: 129–132.

————, and H. L. Shands. 1938. "Fertility as a Factor in Rye Improvement." *Journal of the American Society of Agronomy* 30: 406–418.

Leonard, W. H., and J. H. Martin. 1963. *Cereal Crops.* New York: The Macmillan Company. p. 459.

Morey, D. D., and R. D. Barnett. 1980. "Rye." (In) W. R. Fehr and H. H. Hadley (eds.), *Hybridization of Crop Plants.* American Society of Agronomy and Crop Science Society of America. pp. 523–534.

Muntzing, A. 1951. "Cyto-genetic Properties and Practical Value of Tetraploid Rye." *Hereditas* 37: 17–84.

————. 1954. "An Analysis of Hybrid Vigor in Tetraploid Rye." *Hereditas* 40: 265–277.

Nuttonson, M. Y. 1958. *Rye-Climate Relationships and the Use of Phenology in Ascertaining the Thermal and Photo-Thermal Requirements of Rye.* Washington D.C.: American Institute of Crop Ecology. 219 pp.

O'Mara, J. G. 1946. "The Substitution of a Specific *Secale cereale* Chromosome for a Specific *Triticum vulgare* chromosomes." *Genetics* 32: 99–100.

Pfahler, P. L. 1968a. "Heterosis and Homeostasis in *Secale montanum* Guss. × *S. cereale.* L. Hybrids. I. Vegetative Growth and Development." *Crop Science* 8: 663–666.

————. 1968b. "Heterosis and Homeostasis in *Secale montanum* Guss. × *S. cereale* L. Hybrids. II. Reproductive Characteristics." *Crop Science* 8: 667–670.

Purvis, O. N. 1939. "Studies in Vernalization of Cereals. IV. The Inheritance of the Spring and Winter Habit in Hybrids of Petkus Rye." *Annals of Botany* N.S. 3: 719–730.

Qualset, C. O., and P. E. Hoskinson. 1966. "A Germplasm Source Population of Rye." *Crop Science* 6: 219.

Smith, R. E., and T. M. MacIntyre. 1970. "The Feeding of Rye to Growing Chickens." *Canadian Journal of Animal Science* 50: 107–113.

Starzycki, S. 1976. "Diseases, Pests, and Physiology of Rye." (In) W. Bushuk (ed.), *Rye: Production, Chemistry, and Technology.* St. Paul, Minn.: American Association of Cereal Chemists. pp. 27–61.

Stutz, H. C. 1972. "On the Origin of Cultivated Rye." *American Journal of Botany* 59: 59–70.

Vavilov, N. I. 1917. "On the Origin of Cultivated Rye." *Bulletin of Applied Botany* 10: 561–590.

Vincent, A. 1972. "Relationships Between Vernalization, Hardening and Phytochrome." *Proceedings of a Colloquium on the Winter Hardiness of Cereals.* Agriculture Research Institute. Hungarian Academy of Science. pp. 31–60.

Wells, D. G., and A. R. A. Mahdawi. 1953. "Pollination Control in Rye." *Agronomy Journal* 45: 379–380.

Wieringa, G. W. 1967. *On the Occurrence of Growth Inhibiting Substances in Rye.* Wageningen, The Netherlands: Institute for Storage and Processing of Agricultural Procedure. Publication No. 156. 68 pp.

20
Triticale

ORIGIN, DEVELOPMENT, AND POTENTIAL

Unlike other cereal crops, which emerged through natural hybridization processes and through selection pressure by evolutionary forces over thousands of years, triticale was synthesized by humans. Under protective laboratory conditions or in carefully tended field plots, the law of the survival of the fittest was not allowed to screen triticale in the normal evolutionary sense; instead, the human being acted as a director of evolution. Because triticale is the first man-made crop, human experience in directing evolution was lacking. Based on scientific knowledge, however, shortcuts in the evolutionary process were attempted—and with remarkable success. What took 10,000 years to achieve in other cereals was accomplished in triticale in one one-hundreth of the time, essentially 100 years.

The work of Charles Darwin on evolution and Gregor Mendel on genetics in the second half of the nineteenth century revealed that genetic variability could be developed by hybridization. The extent of limitations on crossing was not fully known, but instinctively, plant breeders envisioned that combining the highly desirable vigor and winter hardiness of rye with the agronomic and quality aspects of wheat could result in a useful new species, provided such a cross could be made.

Unlike a cross between two plants of the same species, where half the chromosomes from each parent pair to provide new genetic combinations, triticale is synthesized by combining the genomes of wheat and rye. Triticale possesses the chromosome complements of both parental species. Such a combination is not easy to achieve, because wheat and rye are separate species, belong to different genera, and as a rule are separated by a strong barrier of incompatibility. Wheat × rye hybrids normally are completely sterile, but spontaneous or induced doubling of the chromosome number results in partially fertile amphiploids. When crossing is achieved, it is when wheat is the female parent. When rye is the female partner in the cross, seeds are produced very rarely, or not at all.

There can be no doubt that over thousands of years of evolution, wheat-rye combinations occurred. It is not surprising, therefore, that in 1875 Wilson (1875) described to the distinguished Botanical Society of Edinburgh a naturally occurring wheat-rye hybrid—which, however, was sterile. Perhaps Wilson's report engendered the vision of developing a new crop that combined wheat and rye, and perhaps scientists attempted numerous crosses that produced only sterile progeny. But by increasing the chances, a very infrequent event occurred: a triticale with partial seed fertility was found by the German scientist

Rimpau and was described in 1890 (Zillinsky and Borlaug, 1971). Fertility resulted from the spontaneous doubling of the chromosome number. The first known triticale that could reproduce itself had come into existence.

The partially fertile triticale described by Rimpau from a direct wheat × rye cross later came to be known as a **primary triticale.** Years later, the plant described by Rimpau was shown by the Swedish worker, Muntzing, to be octoploid ($2n = 8\times = 56$) arising from a hexaploid wheat ($2n = 6\times = 42$) and rye ($2n = 14$).

For many years triticale remained a theoretical curiosity, because new primary triticale plants could not be produced without a technique to effect chromosome doubling. With the discovery of colchicine for chromosome doubling in 1937 (Eigsti, 1938) and of effective embryo culture techniques in the 1940s, a new era for the development of commercial triticale began.

The results of basic research work that extended into the late 1930s were summarized by Briggle (1969), Larter (1974), Lorenz (1974), Muntzing (1939, 1974, 1979), and Zillinsky (1973, 1974). In essence, the initial octoploid triticale plants were partially sterile, had shriveled seed, and exhibited chromosomal irregularities such as aneuploidy and meiotic disturbances as a result of disharmony between the wheat and rye chromosomes. These early triticale plants appeared to offer little commercial potential, and much of the work was of an academic and cytological nature. Understandably, when progress was so limited, interest in triticale waned.

Muntzing in Sweden probably was the first to see the potential of triticale as a new crop species, and he devoted 32 years to the development of octoploid triticale (Muntzing, 1979). Muntzing recognized the high degree of winter hardiness and the ability to grow well on sandy and acid soils, and no doubt he was attracted by the large kernels. Perhaps Muntzing viewed the partial sterility, shriveled seed, and tendency to lodge as a challenge, for he engaged in a program to improve triticale through hybridization and selection. The first triticale evaluated by Muntzing (1939) yielded only 50% of the yield of wheat. In 18 years of effort, Muntzing's best octoploid triticale yielded 90% of the bread wheat yield, but still considerable pessimism prevailed about the commercial value of amphiploids.

Observations on superior meiotic stability and fertility of hexaploid ($2n = 6\times = 42$) triticale relative to octoploid triticale led to the realization that 42 chromosomes were preferred over 56 chromosomes (Sanchez-Monge, 1959; Sisodia and McGinnis, 1970). The first hexaploid triticale was produced about 1948 from *Triticum durum* and rye. Durum wheat contributes the A and B genomes, so that hexaploid triticale lacks the D genome contributed by hexaploid wheat. Shebeski (1958) postulated that the ABR genomic constitution from wheat and rye should result in greater productivity than when the D genome is present.

Hexaploid triticale did not solve all the problems of triticale development. Incomplete fertility remained a problem, but hexaploid triticale did offer more vigor, higher tillering capacity, and greater cytological stability. Improved hexaploid fertility was achieved by crossing octoploid with hexaploid triticale and then selecting for fertile recombinants (Pissarev, 1966; Kiss, 1965). This step, shown in Figure 20-1, did improve fertility, but it took breeders another 25 years and a number of significant discoveries to achieve commercial success for triticale.

Improved fertility and yield potential of hexaploid relative to octoploid triticale resulted in a renewed interest in the early 1950s. In the United States, O'Mara (1948) began a breeding and selection program with hexaploid triticale. In 1954, a large-scale Canadian triticale breeding program was initiated at the University of Manitoba, Winnipeg, under the direction of Drs. L. H. Shebeski and B. C. Jenkins. In the initial stages, both the United States and Canadian programs were devoted to the introduction, synthesis, and evaluation of primary hexaploid triticale.

The development of **secondary hexaploid triticale** provided a second method of deriving the 6× triticale, which proved to be complementary by way of introducing and diversifying the maximum amount of wheat-rye germplasm available. Most triticale programs throughout the world have combined primary and secondary triticale development (Larter, 1974). Muntzing

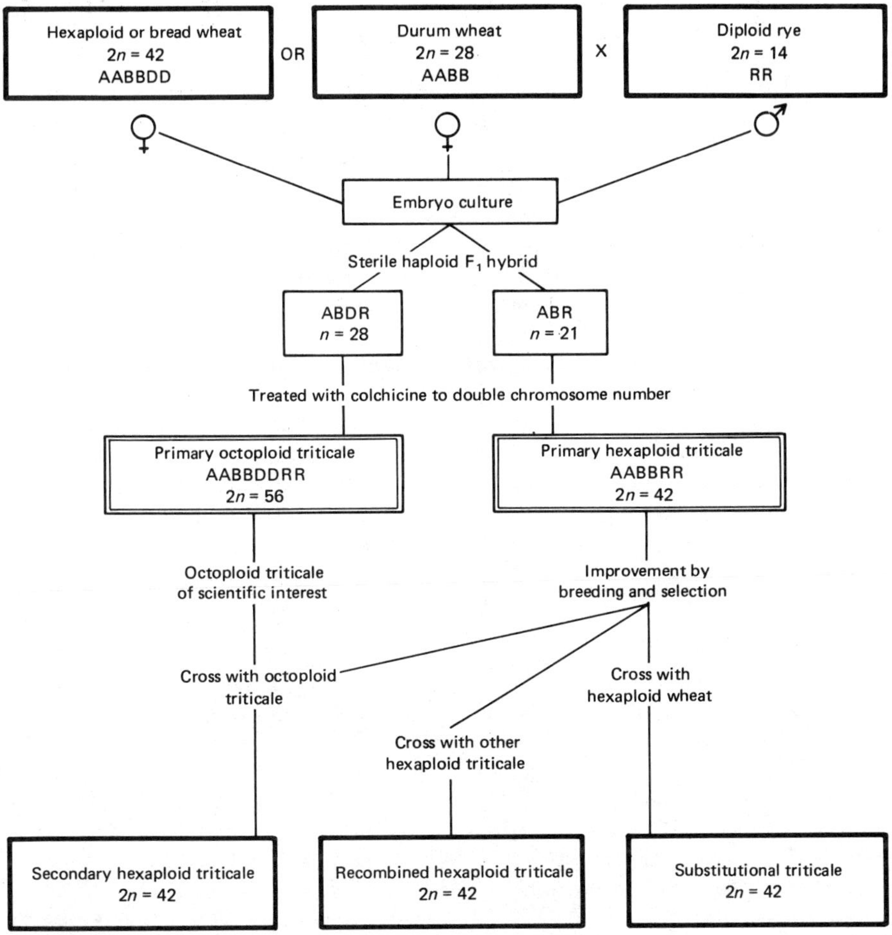

Figure 20–1. Schematic presentation of the steps in the development of triticale. Research emphasis has changed over the years from work on octoploid to primary hexaploid triticale to combined efforts on primary hexaploid and secondary hexaploid forms, recombined hexaploid and substitutional triticale cultivars.

(1979) suggested that a third class of triticale be recognized—that of secondary substitutional triticale in which specific rye chromosomes are replaced by **homoeologous chromosomes** from the D genome of wheat. The sequence of primary, secondary, and substitutional triticale developed to produce octoploid or hexaploid triticale is shown in Figure 20-1.

By 1968, the performance of hexaploid triticale had improved so much that grain yields were approaching bread wheat yields in Manitoba. In 1969, Rosner, the first hexaploid spring triticale to be licensed in Canada, was released (Larter et al., 1970). This step recognized triticale as a new crop of commerce in Canada and served to establish a norm for future improvement.

The release of Rosner by no means signaled the elimination of aneuploidy, sterility, shriveled seed, or lodging problems. Considerable room for improvement existed. To undertake such improvement the most intensive breeding program ever to be mounted for triticale was established in 1964 (Larter, Tsuchiya, and Evans, 1969; Zillinsky and Borlaug, 1971). It involved an exchange of material between CIMMYT in Mexico and the University of Manitoba, and its objective was to develop hexaploid triticale as a new commercial field crop competitive with other cereals. This action, along with the establishment of an international testing program in 1966 in which numerous countries around the world cooperated, raised triticale to international status and provided an avenue for worldwide testing, selection, and exchange of material.

The CIMMYT-University of Manitoba breeding-exchange program involved selection opportunities at three climatically different areas. The winter triticale nursery in the Yaqui Valley in Sonora District in northwestern Mexico is at 28° N latitude and 35 m (115 ft) elevation; the summer nursery near Mexico City is at 18.5° N latitude and 2600 m (8530 ft) elevation; and the University of Manitoba summer nursery is at 50° N latitude and 230 m (755 ft) elevation. These three contrasting test sites added a new dimension to triticale testing and development, resulting in triticale cultivars with reduced daylength sensitivity. An important step was the speeding up of progress by testing and selecting at more than one site and producing two generations annually.

The results of the international testing program were both encouraging and discouraging. Testing under widely different conditions revealed that daylength-insensitive cultivars produced few tillers and short spikes under long-day conditions. Triticale suffered from high temperatures during vegetative growth. Under high elevation and cool temperatures, triticale performance was competitive with or superior to that of wheat. International testing revealed the manifold disadvantages of triticale and led to a clear definition of objectives.

A major advance in triticale improvement was realized with the development in Mexico of a highly fertile cultivar of hexaploid triticale called Armadillo (Zillinsky and Borlaug, 1971). Armadillo was unique among triticale cultivars because of its increased fertility, improved test weight, better yield, insensitivity to daylength, earliness, dwarf gene, and good nutritional quality. Transfer of these heritable traits into secondary hexaploid triticale cultivars resulted in yield increases of 50% to 60% above those obtained from triticale cultivars grown before the 1968–1969 season.

Although substantial improvement was made with Armadillo, further progress was needed to improve endosperm development, to increase disease resistance, to reduce lodging, shattering, and preharvest sprouting, and to overcome deficiencies in tillering, adaptation, and nutritional quality. Such a multiplicity of problems was tackled by increasing genetic diversity. Triticale had lacked the innumerable generations of natural selection under which other cereal crops had evolved. Until 1971, development had been based on a very narrow genetic base. For example, Wolff (1976) reported that as late as 1972 only about 30 primary hexaploids were available at CIMMYT. As outlined in Figure 20-1, three groups of crosses were made to expand the gene pool:

1. Between Armadillo and other hexaploid triticale.
2. Between Armadillo and octoploid triticale.
3. Between Armadillo and dwarf Mexican wheat both hexaploid and tetraploid.

A promising new cultivar named Cinnamon was selected from a cross between a primary octoploid triticale and Armadillo, carried a second dwarfing gene and increased lodging resistance.

Gene-pool enrichment was achieved by crossing the spring triticale cultivars with winter wheat, winter rye, and winter triticale in a cooperative effort with breeders in Canada, the United States, Sweden, Great Britain, Germany, Hungary, and other countries. Environmental conditions inducive to differential winterkill, whereby nonhardy cultivars are eliminated but hardy cultivars survive, occur frequently at Guelph, Canada. In 1980, a winter hexaploid triticale was licensed in Canada named OAC Wintri (Poysa, Knoblauch, and Reinbergs, 1981). In tests over eight station years, OAC Wintri had a 4% to 6% yield advantage over recommended winter wheat cultivars, had 11% higher winter survival, and had superior resistance to powdery mildew, leaf rust, and septoria. Tall vegetative growth has restricted the widespread acceptance of this cultivar, but it may have value as both a grain and forage crop. Lodging in OAC Wintri and other triticale cultivars is a widespread problem.

In the final analysis, the economic production of grain yield will determine the success of triticale. In the earliest trials in Mexico, triticales had failed to yield even half as much grain as the locally adapted wheat. Incorporation of the Armadillo genotype substantially increased yields, but not sufficiently to make them competitive with the best commercial semidwarf Mexican wheat cultivars. For example, the two best cultivars with the Armadillo background averaged 4492 kg/ha (4010 lb/ac), compared to an average wheat yield of 5417 kg/ha (4837 lb/ac). With the introduction of the two-gene dwarf secondary hexaploid Cinnamon in 1971, an advance in yield was made by reduced lodging. In the 1972–1973 CIMMYT yield trials in Sonora District triticale yielded 8000 kg/ha (7143 lb/ac), a yield comparable to the best wheat yields. In 1973–1974, 150 of the 600 cultivars tested yielded over 7000 kg/ha (6250 lb/ac).

Matching the yield of wheat may not assure the success of triticale. Abnormal endosperm formation with an associated wrinkled coat, deep crease, and a lackluster seed has made triticale unattractive to farmers, processors, and consumers. Test weight of triticale ranged from 58 to 72 kg/hl compared to 80 kg/hl for wheat. A standard test weight for triticale is 76 kg/hl. Progress toward this goal is being achieved by increasing the degree of selection pressure to find segregating populations that combine desirable agronomic traits and satisfactory seed quality.

From this historical review it is evident that the development of triticale does not represent the work of a single scientist, a single research unit, or even a single country. Rather, progress in triticale represents the collective contributions from worldwide efforts in countries including Sweden, Spain, Hungary, Russia, Japan, the United States, China, Mexico, and Canada. Progress over the past 115 years, the countries involved, and a description of the contributions made are summarized in Table 20-1.

The evidence of the 1970s suggests that an intense effort will be required to overcome the

Table 20–1
Chronological Sequence of Development and Progress in Triticale over the First 115 Years

Date	Location	Observation
1875	Scotland	First known wheat × rye cross, resulting in a sterile plant, reported by A. S. Wilson.
1888	Germany	First fertile wheat × rye hybrid achieved by W. Rimpau.
1918	Russia	Thousands of wheat × rye hybrids developed at Saratov research station. F_1 plants produce seeds from which true-breeding, fairly fertile, phenotypically intermediate hybrids are derived.
1935	Germany	Name triticale appears in scientific literature.
	Sweden	A. Muntzing begins intensive work on octoploid, triticale; discovers mechanism of spontaneous fertility in wheat × rye hybrids.

(Continued)

Table 20–1 (Continued)

Date	Location	Observation
1937	France	Colchicine technique for doubling chromosomes of sterile hybrids developed; production of fertile triticale made routine.
1940s	?	Embryo culture technique developed for growing hybrid embryos from seeds with malformed endosperm.
1954	Canada	University of Manitoba begins first North American effort to develop triticale as a commercial crop based on primary triticale collection.
1964	Mexico	CIMMYT and University of Manitoba establish cooperative program on triticale.
1968	Mexico	Armadillo cultivar with improved fertility, one drawfing gene, and superior plant type spontaneously appears and becomes progenitor for triticale around the world.
1968–1969	Hungary	Kiss releases two secondary hexaploid triticale cultivars for commercial production.
	Spain	Hexaploid triticale developed by Sanchez-Monge released for commercial production.
1969	Canada	Rosner cultivar becomes first North American triticale released for general use.
1971	Mexico	First two-gene drawf, Cinnamon, used to increase lodging resistance in triticale.
1972	Mexico	Efforts directed toward broadened genetic base for triticale, reduced seed shriveling, and high fertility.
1973	Mexico	Yield trials in Mexico show that the best triticales yield as well as top wheat cultivars.
1974	Mexico	Hectoliter weight averages 72 kilograms, up from 68 kilograms in 1970. Yield of 7000 kg/ha (6250 lb/ac) is obtained. The five highest-yielding triticles in trials at 47 worldwide locations outyield the best bread wheat cultivar by 15%.
1981	Canada	OAC Wintri is the first winter hexaploid triticale licensed in Canada.
1990	World	Triticale is predicted to start to compete seriously with the bread wheats as one of the world's most important food crops (Shebeski, as quoted by Wolff, 1976, p. 14).

Source: Adapted from Wolff (1976).

shortcomings of triticale, but that progress is possible. Compared to wheat, triticale is a competitive crop. Compared to rye, it is superior. The future of triticale was predicted by L. H. Shebeski in 1973 (Wolff, 1976, p. 14) when he said;

> In my opinion, over the next 15 years, yields of triticale will improve much more rapidly than those of wheat and should plateau at a level approximately 50 percent higher than those of wheat. This is no idle speculation. . . . With rapidly expanding programs and a quickly widening genetic base, with improved fertility and seed density, with improving world-wide cooperation, the improvements over the next 15 years will greatly surpass all improvements that so far have been attained.

Shebeski concluded that by 1990 (Wolff, 1976, p. 14), "triticale will have begun to compete seriously with the bread wheats as one of the world's most important food crops."

CLASSIFICATION AND TERMINOLOGY

The common name *triticale* represents a contraction of the two generic names *Triticum* (wheat) and *Secale* (rye). The name triticale is widely accepted to designate all allopolyploids derived from crosses between wheat and rye, including octoploid, hexaploid, and tetraploid forms and both primary and secondary cultivars.

The species name for triticale is × *Triticosecale* Wittmack, which conforms to regulations of the International Code of Botanical Nomenclature. This system of naming and the × *Triticosecale* Wittmack name have become widely accepted, and an earlier proposal by Larter et al. (1970) to name the hexaploid form *Triticale hexaploide* Lar. was rejected. Rejected also was the proposal of Baum (1971) to give a species name in accordance with the parental wheat and rye species involved in the synthesis of the amphiploid; e.g., a cross of *T. turgidum* and *Secale cereale* would be named *Triticale turgidocereale*. Baum's system lacked the flexibility needed to name hybrid combinations arising as a secondary triticale.

Because triticale and wheat cross readily, there appears to be little justification for the use of triticale as a generic name, in which case it would be spelled with a capital T and italicized. The use of *Triticale* as a generic name introduces confusion by having the same word as the scientific and common name. The suggested common name "wye" was not accepted.

The name × *Triticosecale* was credited to Wittmack, who applied the name in 1899 to intergeneric fertile hybrids from wheat × rye combinations (Baum, 1971).

Triticale cultivars differing in chromosome number can be simply indicated by their chromosome number, e.g., octoploid ($22 = 8X = 56$), hexaploid ($22 = 6X = 42$), or tetraploid ($2n = 4X = 28$). Crosses between diploid wheat and rye have not been successful (Sodkiewicz, 1984), but stable tetraploid triticale can be selected from crosses involving hexaploid triticale and diploid rye.

BOTANY OF TRITICALE

Morphologically, triticale most closely resembles its wheat parent but exhibits more vigorous growth characteristics. Triticale may be either spring or winter in growth habit in accordance with parental materials used.

The inflorescence of triticale is a spike, resembling the wheat spike more closely than rye (Figure 20-2). Although not evident in Figure 20-2, the spike of triticale is often considerably larger than that of wheat or rye. Like both parents, the spike of triticale is composed of a series of spikelets arranged alternately on each side of the rachis. The number of spikelets ranges from 30 to 40 per spike. Each spikelet consists of four to eight florets, of which usually only three are fertile.

Each spikelet is subtended by two glumes (outer glumes) of chaff. In turn, the lemma and palea enclose each floret in the spikelet. The lemma commonly terminates in the form of an awn 7 to 10 cm (3 to 4 in.) long. Awns vary in length among cultivars. Triticale is a free-threshing cereal, which means that the lemma

Figure 20–2. Spike comparisons with triticale (right) and the two parental materials of this new species, wheat (left) and rye (center). Generally, triticale resembles wheat more closely than rye and is often larger in size than either wheat or rye spikes. Triticale is a self-pollinated crop, although one of its parents, rye, is a cross-pollinated species. (Courtesy of Ontario Ministry of Agriculture and Food)

and palea do not adhere to the kernel and are readily removed by threshing action.

Three anthers and a bilobed stigma are present in each floret. Triticale is a self-pollinating species, with pollen being released within the floret. Approximately 5% out-crossing was reported among spring triticale cultivars at Winnipeg, Canada (Yeung and Larter, 1972), but additional observations are needed before this value can be accepted as a textbook fact. In new amphiploids and early-generation hybrids, a much higher percentage of out-crossing was observed (Larter and Gustafson, 1980), varying with climatic conditions.

The level of out-crossing is of interest, as it is related to hybrid triticale development. To date, very little attention has been directed to hybrid potential in this crop. Yeung and Larter (1972) considered the extent to which triticale can disseminate viable pollen to a surrounding female plant population. The collective pollen load can be evaluated by (1) the number of anthers per unit area and the number of pollen grains per anther, (2) the degree of anther extrusion, and (3) the duration of anthesis. The study was based on three hexaploid triticale cultivars, which were compared to the common bread wheat cultivar, Manitou. The anthers of triticale were longer than those of wheat and released 15,000 to 21,000 pollen grains per anther compared to about 8000 per anther for wheat. In the triticale cultivar Rosner, 89% anther extrusion occurred, compared to 70% for Manitou. The period of anthesis in triticale varied among cultivars but generally was longer than in wheat and therefore more inducive to out-crossing. This was verified by additional study. Using a triticale strain carrying a dominant marker gene, 50% seed set was obtained at a distance of 12 m (39 ft) leeward of the pollen source. Some seed set occurred at a distance of 30 m (98 ft). From the genetic variation observed in triticale, Yeung and Larter (1972) concluded that selection for prolonged floral opening and unrestricted anther extrusion is possible and that pollen load would not be a limiting factor in the production of hybrid seed. Sodkiewicz (1984) reported higher fertility with cross-pollination in hexaploid triticale produced from diploid wheat × rye.

Anthesis normally begins in the central portion of the spike when the spike has completely emerged from the leaf sheath. Although anthers are extruded from the florets soon after anthesis, the main pollen load may be dumped within a floret, and pollen viability declines rapidly after release from the anther. Viability may be lost within five minutes at 20°C (68°F) (Larter and Gustafson, 1980) and more rapidly at higher temperatures.

The nonshattering kernel of triticale most closely resembles that of wheat (Figure 20-3). As evident from the photograph, partial shriveling of the seed is a common problem. Considerable improvement in reducing this problem has occurred, but further improvement is desirable. Shriveled seed results in lower volume weight, poor milling yield, and unattractive visual appearance. Folds and flakes of outer pericarp material detract considerably from the appearance of the grain and mask the normal yellowish-brown color. Shriveled kernels have a large space at the head of the crease where the endosperm cells have failed to develop. The aleurone layer is irregular in shape, with cells varying in size and wall thickness. When shriveling occurs, irregular invagination of the outer meristematic layer takes place (Simmonds, 1974), distorting the aleurone layer and giving the seed a rough and unattractive appearance. The crease of triticale varies in depth among cultivars.

Triticale seed can be used as human food or livestock feed. A summary of work on nutritional quality (Bushuk and Larter, 1980) has shown that triticale grain is superior to wheat as food for humans and feed for livestock. The protein content of triticale ranged from 11.76% to 22.50%, levels higher than those of bread wheat. Lysine content for triticale varied from 2.55% to 3.74%, intermediate between that of wheat and rye (Villegas, McDonald, and Gilles, 1968). Feeding trials have shown the superior biological value of triticale over wheat. As a source of energy, triticale can replace wheat, barley, corn, or sorghum grain.

Where less than optimum feed performance has been obtained, decreased feed intake has often been observed. This may be attributed to the presence of ergot, the method of processing, or the presence of alkyl-resorcinal (Allee, 1974; Larter, 1974).

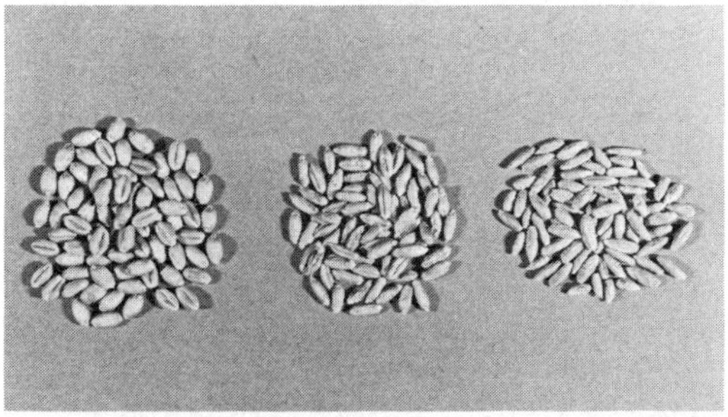

Figure 20–3. Comparison of triticale kernels (center) with wheat (left) and rye (right). Triticale kernel shape is wheatlike and is usually yellowish-brown in color. The color of triticale kernels may be masked by folds and flakes of outer pericarp material which can detract considerably from the appearance of the grain. Shriveled seed has been a problem throughout the development of triticale, and although excellent progress has been made, further improvement is desirable. Triticale kernels can vary from partially shriveled or wrinkled to fully plump. (Courtesy of Ontario Ministry of Agriculture and Food)

Triticale cultivars are often one or two weeks later-maturing than wheat. Direct-combine harvesting is possible when the grain has reached a safe storage level of 14% moisture. Cultivars with abundant vegetative growth and susceptibility to lodging may require swathing.

Kernel density of triticale was reported by Sinha and White (1982) to be about 20% less than that of wheat and 15% less than that of rye. Less vitreous triticale kernels are susceptible to premature germination, especially under moist, humid conditions.

Some triticale cultivars are difficult to thresh, and special attention must be given to proper setting of the combine to avoid loss of grain or cracking of kernels when removing the grain from the spikelets and the glumes from the kernels.

PRODUCTION PRACTICES

It is generally assumed that, because triticale is the product of combining wheat and rye, cultural practices adapted to wheat production will be suitable for triticale. Probably such an approach will not limit the yield performance of triticale, but the full genetic potential may not be realized. Optimum production practices for triticale have not yet been firmly established by experimentation.

A delay in the date of seeding of spring triticale reduced grain yield to a greater extent than for wheat (Larter, Kaltsikes, and McGinnis, 1971). When seeding date was delayed from April 20, the earliest date possible at Winnipeg, Manitoba, to May 5, the general date of seeding for the area, triticale yields were reduced 16% compared to 9% for wheat. In terms of comparative yield potentials of wheat and triticale, delay in seeding reduced the yield advantage of triticale over wheat from 34% to 24%. Early planting of spring triticale appears a critical aspect in production.

For winter triticale, the date of seeding varies with the locality and the preceding crop. Until more extensive research proves otherwise, the suggested seeding dates for winter triticale should coincide with those for winter wheat. Expressed otherwise, winter triticale should be sown at least four weeks before the average date for the end of the growing season in areas with little snow cover and where low temperatures are a major cause of winterkill. In areas where a heavy snow cover occurs and where snow-mold disease losses occur, seeding should occur at least seven weeks before the average date for the end of the growing season. Very early seeding of winter triticale could permit excessive fall growth, resulting in smothering and decay of the abundant vegetative growth or damage by the build-up of Hessian fly.

The optimum rate of seeding for triticale was 100 kg/ha (90 lb/ac) as compared to 68 to 102 kg/ha (62 to 92 lb/ac) recommended for wheat in western Canada (Larter, Kaltsikes, and McGinnis, 1971). As seeding rate increased, kernel weight was reduced. Both date and rate of seeding tests were conducted over a two-year period, 1969 to 1970, for Winnipeg, Manitoba.

In the southeastern United States the optimum rate of seeding was reported to be 75 and 100 kg/ha (67 and 90 lb/ac) for grain and forage production, respectively, when planted in 12.5-cm (5-in.) row spacings (Bishnoi, 1980) when tested over a two-year period, 1975 and 1976. A higher seeding rate for triticale over wheat and rye was justified because of the larger, more shriveled seed of triticale. Clipping effects on triticale and wheat grain yields were inconsistent, but they reduced rye yield considerably.

Winter triticale should be sown at a depth of 2.5 to 5 cm (1 to 2 in.), depending on soil moisture levels. Shallow late seeding of winter triticale can result in winter damage due to heaving.

Soil-fertilization studies with triticale have not resulted in firm recommendations because of the inability of triticale to tolerate the optimum level of fertilization for wheat without lodging. Where lodging is a problem, a total of 80 kg (71 lb) is recommended. This is about 10% less than the recommended rate for wheat.

Fertilizer applications should be based on soil tests along with information on previous crops and management. Phosphate and potash recommendations are similar to those for other cereals.

PHYSIOLOGY

Hexaploid triticale was found to have the highest net photosynthesis and the best water-use efficiency when compared to durum, common bread wheat, and octoploid triticale from observations in France (Planchon, 1979).

In studies on temperature effects Szabo (1972) observed that at 5°C (41°F) germination under laboratory conditions began on the ninth or tenth day; at 10°C (50°F) on the sixth day; at 15°C (39°F) and over on the third day. At 0°C (32°F) no seeds germinated. The optimum germination temperature was about 20°C (68°F). Differences among cultivars were observed. The cardinal points of germination were given by Szabo (1972) as follows:

Optimum 20°C (68°F).
Maximum 35°C (95°F).
Minimum 5°C (41°F).

Dormancy of triticale appeared to be relatively short, owing to the high enzyme activity of the grain (Szabo, 1972). Octoploid triticale has a tendency to sprout at harvest time under moist, humid conditions (Muntzing, 1979).

ADAPTATION AND PERFORMANCE

Centuries of evolutionary pressure and years of plant breeding and selection effort have resulted in wide adaptation among the major cereal crops of the world. Worldwide testing programs have revealed that triticale is notoriously poor in adaptation, but efforts to broaden the gene base are improving adaptation. When triticale is subjected to changes in latitude, elevation, temperature, daylength, fertility, moisture levels, and probably other factors, it exhibits a crippling intolerance to these changes. The CIMMYT breeders, Zillinsky and Borlaug, attributed this poor adaptation to a lack of evolutionary adaptation (Wolff, 1976). To make up for this lack of evolutionary development, it will be necessary to establish populations as genetically diverse as possible and have them grown and selected in various environments around the world. The development of cultivars adapted to specific local environments will eventually indicate the potential this crop has in diverse locations. By bringing the best and most widely adapted cultivars together and rehybridizing them, diverse, widely adapted material may eventually be produced. The system of international testing and exchange of material is essential in this regard.

Triticale, like its rye parent, is well adapted to the poor soil areas of the world. In Sweden and Spain, triticale has replaced rye to a substantial degree on sandy soils. Under similar conditions in the high plateau region of Mexico, triticale has outyielded wheat.

Aluminum is an important growth-limiting factor for plants in many acid soils below *p*H 5.0, and it can limit growth at *p*H levels as high as 5.5. In certain acidic tropical soils characterized by seasonal heavy rainfall and often by aluminum and copper toxicities, triticale has exhibited an advantage in adaptation over wheat. Two of three triticale cultivars tested (Aniol, Hill, and Larter, 1980) were highly tolerant to aluminum. Mugwira et al. (1981) reported large differences in the aluminum tolerances of triticale obtained from different geographic regions and suggested that selection for this attribute might be fruitful.

Some triticale cultivars are well adapted to cultivation under cool growth temperatures found at high elevations of tropical mountain areas. Triticale has shown higher tolerance to frost during the early vegetative stage than wheat.

In Colombia, Ethiopia, Northern India, and Brazil, triticale is well adapted and has outyielded wheat, especially on acid soils.

In areas where cereal diseases restrict successful wheat production, the resistance of triticale to smuts at high elevations provides an advantage. Triticale is not free of disease, however. Symptoms of common cereal diseases can be observed wherever the crop is grown. Evidence of generalized disease resistance in triticale offers new possibilities for a more stable type of resistance against leaf rust in both hexaploid and octoploid triticale (Fuentes, 1973).

To be commercially competitive with other cereal grains, it is probably not sufficient that triticale just be equal to existing grain species in agronomic and quality characteristics. Triticale has been accepted because of its specific adaptation to acidic, droughty, or other extreme conditions under specific conditions. Universal adaptation will occur if it excels in at least one major attribute.

The chapter on octoploid triticale may not be closed as yet. As a rule, octoploid triticale has good winter hardiness, grows well on light soils, and has large kernels. In general, however, octoploid triticale has not reached a level of development sufficient for commercial release. An exception to this statement is found on the Yunnan-Kweichow plateau at 2000 m (6562 ft) in southwestern China, where octoploid triticale started to be grown commercially in 1973 (Muntzing, 1979). In this area of poor soil and high altitude, octoploid triticale clearly is superior to wheat or rye.

The expanded effort on winter triticale promises to introduce new sources of genetic diversity from winter rye and winter wheat that may contribute to wider adaptation.

CONCLUSIONS

The advancing of triticale from a newly developed species with numerous cytological problems to the status of a world crop is a notable achievement. It has meant compressing thousands of years of evolution into mere generations. This was achieved by painstaking breeding and selection against seemingly insurmountable problems.

Triticale was synthesized by combining wheat and rye genomes, making a total of fcur, A, B, D, R. Genomic reconstruction using these four genomes offers a substantial number of different combinations. Perhaps the most desirable recombination has not been produced.

Triticale has been viewed as a crop to be grown in areas unsuited for wheat, just as sorghum was once viewed as a crop to be grown in areas unsuited for corn. It is expected that initially triticale may play such a secondary role.

Progress in overcoming the considerable problems of early triticale cultivars, which has increased yield performance to a level comparable to that of bread wheats, has already started to push triticale into a greater role. There will be a continuing effort to improve hexaploid triticale. Perhaps combining hexaploid triticale with hexaploid wheat-Agropyron hybrids may lead to triticale cultivars with superior forage-production capacity (Jenkins, 1974).

Throughout its history, triticale has been a crop of future promise. In the mid 1980s, triticale in many areas of the world remains a promise on the verge of fulfillment. Progress confidently can be expected.

Landholding patterns and increased population pressure will continue to force people to attempt crop production on marginal areas where rainfall is unreliable and poorly distributed and the danger of drought is great. Food requirements will continue to increase. Necessity may result in greater adaptation of triticale to geographically limited areas.

Triticale is no longer a crop for the future, for production on a world scale has begun. The extent to which this crop is adopted will reflect the progress made. Very few studies have been conducted on heritable traits; cytological problems have not been solved; production techniques mainly reflect those of its wheat parent; quality differs from that of wheat, and perhaps unique milling and baking procedures are needed. The rate at which the crop expands will reflect the degree of progress in finding answers. Opportunities for progress in triticale are abundant.

Questions

1. *a.* List the strengths and weaknesses of triticale in comparison with wheat and rye.
   b. Have the original objectives of combining the best features of wheat and rye been achieved?
2. What are the common disease problems of triticale? (See Zillinsky, 1973.)
3. Genetic engineering and biotechnology promise exciting new genetic combinations in the future. Based on the record of development for triticale, what hope for rapid and dramatic success exists for combining widely different genetic forms either by combining different species or by adding chromosomes or specific genes to another species?
4. *a.* Review the changing status of triticale in the world in terms of yield performance and area of production. Note that Bushuk and Larter (1980) estimated world production to be 150,000 to 170,000 ha (370,000 to 420,000 ac), excluding the People's Republic of Cina. Production in major countries was as follows:

USSR	60,000 ha (148,000 ac)
USA	50,000 ha (124,000 ac)
Canada	30,000 ha (74,000 ac)
Argentina	10,000 ha (25,000 ac)
Hungary	10,000 ha (25,000 ac)

   b. Comment on the 1973 prediction of Shebeski that by 1990 "triticale will have begun to compete seriously with the bread wheats as one of the world's most important crops."
5. Triticale lacks wide adaptation because natural evolutionary forces have been limited. Are cereal crops such as wheat, rye, corn, and so on undergoing evolutionary processes today?

6. Speculate with reasons on the amount of heterosis that might be expected from F_1 hybrid triticale.
7. To a limited extent, triticale is a commercial crop. Speculate on the logistical problems that arise when a new crop enters commercial channels.
8. What problems are associated with the development of winter triticale in comparison with that of spring triticale?
9. *a.* Consider the different impact of human-directed evolution with that exerted by the law of the survival of the fittest on the development of a species in terms of:
   (1) Adaptation.
   (2) Competitive ability.
   (3) Reproductive ability.
   (4) Yield performance.
   (5) Success in terms of agronomic performance.

   b. Comment on the concept of plant breeders acting as "directors of evolution" in terms of a science or an art.

References

Allee, G. L. 1974. "Triticale as a Feed." (In) C. C. Tsen (ed.), *Triticale: First Man-Made Cereal.* St. Paul, Minn.: American Association of Cereal Chemistry. pp. 272–279.

Aniol, A., R. D. Hill, and E. N. Larter. 1980. "Aluminum Tolerance of Spring Rye Inbred Lines." *Crop Science* 20: 205–208.

Baum, B. R. 1971. "The Taxonomic and Cytogenetic Implication of the Problem of Naming Amphiploids of *Triticum* and *Secale.*" *Euphytica* 20: 302–306.

Bishnoi, U. R. 1980. "Effect of Seeding Rates and Row Spacing on Forage and Grain Production of Triticale, Wheat and Rye." *Crop Science* 20: 107–108.

Briggle, L. W. 1969. "Triticale—A Review." *Crop Science* 9: 197–202.

Bushuk, W., and E. N. Larter. 1980. "Triticale: Production, Chemistry, and Technology." (In) *Advances in Cereal Science and Technology.* Vol. III. St. Paul, Minn.: American Association of Cereal Chemists. pp. 115–157.

Eigsti, O. 1938. "A Cytological Study of Colchicine Effects in the Induction of Polyploidy in Plants." *Proceedings of the National Academy of Science USA* 24: 56–63.

Fuentes, S. F. 1973. "Triticale Diseases." *Triticale Breeding and Research at CIMMYT. A Progress Report.* International Maize and Wheat Improvement Center. Mexico. Research Bulletin No. 24: 34–38.

Jenkins, B. C. 1974. "Hexaploid Triticale: Past, Present and Future." (In) C. C. Tsen (ed.), *Triticale: First Man-Made Cereal.* St. Paul, Minn.: American Association of Cereal Chemists. pp. 56–61.

Kiss, A. 1965. "Improvement of the Fertility of Triticale." *Acta Agronomica Hungaricae* 14: 189–201.

Larter, E. N. 1974. "A Review of the Historical Development of Triticale." (In) C. C. Tsen (ed.), *Triticale: First Man-Made Cereal.* St. Paul, Minn.: American Association of Cereal Chemists. pp. 35–52.

Larter, E. N., et al. 1970. "Rosner, A Hexaploid *Triticale* Cultivar." *Canadian Journal of Plant Science* 50: 122–124.

Larter, E. N., and J. P. Gustafson. 1980. "Triticale." (In) W. R. Fehr and H. H. Hadley (eds.), *Hybridization of Crop Plants.* Madison, Wis.: American Society of Agronomy and Crop Science Society of America. pp. 681–694.

Larter, E. N., P. J. Kaltsikes, and R. C. McGinnis. 1971. "Effects of Date and Rate of Seeding on the Performance of Triticale in Comparison to Wheat." *Crop Science* 11: 593–595.

Larter, E. N., T. Tsuchiya, and L. E. Evans. 1969. "Breeding and Cytology of Triticale." *Proceedings of the Third International Wheat Genetics Symposium.* Canberra, Australia. pp. 213–221.

Lorenz, K. 1974. "The History, Development, and Utilization of Triticale." *CRC Critical Reviews in Food Technology* (Chemical Research Company Press, Inc.) 5: 175–280.

Mugwira, L. M., et al. 1981. "Aluminum Tolerance of Triticale and Wheat Cultivars Developed in Different Regions." *Agronomy Journal* 73: 470–475.

Muntzing, A. 1939. "Studies on the Properties and the Ways of Production of Rye-Wheat Amphidiploids." *Hereditas* 25: 387–430.

———. 1974. "Historical Review of the Development of Triticale." (In) *Triticale*. Proceedings of the International Symposium. El Batan, Mexico: International Development Research Center. (IDRC). No. 024e. pp. 13–30.

———. 1979. *Triticale. Results and Problems*. Advances in Plant Breeding, Supplement 10 to the Journal of Plant Breeding. Berlin and Hamburg: Published by Paul Parey. 103 pp.

O'Mara, J. G. 1948. "Fertility in Allopolyploids." *Records of the Genetic Society of America* 17: 52.

Pissarev, V. 1966. "Different Approaches in Triticale Breeding." *Proceedings of the Second International Wheat Genetic Symposium*. Lund, Sweden. 1963. *Hereditas Supplement* 2: 279–290.

Planchon, C. 1979. "Photosynthesis, Transpiration, Resistance to CO_2 Transfer, and Water Efficiency of Flag Leaf of Bread Wheat, Durum Wheat and Triticale." *Euphytica* 28: 403–408.

Poysa, V. W., C. J. Knoblauch, and E. Reinbergs. 1981. "OAC Wintri Winter Triticale." *Canadian Journal of Plant Science* 61: 995–996.

Sanchez-Monge, E. 1959. "Hexaploid Triticale." *Proceedings of the First International Wheat Genetics Symposium*. University of Manitoba, Canada. pp. 181–194.

Shebeski, L. H. 1958. "Speculations on the Impact of the D Genome." *Proceedings of the First International Wheat Genetics Symposium*. Winnipeg, Manitoba. pp. 237–241.

Simmonds, D.H. 1974. "The Structure of the Developing Kernel and Mature Triticale Kernel." (In) C. C. Tsen (ed.), *Triticale: First Man-Made Cereal*. St. Paul, Minn.: American Association of Cereal Chemists. pp. 105–121.

Sinha, R. N., and N. D. G. White. 1982. "Moisture Relations of Stored Triticale and Its Susceptibility to Infestation by Mites and Infection by Microflora." *Canadian Journal of Plant Science* 62: 351–360.

Sisodia, N. S., and R. C. McGinnis. 1970. "Importance of Hexaploid Wheat Germ Plasm in Hexaploid Triticale Breeding." *Crop Science* 10: 161–162.

Sodkiewicz, W. 1984. "Amphiploid *Triticum monococcum*. × *Secale cereale* L. (AARR)—A New Form of Tetraploid Triticale." *Cereal Research Communications* 12(1): 35–40.

Szabo, L. Gy. 1972. "The Germination Physiology of Triticale." *Acta Agronomica Academiae Scientarum Hungaricae* 21: 219–222.

Villegas, E., C. E. McDonald, and K. A. Gilles. 1968. *Variability in the Lysine Content of Wheat, Rye, and Triticale Proteins*. International Maize and Wheat Improvement Center, Mexico. Research Bulletin No. 10. 32 pp.

Wilson, A. S. 1875. "On the Fertilization of Cereals." *Transactions and Proceedings of the Botanical Society of Edinburgh* 12: 237–242.

Wolff, A. 1976. "Wheat × Rye = Triticale." *CIMMYT Today*. No. 5. International Maize and Wheat Improvement Center, Mexico. 15 pp.

Yeung, K. C., and E. N. Larter. 1972. "Pollen Production and Dissemination Properties of Triticale Relative to Wheat." *Canadian Journal of Plant Science* 52: 569–574.

Zillinsky, F. J. (ed.). 1973. *Triticale Breeding and Research at CIMMYT. A Progress Report*. International Maize and Wheat Improvement Center, Mexico. Research Bulletin No. 24. 84 pp.

Zillinsky, F. J. 1974. "The Development of Triticale." *Advances in Agronomy* 26: 315–348.

———, and N. E. Borlaug. 1971. *Progress in Developing Triticale as an Economic Crop*. International Maize and Wheat Improvement Center, Mexico. Research Bulletin No. 17. 27 pp.

21
Pearl Millet

A PERSPECTIVE

Pearl millet is the traditional crop of the semiarid tropics, where it is the staple food for millions. The crop commonly is grown under the most difficult farming conditions, including those in drought-stricken areas where soil fertility is low and where food supplies are dependent on rainfall.

Relative to other cereal grains, pearl millet ranks sixth in importance, following wheat, rice, corn, barley, and sorghum. In India, a country approaching a population of one billion, pearl millet is the fourth most important cereal after rice, wheat, and sorghum. Together, India and Africa produce 93.2% of world production (FAO, 1982).

At one time sorghum had approximately the same distribution as that of pearl millet, but research efforts have extended the range of sorghum distribution substantially. The implication is that a concentrated research effort on pearl millet could contribute to an increased area of adaptation, higher yield performance, and possibly a greater food-producing role in the future. Indeed, as crop production is stretched into marginal land areas, considered to be of little agricultural value because of low moisture and low fertility, pearl millet can be grown, and it warrants a more concentrated research effort. The International Crops Research Institute for the Semi-Arid Tropics (ICRISAT) (Chapter 14) was established at Hyderabad, India, in 1972 to conduct international improvement programs for crops, including pearl millet. The identification of stable stress resistance and wide adaptability has received major attention along with identification of high yield potential. Within a decade, ICRISAT contributed substantially to pearl millet improvement, focusing attention on this crop that may lead to increased research and prestige. Rachie and Majmudar (1980) reviewed the work of ICRISAT since its establishment (pp. 220–231).

In the past pearl millet largely has been overlooked as a crop, despite its agricultural importance. In terms of crop physiology, genetics, cytogenetics, and crop production, much less research information is available than for other cereal crops. Jauhar (1981) gave four reasons why pearl millet largely has been overlooked:

1. Pearl millet has been relegated largely to use in droughty, infertile areas and has come to be viewed as a poor man's crop, lacking prestige and appeal.
2. A lack of prestige and appeal has resulted in pearl millet's being viewed as a crop of secondary

importance. As a result it has not competed for research funding with other crops such as corn, wheat, or rice.

3. Its potential and challenge as a research tool was not appreciated until the early 1980s.
4. Confusion and controversy on classification and nomenclature since the mid-1700s, whereby pearl millet has been treated as a member of six different genera, may have had an adverse impact on research.

Pearl millet is an extremely variable species and is regarded as a dual-purpose crop. Its forage potential, outlined in Chapter 8, gives the genus *Pennisetum* the highest potential of all the millets and miscellaneous cereals. In hot, dry areas of Australia and in the southern part of the United States, pearl millet is recognized as a valuable forage and grain crop with considerable potential.

The attributes and deficiencies of pearl millet are described briefly in the sections that follow.

POSITIVE ASPECTS

1. Pearl millet has the efficient C_4 photosynthetic pathway found in corn and sorghum (refer to Chapter 2, outlining the superior features of C_4 over C_3 plants).

2. Of the eight major cereals, millet has the highest water-use efficiency (Table 2-5). Eight cultivars had an average transpiration ratio of 280, compared to the next lowest ratio found in sorghum of 305, and nearly 2½ times better than that of rice. Expressed otherwise, millet produced 3.64 units of dry matter for every 1000 units of water consumed. The comparable figure for rice was 1.47 units. The reasons for such high water-use efficiency was believed by Jauhar (1981) to be related to fewer stomata, or to stomatal closure under conditions of moisture stress, even in the middle of the day (Begg et al., 1964).

3. A high water-use efficiency coupled with the C_4 photosynthetic pathway means that pearl millet has a remarkable potential to grow under hot, dry agricultural conditions. In sub-Saharan Africa, harvests of pearl millet are obtained with as little as 25 cm (9.8 in.) of annual rainfall (Brunken, 1977). Generally, pearl millet is grown in areas of Asia and Africa with an annual rainfall of 43 to 50 cm (17 to 20 in.). A crop with this drought and heat tolerance has an enormous potential to help fill the food gap of semiarid tropical areas of the world.

For the entire Indian subcontinent, Africa, and Asia, pearl millet is an extremely important cereal food crop, and it has potential for expanding production into areas considered too dry and infertile for sorghum or corn.

4. Although pearl millet is grown extensively on poor, depleted soils, it responds extremely well to fertilizer nutrients. Economic responses to nitrogen fertilizer applications of 448 kg/ha (400 lb/ac) for forage production (Hart and Burton, 1965) and 120 kg/ha (107 lb/ac) for grain yield have been demonstrated when other factors do not limit yield. The crop has a remarkable ability to extract fertilizer nutrients from soils of low fertility.

5. The distinctly protogynous nature (described in Chapter 9) of pearl millet makes this crop ideal for hybrid development.

6. The response of pearl millet to heterosis is outstanding. The discovery of cytoplasmic male sterility in the United States (Burton, 1958) and in India (Athwal, 1966) resulted in the rapid development and adaptation of hybrid pearl millet. F_1 hybrids in India have yielded nearly twice as much as the best standard cultivars (Burton and Powell, 1968).

7. The high level of heterosis obtained over open-pollinated cultivars makes pearl millet an ideal crop for research directed toward an understanding of heterosis. Pearl millet has a low chromosome number and has genetic diversity equal to that of corn and sorghum, but none of this genetic diversity has been mapped. An opportunity and challenge exists with pearl millet to understand the important phenomenon of heterosis and to use the diverse genetic base and broad background of pearl millet for further improvement in hybrid vigor.

8. For the taxonomist and cytologist, pearl millet offers a fascinating and as yet unraveled array of evolutionary problems and phylogenetic

relationships. The crop has a low number ($2n = 14$) of chromosomes but of large size, which offers and attractive cytological tool for observing chromosome pairing and **chiasma** frequencies.

9. The low chromosome number and outbreeding nature of pearl millet makes it a promising crop for polyploid breeding. Polyploidy in pearl millet may stand in contrast to that in other grain crops, such as barley and rye, where most of the artificially induced polyploids have not been of direct practical value.

10. Haploids have been reported to occur spontaneously and can be isolated in the progeny of trisomics, or else can be induced artificially by treating pearl millet seeds with mutagenic agents (Jauhar, 1970 and 1981). In addition to their value in fundamental research, haploids are useful to achieve homozygosity and the acceleration of a breeding program.

11. The genetic variability found in over 140 species of this genus offers a vast potential for improvement through breeding and selection. The many subspecies of *Pennisetum* are interfertile with pearl millet and form what Harlan and de Wet (1971) called a "primary gene pool" or essentially a single, composite reproductive unit. The enormity of such a gene pool offers considerable scope for improvement through plant breeding and selection.

12. Equally exciting as the opportunities for intraspecific crosses are those for interspecific crosses. Although such crosses are difficult to make and may produce sterile progeny, colchicine-induced chromosome doubling may result in the successful development of an amphiploid. For example, a cross between pearl millet and its tetraploid counterpart, napier grass (*Pennisetum purpureum*), to produce a triploid plant produces sterile progeny, but with high F_1 vigor for forage production. Vegetative propagation of this triploid plant is too restrictive, and to be commercially acceptable it must be propagated by seed.

13. Dwarfing genes are available (Burton and Fortson, 1966) that have not been exploited to more effectively distribute photosynthate into grain to maximize yields. Dwarf hybrids, however, have helped reduce lodging problems, especially under high-nitrogen fertility conditions. For mechanized harvesting, dwarf stature is most desirable. Dwarfing may be a key attribute for the expansion of pearl millet into mechanized areas of the world.

14. Pearl millet is the traditional crop of the semiarid tropics, which encompasses much of the developing world. The crop offers a means of conveying research findings to the traditional farmer in these areas.

The high level of heterosis, the ease of hybrid production, the low seeding rate, and the need to increase yield may prove to be the ideal combination needed to introduce F_1 hybrid production to all traditional farmers.

15. Pearl millet is a dual-purpose crop, but perhaps more than 95% is grown primarily for grain. In the southern United States, pearl millet is regarded as the highest-yielding summer annual forage species (Hart and Burton, 1965). Pearl millet is quick to establish, recovers quickly after grazing or clipping, and produces highly palatable and nutritious forage. These features make pearl millet a substitute for a failed cereal crop to provide either forage or grain. Pearl millet cultivars are available with a relatively short growing season.

Most of the research work in Africa and Asia has been directed to the grain potential of pearl millet. Hybrid development and results of trials in other areas of the world on forage production have provided new emphasis and interest in improving the fodder-producing aspects in Israel, South Africa, Australia, Italy, Pakistan, India, and the United States. Breeding and selection offers promise to improve forage production and quality.

In third-world countries, pearl millet stalks are used for fencing, screens, building materials, and fuel.

16. The possibility exists in pearl millet for the development of a perennial crop through interspecific hybridization. If a perennial form of pearl millet can be produced, it may have value in dryland areas where cold hardiness is of no consequence. Perennial pearl millet does not face the winter-survival challenge required of other cereals grown at high latitudes.

17. For the entire Indian subcontinent, Africa, and Asia, pearl millet is an extremely important food crop. The potential of pearl millet in dryland areas in other parts of the world is vast. Climatic changes, depleting soil water reserves, and problems with salinization from irrigation may mean an expansion in the geographic distribution of pearl millet.

Although its attributes and potential are impressive, and though it is an ancient crop, in many respects pearl millet can be regarded as a frontier crop. Its full geographic potential may not have been realized; it has not been the object of concentrated research; yet it appears to have an enormous potential.

Problems also exist with this crop, and these are described briefly below.

PROBLEM AREAS

1. Pearl millet is the lowest-yielding of the cereal crops. Results from 32 test locations in 11 countries in 1976 showed an average yield of 1876 kg/ha (1675 lb/ac) with a range of 804 to 4221 kg/ha (718 to 3769 lb/ac) (Rachie and Majmudar, 1980, p. 229).

Low yields may be a reflection of the harsh conditions under which pearl millet is grown. Other cereals, however, are frequently grown under harsh conditions at high latitudes that include cool, short seasons. World pearl millet grain-yield comparisons (Table 21-1) reveal that the major cereal crops generally outyield pearl millet (see also Tables 14-5 and 14-6).

It may be that agriculture in the pearl millet producing areas of Africa and Asia has not been made a priority, and that breeding, production, crop physiology, genetics, and cytogenetic studies are lacking.

2. A lack of research may be expressed as deficiencies in many aspects of the production system of pearl millet. Although measurements are limited, the harvest index of pearl millet is very low. Ten pearl millet cultivars, four F_1 hybrid and six regular cultivars, grown under irrigated and nonirrigated conditions near Manhattan, Kansas, in the United States had an average harvest index of 14.1 (Table 21-2). Harvest index values ranged from 5.6 to 19.0 (Singh, Kanemasu, and Singh, 1983). A review of Chapter 4 and Tables 4-4, 4-5, and 4-6 shows a range of harvest index values for wheat from 33 to 52. Values can be increased in dwarf cultivars

Table 21–1

World Pearl Millet Grain-Yield Comparisons with Other Cereals, Averaged over a Three-Year Period, 1979–1981, and Expressed in kg/ha (lb/ac)

Crop	*World*	*Africa*	*Asia*	*Europe*	*USSR*	*Developed*	*Developing*
Pearl millet	662 (591)	605 (540)	706 (630)	1627 (1453)	586 (524)	596 (532)	666 (595)
Corn	3252 (2904)	1310 (1196)	2289 (2043)	4612 (4118)	2857 (2551)	5293 (4726)	1903 (1699)
Sorghum	1386 (1237)	720 (643)	963 (860)	3624 (3236)	1138 (1016)	3321 (2965)	1071 (957)
Wheat	1888 (1686)	1096 (979)	1682 (1502)	3586 (3202)	1549 (1383)	2072 (1850)	1622 (1448)
Rice	2767 (2471)	1766 (1576)	2815 (2513)	5062 (4520)	3967 (3542)	5243 (4681)	2680 (2393)
Barley	1965 (1755)	844 (753)	1494 (1334)	3378 (3016)	1342 (1198)	2134 (1905)	1297 (1158)
Oat	1682 (1501)	486 (434)	1476 (1381)	2731 (2438)	1254 (1120)	1710 (1527)	1275 (1139)
Rye	1612 (1440)	230 (205)	1315 (1174)	2327 (2078)	1200 (1071)	1657 (1479)	1240 (1107)

Source: FAO Production Yearbook 1981 (FAO, 1982).

Table 21-2
Harvest Index Values for Both Irrigated and Nonirrigated Pearl Millet Cultivars Grown near Manhattan, Kansas, in the 1980 Season

Cultivar	*Harvest Index Values*		*Plant Height*	
	Nonirrigated	*Irrigated*	*cm*	*in.*
F_1 hybrids:				
2094 × 7024	11.5	16.8	82	32
2221 × 7024	16.8	19.0	77	30
2094 × 4104	10.0	17.1	95	37
2221 × 4104	14.9	13.9	105	41
Average	13.3	16.7	89.8	35
Reg. cultivars:				
HMP 1700	17.3	18.6	86	34
HMP 550	16.7	17.7	98	39
Senegal bulk	11.8	10.8	123	48
HMP 600	17.4	15.9	135	53
Serere-3A	9.5	11.0	181	71
HMP 559	6.3	5.6	219	86
Average	13.2	13.3	140.3	55.2

Source: Singh, Kanemasu, and Singh (1983).

to a high of possibly 60%. Short pearl millet cultivars had the highest harvest index values (Table 21-2), but they are desperately low compared to those of other cereal grains.

3. Generally the protein concentration in pearl millet seed is low and deficient in lysine, valine, and histidine (see Chapter 12). Improvement is possible, but dramatic increases parallel to those in other grain crops have not been achieved.

4. Pearl millet is regarded as a dual-purpose crop, and often the same cultivars are used for grain or forage. Improvement in grain yield and harvest index may be achieved by an approach that concentrates on pearl millet cultivar development for either grain or forage.

5. A systems approach that attempts to optimize production of pearl millet has not been undertaken, possibly because it is viewed as a crop capable of producing with minimal inputs.

CLASSIFICATION AND ORIGIN

The term millet is broadly applied to over 140 species belonging to the genus *Pennisetum*. The generic name is a compound form of the Latin words meaning feather and bristle, a term that describes the floral aspects of the spike of some millet species.

The phylogenetic relationships among the millets are challenging, because species have chromosome numbers as multiples of $x = 5, 7, 8$, and 9, excluding aneuploids. Phylogenetic confusion has resulted in several different scientific names being assigned to pearl millet. Burton and Powell (1968) confessed to changing their terminology from *Pennisetum glaucum* to *Pennisetum typhoides* and then to *Pennisetum americanum* (L.) K. Schum (Burton, 1977). Rachie and Majmudar (1980) used the species name of *Pennisetum typhoides* in the text, although *Pennisetum americanum* is used in the Foreword. Rachie and Majmudar (1980) listed 19 synonyms for *Pennisetum typhoides*, 17 common English names plus a number of local names.

The common names of pearl millet include pearl, bulrush, cattail, or spiked millet. In Hindu the term is bajra (Reddy and Reddy 1978; Shekhawat, Chundawat, and Shrotriya 1970). It is kambu in Tamil, dukhn in Arabic, and mil a chandelles in west North Africa. A total of 55 names were listed by Rachie and Majmudar (1980) that are used around the world.

Five species names have been used for cultivated pearl millet:

1. *Pennisetum typhoideum* L. C. Rich.
2. *P. typhoides* (Burm. f.) Stapf and C. E. Hubbard.
3. *P. spicatum* (L.) Koern.
4. *P. americanum* (L.) K. Schum.
5. *P. glaucum* (L.) R. Br.

In a comprehensive review of the subject Terrell (1976) declared the correct name as

Pearl millet—*Pennisetum americanum* (L.) Leeke

and stated that the combination is not attributed to K. Schumann as listed above.

Pearl millet originated in West Africa, as described in Table 3-5, but some controversy exists with regard to the specific region of Africa. On the basis of present-day distributions, Brunken, de Wet, and Harlan (1977) suggested that the Sahel zone of West Africa appears to be the original home of pearl millet. The Sahel zone stretches from Western Sudan to Senegal. It must be recognized that selection by humans directed its evolutionary development into several morphologically diverse forms. Diversity was aided by its cross-pollinated character.

ADAPTATION

Pearl millet is best adapted to the tropics, but it does well also in hot areas of the temperate zone. The crop is found on every continent of the world and will produce grain on sandy or rocky soils that are too acid, too dry, and too infertile for sorghum or corn. Under favorable environmental conditions pearl millet has the capacity to equal or exceed corn or sorghum in forage yield, and possibly to compete in grain yield.

In Africa, China, and India, pearl millet is grown primarily as a grain crop for human food purposes. In areas too hot and dry for other cereals, pearl millet forms the principal food-producing grain crop.

Cardinal temperatures for germination, growth, flowering, and seed set have not been established. Pearl millet has been known to flower and set seed at temperatures between 25° and 45°C (77° and 113°F) (Burton, 1980).

BOTANY OF PEARL MILLET

Pearl millet is a robust, very rapid growing, erect, tropical, annual, cross-pollinated bunchgrass. The crop can be grown for grain but will give high yields of good-quality forage when properly managed. The forage is free of cyanoglucosides or hydrocyanic acid at all stages of growth. Pearl millet is, along with corn and sorghum, a large-grain crop.

Pearl millet is an extremely variable species. Culms are solid and pithy and, until the discovery of dwarf forms, plants were described as being 2 to 5 m (6.5 to 16 ft) in height. Dwarfs 1 to 1½ m (39 to 60 in.) in height were described by Burton and Fortson (1966). Burton (1977) described modern pearl millet cultivars as ranging from 0.5 to 3 m (1.6 to 10 ft) tall. Culms may be thick or slender, simple or branched. Plant parts such as the sheath, leaves, and nodes may be smooth or hairy and range in color from green, purple, and red to golden yellow.

Plants usually have six to 12 internodes with a leaf sheath arising from each mode, nine internodes being most common. The initial above-ground internode is the shortest, the uppermost or peduncle the longest. In addition to the above-ground internodes there is a group of very closely spaced internodes underground, giving rise to primary tillers.

Leaf sheaths are split and have a prominent ligule (5 mm) at the juncture of the leaf sheath and blade. Leaf blades are up to 1.5 m (5 ft) long and 7 cm (2.8 in.) wide and are long and pointed with small sawlike teeth on the margins. Leaves have a prominent midrib, often white and often pubescent throughout.

The inflorescence of pearl millet consists of a single, terminal, dense, cylindrical, spikelike ear somewhat tapering toward the tip. The head is mostly unbranched, although forking or branching may occur when an interspecific cross is made with napier grass. Botanically, the head is a false spike and ranges in length from 2.5 to 205 cm (1 to 80 in.) and in width from 0.8 to 5.5 cm (0.3 to 2.2 in.). A rosette of bracts consisting of bristles and spikelets united at the base and known as an **involucre** subtends a flower cluster arising from the central rachis. Involucre bracts may drop off or persist at maturity. The involucre itself is borne on stalks about 2.5 cm (1 in) in length and exhibits fine hairs to fingerlike projections or bristles. One to nine fertile spikelets, 3 to 7 mm long and borne on 2-mm-long pedicels or rachillae, are present in each involucre. A single spikelet is really a secondary spike having one upper and one lower floret. The lower floret is staminate or sterile; the upper perfect or fertile floret has three anthers and a pistil with two feathery stylar branches enclosed between the lemma and palea. The anthers are large enough for effective cross-pollination.

Pearl millet is cross-pollinated because of its protogynous nature, but it may be self-pollinated if heads are enclosed in pollen-proof bags prior to style emergence. Under natural conditions one tiller head that reaches anthesis before other tillers on the same plant can pollinate late tillers, thereby resulting essentially in self-pollination. Burton (1974) observed self-pollination as high as 31%. On large heads, later-emerging stigmas may be pollinated by anthers on the same head, as pollen is shed over a four- to seven-day period on any one head. Stigmas remain receptive for three days, and pollination is accomplished primarily by wind, although Leuck and Burton (1966) observed occasional cross-pollination by insects.

In protogynous species, the stigma matures before the anthers of the same flower. Generally, one day after the sigmas have emerged the anthers start emerging from the center of the head toward the tip. Anthers emerge in two distinct waves. The first wave involves the perfect flowers; the second wave includes the staminate flowers. Under arid conditions in Rajasthan, India, first stamens appeared two to three days after the stigma had emerged, and at a temperature of 25°C (77°F) it took two days to complete stigma emergence on one head (Bhatnagar and Kumar, 1960). The length of the head was a factor in determining the total time required (Burton and Powell, 1968). The time lag between stigma and anther extrusion is influenced by temperature and is speeded up or delayed by temperatures above or below 25°C (77°F).

Pearl millet anthers have a tuft of fine hair on the tip of each. The function of these hairs is uncertain, but they may reduce the speed of anther release. Protogyny occurs in species with and without lodiculae despite a popular view that the absence of lodiculae is associated with protogyny.

Pollen of pearl millet normally remains viable for an extended period of up to a few hours, perhaps because of its double wall structure (Jauhar, 1981). Under field conditions pollen remained viable for seven hours, but was most effective for up to five hours. Pollen stored at 4.5°C (40°F) for three weeks gave 80% as good seed set as fresh pollen (Cooper and Burton, 1965).

Flowering on plants with many tillers occurred over a seven- to 21-day period (Rachie and Majmudar, 1980) under irrigated conditions and for a 12-day period under rain-fed conditions. Stigmas emerge when mature, regardless of whether or not the head has emerged from the leaf sheath, thereby restricting seed set because of a lack of pollination. Anthesis starts from the upper third of the head and proceeds toward the base.

Apomixis (review Chapter 5) commonly occurs in several species of Pennisetum, especialy in higher polyploids (Jauhar, 1981). A cross between sexual pearl millet and an apomictic millet, fountain grass (*Pennisetum setaceum*) produced an apomictic hybrid (Hanna, 1979). Apomixis generally is under simple genetic control, and if it can be transferred to hybrids with desired heterozygosity and gene combinations, it could fix heterosis and be a very desirable feature (Hanna and Powell, 1973). *Pennisetum orientale* also has apomictic genes, perennial growth habit, pest resistance, and drought tolerance. A cross with *P. americanum* produced a facultative apomict (Hanna and Dujardin, 1982).

Seeds range in color from light gray, deep gray, and pearly amber to deep yellow and purple (Athwal, 1966). In addition, Burton and Powell (1968) observed white- and brown-seeded types. Most commonly, seeds are yellowish-gray with a reddish tinge on the embryo.

Seed is smaller than corn, but size is highly variable (Figure 21-1). Seeds are typically **obovoid,** 3 to 4 mm long and 2.25 mm wide. Seed protrude from the lemma and palea at maturity, making them susceptible to damage from birds perched on the sturdy head. Damage may be reduced by using bristled cultivars. At maturity, the spike changes from green to brownish. To achieve uniformly dry grain on individual heads and among early and late tillers, swathing may be necessary for combine harvesting. The great majority of the crop in the world, however, is harvested by hand, by cutting mature heads or cutting and **stooking** the entire plant. Periodic pickings reduce field damage from birds and weather. Harvested heads may be dried, stored, and threshed when needed. Dried grain can be stored readily.

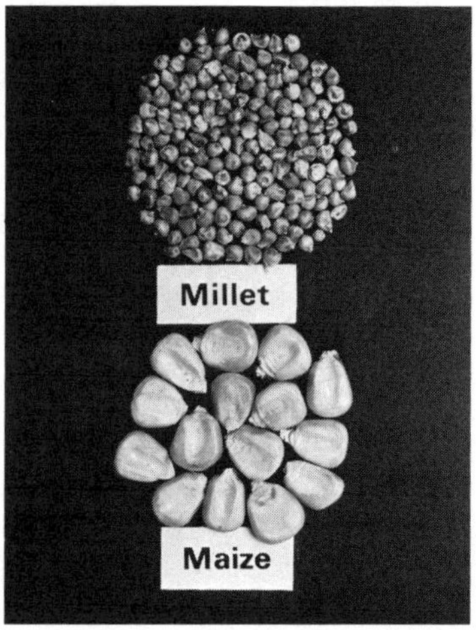

Figure 21–1. Pearl millet seed is smaller than that of maize or corn as evident in this photo. Seed is obovoid or egg-shaped with the narrow end at the base. Seed varies in size and color, is typically 3 to 4 mm long and 2.25 mm wide, and is frequently yellowish-gray in color with a reddish tinge on the embryo. (Courtesy of the International Crops Research Institute for the Semi-Arid Tropics (ICRISAT))

GROWTH AND DEVELOPMENT

Seeding development normally is slow for the first month, because much of the photosynthate is directed to root and tiller development. During this slow-growth period, the plant lacks competitive ability with weeds, making weed control an important production input. Temperatures of 2° to 3°C (32° to 37°F) can damage seedlings. The initial slow growth is followed by rapid growth of leaves and stems. Pearl millet is a determinate plant and reaches its full height after completion of flowering (Figure 21-2).

The primary or radicle root supports the plant until two or three leaves exist, and then the adventitious or secondary roots develop from the basal nodes. Nodes above the soil produce thick, strong roots which serve as prop roots to reduce root lodging. As tillering occurs, whorls of nodal roots arise from each tiller, thereby providing a degree of independence from the main shoot and adding to the volume of roots capable of securing moisture and plant nutrients.

Roots normally penetrate about 1 m (39 in.) into the soil but can penetrate much deeper, depending on moisture conditions and plant population. Main secondary roots are coarse and woody, lateral roots thin and fibrous.

One or more primary tillers arise from the main stalk at or below ground surface and attain about the same height as the main tiller. Secondary tillers may arise from primary tillers and can reach the same height as primary tillers but may have smaller heads.

When daylengths are 12 hours or more pearl millet may fail to flower, making it a short-day crop (Burton, 1965). Most pear millet cultivars are sensitive to photoperiod, and 86% of the introductions from Nigeria and Bourkina Fasso (formerly Upper Volta) were found to be facultative short-day plants that may or may not flower under short-day conditions. The remainder were obligate short-day plants and could reach reproductive development only under a photoperiod of 12 hours or less (Burton,

Figure 21–2. An intermediate-height pearl millet cultivar growing in a semiarid region. Commercial cultivars range in height from 0.5 to 3 m (1.5 to 10 ft) and are capable of tillering. Because the false spike resembles that of the swamp bulrush, the crop is also known as cattail or bulrush millet. Tillers arise from basal nodes and produce stalks similar in height to the main shoot but often with a smaller head. Pearl millet is a major food crop in hot, dry areas of the world and produces grain yields that vary from 200 to 4000 kg/ha (180 to 3500 lb/ac). (Courtesy of the International Crops Research Institute for the Semi-Arid Tropics (ICRISAT))

1965). Facultative short-day cultivars may become obligate short-day plants if exposed to 16-hour days (Begg and Burton 1971; Burton, 1966a and b).

The phenomenon of photoperiodism is under genetic control, and, although it may restrict grain development under long-day conditions, it can be useful for forage production. Burton and Powell (1968), for example, showed that photoperiod-sensitive cultivars were leafier and had a better seasonal distribution of forage production under long-day conditions. Photoperiodism therefore improved forage qualities. Photoperiod-sensitive cultivars planted for seed production in areas of the United States grow very tall and fail to mature seed before frost. The very low harvest index values may reflect excessive vegetative growth because of a high degree of photoperiod sensitivity. Less sensitive cultivars coupled with dwarf stature may improve harvest index values. Early maturity, photoperiod insensitivity, and an improved harvest index value are desirable characteristics for grain production. These traits could extend the range of adaptation, permit more than one crop per season, or enable mature grain in arid regions where late-maturing cultivars would fail.

A photoperiod-insensitive gene reported by Burton (1981a) has made it possible to develop very early maturing cultivars. When these cultivars were tested at Tifton, Georgia, in the United States, they flowered in 45 to 55 days regardless of season planted, compared to cultivars that flowered in 75 to 85 days if planted in May and 55 to 65 days if planted in August.

SEED DORMANCY AND LONGEVITY

Most pearl millet cultivars produce seed that is dormant for several weeks after they mature (Anderson, 1958; Burton, 1969). Dormancy breakdown rate is a function of temperature. Prolonged seed dormancy may be desirable for seed longevity but can hinder a breeding program designed to grow several generations a year. Burton (1969) found that seeds stored in air-tight containers at 0° to 6°C (32° to 43°F) extended the dormancy period for more than 12 months. To break dormancy, seeds can be soaked at 20° to 25°C (68° to 77°F) for one hour in a water solution of 1% 2-chloroethanol plus 0.5% sodium hypochlorite (Burton, 1969).

High temperatures and moist conditions, often found in the tropics, are not inducive to good seed-storage conditions. Unreliable seed storage of expensive hybrid seed of corn or sorghum has restricted the use of hybrids. Retention of good quality and high germinating capacity is essential. Pearl millet is noted for seed longevity. At Tifton, Georgia, seeds dried to less than 12% moisture, enclosed in airtight containers, and stored at 5°C (41°F) had 70% germination after 26 years (Burton, 1976). If stored in sealed containers at 5% to 7% moisture at the proper temperatures, seed storage of 100 years is possible (Burton, 1976).

Under tropical conditions, seed-storage periods are much shorter (Rachie and Majmudar, 1980). Sonavane (1929) reported a drop in germination from 94.5% to 62.2% in five years. Generally, pearl millet seed can be stored for up to 24 months under tropical conditions without appreciable deterioration.

PRODUCTION ASPECTS

Cultural practices for pearl millet are strongly influenced by economic considerations, such as the amount of nitrogen fertilizer applied, pesticide inputs, irrigation water, and the cost of these inputs relative to the price of the product. Pearl millet grain is consumed within the regions of production and almost never enters international trade channels. Recommended practices therefore are often highly specific to a region, its environment, and the local value of the grain.

Frequently grain production occurs under conditions of few additional inputs, including reliance on native fertility, natural rainfall, and little or no attempt at pest control. Under these conditions it may not be economically advisable to use more expensive F_1 hybrid seed, as its full genetic potential may not be expressed. These factors may account for the low yields shown in Table 21-1.

Introduction of superior cultivars with improved lodging resistance, nitrogen responsiveness, dwarf stature, or photoperiod insensitivity may require changes in the total production system and may render previous systems obsolete.

The potential for improving pearl millet yields by management was demonstrated by trials in Nigeria (Nwasike, Baker, and Egharevba, 1982), where most millet is grown mixed with sorghum. When grown alone, yields of 3000 to 4000 kg/ha (2680 to 3570 lb/ac) are possible and are eight to ten times area yields. Plant density and sowing date were considered the most important factors. Density varied from 20,000 plants/ha in the drier Sahel areas to 60,000 plants/ha (8100 to 24,300 plants/ac) in higher-rainfall areas. Nigerian cultivars are only slightly responsive to fertilizers, and only 12 to 13 kg/ha (11 to 12 lb/ac) of nitrogen and phosphorus are recommended. Area yields are low, not only because of competition in mixtures, but because tall cultivars with very low harvest index values are grown, since stover residues are desired.

Pearl millet grain yields in excess of 5000 kg/ha (4500 lb/ac) generally were regarded as unattainable under normal circumstances, but F_1 hybrids made such yields consistently possible with adequate inputs and outyielded the open-pollinated cultivars by 75% to 100% (Rachie, 1966). Intensive systems reviewed by Rachie and Majmudar (1980) have increased yields 50% to 300% over local practices.

General production practices include a recognition that pearl millet requires warm soil for germination and, because it is very susceptible to frost damage, must be seeded after all danger of frost has passed. Seed generally is planted in rows 51 to 107 cm (20 to 42 in.) apart, with plants 10 to 15 cm (4 to 6 in.) apart in the row. Seeding rates are low and vary from 5.5 to 6.5 kg/ha (5 to 6 lb/ac) for grain and 11 to 22 kg/ha (10 to 20 lb/ac) for forage. Shallow planting 1.25 cm (½ in.) deep is recommended (Martin, Leonard, and Stamp, 1976).

The uneven ripening problem can be facilitated by swathing when heads are one-half to two-thirds ripe, followed by combine harvesting when seeds have dried to a storable level.

Several reports reviewed by Jauhar (1981) have shown that nitrogen is a major factor limiting grain production. Rather dramatic increases of 26% to 96% resulted when 20 to 40 kg N/ha (35 lb N/ac) were applied to rain-fed pearl millet trials. Reddy and Reddy (1978) reported that a steady grain-yield increase with an improved hybrid occurred up to 200 kg N/ha (179 lb N/ac).

Forage yields also have shown a steady positive response to nitrogen fertilizer (Hart and

Burton, 1965; Shekhawat, Chundawat, and Shrotriya, 1970) with up to 400 kg N/ha (357 lb N/ac). Crude protein also increased.

Moderate phosphorus levels are required for pearl millet, and levels of 20 to 40 kg P_2O_5/ha were recommended for northern Australia (Norman and Begg, 1968). A higher level of potash is required than for phosphorus, and Burton (1966a) recommended a complete fertilizer having a 4 : 1 : 2 ratio of nitrogen, phosphorus, and potassium, respectively.

Maximum plant response to fertilizer appeared to be at a *p*H of 6.5 (Yuan, Robertson, and Littell, 1978).

CROP PHYSIOLOGY

As in other crops, the yield of pearl millet is produced from photosynthesis and is influenced by the leaf-area index, the duration, and the partitioning of photosynthate into grain or stover (harvest index). Some of the factors affecting photosynthesis are shown in Tables 21-3 and 21-4 for pearl millet grown at 68 kg N/ha (61 lb N/ac) in rows 76 cm (30 in.) apart and with 37.9 cm (14.9 in.) of irrigation water applied in six applications (Singh, Kanemasu, and Singh, 1983). Height ranged from 77 to 219 cm (30 to 86 in.); LAI ranged from 3.6 to 5.6; the vegetative growth made up 60.7% of the total days from emergence to maturity, leaving only 39.3% for the important reproductive development period.

BREEDING

The success in producing triticale by combining two different genera has stimulated interest in comparable wide crosses. Attempts to combine pearl millet with sorghum appear to be restricted by failure of the pollen-tube development in the **gynoecium** (Heslop-Harrison, 1982; Reger and James, 1982). Detailed studies of external and internal forces affecting pollen-tube growth in the gynoecium revealed that pollen-tube development was more extensive when four-day-old pearl millet stigmas were pollinated (Reger and James Sprague, 1983). An understanding of the factors affecting pollen-tube growth might lead to solution of the problem of a successful cross.

The possibility that tropical grasses such as pearl millet might fix nitrogen in the presence of a nitrogen-fixing bacterium, *Spirillum lipoferum*, was raised by Allaby (1977). Pearl millet can fix up to 42 kg N/ha (37.5 lb N/ac) in the presence

Table 21-3
Characteristics of Ten Pearl Millet Cultivars Used in Yield and Water Relations Studies near Manhattan, Kansas, in 1980

Cultivar	*Maximum LAI Observed*	*Vegetative Growth (50% flowering), Days*	*Duration of Reproductive Development, Days*	*Days to Maturity*
F_1 hybrids:				
2094 × 7024 (1)	4.5	47	37	84
2221 × 7024 (2)	4.1	46	30	76
2094 × 4104 (3)	5.1	52	32	84
2221 × 4104 (4)	3.6	49	27	76
Reg. cultivars:				
HMP 1700 (5)	4.3	49	27	76
HMP 550 (6)	4.7	46	38	84
Senegal bulk (7)	4.2	57	34	91
HMP 600 (8)	2.9	46	30	76
Serere-3A (9)	4.0	52	32	84
HMP 559 (10)	5.6	52	34	86

Source: Singh, Kanemasu, and Singh (1983).

Table 21–4

Stover- and Grain-Yield Comparisons in kg/ha (lb/ac) and the Percentage Change of Each under Irrigated and Nonirrigated Conditions of Ten Pearl Millet Cultivars Listed in Table 21–3 and Grown near Manhattan, Kansas, in the 1980 Season

	Stover Yield			*Grain Yield*		
Cultivar	*Nonirrig.*	*Irrig.*	*Percent Change*	*Nonirrig.*	*Irrig.*	*Percent Change*
1	9,810	21,300	+217	1,270	4,300	+339
	(8,759)	(19,018)		(1,134)	(3,839)	
2	7,640	16,700	+219	1,540	3,930	+255
	(6,821)	(14,911)		(1,375)	(3,509)	
3	12,410	17,370	+140	1,380	3,590	+260
	(11,080)	(15,509)		(1,232)	(3,205)	
4	10,980	20,630	+188	1,920	3,340	+174
	(9,804)	(18,420)		(1,714)	(2,982)	
5	9,040	16,430	+182	1,890	3,750	+198
	(8,071)	(14,760)		(1,688)	(3,348)	
6	9,040	15,120	+167	1,810	3,250	+180
	(8,071)	(13,500)		(1,616)	(2,902)	
7	9,970	23,680	+238	1,330	3,240	+244
	(8,902)	(21,143)		(1,188)	(2,893)	
8	8,850	13,870	+157	1,860	2,620	+141
	(7,902)	(12,384)		(1,661)	(2,339)	
9	11,490	20,300	+177	1,200	2,520	+210
	(10,259)	(18,125)		(1,071)	(2,250)	
10	16,150	21,510	+133	1,090	1,280	+177
	(14,420)	(19,205)		(973)	(1,143)	
Average	10,538	18,530	+175	1,529	3,343	+219
	(9,409)	(16,545)		(1,365)	(2,985)	

Source: Singh, Kanemasu, and Singh (1983).

of the bacterium. Jauher (1981) suggested that napier grass (*P. purpureum*) has shown root-associated nitrogenase activity and may be useful in evolving nitrogen-fixing pearl millet. This area of development does indeed merit further investigation.

Forage quality and dwarf stature offer a breeding challenge. One dwarfing gene reduced plant height by 50%, but, because the plant was brachytic and had the same number of leaves, the leaf percentage increased from 54% to 81%. Although total dry-matter yield declined, animal intake increased by 20%, and average daily gain increased by 20% to 50% (Burton et al., 1969).

Decades of intensive, worldwide breeding efforts in cereal crops other than pearl millet have produced remarkable results. This experience, applied to pearl millet, can result in relatively rapid progress, because the genetic variability of pearl millet is so great. At a plant-breeding symposium at Ames, Iowa, Burton (1981b) described the key elements of plant breeding as follows:

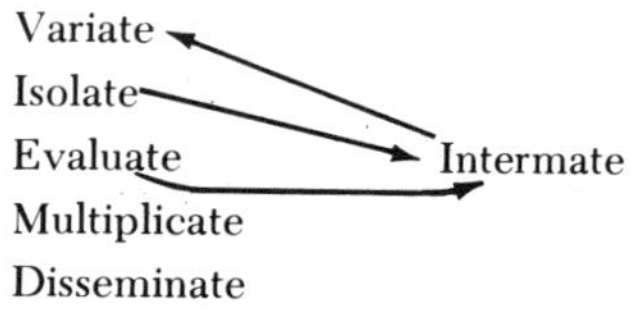

These action verbs sum up modern-day plant breeding.

CONCLUSIONS

If the goal of eradicating hunger and malnutrition from the world is to be achieved, food and agricultural development must be given top priority in all the major food-producing areas and on all crops. Pearl millet is an important crop whose full potential has not been realized or investigated. Breeding, crop production, crop physiology, and genetic studies need to be incorporated into the activities of world scientists. The efforts at ICRISAT are an initial step in giving pearl millet the international recognition the crop deserves.

Just as rye and triticale offer potential to increase food output by increasing the productivity of marginal lands at high latitudes, pearl millet offers a parallel promise of increasing productivity on marginal lands in the tropics.

References cited in this chapter are by no means exhaustive, but instead they represent a cross section of up-to-date findings pertinent to a number of areas. It is immediately apparent from the literature that the work of Glenn W. Burton forms a significant part of the scientific literature on pearl millet. Of the 46 references quoted in this chapter, 18 were authored or co-authored by Dr. Burton. Jauhar (1981) dedicated his article to Dr. Burton out of admiration for his pioneering work on pearl millet, acknowledged its significant contribution to the genetic improvement of the crop.

Questions

1. Comment on apomixis under the following headings:
   a. Obligate apomicts, with absolutely no sexuality, are blind alleys from the evolutionary standpoint.
   b. Since plant breeding is, in fact, man-made evolution, apomixis constitutes a breeder's despair.
   c. Apomixis can help fix heterosis and thus can be a boon to a breeder.

2. a. Account for the geographic location of pearl millet production.
   b. Is there the possibility that the area seeded to pearl millet could be expanded?

3. List the potential opportunities pearl millet offers plant breeders to improve the species.

4. Compare the range in harvest index values found among the cereal crops.

5. Suggest reasons why world average yields of pearl millet are low in Table 21-1.

6. How might the harvest index of pearl millet be improved?

7. a. Consider the potential of cultivar no. 4 (2221 × 4104) in Table 21-3 in terms of increased yield by improved management. Give reasons for your recommendations.
   b. How might this cultivar be improved?

8. Consider pearl millet in terms of yield stability and yield ability compared to other cereal grains.

9. Develop, with reasons, an ideotype for pearl millet similar to that suggested by Donald (1968) for wheat.

10. Item 4 of problems with pearl millet suggests that the concept of a cultivar's serving as either grain or forage is incorrect and that concentrating on cultivar development for one or the other but not both in any cultivar would offer more potential for improvement. Comment, with reasons, on the merits of this suggestion.

References

Allaby, M. 1977. *World Food Resources: Actual and Potential.* London: Applied Science Publishers Limited. p. 290

Anderson, A. M. 1958. "A Preliminary Study of Dormancy in Brown-top and Cattail Millets." *Proceedings of the Association of Official Seed Analysts* 48: 85-92.

Athwal, D. S. 1966. "Current Plant Breeding Research with Special Reference to *Pennisetum*." *Indian Journal of Genetics and Plant Breeding*. Symposium Supplement 26A: 73-85.

Begg, J. E., et al. 1964. "Diurnal Energy and Water Exchanges in Bulrush Millet in an Area of High Solar Radiation." *Agricultural Meteorology* 1: 294-312.

Begg, J. E., and G. W. Burton. 1971. "Comparative Study of the Genotypes of Pearl Millet under a Range of Photoperiods and Temperatures." *Crop Science* 11: 803-805.

Brunken, J. N. 1977. "A Systematic Study of *Pennisetum* sect. *Pennisetum* (*Graminae*)." American Journal of Botany 64: 161-176.

———, J. M. J. de Wet, and J. R. Harlan. 1977. "The Morphology and Domestication of Pearl Millet." *Economic Botany* 31: 163-174.

Bhatnagar, M. P., and K. Kumar. 1960. "Anthesis Studies in Rajasthan Bajra (*Pennisetum typhoideum*)." *Indian Journal of Agricultural Science* 30: 185-195.

Burton, G. W. 1958. "Cytoplasmic Male Sterility in Pearl Millet (*Pennisetum glaucum*) (L.) R. Br." *Agronomy Journal* 51: 479-481.

———. 1965. "Photoperiodism in Pearl Millet, *Pennisetum typhoides*." *Crop Science* 5: 333-335.

———. 1966a. "Pearl Millet Breeding." *African Soils* 11: 39-42.

———. 1966b. "Photoperiodism in Pearl Millet, *Pennisetum typhoides*, Its Inheritance and Use in Forage Improvement." *Proceedings of the 10th International Grassland Congress*. pp. 720-723.

———. 1969. "Breaking Dormancy in Seeds of Pearl Millet, *Pennisetum typhoides*." *Crop Science* 9: 659-664.

———. 1974. "Factors Affecting Pollen Movement and Natural Crossing in Pearl Millet." *Crop Science* 14: 802-805.

———. 1976. "Gene Loss in Pearl Millet Germplasm Pools." *Crop Science* 16: 251-255.

———. 1977. "Fertile Sterility Maintainer Mutants in Cytoplasmic Male Sterile Pearl Millet." *Crop Science* 17: 635-637.

———. 1980. "Pearl Millet." (In) W. R. Fehr and H. H. Hadley (eds.), *Hybridization of Crop Plants*. Madison, Wis.: American Society of Agronomy and Crop Science Society of America. pp. 457-469.

———. 1981a. "A Gene for Early Maturity and Photoperiod Insensitivity in Pearl Millet." *Crop Science* 21: 317-318.

———. 1981b. "Meeting Human Needs Through Plant Breeding: Past Progress and Prospects for the Future." (In) K. J. Frey (ed.), *Plant Breeding II*. Ames, Iowa: The Iowa State University Press. pp. 433-465.

———, and J. C. Fortson. 1966. "Inheritance and Utilization of Five Dwarfs in Pearl Millet (*Pennisetum typhoides*) Breeding." *Crop Science* 6: 69-72.

Burton, G. W., and J. B. Powell. 1968. "Pearl Millet Breeding and Cytogenetics." *Advances in Agronomy* 20: 49-89.

Burton, G. W., et al. 1969. "Effect of the d_2 Dwarf Gene on the Forage Yield and Quality of Pearl Millet." *Agronomy Journal* 61: 607-612.

Cooper, R. B., and G. W. Burton. 1965. "Effect of Pollen Storage and Hour of Pollination on Seed Set in Pearl Millet, *Pennisetum typhoides*." *Crop Science* 5: 18-20.

Donald, C. M. 1968. "The Breeding of Crop Ideotypes." *Euphytica* 17: 385-403.

FAO. 1982. *FAO Production Yearbook 1981*. Vol. 35. Food and Agriculture Organization of the United Nations. FAO Statistics Series No. 40.

Hanna, W. W. 1979. "Interspecific Hybrids Between Pearl Millet and Fountaingrass." *Journal of Heredity* 70: 425-427.

———, and J. B. Powell. 1973. "Stubby Head, an Induced Facultative Apomict in Pearl Millet." *Crop Science* 13: 726-728.

Hanna, W. W., and M. Dujardin. 1982. "Apomictic Interspecific Hybrids Between Pearl Millet and *Pennisetum orientale* L. C. Rich." *Crop Science* 22: 857-859.

Harlan, J. R., and J. M. J. de Wet. 1971. "Toward a Rational Classification of Cultivated Plants." *Taxon* 20: 509-517.

Hart, R. H., and G. W. Burton. 1965. "Effect of Row Spacing, Seeding Rate, and Nitrogen Fertilization on Forage Yield and Quality of Gahi-1 Pearl Millet." *Agronomy Journal* 57: 376-378.

Heslop-Harrison, J. 1982. "Pollen-stigma Interaction

and Cross Incompatibility in the Grasses." *Science* 215: 1358–1364.

Jauhar, P. P. 1981. *Cytogenetics and Breeding of Pearl Millet and Related Species.* New York: Alan R. Less, Inc. 289 pp.

———. 1970. "Haploid Meiosis and Its Bearing on the Phylogeny of Pearl Millet, *Pennisetum typhoides* Stapf and Hubb." *Genetica* 41: 532–540.

Leuck, D. B., and G. W. Burton. 1966. "Pollination of Pearl Millet by Insects." *Journal of Economic Entomology* 59: 1308–1309.

Martin, J. H., W. H. Leonard, and D. L. Stamp. 1976. *Principles of Field Crop Production.* 3d ed. New York: The Macmillan Company. pp. 570–572.

Norman, M. J. T., and J. E. Begg. 1968. "Bulrush Millet (*Pennisetum typhoides* (Burm) S. and H.) at Katherine, N.T.: A Review." *Journal of the Australian Institute of Agriculture Science* 34: 59–68.

Nwasike, C. C., E. F. I. Baker, and P. N. Egharevba. 1982. "The Potential for Improving Millet (*Pennisetum typhoides* (Burm), Stapf and Hubb.) in Farming Systems of the Semi-Arid Areas of Nigeria." *Agriculture and Environment* 7: 15–21.

Rachie, K. O. 1966. "Sorghum and Millet Hybrids in India." *Span* 9: 49–53.

———, and J. V. Majmudar. 1980. *Pearl Millet.* University Park and London: The Pennsylvania State University Press. 307 pp.

Reger, B. J., and J. James. 1982. "Pollen Germination and Pollen Tube Growth of Sorghum When Crossed to Maize and Pearl Millet." *Crop Science* 22: 140–144.

Reger, B. J., and J. James Sprague. 1983. "Pearl Millet and Sorghum Pollen Tube Growth in Pearl Millet Gynoecia of Different Ages." *Crop Science* 23: 931–934.

Reddy, M. R., and B. V. R. Reddy. 1978. "Response of HB-4 and D-174 Bajras to Different Levels of Nitrogen and Row Spacings." *Indian Journal of Agronomy* 23: 385–386.

Shekhawat, G. S., G. S. Chundawat, and G. C. Shrotriya. 1970. "Note on the Response of Bajra Varieties to Levels of Nitrogen on Sandy Soil of Rajasthan." *Indian Journal of Agronomy* 15: 193–194.

Singh, P., E. T. Kanemasu, and P. Singh. 1983. "Yield and Water Relations of Pearl Millet Genotypes under Irrigated and Nonirrigated Conditions." *Agronomy Journal* 75: 886–890.

Sonavane, K. M. 1928. "Longevity of Crop Seeds." *Agricultural Journal of India* 23: 271–276.

Terrell, E. E. 1976. "The Correct Names for Pearl Millet and Yellow Foxtail." *Taxon* 25: 297–304.

Yuan, T. L., W. K. Robertson, and R. C. Littell. 1978. "Soil *p*H Adjustment by the Double-buffer Method for Pearl Millet (*Pennisetum typhoides* (Burm) Stapf and Hubbard) on Ultic Haplaquod." *Proceedings of the Soil and Crop Science Society of Florida* 37: 208–212.

22

Barley and Oat

THE CHANGING SCENE

In some respects it is surprising that oat and barley crops continue to play major roles in world agriculture. In areas where the physiologically superior C_4 crops can be grown, oat and barley crops are not competitive. In cool, short-season areas where C_4 crops are not adapted, the heavier-seeded wheat and triticale can be grown. Yields of oat and barley crops include a low-quality, adhering hull.

Perhaps tradition can account for the continued importance of oat and barley. It is said that the human race adheres to traditional food habits with an irrational rigidity. Immigrants to a country quickly adopt the life style and customs of the new country, except for food habits. Generations later, traditional foods of their homeland continue to be served, especially on festive or commemorative occasions.

Production and use of oat and barley have a long tradition among Old World farmers. At one time, oat and barley crops possibly were the most important cereal crops on earth. As European immigrants settled in various parts of the globe, they carried seed and traditional food dishes with them. A well-established tradition is the use of barley for the production of malt and for brewing. An alcoholic beverage could be made from any of the cereal grains. Barley, however, is the choice grain for this purpose, because it has served this role for centuries.

Barley and oat were not the first choice of neolithic farmers. Heiser (1973, p. 68) noted that wheat formed the basis of civilization in the Near East, rice in the Far East, and corn in the Americas. It was the shift in soil salinity along the Tigris, Euphrates, and Nile rivers that brought oat and barley production into prominence, as both are more salt-tolerant than wheat. It was here that barley was adopted for brewing and that oat became the accepted animal feed, especially for horses.

Once established, the tradition of oat and barley production continued. Harlan (1975, p. 187) noted that barley held a dominant position during classical Greek times. Barley was described as the food of the poor, a situation that described everyone but an aristocratic minority. Barley was described also as the ration of the soldier.

Although the record of cereal production during Greek and Roman times is skimpy, Coffman (1961) noted frequent references to the oat crop. Without science, grain yields were not high. The oat crop can be produced under less-than-ideal production practices. In England, the Dark Ages were not noted for agricultural advances or production. Kerridge (1973) recounted agricultural activities in England for the period 1560 to 1760 and found that oat and barley crops played a dominant role in English agriculture and undoubtedly in Continental Europe as well. Low fertility, inadequate cultural practices, exhausted soils, late seedings, weeds, diseases, and insects were the scourge of agriculture. Kerridge (1973, p. 80) noted that "poor as the soils were, by dint of folding sheep upon them, they could be forced to yield good crops of rye, barley and oats." Yields undoubtedly were not high, but the reliability of oat and barley production was of greater consequence than widely fluctuating yields of other crops that had perhaps a genotype capable of higher yield under favorable conditions. Usually, aspects other than genotype restricted yield. Kerridge (1973, p. 107) noted that a ten- to twentyfold increase of harvest over seed was all that could be expected in the common fields.

Oat and barley grain can be fed to livestock with ease and is associated with the presence of the hull on the grain. The oat groat is soft, compared to hard wheat and other grains, and can be fed whole or rolled. Rolling is achieved with a minimum of mechanical effort, and rolled oat can be fed safely to sheep, lambs, calves and all other ruminants, brood mares, foals, draft horses, chickens, and essentially all classes of livestock. Of the cereal grains, oat seed contains the highest quality and quantity of protein. It possibly was a world dependent on horsepower that elevated oat grain to a position of respect and prominence among the feed grains.

What horse feed did for oat grain, the malting and brewing industry did for barley. Kerridge (1973, p. 100) marveled at the volume of barley that was processed into beer and mused, "Think of what was needed for Northdown barley to be brewed into beer." A quote from an old English folk song best exemplifies the status of barley in medieval England (Beaven, 1947 in the foreword):

> John Barleycorn's the choicest grain
> That e'er was sown on land.

The pride associated with barley production is apparent in the half-century of work in England of one of its early scientists, E. S. Beaven (Beaven, 1947), yet he deplored the relatively small economic advance in barley progress during his 84-year life span.

Not until the industrial revolution, when bread and sandwiches became the food of the factory worker, did wheat start to regain its lost prominence. Although bread came to be recognized as "the staff of life," wheat bread was not always available. Kerridge (1973, p. 158) noted that oat cakes and barley **bannock** were common alternatives to wheat flour, along with rye or a mixture of wheat and rye called maslin.

Settlers to the New World did not abandon centuries of oat and barley culture and tradition for the unknown corn crop. Mixed oat and barley was a more reliable crop than an untested full-season corn crop that preferred high fertility. Danhof (1969, p. 272) noted the frequency of the oat crop in farming systems in the northern United States between 1820 and 1870. It took the revolutionary impact of hybrid corn to replace some of the oat and barley production.

During the era of the binder-reaper, oat and barley straw was softer and more palatable than the coarser, more brittle wheat straw. Oat and barley straw was preferred for bedding or as a low-energy ruminant feed. The need for straw prompted the production of a considerable area of oat and barley.

If oat and barley crops were the most important crops on earth prior to the development of science, they have slipped substantially in the past 200 years. In terms of production, barley ranks fourth and oat sixth (Tables 1-1 and 1-2). In terms of area harvested, the ranking is fourth and seventh for barley and oat, respectively. Expressed otherwise, during the three-year period 1979–1981, barley accounted for 10.1% and oat for 2.8% of total world output of cereal grains. In the ten-year period from 1969–1971 to 1979–1981, oat production decreased 20.9%.

THE FUTURE OF OAT AND BARLEY PRODUCTION

Oat or barley production, but not both, may continue to be of importance in the future. A

20.9% decrease in oat production was accompanied by a 15.6% increase in barley production (Tables 1-1 and 1-2). The combine harvester era and changes in livestock management have reduced the use of barley and oat straw for bedding or supplementary feed. In most developed areas of the world the era of the draft horse has passed. Growing an oat and barley mixture does not maximize yield, does not provide a consistent feed product, and increasingly is viewed as outdated. Oat grain with 25% to 30% hull is too light to enter into world trade. Higher yields can be produced with other crops that are hybrid, refined, and have the C_4 photosynthetic pathway.

The competitive yields for barley (Table 22-1), its dominant role in the malting and brewing industry, scientific advances such as haploidy (refer to Chapter 5) and uniculm development, a highly variable gene pool, and a strong traditional background may mean that barley will replace some of the area seeded to the oat crop in the future.

If the decline in oat production is to be stopped, an increase or the promise of an increase must be found. Little promise exists for a striking advance. Since 1900, genetic gains in oat yields were estimated at 9% to 14% (Browning, Frey, and Grindeland, 1964; Langer, Frey and Bailey, 1978) compared to 35% to 100% increases in other crops over the same time period (Rodgers, Murphy, and Frey, 1983). In a review article on oat improvement, Lawes (1971) revealed no promise for a major advance.

Study of a small 9% increase in 66 oat cultivars, developed over the 40-year period 1935 to 1975, revealed a small burst in yield between 1932 and 1942 (Langer, Frey, and Bailey, 1978). These workers suggested that oat germplasm originally introduced from northern Europe to the midwestern United States may have contained little genetic variability for grain yield. Wide crosses were responsible for the 9% yield advance. A similar low yield increase with little promise for further advance was discerned when genetic relationships and grain yield of oat cultivars released in the north-central and northeastern United States between 1941 and 1980 were examined (Rodgers, Murphy, and Frey, 1983).

Likewise, using 52 oat cultivars representing a wide range of genetic backgrounds, Murphy, Long, and Nelson (1982) noted that modern oat cultivars showed only minor advances over older cultivars. One advance noted was an increase in root development. Unlike other cereal grains, plant height was reduced only slightly over the time span examined.

Still another critical review of oat progress, based on grain yield and associated agronomic and physiological traits of nine oat cultivars released since 1923 and commercially important in Minnesota, produced discouraging results (Wych and Stuthman, 1983). All oat cultivars were grown in common environments for three years. Grain yields over the 58-year period increased from 2240 to 3130 kg/ha (2000 to 2795 lb/ac) or 39.7%, but height remained unchanged although lodging was reduced. Biological yield increased from 8.1 to 9.1 tonnes/ha (3.6 to 4.1 tons/ac), but harvest index was discouragingly low and increased only from 27% to 34%.

The objective of these detailed reviews on oat improvement in the twentieth century was to determine productive areas in which to concentrate breeding efforts to increase grain yields.

Table 22–1
Comparative Performance of Scout Hullless Barley in 1979 to 1981

Cultivar	Grain Yield		Days to Maturity	Test Wt. (kg/hl)	Height	
	kg/ha	lb/ac			cm	in.
Hulled	4575	4085	95.3	65.3	72	28
Hullless (Scout)	3923	3503	92.9	71.7	77	30

Source: Rossnagel, Harvey, and Bhatty (1983).

The conclusion reached was that yield advances have been slow and unexciting and that no single outstanding approach can be identified to ensure a rapid or significant future advance. The studies suggest that genetic variability for yield advances in the oat crop is limited. Conventional breeding techniques do not promise opportunity for major advances.

Oat production will not cease. It is possible that the crop may be the first traditional feed grain to shift to predominant use as a food grain because of its superior nutritional value. In short-season, cool, long-day areas, oat production may continue for some time. A clearer perspective on oat and barley is offered in the next section.

A PERSPECTIVE ON BARLEY AND OAT

Combining the discussion of oat and barley crops in one chapter provides a convenient method of emphasizing their strengths and weaknesses. The reader should not infer any especially strong relationship between the two species. Wheat and rye are the cultivated cereals most closely grouped in the same tribe, *Triticeae;* the oat species belongs to the tribe *Aveneae.* Other reasons for considering the two crops together include cultural practices, area of adaptation, the past record, and end use. In these respects the following points need emphasis:

1. Both crops are **coarse grains,** with the lemma and palea normally adhering to the caryopsis after threshing. The oat grain usually is comprised of 25% to 30% hull with over 30% crude fiber. The barley grain has 12% to 14% hull. Hullless cultivars exist in both species, but they are not grown extensively. Hullless cultivars may not be desired, because they lack the traditional hull and fiber content which makes them safe to feed, they appear to yield less volume, and hullless oats are more difficult to store. The hull of oats can be removed mechanically to achieve the desired product for food. Barley with hulls is preferred for malting purposes, as the hulls provide a protective covering during germination and handling as well as providing a natural filter.

2. The two crops are traditional feed crops with a higher fiber content than other cereal grains. On a dry-weight basis both crops have about a 13% protein content (Appendix Table D). Hullless oat is equal to either wheat or corn as an energy feed. The values of oat and barley crops for livestock feed, either as grain, forage, or by-products of industry, were presented in Chapter 8.

3. Both crops serve specialty food markets. The value of barley for malt production is accepted widely, and its advantages in this respect were outlined in Chapter 12.

Removal of the oat hull or the production of hullless oat provides a product that can be rolled or milled into a number of attractive foods. Oat groats offer the advantage of a balanced amino acid composition relative to other cereals and an oil content ranging from 2.5% to 10%, giving them a high energy value.

4. The two crops have been and continue to be grown extensively in mixtures of varying proportions for producing livestock feed grains. Mixtures are viewed by producers as a means of increasing crop stability on variable soil conditions and from year to year. It is hoped that cooperation of the two genotypes will produce a higher yield as well as a feed product heavier than oat grain. Basic principles of competition were considered in Chapter 8. Generally, a mixture of oat and barley crops will result in a lower yield than the highest-yielding component of the mixture, owing to interference between the two crops.

5. Both are temperate-zone crops and are seldom grown in the warmer portions of either the Northern or Southern Hemisphere above or below 30° latitude. Both crops, however, are grown on every continent, making their best growth in cool, moist regions and on a wide range of soil types as long as they are well drained and reasonably fertile. Coffman (1961, p. 1) found that the greatest area of oat production in North America occurred when June temperatures were below 18°C (65°F).

Few cereals can be grown under such diverse cultural and cropping practices as barley and oat. The oat crop can tolerate a *p*H of 4.5, the barley crop 6.0. Barley is the most tolerant to saline conditions (Chapter 7), and improvement is possible because of genetic variation.

6. Both crops have the inefficient C_3 photosynthetic pathway, which accounts for their adaptation to cool temperatures at high latitudes. C_3 cereal crops require more moisture to produce a given unit of dry matter (refer to Table 2-5) than C_4 crops. Both crops are vulnerable to injury by hot, dry weather, particularly during reproductive development.

Recommendations call for seeding spring cereals as soon as the soil is dry enough to be prepared. Early seeding ensures efficient use of available soil moisture, allows the crop to escape damage by diseases that often develop to epidemic proportions late in the season, and allows escaping midsummer heat and drought. Germination will occur at temperatures of 3° to 5°C (37° to 41°F), but seedlings will emerge faster at temperatures of 5° to 10°C (41° to 50°F).

Although barley can be sown later than the oat crop and still mature before fall frosts, yield may be reduced by the adverse response to high temperatures associated with C_3 crops. Wych, McGraw, and Stuthman (1982) stressed the sensitivity of the oat crop to hot or dry weather.

7. Both crops, as well as spring wheat, can withstand very low temperatures during the seedling or tillering stages. Low-temperature damage will depend on intensity and duration of the freezing period, but a temperature of about −8°C (17.6°F) is required to kill the leaves of seedlings (Anonymous, 1983). Spring temperatures following seedling emergence are seldom this low, and, if leaves are damaged, recovery is usually complete, because the growing point is below the soil surface and thus is protected from damage.

The risk of frost damage should not serve as a deterrent to early seeding. The use of early-maturing cultivars will not avoid damage from very late spring frosts, but they happen very rarely. Early seeding may help ensure physiological maturity before fall frosts in short-season areas at high latitudes or altitudes.

Susceptibility to frost damage increases with increasing plant development, but the likelihood of severe frosts decreases as the season advances. If freezing does occur, the developing sexual organs may be damaged. With the onset of flowering, a frost of 0° to −2°C (32° to 28.4°F) is sufficient to damage the young florets, resulting in partial or even complete sterility in the head (Anonymous, 1983). Frost-damaged crops can be salvaged by harvesting them for hay or silage.

Barley and oat crops grown at high latitudes are most frequently damaged in the adult stage by fall frosts. Symptoms of the damage are seldom obvious, as the leaves apparently are unharmed, but the germ of the soft or immature kernel may be killed. Immature kernels do not develop further after freezing, and yield loss depends directly on the stage of filling, with losses decreasing with advancing maturity. At harvest, frozen grain has a dull greenish color, lacks luster, but is suitable for feed. Frozen grain will not germinate and is useless for seed or malting purposes.

8. Barley and oat crops have the poorest lodging resistance of the cereals. The potential lodging hazard precludes intensive management such as high levels of fertilizers and irrigation water. Lodging results in direct yield losses by reducing photosynthesis and in indirect yield losses by promoting conditions inducive to mold and disease, and it increases harvesting difficulty and losses (review Chapter 2).

9. Although genetic sources of dwarf barley and oat exist, most dwarfs lack acceptable agronomic traits and in general have not become commercially successful. Increases in straw strength are often taken as a signal by producers to increase production inputs, and lodging reoccurs.

Problems with dwarf cultivars include inadequate emergence of the head from the upper sheath or boot, which results in poor seed set. Other problems with dwarf barley include late maturity, disease susceptibility, and brittle straw. One or more of these deficiencies can limit the success of dwarf cultivars. For plant height-yield relationships review Chapter 4.

10. Both crops have a winter form which can be produced in favored areas, as they are not as hardy as winter wheat, rye, or triticale. The

winter oat crop is the least winter-hardy of the cereals. Details of low-temperature injury are outlined in Chapter 7. Pfeifer and Kline (1960) found that if the soil temperature at the 2.5 cm (1 in.) depth dropped to −12°C (11°F), winter oat cultivars normally were killed. At −5°C (22°F) the more hardy survived. The area of winter oat production lies to the south of the zone of −5°C (21°F) average temperature for December to February.

11. Both oat and barley are long-day crops, and, true to the Law of Homologous Series (refer to Chapter 3), genes for daylength insensitivity have been found. Cultivars developed at high latitudes from a narrow gene base and without benefit of a winter nursery often fail to flower normally at lower latitudes, because daylengths are too short. This prohibits growing a winter increase generation in southern locations and slows the breeding programs to increase yield. Incorporation of genes for daylength insensitivity means that cultivars tend to flower too early to obtain high yields, but selection for adaptation is possible.

Generally, the most productive cultivars are those that have been tested, and selected for a specific area. Few oat and barley cultivars are adapted coast to coast. The transfer of cultivars for commercial use from one area to another is limited.

Early seeding allows long-day cereal crops to take advantage of the long days of spring. The closer the reproductive period coincides with the long photosynthetically active days of early spring, the greater the potential yield.

Under conditions of delayed seeding, contrary to expectations, an early-maturing cultivar should not be used. Sensitivity to long days compresses the differences in maturity between late- and early-maturing cultivars. If an early-maturing cultivar is seeded late, reproductive growth may be stimulated before leaf-area development and tillering are adequate, resulting in low yields. A relatively large difference in maturity between two cultivars will be reduced by late seeding, and therefore a cultivar normally considered as late-maturing will develop more suitably under late-seeding conditions than an early-maturing cultivar.

12. Because oat and barley cultivars exist that mature in as few as 80 frost-free days, these crops can be grown in short-season areas. The number of days required from seeding to harvest will vary considerably, depending on location, season, and date of seeding. The higher the latitude, the longer the daily duration of photosynthesis. Oat or barley crops grown at 60° N latitude, for example, are subjected on June 21 to a daylength six minutes short of 19 hours. If grown at 40° N latitude, the daily duration of photosynthesis is four hours less (refer to Table 2-4).

13. Both crops can be harvested in a similar manner and when the grain contains less than 15% moisture. When kernel moisture content is uniformly 14.5% or less, direct-combine harvesting may be most economical, compared with swathing before harvesting. Swathing may be preferred if the crop is excessively weedy, since green, moist seeds or plant parts may heat in storage. Swathing may be preferred if the crop has late secondary tillers, if the field is variable and has uneven maturity, if plants have been frosted, or if less-than-ideal weather conditions prevent rapid drying.

Lodged fields should be swathed at or near maturity to encourage uniform and rapid drying before being threshed. Swathers should be equipped with extended cutter-bar guards to lift the lodged grain before it is cut. Lodged grain matures, ripens, and dries more slowly than nonlodged grain, because the soil dries more slowly under the blanketlike covering of lodged grain.

14. Both oat and barley crops are predominantly self-pollinated. Natural out-crossing was believed to be of the magnitude of less than 0.2% and 0.5% for barley and oat crops, respectively (Starling, 1980; Brown, 1980). Florets of both crops have three stamens, a pistil consisting of a two-lobed ovule and a two-branched feathery stigma, two lodiculae at the base, and reproductive parts enclosed in a single floret.

15. Under most conditions, barley is a more competitive crop than oat. Where herbicide weed control is not adequate for direct seeding of forage seedlings, oat and barley crops are commonly grown as a companion crop (see

Chapter 2). When a cereal companion crop is used, the oat crop is preferred. Generally, oat cultivars have a lower tillering capacity than barley.

16. Hybrid development remains of potential value in both crops but is restricted by a number of factors which limit its commercial use (review Chapter 9).

17. Minor problems exist with both crops. In oat, the oil content is high and partially compensates for the energy reduction associated with the high hull content. The unsaturated fats in the oat groat oxidize readily, however, causing deterioration in storage. When hullless oat groats are produced, handling results in bruising, and the oil-rancidity problem is aggravated.

Before being fed to any class of livestock, newly harvested barley, whether dry or high-moisture grain, should be stored about four to six weeks to allow it to undergo a sweat period. Serious losses in cattle, pigs, and poultry have been attributed to feeding barley grain that has not undergone the sweat period. Apparently, compounds toxic to animals undergo chemical changes that render them nontoxic. Processing newly harvested barley for feed does not eliminate the problem.

ORIGIN AND CLASSIFICATION

The time and place of the origin of barley and oat are lost in antiquity, and the best that can be provided is an educated guess based on cytogenetics and archaeological records. Table 3-5 listed the Near East as the origin for common barley and Northern Europe for the cultivated oat. Discussions of the evidence on the origin for the cultivated oat were provided by Rajhathy and Thomas (1974) and Coffman (1961) and for barley by Harlan (1968 and 1975). Possibly a two-row wild barley species with a brittle rachis and known as *Hordeum spontaneum,* which crosses readily with cultivated barley, was the wild progenitor of our cultivated barley species. Harlan (1975, p. 179) suggested that fully domesticated barley was probably hulled and two-rowed.

Cultivated barley has the binomial Latin name *Hordeum vulgare* L., is a diploid, with $2n = 2X = 14$ or $n = 7$ chromosomes. Briggs (1978, p. 76–88) reviewed various barley classifications that have become muddled by the adoption of different criteria on which to base a classification. Divisions have emerged that increased knowledge of experimental genetics has made untenable. Of interest is the barley species *Hordeum bulbosum,* described in Chapter 5, which is used to produce haploid barley embryos.

For a crop as important as barley, little is known about its relatives, and species relationships within the genus are practically unexplored (Harlan, 1968).

Likewise the origin of oat is obscure. Like rye, the oat plant may have persisted as an impurity in barley or wheat crops before becoming cultivated as a crop. This may not have happened until the crop was widespread in northern Europe. Climatic conditions favorable to oat may have led to the conclusion that oat evolved in northern Europe.

Two cultivated forms of oat exist: *Avena sativa* L. and *Avena byzantina* C. Koch with $2n = 6X = 42$ chromosomes, a hexaploid. Rajhathy and Thomas (1974) have classified oat in the genus *Avena,* which forms a polyploid series of diploids, tetraploids, and hexaploids with a basic chromosome number of seven. The genus has been classified (Rajhathy and Thomas, 1974) into 19 taxonomic species, of which ten are diploid ($2n = 14$), five are tetraploid ($2n = 28$), and four are hexaploid ($2n = 42$). Two wild oat species, *Avena fatua* L. and *Avena sterilis* L., are hexaploid and are readily crossed with the common cultivated oat species *A. sativa* and *A. byzantina* and produce fertile progeny. Wild species possess disease-resistance and dormancy factors. Crosses with diploid and tetraploid species are complicated, difficult to achieve, or may carry undesirable weedy characteristics (McMullen, Phillips, and Stuthman, 1982). Until problems can be overcome, the gene-pool resources of breeding value are restricted to the hexaploid forms.

BOTANY OF BARLEY

To illustrate some aspects associated with the cereal grasses, several morphological features of barley were described in Chapter 3. A condensed version of that discussion is provided here.

The inflorescence of barley is a spike with alternating rachis nodes and three spikelets at each node. The shorter the internodes, the more compact the ear (Figure 22-1). Each spikelet normally consists of a pair of narrow glumes and one floret. The central floret is the largest, with laterals ranging from fully fertile to **vestigial,** thereby forming six-row, two-row, or intermediate spikes. In six-row cultivars, lateral spikelets are fertile; in two-row cultivars they are sterile. Crosses between six-row and two-row cultivars readily give rise to true-breeding intermediate forms have awnless lateral florets of variable fertility. At least two loci are involved, with an allelic series at each controlling various intermediate forms. The change from two-row to six-row is achieved from a single recessive mutation (Harlan, 1975, pp. 129–130). Six-row cultivars usually have from 25 to 60 kernels and two-row cultivars from 15 to 30 kernels per spike. Two-row cultivars are noted for a high tillering ability and, although they have fewer seeds per spike, yield competitively with six-row barley cultivars. Cyclical preferences for two- or six-row barley cultivars occur. In the 1980s world demand for malt has shifted to high-quality, two-row barley (Anonymous, 1983).

Figure 22–1. A six-row, awned barley spike with short rachis internodes resulting in a compact, dense ear. The rachis may vary in length from 2.5 to 12.7 cm (1 to 5 in.) and may carry 25 to 60 kernels in six-row cultivars or 15 to 30 kernels in two-row cultivars. Awns are an extension of the lemma and may be barbed or smooth. If barbed, the teeth are angled upward. (Courtesy of Ontario Ministry of Agriculture and Food)

Barley spikes may be awned or awnless. If awns are present, they are an extension of the lemma and may be barbed or smooth. Awns may be fragile, tough, or **dehiscent.** Lemma appendages may vary from awnless, awned, or awnletted to a number of hood shapes (Briggs, 1978, p. 61). Normally the lemma and palea adhere tightly on the caryopsis, although hullless cultivars exist.

Like other cereal grasses, barley has both primary and secondary (seminal and adventitious) roots, a cylindrical stem or culm with hollow internodes, and five to seven solid nodes or joints. The internodes are shortest at the base of the plant and increase in length progressively toward the top. Total length varies with cultivar and environment from 10 to 20 cm (4 to 8 in.) in brachytic types to 152 cm (50 in.) in tall cultivars.

A split leaf sheath arises from each node and carries a flat, leaf blade; the blades are borne alternately on opposite sides of the stem. Leaves consist of a sheath, ligule, auricle, and blade.

The terminal spike may be held upright or in a nodding position, and the angle is a factor influencing kernel weathering and quality. In barley-growing areas where humid conditions occur during grain maturation, kernel weathering is a serious problem. Maltsters and brewers avoid using weathered and stained barley if nonweathered barley is available. Studies by Brinkman and Luk (1979) established that a nodding angle greater than 90° from vertical contributed to kernel brightness in moist conditions. They suggested that barley breeders consider spike nodding as a selection criterion. The nodded-spike characteristic should be transferable in barley-breeding programs without unusual difficulty (Miller and Brinkman, 1983).

The number of stems per plant as a result of tillering is influenced by the density of the stand, competition for growth factors, cultivar (two-row versus six-row), and the environment. At normal seeding rates a single plant may average 1.5 to 5 heads. Under thin stands and favorable conditions tiller number may be several times greater. Single-stem or uniculm mutant forms exist.

OAT BOTANY

The inflorescence of the oat plant is a panicle (Figure 22-2), composed of a central loose, open rachis with five to seven nodes, from which arise branches bearing spikelets. Each lateral branch terminates in a single apical spikelet. Other spikelets are borne on second- or third-order branches. Each panicle may have 20 to 50 spikelets.

Spikelets consist of several florets enclosed in two empty glumes, with the tip of one glume extending slightly above the other. One species, *Avena longiglumis* Durieu, has very long subequal glumes. Florets within each spikelet are arranged alternately upon a central axis, the rachilla, and usually the two basal florets are fertile.

Figure 22–2. A panicle of the oat plant with five whorls of branches spreading equilaterally on either side of the rachis. Lateral branches arise from nodes on the open rachis and terminate in a single apical spikelet. (Courtesy of the Canadian Seed Growers' Association)

Except in hullless oats, the lemma and palea normally adhere to the kernel or caryopsis. Wild oat forms usually have a strongly developed, twisted awn that is bent like a knee and arises as an extension of the midrib of the lemma. In the cultivated oat, an awn may be present only on the lower grain of the spikelet and usually is small, untwisted, curved or straight. The twisted awn responds to environmental changes by untwisting like a wound-up spring, propelling the seed across the soil surface or helping to bury the seed in loose soil.

The oat groat or caryopsis is spindle-shaped with a crease or furrow on one side. It is usually buff colored and covered with moderately long, fine, silky hairs.

The lemma of the oat kernel may vary in color from white, yellow, and gray to red or black.

The kinds of plant parts produced, the position of the leaves, the structure of roots, and the tillers are similar to those in barley and other cereal grasses and were described in detail by Bonnett (1961). The oat stem internode is hollow, however, and speculation exists that **vascular bundle** capacity may limit transport and panicle yield (Housley and Peterson, 1982; Peterson, Housley, and Luk, 1982). These workers concluded that the number and size of vascular bundles develop in concord with the number of spikelets initiated and that some other restriction may limit oat yield.

OAT AND BARLEY PRODUCTION

If oat and barley crops can be justified in a cropping system, they deserve to be treated as first-rate crops in order to produce yields that are competitive. To produce high yields requires that all parts of the production system be considered. Major aspects include proper seeding date, seeding depth, and seeding rate, an adequate and balanced fertility program, and pest control.

Production systems are not fixed packages to be used year after year. The correct seeding rate depends on the area of production, the cultivar, the intended end use of the crop, and fertility. Where moisture is plentiful throughout the season, a higher seeding rate may be used than in areas of limited rainfall. Large-seeded, two-row barley cultivars should be seeded at a slightly higher rate than average-sized, six-row cultivars. Seeding rates for barley in Western Canada vary from 40 to 100 kg/ha (35 to 90 lb/ac). For oat, the average rate of seeding in North America is about 90 kg/ha (80 lb/ac).

It generally is better to plant on the thin side than to plant too much seed. Heavy seeding can result in lightweight seed. Seed growers multiplying expensive and scarce seed of a new cultivar should sow at a low seeding rate (25 to 30 kg/ha or 22 to 28 lb/ac) and possibly in rows 38 cm (15 in.) apart.

Increasing the seeding rate is not a satisfactory method of compensating for seed with poor germination, as such seed may produce weak seedlings subject to disease and resulting in low yield. It is recommended that pedigree seed be purchased to assure high quality.

The usual row width for oat and barley crops is 15 to 20 cm (6 to 8 in.) apart, as determined by mechanical seeding equipment. Under conditions of row spacings wider than 20 cm (8 in.), sunlight interception may be inadequate, weed growth may flourish, and swath support becomes a problem.

Plant density affects such factors as tillering, head size, harvest index, and grain yield. Ten barley cultivars all showed similar density responses when seeded at 1.6, 6.2, 25, 100, and 400 plants/m^2 (1.5, 5.7, 23, 92, 369 plants/yd^2) (Baker and Briggs, 1982).

Where there is adequate moisture for germination, shallow seeding is recommended. Shallow seeding encourages rapid emergence and commencement of photosynthesis and reduces the incidence of root rot organisms[1] (Duczek and Piening, 1982). Seeding depth for oat and barley crops should not exceed 5 cm (2 in.).

If hullless oat and barley cultivars are produced, special care may be required. The exposed embryos of the naked kernels are damaged more readily, and germination for seed purposes may be reduced. Hullless oat cultivars do not store as well as conventional oat seed, and this may be associated with the fine hairs on the kernels, which provide an ideal environment for fungi that cause rapid spoilage. Storage is safest at a moisture level of less than 12.5%. High oil content can lead to rancidity and off-flavors. Yields of hullless oat and barley cultivars are as high as those of normal cultivars if allowance is made for hulls (Rossnagel, Harvey, and Bhatty, 1983). Performance aspects of a two-row hullless barley tested over approximately 50 station years in the Canadian prairies are shown in Table 22-1.

A possible switch from producing feed to producing milling oat for food purposes will require increased attention to production aspects. A milling oat is a high-quality product that equals or exceeds well-defined specifications of plumpness, kernel weight, freedom from foreign materials, freedom from discoloration, and off-flavors. Any recommended cultivar, hulled or hullless, is capable of producing a milling oat, but attention should be directed to reducing lodging, controlling disease, and harvesting promptly to reduce weathering and discoloration. Grain drying may be advisable, especially of a hullless cultivar, to reduce the grain to a safe storage level. Clarke et al. (1982) reported higher grades with artificial drying when rain occurred during the harvest period.

Choice of cultivar is an important consideration in producing a crop. For many areas, performance trials provide the best information on cultivar performance. Grain-yield performance is usually the major criterion for cultivar recommendations, as it reflects lodging and disease resistance and physiological factors associated with yield production. On-farm testing on a field scale provides producers with observations at a specific location. Table 22-2 compares six performance characteristics of ten oat cultivars. Harvest index values ranged from 40.2 to 48.9.

---

[1]Root rot organisms—*Cochliobolus sativas*, *Helminthosporium sativum*, and *Fusarium* spp. (*F. culmorum*).

Table 22–2
Comparisons for Six Traits among Ten Oat Cultivars Grown at Three Locations in Iowa in 1973 and 1974

Cultivar	*Grain Yield*		*Total Plant Biomass*		*Plant Height*		*Harvest Index, %*	*Heading Date, No. Days after May 31*
	kg/ha	*lb/ac*	*kg/ha*	*lb/ac*	*cm*	*in.*		
X434-II	3600	3214	7300	6518	95	37	48.9	16.6
Pettis	3500	3125	7300	6518	99	39	48.4	17.0
Clintford	3400	3036	7400	6607	91	36	45.6	18.7
Grundy	3400	3036	7100	6339	91	36	47.5	18.8
Portal	3300	2946	7500	6696	101	40	45.4	22.4
O'Brien	3200	2857	7100	6339	102	40	45.9	21.1
Otee	3200	2857	7300	6518	91	36	43.2	20.8
Garland	3000	2679	6700	5982	94	57	44.2	22.0
Tippecanoe	2900	2589	7200	6429	94	37	40.2	19.0
Dal	2800	2500	7000	6250	98	39	39.4	24.8

Source: Rosielle and Frey (1977).

Likewise, production aspects for malting barley must be considered, as cultural practices may determine whether the barley crop will meet the requirements for malting grades. Excessive nitrogen fertilizer may result in an undesirably high protein level. Skinned, cracked, or broken kernels are undesirable. Grain should be plump, uniform in size, mature, and have a high level of germination.

DORMANCY OF DORMOAT

Seeds of many species have a built-in form of dormancy to prevent the seeds from germinating in the fall when they fall from the seed to the plant. A short period of dormancy in cereals after maturity is desirable to help prevent sprouting in the head. The wild oat, *Avena fatua* L., has a form of dormancy that allows it to germinate in the spring. Oat and barley growers have attempted late fall seeding, just before freeze-up, to try and duplicate the annual cycle of the very well adapted wild oat plant. Seeding conventional oat and barley cultivars in this manner is seldom advantageous.

Burrows (1964) proposed a scheme to transfer the dormancy factor from wild oat into high-yielding modern cultivars. The resulting cultivars, called "dormoat," can be seeded in the fall to germinate very early in the spring. When properly handled (Burrows, 1970; Andrews and Burrows, 1972), dormoat was able to utilize the favorable growing conditions of early spring, mature early in the season, and escape disease, drought, and heat that often reduce the yield of late-seeded oat crops. Difficulties include volunteer oat plants in subsequent crops and slow spring germination if seeded too deeply, as unworked soil is slow to warm up.

FATUOIDS

Fatuoids, or false wild oat, occasionally appear in oat cultivars, possibly as a result of natural crossing, gene mutation, or **chromosome aberration** (Jensen, 1961, p. 132). Fatuoids are undesirable because they shatter as wild oat shatters at maturity. Fatuoids exhibit many of the characteristics of the cultivar in which they are found, except for a basal cavity on the seed, or so-called sucker mouth, dense and often very long hairs or bristles on the base of the lemma as well as on the rachilla segment, and a long, twisted, knee-bent awn on all forets of the spikelet (Figure 22-3). Stanton (1936) suggested that the occurrence of fatuoids might be considered a provi-

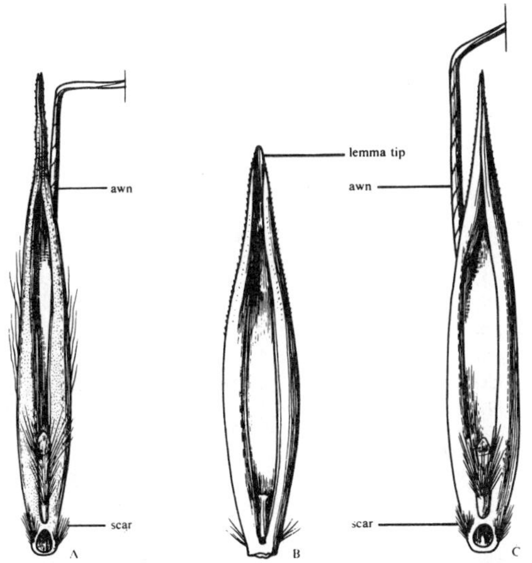

Figure 22–3. Typical seeds of wild oat (*Avena fatua* L.) (A), common oat (*Avena sativa* L.) (B), and a fatuoid oat (C). Fatuoids arise spontaneously in oat cultivars and resemble wild oat in that they have a typical basal cavity or sucker mouth, a prominent, twisted, knee-bent awn, and long hairs or bristles on the base of the lemma and on the rachilla segment. At maturity fatuoids shatter their seed similar to that of wild oat. (Courtesy of the Canadian Seed Growers' Association)

sion of nature to return cultivated oat to wild forms, thus making them self-propagating in case, through some disaster, the cultivated forms were no longer in the hands of human beings.

WILD OAT

Wild oat (*Avena fatua* L.) is an important weed problem in many small-grain producing regions of the world and is well adapted to grow and persist under the wide range of climatic soil conditions in these areas. The wide distribution of wild oat indicates considerable genetic diversity in the population, resulting in wide adaptation. Bowden (1971) described wild oat as one of the main problems of cereal production. The wild oat problem has increased as dwarf cultivars of wheat and other cereals have been introduced (Jordan, 1977). Wild oat causes losses directly through competition, resulting in decreased crop yields. Such losses occur early in the growing season, attaining a maximum before the crop is 45 to 50 days old.

Cultural attempts at wild oat control involve early seedbed preparation to encourage the wild oat seed germination followed by a second cultivation to remove germinated seedlings. Selective herbicidal weed control gives excellent control, but repeated use of single herbicide could result in the development of a tolerant wild oat population (Miller, Nalewaja, and Mulder, 1982). The variation found in wild oats indicates a potential within the natural population to become more weedy than others. Use of herbicides may favor evolutionary shifts to new aggressive types. Wild oat seed that is incorporated into the soil may not receive sufficient oxygen or other factors required for germination and may remain dormant for years until brought to the surface by tillage. A contained weed seed population achieved by shallow or zero-tillage offers some control of this weed (Bubar, 1983).

In addition to yield losses, wild oat causes serious losses from dockage. The seed is easily identified by its black lemma, sucker mouth, twisted bent awn, and pubescence. Cereal production costs are increased by additional tillage costs. After-harvest tillage to stimulate growth and allow subsequent destruction of wild oat is ineffective because of dormancy factors of newly produced seed. Cultural practices that delay spring seeding reduce yields, and chemical applications to control wild oat weed plants are energy-intensive and expensive.

CONCLUSIONS

Reliable performance, ease of production, acceptability as a feed, use as food, preference of barley for malt and brewing, adaptability to short-season, cool conditions with long-day photoperiods, and traditional conditioning have maintained the status of oat and barley as major grain crops. Although a position in the annals of history is secure, their future is less certain. The immediate future of barley appears to be assured

Table 22–3
Major Oat-Producing Areas of the World, Showing Production and Performance Changes over a Ten-Year Period.

	Area Harvested, Thousands of Hectares (Thousands of Acres)		Yield, kg/ha (lb/ac)	
Area	1969–1971	1979–1981	1969–1971	1979–1981
World	30,358 (74,984)	25,898 (63,968)	1,795 (1,603)	1,681 (1,501)
USSR	9,394 (23,203)	12,160 (30,035)	1,488 (1,329)	1,254 (1,120)
Europe	7,240 (17,883)	5,226 (12,909)	2,478 (2,213)	2,731 (2,438)
United States	7,051 (17,416)	3,742 (9,244)	1,893 (1,690)	1,929 (1,722)
Canada	2,834 (7,000)	1,603 (3,960)	1,943 (1,735)	1,988 (1,775)
Asia	1,474 (3,641)	782 (1,932)	997 (890)	1,476 (1,318)
Oceania	1,409 (3,480)	1,282 (3,166)	1,018 (909)	1,115 (995)

Source: *FAO Production Yearbook 1981* (FAO, 1982).

by the fact that world area of production in the ten-year period 1969–1971 to 1979–1981 has increased 20% along with a small yield increase of 4.8% (FAO, 1982).

In contrast, world oat production is declining (Table 22-3). Worldwide, the area of production has decreased 14.7% and yield per unit area has declined 6.3%. The six areas of the world in Table 22-3 account for 95.7% of total production, and five of these six have shown a decline in area of production. Yield increases have occurred in all areas except the USSR, where a 29.4% increase in area grown has occurred.

The reasons for this decline are evident from this chapter. In addition, the problems of the oat crop are not being addressed at any of the International Research Institutes.

Economic factors in competition with other cereal grains along with market forces will determine the future of the oat crop. Traditional farming methods are changing, and few things remain static. A changing agricultural scene is causing major changes in world oat and barley production.

Questions

1. Plot trends in yield and area of production for both oat and barley crops for a longer time period than presented in this chapter. Consider whether:
   a. The decline in the importance of the oat crop is a short-term or a longer-term phenomenon.
   b. Barley yields have only recently started to improve.
   c. Changes in oat and barley production aspects follow any cyclical trend.
2. Compare the strengths and weaknesses of the oat crop with those of other cereal grain crops that might replace it.
3. *a.* Based on observations in Questions 1 and 2 above, and from information in this chapter, what conclusions do you reach regarding the future production of oat as a major world cereal grain?
   b. How important do you believe the oat crop will be 50 years from now?
4. Conduct an analysis for barley similar to that for the oat crop in Questions 1 and 2. What conclusions do you reach regarding the future of barley as a major world cereal grain?
5. Compare the advantages and disadvantages of the oat and barley crops.

6. Suggest changes, if any, that might help restore the importance of oat as a major world cereal crop.
7. Based on the comparative data shown in Table 22-2, which oat cultivar would you advise a producer to select? Give reasons for your decision.
8. Oat and barley are long-day crops.
   a. How would you produce a plot of these crops outdoors in the tropics for breeding and observation?
   b. What is the importance of having daylength insensitivity in these crops from a production standpoint?
   c. How do you suppose a daylength-insensitive cultivar of oat or barley might perform under tropical or semitropical conditions? Explain your answer.
9. Clearly explain why an early-seeded oat or barley crop is likely to outyield a late-seeded crop under otherwise comparable field conditions.
10. What management changes might you consider to maximize grain yield in oat or barley production under early- and late-seeding conditions?

References

Andrews, C. J., and V. D. Burrows. 1972. "Germination Response of Dormoat Seeds to Low Temperature and Gibberellin." *Canadian Journal of Plant Science* 52: 295–303.

Anonymous. 1983. *Barley Production in Alberta.* Alberta Agriculture, Agdex 114/20-1. 36 pp.

Baker, R. J., and K. G. Briggs. 1982. "Effects of Plant Density on the Performance of 10 Barley Cultivars." *Crop Science* 22: 1164–1167.

Beaven, E. S. 1947. *Barley. Fifty Years of Observation and Experiment* London: Duckworth. 394 pp.

Bonnett, O. T. 1961. "Morphology and Development." (In) F. A. Coffman (ed.), *Oats and Oat Improvement.* Madison, Wis.: American Society of Agronomy. pp. 41–74.

Bowden, B. A. 1971. "Wild Oat: Menace of the Canadian Prairies." *Span* 14 (3): 139–141.

Briggs, D. E. 1978. *Barley.* London: Chapman and Hall. 612 pp.

Brinkman, M. A., and T. M. Luk. 1979. "Relationship of Spike Nodding Angle and Kernel Brightness under Simulated Rainfall in Barley." *Canadian Journal of Plant Science* 59: 481–485.

Brown, C. M. 1980. "Oat." (In) W. R. Fehr and H. H. Hadley (eds.), *Hybridization of Crop Plants.* Madison, Wis.: American Society of Agronomy and Crop Science Society of America. pp. 189–202.

Browning, J. A., K. J. Frey, and R. L. Grindeland. 1964. "Breeding Multiline Oat Varieties for Iowa." *Iowa Farm Science* 18 (8): 5–8.

Bubar, J. S. 1983. "Growth Habit and Control of Wild Oats." Agriculture Canada Publication No. 1983-1E.

Burrows, V. D. 1964. "Seed Dormancy, A Possible Key to High Yields of Cereals." *Agricultural Institute Review* 19: 33–35.

———. 1970. "Yield and Disease—Escape Potential of Fall-Sown Oats Possessing Seed Dormancy." *Canadian Journal of Plant Science* 50: 371–377.

Clarke, J. M. et al. 1982. "Effect of Kernel Moisture Content at Harvest and Windrow vs. Artificial Drying on Quality and Grade of Oats." *Canadian Journal of Plant Science* 62: 845–854.

Coffman, F. A. 1961. "World Importance and Distribution." "Origin and History." (In) F. A. Coffman (ed.), *Oats and Oat Improvement.* Madison, Wis.: American Society of Agronomy. Chaps. I and II, pp. 41–74.

Danhof, C. H. 1969. *Change in Agriculture: The Northern United States, 1820–1870.* Cambridge, Mass.: Harvard University Press. 322 pp.

Duczek, L. J., and L. J. Piening. 1982. "Effect of Seeding Depth, Seeding Date, and Seed Size on Common Root Rot of Spring Barley." *Canadian Journal of Plant Science* 62: 885–891.

FAO. 1982. *FAO Production Yearbook 1981.* Vol. 35. Rome, Italy: Food and Agriculture Organization of the United Nations. FAO Statistics Series No. 40.

Harlan, J. R. 1968. "On the Origin of Barley." (In) *Barley: Origin, Botany, Culture, Winterhardiness,*

Genetics, Utilization, Pests. Agriculture Handbook No. 38. Agriculture Research Service, United States Department of Agriculture. pp. 9–31.

———. 1975. *Crops and Man.* Madison, Wis.: American Society of Agronomy and the Crop Science Society of America. 295 pp.

Heiser, C. B., Jr. 1973. *Seed to Civilization: The Story of Man's Food.* San Francisco: W. H. Freeman and Company. 243 pp.

Housley, T. L., and D. M. Peterson. 1982. "Oat Stem Vascular Size in Relation to Kernel Number and Weight. I. Controlled Environment." *Crop Science* 22: 259–263.

Jensen, N. F. 1961. "Genetics and Inheritance in Oats: Inheritance of Morphology and Other Characters." (In) F. A. Coffman (ed.), *Oats and Oat Improvement.* Madison, Wis.: American Society of Agronomy. pp. 125–206.

Jordan, D. 977. "A Decade of Wild Oats." *Span* 20 (1): 21–24.

Kerridge, E. 1973. *The Farmers of Old England.* London: George Allen and Unwin Ltd. 180 pp.

Langer, I., K. J. Frey, and T. B. Bailey. 1978. "Production Response and Stability Characteristics of Oat Cultivars Developed in Different Eras." *Crop Science* 18: 938–942.

Lawes, D. A. 1971. "Oat Improvement—Recent Research and Developments." *Field Crop Abstracts* (Review Article) 24: 203–215.

McMullen, M. S., R. L. Phillips, and D. D. Stuthman. 1982. "Meiotic Irregularities in *Avena sativa* L./*A. sterilis* L. Hybrids and Breeding Implications." *Crop Science* 22: 890–897.

Miller, J. P., and M. A. Brinkman. 1983. "Inheritance of Spike Nodding Angle in Spring Barley" *Crop Science* 23: 688–691.

Miller, S. D., J. D. Nalewaja, and C. E. G. Mulder. 1982. "Morphological and Physiological Variation in Wild Oat." *Agronomy Journal* 74: 771–775.

Murphy, C. F., R. C. Long, and L. A. Nelson. 1982. "Variability of Seedling Growth Characteristics Among Oat Genotypes." *Crop Science* 22: 1005–1009.

Peterson, D. M., T. L. Housley, and T. M. Luk. 1982. "Oat Stem Vascular Size in Relation to Kernel Number and Weight. II. Field Environment." *Crop Science* 22: 274–278.

Pfeifer, R. P., and J. P. Kline. 1960. "A Major Cause of Winter-kill of Winter Oats." *Agronomy Journal* 52: 621–623.

Rajhathy, T., and H. Thomas. 1974. *Cytogenetics of Oats (Avena L.).* Miscellaneous Publications of the Genetics Society of Canada. No. 2. Ottawa. 90 pp.

Rodgers, D. M., J. P. Murphy, and K. J. Frey. 1983. "Impact of Plant Breeding on the Grain Yield and Genetic Diversity of Spring Oats." *Crop Science* 23: 737–740.

Rosielle, A. A., and K. J. Frey. 1977. "Inheritance of Harvest Index and Related Traits in Oats." *Crop Science* 17: 23–28.

Rossnagel, B. G., B. L. Harvey, and R. S. Bhatty. 1983. "Scout Hullless Barley." *Canadian Journal of Plant Science* 63: 751–752.

Stanton, T. R. 1936. "Superior Germ Plasm in Oats." (In) *Yearbook of Agriculture.* Washington, D. C.: United States Department of Agriculture. pp. 347–414.

Starling, T. M. 1980. "Barley." (In) W. R. Fehr and H. H. Hadley (eds.), *Hybridization of Crop Plants.* Madison, Wis.: American Society of Agronomy and Crop Science Society of America. pp. 189–202.

Wych, R. D., and D. D. Stuthman. 1983. "Genetic Improvement in Minnesota-Adapted Oat Cultivars Released Since 1923." *Crop Science* 23: 879–881.

Wych, R. D., R. L. McGraw, and D. D. Stuthman. 1982. "Genotype × Year Interaction for Length and Rate of Grain Filling in Oats." *Crop Science* 22: 1025–1028.

23
Wild Rice

INTRODUCTION

Wild rice does not rank in importance with the major world cereal crops described in this book, but it is included because it offers modern humans an opportunity to participate in domesticating a wild crop. An increasing demand for wild rice grain, an attractive price, selection of cultivars with reduced grain shattering, and perhaps recognition of the potential associated with a new crop have led to increased interest in this crop. The time has come to move wild rice out of its native habitat of shallow lakes and streams, where it has existed for centuries, into a system of controlled commercial production. Commercial wild rice production is in its infancy, but the agricultural possibilities have been established.

Wild rice, *Zizania palustris* L., also known as water oat, is native to North America. Although it had been a food source for Indian tribes, early explorers, fur traders, and trappers (Steeves, 1952; Stichney, 1896), no effort was made to domesticate the crop. Present attempts at commercial production lack the test of time. The harvesting procedures involved are essentially a reenactment of the ancient steps taken by early peoples undergoing the transition from nomadic hunters to agriculturists. Much of the present wild rice harvest in Canada is done by native peoples from natural stands using traditional techniques. To be successfully domesticated, the crop must be transformed from one adapted to reseeding and survival on its own to one modified to meet commercial production methods. Domestication represents a major transformation: a shift away from the seed-shattering habit so that mechanical harvesting can be accomplished. An extended seed dormancy and underwater seed storage present challenges to commercial production.

Wild rice has been harvested by native Indians for centuries, and these people seek exclusive right to harvest the crop. A "Wild Rice Harvesting Act" was passed by the government of Ontario in 1960, which legislates that harvest of indigenous stands is legal only with a license. Government funding, although limited, has been directed to aid native people in harvesting and in erecting water-control structures and processing facilities. As yet, little government funding has been directed to plant breeding and development or to commercial production studies in controlled water paddies.

Wild rice domestication technology and commercial production truly are in their infancy.

A PERSPECTIVE ON WILD RICE

Shatter-resistant cereal crop plants were found and improved centuries ago. In wild rice, a major step toward domestication occurred with the discovery in 1963 of a cultivar with reduced grain shattering, which became known as Algot Johnson after the grower who selected it (Elliot and Oelke, 1977; Elliot and Perlinger, 1977). Shattering losses still occurred, but the use of a more shatter-resistant cultivar permitted producers to use harvesting equipment similar to that for cultivated rice, *Oryza sativa*, and stimulated interest in commercial production (Oelke et al., 1973). Delay in seed abscission allows for destructive machine harvesting of a near-full crop in a single sweep.

Unlike triticale, in which the feed or food potential needs to be determined, wild rice is a highly marketable resource (Figure 23-1). It is recognized as an epicurean delight. After processing, its nutty flavor and chewy texture make it a prized gourmet delicacy throughout the world. An inflated price has kept wild rice out of the mass market, so that it is virtually unknown to the average consumer. Perhaps an increase in production per unit area and a lowered price might enhance the appeal and result in increased consumption. Characteristically the high price has resulted in the marketing of wild rice in the gourmet section of food stores, in specialty shops, or in blends with common rice. A wild rice marketing board may be needed to determine a suitable price, to undertake the giant task of education and promotion, and to effectively distribute and market the product nationally and internationally.

Figure 23–1. The mature caryopsis of wild rice is about 16 mm long and 1.5 mm in diameter and dark brown to black in color. The embryo or germ extends the entire length of the grain along the grooved side. The slender grain must be threshed and handled carefully to prevent breaking. Following harvest, the grain is dried and then parched in ovens to remove any moisture. Cooking transforms the hard grains into fluffy, brown-gray curls prized by gourmets for their distinctive, nutty flavor and chewy texture. Wild rice swells to four times its volume when cooked. (Courtesy of Ontario Ministry of Agriculture and Food)

From a nutritional standpoint, wild rice compares favorably with other cereal grains. Total protein, using a recommended 5.7 multiplication factor for converting Kjeldahl nitrogen to protein, was 13.8% (Watts and Dronzek, 1981). Amino acid composition varied little among the 12 samples of Canadian-grown wild rice analyzed. Threonine and lysine were the most limiting amino acids. Total carbohydrate content was 82.8%, ash 1.7%, and crude fiber 1.2%. These values are similar to those of other cereal grains, and direct comparisons can be made with values in Appendix Table D.

Plant-breeding programs have been initiated with the objective of producing domesticated cultivars suited to commercial paddy rice production (Elliott, 1976; Elliott, 1980; Everett and Stucker, 1983; Woods and Gutek, 1974). In its native state, wild rice is not suitable for production under conventional, modern agronomic practices (Harlan, de Wet, and Price, 1973). This is not solely because of its seed-shattering habit, but because of asynchronous tiller maturity, seed-storage problems to allow spring seeding to take place in paddy fields, the soft, fragile nature of the long, slender seed, earliness, plant height, and yielding ability. The speed of improvement depends on the variation that is readily available for rapid exploitation. The amount of variation for four agronomic traits based on 100 randomly selected plants in this heterozygous species is shown in Table 23-1 (Foster and Rutger, 1980).

Domestication, production, breeding and selection, cultural practices, pest-management systems, and a score of associated promising and problem attributes offer a new and largely unexplored potential to develop a food crop (Zarkin, 1973). The reward for efforts devoted to wild rice is an industry that could use

Table 23–1
Mean and Range of Four Agronomic Wild Rice Characters Studied at Davis, California, in 1973

Character	Mean	Range
Heading days (days from June 1)	49.8	40.0 to 56.0
Plant height in cm (in.)	139.7 (55)	95.6 to 175.9 (37.6 to 69.3)
Tiller number	11.3	4.3 to 17.5
Seed length (mm)	13.2	11.5 to 15.1

Source: Foster and Rutger (1980).

economically depressed, essentially nonproductive swampland regions. Wild rice offers potential to increase world food output on continually flooded and agriculturally limited areas (van der Meer, 1983). Through genetic engineering, desirable genes in the wild rice crop may prove to be useful in existing cereal crops, if they can be successfully transferred. If wild rice is to be exploited fully and intelligently, an understanding of the crop is essential.

As early as 1952, concern was expressed about the dwindling natural stands of wild rice (Nickels, 1952). The carp,[1] a coarse, omnivorous fish that favors shallow, warm waters where wild rice grows well, may be partially responsible. These fish dislodge plants, consume seed, and stir up the mud sediment. Carp entered Ontario waters early in the 1900s and have continued to spread and increase rapidly.

Acid rain may be a major threat to indigenous stands of wild rice. Wild rice succeeds best in water with a *p*H between 7.5 and 8.5. Thousands of lakes in North America already have been destroyed by acidity from this insidious killer (Demitor, 1982).

Centers of origin and detailed surveys on the distribution of wild rice have not been made. Wild rice occurs over a large portion of North America, but there are areas where wild rice always has been absent and other areas where it is sparsely scattered. This has led to a general conclusion that the ecological range of wild rice is limited. In support of this view, efforts to introduce the crop to England were unsuccessful. On the other hand, intentional plantings into outlying regions in North America have been successful and have performed better than in the area from which it was introduced. Such evidence indicates that the potential range is very great, probably encompassing suitable conditions within the temperate and tropical portions of all continents (Dore, 1969).

Possible reasons for a seemingly narrow distribution include unsuitable habitats, certain biological barriers, or physiographic limitations. Perhaps wild rice migrates slowly and has not reached the ultimate limits of its range. Flowing water is the prime means of dissemination of seeds, but plants cannot move upstream. Perhaps native people intentionally or otherwise spread wild rice seed, but only fresh grains from newly harvested plants would be viable. Only recently have conservationists, interested in providing food for wildlife, made a conscious effort to introduce wild rice into remote wilderness areas and encourage its production.

Threats from pollution, acid rain, carp, cottagers destroying wild rice stands along the shoreline in front of their cottage, and power boats churning the water may individually or collectively threatened the native stands of the wild rice resource. Just as, among the other major cereal grains, cultivars have been collected and preserved in seed-storage units, so too should action be taken to assemble the genetic variability found in wild rice. Elliott (1980) indicated he had collected a wild rice germplasm pool consisting of 89 genotypes assembled from Canada and the state of Minnesota in the United

[1]Carp—*Cyprinus carpio* L.

States. Perhaps international efforts between Canada and the United States are needed to ensure a more complete collection of genotypes.

CLASSIFICATION

Most reports refer to wild rice as *Zizania aquatica* L., based on the classification by Fassett (1924). All wild rice found in North America has a chromosome number of $2n = 30$ and should not be confused with wild forms of *Oryza*. A species of wild rice found in China and northeast Asia has a chromosome number of $2n = 34$ and was recognized as *Zizania latifolia*. Fassett (1924) also recognized a wild rice growing in a small area of Texas as *Zizania texana* (Hitchc.).

On the basis of morphological differences, Dore (1969) split *Zizania aquatica* into two species: *Zizania aquatica* and *Zizania palustris* L. Elliott (1980) reported that W. A. P. Emery was able to make crosses and reciprocal crosses among the three North American species. He suggested that all belong to *Zizania palustris* L.

Advances in science have resulted in new insights and understandings as to what constitutes a species within the genus *Zizania*. Generally, *Zizania palustris* L. is becoming the recognized name to identify the wild rice species (Dore, 1975; Elliott and Perlinger, 1977; Oelke and Albrecht, 1980).

The indigenous wild rice of North America morphologically bears little resemblance to the cultivated rice, *Oryza sativa*. Wild rice belongs to the grass tribe *Zizanieae*, and cultivated rice to the tribe *Oryzeae*. Wild rice has 15 chromosome pairs, cultivated rice 12 pairs. Reeder (1957), however, concluded that studies of the embryos suggest the two species are closely related. Weir and Dale (1960) found a close similarity in the anatomy of leaves as well as in other anatomical features. Both species have six stamens, a situation unique among Canadian grasses. Perhaps some similarities are to be expected, just as certain similarities exist among the other cereal grasses.

REPRODUCTION

Wild rice is an annual grass which ranges in height from 150 to 300 cm (5 to 10 ft). Plants are monoecious, with unisexual florets borne on a terminal panicle 20 to 25 cm (8 to 10 in.) in length. The pistillate or female florets are borne on the upper portion of the panicle, are held upright on short stiff pedicels, and are held closely to the panicle rachis.

The staminate or pollen-producing florets hang on threadlike pedicels from the spreading panicle branches (Figure 23-2) and drop off a day or two after they shed their pollen. The unisexual nature of the florets makes wild rice well adapted to cross-pollination, but wild rice can set seed by self-pollination when panicles are bagged or grown in isolation. On the inflorescence, at the junction of pistillate and

Figure 23–2. A wild rice inflorescence or panicle. The monoecious wild rice plant bears unisexual flowers on the same head. The upper portion bears pistillate florets that are held upright on short, stiff pedicels close to the panicle rachis. The staminate florets hang pendulously on the threadlike pedicels from the spreading branches on the lower portion of the panicle. After pollen is shed, the staminate florets drop off. Seeds are borne in the pistillate flowers only. (Courtesy of Ontario Ministry of Agriculture and Food)

staminate spikelets, sometimes there are florets with well-developed stamens and ovaries. These florets may be self-pollinated and may produce seed.

Maturation proceeds from the tip downward, and each inflorescence lasts for a number of days. The staminate florets begin to shed their pollen just as the last pistillate florets have been pollinated or reach anthesis. Self-pollination is therefore unlikely in unisexual florets.

The two outer glumes that define a spikelet in other grasses are lacking in staminate spikelets, and in the pistillate spikelets they appear as rudimentary cup-shaped bracts.

Dore (1969) noted that all pistillate (female) florets on the panicle do not produce grain and reported that the amount of sterility may be as high as 80%. The cause of such low and variable seed set is not known, but it may be associated with factors influencing both pollination and fertilization. Hot weather with low humidity at the time of flowering may shrivel the stigmatic hairs on the bilobed stigma. Wet, rainy weather may reduce bursting of the stamens and pollen drift. Perhaps the very first pistillate florets may not get pollinated if stamens are not exposed until a few days later. Secondary tillers with an asynchronous tillering habit and produced late in the season have less chance of being pollinated.

The lemma and palea of pistillate florets are interlocked and do not spread widely apart, as do the staminate flowers. Pistillate flowers separate at the base and provide a slit only wide enough to allow the stigmas to protrude.

GROWTH AND DEVELOPMENT

In the spring, seeds in the mud, gravel, sand, or ooze material at the bottom of shallow lakes or rivers germinate and produce two or three long, soft, ribbonlike leaves. These submersed leaves may be seen streaming out in the water with the current. Neither the upper nor the lower epidermis of submerged leaves contains stomata or chloroplasts. These soft leaves soon decay as upper leaves are formed.

As plant growth reaches the surface, one or two limp, floating leaves per plant are produced that are long, lie flat on the water, and effectively cover the water surface to intercept sunlight. These second leaves are juvenile leaves and are part of the vegetative growth period. It seems necessary for these leaves to reach the surface before a flowering stem can develop. Floating leaves have an upper surface which is never wet. A cuticle prevents the surface from becoming wet, as water tends to ball up and roll off. No stomata are present on the undersurface of floating leaves. Airspaces are found on either side of the midrib, which is often a little nearer to one margin than to the other. Bulliform cells are a feature of the upper surface of floating leaf blades.

A third type of aerial leaves develop which are stiff and upright. Bulliform cells are present in aerial leaves. The total number of leaves in each category varies with water depth and the total height of the plant. In water 60 cm (2 ft) deep, total plant height may be 120 cm (4 ft) from base to flower heads. In deeper water the internodes elongate.

Plants bear four to six leaves which are 1 to 5 cm (0.39 to 2 in.) in width and up to 65 cm (26 in.) long. The leaf sheath clasps the stem, and the edges, which are often streaked with blue, overlap, so that the stalk is completely covered from the node where the leaf arises almost to the node above. The nodes are covered with dense, fine hairs.

The stem between the internodes is hollow but is divided at various intervals by partitions, or **diaphragms** (Figure 23–3). These partitions lack vascular and strengthening tissue, are very porous, and allow for gas movement up and down the stem. This is an important feature to allow aeration to the base of the stem, if the roots are to grow actively under submerged conditions.

Wild rice stems have a very weak physical structure. Air spaces interspersed throughout the living tissues, even root tissues, make the whole plant particularly buoyant. Stems and plants could not support themselves well in an upright position in the water if it were not for their

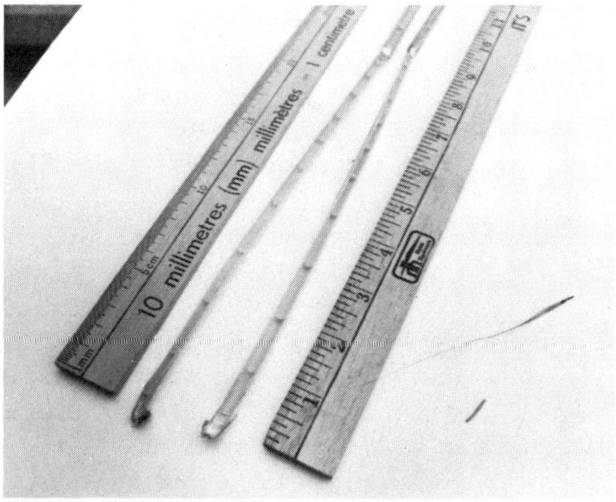

Figure 23–3. The culm of wild rice is hollow and divided into nodes and internodes. The internodes are divided by transverse diaphragms or pseudonodes, as shown in this split culm. These partitions divide the stem into watertight compartments so that the plant will float, even if a section of the stem is damaged or punctured. Internode lengths of the plants in this photo are about 25 cm (10 in.) apart, and airtight partitions are about 2.5 cm (1 in.) apart. (Courtesy of Ontario Ministry of Agriculture and Food)

acquired buoyancy. Uprooted plants float quickly to the surface and never regain a roothold.

Tillering or branching sometimes occurs, particularly in isolated plants and usually at the first or second node, 40 to 90 cm (15 to 35 in.) from the soil surface (Figure 23–4), depending on water depth.

The root system is fibrous and spreading. Roots are clustered at the base of the stem, grow diagonally into the mud, and anchor the plant against the lifting force of waves and currents. Wild rice roots do not have the true microscopic and unicellular root hairs that are present on land plants. Instead the roots are straight, spongy, thick, and white.

Flowering occurs from the middle of July through August in Ontario, Canada, where wild rice occurs as far north as 53° N latitude.

During grain development the caryopsis undergoes color changes. In the dough stage, the grain is light green, changes to an olive brown, and finally at maturity becomes dark brown or black. Seeds are shed in the firm-dough stages.

The shape of the seed, with its heavy base, the enveloping thin lemma and palea, and the bristle attached to the lemma (Figure 23-5), results in dropped seed plummeting directly to the bottom. The seed become firmly lodged in the flooded soil and is not carried any distance by the current. The spread of wild rice is slow, but seeds are not swept into deep water and lost.

Because grain falls from the mother plant probably as soon as physiological maturity occurs, a single harvest will miss many of the grains. The traditional method of harvesting employed by Indians is to press plants against the gunwales of a canoe with a paddle (Figure 23-6), thereby leaving the plant intact and undamaged for a subsequent harvest. Only a few mature grains may be collected at one time, and repeated collections from shattering cultivars are required.

Figure 23–4. Wild rice plants pulled out by the roots and placed beside meter sticks reveal a height of about 2.5 m (8.2 ft). Branches or tillers are shown arising from the second node. The long, ribbonlike leaves are apparent. These plants bear mature seed and were pulled in late August from a lake in northern Ontario. (Courtesy of Ontario Ministry of Agriculture and Food)

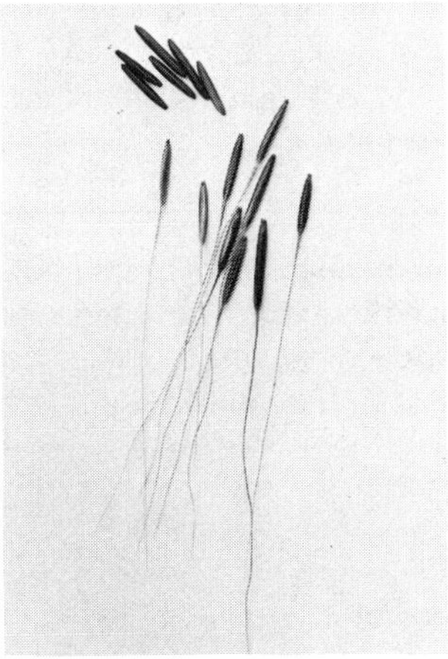

Figure 23–5. Wild rice kernels have a papery thin lemma and palea completely surrounding the caryopsis. The tip of the lemma tapers off to become an awn. The black, cylindrical naked grain is shown. (Courtesy of Ontario Ministry of Agriculture and Food)

The depth of water in which wild rice grows may range from 15 cm (6 in.) to 110 cm (42 in.), but water depth of 210 cm (84 in.) is possible if the depth is fairly constant. Wild rice seldom grows where variations in water level annually exceed 90 cm (35 in.). Flooding may uproot the plant if in the floating leaf stage.

The ideal habitat for wild rice is in the shallow waters of lakes and ponds (Figure 23-7), or along the borders of sluggish streams.

SEED DORMANCY

Because wild rice is an annual plant, it is dependent on reseeding each year. For species survival in a climate subject to the extremes of temperature found in Canada, some degree of dormancy is essential. To prevent premature sprouting in the fall and to permit survival

Figure 23–6. Seeds of indigenous wild rice shatter when mature and can be harvested several times if nondestructive methods are used. The traditional method of harvesting is to press the panicles against the gunwales of a canoe or boat and dislodge the seed into the bow of the boat. Wild rice is an annual, but because so much seed is dropped, the stand for the following year is assured. (Courtesy of Ontario Ministry of Agriculture and Food)

Figure 23–7. Wild rice stands are found in protected coves and shallow bays of lakes and rivers, just outside the zone of cattails and among water lilies. The establishment of commercial wild rice production may not mean that rice becomes transferred entirely to farm paddy fields. Native stands, shown in this photo, may serve as sources of plant-breeding material, as an area for cross-pollination, and for selection purposes. New stands may be established of selected cultivars for breeding or for food purposes and to attract and provide a food source for wildlife, including deer, moose, and muskrat and wild fowl such as ducks and geese. Wild rice is able to effectively utilize areas that are otherwise of no commercial agricultural value. (Courtesy of Ontario Ministry of Agriculture and Food)

through the winter under ice, seeds need a minimum dormancy period of approximately eight months (September to May). In addition, when mature seeds drop from the adult plant in the fall, they go right to the bottom and penetrate the mud, which is presumably low in oxygen. Hence for seeds to germinate in spring, it is reasonable to expect that some selective adaptation has occurred that favors loss of dormancy under low-oxygen conditions.

Seeds fully ripened on the stalk have a definite dormancy of three months or longer (Simpson, 1966), even if temperature, moisture, and substrate are satisfactory for growth. Seeds require an after-ripening period involving freezing or near-freezing temperatures before the embryo will break dormancy.

When seed is stored dry, germinability is lost rather rapidly, and after one month germination can be reduced to less than 50%. Simpson (1966) reported that air-drying resulted in a detectable decrease in germination within 14 days, and within 90 days viability was almost completely lost.

Wild rice seed obtained from several lakes in northern Saskatchewan had a six-month dormancy period when stored at 1° to 3°C (34° to 37°F). Under natural conditions, not all seeds germinate after six months. Evidence exists that in a considerable quantity of seed, dormancy may delay germination for at least 18 months (Steeves, 1952).

To facilitate plant-breeding programs, means of breaking dormancy are desirable. Simpson (1966) tried pricking the seeds to allow moisture to penetrate the impermeable pericarp. This reduced the dormancy period but did not eliminate it. Freezing seeds to −20°C (−4°F) killed the embryos.

Ultrasonic vibration appeared to reduce dormancy, but material tested had been after-ripened three months (Halstead and Vicario, 1969).

LaRue and Avery (1938) grew excised embryos on agar culture successfully if the seeds were mature but "unripe," a vague and uncertain term.

Woods and Gutek (1974) reported successful germination of mature seeds during the commencement of the dormancy period by scraping the pericarp from above the embryo with a scalpel. Although this method is cumbersome and practical only on a small scale, it did allow germination in petri plates within two weeks and allowed for three or more generations to be grown in one year. This method, however, reduced seedling vigor and may result in a low survival rate. These observations were verified by Cardwell, Oelke, and Elliot (1978).

Mechanical scarification to facilitate moisture penetration through the impervious pericarp gives germination percentages of 44% to 60%. Best results were obtained from seeds germinated in water (Oelke and Albrecht, 1978).

Soaking seed for seven hours in 43% ethanol was reported by Oelke and Albrecht (1980) to increase germination from 15% with no treatment to 61%.

PRODUCTION

Viable seed that will germinate quickly following spring seeding into man-made rice paddies is a prime requirement. Seed can be overwintered by storing it in water at 0.5° to 2°C (33° to 36°F).

Seeding can be achieved by planting into dry ground and then flooding, or by flooding and then seeding from the air. About 45 kg/ha (40 lb/ac) of seed are required to obtain a stand, but after the first year the paddy will be reseeded from shattered seed.

Yields are not high. Native stands harvested with traditional methods yield only 55 to 85 kg/ha (50 to 75 lb/ac). Under field paddy conditions yields of up to 1200 kg/ha (1000 lb/ac) of unprocessed rice at about 40% moisture are possible. It takes about two to three units of green rice to produce a ready-to-cook wild rice. Harvest efficiency is reduced by uneven ripening over a ten- to 14-day period, and only 350 to 700 kg/ha (325 to 700 lb/ac) are obtained with nonshattering cultivars that are machine-harvested.

Optimum water depth, plant population, fertilizer rates and analysis, thinning techniques, and the best harvest time need to refined by research and experience.

Problems with wild rice culture include birds, especially blackbirds (although ducks will be attracted, they often eat fallen seed), wildlife such as muskrats and deer, fluctuating water levels, windstorms, insects, and diseases. Management is complicated by dropped seed that can remain viable for years. The result may be overly thick stands that need thinning. The process of changing a field paddy to an improved cultivar is complicated by volunteer plants that out-cross with the new cultivar, resulting in plants with new genetic combinations in subsequent years.

The value of the straw as forage should not be overlooked as it has nutritional value even after the grain is mature. Removal of wild rice residues helps control pests. A practical method of doing this, and at the same time encouraging reseeding of the crop for the next year, remains to be developed.

Processing of the grain appears to be more an art than a science. The first step is drying the green rice by spreading it thinly over a large area. The purpose of drying is to reduce the moisture content to prevent mold and decay. Drying also is associated with curing, a term that might seem to imply a mellowing period, and must be done gradually. Gradual drying allows the grain to become firm and black.

Parching or roasting is achieved with low heat to drive off all residual moisture. Parched grain may be cleaned, graded, and packaged for marketing.

Future wild rice production may not all be transferred to farm paddy fields. Stands may be established in lakes and streams for cross-pollination and selection purposes. As wild rice production aspects become better known, plantings may be undertaken. New stands may be established by casting freshly harvested, viable seed on the water surface near the shore from a boat. Shore seeding is less effective but can be aided by making mudballs containing a few seeds each and casting them into the water.

The water in protected coves and shallow bays of lakes, just outside the zone of cattails and among the water lilies, is a good place to plant.

Wild rice does not compete well with other vegetation and will not grow in the shade of over hanging trees. Neither saltwater marshes and lagoons along the seacoast nor small bog-margined lakes are suitable for wild rice production.

CONCLUSIONS

Even before becoming a domesticated crop, wild rice grain is well established as a food delicacy. The process of domestication has only begun, and an opportunity exists for participation in developing a historic crop. Whether wild rice ever ranks as a major world grain will depend on the success of the domestication process.

Along with commercial importance, wild rice may be of genetic value in expanding the food potential of poorly drained swamplands at high latitudes, providing additional and exciting variability to food, and possibly providing specific genes for transfer to other crops such as cultivated rice. Wild rice is a cross-pollinated crop, a desirable attribute in an era of hybrid crop development.

Learning to utilize the features exhibited by wild rice presents students of grain crops with a challenging opportunity. It provides an opportunity to participate in and record the process of domestication that in other world crops took place centuries ago. Perhaps wild rice will clarify some steps that have been missing in our reconstruction of the processes of domestication.

QUESTIONS

1. What steps do you consider essential in the successful domestication of wild rice?
2. Wild rice is monoecious. Clearly determine the meaning of this term and state what other cereal grain crops are monoecious.
3. *a.* What may be the greatest deterrent to domestication of wild rice?
   b. Why was wild rice not domesticated earlier?
4. Consider the various aspect of marketing.
   a. Should wild rice be promoted as an epicurean delight or for the mass market?
   b. Should production be controlled so as to keep the price high, or should a "laissez-faire" attitude be adopted?
   c. What sort of publicity should be undertaken?
5. What might be the primary objectives of a wild rice breeding program?
6. List features of the wild rice plant that you consider valuable, and explain why.
7. Speculate on the spread of wild rice in terms of:
   a. Limitations as compared to other crops.
   b. How it was achieved.
   c. Whether a center of origin could be found, and how the search could be undertaken.
8. Heiser (1973), in a brief statement on wild rice (p. 89), considered that the introduction of mechanical means to harvest native stands of the crop could have proved disastrous if their use had gone unchecked. Consider and explain the validity of this observation.
9. What production aspects need to be researched and resolved to ensure the commercial production of wild rice in field paddies?

REFERENCES

Cardwell, V. B., E. A. Oelke, and W. A. Elliot. 1978. "Seed Dormancy Mechanisms in *Zizania aquatica* L." *Agronomy Journal* 70: 481–484.

Demitor, P. R. 1982. "The Silent Killer: Acid Rain." *Sportsmen Magazine*. 1982. pp. 6–10.

Dore, W. G. 1969. *Wild Rice*. Canada Department of Agriculture Publication No. 1393. 84 pp.

———. 1975. "Rapid Evolution of a Cultigen Cereal." *Abstracts of Technical Papers*. Eastern Section Meeting, Canadian Society of Agronomy, and Northeastern Branch Meeting, American Society of Agronomy. University of Guelph. p. 37.

Elliott, W. A. 1976. *Wild Rice Breeding Research*. Progress Report of Wild Rice Research. St. Paul, Minn.: Minnesota Agriculture Experiment Station. pp. 51–64.

———. 1980. "Wild Rice." (In) W. R. Fehr and H. H. Hadley (eds.), *Hybridization of Crop Plants*. Madison, Wis.: American Society of Agronomy and Crop Science Society of America. pp. 721–731. pp. 721–731.

———, and E. A. Oelke. 1977. "New Era for Wild Rice." *Crops and Soils* 29 (9): 8–11.

———, and G. J. Perlinger. 1977. "Inheritance of Shattering in Wild Rice." *Crop Science* 17: 851–853.

Everett, L. A., and R. E. Stucker. 1983. "A Comparison of Selection Methods for Reduced Shattering in Wild Rice." *Crop Science* 23: 956–960.

Fassett, N. C. 1924. "A Study of the Genus *Zizania*." *Rhodora* 25: 153–160.

Foster, K. W., and J. N. Rutger. 1980. "Genetic Variation of Four Traits in a Population of *Zizania Aquatica*." *Canadian Journal of Plant Science* 60: 1–4.

Halstead, E. H., and B. T. Vicario. 1969. "Effects of Ultrasonics on the Germination of Wild Rice (*Zizania aquatica*)." *Canadian Journal of Botany* 47: 1638–1640.

Harlan, J. R., J. M. J. de Wet, and E. G. Price. 1973. "Comparative Evolution of Cereals." *Evolution* 27: 311–325.

Heiser, C. B., Jr. 1973. *Seed to Civilization: The Story*

of Man's Food. San Francisco: W. H. Freeman and Company. 243 pp.

La Rue, C. D., and G. S. Avery. 1938. "The Development of the Embryo of *Zizania aquatica* in the Seed and in Artificial Culture." *Torrey Botanical Club Bulletin* 65: 11–21.

Nickels, N. 1952. "Is the Wild Rice Crop Doomed?" *Rod and Gun* 54(11): 33.

Oelke, E. A., et al. 1973. *Commercial Production of Wild Rice*. University of Minnesota Agriculture Extension Service, Extension Folder 284.

Oelke, E. A., and K. A. Albrecht. 1978. "Mechanical Scarification of Dormant Wild Rice Seed." *Agronomy Journal* 70: 691–694.

———. 1980. "Influence of Chemical Seed Treatments on Germination of Dormant Wild Rice Seeds." *Crop Science* 20: 595–598.

Reeder, J. R. 1957. "The Embryo in Grass Systematics." *American Journal of Botany* 44: 756–758.

Simpson, G. M. 1966. "A Study of Germination in the Seed of Wild Rice (*Zizania aquatica*)." *Canadian Journal of Botany* 44: 1–9.

Steeves, T. A. 1952. "Wild Rice—Indian Food and a Modern Delicacy." *Economic Botany* 6: 107–142.

Stickney, G. P. 1896. "Indian Uses of Wild Rice." *American Anthropologist* 9: 115–121.

Van der Meer, J. P. 1983. "The Domestication of Seaweeds." *BioScience* 33: 172–176.

Watts, B. M., and B. L. Dronzek. 1981. "Chemical Composition of Wild Rice Grain." *Canadian Journal of Plant Science* 61: 437–446.

Weir, C. E., and H. M. Dale. 1960. "A Developmental Study of Wild Rice, *Zizania aquatica* L." *Canadian Journal of Botany* 38: 719–739.

Woods, D. L., and L. H. Gutek. 1974. "Germinating Wild Rice." *Canadian Journal of Plant Science* 54: 423–424.

Zarkin, D. A. 1973. "Wild Rice, Historic Commodity Under Study." *Minnesota Science* 29 (1): 13–14.

Glossary

Abscisic acid. A plant hormone that inhibits rather than stimulates plant growth.

Accessions. Individual plant entries in world collections.

Acclimatization. A plant response to stress conditions to increase tolerance to a specific environmental variable such as low temperature (see *Adaptability*).

Active collections. In gene banks, these are the working collections from which seed may be drawn as needed for distribution for evaluation and plant breeding. Active collections are considered as a medium-term form of storage (see *Base collections*).

Actual evapotranspiration. The actual water loss that occurs from a field in which a crop is growing, calculated from meteorological parameters.

Adaptation. In plants, a modification in the structure or function of plant processes to fit minor environmental changes (see *Acclimatization*).

Adventitious. Secondary roots or shoots arising from an internode other than the expected or usual source. Also known as coronal or permanent roots.

Aeciospores. Wind-borne spores released from barberry leaves that cause the initial infection of stem rust in cereals and grasses.

Aecium (pl. **aecia**). An orange-yellow lesion that forms on the underside of the barberry leaf and releases aeciospores of the stem rust fungus.

Aerenchyma. Thin-walled cells of plant tissue with large, air-filled intercellular spaces, found in roots and stems of rice and other aquatic plants.

Agroecosystem. A comprehensive approach to factors associated with agriculture and ecology and their interactions.

Albinos. Mutant forms of plants showing degrees of chlorophyll deficiency ranging from complete to banding.

Aleurone. A layer of cells surrounding the endosperm of cereal grain.

Alkyl-resorcinol. A naturally occurring animal-growth-inhibitory substance concentrated in the oil fraction of rye seed that imparts a bitter flavor and reduces feed intake when fed in large quantities to livestock.

All purpose wheat flour. Flour produced from hard spring wheat and suited for bread making because of its gluten-forming protein.

Alleles. Pairs of genetic factors found at similar points on homologous chromosomes that are inherited in alternative pairs. An allele is an alternative form of a gene.

Allelopathy. Release of growth-inhibiting substances from higher plants.

Allogamy. The process of cross-pollination whereby pollen from one plant lands on the female organ of another plant. Cross-pollination results in heterozygous plants.

Amino acids. Basic organic compounds that are the building blocks for proteins such as lysine and methionine.

Amphiploids. Fertile hybrids that are diploid for two genomes, each from a different species.

Anaerobic. Existing in the absence of free oxygen; the opposite of aerobic.

Aneuploidy. Having a chromosome number other than an exact multiple of the basic number.

Angiosperms. Seed-bearing plants in which the female gamete is protected within an enclosed ovary.

Annuals. Plants that complete their life cycle from seed in one year.

Anthesis. A developmental stage in flowering at which anthers rupture, pollen is shed, and the floret is in full bloom.

Apical dominance. In plants, the inhibition of lateral buds by high levels of auxins produced in the lead shoot or apical meristem.

Apomixis. A general term for several methods of nonsexual reproduction that involve some or most of the apparatus of sexual reproduction.

Aracicides. Chemicals toxic to mites and ticks.

Asexual reproduction. The vegetative production of a new plant that does not involve meiosis and the union of gametes.

Auricles. Clasping appendages arising at the junction of the leaf blade with the sheath. In barley, auricles are large and clasping.

Autogamy. The process of self-pollination whereby pollen lands on the female organ of the same plant. Self-pollination leads to homozygous plants.

Auxins. A group of plant hormones that influence cell enlargement and elongation, photo- and geotropism, apical dominance, abscission of plant parts, flower initiation and development, root initiation, and seed germination.

Available water. Soil water useful to plants.

Azolla. A water fern plant that enhances nitrogen fixation when grown in association with blue-green algae in flooded rice fields.

Backcross method. A system for incorporating a desirable genetic trait into an otherwise acceptable cultivar through a system of repeated crossing onto a recurrent F_1 selection, using the same nonrecurrent parent having a desirable trait.

Bacteria. Single-celled microscopic plants that lack chlorophyll and that reproduce by splitting in two.

Bannock. A flat cake often made of oatmeal or barley meal and cooked on a griddle.

Barren plants. Plants that fail to produce a normal ear. In corn, this is a major limitation to efficient conversion of light energy into grain when grown at high plant densities. Barrenness is a physiological response of the corn plant to inter- and intraplant competition for factors such as mineral nutrients, soil moisture, and light at high plant densities.

Base collections. In gene banks, this is the primary or long-term seed source, in which germination is renewed only after long time intervals (see *Active collections*).

Basidiospores. Haploid sexual spores arising from telespores and which serve to carry the stem rust fungus from overwintering material to the alternate host, barberry.

Beriberi. A vitamin B_1 deficiency disorder resulting in neurologic symptoms, cardiovascular abnormalities, and edema.

Biennial. A plant that requires two years to complete its life cycle from seed to seed. Food stored in the first season is used in the second season to produce flowers and seed.

Binomial system. A biological classification method whereby each species is indicated by two names—first the genus and second the species name.

Biochemistry. The chemistry of living organisms.

Biological buffer. A living system that removes contaminants before food is used for human consumption.

Biological value. An index of the portion of absorbed nitrogen retained in the body for growth or maintenance or both.

Biomass. Renewable organic matter formed through the process of photosynthesis. Biomass includes all food and feed products, plant and animal residues, processing by-products, tree and forest products, algae, and fresh- and saltwater plants.

Biotechnology. The application of engineering and technological principles to the manipulation of genetic materials.

Biotic factors. All organisms ranging from soil microorganisms to insects, animals and humans that directly or indirectly influence plants.

Biotype. Strain or race of a biological organism differing in its ability to survive or attack in a specific environment (see *Race*).

Bivalents. A pair of homologous chromosomes.

Blends. Combinations of two or more cultivars of the same species together for commercial production without regard to genetic differences or similarities.

Brachytic. Refers to short plants having the same number of nodes and internodes as taller counterparts. The length of each internode is reduced in length.

Bran. The outer layers of a cereal grain removed in milling, consisting of the pericarp, the testa, and the aleurone layer.

Bread flour. Flour milled from hard wheats for

commercial bakers and with high gluten-forming properties.

Breeder seed. Cultivar seed that is recognized by a regulatory body as having been produced by a plant breeder in a manner maintaining phenotypic characteristics of the cultivar and that provides the sources for the initial and recurring increase of seed of the pedigreed classes.

Breeders' Rights. Legislation giving the equivalent of patent rights for plant cultivars. The effect is to give the developer of a cultivar sole legal possession of that cultivar and a legal basis for compensation for its use by others. Breeders' Rights may be referred to as "plant cultivar protection."

Brewing process. The production of an alcoholic beverage using the readily fermentable sugars contained in the malt along with hops to provide flavor and preservation.

Bulk density (soil). The weight of a centimeter of soil as compared to a cubic centimeter of water and expressed as grams per cubic centimeter.

Bulk method. A plant-breeding system involving the growing of segregated genertions of self-pollinated crops in a randomly mixed block. Seed for the subsequent generation may be obtained by mass selection or a random sample. Individual plant selection is not conducted until the F_6 or later generations.

Bulliform. A specialized type of plant cell found in some cereal grasses and involved in tissue contraction.

Bunded. Leveed, dyked, or embanked to contain water and regulate its depth for lowland rice production.

Carbohydrate. Organic compound composed of carbon, hydrogen, and oxygen. Carbohydrates include sugars, starches, and celluloses.

Carbon-nitrogen ratio. A weight ratio of organic carbon to total nitrogen, obtained by dividing the percentage of carbon by the percentage of total nitrogen.

Cardinal temperatures. Specific temperatures at which physiological plant processes are at the minimum, optimum, and maximum levels.

Caryopsis. A dry, small fruit that does not shatter or break open and that has a single seed with such a thin, closely adhering pericarp that a single body, a grain, is formed.

Cation exchange capacity. The total number of cations that can be interchanged between the positively charged cations in solution and the cations of any surface-active material such as clay, root surfaces, or humus.

Celiac disease. In humans, an intestinal malabsorption of fat and an inability to digest cereal gluten, resulting in a complex of symptoms. Also known as sprue syndrome.

Cereal grains. Grasses cultivated primarily for their edible seeds or grains.

Certified seed. The approved progeny of breeder, select, foundation, or registered seed produced by standards determined by the Seed Growers' Association and graded and sealed by authorized inspectors.

Chaff. Fragments of straw including the glume and hull removed from cereal grains in threshing or processing.

Chapatti. An unleavened bread made from a stiff dough of flour and water and cooked on a hot, flat plate into a round, flat food. Typically from India.

Check cultivar. A standard reference cultivar for test comparisons.

Check-row planting. Spacing of plants in hills equally in all directions, often 107 cm (42 in.) apart and with three or four plants per hill.

Chiasma. Crossing over resulting in an interchange at corresponding points between homologous chromosomes.

Chlorophyll. The green pigmentation in the chloroplasts of plant cells necessary for photosynthesis.

Chloroplasts. Specialized structures found in plant cells containing chlorophyll, the site of photosynthesis.

Chromatids. Half chromosomes, resulting from longitudinal division during meiosis and mitosis.

Chromatography. A process of separation of materials into bands of different colors when a solution of dyes is dropped upon blotting paper.

Chromosome. A highly organized body carrying genes and contained in the nucleus of a cell.

Chromosome aberration. The lack of a complete chromosome complement.

Clavate. A term to describe a spike that is compacted and enlarged at the top, giving a club-shaped appearance.

Climate. Average long-term temperature, moisture, light, and wind (see *Weather*).

CO_2 flux. The transfer of CO_2 to or away from the ground.

Coarse grains. A connotation given to certain cereal crops that are considered common, crude, ordinary, or of inferior quality because of hull, taste, or other features. Coarse grains are considered livestock feed.

Cob. The central axis of an ear of corn on which kernels develop.

Code. A particular sequence of specific chemicals containing genetic information.

Colchicine. An alkaloid that is used to induce polyploidy through the mechanism of preventing spindle fibers from pulling the duplicated chromosomes apart in cell division.

Coleoptile. A membranous cover over the shoot apex of cereal crops that protects the plumule as it grows from the seed to the soil surface.

Combining ability. The relative ability of a parent to transfer desirable performance to its progreny.

Companion crop. A cereal crop grown in association with forage seedlings to act as a cover crop to suppress weeds.

Compensation point. In plants, the level of photosynthetically active radiation or CO_2 required to equal respiration losses.

Competition. In plants, when the immediate supply of a single necessary factor falls below the combined demands of the plant or plants, competition occurs. Competition may result in a retardation of plant growth.

Concentrates. Feeds low in fiber content and of high digestibility, such as the grains of cereal crops, and components of grains such as bran, gluten, or grain by-products having a high feed value, such as brewers' grains.

Cooperation. In plants, when the yield of a mixture exceeds the yield of the highest yielding component of that mixture.

Corn heat units (CHU). Values based on the relationships between day (maximum) and night (minimum) temperatures and the rate of development to maturity for corn. Both corn hybrids and geographic regions are rated in the same terms, thereby providing a basis for matching hybrids with a suitable maturity rating to a specific location.

Cotyledon. The single leaf contained in the seed of cereal crops.

Couscous. A food made by steaming durum flour or semolina.

Critical leaf area. A leaf area above which detrimental mutual shading reduces net photosynthesis rate.

Critical planting date. The date beyond which a crop cannot be expected to mature.

Crop ecology. The study of the relationships between crop plants and their environment.

Crop rotation. Growing of crops in a regularly scheduled sequence on the same land area, as contrasted to the continuous culture of one crop or the growing of different crops in haphazard order.

Cross-pollinated. Fertilized by pollen from another plant.

Crossing over. The exchange of corresponding segments between two paired, homologous chromosomes, resulting in an exchange of genes and producing a new genetic combination differing from those that were provided by the parents.

Cryopreservation. Seed storage at −196°C, the temperature of liquid nitrogen.

Culm. A commonly used name for the stem of the grass plant, consisting of nodes and usually hollow internodes.

Cultivars. A contraction of the words *cultivated varieties*, an inclusive term for lines, hybrids, selections, or varieties of crops. To be classed as such, a cultivar must be distinct from other cultivars of the same species.

Cultural energy. Energy from human and animal labor, and from fossil fuels used in making and operating machines, in all aspects of food production, transportation, processing, and distribution.

Cuticle. A layer over the epidermal cells of plant organs and impermeable to water.

Cytokinins. A group of plant hormones that promote primarily cell division as well as cell enlargement, differentiation, dormancy, and retardation of leaf senesence.

Cytology. A study of cell structure and components in relation to their function.

Cytoplasm. The contents of a cell, excluding the nucleus.

Cytoplasmic factors. Inheritance-determining factors outside of the nucleus and hence not associated with the genes.

Cytoplasmic inheritance. The transfer of traits determined outside of the nucleus by factors contained in the cytoplasm or cell material.

Cytoplasmic male sterility. Sterility due to factors that are carried only through the female and are not diluted or lost in successive generations of reproduction.

Deep-water or floating rice. Rice plants that can elongate rapidly when flood waters rise and that have adventitious roots at submerged nodes which allow them to obtain nutrients and reestablish themselves if dislodged by flooding water.

Dehiscent. Breaking free or open.

Deletion. The loss of a chromosome segment involving one or more genes.

Detasseling. The removal of the pollen-bearing tassel from each plant of the female seed parent at the proper stage of growth but before pollen is produced.

Determinate. Of plants, exhibiting clearly defined vegetative growth and reproductive development stages.

Developed countries. The more industrialized nations, such as the United States, all of Europe, the USSR, Japan, Australia, New Zealand, and Canada.

Developing countries. Nations in the early stages of economic development and characterized by low per-capita incomes. Also called the "third world," "less-developed countries (LDCs)" or "underdeveloped countries." About three-fourths of the world's population can be classed in this group.

Devernalization. A high-temperature response in plants, whereby fully vernalized plants revert back to the original nonflowering condition.

Diallel crosses. Progeny produced by crossing three or more parents in all possible combinations.

Diaphragm. A sheet of tissue forming a wall and separating cavities or sections in plants or animals.

Dicotyledonous. Having two colyledons or two sections of a seed that forms two cotyledonary leaves at the first node.

Differentiation. The formation of thick-walled or specialized tissues or substances during growth and development.

Dioecious. Having the male and female organs on separate plants.

Distichous. Having two ranks of single leaves borne in alternate positions on the stem.

DNA. Deoxyribonucleic acid, the material of inheritance, arranged in specific sequence to form a chromosome.

Dominant. A gene that influences a specific character is said to be dominant when it dominates a contrasting gene at the same allele in the development of the character (see *Recessive*).

Dormancy. A condition when healthy, morphologically ripe seed does not germinate or has a delayed response when placed under optimum conditions of moisture, temperature, and light.

Double cropping. Growing two crops a year in sequence.

Double fertilization. A phenomenon in angiosperms whereby one male nucleus unites with the egg nucleus to form a zygote which develops into the embryo. A second male nucleus unites with two polar nuclei in the embryo sac to form the endosperm.

Drought hardiness. The capacity of plants to survive desiccation.

Drought resistance. In plants, overall suitability for cultivation under dry conditions.

Ear. The seed-bearing organ of the corn plant.

Ecology. The study of reciprocal relations among plants and animals and their environment.

Electrophoresis. The movement of colloidal particles through a fluid under the action of an electric field.

Emasculation. The removal of anthers from a flower before pollen is shed, a preliminary step in hybridization to prevent self-pollination.

Embryo. The rudimentary plantlet contained in a seed.

Empirical. Based on observation and experience without a scientific understanding.

Endosperm. Food-storage material in a seed.

Environment. The sum total of a number of variables including climate, soil, physiographic, and biotic factors.

Enzymes. Substances produced from a gene that control or regulates cell functions and hence the entire organism. An enzyme may catalyze and initiate a biochemical reaction.

Ethylene. A gaseous plant hormone that influences flower initiation and fruit ripening.

Eucarpia. Acronym for the European Association for Research on Plant Breeding.

Evapotranspiration. The combined loss of water from an area through evaporation and transpiration.

Exponential rate. Used to indicate variation in which one variable factor depends upon another, resulting in an ever-increasing rate.

F_1. The first-generation progeny from a cross. F_1 indicates the first filial generation.

F_2. The second filial generation, produced from selfing or crossing of the F_1. Subsequent filial generations are F_3, F_4, F_5, and so on.

Facultative cereals. Plants that are flexible in their development and are able to survive under variable environmental conditions.

FAO. The Food and Agricultural Organization of the United Nations, an autonomous agency within the United Nations system. FAO's main goals are to raise levels of nutrition, improve production and distribution of food and agricultural products, and better the living and working conditions of rural peoples. Its ultimate purpose is to achieve freedom from hunger—food security—for all people in the world.

Farina. An unrefined or coarsely ground form of flour made from cereal grains.

Feed. All plant products consumed by livestock. Those which are bulky and with low feed value are roughages, and those which have high digestibility and a low content of crude fiber are concentrates.

Feedlot. A production unit involving a concentration of animals and produced on an area insufficient to sustain plant growth.

Field capacity (water). The amount of moisture remaining in soil after gravitational water has drained away. This is the maximum amount of water that can be held by capillary action in the soil and be available to plants.

Flag leaf. The final, uppermost leaf to develop in a cereal plant.

Floret. The stamens, pistil, and lodiculae enclosed by the lemma and palea.

Forages. Those plants grown primarily for feed and in which all or nearly all of the top growth is harvested. Cereal grain crops used for silage, hay, or pasture, and of which the stems, leaves, and floral parts are fed intact, are called forages.

Foundation seed. The approved progeny of breeder or select seeds produced by recognized growers. The seed is graded and sealed by a federal government inspector.

Frost damage. Estimated amount of flower sterility in a cereal crop resulting from freezing spring temperatures.

Fungi. Microscopic plants consisting of a vegetative structure called a mycelium, lacking chlorophyll and conductive tissue, and reproducing by spores.

Fungicides. Chemicals toxic to fungus disease organisms.

Fusiform. A term describing a spike that is widest in the middle and tapers to both its tip and base.

Futures contract. An agreement for the purchase or sale of a commodity, to be delivered at a specified future time and at a definite price.

Gametes. Reproductive cells (pollen or ovary) containing the haploid number of chromosomes. The diploid chromosome number is restored at fertilization.

Gamma rays. Short-wavelength electromagnetic radiation emitted from nuclei of radioactive isotopes.

Gene. A unit of heredity occupying a fixed chromosome locus.

Genetic diversity. Variations in genotypes resulting from hybridization and selection under the law of the survival of the fittest over many millennia.

Genetic vulnerability. Refers to pest susceptibility when large geographic regions are planted with the same crop cultivar or with cultivars having the same genes or cytoplasmic factors.

Genetics. The science of heredity.

Genome. The sum of all the genes carried by a single gamete.

Genotype. The genetic makeup of an individual.

Genus (pl. **genera**). A taxonomic category that includes groups of closely related species. A group of closely related species.

Geotropic response. Growth or movement of a plant part with respect to gravitational forces. Vegetative plant parts are usually negatively geotropic and grow upward; roots are positively geotropic and grow downward.

Germplasm. In crops, refers to plant materials that serve as a basis of crop improvement or related research; a reservoir of genes.

Gibberellins. A group of plant hormones that dramatically stimulate plant growth by cell division, cell elongation, or both.

Gigas. Exhibiting large-sized characteristics.

Glabrous. Smooth-surfaced and free of hair or trichomes.

Glassine bag. A glazed, semitransparent paper that withstands outdoor conditions and is used to protect emasculated plant parts from cross-pollination.

Glumes. Two bracts found at the base of a grass spikelet.

Gluten. A mixture of proteins that results in dough elasticity and cohesiveness and is found most abundantly in wheat and rye.

Grain. A single seed or kernel called a caryopsis; a collective term for cereal crops; cereal seeds in bulk. In botanical terms, the "seed" of cereals is a dry fruit with ovary walls fused to the seed.

Green revolution. The development and successful introduction of high-yielding semidwarf wheat and rice cultivars in Africa, Asia, and Latin America.

Greenhouse effect. An increase in carbon dioxide in the earth's atmosphere that has a warming effect, as it restricts reradiation of sunlight to the outer atmosphere.

Guard cells. Modified epidermal cells that surround stomatal openings on the leaf surfaces. Guard cells respond to turgidity and open or close the stomatal openings as they expand or contract.

Gymnosperm. A seed plant with seeds not enclosed in an ovary wall.

Gynoecium. Collective name for the stigma, style, and ovary.

Haploid. Having a single set of unpaired chromosomes in each nucleus.

Haploid method. A procedure in plant breeding for obtaining embryos with half the chromosome number, followed by doubling with colchicine to produce a homozygous plant.

Harvest index. Grain weight as a percentage of total above-ground dry weight at maturity.

Hedging. Offsetting the risk of loss from market fluctuations by purchase or sale of commodity futures.

Herbicides. Chemicals toxic to weeds.

Heredity. The transmission of plant characteristics from parents to their offspring.

Heritability. A measure of the effect of genetics on phenotype.

Heterosis. A term synonymous with hybrid vigor. Defined as an increase in vigor in the first-generation progeny following a cross of different inbred lines, cultivars, or species.

Heterozygous. When a single pair of genes at one allele on homologous chromosomes differ, the gene pair is heterozygous. A plant is heterozygous when gene pairs at one or more alleles are unlike and progeny segregate.

High-moisture grain. Grain crops harvested with a moisture level of about 25% to 30% and stored without drying in an airtight silo in a semiensiled state.

Hills. In corn, refers to closely clustered groups of plants, in contrast to row planting. North American Indians used hill-planting methods to avoid cultivating the entire cropped area; early corn producers used hill plantings to allow multidirectional cultivation to reduce hand weeding within the row.

Holocoenotic. Refers to the total impact on living organisms of many complex environmental factors acting together.

Homoeologous chromosomes. Chromosomes with similar genetic function and from a common evolutionary source, but which no longer pair at meiosis because they have undergone genetic diversity.

Homologous chromosomes. The two members of a chromosome pair.

Homologous variation. The expression of similar traits in different species as a result of a common ancestor (see *Parallel variation*).

Homozygous. When a single pair of genes at one allele on homologous chromosomes are alike, the gene pair is homozygous. A plant is homozygous when all genes on homologous chromosomes are alike. Homozygosity can be achieved by doubling the haploid chromosome number or through inbreeding (self-pollination) so that cell progeny are identical to the parent plant (see *Heterozygous*).

Hooded. Refers to awns on glumes that terminate in various winglike structures.

Hull. The outer covering of a fruit or seed. In cereal crops, hulls may be removed freely as in wheat or adhere as in barley (see *Pericarp*).

Humus. A form of soil organic matter that is somewhat resistant to further rapid decomposition and has a carbon : nitrogen ratio of 10 : 1.

Hybrid. The first-generation progeny of a cross between two different plants of the same species. A hybrid may combine characteristics derived from the two parent stocks and may be more desirable than either parent.

Hybrid vigor. (See *Heterosis.*)

Hybridization. The crossing of genetically unlike individuals to obtain cultivars with new genetic recombinations.

Hydrocyanic acid. A highly poisonous compound, HCN or hydrogen cyanide, also called prussic acid.

Hydrophyte. Plant capable of living under wet soil conditions.

Hydroponics. The practice of growing plants without soil by using nutrient solutions alone or with an artificial medium or in sand or gravel.

Imperfect flower. A floret lacking either stamens (pistillate) or a pistil (staminate).

Inbreeding. Crossing closely related individuals or crossing a bisexual plant with itself, called selfing, which results usually in a decline in vigor.

Independent assortment. Genetic segregation of nonallelic genes completely at random.

Indeterminate. Refers to plants in which vegetative growth or reproductive development stages overlap.

Inflorescence. The flowering part of a plant.

Insecticides. Chemicals toxic to insects.

Interference. Hardships caused by the proximity of neighboring plants when the immediate supply of a single necessary factor falls below the combined demands of the plants. The word *competition* is often used to express interference.

Internode. The region between two nodes of a plant stem. In cereal crops the leaf sheath encases much of the internode.

Inversion. Refers to a rearrangement of a group of genes in a chromosome in such a way that their order is reversed.

In vitro. Conducted outside a living organism, in contrast to *in vivo,* which is within a living organism.

Involucre. Protective group of bracts enveloping sexual parts of a young inflorescence.

Isoline. Identical in all genetic traits but one.

Jointing. In cereals, refers to the period when nodes and internodes are developed during stem elongation.

Joints. The solid or nodal portion separating the internodes of a cereal stem and from which the leaf sheath, adventitious roots, or tillers arise.

Juvenile phase. An early stage in the development of a plant when tillers develop, internodes elongate, and the plant is insensitive to external stimuli.

Kjeldahl method. An analysis of organic compounds to determine the quantitative nitrogen content by interaction with concentrated sulfuric acid and by distillation of ammonia from the NH_4SO_4 formed.

Kneeing ability. A positive geotropic response whereby the floating leaves and stems of deep-water or floating rice become erect.

Kwashiorkor. A protein-deficiency syndrome in infants and young children, found mainly in the tropics.

Lanceolate. Having a slender or spear-headed shape.

Land races. Traditional cultivars with sufficient genetic integrity to be morphologically identifiable and differing in adaptation and to cultural practices, but genetically variable.

Large grains. Corn and sorghum are described as large grains because of their large plant size relative to small grains such as wheat and rice.

Lateritic soil. Soil that hardens irreversibly into rock-like material when exposed and dried by sunlight. The name is obtained from the Latin word for brick. The soil is high in iron and aluminum.

Law of diminishing returns. The increase in any crop produced by a unit increment of a deficient factor is proportional to the decrement of that factor from the maximum (as stated by E. A. Mitscherlick, a German soil scientist, in 1909). In numerical terms, it follows that if one unit of input produces half the maximum yield, then two units will produce 75% of the maximum yield, three units will give 87.5% of the maximum, and so on.

Leaf-area index (LAI). The ratio of leaf area to land area. An LAI of 6 means that on each unit of land area there are six units of leaf area. One side of the leaf lamina only is considered.

Leavened. A dough product that has undergone fermentation by the addition of yeast or baking powder to lighten the dough.

Lemma. The lower or dorsal bract of a grass spikelet enclosing the caryopsis. In wheat the lemma is readily removed at threshing, and in oat and barley it adheres.

Light flecking. A phenomenon favorable to photosynthesis, in which light utilization is increased by leaf movement that allows light to penetrate the leaf canopy.

Ligule. A membranous fringe on the inner side of a leaf at the top of the sheath.

Linked genes. A situation in which genes are linked or close to one another so that independent assortment of nonallelic genes is hindered.

Loci (sing. **locus**). The position of a gene on a chromosome relative to other genes on that chromosome.

Lodging. The displacement of the stems of crops from their upright position.

Lodiculae. Small glandlike organs in grasses and cereals located between the ovary and the surrounding glumes. Swelling of the lodiculae results in opening of the glumes.

Long-day plants. Plants that require a short night or dark period for reproduction development. Although it is the length of night that influences reproductive development, the original classification of long-day, short-day, and day-neutral plants persists (see *Short-day plants*).

Lowland rice. Rice grown in fields flooded by irrigation or rain water. The water may be contained by bunds.

Major elements. Nitrogen, phosphorus, potassium, calcium, magnesium, and sulphur, which are essential to plant growth and are used in relatively large amounts (see *Minor elements*).

Male sterility. An inherited factor, useful in hybrid seed production, that prevents viable pollen from being produced.

Malt. A grain product in which the starch in the endosperm has been converted by enzymes to sugar through the germination process and from which the dried rootlets have been removed.

Mass selection. Selection of individual plants with desirable phenotypes from a large population of plants grown in a field or in bulk.

Meiosis. Nuclear division in which the diploid ($2n$) chromosome number is reduced to the haploid ($1n$) chromosome number. Genetic segregation accompanies meiosis.

Meristem regions. Areas in the plant where cells are rapidly dividing.

Mesocotyl. The subcrown internode of a grass seedling.

Metamorphosis. In insects, the transformation from young to adult form.

Microclimate. The climate at the soil surface, in the soil, at the surface of a leaf, or within a crop canopy, as contrasted to the climate of a more extensive field or region.

Minor elements. Elements such as iron, manganese, zinc, copper, boron, and molybdenum, which are required in cereal crop physiology in small or trace amounts (see *Major elements*).

Mitochrondria (sing. **mitochondrium**). Rodlike or spherical bodies found in the cytoplasm of all cells and containing enzymes needed for respiration. Mitochondria are regarded as the powerhouse of the cells because they are a primary source of energy needed in all life processes.

Mitosis. Nuclear division in which chromosomes duplicate and divide to produce two nuclei that are identical with the original nucleus.

Mixed grain. The growing of two or more crops of different cereal grain species together, primarily for feed production. Mixed grain often refers to an oat and barley mixture in a 50 : 50 ratio or a 65 : 35 ratio of oat to barley on a weight basis. Spring wheat may be added to an oat-barley mixture.

Monocotyledonous. Having a single cotyledon or leaf at the first node of the lead shoot or stem.

Monoculture production. Refers both to the production of a single species on the same field year after year, and to the production of a single crop, often the same cultivar, over a wide geographical area.

Monoecious. Having separate male and female flowers on the same plant.

Monsoon. A cyclic weather pattern involving rain and wind.

Morphology. The form, structure, and development of plants.

Multilines. Blends or mixtures of cereal cultivars of one species that are genetically very similar but differ in specificity of resistance to strains of a disease.

Multiple alleles. A series of alleles at similar loci on homologous chromosomes of related species.

Multiple cropping. The production of more than one crop in a given area in one growing season or year.

Mutagens. Substances that are known to induce genetic changes or mutations.

Mutation. A heritable change resulting from a sudden modification in an individual gene or chromosome.

Mycelium (pl. **mycelia**). A mass of threadlike filaments (hyphae) of a fungus comprising the main body.

Mycoplasmas. Microscopic, unicellular organisms similar to but smaller than bacteria and lacking rigid cell walls.

Nematicides. Chemicals toxic to nematodes.

Nematodes. Microscopic, eellike worms that attack and parasitize plant roots, stems, foliage, and inflorescences and which reproduce by eggs which hatch into larvae.

Net assimilation rate. Total rate of photosynthesis minus respiration.

Nicking. Synchronization of the receptivity of the male-sterile plant to the maximum pollen load of the pollinator for cross-pollination in hybrid seed production.

Night soil. Human excreta.

Nitrate reductase. An enzyme found in plant leaves and tissues that is involved in the first steps of nitrate fertilizer reduction and influences the levels of amino acids and proteins in plants.

Nubbin. A small, poorly developed corn ear.

Obovoid. Egg-shaped with the narrow end at the base.

Offal. Refuse in general, but associated especially with the parts of a butchered animal that are considered inedible by humans.

OPEC. Acronym for "Organization of Petroleum Exporting Countries," a cartel formed in 1972 that resulted in rapid and dramatic price increases of fossil fuels.

Optimum leaf area. The leaf area at which dry-matter production is maximized.

Organic farming. A production system that avoids or largely excludes the use of synthetically compounded fertilizers, pesticides, growth regulators, and livestock feed additives. To the maximum extent feasible, organic farming systems rely upon crop rotations, crop residues, animal manures, legumes, green manures, off-farm organic wastes, mechanical cultivation, mineral-bearing rocks, and aspects of biological pest control to maintain soil productivity and tilth, to supply plant nutrients, and to control insects, weeds, and other pests (USDA Report, 1980).

Organic matter (soil). Plant and animal residues at various stages of decomposition expressed as a percentage. Organic-matter content can be estimated by multiplying percentage of carbon by 1.724.

Ovary. The basal, enlarged part of the pistil in which seed is found.

Oxalic acid. A bitter, poisonous acid.

Palea. A thin, papery inner bract of a floret lying next to the caryopsis in grasses. In wheat, the palea is removed at threshing, and in oat and barley it adheres.

Panicle. A branched inflorescence such as in oat or sorghum in which spikelets are borne on subdivided branches arising from a main axis.

Parabolic. Umbrella- or arch-shaped.

Parallel variation. Independent but similar evolutionary development resulting from a common ancestor (see *Homologous variation*).

Particle density. The weight of soil per unit volume of soil solids without pore space.

Paste products. Food items such as spaghetti, macaroni, and noodles manufactured from durum wheat, often referred to as pasta products.

Pastry flour. Cake or cookie flour, milled from soft wheat. This flour is not suitable for bread making because it contains very little gluten-forming protein.

Pathogenic. Capable of causing disease.

Pedicel. The stalk of an individual flower of an inflorescence.

Pedigree method. A system of plant breeding based on selection of individual plants during segregating generations of a cross. Plants in each generation are treated as individuals, and the pedigree of each selection is recorded.

Pedigreed status. Seed has pedigreed status when the crop from which it is derived meets the requirements of the Seed Growers' Association and a crop certificate has been issued, or, in the case of foreign seed, when it is certified by a recognized foreign certification agency.

Peduncle. Stalk of an inflorescence.

Pellagra. A disease caused by nicotinic acid deficiency. Symptoms include inflammation of the soft tissues of the mouth, skin lesions, diarrhea, and disorders of the central nervous system.

Perennation. Survival from year to year, considered in annual cereal crops as an evolutionary remnant from a crop that persisted over several years by regenerating new growth from tillers.

Perennial. A plant that lives for three or more years.

Perfect flower. A bisexual flower that contains both stamens and a pistil.

Pericarp. The outer covering or hull that develops from the ovary wall.

Pesticide. A chemical to control weeds, diseases, insects, and other pests.

***p*H.** A numerical value of the concentration of hydrogen ions in solution. A *p*H of 7 is neutral, values less than 7 are acidic, and values greater than 7 and up to 14 are alkaline. A value of 6.5 is five times as acidic as *p*H 7.

Phenotype. The external physical appearance.

Phloem. Conducting tissue within a plant to transport the products of photosynthesis to other plant parts.

Photon. A quantum of light.

Photoperiod. The duration of the light period between sunrise and sunset, including the twilight period.

Photoperiodism. Reproductive response to the relative lengths of light and dark periods in a day.

Photorespiration. A form of respiration stimulated by light, found in C_3 plants, and having no known essential function.

Photosynthesis. The process that transforms carbon dioxide (CO_2) into food—the basis of all crop yield and the essence of agriculture.

Photosynthetically active radiation (PAR). Light energy within the spectral wavelength of 0.4 to 0.7 microns or 400 to 700 nanometers.

Phylogeny. The evolutionary development of a species or genus.

Physiographic. Refers to environmental factors related to the structure and behavior of the earth's surface, such as topographic features of elevation and slope and processes of silting and erosion (wind and water processes).

Physiological maturity. The point in grain development at which no further increase in dry weight takes place.

Physiologic forms. Groups within species that are similar morphologically but differ in characteristics such as their ability to cause disease.

Physiology. Study of the processes that occur in living organisms.

Pistil. The seed-bearing organ of a flower, comprised of a stigma, style, and ovary.

Placentochalazal region. The area at the base of the seed and the maternal plant.

Plant architecture. The number, size, shape, structure, arrangement, and display of particular plant parts.

Plant breeding. An organized attempt to produce progressively better-adapted plants.

Plant growth regulators. Nonnutrient chemicals that do not occur naturally in plants but when applied to plants can influence their growth and development.

Plant hormones. Naturally occurring plant substances that control plant activities.

Plasmids. Tiny, independent molecular entities associated with the mitochondria (energy-producing bodies) in the cytoplasm of plants.

Plastic responses. The ability of plants to respond (within limits) to competitive stress such as seeding rate.

Plumule. The first leaf in an embryo that emerges as the first leaf shoot from the coleoptile.

Polygenes. Numerous genes that influence a single trait, with each having a minor impact.

Polymerized. A chemical term pertaining to a compound of high molecular weight formed from smaller molecules of the same substance.

Polymorphism. Showing a great deal of variability in phenotype and genotype.

Polyploid. Having a cell nucleus that contains more than the $2n$ chromosome number.

Pore space. In a bulk volume of soil, the space not occupied by soil particles, expressed as a percentage.

Potential evapotranspiration rate. The potential water need computed by means of a formula (Thornwaite) from the mean temperature modified by a length-of-day factor.

Primary triticale. The immediate progeny produced from a wheat-rye cross followed by chromosome doubling.

Protogyny. A mechanism to prevent self-pollination. The female reproductive structures mature before the male structures.

Protoplast. Cell contents without the cell wall.

Prussic acid. A name sometimes used for hydrocyanic acid, a poisonous compound.

Pubescence. Downiness or soft hair or trichomes on plant parts.

Puddling. The destruction of the granular structure of the soil associated with cultivation when the soil is wet.

Punnett square. A method for determining all possible combinations for male and female reproductive cells or gametes.

Pycnia (sing. **pycnium**). A flask-shaped vessel on the upper surface of the alternate host of the stem rust fungus, which serves as the site of genetic recombinations of the rust to develop races and which forms from basidiospores.

Pycniospores. Haploid spores produced from the pycnium on the barberry leaf in a stage allowing for genetic recombinations of the stem rust fungus.

Quadrivalent. A union of four homologous chromosomes at meiosis.

Qualitative genetics. The inheritance of those differences among individuals that are characterized by discrete classes.

Quantitative genetics. The inheritance of those differences among individuals that are expressed in terms of degree rather than kind.

Races. Groups of organisms within a species. Races of pests are distinguished by their virulence.

Rachilla. The axis or little rachis of a spikelet.

Rachis. The central axis of the inflorescence of a cereal crop plant, such as the spike of wheat or barley.

Ramose. A lax inflorescence having lateral branches or divisions.

Ratoon cropping. Harvesting for grain the regrowth arising from basal buds on the stem or crown.

Recessive. A gene which in the presence of a contrasting dominant gene does not express itself in a specific character (see *Dominant*).

Recombinant DNA. The insertion of DNA containing genetic information into an organism or cell using in vitro techniques.

Recurrent parent. The parent onto which the F_1 repeatedly crossed.

Reduction division. A nuclear division whereby the diploid chromosome number is reduced to the haploid number. Also known as *meiosis*.

Registered seed. The approved progeny of breeder, select, or foundation seed produced by standards determined by the Seed Growers' Association and graded and sealed by a federal government inspector.

Relative humidity. The amount of water vapor in the atmosphere expressed as a percentage of that which could be held at saturation at a particular temperature and pressure.

Replications. Repetition of treatments in experiments that allows for statistical analysis.

Respiration. A process involving the use of oxygen, the giving off of carbon dioxide, and the release of energy for the growth and survival of a living organism.

RNA. Abbreviation of ribonucleic acid, an intermediate chemical that translates genetic information into action. In contrast, DNA (deoxyribonucleic acid) transmits genetic information from one generation to the next.

Rodenticides. Chemicals toxic to rodents.

Rogues. Undesirable plants growing in a field of morphologically uniform plants. A rogue may arise through genetic change within a cultivar, from intercrossing, or from mechanical mixtures. Rogues are removed to ensure the production of high-quality and uniform pedigreed seed.

Rotation. See *Crop rotation*.

Roughage. Coarse, bulky feed high in crude fiber content and of relatively low digestibility and feeding value, such as stover, straw, silage, and hay.

Saline soil. A soil containing sufficient soluble salts to impair crop productivity.

Sclerotia (sing. **sclerotium**). A dense, compact mass of mycelium capable of remaining dormant for an extended period.

Scutellum. A shield-shaped body covering the embryo of a grass seed.

Secondary hexaploid triticale. The progeny produced from crosses between octoploid × hexaploid triticale or between hexaploid triticale × hexaploid wheat.

Seed vigor. The overall ability of seeds to perform well in the field.

Segregation. In plants, refers to progeny having contrasting genetic features from those of the parents.

Select seed. The approved progeny of breeder or select seed produced by recognized growers in a manner to ensure its specific genetic identity. The production of select crops from select seed is limited to several generations.

Self-pollinated. Fertilized by pollen from the same plant.

Seminal. Arising from the seed.

Semolina. The end product of the ground-up endosperm of durum wheat, consisting of particles of a certain size and used to produce pasta or paste products.

Sessile. Without a stalk.

Short-day plants. Plants that require a long night or dark period for reproductive development (see *Long-day plants*).

Silage. Succulent components of plants preserved by fermentation products in a compact and airtight mass.

Small grains. A loose classification of cereals based on small plant structure and including wheat, rice, barley, oat, triticale, and rye.

Sodic soil. A soil containing sufficient sodium to interfere with the growth and productivity of most crop plants.

Soil. A major environmental aspect having a number of properties (profile, structure, texture, moisture, and air), chemical (clay minerals, base exchange properties, *p*H, anions, organic compounds), and biotic (flora, fauna, litter and humus, antibiotic effects).

Soil depletion. The reduction in crop productivity due to a change in soil properties.

Soil deterioration. The reduction in agricultural value due to soil erosion or other factors that reduce its productivity.

Soil fertility. The quality that enables a soil to provide the proper compounds, in the proper amounts, and in the proper balance for the growth of selected crops when other environmental factors are favorable.

Soil productivity. The capability of a soil to produce a specified plant or sequence of plants under a specified system of management.

Soil strength. The combined effects of soil bulk density, moisture, texture, organic matter, and management, measured in terms of mechanical impedance to soil penetration by a rigid probe and expressed in bars.

Somatic cell. Nonreproductive or vegetative tissue cell that is developed through mitosis and will not undergo meiosis.

Sour dough. Fermented dough kept from one baking to the next to start fermentation instead of beginning each time with fresh yeast.

Species. A taxonomic category of a group of individuals having a common ancestry, which cross or interbreed freely, and which have similar structure and behavior.

Spike. An unbranched inflorescence, such as wheat or barley, in which spikelets are sessile on the rachis.

Spikelet. The unit of inflorescence in grasses encased by two glumes and consisting of one or more florets.

Spindle fibers. Fiberlike structures that appear to pull the chromosome halves apart during cell division.

Spores. Reproductive units in fungi and lower plant forms.

Stamens. Pollen-bearing organs of a flower.

Stomatal openings. Apertures in the epidermis of a leaf, bounded by two guard cells which regulate the size of the opening. Stomata (plural form) allow for free exchange of carbon dioxide and oxygen between the leaf and the outside areas in the processes of photosynthesis and respiration and are openings through which water vapor passes in transpiration.

Stooking. Placing bound grain sheaves in a propped upright position in the field to facilitate drying, also known as shocking.

Stover. Dried corn, sorghum, and pearl millet stalks and leaves after the grain is removed.

Straw. Dried residue of the fine-stemmed cereal crops from which the grain has been removed.

Suckering. The development of undesirable tillers or shoots arising from the basal node or leaf axils of plants.

Summer fallowing. Not cropping but cultivating a field with the purpose of replenishing soil moisture.

Swathing. Cutting and windrowing grain that has reached physiological maturity but is too high or has too uneven a moisture content to allow for uniform drying before combine harvesting.

Symbiotic relationship. A situation where two different organisms live in close association and depend on each other for their mutual benefit.

Synapse. Pairing by conjugation of homologous chromosomes.

Synthetic cultivars. Refers to forage cultivars produced by the combination of selected plants through normal cross-pollination.

Systemic. A chemical absorbed into the plant system and carried along with the transpiration stream, thus spreading throughout the plant rather than remaining localized.

Tapioca. A starchy food normally produced from cassava for use in puddings or as a thickener. A substitute form can be produced from the starch of corn.

Taxa (sing. **taxon**). A general term for a taxonomic classification, whatever its rank.

Teliospores. Resting spores produced from telia borne on cereals and grasses following the overwintering stage of the rust and smut fungi.

Testa. A seed coat formed from the inner parts of the ovule which serves as a semipermeable membrane under the pericarp.

Test weight. Volume weight of clean grain expressed in kilograms per hectoliter.

Thermal neutrons. Neutrons whose atomic or molecular energy has been reduced by the temperature-generating process of scattering and collisions with other particles.

Thermoperiodism. Plant responses to temperature cycles.

Tillering. Secondary stems of grasses arising from a basal node and capable of producing reproductive organs (See *Suckering*.)

Tissue culture. The development of an entire organism from a single somatic cell.

Tortillas. An unleavened bread usually prepared from corn meal and cooked on a hot, flat plate into a round, flat food. Typically from Mexico.

Totipotency. The regeneration of a whole plant from a single cell or protoplast.

Trace elements. (See *Minor elements*.)

Tramlines. Unseeded, equispaced tracks established in a grain field at seeding time to provide a travel pass for tractor operations, such as spraying and fertilizing, which are carried out after the crop is established.

Translocation. In genetics, refers to a change in position of a segment of a chromosome to another part of the same chromosome or of a different chromosome. In physiology, refers to the movement of a substance in a plant from its point of origin to where it is utilized.

Transpiration. The loss of water vapor from aerial parts of plants, chiefly through leaf stomata.

Transportation ratio. The weight of water absorbed by the plant during its growth period relative to the weight of dry matter harvested. Approximately 99% of the water absorbed by a plant is transpired.

Trichome. A leaf hair arising from an epidermal cell. Trichomes are a form of pubescence.

Triploid. Having three haploid sets of chromosomes. If the embryo is triploid, shriveled seed or sterility may result.

Trisomic. A plant with an extra chromosome over and above the usual diploid number.

Trivalent. Formed by a union of three homologous chromosomes at meiosis.

Uniculm. A plant having only one shoot and lacking secondary tillers.

Upland rice. Rice grown like wheat or other cereals in nonbunded fields.

Urediospores. Asexual spores produced from uredia borne on cereals and grasses as a stage of multiplication of the rust fungi and serving to spread the disease.

Vapor-pressure deficit. The difference between the actual water-vapor pressure and that needed to saturate the air at the same temperature.

Vascular bundle. Longitudinal strand of conducting (vascular) tissue consisting essentially of xylem and phloem and occurring in veins of leaves or scattered throughout stem tissue.

Vector. An agent for transferring a gene from one plant to another.

Vernalization. The exposure to certain conditions of cold temperature and photoperiod needed by winter cereals as a prerequisite to developing reproductive organs.

Vestigial. A biological structure that is rudimentary or imperfectly developed and has no useful purpose.

Virulence. The ability of a pest organism to overcome the defenses of a host.

Viruses. Particles consisting of nucleic acid contained in a protein covering.

Vitreous. Having a hard, transparent, glasslike structure. In hard bread or macaroni wheat this is a desirable characteristic of the endosperm.

Volcanic tuff. A geological term referring to small rock fragments produced from disintegrated volcanic material and usually stratified.

Weather. Short-term variations in climatic variables (see *Climate*).

Whole grain flour. A product containing bran and having a low percentage of gluten. When used to make bread, it produces a firm, heavy, close-textured loaf.

Whorl. A circular arrangement of plant leaves.

Winter annual. A plant that develops a seedling stage in the fall, becomes vernalized over the winter, and completes vegetative and reproductive growth the following season.

Winter hardiness. The ability of the plant to withstand environmental conditions during the winter months, including those of soil and climate, without damage or subsequent reduction in yield.

X-rays. Penetrating electromagnetic radiations having wavelengths shorter than those of ultraviolet or visible light.

Xanthophyll. A yellow carotenoid pigment found in chloroplasts and found also in plastids in other plant parts where chlorophyll is absent.

Xenia. Changes due to the effect of foreign pollen in visible characters of the endosperm. Crossing two widely different corn cultivars provides classical examples, owing to the phenomenon of double fertilization in the seed plants.

Xerophytic. Capable of living under extremely dry soil conditions.

Xylem. Conducting tissue in plants for transporting to the other plant parts water and nutrients absorbed by the root.

Yellow berry. A phenomenon occurring in wheat and triticale, whereby a maldeveloped endosperm results in kernels with a yellowish color, caused by a lower protein content and adversely affecting milling and baking quality.

Yield ability. Of a genotype, a very high yield capacity under ideal growing conditions.

Yield stability. Of a genotype, the capability of producing a reliable or consistent but not necessarily a high yield under conditions ranging from ideal to stress.

Zero-tillage. A system in which a crop is planted with minimum soil disturbance directly into an untilled stubble remaining from the previous crop and in which chemicals are used to control unwanted vegetation.

Zygote. The fertilized ovary, before it undergoes further differentiation and mitotic division to develop into the embryo.

APPENDIX A

Crop Names and Chromosome Numbers

Table A–1
Names and Chromosome Numbers for Selected Field Crops and Some Relatives

Common Name	*Botanical Name*	*Chromosome (Gamete) Number*
Amaranth	*Amaranthus hybridus* L.	16
	spp. *A. hypochondriacus* (L.) Thell.	16
Banana	*Musa paradisiaca*	14
Barley	*Hordeum vulgare* L.	7
Six-row	*H. sativum; H. hexastichon*	7
Two-row	*H. distichon; H. deficiens*	7
Awnless	*H. intermedium*	7
Tetraploid	*H. bulbosum*	14
	H. jubatum	14
Wild (two-row)	*H. spontaneum*	
Bean		
Broadbean	*Vicia faba* L.	6
Faba	*Faba vulgaris* Moench	6; 7
Field, dry or common	*Phaseolus vulgaris* L.	11
Buckwheat	*Fagopyrum esculentum* (*F. sagittatum Gilib.*)	8
Canola (see rapeseed)		
Castor bean	*Ricinus communis* L.	10
Chickpea	*Cicer arietinum* L.	7; 8
Coconut	*Cocos mucifera* L.	
Cassava (mandioca, yuca, tapioca, manioc)[a]	*Manihot esculenta* Crantz *M. utilissima*	18; 36
Corn	*Zea mays* L.	10
Dent	*Z. mays indentala*	10
Flint	*Z. mays indurata*	10
Flour	*Z. mays amylacea*	10
Sweet	*Z. mays saccharata*	10

Table A-1 (Continued)

Common Name	Botanical Name	Chromosome (Gamete) Number
Pod	*Z. mays tunicata*	10
Pop	*Z. mays praecox*	10
Perennial wild forms	*Z. perennis*[b] (tetraploid)	20
(teosinte)	*Z. diploperennis* (diploid)	10
Flax (seed)	*Linum usitatissimum* L. (The genus contains over 100 species.)	15
Guinea grass	*Panicum maximum* Jacq.	21
Lentil	*Lens esculenta* Moench (*L. culinaris* Medik.)	7
Manna grass	*Glyceria fluitans* R. Br.	20
Millet (cultivated pearl)	*Pennisetum americanum* (L.) Leeke (*P. typhoides* (Burm.) Stapf et Hubb.) (*P. typhoideum* Rich.) (*P. spicatum* (L.) Koern.) (*P. glaucum* (L.) R. Br.) (*P. americanum* (L.) K. Schum.)	7
Oat		
Common	*Avena sativa* L.	21
Red	*A. byzantina* C. Koch.	21
Hullless	*A. nuda*	21
Side	*A. sativa orientalis*	21
Wild	*A. fatua*	21
Wild	*A. sterilis*	21
Pea—field or dry	*Pisum arvense* L. (*P. sativum* L.)	7
Peanut	*Arachis hypogaea* L.	20
Valencia	*A. hypogaea fastigiata*	20
Virginia and runner	*A. hypogaea hypogaea*	20
Spanish	*A. hypogaea vulgaris*	20
Potato	*Solanum tuberosum* L.	12; 24
Rice (common)	*Oryza sativa* L.	12
(Bulu type)[c]	*O. sativa bulu*	12
(Indico type)[c]	*O. sativa indica*	12
(Japonica type)[c]	*O. sativa japonica*	12
(African)	*O. glaberrima* Steud.	12
Rye	Secale cereale L.	7
Safflower	*Carthamus tinctorius* L.	12
Sesame	*Sesamum indicum* L.	13; 32
Sorghum	*Sorghum bicolor* L. Moench	10
Soybean	*Glycine max* (L.) Merrill (*Soya max* (Moench) F. J. Herm.)	20
Sudan grass	*Sorghum sudanense* (Pipa) Stapf (usually considered a cultivar of sorghum)	10
Sugar beet	*Beta vulgaris* L.	9

(Continued)

Table A-1 (Continued)

Common Name	Botanical Name	Chromosome (Gamete) Number
Sunflower	*Helianthus annuus* L	17; 34; 51
Sweet potato	*Ipomea batatas* (L.) Lam.	45
Taro	*Colocasia esculenta* (L.) Schott. (*C. antiquorum*).	21
Tobacco	*Nicotiana tabacum* L.	24
Triticale	×*Triticosecale* Wittmack (ABR genomes)	21
Wheat (common)	*Triticum aestivum* L. em. Thell. (*T. sativum; T. vulgare* Vill.)	21
Wild einkorn	*T. aegilopoides*	7
Einkorn	*T. monococcum*	7
Wild emmer	*T. dicoccoides*	14
Emmer	*T. dicoccum*	14
Macaroni wheat	*T. durum*	14
Persian wheat	*T. persicum*	14
Rivet wheat	*T. turgidum*	14
Polish wheat	*T. polonicum*	14
Timopheevi	*T. timopheevi*	14
Shot wheat	*T. sphaerococcum*	21
Club wheat	*T. compactum*	21
Spelt	*T. spelta*	21
Macha wheat	*T. macha*	21
Wild rice		
(Small-seeded)	*Zizania aquatica* L.	15
(Large-seeded)	*Z. palustris* L.	15
Asian	*Z. latifolia*	17
	Z. texana (found in Texas)	15

[a]Alternate or synonymous names are given in parentheses. The genus *Manihot* contains more than 100 species. Botanical names vary because of new knowledge of the species gained through taxonomic studies.
[b]Considered a close relative of corn and included in the same genus.
[c]These three types were initially thought to be subspecies of *O. sativa* but are now considered to be ecogeographic races.

APPENDIX B
Metric Information

DEVELOPMENT OF SI UNITS

Throughout the centuries many measurement systems were developed. Local adaptations and customs led to the evolution of so many variations that a rational structure was lost. The imperial system, based on such measurements as the yard, gallon, and pound, is an example of poorly correlated units.

In 1791, France adopted a metric system to bring order to measurements. This metric system was effective, popular, and logical, but continued evolution led to versions that caused unnecessary congestion.

In 1960, the International System of Units (SI) was established to replace all former systems of measurement, including former versions of the metric system. About 98% of the world's population live in countries that have adopted or are changing to SI, the most up-to-date system of measurement.

SI usage is not just a change for the sake of change; it is an essential step to a universal language of measurement. The worldwide importance of cereal crops, especially as manifested in the international marketing system, emphasizes the need for a globally common and acceptable system of weights and measures.

The simplicity of the system is a key point in its adaptation; no mathematical factors other than "one" are involved in the formation of derived units.

STRUCTURE OF SI

Only seven base units exist in SI, and all other units evolve from these primary units (Table B-1).

Derived units are formed algebraically from the seven base units. Some derived units have special names and symbols which, like those of the base units, are used in place of their full expression (Table B-2).

SI units internationally and nationally approved are shown in Table B-3.

PREFIXES

The decimal structure is derived from a system of prefixes; multiples and submultiples of SI units are explained as follows:

The prefix kilo means (1×10^3); hence kilojoule, kilometer, and kilogram mean 1000 joules, 1000 meters, and 1000 grams, respectively (Table B-4).

Table B–1
Base Units for SI

Physical Quantity	*Base Unit*	*SI Symbol*
Length	meter[a]	m
Mass[b]	kilogram	kg
Time	second	s
Amount of substance	mole	mol
Thermodynamic temperature	kelvin	K
Electric current	ampere	A
Luminous intensity	candela	cd

[a]Spelling variations are found worldwide. In Canada, the accepted spelling is metre and litre; in the United States, meter and liter. Whichever spelling is used, the terms are synonymous.
[b]Mass, as unlike weight, does not depend on local gravitational force and is not subject to differences caused by elevation.

Table B–2
Some Derived SI Units

Quantity and Derivation	*Name of Unit*	*SI Symbol*
Force	newton	N
Pressure (N/m^2)	pascal	Pa
Energy ($N \times m$)	joule	J
Power (J/s)	watt	W
Electric charge ($A \times s$)	coulomb	C
Electric potential (V/A)	volt	V
Capacitance (C/V)	farad	F
Conductance (A/V)	siemens	S
Magnetic flux ($V \times s$)	weber	Wb
Luminous flux ($cd \times s$)	lumen	lm
Illuminance (lm/m^2)	lux	lx
Temperature (°K)	degree Celsius	°C
Activity (of a radionuclide)	becquerel	Bq

Table B–3
Units Permitted for Use with SI

Name of Unit	*Symbol*	*Quantity*
Minute	min	time
Hour	h	time
Day	d	time
Year	a	time
Hectare ($100\ m \times 100\ m$)	ha	area
Liter	L or l	volume
Hectoliter	hL	volume
Tonne (metric ton)	t	mass

Table B–4
Some Common Prefixes

Prefix	*Symbol*	*Means multiply by*	*Factor*
mega	M	1,000,000	10^6
kilo	k	1,000	10^3
hecto	h	100	10^2
deca	da	10	10^1
deci	d	0.1	10^{-1}
centi	c	0.01	10^{-2}
milli	m	0.001	10^{-3}
micro	μ	0.000001	10^{-6}

METRIC STYLE USAGE

SI symbols are written in lower-case letters, except when the name of the unit is derived from proper name. For example:

gram = g
newton = N
joule = J
watt = W
kilowatt = kW

The liter is the only unit that can deviate from this rule:

liter = L or l

SI symbols are never pluralized. For example:

10 g; 1200 kg

SI symbols are never followed by a period except at the end of a sentence.

THE LITER

The coherent SI unit is the cubic meter. The volume of water contained in a cube with length, width, and depth of 1 meter (m) is 1000 kilograms (kg) or 1 tonne (t).

One tonne (t) of water contains 1000 liters of water. One liter (L) weighs 1 kilogram. There are 1000 cubic centimeters in a liter. Concentrations of solutions are expressed in terms of the correct SI symbol, the milliliter (mL), rather than the cubic centimeter (cc).

All measurements of length, area, volume, capacity, and mass are interrelated and are based on the meter.

THE JOULE

The joule is the unit of work done by a force of 1 newton (N) when its point of application is displaced 1 meter (m) in the direction of the force (J = N × m). The thermochemical calorie should be defined solely in terms of the joule. The calorie is a metric unit but is not an SI unit.

The conversion factor from calorie to joule is 4.184.

CONVERSION

Conversion of units of accepted measure of SI does not simply mean replacing the imperial measure by the exact SI measurement (soft conversion; e.g., 1 ft = 30.48 cm). Conversion means the adoption of SI standards that assign convenient rounded measurements to each unit (hard conversion; e.g., 30 cm replaces 1 ft).

Table B-5 (p. 490) shows factors for converting non-SI units to SI units.

Table B-6 shows factors for converting selected SI units to non-SI units.

Table B–6
Factors for Converting Some SI Units to Non-SI Units

SI Units: Multiply	*By*	*Non-SI Units: To Obtain:*
Linear:		
centimeter	0.39	inch
meter	3.28	foot
kilometer	0.62	mile
Area:		
square centimeter	0.15	square inch
square meter	1.15	square yard
square kilometer	0.39	square mile
hectare	2.47	acre

Table B–5
Factors for Converting Some Non-SI Units to SI Units

Non-SI Units: Multiply	*By*	*SI Units: To Obtain*
acre	0.405	hectare
Angstrom unit	0.1	nanometer
atmosphere	0.101	megapascal
bar	10^{-1}	megapascal
British thermal unit	1.05×10^3	joule
calorie	4.184	joule
calorie per square centimeter minute (irradiance)	698	watt per square meter
calorie per centimeter (langley)	4.19×10^4	joule per square meter
centimeter per day (elongation rate)	0.116	micrometer per second
cubic foot	0.0283	cubic meter
cubic inch	16.4	cubic centimeter
curie	3.7×10^{10}	becquerel
degree (angle)	1.75×10^{-2}	radian
foot	0.305	meter
gallon (U.S.)	3.78	liter
gallon per acre	9.35	liter per hectare
inch	2.54	centimeter
mile	1.61	kilometer
mile per hour	0.477	meter per second
ounce	28.4	gram
ounce (fluid)	2.96×10^{-2}	liter
pint (liquid)	0.473	liter
pound	453.6	gram
pound per acre	1.12	kilogram per hectare
Volume:		
cubic centimeter	0.06	cubic inch
cubic meter	35.31	cubic foot
	1.31	cubic yard
Capacity:		
liter	0.035	cubic foot
hectoliter	22	gallon
	2.5	bushel
Weight:		
gram	0.04	ounce
kilogram	2.2	pound
tonne	1.1	short ton
Agricultural:		
liter per hectare	0.089	gallon per acre
	0.357	quart per acre
	0.71	pint per acre
milliliter per hectare	0.014	fluid ounce per acre
tonne per hectare	0.45	ton per acre
kilogram per hectare	0.89	pound per acre
gram per hectare	0.014	ounce per acre
plant per hectare	0.405	plant per acre

(*Continued*)

Table B–5 (Continued)

To Find	*Divide by*	*When You Know*
pound per cubic foot	16.02	kilogram per cubic meter
pound per cubic inch	2.77×10^4	kilogram per cubic meter
pound per square foot	47.9	pascal
pound per square inch	6.90×10^3	pascal
quart (liquid)	0.946	liter
quintal (metric)	100	kilogram
square foot	0.0929	square meter
square inch	645	square millimeter
square mile	2.59	square kilometer
temperature (°F − 32)	0.550	temperature °C
ton (2000 lb)	907	kilogram

APPENDIX C

Institutions Supported by the Consultative Group on International Agricultural Research (CGIAR)

CIAT	Centro Internacional de Agricultura Tropical Apartado Aereo 6713 Cali, Colombia
CIMMYT	Centro Internacional de Mejoramiento de Maiz y Trigo Londres 40 Mexico 6, D.F., Mexico
CIP	Centro Internacional de la Papa Apartado 5969 Lima, Peru
IBPGR	International Board for Plant Genetic Resources Food and Agriculture Organization of the United Nations Via delle Terme de Caracalla 00100 Rome, Italy
ICARDA	International Center for Agricultural Research in the Dry Areas P.O. Box 114/5055 Beirut, Lebanon
ICRISAT	International Crops Research Institute for the Semi-Arid Tropics Patancheru P.O. Andra Pradesh 502 324, India
IFPRI	International Food Policy Research Institute 1776 Massachusetts Avenue, N.W. Washington, D.C. 20036, U.S.A.
IITA	International Institute of Tropical Agriculture P.O. Box 5320 Ibadan, Nigeria
ILCA	International Livestock Center for Africa P.O. Box 5689 Addis Ababa, Ethiopia
ILRAD	International Laboratory for Research on Animal Diseases P.O. Box 30709 Nairobi, Kenya

IRRI	International Rice Research Institute P.O. Box 933 Manila, Philippines
ISNAR	International Service for National Agricultural Research P.O. Box 93375 2509 A.J. The Hague, The Netherlands
WARDA	West African Rice Development Association E. J. Roye Memorial Building P.O. Box 1019 Monrovia, Liberia

Two programs not totally funded by CGIAR and not included in the system are:

AVRDC	Asian Vegetable Research and Development Center Tainan, Taiwan
IFDC	International Fertilizer Development Center Alabama, U.S.A.

APPENDIX D

Feed Composition of Cereal Crop Products

Product	Basis	Dry Matter, %	Ash,[a] %	Crude Fiber,[b] %	Crude Protein,[c] %
Barley straw	As fed	86.9	6.0	36.2	3.6
	Dry	100.0	6.9	41.6	4.1
Barley grain	As fed	89.0	3.0	5.3	11.6
	Dry	100.0	3.4	6.0	13.0
Corn stover	As fed	84.4	6.2	28.2	5.7
	Dry	100.0	7.3	33.4	6.8
Corncobs (ground)	As fed	89.8	1.6	31.1	2.8
	Dry	100.0	1.8	34.6	3.1
Corn silage	As fed	27.8	1.9	7.6	2.2
	Dry	100.0	7.0	27.2	8.0
Corn ear silage	As fed	43.4	2.1	5.1	3.8
	Dry	100.0	4.9	11.8	8.8
Corn bran	As fed	88.7	1.9	9.6	8.0
	Dry	100.0	2.2	10.9	9.1
Corn and cob meal	As fed	85.4	1.5	7.1	8.0
	Dry	100.0	1.7	8.3	9.3
Corn hominy feed (grits)	As fed	89.2	2.7	4.4	10.6
	Dry	100.0	3.0	5.0	11.9
Corn starch	As fed	90.4	0.2	0.2	0.6
	Dry	100.0	0.2	0.2	0.7
Corn gluten meal	As fed	91.0	3.3	4.6	39.0
	Dry	100.0	3.6	5.1	42.9
Corn (shelled) (yellow dent)	As fed	86.0	1.1	2.0	8.8
	Dry	100.0	1.3	2.3	10.2
Brewer's grains (dried)	As fed	91.0	3.8	14.7	25.8
	Dry	100.0	4.2	16.1	28.3

(Continued)

Product	*Basis*	*Dry Matter, %*	*Ash,*[a] *%*	*Crude Fiber,*[b] *%*	*Crude Protein,*[c] *%*
Brewer's grains (wet)	As fed	23.8	1.1	3.8	5.5
	Dry	100.0	4.8	16.1	23.0
Distiller's grains (dried)	As fed	92.5	1.6	12.8	27.4
	Dry	100.0	1.7	13.8	29.6
Millet hay (foxtail)	As fed	85.8	7.0	25.0	8.5
	Dry	100.0	8.1	29.2	9.9
Oat hay	As fed	90.7	7.5	27.9	7.7
	Dry	100.0	8.2	30.8	8.5
Oat hulls	As fed	92.7	6.0	29.8	3.8
	Dry	100.0	6.5	32.2	4.3
Oat straw	As fed	88.6	6.8	36.3	3.8
	Dry	100.0	7.6	41.0	4.3
Oat grain	As fed	88.9	3.4	10.6	11.7
	Dry	100.0	3.8	11.9	13.2
Oat groats	As fed	90.7	2.2	2.6	16.5
	Dry	100.0	2.4	2.8	18.2
Rice hulls	As fed	92.4	18.4	41.1	2.8
	Dry	100.0	19.9	44.5	3.1
Rice bran	As fed	90.7	12.1	11.2	13.0
	Dry	100.0	13.3	12.4	14.3
Rye fresh forage	As fed	20.6	2.1	5.2	4.3
	Dry	100.0	10.3	25.2	20.7
Rye grain	As fed	88.2	1.7	2.0	11.3
	Dry	100.0	2.0	2.3	12.8
Sorghum silage	As fed	30.7	2.3	7.3	3.0
	Dry	100.0	7.3	23.8	9.6
Sorghum grain	As fed	88.5	2.1	2.3	8.9
	Dry	100.0	2.4	2.6	10.0
Sorghum hay (Sudan grass)	As fed	89.6	8.6	27.5	8.7
	Dry	100.0	9.6	30.7	9.6
Wheat straw	As fed	87.8	6.3	38.3	3.2
	Dry	100.0	7.2	43.6	3.7
Wheat bran	As fed	88.7	5.9	8.6	15.7
	Dry	100.0	6.7	9.7	17.8
Wheat shorts	As fed	87.1	4.0	6.3	15.8
	Dry	100.0	4.5	7.3	18.1
Wheat middlings	As fed	87.4	2.5	2.5	15.5
	Dry	100.0	2.8	2.8	17.7
Wheat grain	As fed	88.9	1.9	2.5	11.9
	Dry	100.0	2.1	2.8	13.4

Source: A. E. Cullison, *Feeds and Feeding* (Reston, Va.: Reston Publishing Company, Inc., 1982), 3d ed.

[a]*Ash*. Analysis includes the inorganic or mineral components of a feed.

[b]*Crude fiber*. Those materials in a feed that are of low digestibility, such as cellulose, hemicellulose, and lignin.

[c]*Crude protein*. Includes all materials containing nitrogen and assumes them to be proteins. Protein is comprised of an average of 16% nitrogen. The word protein usually means crude protein or "N × 6.25."

Index

Bold face figures refer to pages where references are given in full.

AUTHOR INDEX

A

B

C

D

S

T

SUBJECT INDEX

D

E

F

Q

R

T

Z